Results and Problems in Cell Differentiation

Volume 71

Results and Problems in Cell Differentiation is an up-to-date book series that presents and explores selected questions of cell and developmental biology. Each volume focuses on a single, well-defined topic. Reviews address basic questions and phenomena, but also provide concise information on the most recent advances. Together, the volumes provide a valuable overview of this exciting and dynamically expanding field. This series is indexed in PubMed.

Malgorzata Kloc • Ahmed Uosef
Editors

Syncytia: Origin, Structure, and Functions

Springer

Editors
Malgorzata Kloc
Houston Methodist Research Institute and
The Houston Methodist Hospital
Houston, TX, USA

Ahmed Uosef
Houston Methodist Research Institute and
The Houston Methodist Hospital
Houston, TX, USA

ISSN 0080-1844 ISSN 1861-0412 (electronic)
Results and Problems in Cell Differentiation
ISBN 978-3-031-37935-2 ISBN 978-3-031-37936-9 (eBook)
https://doi.org/10.1007/978-3-031-37936-9

This Springer imprint is published by the registered company Springer Nature Switzerland AG
The registered company address is: Gewerbestrasse 11, 6330 Cham, Switzerland

Preface

The volume reviews the current knowledge and hypotheses on the evolution, origin, and function of the germline, embryonic, and specialized somatic syncytia and the formation and role of pathogen-induced and cancer-related syncytia. The first part of this volume describes the structure and function of germline syncytia and presents hypotheses on their ancient origin and evolution. It also describes the unusual features of insect somatic and germline syncytia, including accessory nuclei and spermatid bundles. The second part is devoted to the syncytia in insect embryogenesis and organogenesis, trophoblast and mammalian placenta formation, and the development of the carnivorous plant Utricularia. The third part reviews cell fusion in fungi and the consequences on multinucleate cellular compartments for protein synthesis and exemplifies how the genomic changes accumulated in the microorganisms living in extreme environments could affect the evolution of uninuclear and multinuclear cells. It also discusses the role of somatic cell fusion in host defense and adaptation, and the origin and function of specialized somatic syncytia, such as osteoclasts and skeletal muscle. The fourth part describes pathogen (virus and nematode)-induced syncytia in humans, animals, and plants and reviews the difference between fusion with the plasma membrane versus endocytosis and virus infectivity. It also describes current achievements and future directions in mathematical modeling of the formation of virally induced syncytia. The fifth part is devoted to the phenomenon of cell fusion in cancer and the role of syncytia and syncytial hybrid cells in pathological conditions. This illuminating volume should inspire scientists to study cellular and molecular aspects of syncytiogenesis in animals and plants and its relevance to the development and progression of diseases.

Houston, TX, USA — Malgorzata Kloc

Book Abstract

This book combines the most current information on the evolution, origin, structure, and functions of germline and somatic cell syncytia during embryogenesis and organogenesis. It also reviews pathogen-induced syncytia and the role of syncytial cells in cancer development. The book covers the following subjects: the germline syncytia, evolution, function, and structure; the syncytia in embryogenesis and development; the role of somatic cell fusion in fungi, specialized somatic tissues, host defense, and adaptation; the virus- and parasite-induced syncytia, and the syncytia and circulating hybrid cells in cancer and other pathological conditions, and discusses how the genomic adaptations of microorganisms to extreme habitats may prompt evolution of mononuclear and multinuclear/syncytial cells. This book should provide the audience with a new perspective on the role of syncytia in embryogenesis, organogenesis, adaptation, host defense, and development of specialized tissues, and the importance of syncytia for pathogen–host interactions and cancer development.

Contents

Part I
Germline Syncytia, Evolution, Function, and Structure

Chapter 1
The Ancient Origin and Function of Germline Cysts

Allan C. Spradling

Abstract Gamete production in most animal species is initiated within an evolutionarily ancient multicellular germline structure, the germline cyst, whose interconnected premeiotic cells synchronously develop from a single progenitor arising just downstream from a stem cell. Cysts in mice, Drosophila, and many other animals protect developing sperm, while in females, cysts generate nurse cells that guard sister oocytes from transposons (TEs) and help them grow and build a Balbiani body. However, the origin and extreme evolutionary conservation of germline cysts remains a mystery. We suggest that cysts arose in ancestral animals like *Hydra* and *Planaria* whose multipotent somatic and germline stem cells (neoblasts) express genes conserved in all animal germ cells and frequently begin differentiation in cysts. A syncytial state is proposed to help multipotent stem cell chromatin transition to an epigenetic state with heterochromatic domains suitable for TE repression and specialized function. Most modern animals now lack neoblasts but have retained stem cells and cysts in their early germlines, which continue to function using this ancient epigenetic strategy.

1.1 Animals Evolved Special Germline Cells to Propagate a Complex Multicellular System

Meiosis and sexual reproduction were key eukaryotic innovations established in single-celled protists long before the advent of metazoans. In animals, however, sexual reproduction was further modified by evolving a special cell type, germ cells, to protect the larger genome necessitated by multicellularity and to generate male and female gametes. Following fertilization, animals undergo a process of embryonic development to generate multiple somatic cell types that usually survive only

A. C. Spradling (✉)
Carnegie Institution for Science/Howard Hughes Medical Institute, Baltimore, MD, USA
e-mail: spradling@carnegiescience.edu

M. Kloc, A. Uosef (eds.), *Syncytia: Origin, Structure, and Functions*, Results and Problems in Cell Differentiation 71,
https://doi.org/10.1007/978-3-031-37936-9_1

one generation. Somatic cells facilitate germ cell function and promote sophisticated structures and survival strategies (Denis and Lacroix 1993; Strome and Lehmann 2007; Woodland 2016).

Limiting meiosis and long-term reproduction to germ cells represents a highly successful specialization but one that imposes stringent developmental and evolutionary challenges (Bergero et al. 2021). Each generation of reproductive cells must minimize and repair damage to their genomes and also reverse changes affecting the abundance and function of every other vital cellular organelle and substructure. For an animal species to persist long enough to enter the fossil record, hundreds of thousands of generations are required, an impossible task unless germ cells on average in each generation manage to fully repair damage to both genome and cytoplasm. In addition to the burden of maintaining cellular immortality, the germ cell's position as cellular and genomic gatekeeper ensures they will need to confront an endless succession of genomic and cellular parasites (review Russell et al. 2019).

Germ cells across the animal kingdom share a large set of core germ cell genes (CGGs) such as Piwi, Vasa, Nanos, Tudor, and more than 20 others that help them accomplish these difficult tasks (Fierro-Constaín et al. 2017). Some CGG genes, such as those encoding cytoplasmic piwi-clade Argonautes, Vasa, and Tudor domain proteins, function in the cytoplasmic piRNA pathway to slice up transposable element transcripts in perinuclear aggregates known as nuage (Lim and Kai 2007; Czech et al. 2018). Nuclear Argonautes like Piwi use piRNA guides (or individual targeting proteins) to zero in and silence TE and repetitive element transcription by causing nearby chromatin proteins to undergo SUMO modification by PIAS1/Suvar2–10, followed by H3K9methylation by Setdb1/Egg (Seah et al. 2019; Ninova et al. 2020; review: Onishi et al. 2021).

Germ cells employ other CGGs to regulate translation (Mercer et al. 2021), possibly to avoid compromising pluripotency by expressing transcription factors involved in somatic cell development. Frequently, germ cell mRNAs along with translation factors and regulatory proteins are packaged into phase-separated particles known as germ granules that allow their activity to be activated or repressed based on modifications of granule proteins and by alterations in their subcellular location (review: Seydoux and Braun 2006; Chiappetta et al. 2022). Other germ cell tasks, such as those involving organelle maintenance and preserving germline immortality, are just beginning to be understood (Lieber et al. 2019; Chen et al. 2020; Samaddar et al. 2021).

1.2 Animals Expanded the Protist Meiotic Program by Adding Germline Cysts

Single-celled eukaryotes undergo an indefinite number of mitotic divisions before they enter meiosis, in response to environmental factors (Fig. 1.1a–c). For example, in *Saccharomyces cerevisiae*, nitrogen starvation or another deficit promotes

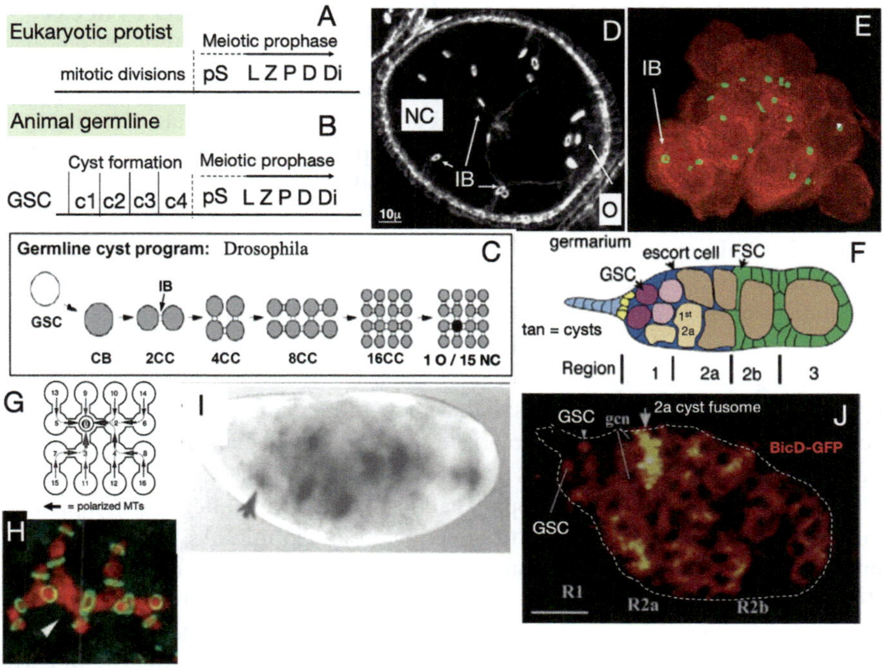

Fig. 1.1 Animal germ cells added cysts between the GSC and meiosis. (**a–b**) Cell cycle and meiotic entry in eukaryotic protists (**a**) compared to animal germlines (**b**), showing the germline cyst-forming divisions c1–c4 downstream from the germline stem cell (GSC). Vertical dashed line indicates onset of meiosis: *pS* premeiotic S phase, *L* leptotene, *Z* zygotene, *P* pachytene, *D* diplotene, *Di* diakinesis/dictyate. (**c**) The germline cyst program in Drosophila. *CB* cystoblast, *nCC* n-cell cyst, *IB* intercellular bridge, *NC* nurse cell, *O* oocyte. (**d**) A 16-cell Drosophila cyst in a growing follicle; **c**, **d** from de Cuevas et al. (1996). (**e**) An early mouse germline cyst. From Lei and Spradling (2016). (**f**) Drawing of a Drosophila germarium showing subregions 1–3. FSC follicle stem cell, forming cysts (light pink), earliest meiotic cysts (light tan), later cysts (tan). (**g**) Diagram of 16CC cyst fusome polarity (arrows); microtubule (MT) minus ends (arrowheads) focus on the initial cell, which becomes the oocyte (O). (**h**) Fusome (red) from a 16CC, green = IBs. The large fusome compartment indicated by the arrowhead is in the oocyte; from de Cuevas and Spradling (1998). (**i**) *hts* RNA is localized to the oocyte beginning early in region 2a, from Spradling et al. (1997). (**h**) BicD associates with the fusome beginning in early region 2a 16CCs (arrow). BicD-GFP (red), gcn germ cell nucleus, reproduced and adapted with permission from Paré and Suter (2000)

meiotic entry. Animal germlines, in addition to evolving CGGs, generated a novel cellular feature, the germline cyst, which always develops just after a germline stem cell (GSC) or primordial germ cell and just preceding entry into meiosis. Germline cysts are built from a single progenitor germ cell, ensuring that their cell's genomes are virtually identical, thereby empowering cyst cells to act altruistically. The reason germline cysts arose and have been conserved in animal germlines remain a major mystery, but one that has attracted surprisingly little study.

1.3 Cysts Are Ancient and Are Built by a Conserved Process

The ubiquitous presence of germline cysts in male germlines argues strongly that germline cysts arose at the time of animal origins. Similarly, female germline cyst production supports oocyte development in phylogenetically diverse species, including holometabolous insects (Table 1.1I) and mammals (Table 1.1O) (reviews: Telfer 1975; Pepling et al. 1999; Matova and Cooley 2001; Lu et al. 2017; Spradling et al. 2022). Like cysts in Drosophila and many other species, mouse cysts (Lei and Spradling 2016; Niu and Spradling 2022) develop into nurse cells as well as oocytes.

Studies in Drosophila still provide most of what we know about how cysts become polarized, which allows them to specify the oocyte, while the remaining cells, baring cyst breakage, become nurse cells. The fusome ("spindle-derived"), a cyst organelle built from mitotic spindles, new microtubules (MT), spectraplakin, actin, and ER vesicles, becomes polarized with MT minus ends focused on the initial cell (the future oocyte) as cysts develop (reviews, Hinnant et al. 2020; Spradling et al. 2022; Fig. 1.1g, h). In Drosophila, almost immediately after cyst completion, microtubule motor activity starts to directionally move mRNAs from nurse cells into the oocyte, where they become enriched beginning at the onset of meiosis (Fig. 1.1i, arrow; Spradling et al. 1997). Enrichment is blocked by mutations disrupting the fusome, minus end-directed microtubule transport proteins like Dynein, the mRNA adaptor Egl, or the dynactin component BicD. Consistent with early transport, high levels of BicD are found associated with the fusome, the site of polarized transport, in the first region 2a cysts (Fig. 1.1j, arrow; from Paré and Suter 2000). Early cyst differentiation in Drosophila limits meiotic prophase entry to just the oocyte and its initial daughter, whereas in most insects, all cyst cells typically enter meiosis (Büning 1994). In the mouse, most cyst cells initially enter meiosis, and even the 80% fated as nurse cells behave like future oocytes until they are sequentially activated for transfer (Niu and Spradling 2022). By the time the follicle forms, however, nurse cells in both species have transferred centrosomes and organelles into the oocyte to establish a new microtubule organizing center (MTOC) and formed a Balbiani body (Bb) (Cox and Spradling 2003; review Spradling et al. 2022).

Other large animal groups modified the fate of cysts to generate telotrophic ovaries (Büning 1994; Świątek and Urbisz 2019; Kloc 2019), including nematodes, annelids, Heteroptera, and polyphagous beetles. These groups build syncytial ovaries that transport trophic material from distal nurse cells to oocytes through a common cytoplasmic core or individual trophic cords (Büning 1994; Gottanka and Büning 1993; Seidel et al. 2018). How the original cyst program is altered in telotrophic groups remains incompletely understood. Relatively few changes may be required, as *Rhabditis SB347*, a hermaphroditic nematode related to *C. elegans*, uses conventional germline cysts to produce sperm throughout adulthood (McCaig et al. 2017). Some animals generate oocytes in panoistic ovaries that only use cysts in early developmental stages (see Büning and Sohst 1988; Gottanka and Brining 1990) or that do not generate cysts at all in the female germline (e.g., Tworzydlo et al. 2014). The true panoistic condition in the clearest cases appears to have resulted from secondary loss of cysts during evolution (Büning 1994). Nonetheless, animals whose female gametes utilize or are likely to utilize cysts and nurse cells can be seen to span most of animal diversity (Fig. 1.2).

Germline cysts are formed by stereotyped cell cycle programs whose detailed characteristics have been conserved in multiple animal groups (Telfer 1975; King 1970; Büning 1994). These features include the following: (1) A single cyst progenitor known as a "cystoblast" divides multiple times with incomplete cytokinesis, to generate a persistent cluster of interconnected daughter germ cells. (2) Cyst divisions are synchronous, at least initially, biasing cyst cell numbers to powers of two. (3) Cytokinesis is blocked by mitotic spindle remnants, presumably the midbody. (4) Rather than abscising, the arrested furrow converts into an intercellular bridge of characteristic appearance (Fawcett et al. 1959; Price et al. 2022). (5) Cyst interconnections allow the sharing of cytoplasmic materials between cyst cells. (6) Cyst forming divisions may generate a persistent, asymmetric fusome (Fig. 1.1d, e).

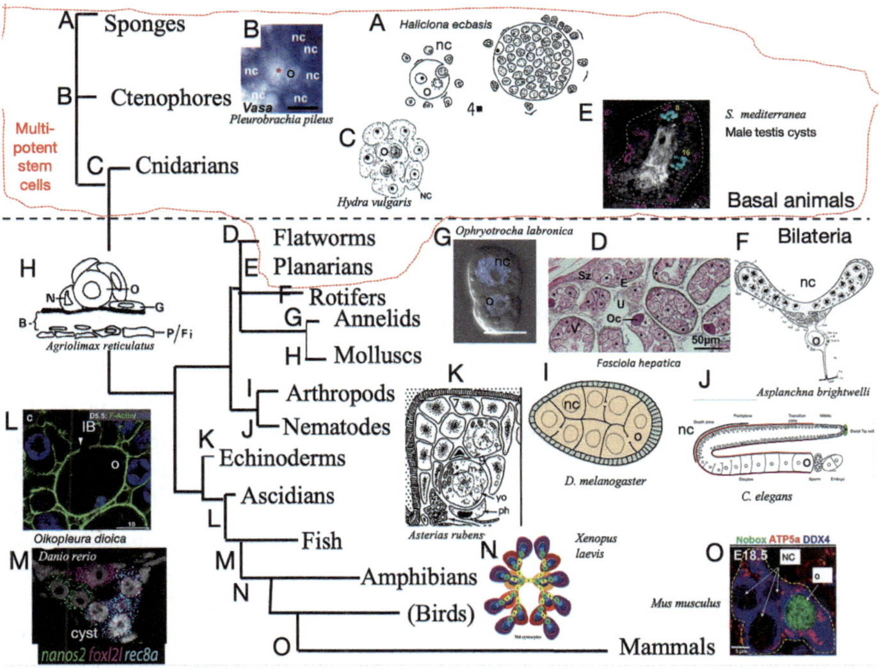

Fig. 1.2 Phylogenetic diversity of animals with nurse cells and oocytes. A simplified phylogenetic tree of major animal phyla indicated by name and letter (A–O). An image of a cyst or of an oocyte with associated nurse cells from that group is also shown. o oocyte; nc nurse cell. Basal animals are above the dashed line, and bilateral animals are shown below. Red dashed lines indicate phyla with multipotent stem cells (neoblasts). See Table 1.1 for image citations and further information

Table 1.1 Groups and species in Fig. 1.2

Group	Species	Citation (see References)
A. Sponges	*Haliclona ecbasis*	Reprinted from Fell (1968) with permission from Wiley. Choanocyte (stem cells) generate oocytes that engulf NCs as shown. Clonal origin of NCs plausible but unproven
B. Ctenophores	*Pleurobrachia pileus*	Reprinted from Alié et al. (2011) with permission from Elsevier. Clonal origin of NCs plausible but unproven
C. Cnidaria	*Hydra vulgaris*	Reprinted from Alexandrova et al. (2005) with permission from Elsevier. One of many 8-cell cysts depicted prior to competition, fusion, and oocyte selection. Note intercellular bridges
D. Flatworms	*Fasciola hepatica*	Reprinted from Hanna et al. (2016) with permission from Elsevier. Each egg contains approximately 30 vitelline cells (V) and a single oocyte (Oc). Vitelline cells derive from separate stem cell
E. Planaria	*S. mediterranea*	Reprinted from Issigonis et al. (2022) Fig. 1.3 under CC license. 8-cell male cysts are shown, which give rise to 32 sperms. Use of cyst to generate yolkless oocytes plausible but unproven. Oocytes later fuse with yolk cells generated from independent stem cells during transport through the oviducts
F. Rotifers	*Asplanchna brightwelli*	Reprinted from Bentfeld (1971) with permission from Springer Nature. Oocytes individually connected by intercellular bridges to syncytial tropic chamber of nurse cells
G. Annelids	*Ophryotrocha labronica*	Reprinted from Brubacher and Huebner (2011) with permission from Elsevier. Initial cyst fragments to produce 2-cell cysts (shown) with 1 NC and 1 oocyte
H. Mollusk	*Agriolimax reticulatus*	Reprinted from Hill (1977) with permission from Springer Nature. An oocyte (O) with nurse cell (N) is shown
I. Arthropods	*D. melanogaster*	Reprinted from Frydman and Spradling (2001). All oocytes develop from cyst with 15 nurse cells
J. Nematodes	*C. elegans*	Reprinted from Gartner et al. (2008). Under CC license. Germ cells develop from GSCs, lineage not yet mapped, in telotrophic gonad, with most serving as nurse cells
K. Echinoderms	*Asterias rubens*	Reprinted from Schoenmakers et al. (1981) with permission from Springer Nature. A cell nest with a young oocyte (yo) is shown
L. Tunicate	*Oikopleura dioica*	Reprinted from Ganot et al. 2007 with permission from Elsevier. Developing oocyte shown with intercellular bridge (IB) connecting to polyploid nurse cells within the syncytial coenocyst
M. Fish	*Danio rerio*	Reprinted from Liu et al. (2022) under a CC attribution license. Early developing ovarian germline cysts expressing Foxl2 and rec8a. See also Mytlis et al. (2022)
N. Amphibian	*Xenopus laevis*	Reprinted from Kloc et al. (2004) with permission from Elsevier. A 16-cell germline cyst consisting of an unknown number of oocytes and nurse cells

(continued)

Table 1.1 (continued)

Group	Species	Citation (see References)
O. Mammal	*Mus musculus*	Reprinted from Niu and Spradling (2022). A germline cyst is shown containing one oocyte (o) and several nurse cells (nc). Oocyte-specific expression of the Nobox gene is already evident prior to completion of cyst breakdown

Note: for complete details of sources, see References

1.4 Early Metazoan Animals Also Use Cellular Cysts Downstream from Stem Cells

Where did the intricate germline cyst program originate and how has it managed to persist more than 500 million years during the metazoan animal era? Studies of simple animals, such as hydra, planaria, and parasitic flatworms provide fascinating insights into this issue (Newmark and Sánchez 2022; Reddien 2022; Rozario et al. 2022). These basal animals are thought to resemble early metazoan animals, and they display exceptional abilities at regeneration and somatic propagation, but little if any indication of aging. This group contains multipotent stem cells, "neoblasts," that are not widespread in most animal groups today. Some neoblasts are able to generate both germline and somatic cell types, or rarely all cell types, as well as downstream stem cells with more limited outputs (Raz et al. 2021). Neoblasts are responsible for the extensive regenerative capacity of these species and for their ability to propagate somatically for extended periods by budding or fragmenting. Neoblasts express the same CGG genes as bilaterian animal germ cells (Reddien et al. 2005; Alié et al. 2011; Reddien 2018; Siebert et al. 2019). Moreover, many types of neoblasts begin differentiation by generating cysts of interconnected cells that are similar to germline cysts in Drosophila. Thus, animal germline cysts may have arisen originally within animals whose lifestyle was based heavily on flexible, multipotent stem cells.

1.5 Multipotent Stem Cells Use CGGs to Control Transposable Elements

Study of *Hydra* stem cells and planarian neoblasts have provided insight into their properties (David 2012; Zeng et al. 2018; Siebert et al. 2019; Issigonis et al. 2022). For example, *Hydra* interstitial stem cells (ISCs) produce gland cells, neuroblasts generating neural cells, nematoblasts producing nematocysts, and two germ stem cell types supporting either sperm or oocyte development (Bosch and David 1987; Nishimiya-Fujisawa and Kobayashi 2012; David 2012). CGG genes in hydra, like in animal germ cells, regulate transposon activity. The *Hydra* Piwi proteins Hywi and Hyli are strongly expressed in ISCs, GSCs, and oogenic cells (Lim et al. 2014). ISC-derived germ cells contain nuage containing these Piwi proteins, where they likely support a ping-pong cycle. Both Hywi and Hyli contain symmetrical dimethyl-modified arginine residues that are known to aid in Piwi localization to nuage and that support piRNA loading. In planaria, both Piwi-1 and Piwi-2 genes are required for oocyte production (Kimoto et al. 2021).

The Piwi proteins carry out similar functions in somatic stem cells and their downstream derivatives, as shown by studies using *Hydra* lacking the interstitial cell linage (Teefy et al. 2020) and also in the sea anemone, *Nematostella vectensis* (Praher et al. 2017). Somatic piRNAs predominantly target transposable elements, and knockdown of *hywi* in epithelial cells increases TE expression levels in somatic cells. Functions of CGGs including Piwi proteins were likewise found in both somatic and germline multipotent stem cells in the ctenophore *Pleurobrachia pileus* (Alié et al. 2011). These studies show that in basal animals, *piwi* genes and piRNAs function to control TEs in both germline and somatic stem and downstream cells, as well as other possible tasks.

1.6 *Hydra* ISCs Give Rise to Somatic Cnidoblast Cells Using a Pathway Like the Germline Cyst

If basal neoblast stem cells and their downstream cysts were predecessors of the animal germline cyst, then many of the details of cyst formation should match the properties deduced in previous germ cell studies. *Hydra* ISCs produce nematocysts, a major cell type used for food capture, using cysts similar to those GSCs use to produce gametes. ISCs divide asymmetrically with compete cytokinesis to self-renew while generating a founder nematoblast. The nematoblast then divides 2, 3, or 4 times synchronously and with incomplete cytokinesis, to generate interconnected cysts of 4, 8, or 16 nematoblasts joined by intercellular bridges (IBs). Although the exact pathway of interconnections was not reported, drawings suggest that nematoblast cysts are organized in a largely linear manner, with most cells having two IBs.

Electron microscopic studies of the nematoblast cysts revealed that cytokinesis is blocked by bundles of filaments that appear to derive from mitotic spindle remnants (Fawcett et al. 1959). As development proceeds, the filaments gradually disappear from the arrested furrows, which convert into intercellular bridges indistinguishable in appearance from those in animal germ cells. Cytological evidence of cytoplasmic sharing also becomes evident, since advanced nematoblast cyst cell bridges show evidence that endoplasmic reticulum likely passes between cells (Fawcett et al. 1959). ER is also found in the Drosophila fusome, which passes through the germline cyst. While there are four recognized nematocyte subtypes, each cyst only produces a single type (David 2012). Thus, current knowledge of cyst formation in basal animals is consistent with and supports the idea that animal germline cysts are derived from these precursors.

1.7 Germline Cysts in *Hydra* and *Planaria* Downstream from Male GSCs Also Parallel Their Function in Advanced Animal Testes

In *Planaria*, neoblasts and developing germ cells express and require multiple CGG genes and many other genes that are associated with germ cell development in diverse animals (reviewed in Issigonis and Newmark 2019). Neoblasts and germ cells display distinctive germ cell characteristics such as a chromatoid body, nuage, and gene expression associated with germ granule production in other systems. Neoblasts that are specialized as male germline progenitors generate spermatogonial cysts (Farnesi et al. 1977; Issigonis and Newmark 2019). Cyst progenitor cells undergo three rounds of synchronous division with incomplete cytokinesis to generate eight cell spermatogonial cysts. They subsequently enter meiosis and undergo synchronous spermiogenesis to form cysts of 32 spermatids. *Hydra* male germline development takes place in cysts of 32 interconnected cells (Littlefield 1985; Munck and David 1985).

1.8 Female GSCs Develop into Clusters That Generate Oocytes and Either Differentiate or Acquire Nurse Cells

Female germline cysts in basal animals support the production of oocytes that can reach very large size (David 2012). *Hydra* contain a subset of ISCs that are restricted to oocyte production (Littlefield 1991) and that produce both oocytes and nurse cells. In *Hydra vulgaris*, eight cell cysts are initially formed in which two central cells containing three IBs soon become larger and show a distinctive nuclear morphology compared to six outer cells with one or two IBs (Miller et al. 2000; Alexandrova et al. 2005). Female germ cells enter meiosis in the cyst stage and occupy a large "egg patch" containing many of the initial interconnected germ cell groups that compete with each other as they rapidly grow and fuse. Only one of the central cells from one cyst continues to expand and ultimately differentiate as an enormous oocyte, while all the other cells, numbering in the thousands, are engulfed in the oocyte cytoplasm as nurse cells.

Following fusion with the oocyte cytoplasm, nurse cells transfer their cytoplasmic contents to the oocyte. However, their nuclei and nuclear envelope remain largely intact even as their DNA is cleaved to a smaller size (Technau et al. 2003). In some species, only after the oocyte is fertilized are the nurse cell nuclei expelled and turn over outside the embryo. Thus, like nurse cells in Drosophila and mouse, hydra nurse cells remain for an extended period as small remnant cells, consisting largely of a partially degraded nucleus. Further study will be required to determine if they turn over by a process of "developmental cell death" mediated by somatic cells (Lebo and McCall 2021), like mouse and Drosophila remnant nurse cells.

Developing female germ cells in *Planaria* interact extensively with somatic cells, and eggs are nourished by a separate lineage of yolk cells (Khan and Newmark 2022; Issigonis et al. 2022). This style of yolk production is characteristic only of planarians and parasitic flatworms and is associated with gonad-like yolk glands that supply yolk cells to oocytes as they move along the oviduct. Stem cells likely support yolk cell production, but these have not been characterized.

1.9 Cysts May Be Needed to Synchronize Downstream Cells and Modify Their Cell Cycles

In Drosophila, cyst mitotic synchrony depends on regulated cyclin degradation on the fusome (Lilly et al. 2000; Ohlmeyer and Schüpbach 2003). The fact that a physical structure shared between all cyst cells catalyzes cell cycle progression probably explains the precise synchrony observed in germline cysts. Most but not all insect groups with cysts are thought to contain a fusome-like structure (Büning 1994). It would be worthwhile to study whether cyst-based cell cycle synchrony and somatic cell development downstream of neoblasts are equally precise and to investigate their structural bases.

1.10 Drosophila Female GSC Daughters Become Epigenetically Modified Within Cysts Like Pre-blastoderm Nuclei in the Syncytial Embryo

The chromatin state of one broadly potent stem cell, the Drosophila female GSC, was recently found to be largely open and free of repressive chromatin (DeLuca et al. 2020; Pang et al. 2023). HP1c levels, a marker of H3K9me2/3 binding and of heterochromatin, are lower in GSCs and mitotic germline cysts compared to older germ cells (Fig. 1.3a). Chip-seq studies using purified GSC DNA showed that the GSC genome contains very little Polycomb-related H3K27me3 repressive marks, even within Polycomb domains (Fig. 1.3b, GSC 2C). This open state contrasts with somatic follicle cells (Fig. 1.3b, FC 2C, 16C) or follicular nurse cells (Fig. 1.3b, NC 512C).

The open nature of GSC chromatin was functionally validated using heat shock-GFP (hs-GFP) reporters inserted at more than 100 diverse, known genomic sites defined by MiMIC transposons. Some hs-GFP insertions at repeat-rich pericentromeric sites could be activated in GSCs and germline cysts, but not in germ cells following meiotic entry (Fig. 1.3a‘). This suggests that expression-repressing H3K9me3-rich heterochromatin forms prior to the onset of meiosis, likely through the action of Setdb1/Egg (Clough et al. 2014). Repressing repeat-rich sequences in heterochromatin would reduce non-allelic recombination that might contribute to

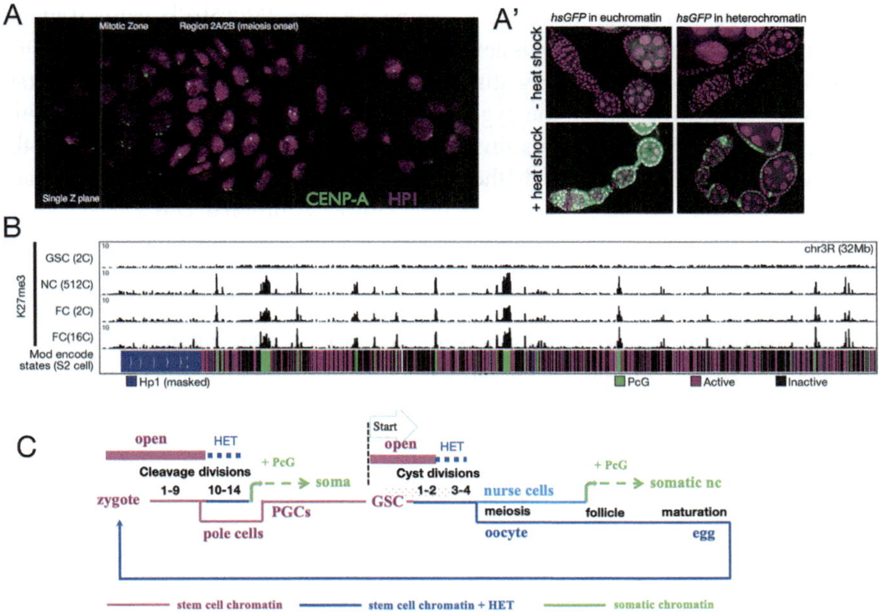

Fig. 1.3 Chromatin changes in germline cysts and germ cells downstream from the GSC resemble changes from the zygote during early embryonic development. (**a**) Immunofluorescence staining of the H3K9me3-binding heterochromatin protein HP1c (magenta) shows relatively low levels of heterochromatin in the stem cell and mitotic (cyst forming) region of a Drosophila germarium (see Fig. 1.1f). (**a′**) Functional analysis of chromosome repression using hsp70-GFP reporter genes. Examples of GFP expression among over 100 analyzed reporter insertions located either in euchromatin (left) or in pericentric heterochromatin (right) with or without heat treatment are shown (Pang et al. 2023). (**b**) Chip-seq results showing levels of H3K27me3 on Drosophila chromosome 3R, comparing germline stem cells (GSC), nurse cells (NC 512c), and follicle cells (FC 2c, 16c). The type of chromatin present in each chromosome subregion is shown by a color code below, including centromeric heterochromatin enriched in Hp1 (blue), Polycomb (PcG) domains (green), active chromatin (pink), and black chromatin (black) as described in DeLuca et al. (2020). (**c**) Summary of the Drosophila female germ cell generational cycle, diagramed to indicate when chromatin exists in an open state (pink, open), when heterochromatin is forming (blue, HET), and when Polycomb domains become repressive (green dashed lines) as somatic cells are specified. Two repeating cycles of stem cell chromatin differentiation occur each generation. The pre-blastoderm embryo undergoes 14 syncytial mitotic cycles downstream from the zygote, giving rise to pole cells at cycle 9 and somatic cells beginning at cycle 14 (cellular blastoderm). Similarly, four cyst-forming cell cycles downstream from an adult GSC lead to heterochromatin formation, oocyte, and nurse cell specification and to nurse cell PcG repression and somatic differentiation after follicle formation. Above, diagrams illustrate the germ cell chromatin state during these periods. **b**, **c** modified from DeLuca et al. (2020; Pang et al. 2023)

the sterility of *egg* germline clones (Clough et al. 2014). Transgenes in Polycomb domains express GFP following heat shock in GSCs but are not activated after follicle formation in polyploid nurse cells, a change that depends on the Polycomb H3K27me3 methylase E(z) (DeLuca et al. 2020). Thus, Drosophila nurse cells after follicle formation differentiate as a somatic cell (but only after nurse cells would have turned over in most groups), contradicting the view that germ cells sequestered by germ plasm in an early embryo never give rise later to somatic cells.

A similar but more thoroughly studied transition from an undifferentiated state takes place downstream from the zygote in the preblastoderm embryo. The zygote undergoes 14 rapid, synchronous divisions in a syncytial state. At division 9, the pole cells separate and establish the embryonic germline (Seydoux and Braun 2006). The remaining nuclei undergo four slower additional cell cycles before establishing the cellular blastoderm. Prior to completing the last divisions, a series of steps establishes heterochromatin (Seller et al. 2019). Subsequently, during early embryonic development, many cells repress Polycomb domains as part of somatic cell differentiation. Thus, germ cells downstream from the female GSCs and the Drosophila zygote both undergo heterochromatin formation, followed by somatic differentiation (Fig. 1.3a). In both instances, these changes occur in the syncytial context of a germline cyst or syncytial embryo.

1.11 Cysts and Syncytia May Have Evolved to Limit Germline Parasites

The ubiquity of germline cysts in male germlines is thought to result from the high vulnerability of this process, including its haploid stages, to selfish elements. Cysts are likely to suppress meiotic drive elements by ensuring product sharing within cysts of developing sperm (Braun et al. 1989). They also may share small RNA-based defenses synthesized during the primary spermatocyte stages against drive elements.

The primary function of cysts or syncytia is likely to be defending early differentiating cells from a different threat-TE activation. The germline movement of the P element has been studied in individual males in a manner that allows "clusters" of identical mutants caused by premeiotic transposition to be identified. Transposition prior to or at the GSC state would produce a cluster of identical mutations in at least 5% of sperm, whereas transposition during cyst stages would generate smaller clusters and single insertions thereafter. In a P element mutagenesis screen involving more than 15,000 male parents, clusters of 2 identical insertions were observed in about 5% of 8276 transposition events, but clusters larger than 2 were not detected (Karpen and Spradling 1992; Pang et al. 2023). Thus, P elements rarely transpose in germ cells from the zygote to the GSC stage but start to move during the final cyst divisions.

This suggests that potent stem cells such as neoblasts and germline stem cells utilize a novel mechanism to repress transposon activity in addition to employing CGG genes encoding piRNA pathway genes. This currently unknown mechanism allows such stem cells to be highly resistant to transposon activity even without the high heterochromatin levels that are characteristic of more differentiated cells. We suggest that this stem cell-specific system is capable of even higher levels of suppression than are observed in downstream cells, consistent with the ability of potent stem cells to propagate genomes over evolutionary time scales (Pang et al. 2023). The system is likely to be incompatible with differentiation into either gametes or most somatic cells, explaining the transitions observed in Fig. 1.3c.

Moreover, the chromatin changes taking place downstream from potent stem cells that are needed to establish heterochromatin-mediated TE protection may temporarily pose significant vulnerability to TE activation. During early embryogenesis, a period where chromatin is transitioning from the pluripotency of the zygote, some TEs, such as the I element retrotransposon, are capable of increased movement (Lavige and Lecher 1982). In both Drosophila and mammals, early embryogenesis is a time of increased transposon transcription (Scherer et al. 1982; Percharde et al. 2018), and in mammals exit from the pluripotent state likely also involves changes in transposon mRNA methylation (Chen et al. 2021). Thus, during the process of epigenetic remodeling downstream from GSCs and zygotes, the potential of TEs to increase their activity may have led to the selection and maintenance of a cyst or syncytial stage that provide more effective defenses to mitigate this threat. If unrestrained, either P or I elements can kill germ cells and cause sterility.

How would the sharing of materials between cells enhance TE suppression under these conditions? Sharing of defensive molecules such as piRNAs among all the interconnected cyst or syncytial cells is likely to provide a stronger and more diverse TE and parasite defense than if each cell defended itself independently. During an epigenetic transition, TE movement is elevated, but only a minority of cells will experience any individual new insertion. The production and sharing of primary piRNAs by a few cells in the cluster might allow the other cells to jumpstart the ping pong cycle to gain a head start in defense against the active element. The ability to share new piRNAs targeting other genome regions altered by element activity, for example, repeat copy number changes or local chromosome rearrangements, might help all the cells react faster and more effectively by targeting the most critical endogenous elements, genes, and chromatin regions for heterochromatization. When TEs become activated, including previously inactive elements, the identity and location of the threat may be difficult for a cell to identify accurately. Because of the genetic similarity of cystocytes, vulnerable sites in one cell are likely to require defense in others. Thus by mechanisms known and unknown, groups of interconnected cells likely provide an advantage that has proved sufficiently valuable to conserve germline cysts and equivalent mechanisms throughout the history of animal life (Pang et al. 2023).

1.12 Concluding Thoughts

Drosophila and many other animal lifecycles have two points of epigenetic simplicity followed by genomic specialization and production of new cell types (Fig. 1.3c). However, even in animals like Drosophila, GSC and zygote chromatin probably differ in ways that assist them to play different roles. GSCs must prepare a new generation of immortal germ cells, early assistance to patterning the embryo, and, at least in higher insects, engender one simple somatic cell type, the differentiated follicular nurse cell. In contrast, zygotes lay the epigenetic groundwork to generate hundreds of different cell types and diverse tissues. These somatic cells will build all the structures on which the survival of another animal generation depends, including a complex nervous system.

Is the amazing lifestyle of basal animals based on totipotent and multipotent stem cells confined to the members of those groups that continue to thrive today? As summarized here, it seems likely that their legacy has been preserved in germ cell lineages throughout animals, which continue to maintain cysts and syncytial states. However, when responding to tissue damage, some somatic cells can diversify their output and, at least in a few species, even produce new germ cells, like somatic neoblasts. Neuroblasts that generate complex programs of neural and glial cell production might also still use mechanisms derived from basal animals. Overall, however, most animals seem to have left the Eden-like state where basal animals lived simple lives of near immortality. By gaining the capacity for great complexity, some species have even learned to contemplate whether evolution has done them a favor by this transition.

Acknowledgments We thank Dr. Steven DeLuca and Haolong Zhu who carried out the studies in Fig. 1.3a and a′ (DeLuca et al. 2020; Pang et al. 2023). We thank Drs. Qi Yin, Bhawana Bhawana, and Asya Davidian of the Spradling lab for helpful comments on the manuscript.

References

Alexandrova O, Schade M, Böttger A, David CN (2005) Oogenesis in Hydra: Nurse cells transfer cytoplasm directly to the growing oocyte. Dev Biol 281:91–101

Alié A, Leclère L, Jager M, Dayraud C, Chang P, Le Guyader H, Quéinnec E, Manuel M (2011) Somatic stem cells express Piwi and Vasa genes in an adult ctenophore: ancient association of “germline genes” with stemness. Dev Biol 350:183–197. https://doi.org/10.1016/j.ydbio.2010.10.019

Bentfeld ME (1971) Studies of oogenesis in the rotifer, Asplanchna. I. Fine structure of the female reproductive system. Z Zellforsch Mikrosk Anat 115(2):165–183. https://doi.org/10.1007/BF00391123

Bergero R, Ellis P, Haerty W, Larcombe L, Macaulay I, Mehta T, Mogensen M, Murray D, Nash W, Neale MJ, O’Connor R, Ottolini C, Peel N, Ramsey L, Skinner B, Suh A, Summers M, Sun Y, Tidy A, Rahbari R, Rathje C, Immler S (2021) Meiosis and beyond – understanding the mechanistic and evolutionary processes shaping the germline genome. Biol Rev Camb Philos Soc 96:822–841

Bosch TCG, David CN (1987) Stem cells of Hydra magnipapillata can differentiate into somatic cells and germ line cells. Dev Biol 21:182–191

Braun RE, Behringer RR, Peschon JJ, Brinster RL, Palmiter RD (1989) Genetically haploid spermatids are phenotypically diploid. Nature 337:373–376. https://doi.org/10.1038/337373a0

Brubacher JL, Huebner E (2011) Evolution and development of polarized germ cell cysts: new insights from a polychaete worm, Ophryotrocha labronica. Dev Biol 357(1):96–107. https://doi.org/10.1016/j.ydbio.2011.06.026. Epub 2011 Jun 25. PMID: 21726546

Büning J (1994) The insect ovary: ultrastructure, previtellogenic growth and evolution. Chapman and Hall, London

Büning J, Sohst S (1988) The flea ovary: ultrastructure and analysis of cell clusters. Tissue Cell 20:783–795. https://doi.org/10.1016/0040-8166(88)90023-7

Chen Z, Wang ZH, Zhang G, Bleck CKE, Chung DJ, Madison GP, Lindberg E, Combs C, Balaban RS, Xu H (2020) Mitochondrial DNA segregation and replication restrict the transmission of detrimental mutation. J Cell Biol 219:e201905160

Chen C, Liu W, Guo J, Liu Y, Liu X, Liu J, Dou X, Le R, Huang Y, Li C, Yang L, Kou X, Zhao Y, Wu Y, Chen J, Wang H, Shen B, Gao Y, Gao S (2021) Nuclear m^{6}A reader YTHDC1 regulates the scaffold function of LINE1 RNA in mouse ESCs and early embryos. Protein Cell 12:455–474. https://doi.org/10.1007/s13238-021-00837-8

Chiappetta A, Liao J, Tian S, Trcek T (2022) Structural and functional organization of germ plasm condensates. Biochem J 479:2477–2495. https://doi.org/10.1042/BCJ20210815

Clough E, Tedeschi T, Hazelrigg T (2014) Epigenetic regulation of oogenesis and germ stem cell maintenance by the Drosophila histone methyltransferase Eggless/dSetDB1. Dev Biol 388:181–191. https://doi.org/10.1016/j.ydbio.2014.01.014

Cox RT, Spradling AC (2003) A Balbiani body and the fusome mediate mitochondrial inheritance during Drosophila oogenesis. Development 130:1579–1590. https://doi.org/10.1242/dev.00365

Czech B, Munafò M, Ciabrelli F, Eastwood EL, Fabry MH, Kneuss E, Hannon GJ (2018) piRNA-guided genome defense: from biogenesis to silencing. Annu Rev Genet 52:131–157. https://doi.org/10.1146/annurev-genet-120417-031441

David CN (2012) Interstitial stem cells in Hydra: multipotency and decision-making. Int J Dev Biol 56:489–497. https://doi.org/10.1387/ijdb.113476cd

de Cuevas M, Spradling AC (1998) Morphogenesis of the Drosophila fusome and its implications for oocyte specification. Development 125:2781–2789. https://doi.org/10.1242/dev.125.15.2781

de Cuevas M, Lee JK, Spradling AC (1996) alpha-spectrin is required for germline cell division and differentiation in the Drosophila ovary. Development 122:3959–3968. https://doi.org/10.1242/dev.122.12.3959

DeLuca SZ, Ghildiyal M, Pang LY, Spradling AC (2020) Differentiating *Drosophila* female germ cells initiate Polycomb silencing by regulating PRC2-interacting proteins. elife 9:e56922. https://doi.org/10.7554/eLife.56922

Denis H, Lacroix JC (1993) The dichotomy between germ line and somatic line, and the origin of cell mortality. Trends Genet 9:7–11. https://doi.org/10.1016/0168-9525(93)90065-P

Farnesi RM, Marinelli M, Tei S, Vagnetti D (1977) Ultrastructural research on the spermatogenesis in Dugesia lugubris. SL Riv Di Biol 70:113–136

Fawcett DW, Ito S, Slauterback D (1959) The occurrence of intercellular bridges in groups of cells exhibiting synchronous differentiation. J Biophys Biochem Cytol 5:453–460. https://doi.org/10.1083/jcb.5.3.453

Fell PE (1968) The involvement of nurse cells in oogenesis and embryonic development in the marine sponge, *Haliclona ecbasis*. J Morph 127:133–150

Fierro-Constaín L, Schenkelaars Q, Gazave E, Haguenauer A, Rocher C, Ereskovsky A, Borchiellini C, Renard E (2017) The conservation of the germline multipotency program, from sponges to vertebrates: a stepping stone to understanding the somatic and germline origins. Genome Biol Evol 9:474–488

Frydman HM, Spradling AC (2001) The receptor-like tyrosine phosphatase lar is required for epithelial planar polarity and for axis determination within drosophila ovarian follicles. Development 128:3209–3220. https://doi.org/10.1242/dev.128.16.3209

Ganot P, Bouquet JM, Kallesøe T, Thompson EM (2007) The Oikopleura coenocyst, a unique chordate germ cell permitting rapid, extensive modulation of oocyte production. Dev Biol 302(2):591–600. https://doi.org/10.1016/j.ydbio.2006.10.021

Gartner A, Boag PR, Blackwell TK (2008) Germline survival and apoptosis. WormBook Sep 4:1–20. https://doi.org/10.1895/wormbook.1.145.1

Gottanka J, Büning J (1990) Oocytes develop from interconnected cystocytes in the panoistic ovary of Nemoura sp. (Pictet) (Plecoptera: Nemouridae). Int J Insect Morphol Embryol 19:219–225

Gottanka J, Büning J (1993) Mayflies (Ephemeroptera), the most "primitive" winged insects, have telotrophic meroistic ovaries. Rouxs Arch Dev Biol 203:18–27

Hanna RE, Moffett D, Forster FI, Trudgett AG, Brennan GP, Fairweather I (2016) Fasciola hepatica: a light and electron microscope study of the ovary and of the development of oocytes within eggs in the uterus provides an insight into reproductive strategy. Vet Parasitol 221:93–103. https://doi.org/10.1016/j.vetpar.2016.03.011

Hill RS (1977) Studies on the ovotestis of the slug agriolimax reticulatus (Müller). 2. The epithelia. Cell Tissue Res 183(1):131–141. https://doi.org/10.1007/BF00219997

Hinnant TD, Merkle JA, Ables ET (2020) Coordinating proliferation, polarity, and cell fate in the *Drosophila* female germline. Front Cell Dev Biol 8:19

Issigonis M, Newmark PA (2019) From worm to germ: germ cell development and regeneration in planarians. Curr Top Dev Biol 135:127–153. https://doi.org/10.1016/bs.ctdb.2019.04.001

Issigonis M, Redkar AB, Rozario T, Khan UW, Mejia-Sanchez R, Lapan SW, Reddien PW, Newmark PA (2022) A Krüppel-like factor is required for development and regeneration of germline and yolk cells from somatic stem cells in planarians. PLoS Biol 20:e3001472. https://doi.org/10.1371/journal.pbio.3001472

Karpen GH, Spradling AC (1992) Analysis of subtelomeric heterochromatin in the Drosophila minichromosome Dp1187 by single P element insertional mutagenesis. Genetics 132:737–753. https://doi.org/10.1093/genetics/132.3.737

Khan UW, Newmark PA (2022) Somatic regulation of female germ cell regeneration and development in planarians. Cell Rep 38:110525. https://doi.org/10.1016/j.celrep.2022.110525

Kimoto C, Nakagawa H, Hasegawa R, Nodono H, Matsumoto M (2021) Co-localization of DrPiwi-1 and DrPiwi-2 in the oogonial cytoplasm is essential for oocyte differentiation in sexualized planarians. Cells Dev 167:203710. https://doi.org/10.1016/j.cdev.2021.203710

King RC (1970) Ovarian development in Drosophila melanogaster. Academic Press

Kloc M (2019) The Rove Beetle Creophilus maxillosus as a model system to study asymmetric division, oocyte specification, and the germ-somatic cell signaling. Results Probl Cell Differ 68:217–230. https://doi.org/10.1007/978-3-030-23459-1_9

Kloc M, Bilinski S, Dougherty MT, Brey EM, Etkin LD (2004) Formation, architecture and polarity of female germline cyst in Xenopus. Dev Biol 266(1):43–61. https://doi.org/10.1016/j.ydbio.2003.10.002. PMID: 14729477

Lavige JM, Lecher P (1982) Mttoses anormales dans les embryons a developpement bloque dans le systeme I-R de dysgenesie hybnde chez Drosophila melanogaster. Brol Cell 44:9–14

Lebo DPV, McCall K (2021) Murder on the ovarian express: a tale of non-autonomous cell death in the Drosophila ovary. Cell 10:1454–1473. https://doi.org/10.3390/cells10061454

Lei L, Spradling AC (2016) Mouse oocytes differentiate through organelle enrichment from sister cyst germ cells. Science 352:95–99

Lieber T, Jeedigunta SP, Palozzi JM, Lehmann R, Hurd TR (2019) Mitochondrial fragmentation drives selective removal of deleterious mtDNA in the germline. Nature 570:380–384

Lilly M, de Cuevas M, Spradling AC (2000) Cyclin A associates with the fusome during germline cyst formation in the Drosophila ovary. Dev Biol 218:53–63

Lim AK, Kai T (2007) Unique germ-line organelle, nuage, functions to repress selfish genetic elements in *Drosophila melanogaster*. Proc Natl Acad Sci 104(16):6714–6719. https://doi.org/10.1073/pnas.0701920104

Lim RS, Anand A, Nishimiya-Fujisawa C, Kobayashi S, Kai T (2014) Analysis of Hydra PIWI proteins and piRNAs uncover early evolutionary origins of the piRNA pathway. Dev Biol 386:237–251. https://doi.org/10.1016/j.ydbio.2013.12.007

Littlefield CL (1985) Germ cells in Hydra oligactis males. I. Isolation of a subpopulation of interstitial cells that is developmentally restricted to sperm production. Dev Biol 112:185–193

Littlefield CL (1991) Cell lineages in Hydra: isolation and characterization of an interstitial stem cell restricted to egg production in Hydra oligactis. Dev Biol 143:378–388

Liu Y, Kossack ME, McFaul ME, Christensen LN, Siebert S, Wyatt SR, Kamei CN, Horst S, Arroyo N, Drummond IA, Juliano CE, Draper BW (2022) Single-cell transcriptome reveals insights into the development and function of the zebrafish ovary. elife 11:e76014. https://doi.org/10.7554/eLife.76014

Lu K, Jensen L, Lei L, Yamashita YM (2017) Stay connected: a germ cell strategy. Trends Genet 33:971–978. https://doi.org/10.1016/j.tig.2017.09.001

Matova N, Cooley L (2001) Comparative aspects of animal oogenesis. Dev Biol 231:291–320

McCaig CM, Lin X, Farrell M, Rehain-Bell K, Shakes DC (2017) Germ cell cysts and simultaneous sperm and oocyte production in a hermaphroditic nematode. Dev Biol 430:362–373. https://doi.org/10.1016/j.ydbio.2017.08.010

Mercer M, Jang S, Ni C, Buszczak M (2021) The dynamic regulation of mRNA translation and ribosome biogenesis during germ cell development and reproductive aging. Front Cell Dev Biol 9:710186

Miller MA, Technau U, Smith KM, Steele RE (2000) Oocyte development in Hydra involves selection from competent precursor cells. Dev Biol 224:326–338

Munck A, David CN (1985) Cell proliferation and differentiation kinetics during spermatogenesis in Hydra carnea. Rouxs Arch Dev Biol 194:247–256

Mytlis A, Kumar V, Qiu T, Deis R, Hart N, Levy K, Masek M, Shawahny A, Ahmad A, Eitan H, Nather F, Adar-Levor S, Birnbaum RY, Elia N, Bachmann-Gagescu R, Roy S, Elkouby YM (2022) Control of meiotic chromosomal bouquet and germ cell morphogenesis by the zygotene cilium. Science 376:eabh3104. https://doi.org/10.1126/science.abh3104

Newmark PA, Sánchez AA (2022) Schmidtea happens: re-establishing the planarian as a model for studying the mechanisms of regeneration. Curr Top Dev Biol 147:307–344. https://doi.org/10.1016/bs.ctdb.2022.01.002

Ninova M, Chen YA, Godneeva B, Rogers AK, Luo Y, Fejes Tóth K, Aravin AA (2020) Su(var)2-10 and the SUMO pathway link piRNA-guided target recognition to chromatin silencing. Mol Cell 77:556–570. https://doi.org/10.1016/j.molcel.2019.11.012

Nishimiya-Fujisawa C, Kobayashi S (2012) Germline stem cells and sex determination in Hydra. Int J Dev Biol 56(6–8):499–508. https://doi.org/10.1387/ijdb.123509cf

Niu W, Spradling AC (2022) Mouse oocytes develop in cysts with the help of nurse cells. Cell 185:2576–2590

Ohlmeyer JT, Schüpbach T (2003) Encore facilitates SCF-Ubiquitin-proteasome-dependent proteolysis. Development 130:6339–6349

Onishi R, Yamanaka S, Siomi MC (2021) piRNA- and siRNA-mediated transcriptional repression in Drosophila, mice, and yeast: new insights and biodiversity. EMBO Rep 22:e53062. https://doi.org/10.15252/embr.202153062

Pang L-Y, DeLuca S, Zhu H, Urban JM, Spradling AC (2023) Chromatin and gene expression changes during female Drosophila germ cell development illuminate the biology of highly potent stem cells. elife, in press. See BioRxiv. https://doi.org/10.1101/2023.05.31.543160

Paré C, Suter B (2000) Subcellular localization of Bic-D::GFP is linked to an asymmetric oocyte nucleus. J Cell Sci 113(Pt 12):2119–2127. https://doi.org/10.1242/jcs.113.12.2119

Pepling ME, de Cuevas M, Spradling AC (1999) Germline cysts: a conserved phase of germ cell development? Trends Cell Biol 9:257–262

Percharde M, Lin CJ, Yin Y, Guan J, Peixoto GA, Bulut-Karslioglu A, Biechele S, Huang B, Shen X, Ramalho-Santos M (2018) A LINE1-nucleolin partnership regulates early development and ESC identity. Cell 174:391–405.e19. https://doi.org/10.1016/j.cell.2018.05.043

Praher D, Zimmermann B, Genikhovich G, Columbus-Shenkar Y, Modepalli V, Aharoni R, Moran Y, Technau U (2017) Characterization of the piRNA pathway during development of the sea anemone Nematostella vectensis. RNA Biol 14:1727–1741

Price, K, Dyuthi L, Tharakan M, Cooley L (2022) Evolutionarily conserved midbody reorganization precedes ring canal formation during gametogenesis. bioRxiv 2022.06.03.494691; https://doi.org/10.1101/2022.06.03.494691

Raiders SA, Eastwood MD, Bacher M, Priess JR (2017) Binucleate germ cells in Caenorhabditis elegans are removed by physiological apoptosis. PLoS Genet 14:e1007417. https://doi.org/10.1371/journal.pgen.1007417

Raz AA, Wurtzel O, Reddien PW (2021) Planarian stem cells specify fate yet retain potency during the cell cycle. Cell Stem Cell 28, 1307–1322.e5. https://doi.org/10.1016/j.stem.2021.03.021

Reddien PW (2018) The cellular and molecular basis for planarian regeneration. Cell 175:327–345

Reddien PW (2022) Positional information and stem cells combine to result in planarian regeneration. Cold Spring Harb Perspect Biol 14:a040717. https://doi.org/10.1101/cshperspect.a040717

Reddien PW, Oviedo NJ, Jennings JR, Jenkin JC, Sánchez Alvarado A (2005) SMEDWI-2 is a PIWI-like protein that regulates planarian stem cells. Science 310:1327–1330

Rozario T, Collins JJ 3rd, Newmark PA (2022) The good, the bad, and the ugly: from planarians to parasites. Curr Top Dev Biol 147:345–373. https://doi.org/10.1016/bs.ctdb.2021.12.015

Russell SL, Chappell L, Sullivan W (2019) A symbiont's guide to the germline. Curr Top Dev Biol 135:315–351. https://doi.org/10.1016/bs.ctdb.2019.04.007

Samaddar M, Goudeau J, Sanchez M, Hall DH, Bohnert KA, Ingaramo M, Kenyon C (2021) A genetic screen identifies new steps in oocyte maturation that enhance proteostasis in the immortal germ lineage. elife 10:e62653. https://doi.org/10.7554/eLife.62653

Scherer G, Tschudi C, Perera J, Delius H, Pirrotta V (1982) B104, a new dispersed repeated gene family in Drosophila melanogaster and its analogies with retroviruses. J Mol Biol 157:435–451. https://doi.org/10.1016/0022-2836(82)90470-3

Schoenmakers HJ, Colenbrander PH, Peute J, van Oordt PG (1981) Anatomy of the ovaries of the starfish Asterias rubens (Echinodermata). A histological and ultrastructural study. Cell Tissue Res 217(3):577–597. https://doi.org/10.1007/BF00219366

Seah MKY, Wang Y, Goy PA, Loh HM, Peh WJ, Low DHP, Han BY, Wong E, Leong EL, Wolf G, Mzoughi S, Wollmann H, Macfarlan TS, Guccione E, Messerschmidt DM (2019) The KRAB-zinc-finger protein ZFP708 mediates epigenetic repression at RMER19B retrotransposons. Development 146:dev170266. https://doi.org/10.1242/dev.170266

Seidel HS, Smith TA, Evans JK, Stamper JQ, Mast TG, Kimble J (2018) *C. elegans* germ cells divide and differentiate in a folded tissue. Dev Biol 442:173–187

Seller CA, Cho CY, O'Farrell PH (2019) Rapid embryonic cell cycles defer the establishment of heterochromatin by Eggless/SetDB1 in *Drosophila*. Genes Dev 33:403–417. https://doi.org/10.1101/gad.321646.118

Seydoux G, Braun RE (2006) Pathway to totipotency: lessons from germ cells. Cell 127:891–904

Siebert S, Farrell JA, Cazet JF, Abeykoon Y, Primack AS, Schnitzler CE, Juliano CE (2019) Stem cell differentiation trajectories in *Hydra* resolved at single-cell resolution. Science 365:eaav9314. https://doi.org/10.1126/science.aav9314

Spradling AC, de Cuevas M, Drummond-Barbosa D, Keyes L, Lilly M, Pepling M, Xie T (1997) The Drosophila germarium: stem cells, germ line cysts, and oocytes. Cold Spring Harb Symp Quant Biol 62:25–34

Spradling AC, Niu W, Yin Q, Pathak M, Maurya B (2022) Conservation of oocyte development in germline cysts from *Drosophila* to mouse. elife 11:e83230. https://doi.org/10.7554/eLife.83230. PMC9708067

Strome S, Lehmann R (2007) Germ versus soma decisions: lessons from flies and worms. Science 316:392–393

Świątek P, Urbisz AZ (2019) Architecture and life history of female germ-line cysts in clitellate annelids. Results Probl Cell Differ 68:515–551

Technau U, Miller MA, Bridge D, Steele RE (2003) Arrested apoptosis of nurse cells during Hydra oogenesis and embryogenesis. Dev Biol 260:191–206. https://doi.org/10.1016/s0012-1606(03)00241-0. PMID: 12885564

Teefy BB, Siebert S, Cazet JF, Lin H, Juliano CE (2020) PIWI–piRNA pathway-mediated transposable element repression in Hydra somatic stem cells. RNA 26:550–563

Telfer WH (1975) Development and physiology of the oocyte-nurse cell syncytium. Adv Insect Physiol 11:223–319

Tworzydlo W, Kisiel E, Jankowska W, Bilinski SM (2014) Morphology and ultrastructure of the germarium in panoistic ovarioles of a basal "apterygotous" insect, Thermobia domestica. Zoology (Jena) 117:200–206

Woodland HR (2016) The birth of animal development: multicellularity and the germline. Curr Top Dev Biol 117:609–630. https://doi.org/10.1016/bs.ctdb.2015.10.020

Zeng A, Li H, Guo L, Gao X, McKinney S, Wang Y, Yu Z, Park J, Semerad C, Ross E, Cheng LC, Davies E, Lei K, Wang W, Perera A, Hall K, Peak A, Box A, Sánchez-Alvarado A (2018) Prospectively isolated tetraspanin^{+} neoblasts are adult pluripotent stem cells underlying planaria regeneration. Cell 173:1593–1608.e20. https://doi.org/10.1016/j.cell.2018.05.006

Chapter 2
Female Germline Cysts in Animals: Evolution and Function

John L. Brubacher

Abstract Germline cysts are syncytia formed by incomplete cytokinesis of mitotic germline precursors (cystoblasts) in which the cystocytes are interconnected by cytoplasmic bridges, permitting the sharing of molecules and organelles. Among animals, such cysts are a nearly universal feature of spermatogenesis and are also often involved in oogenesis. Recent, elegant studies have demonstrated remarkable similarities in the oogenic cysts of mammals and insects, leading to proposals of widespread conservation of these features among animals. Unfortunately, such claims obscure the well-described diversity of female germline cysts in animals and ignore major taxa in which female germline cysts appear to be absent. In this review, I explore the phylogenetic patterns of oogenic cysts in the animal kingdom, with a focus on the hexapods as an informative example of a clade in which such cysts have been lost, regained, and modified in various ways. My aim is to build on the fascinating insights of recent comparative studies, by calling for a more nuanced view of evolutionary conservation. Female germline cysts in the Metazoa are an example of a phenomenon that—though essential for the continuance of many, diverse animal lineages—nevertheless exhibits intriguing patterns of evolutionary innovation, loss, and convergence.

2.1 Introduction

In sexually reproducing organisms, female gametes provide a cytoplasmic environment that is competent to initiate the development of a complex, multicellular organism. Eggs are therefore complicated cells, and the process of making an egg (oogenesis) is a complicated business—one that is necessary for the ongoing

J. L. Brubacher (✉)
Canadian Mennonite University, Winnipeg, MB, Canada
e-mail: jbrubacher@cmu.ca

M. Kloc, A. Uosef (eds.), *Syncytia: Origin, Structure, and Functions*, Results and Problems in Cell Differentiation 71,
https://doi.org/10.1007/978-3-031-37936-9_2

existence of sexually reproducing species. As such, oogenesis has long been a topic of fascination and insight for biologists.

In animals, the premeiotic stages of gametogenesis often include a syncytial stage, or **cyst**, in which germline precursor cells are interconnected by micrometre-scale channels variously known as intercellular bridges, cytoplasmic bridges, or ring canals. Such syncytia were first described in histological studies of the late nineteenth century and confirmed by electron microscopic observations in the 1950s (reviewed by Dym and Fawcett 1971). They are a near-universal feature of animal spermatogenesis (Pepling et al. 1999; Guo and Zheng 2004; White-Cooper and Bausek 2010) and are also widespread in oogenesis, among species separated by hundreds of millions of years of evolution (Giardina 1901; Gondos 1973; Telfer 1975; Pepling and Spradling 1998). Female germline cysts are often described as conserved, universal, or even ubiquitous features of oogenesis (though a few exceptions may be noted) perhaps because they *are* present in most traditional model species commonly studied by developmental biologists (e.g. *Drosophila melanogaster*, *Caenorhabditis elegans*, *Danio rerio*, *Xenopus laevis*, *Mus musculus*).

The degree of conservation of metazoan female germline cysts has clear implications for efforts to elucidate general functions of such cysts. In this review, I survey key groups of animals, to clarify that while cysts are a common theme in animal oogenesis, they are neither ubiquitous nor as conserved as is often assumed.

2.2 Oogenic Cysts in *Drosophila melanogaster*: The Standard Model

The development and structure of oogenic germline cysts were first described at the ultrastructural level in fruit flies (*Drosophila melanogaster*; Koch et al. 1967), and this species remains the most thoroughly studied example of the phenomenon. It is therefore helpful to begin with a summary of cyst development in female fruit flies, to begin introducing core themes and terminology that recur with varying degrees of conservation and innovation in other taxa.

In flies, embryonic primordial germ cells (PGCs) give rise to germline stem cells (GSC) that persist through adulthood at the distal ends of ovarioles: the functional units of each ovary. GSCs divide asymmetrically to self-renew and produce a germline-cyst precursor: a **cystoblast** (Wieschaus and Szabad 1979; Lin and Spradling 1993). Figure 2.1a summarizes the steps by which a cystoblast generates a 16-cell cyst in *D. melanogaster*. Each undergoes a series of synchronous mitotic divisions in which cytokinesis is incomplete, leaving a syncytium of **cystocytes** interconnected by cytoplasmic bridges. The cystogenic mitoses follow a consistent branching geometry and asymmetrically partition a membranous, spectrin-rich organelle, the **fusome**. Asymmetric distribution of the fusome results in a polarized cytoskeletal network that connects all cystocytes (Deng and Lin 1997). The minus ends of microtubules in this network are localized to one of two cystocytes at the

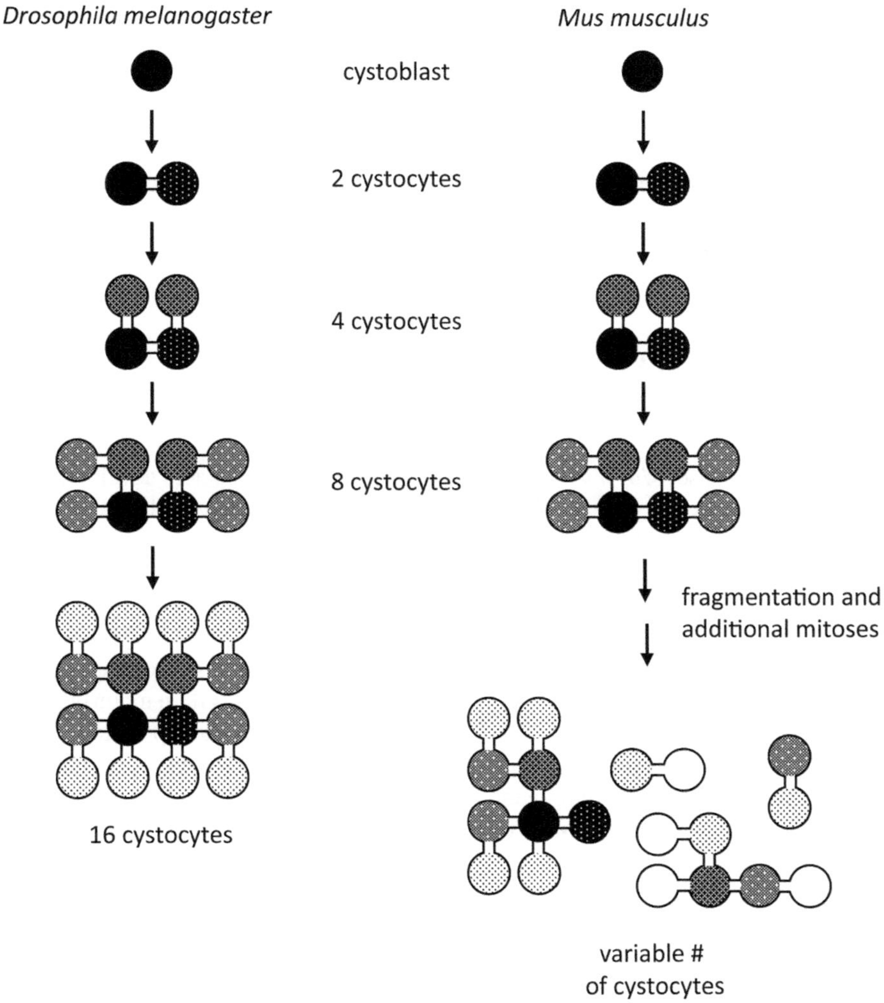

Fig. 2.1 Oogenic cyst formation in fruit flies and house mice. In both species, cystogenic mitoses are initially synchronous, resulting in cysts containing 2^n cystocytes, where n is an integer. (**a**) *Drosophila melanogaster*—Synchrony is maintained in the fruit fly through 4 divisions, resulting in 16-cell cysts with a consistent branching geometry. The cytoskeleton of cysts in *D. melanogaster* is polarized, with microtubule minus ends in a central cell (shaded black) which is the presumptive oocyte. The remaining 15 cells differentiate as nurse cells, become polyploid via incomplete endocycles, and supply the oocyte with macromolecules and organelles. The overall direction of cytoplasmic transport is from lighter-shaded cells to the darker ones. (**b**) In the mouse, cysts begin to fragment after about three cystocyte mitoses, and synchrony of divisions is lost. Within each cyst, the most-central cell accumulates centrosomes, setting up polarized transport of cytoplasmic contents from less-connected (lighter) cystocytes to the most-connected cell in the cyst. The less-connected cells do not become polyploid, but do function as nurse cells, eventually undergoing programmed cell death. Each cyst fragment is thought to produce a single oocyte, which associates with somatic granulosa cells to form a primordial follicle

centre of the branching structure (possessing four intercellular bridges and the largest share of the fusome). This cell is the presumptive oocyte (de Cuevas and Spradling 1998).

The remaining cystocytes differentiate as **nurse cells**, which will undergo genomic amplification via endocycles (Dej and Spradling 1999) and support the early growth and patterning of the oocyte. As the oocyte enters vitellogenesis, nurse cells ultimately undergo a programme of developmental cell death and are scavenged by surrounding phagocytic follicle cells (Lebo and McCall 2021).

2.3 Of Mice and Flies: Apparent Conservation of Female Cysts

In the house mouse, *Mus musculus*, the cyst phase of female gamete development is restricted to late embryogenesis and the perinatal period, between days E10.5 and P5 (Pepling et al. 2007). In key respects, female cyst development in mice is remarkably like that in *D. melanogaster* (Fig. 2.1b). As in flies, murine cyst development begins with a few synchronous, incomplete mitotic divisions, which generate early cysts with up to eight cystocytes (Lei and Spradling 2013). By E12.5, cysts begin to fragment stochastically, and their early mitotic synchrony is lost. Continued asynchronous mitoses combined with death of certain cystocytes result in maturing cysts that do not typically contain a fixed number of cells (Lei and Spradling 2013). Though cyst fragmentation is not a feature of cyst development in flies, other significant aspects of cyst development appear to be well conserved in these two species. As in the fly, mouse cystocytes initially enter meiosis, but then diverge into separate nurse-cell or oocyte fates (Lei and Spradling 2016; Niu and Spradling 2022). Murine nurse cells transfer much of their cytoplasm into presumptive oocytes and then undergo programmed cell death via a molecular mechanism that mirrors that of nurse cells in *D. melanogaster* (Niu and Spradling 2022). Mitochondria, Golgi elements, and centrioles transported from nurse cells, and those endogenous to the presumptive oocyte, coalesce to form a Balbiani body near the oocyte nucleus (Pepling et al. 2007). It is not entirely clear whether functional cytoplasmic connections between cystocytes are *essential* for female fertility in mice, but disrupting cyst formation or stability alters the development of oocytes and impairs fertility (Greenbaum et al. 2009; Soygur et al. 2021; Ikami et al. 2021; Niu and Spradling 2022).

Given such similarities between distantly related animal species, and the appearance of a similar cyst stage in oogenesis of most animal phyla, many authors have proposed that syncytial cysts are an evolutionarily conserved feature of oogenesis. Presumably then, there also must be conserved *functions* for cysts in oocyte development (Pepling et al. 1999; Greenbaum et al. 2011; Lu et al. 2017; Bauer et al. 2021; Spradling et al. 2022).

Certainly, in metazoan *spermatogenesis*, a syncytial cyst stage is universal—or nearly so (Fawcett et al. 1959; Dym and Fawcett 1971; Guo and Zheng 2004; White-Cooper and Bausek 2010). In experimentally tractable species, functional roles for spermatogenic syncytia have been demonstrated, which solve challenges particular to sperm development. One example is the likely benefit of sharing gene products among interconnected spermatocytes and spermatids. In males, spermatogenic cysts persist through and beyond meiosis and therefore permit sharing of gene products between haploid spermatocytes. Such sharing would compensate for the lack of hemizygous genes carried on sex chromosomes in species where males are the heterogametic sex (Braun et al. 1989). More generally, sharing would buffer against stochastic variations in gene expression (Guo and Zheng 2004) and possibly also protect against the production of sperm carrying selfish genes that result in meiotic drive or segregation distortion (LeGrand 1997).

In female animals, however, meiotic progression in oocytes is arrested in prophase I prior to maturation (maintaining diploidy), and cysts do not persist beyond this meiotic arrest. Thus, buffering variable gene expression would be less crucial in oogenesis than in spermatogenesis, and evolutionary conservation of cyst function should therefore derive from different selective pressures. Several functions have been proposed over the years. Perhaps the most obvious is facilitating the resource-intensive growth of oocytes via cytoplasmic transfer from a subset of cystocytes (nurse cells) (Pepling and Spradling 2001; Lei and Spradling 2016). A second hypothesis with experimental support is that cysts play a role in organelle biogenesis and organization. This role also encompasses the selection and transfer of high-quality organelles to presumptive oocytes, with nurse cells (destined for programmed cell death) effectively serving as disposal sites for damaged or dysfunctional organelles and other subcellular structures (Cox and Spradling 2003; Pepling et al. 2007; Lei and Spradling 2016). Third, recent studies have shown that in mice, as in other animals, the development of a cytoskeletal network throughout the cyst is essential for the subsequent organization of mitochondria, Golgi components, and germ granules into a Balbiani body (Niu and Spradling 2022; Spradling et al. 2022). To assess the generality of these hypotheses, a broader view of the conservation and prevalence of female germline cysts are in the animal kingdom is essential. An exhaustive overview would require a book. In what follows, I highlight only a few particularly informative clades.

2.4 Cyst Evolution in Hexapods: A Case Study

2.4.1 Ovariole Diversity in Hexapods

There is more information about female cyst development in hexapods than in any other animal group. Much of the terminology used to describe oogenic cysts arose through studies of these invertebrates. Because our understanding of hexapod phylogeny at the ordinal level is also relatively secure (Misof et al. 2014; Tihelka et al. 2021), this group makes a useful case study to examine the evolutionary history of female germline cysts in animals. As we will see, statements to the effect that female germline cysts are conserved "from *Drosophila* to mice" (Spradling et al. 2022) or "from fruit flies to mammals" (Greenbaum et al. 2011) are difficult to support, as this trait is not even a conserved feature of the female germline among insects themselves. The literature on female germline cysts in hexapods has been thoroughly reviewed in the past (Telfer 1975; Štys and Biliński 1990; Büning 1993, 1994; Heming 2003) and arguably needs to be revisited. While much work has been conducted in the decades since publication of these works, all remain excellent overviews of cyst evolution, whose central insights still largely hold true.

In most insects (though among basal hexapods only in the japygid Diplura), the ovaries are subdivided into functional units called **ovarioles** (Heming 2003). Figure 2.2a provides a schematic view of ovariole structure (top right) and summarizes the broad categories of cysts found in hexapods. Mitotic germline precursors (germline stem cells and/or cystoblasts) and their immediate descendants are found distal to the oviduct, adjacent to the **terminal filament**. This region of the ovariole is known as the **germarium**. As oocytes differentiate and begin vitellogenesis, they move towards the oviduct, are enclosed by an epithelial layer of follicle cells to form **follicles**, and enter the **vitellarium**, where they will complete their development and become chorionated. Brandt (1874) categorized ovarioles as **panoistic** or **meroistic**, depending on whether the germ lineage produces oocytes only or oocytes and nurse cells, respectively. In panoistic ovarioles, each follicle contains an individual oocyte, as no nurse cells are formed in the germarium. If cysts are produced in a panoistic ovariole, these fragment to release oocytes one by one. Meroistic ovarioles were further subdivided into two types (Gross 1901) depending on the organization of nurse cells: **polytrophic**, in which individual cysts (an oocyte plus connected nurse cells) are segregated into follicles as discrete units, or **telotrophic**, in which nurse cells remain in the germarium but stay connected to oocytes via elongated cytoplasmic bridges known as trophic cords (Heming 2003).

Ovariole types reflect the presence or absence, and type, of germline cysts they contain. As such, the terms just described also apply to cyst types. The remainder of Fig. 2.2a schematically illustrates these cyst categories among hexapods. The red-shaded cysts are meroistic, exhibiting linear or branched arrangements, which reflect differences in partitioning of cytoplasmic bridges during cyst-generating mitoses. In many panoistic ovarioles, oocytes are produced directly from germline precursors by complete cytokinesis, and cysts are not formed (blue-shaded cells). This acystic mode of panoism has been termed "primary" panoism, in that it is almost certainly the ancestral condition in insects (Štys and Biliński 1990; see below). Some other panoistic hexapods do produce cysts, however, in which all cystocytes differentiate as oocytes (purple-shaded cyst). Such "secondary" or "neo-" panoism is a derived condition that has independently evolved from meroism in several lineages (Štys and Biliński 1990; Heming 2003).

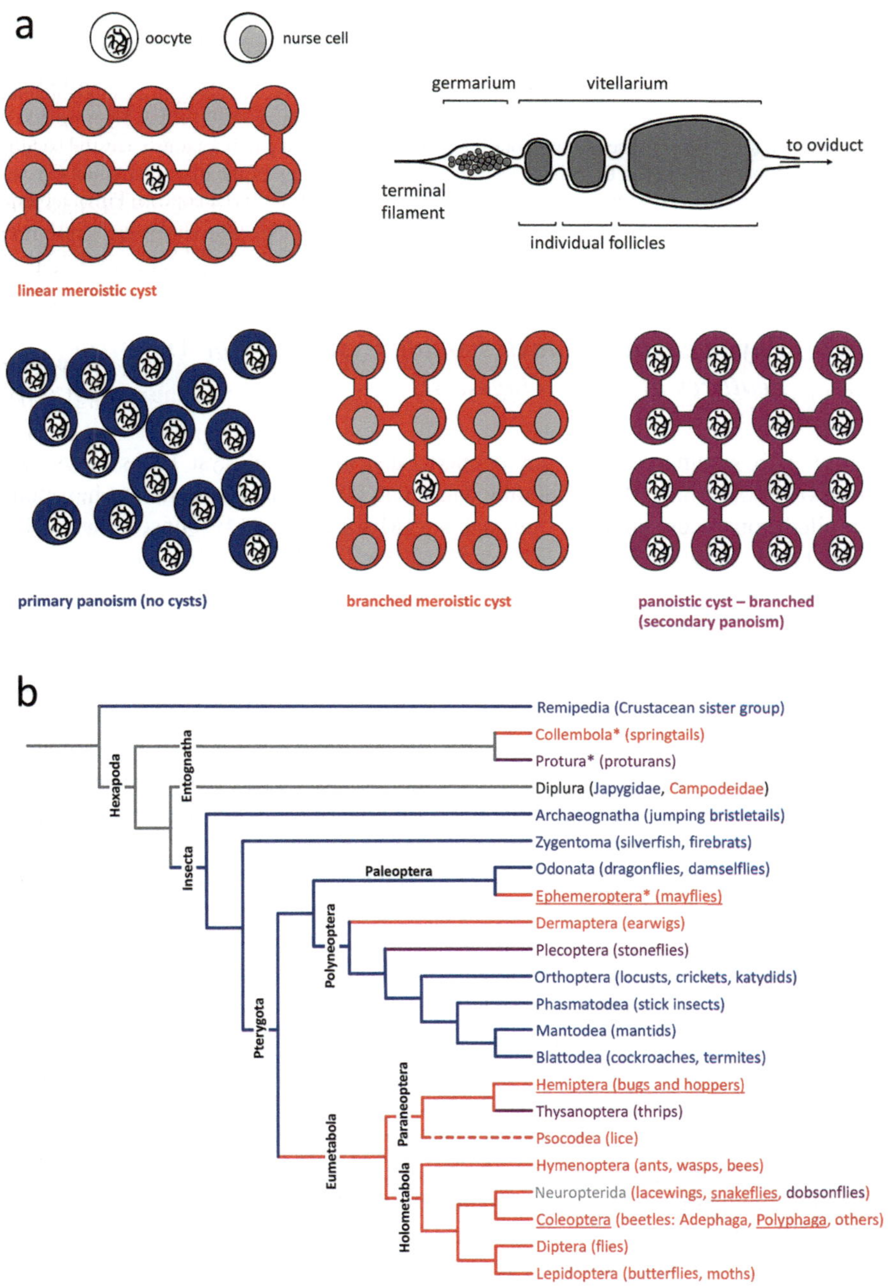

Fig. 2.2 Cyst types and their phylogenetic distribution in the Hexapoda. (**a**, top right) Schematic illustration of an insect ovariole. Grey-shaded areas in follicles represent oocytes for panoistic or telotrophic meroistic ovarioles or cysts for polytrophic meroistic ovarioles. (**a**, top left and bottom) Cyst types, with presumptive oocytes or nurse cells indicated by shading of the nucleus, as indicated in legend at top. Colours of cyst types correspond to panel **b**. (**b**) Cladogram of the hexapods,

(continued)

with Remipedia as an outgroup (Giribet and Edgecombe 2019). Names of major (supraordinal) clades are in black text. Text and branch colours correspond to the most common cyst/ovariole types in a group: blue, primary (acystic) panoistic; plain red, polytrophic meroistic; underlined red, telotrophic meroistic; purple, secondary (cystic) panoistic. Grey text indicates groups in which no cyst type predominates. (References: Štys and Biliński 1990; Büning 1993, 1994; Heming 2003, and see main text). Grey branches and nodes indicate that the most parsimonious ancestral state is unclear. *: Linear rather than branched cyst organization. Topology of the tree is based on Giribet and Edgecombe (2019) (Entognatha, Remipedia); Wipfler et al. (2019) (Paleoptera, Polyneoptera, Paraneoptera); Johnson et al. (2018) (Paraneoptera); Tihelka et al. (2021) (Eumetabola). The Psocodea are grouped with the Paraneoptera by a dashed line to indicate that some analyses (e.g. Johnson et al. 2018) place them as the sister lineage of the Holometabola

2.4.2 Phylogenetic Context: Female Cysts Are Not Ancestral Features of Insect Oogenesis

Figure 2.2b summarizes the phylogenetic distribution of cyst types (Štys and Biliński 1990; Büning 1994; Kubrakiewicz et al. 2012) on a cladogram of hexapods compiled from recent phylogenetic analyses (Misof et al. 2014; Johnson et al. 2018; Wipfler et al. 2019; Giribet and Edgecombe 2019; Tihelka et al. 2021). Remipedia—likely the crustacean sister group to the hexapods (Giribet and Edgecombe 2019)—is included as an outgroup. Branch and text colours in the cladogram correspond to those used to illustrate different cyst types in Fig. 2.2a.

Cyst types in the early-branching hexapods are quite variable: linear (unbranched) meroistic cysts occur in the Collembola and campodeid Diplura (red text, Fig. 2.2b), whereas the Protura have linear panoistic cysts (purple text, Fig. 2.2b), and japygid Diplura exhibit non-cystic (primary) panoism (blue text, Fig. 2.2b). This situation is ambiguous regarding the ancestral mode of oogenesis in the hexapoda overall. Bilinski (1994) proposed that primary (non-cystic) panoism was ancestral to the hexapods, with linear meroism evolving independently in a monophyletic group of Entognathans and subsequent transition to secondary (cystic) panoism in the Protura. This view is somewhat supported by primary panoism in the Remipedia, but the branchiopod crustaceans, which are consistently meroistic (Jaglarz et al. 2014a, b), also have close affinity with the hexapods (Misof et al. 2014; Giribet and Edgecombe 2019). Additionally, current phylogenies propose that the Entognatha are paraphyletic as indicated in Fig. 2.2b (Wipfler et al. 2019; Giribet and Edgecombe 2019) or phylogenetically unresolved (Tihelka et al. 2021), which would imply a more complex story of lineage-specific evolutionary trends. Overall, the plasticity of the existence and type of cysts at the base of the hexapod tree is itself a message about the degree to which these traits are conserved.

Despite the ambiguous situation in basal hexapods, for the Insecta sensu stricto, primary panoism (i.e. acystic oogenesis) is clearly the ancestral state (Büning 1993, 1994; Bilinski 1994; Heming 2003). Among the earlier-branching insects—the Archaeognatha, Zygentoma, Paleoptera, and Polyneoptera—primary (acystic) panoism predominates (Gottanka and Büning 1990) (blue branches and text, Fig. 2.2b). More recent studies have confirmed primary panoistic oogenesis in key lineages (Tworzydlo et al. 2014; Ramos et al. 2020; Sekula et al. 2022). Thus, though cysts are characteristic of the "higher" insects, such as *D. melanogaster* (Diptera), cystic oogenesis in these insects is derived from primary panoism (Fig. 2.2b). Unfortunately, the acystic nature of oogenesis in basally branching insects is obscured in the literature by the fact that, as noted above, the term "panoistic" applies to both cystic and acystic modes of oogenesis, and is often used without further elaboration.

The primary panoistic polyneopteran orders are not as speciose as many eumetabolan orders; thus, one might argue that meroism is intrinsically adaptive—a superior mode of oogenesis. Such a view would be tenuous, however, as acystic oogenesis is used by some of the most successful insects on the planet, in terms of population size and geographic distribution: dragonflies (Odonata), locusts (Orthoptera), cockroaches, and termites (Blattodea). The extant members of these clades represent lineages that have used this mode of oogenesis for some 450 million years, since the initial diversification of the Insecta (Giribet and Edgecombe 2019).

Although there are polyneopteran lineages in which oogenesis includes a cyst stage, these are exceptions rather than the norm. The distinctive characteristics of female cysts in mayflies, earwigs, and stoneflies, respectively, suggest independent evolutionary origins of cysts in each of these orders. Although Fig. 2.2b's summary of the phylogenetic relationships is not universally accepted, plausible alternatives (e.g. paraphyly of the Paleoptera, with Ephemeroptera as the sister group of pterygotes other than Odonata—Sharma 2019; Tihelka et al. 2021) do not alter this conclusion.

In mayflies (Ephemeroptera), oogenic cysts are meroistic, with multiple nurse cells nourishing a single developing oocyte. These cysts are unique to this order, however, in that they are linear (like those of basal hexapods) but have a telotrophic organization—which is otherwise only found in eumetabolan groups (Gottanka and Büning 1993).

Among female earwigs (Dermaptera) examined to date, ovarian follicles contain two-cell meroistic cysts consisting of a single nurse cell and oocyte (Yamauchi and Yoshitake 1982). In "higher" dermapteran families, the two-cell units form directly via a single mitotic division of a progenitor cystoblast (Tworzydło et al. 2010). Basal families exhibit a more complicated process, however, in which eight-cell cysts form via three synchronous mitotic divisions and later fragment into the final two-cell units (Yamauchi and Yoshitake 1982; Tworzydło et al. 2010; Núñez-Pascual et al. 2023). Earlier reviewers differ as to whether meroism arose independently in the Dermaptera (King and Büning 1985) or as a synapomorphy uniting the Dermaptera with the Eumetabola (Štys and Biliński 1990). Given the unique fragmentation of these cysts, and the fact that current molecular phylogenies do not place Dermaptera as a sister group to the Eumetabola (Wipfler et al. 2019; Tihelka et al. 2021), the independent origin of meroism in this lineage seems more likely.

The stoneflies (Plecoptera) produce branched cysts that fragment into smaller clusters, from which individual oocytes detach. Gottanka and Büning (1990) found no evidence of nurse cells in *Nemoura sp.* Heming (2003) suggested that plecopteran panoistic cysts were derived from a meroistic condition, but the original paper and Büning's later reviews (1993, 1994) clearly interpreted them to have arisen independently from the primary panoistic condition. As with the dermapteran cysts then, those of the stoneflies are distinctive enough that it seems most parsimonious to conclude these evolved independently of eumetabolan cysts. The alternative hypothesis, that plecopteran and dermapteran cysts share a common evolutionary origin, would require a secondary "re-loss" of cysts in the remaining polyneopteran orders, as the Plecoptera are a sister to that clade.

2.4.3 Diversity of Cyst Types in Eumetabolan Insects

Meroistic cysts with branching geometry are clearly ancestral to the Eumetabola (Büning and Sohst 1990). Büning (1993) described the "basic type" of such cysts as having similar characteristics to those of *D. melanogaster*—namely, construction from 2^n cystocytes (n = the number of synchronous mitotic divisions that established the cyst) of which one differentiates as an oocyte, while the rest become polyploid nurse cells. While the polytrophic meroistic condition is widespread among the Eumetabola, substantial variations on this theme occur between and within several orders of this group. Neo- or secondary panoism occurs in the Thysanoptera (Pritsch and Büning 1989; Tsutsumi et al. 1995), and also in some families of the Neuroptera, along with fleas (Siphonaptera) and boreid scorpionflies (Mecoptera) (Štys and Biliński 1990; Heming 2003). The latter two orders are not included in Fig. 2.2b.

Telotrophic ovarioles have also evolved from polytrophic ones in several eumetabolan taxa: they are near-universal in the Hemiptera (true bugs, hoppers, and aphids). Among beetles (Coleoptera), the suborder Polyphaga is telotrophic, whereas the Adephaga retain polytrophic ovarioles. Finally, among the closely related orders of the Neuropterida, the Raphidioptera (snakeflies) and alderflies (Sialidae: Megaloptera) possess very similar telotrophic ovarioles, in contrast to the secondarily panoistic (cystic panoistic) dobsonflies and fishflies (Megaloptera: Corydalidae) and the polytrophic meroistic lacewings (Neuroptera). The organization of nurse cells in the germarium differs substantially between the hemipteran, polyphagan, and *Sialis*-type telotrophic ovarioles, most likely reflecting separate evolutionary origins for these types (Büning 1993; Rübsam and Büning 2017). These differences include the mitotic activity and synchrony of nurse cells, the partial or complete union of cysts to form a common mass of nurse cell nuclei, and the ploidy of nurse cells. All are significant modifications of the "basic" meroistic cyst plan described above.

The presence of *Sialis*-type telotrophic ovarioles in the Sialidae (Megaloptera) and Raphidioptera is intriguing, as these taxa do not form a monophyletic clade in recent phylogenetic analyses of the Neuropterida (e.g. Song et al. 2018; Vasilikopoulos et al. 2020). Thus, Heming (2003) proposed four separate evolutionary origins of telotrophy in the Eumetabola. Since that time, however, *Sialis*-type ovarioles have also been described in beetles of the family Hydroscaphidae (suborder Myxophaga) (Büning 2005), along with significant similarities between the early stages of ovariole development in the Sialidae and Polyphaga (Trauner and Büning 2007). Such affinities complicate our understanding of the evolutionary story of telotrophy (Rübsam and Büning 2017). The difficulties involved in clearly determining the relative degree of evolutionary conservation, innovation, convergence, and parallelism from the extant patterns of cyst types in the Eumetabola emphasizes the variability of cyst architecture in this group.

In summary, while cystic oogenesis is clearly a conserved aspect of eumetabolan oogenesis, the specifics of female germline cysts in this group differ among and within major taxa. Variations on the theme include reversions to panoism and elaborations of the polytrophic meroistic pattern into a variety of telotrophic types. When we consider cyst evolution in the hexapods as a whole, the picture that emerges is one of evolutionary plasticity, in which cysts are a robust strategy for egg production, but not one that is essential to oogenesis itself or uniform in its manifestations.

2.5 Diversity of Female Germline Cysts in Other Animal Lineages

2.5.1 Overview

It is difficult to gauge the actual prevalence of female germline cysts across the full range of animal diversity. The early stages of oogenesis, in which cysts occur, may be restricted to developmental stages that have not yet been thoroughly investigated in certain animal lineages. Even where cysts are relatively accessible to examination, confirming their presence requires the observation of intercellular bridges—typically by electron microscopy, which remains a laborious approach that is not well suited to surveying large volumes of tissue or large numbers of specimens. As such, one should not be too eager to interpret the absence of evidence for cysts in a given clade as evidence of their absence. The near-universality of male germline cysts, and their remarkable similarity across disparate taxa, means that the cytological machinery and developmental processes required to build cysts in the germline are available in essentially all animals (Fawcett et al. 1959; Dym and Fawcett 1971; Guo and Zheng 2004; White-Cooper and Bausek 2010). Finally, female germline syncytia have been described in many major animal phyla, including non-bilaterians (Eckelbarger and Hodgson 2021; Chaigne and Brunet 2022). Thus, the preponderance of evidence suggests that oogenic cysts have a long evolutionary history in animals; indeed, the lack of cysts at the root of the insect tree is generally understood to indicate a secondary loss of cysts from an earlier ancestral arthropod lineage (Büning 1993—see in particular the caption for Fig. 5).

However, as the above discussion of hexapods should demonstrate, finding female cysts in a few species of a lineage does not necessarily mean that such cysts are a conserved feature of the group in question, let alone animals overall. Thus, we need to maintain a broad perspective on this phenomenon. A comprehensive overview of female cysts across animal phyla is well beyond the scope of this review, but a few comments on taxa of interest are in order. The pattern of variability evident in hexapods extends to the metazoans overall.

2.5.2 Non-bilaterian Phyla: Sponges and Cnidarians

In their recent overview of invertebrate oogenesis, Eckelbarger and Hodgson (2021) note that while "nurse cells" have been described in various sponge species, in the sponge literature, this term typically refers to cells that are not clonal germline siblings of presumptive oocytes in a meroistic cyst (e.g. Maldonado 2007). For example, in the calcareous sponge *Paraleucilla magna*, oocytes obtain nutrients by phagocytosing vesicles secreted by adjacent follicle-like choanocytes and pinacocytes, which have thus been referred to as nurse cells. However, female germline cysts do not form in this species, though spermatogenic cysts do (Lanna and Klautau 2010). Cytoplasmic bridges have observed to form between oocytes and "nurse cells" in demosponges and presumably play a role in oocyte nourishment. However, these connections appear to arise via transient fusion of non-clonal cells, rather than the formation of a clonal germline syncytium (Kaye 1991; Maldonado and Riesgo 2009). In glass sponges (Hexactinellida) oocytes differentiate from archaeocytes that are already interconnected by cytoplasmic bridges in groups called congeries (Leys and Ereskovsky 2006). The overall pattern for sponges is that female germline cysts—to the extent that they form at all—are of a different sort than those of other Metazoa.

Among some Cnidarians, oogenesis includes syncytial stages produced by fusion of germline cells. For example, in *Hydra vulgaris* (Hydrozoa), presumptive female germ cells differentiate from patches of interstitial cells of the ectodermal layer. Of these germline cells (numbering in the hundreds to thousands per patch), a few dozen receive cytoplasm from the rest via the formation of transient cytoplasmic connections. The sacrificial nurse cells then disconnect, embark on a process of programmed cell death, and are phagocytosed by the growing oocyte. Ultimately, only a handful of oocytes survive. These then fuse to form a single cell, in which one nucleus will persist (Alexandrova et al. 2005). This process shares similarities with meroistic oogenesis in other animals, but the temporary fusions between germ cells are a noteworthy difference. For example, the transience of these intercellular connections would seem to preclude the formation of a stable cytoskeletal network throughout the interconnected cells, which has been proposed as fundamental to the function of meroistic cysts (see introduction in Niu and Spradling 2022).

The process of syncytial oogenesis observed in *Hydra* (Hydrozoa) is not necessarily widespread among cnidarians. Early records of "nurse cells" exist for a variety of species in the Scyphozoa, but more recent ultrastructural work has demonstrated that these are not classical nurse cells, in the sense of being germline cells that connect to oocytes via cytoplasmic bridges (Eckelbarger 1994; Eckelbarger and Hodgson 2021). The situation is largely similar in the Anthozoa. For example, in the emerging cnidarian model species, the starlet sea anemone *Nematostella vectensis*, vitellogenic oocytes arise and begin growth asynchronously in the gastrodermis. They enter vitellogenesis individually, bulging into the mesoglea but retaining contact with specialized groups of gastrodermal cells (trophocytes) known as trophonemata. Despite the close contact between oocytes and the trophocytes, cytoplasmic connections are not established (Eckelbarger et al. 2008). This mode of oogenesis appears to be widespread in anthozoan corals, although trophonemata have not yet been described in all orders (Shikina and Chang 2016; Lauretta et al. 2018).

2.5.3 Annelids

Aside from the hexapods, female germline cysts are best documented in the segmented worms, the Annelida (earthworms, leeches, polychaetes, and relatives). This literature has expanded rapidly in recent years, particularly for the oligochaetes and leeches (Clitellata). The review in this book series by Świątek and Urbisz (2019) gives a detailed overview of female cysts in clitellates, helpfully examining not only structural features, but also developmental details. As with hexapod cysts, there are common traits to be found in those of annelid clades—especially the Clitellata—but cysts are neither universal nor homogeneous in this phylum.

At least one example of acystic, panoistic oogenesis has been described in the basal clitellate *Capilloventer australis* (Świątek et al. 2016). However, meroistic female cysts are the norm among clitellates, and their organization is quite similar (Świątek and Urbisz 2019; Świątek et al. 2020; Urbisz et al. 2020, 2021; Ahmed et al. 2021). Cyst architecture differs notably from that of meroistic insects, in that all cystocytes connect via a single bridge to a common cylindrical core of shared cytoplasm, the **cytophore**. This arrangement is also found in other phyla, including the nematodes (Gibert et al. 1984; Gumienny et al. 1999), oribatid mites (Liana and Witaliński 2012), and the spoon worms (Echiura; annelid relatives of the clitellates; Leutert 1974). Within a typical clitellate cyst, only one or a few cystocytes differentiate as oocytes, with the rest functioning as nurse cells. Nurse cells do not appear to become polyploid, although their DNA content has been quantified in only *Enchytraeus albidus* (Urbisz et al. 2017). Clitellate cysts have been categorized into seven distinct types, though these retain the basic structure just described and vary mostly in the number of cystocytes per cyst (from 16 to thousands), the branching or linearity of the cytophore, and the synchrony or lack thereof in the cystogenic cell divisions (Świątek and Urbisz 2019).

In contrast to the clitellates, polychaete annelids are more variable in their modes of oogenesis. Female germline cysts are less commonly reported in polychaetes than in clitellates, though detailed ultrastructural investigations of oogenesis remain infrequent in this group (Eckelbarger 2005; Eckelbarger and Hodgson 2021). Meroistic cysts have been reported in several species (Anderson and Huebner 1968;

Heacox and Schroeder 1981; Eckelbarger 1992), of which the dorvilleid *Ophryotrocha labronica* has received the most attention (Emanuelsson 1969; Brubacher and Huebner 2009, 2011). The cysts of *O. labronica* are reminiscent of those found in earwigs (see above), in that they consist of two-cell units (a polyploid nurse cell and an oocyte) produced by fragmentation of a larger parental cyst. These parental cysts differ in significant ways from typical polytrophic meroistic cysts of insects, however, in that cystocyte proliferation is asynchronous, cyst size is indeterminate, and oocytes differentiate from peripheral cystocytes rather than centrally located ones (Brubacher and Huebner 2011). In addition to these scattered examples of meroistic cysts, panoistic cysts have been conclusively documented in more than one species of the Nereidae, including the model polychaete *Platynereis dumerilii* (Fischer 1974, 1975; Dhainaut 1984).

2.5.4 Vertebrates

Cyst development in mice is described above and in Fig. 2.1, and similar cysts have been identified in other mammalian species (reviewed in Gondos 1973; see also Lechowska et al. 2012). Aside from mammals, oogenic cysts have been studied in some detail in the African clawed frog *Xenopus laevis* (Kloc et al. 2004), zebrafish *Danio rerio* (Marlow and Mullins 2008), and the Japanese medaka (*Oryzias latipes*; Nakamura et al. 2010). In these non-mammalian cases, cysts are apparently panoistic. However, mice were also considered to be panoistic until quite recently (e.g. Lechowska et al. 2012), as oocytes and nurse cells in the house mouse are indistinguishable by routine examination. It has taken careful, elegant work to demonstrate that most mouse cystocytes function as nurse cells (Lei and Spradling 2016; Soygur et al. 2021; Niu and Spradling 2022). Mouse nurse cells can be identified by their ultimate fate, which is programmed cell death by a non-apoptotic mechanism, similar to that seen among nurse cells in flies (Lebo and McCall 2021; Niu and Spradling 2022).

Spradling et al. (2022) seem to raise the possibility that female germline cysts in *Xenopus* may be meroistic rather than panoistic, implying that studies of apoptosis in the germline of female frogs (Kloc et al. 2004) may have missed cell death by other mechanisms—and therefore the cryptic presence of nurse cells. Kloc and coworkers used the TUNEL assay to detect programmed cell death, which, in mice, seems not to capture programmed death of nurse cells as effectively as other techniques. However, TUNEL does robustly label the developmental cell death of nurse cells in flies. It is noteworthy that Kloc and coworkers not only found very few $TUNEL^+$ cells in *Xenopus* ovaries but also noted no correlation of TUNEL positivity with particular stages of oogenesis and observed that most TUNEL staining was in somatic cells. In particular, $TUNEL^+$ cells were especially rare in older cysts (Kloc et al. 2004). Taken together, these findings are consistent with a genuine lack of programmed cell death in the *Xenopus* oogenic cysts. Similarly, in three species of frogs of the genus *Rana*, degeneration of germ cells was examined histologically in all stages of oogenesis and found to be quite rare (Ogielska et al. 2010).

While Nakamura et al. (2010) noted cell death among cystocytes of the medaka on the basis of condensed (DAPI-stained) chromatin, their study did not quantify this phenomenon or note its correlation with particular stages of cyst development. Thus, it is difficult to assess whether cysts in this species are panoistic or meroistic. Elkouby and Mullins (2017) found cystocyte death to be relatively scarce in zebrafish ovaries and concluded that most likely, cysts in these fish are more similar to (panoistic) *Xenopus* cysts than to those of mice. In certain sturgeons, however, there is some evidence to suggest transfer of cytoplasmic materials among cystocytes, and cystocyte death seems to be relatively frequent—though such assessments have been qualitative. As such, at least some authors have concluded that cysts in these basal, ray-finned fish are meroistic (Żelazowska and Fopp-Bayat 2017).

Overall, female germline cysts are widespread in vertebrates. The technical difficulties involved in studying cyst development in vertebrates (Spradling et al. 2022) and the cytological similarity of presumptive oocytes and nurse cells—even in authentically meroistic species like mice—make it difficult to draw conclusions about the relative abundance of meroism and panoism. It does, however, appear that both types of cysts occur in this phylum, which has implications for the conservation of cyst function, to be discussed below.

2.5.5 *Major Taxa in Which Female Cysts Have Not Been Described*

Notwithstanding the inherent difficulty of proving the *absence* of a phenomenon, there are notable clades of animals in which female germline cysts have not been described, despite substantial research interest in oogenesis (Chaigne and Brunet 2022). These include the acoels (a basal bilaterian order; Laumer et al. 2019) and the diverse, highly successful molluscs. Although Chaigne and Brunet list the Platyhelminthes as a phylum in which female germline cysts *do* occur, this appears to be a case of mistaken identity; the work cited to support that conclusion is a description of meroistic cysts in the polychaete annelid, *Diopatra cuprea* (Anderson and Huebner 1968; Huebner and Anderson 1976). Ironically, these authors list the Platyhelminthes as a phylum in which the existence of spermatogenic cysts is uncertain, though male cysts are well documented in planarians (Franquinet and Lender 1972, 1973; Farnesi et al. 1977; Issigonis and Newmark 2019).

Female germline cysts have not (yet) been described in flatworms despite a substantial amount of ultrastructural investigation (reviewed by Gremigni 1997; Eckelbarger and Hodgson 2021). Beyond the lack of direct visual confirmation of cysts, in a quantitative investigation of the early differentiation of female germ and yolk cells in the planarian *Schmidtea mediterranea*, Issigonis et al. (2022) found no evidence of synchronized mitoses in the ovaries. Conversely, mitotic synchrony was commonplace in the testes, where cysts are known to form. The lack of synchrony in the female germline, though not definitive, is consistent with the absence of female germline cysts in this species.

Chaigne and Brunet also include echinoderms (sea stars, sea urchins, and relatives) as a phylum with female germline cysts, citing Chia (1968) to support this conclusion. While Chia did describe cytoplasmic bridges occurring between vitellogenic oocytes and somatic follicle cells in the sea star *Leptasterias hexactis*, these are not clonal germline cysts, nor does their proposed function of transferring yolk granules match that of "authentic" female cysts. In their extensive survey of invertebrate oogenesis, Eckelbarger and Hodgson (2021) did not note the presence of cysts or intercellular bridges in the Echinodermata. As with other phyla, use of the term "nurse cell" in the older literature on echinoderm oogenesis does not refer to germline cells that are connected to oocytes by cytoplasmic bridges. Careful electron microscopists, who would have been well acquainted with the phenomenon of germline cysts, did not note their presence in studies of other echinoderms (Anderson 1974).

2.6 Concluding Thoughts: The Function(s) of Female Germline Cysts

There is certainly a sense in which female germline cysts are a "conserved" aspect of animal oogenesis, in that they are a recurrent trait in many metazoan lineages. However, the occurrence of this theme in popular model species obscures the significance of acystic oogenesis in the animal kingdom. Furthermore, oogenic cysts are a plastic trait that has been lost and regained in more than one branch of the animal tree. Among other implications, the apparent loss of female germline cysts at the base of the Insecta, and their re-appearance in Eumetabola, means that the similarities between cysts of flies and mice cannot be traced back to a common ancestor without interruption (they are not synapomorphies). Instead, these similarities are the result of evolutionary convergence—or perhaps large-scale parallelism if we take the conserved programme of spermatogenic cyst development as the "raw material" from which female cyst types converged (Abouheif 2008; Pearce 2012).

Convergence suggests that there may be intrinsic selective advantages to particular cyst types in particular situations. Certainly, functional studies of meroistic cysts in flies and mice underline the importance of cysts to female fertility in these species and more broadly. Nevertheless—because convergent or parallel evolution necessarily involves the evolution of similarity from *difference*—the functional roles of cysts in particular lineages cannot be universal.

Above, I noted three hypotheses that have been proposed to explain why cysts are prevalent in animal oogenesis (Niu and Spradling 2022; Spradling et al. 2022):

1. That female cysts support oocyte growth, via cytoplasmic transfer from nurse cells
2. That female cysts facilitate the selection and organization of high-quality organelles
3. That female cysts establish a cyst-wide cytoskeletal network, which plays an essential role in patterning developmental axes in the oocyte

The distinction between meroistic and panoistic female cysts is crucial in assessing the generalizability of the first two of these hypotheses: each depends on the supportive behaviour of nurse cells in oogenesis, and therefore they apply only to meroistic cysts (Lu et al. 2017; Peterson and Fox 2021). The first hypothesis requires the biosynthetic activity of nurse cells. This role is almost self-evident in meroistic cysts, particularly in cases where nurse-cell polyploidy or expansion of the nurse cell population (as in a telotrophic ovariole) increases the output of the nurse cells. However, this "provisioning hypothesis" can be extended beyond the supply of nutrients and organelles to include protective molecules, such as small RNAs that inhibit the activity of transposable elements (see discussion in Niu and Spradling 2022). The second hypothesis requires the death of nurse cells, which makes them a convenient site to sequester both potentially hazardous metabolic activity (Tworzydlo et al. 2020; Urbisz et al. 2022) and damaged cellular materials—such as defective mitochondria (Lieber et al. 2019; Chen et al. 2020).

Panoistic cysts, in which all cystocytes will become viable oocytes, have no nurse cells to provide such services. Thus, though the first two hypotheses may point towards widespread, real benefits of meroistic cysts, they do not do so for cysts *in general*. Furthermore, alternatives to germline cysts have evolved that accomplish these functions in acystic lineages. For example, parasitic platyhelminthes such as tapeworms (an acystic group) have become prodigious egg producers by outsourcing yolk production to a separate organ: the vitellarium (Scholz et al. 2009; Egger et al. 2015). In the acystic, panoistic firebrats (*Thermobia domestica*), oocytes autonomously produce functional Balbiani bodies, from which dysfunctional mitochondria are eliminated (Tworzydlo et al. 2016).

The third hypothesis—that cysts facilitate axis patterning in oocytes—does apply to both panoistic and meroistic cysts. In branching cysts of the type found in flies, mice, and *Xenopus*, regardless of meroism or panoism, cystocyte divisions are asymmetric in their partitioning of pre-existing cytoplasmic bridges and associated cytoskeletal elements, such as the fusome (de Cuevas and Spradling 1998; Kloc et al. 2008; Bilinski et al. 2017; Niu and Spradling 2022). In each *Xenopus* cystocyte, the Balbiani body is asymmetrically located on the side of the cell facing the cytoplasmic bridge that formed in the mitotic division that gave rise to that cell. This asymmetry coincides with the future animal-vegetal axis of the oocyte and embryo (Kloc et al. 2004; Bilinski et al. 2017). Again, however, cysts are not essential for the establishment of developmental axes. Animals with acystic oogenesis also produce polarized oocytes, though it is not yet known whether this initial asymmetry prefigures embryonic axes (Tworzydlo et al. 2017). Furthermore, other animals—even some with oogenic cysts—do not specify embryonic axes until after fertilization, making patterning of the oocyte unnecessary. In *Caenorhabditis elegans* hermaphrodites, for example, oocytes develop in a cryptically meroistic cyst similar to those of clitellate annelids, as noted above (Gibert et al. 1984; Gumienny et al. 1999). However, specification of the embryonic axes is not prefigured in the oocytes of this species, but is instead determined by the location of the centrosome delivered by the sperm at fertilization (Reich et al. 2019).

Certainly, animals have evolved modes of oogenesis in which female germline cysts play central and essential roles. However, oogenic cysts are not as widespread in the animal kingdom as their male equivalents, which are nearly universal and much more uniform in their organization and development (White-Cooper and Bausek 2010). As such, the "seemingly whimsical occurrence" (Telfer 1975) and variability of female cysts over the animal evolutionary tree may be best explained by different degrees of selective pressure on germline cysts. In the male germline, strong selective pressure preserves the cyst-generating developmental programme, which is then available to be co-opted and modified in the female germline—where selection is evidently more relaxed. Going forward, there is much fascinating work to be done and surely more surprising insights to emerge, as we explore which apparently conserved aspects of oogenesis are non-negotiable for the continuation of a species, which are convergent adaptations to particular life history strategies, and which are the result of evolutionary happenstance.

References

Abouheif E (2008) Parallelism as the pattern and process of mesoevolution. Evol Dev 10:3–5. https://doi.org/10.1111/j.1525-142X.2007.00208.x

Ahmed RB, Urbisz AZ, Świątek P (2021) An ultrastructural study of the ovary cord organization and oogenesis in the amphibian leech *Batracobdella algira* (Annelida, Clitellata, Hirudinida). Protoplasma 258:191–207. https://doi.org/10.1007/s00709-020-01560-7

Alexandrova O, Schade M, Böttger A, David CN (2005) Oogenesis in *Hydra*: nurse cells transfer cytoplasm directly to the growing oocyte. Dev Biol 281:91–101. https://doi.org/10.1016/j.ydbio.2005.02.015

Anderson E (1974) Comparative aspects of the ultrastructure of the female gamete. Int Rev Cytol Suppl 4:1–70

Anderson E, Huebner E (1968) Development of the oocyte and its accessory cells of the polychaete, *Diopatra cuprea* (Bosc). J Morphol 126:163–197. https://doi.org/10.1002/jmor.1051260203

Bauer J, Poupart V, Goupil E et al (2021) The initial expansion of the *C. elegans* syncytial germ line is coupled to incomplete primordial germ cell cytokinesis. Development 148:dev199633. https://doi.org/10.1242/dev.199633

Bilinski S (1994) The ovary of Entognatha. In: Büning J (ed) The insect ovary: ultrastructure, previtellogenic growth, and evolution. Chapman and Hall, London, pp 7–30

Bilinski SM, Kubiak JZ, Kloc M (2017) Asymmetric divisions in oogenesis. In: Tassan J-P, Kubiak JZ (eds) Asymmetric cell division in development, differentiation and cancer. Springer International Publishing, Cham, pp 211–228

Brandt A (1874) Uber die Eirohren der *Blatta orientalis* (Periplaneta). Mém Académie Impériale Sci St-Pétersbourg 21:1–30

Braun RE, Behringer RR, Peschon JJ et al (1989) Genetically haploid spermatids are phenotypically diploid. Nature 337:373–376. https://doi.org/10.1038/337373a0

Brubacher JL, Huebner E (2009) Development of polarized female germline cysts in the polychaete, *Ophryotrocha labronica*. J Morphol 270:413–429. https://doi.org/10.1002/jmor.10687

Brubacher JL, Huebner E (2011) Evolution and development of polarized germ cell cysts: new insights from a polychaete worm, *Ophryotrocha labronica*. Dev Biol 357:96–107. https://doi.org/10.1016/j.ydbio.2011.06.026

Büning J (1993) Germ cell cluster formation in insect ovaries. Int J Insect Morphol Embryol 22:237–253. https://doi.org/10.1016/0020-7322(93)90013-Q

Büning J (1994) The Insect Ovary: ultrastructure, previtellogenic growth, and evolution. Springer, Dordrecht

Büning J (2005) The telotrophic ovary known from Neuropterida exists also in the myxophagan beetle *Hydroscapha natans*. Dev Genes Evol 215:597–607. https://doi.org/10.1007/s00427-005-0017-8

Büning J, Sohst S (1990) Ultrastructure and cluster formation in ovaries of bark lice, *Peripsocus phaeopterus* (Stephens) and *Stenopsocus stigmaticus* (Imhof and Labram) (Insecta: Psocoptera). Int J Insect Morphol Embryol 19:227–241. https://doi.org/10.1016/0020-7322(90)90009-E

Chaigne A, Brunet T (2022) Incomplete abscission and cytoplasmic bridges in the evolution of eukaryotic multicellularity. Curr Biol 32:R385–R397. https://doi.org/10.1016/j.cub.2022.03.021

Chen Z, Wang Z-H, Zhang G et al (2020) Mitochondrial DNA segregation and replication restrict the transmission of detrimental mutation. J Cell Biol 219:e201905160. https://doi.org/10.1083/jcb.201905160

Chia F (1968) Some observations on the development and cyclic changes of the oöcytes in a brooding starfish, *Leptasterias hexactis*. J Zool 154:453–461. https://doi.org/10.1111/j.1469-7998.1968.tb01677.x

Cox RT, Spradling AC (2003) A Balbiani body and the fusome mediate mitochondrial inheritance during *Drosophila* oogenesis. Development 130:1579–1590. https://doi.org/10.1242/dev.00365

de Cuevas M, Spradling AC (1998) Morphogenesis of the *Drosophila* fusome and its implications for oocyte specification. Development 125:2781–2789. https://doi.org/10.1242/dev.125.15.2781

Dej KJ, Spradling AC (1999) The endocycle controls nurse cell polytene chromosome structure during *Drosophila* oogenesis. Development 126:293–303. https://doi.org/10.1242/dev.126.2.293

Deng W, Lin H (1997) Spectrosomes and fusomes anchor mitotic spindles during asymmetric germ cell divisions and facilitate the formation of a polarized microtubule array for oocyte specification in *Drosophila*. Dev Biol 189:79–94. https://doi.org/10.1006/dbio.1997.8669

Dhainaut A (1984) Oogenesis in polychaetes: ultrastructural differentiation and metabolism of nereid oocytes. Fortschr Zool 29:183–205

Dym M, Fawcett DW (1971) Further observations on the numbers of spermatogonia, spermatocytes, and spermatids connected by intercellular bridges in the mammalian testis. Biol Reprod 4:195–215. https://doi.org/10.1093/biolreprod/4.2.195

Eckelbarger KJ (1992) Polychaeta oogenesis. In: Harrison FW, Gardiner FW (eds) Microscopic anatomy of invertebrates, Annelida, vol 7. Wiley, New York, pp 109–127

Eckelbarger KJ (1994) Oocyte nutrition in the lower Metazoa: the Scyphozoa. In: Wilson WH, Stricker SA, Shinn GL (eds) Reproduction and development of marine invertebrates. Johns Hopkins University Press, Baltimore, pp 15–28

Eckelbarger KJ (2005) Oogenesis and oocytes. Hydrobiologia 535–536:179–198. https://doi.org/10.1007/s10750-004-4397-y

Eckelbarger KJ, Hodgson AN (2021) Invertebrate oogenesis – a review and synthesis: comparative ovarian morphology, accessory cell function and the origins of yolk precursors. Invertebr Reprod Dev 65:71–140. https://doi.org/10.1080/07924259.2021.1927861

Eckelbarger KJ, Hand C, Uhlinger KR (2008) Ultrastructural features of the trophonema and oogenesis in the starlet sea anemone, *Nematostella vectensis* (Edwardsiidae). Invertebr Biol 127:381–395. https://doi.org/10.1111/j.1744-7410.2008.00146.x

Egger B, Lapraz F, Tomiczek B et al (2015) A transcriptomic-phylogenomic analysis of the evolutionary relationships of flatworms. Curr Biol 25:1347–1353. https://doi.org/10.1016/j.cub.2015.03.034

Elkouby YM, Mullins MC (2017) Coordination of cellular differentiation, polarity, mitosis and meiosis – new findings from early vertebrate oogenesis. Dev Biol 430:275–287. https://doi.org/10.1016/j.ydbio.2017.06.029

Emanuelsson H (1969) Electronmicroscopical observations on yolk and yolk formation in *Ophryotrocha labronica* LaGreca and Bacci. Z Zellforsch Mikrosk Anat 95:19–36. https://doi.org/10.1007/BF00319266

Farnesi RM, Marinelli M, Tei S, Vagnetti D (1977) Ultrastructural research on the spermatogenesis in *Dugesia lugubris* S.L. Riv Biol 70:113–136

Fawcett DW, Ito S, Slautterback D (1959) The occurrence of intercellular bridges in groups of cells exhibiting synchronous differentiation. J Biophys Biochem Cytol 5:453–460. https://doi.org/10.1083/jcb.5.3.453

Fischer A (1974) Stages and stage distribution in early oogenesis in the Annelid, *Platynereis dumerilii*. Cell Tissue Res 156:35–45. https://doi.org/10.1007/BF00220100

Fischer A (1975) The structure of symplasmic early oocytes and their enveloping sheath cells in the polychaete, *Platynereis dumerilii*. Cell Tissue Res 160:327–343. https://doi.org/10.1007/BF00222043

Franquinet R, Lender T (1972) Quelques aspects ultrastructuraux de la spermiogenèse chez *Polycelis tenuis* et *Polycelis nigra* (planaires). Z Für Mikrosk-Anat Forsch 86:481–495

Franquinet R, Lender T (1973) Étude ultrastructurale des testicules de *Polycelis tenuis* et *Polycelis nigra* (Planaries). Evolution des cellules germinales mâles avant la spermiogenèse. Z Für Mikrosk-Anat Forsch 87:4–22

Giardina A (1901) Origine dell'oocite e delle cellule nutrici nel *Dytiscus*. Int Monatsschrift Für Anat Physiol 18:417–484

Gibert MA, Starck J, Beguet B (1984) Role of the gonad cytoplasmic core during oogenesis of the nematode *Caenorhabditis elegans*. Biol Cell 50:77–85. https://doi.org/10.1111/j.1768-322x.1984.tb00254.x

Giribet G, Edgecombe GD (2019) The phylogeny and evolutionary history of arthropods. Curr Biol 29:R592–R602. https://doi.org/10.1016/j.cub.2019.04.057

Gondos B (1973) Intercellular bridges and mammalian germ cell differentiation. Differentiation 1:177–182. https://doi.org/10.1111/j.1432-0436.1973.tb00112.x

Gottanka J, Büning J (1990) Oocytes develop from interconnected cystocytes in the panoistic ovary of *Nemoura sp.* (Pictet) (Plecoptera: Nemouridae). Int J Insect Morphol Embryol 19:219–225. https://doi.org/10.1016/0020-7322(90)90008-D

Gottanka J, Büning J (1993) Mayflies (ephemeroptera), the most "primitive" winged insects, have telotrophic meroistic ovaries. Rouxs Arch Dev Biol 203:18–27. https://doi.org/10.1007/BF00539886

Greenbaum MP, Iwamori N, Agno JE, Matzuk MM (2009) Mouse TEX14 is required for embryonic germ cell intercellular bridges but not female fertility. Biol Reprod 80:449–457. https://doi.org/10.1095/biolreprod.108.070649

Greenbaum MP, Iwamori T, Buchold GM, Matzuk MM (2011) Germ cell intercellular bridges. Cold Spring Harb Perspect Biol 3:a005850. https://doi.org/10.1101/cshperspect.a005850

Gremigni V (1997) The evolution of the female gonad in Platyhelminthes-Turbellaria: ultrastructural investigations. Invertebr Reprod Dev 31:325–330. https://doi.org/10.1080/07924259.1997.9672592

Gross J (1901) Untersuchungen über das Ovarium der Hemipteren, zugleich ein Beitrag zur Amitosenfrage. Z Für Wiss Zool 69:139–201

Gumienny TL, Lambie E, Hartwieg E et al (1999) Genetic control of programmed cell death in the *Caenorhabditis elegans* hermaphrodite germline. Development 126:1011–1022. https://doi.org/10.1242/dev.126.5.1011

Guo G-Q, Zheng G-C (2004) Hypotheses for the functions of intercellular bridges in male germ cell development and its cellular mechanisms. J Theor Biol 229:139–146. https://doi.org/10.1016/j.jtbi.2004.03.010

Heacox AE, Schroeder PC (1981) A light- and electron-microscopic investigation of gametogenesis in *Typosyllis pulchra* (Berkeley and Berkeley) (Polychaeta: Syllidae): II. Oogenesis. Cell Tissue Res 218:641–658. https://doi.org/10.1007/BF00210121

Heming BS (2003) Insect development and evolution. Comstock Pub. Associates, Ithaca

Huebner E, Anderson E (1976) Comparative Spiralian oogenesis—structural aspects: an overview. Am Zool 16:315–343. https://doi.org/10.1093/icb/16.3.315

Ikami K, Nuzhat N, Abbott H et al (2021) Altered germline cyst formation and oogenesis in *Tex14* mutant mice. Biol Open 10:bio058807. https://doi.org/10.1242/bio.058807

Issigonis M, Newmark PA (2019) From worm to germ: germ cell development and regeneration in planarians. Curr Topics Dev Biol 135:127–153. https://doi.org/10.1016/bs.ctdb.2019.04.001

Issigonis M, Redkar AB, Rozario T et al (2022) A Krüppel-like factor is required for development and regeneration of germline and yolk cells from somatic stem cells in planarians. PLoS Biol 20:e3001472. https://doi.org/10.1371/journal.pbio.3001472

Jaglarz MK, Kubrakiewicz J, Bilinski SM (2014a) The ovary structure and oogenesis in the basal crustaceans and hexapods. Possible phylogenetic significance. Arthropod Struct Dev 43:349–360. https://doi.org/10.1016/j.asd.2014.05.003

Jaglarz MK, Kubrakiewicz J, Jędrzejowska I et al (2014b) Ultrastructural analysis of the ovary and oogenesis in Spinicaudata and Laevicaudata (Branchiopoda) and its phylogenetic implications. Zoology 117:207–215. https://doi.org/10.1016/j.zool.2013.12.002

Johnson KP, Dietrich CH, Friedrich F et al (2018) Phylogenomics and the evolution of hemipteroid insects. Proc Natl Acad Sci USA 115:12775–12780. https://doi.org/10.1073/pnas.1815820115

Kaye HR (1991) Sexual reproduction in four Caribbean commercial sponges. II. Oogenesis and transfer of bacterial symbionts. Invertebr Reprod Dev 19:13–24. https://doi.org/10.1080/07924259.1991.9672152

King RC, Büning J (1985) The origin and functioning of insect oocytes and nurse cells. In: Kerkut GA, Gilbert LI (eds) Comprehensive insect physiology, biochemistry and pharmacology 1: embryogenesis and reproduction. Pergamon Press, Oxford, pp 37–82

Kloc M, Bilinski S, Dougherty MT et al (2004) Formation, architecture and polarity of female germline cyst in *Xenopus*. Dev Biol 266:43–61. https://doi.org/10.1016/j.ydbio.2003.10.002

Kloc M, Jaglarz M, Dougherty M et al (2008) Mouse early oocytes are transiently polar: three-dimensional and ultrastructural analysis. Exp Cell Res 314:3245–3254. https://doi.org/10.1016/j.yexcr.2008.07.007

Koch EA, Smith PA, King RC (1967) The division and differentiation of *Drosophila* cystocytes. J Morphol 121:55–70. https://doi.org/10.1002/jmor.1051210106

Kubrakiewicz J, Jaglarz MK, Iliffe TM et al (2012) Ovary structure and early oogenesis in the remipede, *Godzilliognomus frondosus* (Crustacea, Remipedia): phylogenetic implications. Zoology 115:261–269. https://doi.org/10.1016/j.zool.2012.01.001

Lanna E, Klautau M (2010) Oogenesis and spermatogenesis in *Paraleucilla magna* (Porifera, Calcarea). Zoomorphology 129:249–261. https://doi.org/10.1007/s00435-010-0117-5

Laumer CE, Fernández R, Lemer S et al (2019) Revisiting metazoan phylogeny with genomic sampling of all phyla. Proc R Soc B Biol Sci 286:20190831. https://doi.org/10.1098/rspb.2019.0831

Lauretta D, Wagner D, Penchaszadeh PE (2018) First record of a trophonema in black corals (Cnidaria: Antipatharia). Coral Reefs 37:581–584. https://doi.org/10.1007/s00338-018-1682-1

Lebo DPV, McCall K (2021) Murder on the ovarian express: a tale of non-autonomous cell death in the *Drosophila* ovary. Cells 10:1454. https://doi.org/10.3390/cells10061454

Lechowska A, Bilinski SM, Rasweiler JJ et al (2012) Early oogenesis in the short-tailed fruit bat *Carollia perspicillata*: transient germ cell cysts and noncanonical intercellular bridges. Genes N Y N 50:18–27. https://doi.org/10.1002/dvg.20780

LeGrand EK (1997) An adaptationist view of apoptosis. Q Rev Biol 72:135–147. https://doi.org/10.1086/419763

Lei L, Spradling AC (2013) Mouse primordial germ cells produce cysts that partially fragment prior to meiosis. Development 140:2075–2081. https://doi.org/10.1242/dev.093864

Lei L, Spradling AC (2016) Mouse oocytes differentiate through organelle enrichment from sister cyst germ cells. Science 352:95–99. https://doi.org/10.1126/science.aad2156

Leutert VR (1974) Zur Geschlechtsbestimmung und Gametogenese von *Bonellia viridis* Rolando. Development 32:169–193. https://doi.org/10.1242/dev.32.1.169

Leys SP, Ereskovsky AV (2006) Embryogenesis and larval differentiation in sponges. Can J Zool 84:262–287. https://doi.org/10.1139/z05-170

Liana M, Witaliński W (2012) Female and male reproductive systems in the oribatid mite *Hermannia gibba* (Koch, 1839) (Oribatida: Desmonomata). Int J Acarol 38:648–663. https://doi.org/10.1080/01647954.2012.734333

Lieber T, Jeedigunta SP, Palozzi JM et al (2019) Mitochondrial fragmentation drives selective removal of deleterious mtDNA in the germline. Nature 570:380–384. https://doi.org/10.1038/s41586-019-1213-4

Lin H, Spradling AC (1993) Germline stem cell division and egg chamber development in transplanted *Drosophila* germaria. Dev Biol 159:140–152. https://doi.org/10.1006/dbio.1993.1228

Lu K, Jensen L, Lei L, Yamashita YM (2017) Stay connected: a germ cell strategy. Trends Genet 33:971–978. https://doi.org/10.1016/j.tig.2017.09.001

Maldonado M (2007) Intergenerational transmission of symbiotic bacteria in oviparous and viviparous demosponges, with emphasis on intracytoplasmically-compartmented bacterial types. J Mar Biol Assoc U K 87:1701–1713. https://doi.org/10.1017/S0025315407058080

Maldonado M, Riesgo A (2009) Gametogenesis, embryogenesis, and larval features of the oviparous sponge *Petrosia ficiformis* (Haplosclerida, Demospongiae). Mar Biol 156:2181–2197. https://doi.org/10.1007/s00227-009-1248-4

Marlow FL, Mullins MC (2008) Bucky ball functions in Balbiani body assembly and animal–vegetal polarity in the oocyte and follicle cell layer in zebrafish. Dev Biol 321:40–50. https://doi.org/10.1016/j.ydbio.2008.05.557

Misof B, Liu S, Meusemann K et al (2014) Phylogenomics resolves the timing and pattern of insect evolution. Science 346:763–767. https://doi.org/10.1126/science.1257570

Nakamura S, Kobayashi K, Nishimura T et al (2010) Identification of germline stem cells in the ovary of the teleost medaka. Science 328:1561–1563. https://doi.org/10.1126/science.1185473

Niu W, Spradling AC (2022) Mouse oocytes develop in cysts with the help of nurse cells. Cell 185:2576–2590.e12. https://doi.org/10.1016/j.cell.2022.05.001

Núñez-Pascual V, Calleja F, Pardo RV et al (2023) The ring-legged earwig *Euborellia annulipes* as a new model for oogenesis and development studies in insects. J Exp Zool B Mol Dev Evol 340:18–33. https://doi.org/10.1002/jez.b.23121

Ogielska M, Rozenblut B, Augustyńska R, Kotusz A (2010) Degeneration of germ line cells in amphibian ovary. Acta Zool 91:319–327. https://doi.org/10.1111/j.1463-6395.2009.00411.x

Pearce T (2012) Convergence and parallelism in evolution: a neo-Gouldian account. Br J Philos Sci 63:429–448. https://doi.org/10.1093/bjps/axr046

Pepling ME, Spradling AC (1998) Female mouse germ cells form synchronously dividing cysts. Development 125:3323–3328. https://doi.org/10.1242/dev.125.17.3323

Pepling ME, Spradling AC (2001) Mouse ovarian germ cell cysts undergo programmed breakdown to form primordial follicles. Dev Biol 234:339–351. https://doi.org/10.1006/dbio.2001.0269

Pepling ME, de Cuevas M, Spradling AC (1999) Germline cysts: a conserved phase of germ cell development? Trends Cell Biol 9:257–262. https://doi.org/10.1016/s0962-8924(99)01594-9

Pepling ME, Wilhelm JE, O'Hara AL et al (2007) Mouse oocytes within germ cell cysts and primordial follicles contain a Balbiani body. Proc Natl Acad Sci 104:187–192. https://doi.org/10.1073/pnas.0609923104

Peterson NG, Fox DT (2021) Communal living: the role of polyploidy and syncytia in tissue biology. Chromosom Res 29:245–260. https://doi.org/10.1007/s10577-021-09664-3

Pritsch M, Büning J (1989) Germ cell cluster in the panoistic ovary of Thysanoptera (Insecta). Zoomorphology 108:309–313. https://doi.org/10.1007/BF00312163

Ramos S, Chelemen F, Pagone V et al (2020) Eyes absent in the cockroach panoistic ovaries regulates proliferation and differentiation through ecdysone signalling. Insect Biochem Mol Biol 123:103407. https://doi.org/10.1016/j.ibmb.2020.103407

Reich JD, Hubatsch L, Illukkumbura R et al (2019) Regulated activation of the PAR polarity network ensures a timely and specific response to spatial cues. Curr Biol 29:1911–1923.e5. https://doi.org/10.1016/j.cub.2019.04.058

Rübsam R, Büning J (2017) Germ cell proliferation and cluster behavior in ovarioles of *Sialis flavilatera* (Megaloptera: Sialidae) during larval growth. Arthropod Struct Dev 46:246–264. https://doi.org/10.1016/j.asd.2016.11.010

Scholz T, Garcia HH, Kuchta R, Wicht B (2009) Update on the human broad tapeworm (genus *Diphyllobothrium*), including clinical relevance. Clin Microbiol Rev 22:146–160. https://doi.org/10.1128/CMR.00033-08

Sekula M, Tworzydlo W, Bilinski SM (2022) Morphology and ultrastructure of the Balbiani body in the oocytes of closely related bush cricket species. Shared features reveal important aspect of functioning. Zoology 155:126051. https://doi.org/10.1016/j.zool.2022.126051

Sharma PP (2019) Integrating morphology and phylogenomics supports a terrestrial origin of insect flight. Proc Natl Acad Sci USA 116:2796–2798. https://doi.org/10.1073/pnas.1822087116

Shikina S, Chang C-F (2016) Sexual reproduction in stony corals and insight into the evolution of oogenesis in Cnidaria. In: Goffredo S, Dubinsky Z (eds) The Cnidaria, past, present and future: the world of Medusa and her sisters. Springer International Publishing, Cham, pp 249–268

Song N, Lin A, Zhao X (2018) Insight into higher-level phylogeny of Neuropterida: evidence from secondary structures of mitochondrial rRNA genes and mitogenomic data. PLoS One 13:e0191826. https://doi.org/10.1371/journal.pone.0191826

Soygur B, Jaszczak RG, Fries A et al (2021) Intercellular bridges coordinate the transition from pluripotency to meiosis in mouse fetal oocytes. Sci Adv 7:eabc6747. https://doi.org/10.1126/sciadv.abc6747

Spradling AC, Niu W, Yin Q et al (2022) Conservation of oocyte development in germline cysts from *Drosophila* to mouse. eLife 11:e83230. https://doi.org/10.7554/eLife.83230

Štys P, Biliński S (1990) Ovariole types and the phylogeny of hexapods. Biol Rev 65:401–429. https://doi.org/10.1111/j.1469-185X.1990.tb01232.x

Świątek P, Urbisz AZ (2019) Architecture and life history of female germ-line cysts in clitellate annelids. Results Probl Cell Differ 68:515–551. https://doi.org/10.1007/978-3-030-23459-1_21

Świątek P, Płachno BJ, Marchant R et al (2016) Germ-line cells do not form syncytial cysts in the ovaries of the basal clitellate annelid *Capilloventer australis*. Zool Anz – J Comp Zool 260:63–71. https://doi.org/10.1016/j.jcz.2015.12.002

Świątek P, Pinder A, Gajda Ł (2020) Description of ovary organization and oogenesis in a phreodrilid clitellate. J Morphol 281:81–94. https://doi.org/10.1002/jmor.21081

Telfer WH (1975) Development and physiology of the oöcyte-nurse cell syncytium. Adv Insect Physiol 11:223–319. https://doi.org/10.1016/S0065-2806(08)60164-2

Tihelka E, Cai C, Giacomelli M et al (2021) The evolution of insect biodiversity. Curr Biol 31:R1299–R1311. https://doi.org/10.1016/j.cub.2021.08.057

Trauner J, Büning J (2007) Germ-cell cluster formation in the telotrophic meroistic ovary of *Tribolium castaneum* (Coleoptera, Polyphaga, Tenebrionidae) and its implication on insect phylogeny. Dev Genes Evol 217:13–27. https://doi.org/10.1007/s00427-006-0114-3

Tsutsumi T, Matsuzaki M, Haga K (1995) Formation of germ cell cluster in tubuliferan thrips (Thysanoptera). Int J Insect Morphol Embryol 24:287–296. https://doi.org/10.1016/0020-7322(95)00001-K

Tworzydło W, Biliński SM, Kočárek P, Haas F (2010) Ovaries and germline cysts and their evolution in Dermaptera (Insecta). Arthropod Struct Dev 39:360–368. https://doi.org/10.1016/j.asd.2010.05.004

Tworzydlo W, Kisiel E, Jankowska W, Bilinski SM (2014) Morphology and ultrastructure of the germarium in panoistic ovarioles of a basal "apterygotous" insect, *Thermobia domestica*. Zoology 117:200–206. https://doi.org/10.1016/j.zool.2014.01.002

Tworzydlo W, Kisiel E, Jankowska W et al (2016) Exclusion of dysfunctional mitochondria from Balbiani body during early oogenesis of *Thermobia*. Cell Tissue Res 366:191–201. https://doi.org/10.1007/s00441-016-2414-x

Tworzydlo W, Marek M, Kisiel E, Bilinski SM (2017) Meiosis, Balbiani body and early asymmetry of *Thermobia* oocyte. Protoplasma 254:649–655. https://doi.org/10.1007/s00709-016-0978-7

Tworzydlo W, Sekula M, Bilinski SM (2020) Transmission of functional, wild-type mitochondria and the fittest mtDNA to the next generation: bottleneck phenomenon, Balbiani body, and mitophagy. Genes 11:E104. https://doi.org/10.3390/genes11010104

Urbisz AZ, Chajec Ł, Brąszewska-Zalewska A et al (2017) Ovaries of the white worm (*Enchytraeus albidus*, Annelida, Clitellata) are composed of 16-celled meroistic germ-line cysts. Dev Biol 426:28–42. https://doi.org/10.1016/j.ydbio.2017.04.009

Urbisz AZ, Nakano T, Świątek P (2020) Ovary cord micromorphology in the blood-sucking haemadipsid leech *Haemadipsa japonica* (Hirudinida: Arhynchobdellida: Hirudiniformes). Micron 138:102929. https://doi.org/10.1016/j.micron.2020.102929

Urbisz AZ, Martin P, Lagnika M et al (2021) Microorganization of ovaries and oogenesis of *Haplotaxis sp.* (Clitellata: Haplotaxidae). J Morphol 282:98–114. https://doi.org/10.1002/jmor.21285

Urbisz AZ, Chajec Ł, Małota K et al (2022) All for one: changes in mitochondrial morphology and activity during syncytial oogenesis. Biol Reprod 106:1232–1253. https://doi.org/10.1093/biolre/ioac035

Vasilikopoulos A, Misof B, Meusemann K et al (2020) An integrative phylogenomic approach to elucidate the evolutionary history and divergence times of Neuropterida (Insecta: Holometabola). BMC Evol Biol 20:64. https://doi.org/10.1186/s12862-020-01631-6

White-Cooper H, Bausek N (2010) Evolution and spermatogenesis. Philos Trans R Soc B Biol Sci 365:1465–1480. https://doi.org/10.1098/rstb.2009.0323

Wieschaus E, Szabad J (1979) The development and function of the female germ line in *Drosophila melanogaster*: a cell lineage study. Dev Biol 68:29–46. https://doi.org/10.1016/0012-1606(79)90241-0

Wipfler B, Letsch H, Frandsen PB et al (2019) Evolutionary history of Polyneoptera and its implications for our understanding of early winged insects. Proc Natl Acad Sci USA 116:3024–3029. https://doi.org/10.1073/pnas.1817794116

Yamauchi H, Yoshitake N (1982) Origin and differentiation of the oocyte – nurse cell complex in the germarium of the earwig, *Anisolabis maritima* Borelli (Dermaptera: Labiduridae). Int J Insect Morphol Embryol 11:293–305. https://doi.org/10.1016/0020-7322(82)90018-6

Żelazowska M, Fopp-Bayat D (2017) Ovarian nests in cultured females of the Siberian sturgeon *Acipenser baerii* (Chondrostei, Acipenseriformes). J Morphol 278:1438–1449. https://doi.org/10.1002/jmor.20723

Chapter 3
Germline and Somatic Cell Syncytia in Insects

Malgorzata Kloc, Wacław Tworzydło, and Teresa Szklarzewicz

Abstract Syncytia are common in the animal and plant kingdoms both under normal and pathological conditions. They form through cell fusion or division of a founder cell without cytokinesis. A particular type of syncytia occurs in invertebrate and vertebrate gametogenesis when the founder cell divides several times with partial cytokinesis producing a cyst (nest) of germ line cells connected by cytoplasmic bridges. The ultimate destiny of the cyst's cells differs between animal groups. Either all cells of the cyst become the gametes or some cells endoreplicate or polyploidize to become the nurse cells (trophocytes). Although many types of syncytia are permanent, the germ cell syncytium is temporary, and eventually, it separates into individual gametes. In this chapter, we give an overview of syncytium types and focus on the germline and somatic cell syncytia in various groups of insects. We also describe the multinuclear giant cells, which form through repetitive nuclear divisions and cytoplasm hypertrophy, but without cell fusion, and the accessory nuclei, which bud off the oocyte nucleus, migrate to its cortex and become included in the early embryonic syncytium.

M. Kloc (✉)
The Houston Methodist Research Institute, Houston, TX, USA

The Houston Methodist Hospital, Department of Surgery, Houston, TX, USA

Department of Genetics, MD Anderson Cancer Center, The University of Texas, Houston, TX, USA
e-mail: mkloc@houstonmethodist.org

W. Tworzydło · T. Szklarzewicz
Department of Developmental Biology and Invertebrate Morphology, Institute of Zoology and Biomedical Research, Faculty of Biology, Jagiellonian University in Krakow, Krakow, Poland
e-mail: w.tworzydlo@uj.edu.pl; teresa.szklarzewicz@uj.edu.pl

M. Kloc, A. Uosef (eds.), *Syncytia: Origin, Structure, and Functions*, Results and Problems in Cell Differentiation 71,
https://doi.org/10.1007/978-3-031-37936-9_3

3.1 Types and Origin of Syncytia and Giant Cells

Syncytia are a common feature of multicellular organisms under natural and pathological conditions (Kloc et al. 2022a). Syncytia derive from cell fusion (the "real" syncytia) or incomplete divisions of the founder cell (Kloc et al. 2004, 2008; Spradling et al. 2022, in this volume). The fusing cells can be the same (homotypic syncytia) or a different type (heterotypic syncytia) (Oren-Suissa and Podbilewicz 2007; Steinberg et al. 2010). Depending on the behavior of fusing cell nuclei, the syncytia can be divided into heterokaryotic and synkaryotic. In heterokaryotic syncytia, the homotypic or heterotypic nuclei remain separate within the common cytoplasm forming binuclear or multinuclear syncytium (Strom and Bushley 2016). In synkaryotic syncytia, the nuclei fuse resulting in a mononuclear syncytium (Berndt et al. 2013). In some syncytia derived from incomplete divisions, the participating cells differ in the stage of differentiation and/or have a different fate (Kloc 2019; Spradling et al. 2022). Interestingly, in *Drosophila*, the positioning and movement of multiple nuclei within the syncytium are highly coordinated and regulated, and the nuclear movement depends on pulsed cytoplasmic flows coupled with synchronous divisions (Donoughe et al. 2022; Padilla et al. 2022). In contrast, in the cricket *Gryllus bimaculatus*, during blastoderm formation, the nuclear movements do not depend on cytoplasm flow, their directions are asynchronous, and the movement speed and cycle length depend on the local density of nuclei (Donoughe et al. 2022). An interesting phenomenon in insect and other invertebrate oogenesis and early embryogenesis is the formation of accessory nuclei, which bud off the oocyte nucleus (Fig. 3.1a–d). In some insects, they remain attached to the oocyte nucleus, but in others, they migrate to the periphery of the oocyte (Fig. 3.1e) and future embryo. Accessory nuclei contain coilin and small ribonucleoproteins (snRNPs) and are believed to deliver these compounds to the early embryo (Biliński and Kloc 2002; Jaglarz et al. 2008). Syncytia can be either permanent or temporary structures, which eventually cellularize (separate) into individual cells. Some syncytia polyploidize or endoreplicate DNA enhancing transcriptional activity. A separate category is the giant cells. Giant cells can be of syncytial origin or form through the hypertrophy of the cytoplasm combined with the polyploidization or endoreplication of DNA. Like syncytia, giant cells occur both under natural and pathological conditions (Kloc et al. 2022b). Additionally, there are multinuclear coenocytic cells, which derive from sequential nuclear divisions without cytokinesis or fusion (Ondracka et al. 2018; Xu et al. 2022).

3.2 Insect Germline Syncytia

In insects, like in many other invertebrates and vertebrates, female and male germ cells derive from the progenitor cells (Kloc et al. 2004, 2008; Spradling et al. 2022, in this volume; Yoshida 2016). The incomplete divisions (with partial or incomplete

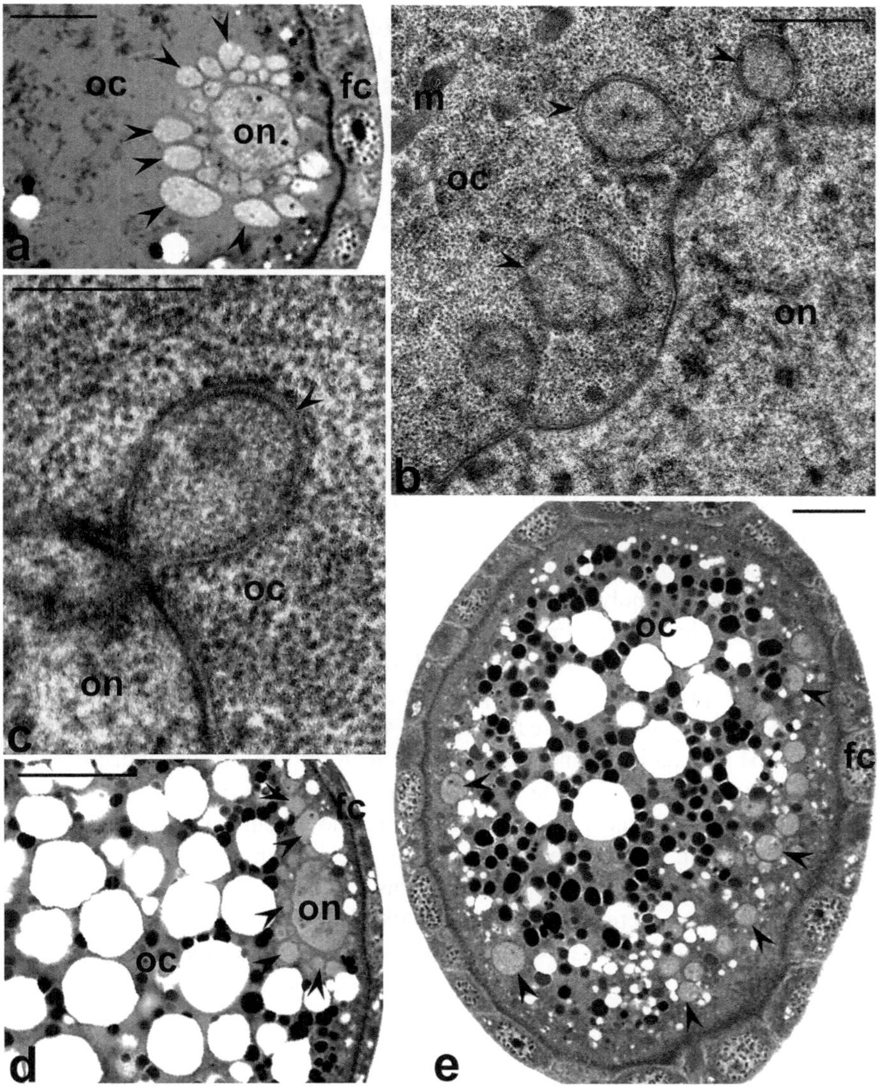

Fig. 3.1 Accessory nuclei in insect oocytes. (**a–e**) *Matsucoccus pini* (Hemiptera, Coccomorpha: Matsucoccidae). (**a**, **b**) Oocyte nucleus surrounded by accessory nuclei. (**c**) Formation of accessory nucleus by budding off the nuclear envelope of oocyte nucleus. (**d**) Accessory nuclei translocate to the cortical ooplasm. (**e**) Accessory nuclei at the periphery of the oocyte. White spheres in the oocyte cytoplasm are lipid droplets. **a**, **d**, **e** Light microscopy, scale bar = 20 μm, **b**, **c** Transmission electron microscopy, scale bar = 1 μm. *Black arrowheads* accessory nuclei, *fc* follicular cells, *m* mitochondrion, *oc* oocyte, *on* oocyte nucleus

cytokinesis) of progenitor cells result in syncytium (cyst or nest) of cells connected by intercellular bridges. The incomplete or complete cytokineses of germline stem cells depend on the ESCRT-III protein Shrub (Mathieu et al. 2022; Matias et al.

2015). After progressing through the consecutive stages of gametogenesis within syncytium, eventually, the germ cells separate to form individual gametes. In some animals, including insects, some of the cells in the germline syncytium do not develop into gametes but form specialized cells which support germ cell/gamete development (Kloc 2019).

3.2.1 Syncytia in Spermatogenesis

Most of the studies on insect spermatogenesis have been done in *Drosophila*. *Drosophila* testes contain two subpopulations of stem cells: germline stem cells (GSCs) and somatic cyst stem cells (CySCs). Divisions of CySCs produce cyst cells, which encapsulate and support the development of germ cells (Zoller and Schulz 2012). Asymmetrical division of GSC results in another GCS and a precursor of the gonioblast. Gonioblast undergoes four divisions with incomplete cytokinesis. The result is the cyst of 16 spermatogonia (SGs) connected by intercellular bridges (ring canals). The signaling between the members of the cyst is coordinated by the fusome (or polyfusome), a membranous/cytoskeletal structure which branches within the bridges (Frappaolo et al. 2022; McKearin 1997). The cyst of 16 spermatogonia enters the S phase, followed by the prolonged G2 phase. Spermatogonia grow in volume, prepare for entry into meiosis, and become spermatocytes. Spermatocytes divide meiotically, with incomplete cytokinesis, forming the cyst of 64 spermatids connected by intercellular bridges. Mitochondria of early-post-meiotic spermatids migrate toward the nucleus and wrap around each other into a giant (6 micrometers in diameter) structure, the nebenkern (Chen and Megraw 2014; Sawyer et al. 2017; Vedelek et al. 2016). The nebenkern provides a structural platform for sperm tail elongation. It contains a testes-specific protein Spermitin (Sprn) of unknown function, which contains Pleckstrin homology-like (PH) domain related to Ran-binding protein 1 (Chen and Megraw 2014). Spermatids enter the final differentiation program, spermiogenesis, to become functional gametes (sperm cells). Spermatid nuclei and cytoplasm shrink, and they lose most organelles. The nebenkern splits in half to surround the elongating axoneme and the acrosome forms. In the final step, the cyst individualizes into separate mature sperm cells, which migrate to the seminal vesicles (Fabrizio et al. 2012, 2020; Frappaolo et al. 2022). The terminal differentiation, elimination of organelles, and individualization of sperm cells involve apoptotic enzymes (caspases) and proteasome proteolytic activity but without cell death (Huh et al. 2004). Caspase activation during this process is facilitated by the F-box protein Nutcracker, which interacts with a Cullin-1-based ubiquitin ligase complex (SCF): Cullin-1 and chaperone protein Skp (Bader et al. 2010). In *Drosophila*, like in other animals, including mammals, during spermatogenesis, nearly a quarter of newly formed spermatogonia are eliminated through the programmed cell death that removes damaged or excess cells. When some of the cells within a 2–16 cell cyst require elimination, the neighboring cyst cells remove them through phagoptosis. Recent studies showed that the phagoptotic ability of the cyst cells is facilitated by the transmembrane phagocytic receptor Draper (Drpr), which is the ortholog of mammalian Jedi/LRP/MEGF10. These cells produce endosomes and lysosomes, which acidify neighboring spermatogonia leading to the fragmentation and degradation of their DNA, which is followed by dead cell removal (Zohar-Fux et al. 2022).

The described above scenario of spermatogenesis differs between different insect groups. For example, the number of divisions and thus spermatids in the cyst differ. Some flies usually have 32 spermatids per cyst; particular species of coccids may have 16, 32, or 64 spermatids; and caddis flies have 128 spermatids, while most butterflies, moths, and beetles have 256 spermatids per cyst (Phillips 1970; Robison 1990).

The most fascinating examples of spermatogenic syncytia are those found in scale insects. These insects produce motile sperm bundles. Each bundle is surrounded by a specialized cell (ensheathing cell) (Fig. 3.2a) derived from the cyst wall. While in the male reproductive tract, the cytoplasm and the nucleus of the ensheathing cell are discarded, and the sperm bundle becomes surrounded only by a double layer of membranes (Fig. 3.2b, c). The spermatozoa lack acrosome, mitochondria, centrioles, and flagellum. Each spermatozoon in the bundle is delineated lengthwise by a sheet of microtubules (Fig. 3.2b, c), which allow the sperm bundle to move in the female reproductive tract. Additionally, the whole bundle can propel forward and rotate through the coordinated motion of its spermatozoa (Robison 1966, 1990; Ross and Robison 1969). Interestingly, in some scale insects, the sperm bundle penetrates the ovariole (i.e., the structural subunit of the ovary—for details, see the chapter by Brubacher in this volume). Sperm bundles enter the specialized cell located at the base of the ovariole (termed vestibular cell) (Fig. 3.2d, e), where they wait for the oocyte to complete its growth (Niżnik 2007; Robison Jr. 1990). It has been suggested that this unusual behavior of scale insect spermatozoa is related to the lack of micropylar openings in egg capsule. Consequently, the sperm needs to penetrate the oocyte before the formation of eggshell (Szklarzewicz et al. 2022).

Interestingly, the lepidopteran insects produce two types of sperm cells: eupyrene sperm cells, which have nuclei and fertilize eggs, and apyrene sperm (parasperm, anuclear sperm), which do not contain DNA. Although the apyrene sperm is smaller than regular sperm, in butterflies and moth, it consists of >85% of all sperm. Both eupyrene and apyrene sperm derive from the bipotential spermatocytes. The eupyrene versus apyrene fate of spermatocyte depends on a hemolymph factor activated before or after pupation. The shift of spermatocyte toward apyrene fate depends on meiotic prophase shortening and inhibition of a meiotic lysine-rich protein synthesis (Friedländer 1997). Although it remains, unknown what is the exact role of apyrene sperm, it seems that it may help in sperm motility and dissociation from the sperm bundles, and/or by filling the spermatheca reduce female receptivity for re-mating (Friedländer 1997; Konagaya et al. 2020). Yang et al. (2022a) showed that eupyrene sperm development depends on methyltransferase BmHen1, which modifies Piwi-interacting RNAs (piRNAs) involved in silencing and control or transposable elements in the germ cells (Tóth et al. 2016). Recently, Yang et al. (2022b) showed that eupyrene spermatogenesis in the silkworm moth *Bombyx mori*

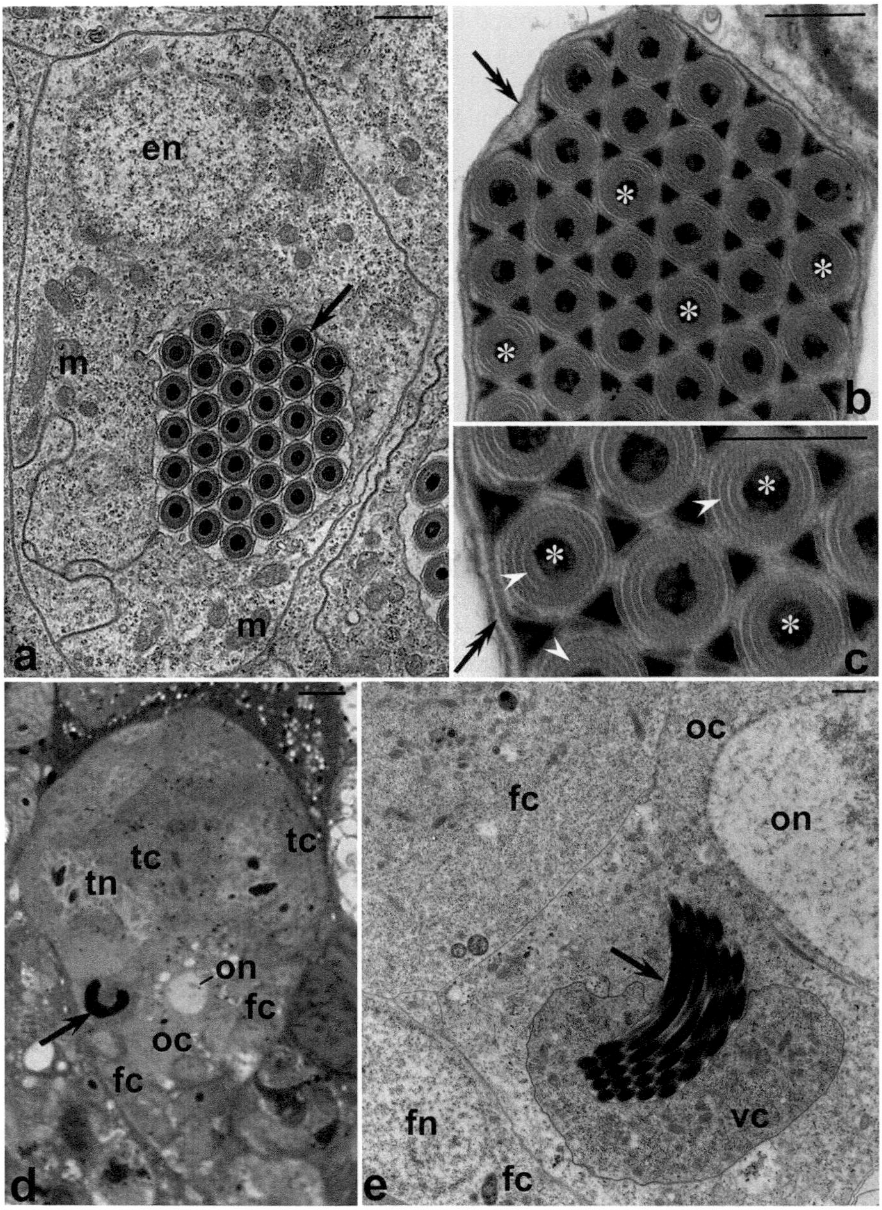

Fig. 3.2 Syncytia in spermatogenesis. (**a**–**e**) *Icerya purchasi* (Hemiptera, Coccomorpha: Monophlebidae). (**a**) Sperm bundle surrounded by a somatic cell. (**b**, **c**) Sperm bundle surrounded by two membranes. (**d**) Previtellogenic ovariole (longitudinal section). Note the vestibular cell surrounding the sperm bundle (arrow). (**e**) Sperm bundle (arrow) inside the ovariole. a-c, e TEM, scale bar = 2 μm, d LM, methylene blue, scale bar = 20 μm. *Arrow* sperm bundle, *white arrowheads* microtubules, *white asterisk* sperm nucleus, *double arrow* double layer of membranes, *en* nucleus of the ensheathing cell, *fc* follicular cells, *fn* follicular cell nucleus, *m* mitochondrion, *oc* oocyte, *on* oocyte nucleus, *tc* trophocyte, *tn* trophopcyte nucleus, *vc* vestibular cell (courtesy of Sylwia Niżnik, Jagiellonian University, Kraków, Poland)

(Bm) depends on the BmPMFBP1 (*Bombyx mori*-specific version of polyamine-modulated factor 1 binding protein 1). Their bioinformatic analysis indicates that BmPMFBP1 is not homologous to mammalian PMFBP1. The BmPMFBP1 deficiency resulted in abnormal eupyrene but normal apyrene sperm bundles and prevented the release of eupyrene sperm bundles. This finding indicates that at least in the silkworm, the development of these two sperm types is regulated by different genes.

3.2.2 Syncytia in Oogenesis

The formation of cysts during insect oogenesis follows a similar pattern to the formation of male cysts (see Sect. 3.2.1). The progenitor cell of female germ cells (termed cystoblast) undergoes a series of synchronous incomplete mitotic divisions leading to the formation of syncytia termed clusters of cystocytes (Fig. 3.3a, b) (Bilinski et al. 2017). The number of cycles of divisions is genetically determined and species-specific, for example, in *Drosophila melanogaster* and most dipterans, the initial cystoblast undergoes four rounds of divisions resulting in the formation of 16 cystocytes (Fig. 3.3c). In butterflies, the cystoblast divides three times which leads to the formation of eight cystocytes, while in certain dermapterans, cystoblast divides only once to generate a simple two-cell cyst (Fig. 3.3d) (Büning 1994; Tworzydło et al. 2010). However, in some carabid beetles and scale insects, the number of cystocytes does not reflect the number of cell divisions. This is probably caused by the loss of synchrony of cystocyte divisions (Jaglarz 1992; Szklarzewicz 1998). As a result of incomplete cytokinesis, the cystocytes are connected by intercellular bridges (for details, see the chapter by Brubacher in this volume). Bridges are filled with amorphous, electron-dense fusomal material. The fusomal material of all bridges fuse, forming a polyfusome, which integrates all the cystocytes within the cluster (Fig. 3.3b) (Büning 1994; Szklarzewicz 1997). According to Storto and King (1989), the polyfusome coordinates the divisions and affects the determination of the oocytes. In the polytrophic ovariole (for the detailed organization of insect ovariole types and their distribution among insect taxa, see the chapter by Brubacher in this volume), in each cluster, only a single oocyte forms, while the remaining cystocytes become trophocytes (i.e., cells responsible for the synthesis of RNA for the future embryo) (Fig. 3.3c, d). In the telotrophic ovariole, even 50% cystocytes may become oocytes. After completion of the differentiation of cystocytes into oocytes and trophocytes, the polyfusome vanishes. Simultaneously, intercellular bridges increase the diameter and thickness of the wall, which is related to their function—the transport of macromolecules (mainly RNAs) and organelles from the trophocytes to the developing oocyte (Büning 1994). The ovarioles retain their syncytial character until the end of the oocyte growth (Fig. 3.3c–e).

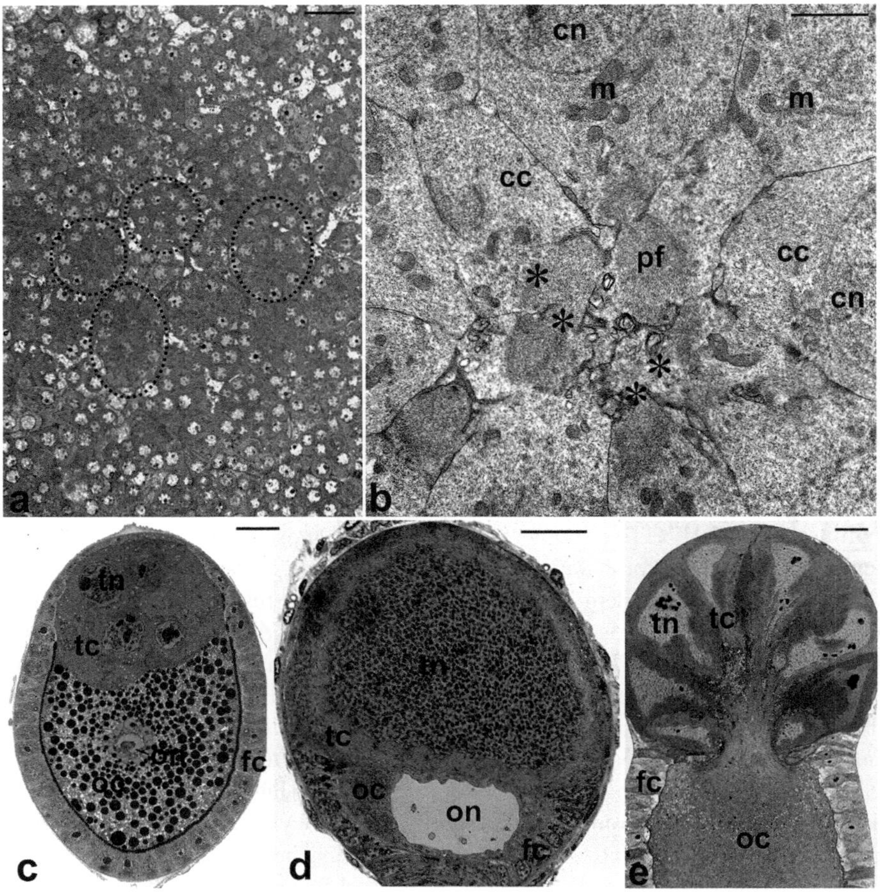

Fig. 3.3 Syncytia in oogenesis. (**a**, **b**) *Orthezia urticae* (Hemiptera, Coccomorpha: Ortheziidae). (**a**) Cysts containing interconnected cystocytes (encircled). (**b**) Polyfusome in the center of the cyst. (**c**) *Tachypeza* sp. (Diptera: Hybotidae). Ovarian follicle in the polytrophic ovariole (longitudinal section) (courtesy of Szczepan Biliński, Jagiellonian University, Kraków, Poland). (**d**) *Doru lineare* (Dermaptera: Forficulidae). Ovarian follicle in the polytrophic ovariole (longitudinal section). (**e**) *Marchalina hellenica* (Hemiptera, Coccomorpha: Marchalinidae). Telotrophic ovariole (longitudinal section). **a**, **c–e** LM, methylene blue, scale bar = 2 μm, **d** TEM, scale bar = 10 μm. *Black asterisk* intercellular bridge, *cc* cystocyte, *cn* cystocyte nucleus, *fc* follicular cells, *m* mitochondrion, *oc* oocyte, *on* oocyte nucleus, *pf* polyfusome, *tc* trophocyte, *tn* trophopcyte nucleus

3.3 Insect Somatic Cell Syncytia

3.3.1 Epithelial Syncytia

In insects, a common strategy to repair injured epithelium is the formation of giant polyploid cells, which contain more than the diploid copy of chromosomes. Usually, the cell size increases in parallel with the polyploidization level. Polyploidization occurs by endoreplication of DNA, cell fusion, or both. The wound-induced polyploidization (WIP), accompanied by cell growth, is a conserved healing strategy in invertebrates and vertebrates. During wound repair, an increased cell size compensates for cell loss. One of the advantages of forming giant polyploid cells instead of increased cell proliferation for tissue repair is lowering the possibility of cancer formation through uncontrolled cell divisions (Bailey et al. 2020; Losick and Duhaime 2021; Nandakumar et al. 2020). In *Drosophila*, the polyploid cells are generated within a few days after mechanical puncture. The post injury endoreplication in the epithelium is regulated by the protooncogene Myc (Bailey et al. 2020). Another pathway involved in polyploidization, cell fusion, and syncytium formation during wound healing is Hippo and its effector Yorkie which regulate cell proliferation and apoptosis (Besen-McNally et al. 2021; Losick et al. 2013). Kakanj et al. (2022) recently showed that during wound healing in *Drosophila*, the formation of the multinucleated syncytium depends on the mTORC1-regulated autophagy.

Another example of giant polyploid epithelial cells is those involved in the production of male pheromone in the goldenrod gall fly *Eurosta solidaginis.* Males of *E. solidaginis* produce an incredibly high amount of sex pheromone. One male emits 70-85 μg of pheromone (tens-hundreds of times more than other insects) within 24 hrs, which requires super large cells for its production. These cells, located in the abdominal pleura, are 60–100 larger than equivalent epithelial cells in females and are the largest animal epithelial cells even recorded (Yip et al. 2021). They are multinucleated (>20 nuclei per cell), and each nucleus contains many nucleoli, indicating increased transcriptional activity. The polyploidization probably occurs via endoreplication (Yip et al. 2021), but there is no information if the cell fusion is also involved in the formation of these giant cells. Recent studies of *Drosophila* brain cells suggest that the increased cell size and ploidy of the long-lived cells, such as neurons and glial cells, can play a protective role against oxidative stress related to the accumulation of age-related damage (Nandakumar et al. 2020). This study also showed that only some of all polyploid cells derive from cell fusion, while the majority arise from the endoreplication of DNA.

In summary, it seems that cell fusion and/or polyploidization of somatic cells can not only increase synthetic capacity but also better protect from cell death.

3.3.2 Multinucleated Giant Cells in Insect Immune Response

The traditional view divides the immune system into innate and adaptive. According to this division, the role of innate immune cells is to prevent pathogen entry and rapid elimination of those already internalized. The innate immune response is characteristic of invertebrates, including insects. It is antigen nonspecific, responds only to a small fraction of pathogens, and is short-lived and evolutionary older. In contrast, the vertebrates, besides innate immunity, developed more sophisticated, adaptive responses, which, through somatic recombination, can respond to nearly all encountered pathogens. Additionally, the adaptive immune cells can memorize pathogen encounters. Such immunological memory allows for rapid and enhanced response to the reoccurring encounter with the same or similar pathogen. Although immunological memory was traditionally viewed as a trait of the adaptive immune system, thus, absent in invertebrates, recent studies showed that the innate immune cells could also memorize pathogen encounters (Kloc et al. 2023). Studies of innate immunological memory showed that some insect species develop lifelong protection against a wide range of pathogens and that pathogen recognition can be highly specific, indicating that insect immune cells can have immunological memory (Cooper and Eleftherianos 2017). A fascinating phenomenon related to immunological memory is transgenerational immune priming (TgIP). In TgIP, immune protection acquired by an insect is transferred to its offspring (López et al. 2014; Moret 2006). Although it is currently unknown what is the mechanism of TgIP, it seems that some unknown factor transferred within the egg transmits memory of pathogen encounter to the developing immune cells of the offspring (López et al. 2014) and/or induces the production of antimicrobial peptides in the hemolymph (Moret 2006).

Insects' body cavity (hemocoel) is filled with hemolymph, containing several types of immune cells (hemocytes). Hemocytes derive from the hematopoietic organ called a lymph gland through the proliferation of the prohemocytes (Grigorian et al. 2011). Hemocytes common for most insect species are plasmatocytes, granulocytes, and coagulocytes. Oenocytoids occur in Coleoptera, Lepidoptera, some Diptera, and some Heteroptera. Spherulocytes and adipohemocytes occur in Lepidoptera and Diptera. The main functions of hemocytes are phagocytosis, encapsulation of parasites too large to phagocytose, wound healing, coagulation, secretion of mucopolysaccharides, and other factors. Hemocytes are necessary for insect immunity. Experimental elimination of hemocytes from *Drosophila* larvae dramatically increases susceptibility to bacterial infection (Rosales 2017).

Discovered recently is a novel subtype of insect hemocytes, the multinucleated giant hemocytes (MGHs). These cells are engaged in the anti-parasitic immune response (Cinege et al. 2020, 2022). For example, in the ananassae subgroup of Drosophilidae, such as *D. ananassae*, *D. pallidosa*, *D. atripex*, *D. pseudoanassae*, *D. bipectinata*, *D. malerkotliana*, and *D. parabipecinata*, the MGHs encapsulate and kill eggs and larvae of parasitotic wasps, but do not participate in phagocytosis. MGHs form through the fusion of parasite-induced mononuclear and multinuclear hemocytes (Márkus et al. 2015). In *Drosophila melanogaster* that does not have MGHs, the encapsulation and killing of the parasite are performed by the lamellocytes (hemocytes formed in response to parasite entry) and melanization, which generates reactive oxygen species (Nappi et al. 2009). In *Drosophila ananassae*, the MGHs develop within 72 hrs after oviposition of wasp eggs; they are highly mobile and express a high level of transcripts encoding hemolysin-like proteins and pore-forming toxins of prokaryotic origin, which may facilitate the elimination of the parasite (Cinege et al. 2022). Studies of MGHs development in drosophilid *Zaprionus indianus* showed that they form both inside and outside of the lymph gland through cell fusion accompanied by nuclear divisions (Cinege et al. 2020). The MGHs have many similarities to mammalian macrophage-derived multinucleated giant cells formed in the inflammatory granulatomas (Kloc et al. 2022a, b; Márkus et al. 2015).

Another example of a giant but non-syncytial mononuclear cell is a mosquito megacyte formed in response to *Plasmodium* infection (Barletta et al. 2022). Megacytes form from the granulocytes in response to the Toll signaling pathway triggered by the *Plasmodium*. Mosquito *Anopheles gambiae* responds to *Plasmodium* infection by recruiting megacytes to the midgut surface. Megacytes release hemocyte-derived micro vesicles (HdMv) that activate the mosquito's complement-like system, destroying the parasites (Anderson 2000; Barletta et al. 2022).

3.3.3 Muscle Cell Syncytia

Cell fusion during development and regeneration of muscle syncytia has been thoroughly studied in invertebrates and vertebrates (Kim et al. 2015; Luo et al. 2022; Petrany and Millay 2019; Rout et al. 2022; Whitlock in this volume). Among insects, *Drosophila*, in which 60% of the genome is conserved with humans, is a well-established model to study muscle formation and regeneration. During embryogenesis the mononuclear myoblasts fuse into multinucleated syncytia (muscle fibers). In the insect's life cycle, muscle syncytia first develop during embryogenesis and second time during metamorphosis. Fusing myoblasts make contact and fuse using the fusogenic synapse (fusion restricted myogenic adhesive structure (FuRMAS), also called a podosome-like structure (PLS)) rich in adhesion molecules surrounding the F actin center (Rout et al. 2022). Fusogenic synapse was discovered in *Drosophila*, but it also occurs in fusing mammalian myoblasts and other cell types. During fusion, one of the fusing partners forms actin-based protrusions, which push on the partner's cell membrane evoking actomyosin-based mechano-response. Forces generated by the pushing and resisting membranes engage the transmembrane proteins (fusogens), which induce the formation of the fusion pores and, eventually, unify the membranes (Brukman et al. 2019; Kim and Chen 2019). *Drosophila*, like other holometabolic insects, have larval and adult stages. During metamorphosis, larval tissues formed during embryogenesis are replaced by adult tissues and organs. The body muscles of *Drosophila* larva are syncytia containing between 4 and 24 nuclei. The larva also has syncytial visceral muscles surrounding the midgut, which contain partially fused binuclear cells and multinuclear syncytia (Kim and Chen 2019). The pro-muscle clusters of the body muscles precursor cells express the bHLH protein Lethal of Scute (L'sc) under the control of the Receptor Tyrosine Kinase (RTK) pathway. The RTK, Ras, and Notch signaling limit the expression of L'sc to one cell only, which becomes the muscle progenitor cell. The muscle progenitor cell divides into two founder cells (FC) or one FC and one adult muscle progenitor cell (AMP). The muscle identity of the progenitor cells depends on the expression of muscle-specific proteins regulated by the cytoplasmic membrane-associated protein Numb that prevents Notch-mediated inhibition. The other cells of the pro-muscle cluster become the fusion-competent myoblasts

(FCMs). The fate of FCMs is determined by the Notch-mediated inhibition and transcription factors Lame duck and Tramtrack. The FCs undergo several rounds of fusion, forming different types of syncytial muscle. FC expressing a LIM-homeodomain protein Apterous forms lateral transverse muscle, while FC expressing myogenic basic helix-loop-helix protein Nautilus forms visceral muscle (Deng et al. 2017).

Adult *Drosophila* contains flight muscle syncytia which have ~1000 nuclei and resemble mammalian skeletal muscle. Flight muscles derive from the myogenic precursor cells located in the wing imaginal disc which is set up during embryogenesis. During metamorphosis, some larval muscles remain and serve as a template for dorsal longitudinal muscles (DLM). DLM fuse with the migrating toward them myoblasts, which are derived from the precursor cells of the wing disc. The flight musculature also contains a subpopulation of satellite cells that regenerate muscle after the injury.

Another subset of adult muscle is the testes muscles, which derive from precursor cells located in the genital imaginal disc. In contrast to the flight muscle, they do not migrate but fuse in situ in the genital imaginal disc (Kim and Chen 2019; Rout et al. 2022). Studies of *Drosophila* testes muscle development showed that the founder cell (FC)-like myoblasts express Dumbfounded (Duf) and Roughest (Rst), and the fusion-competent myoblast (FCM)-like cells mainly express Sticks and stones (Sns) (Kuckwa et al. 2016). Duf, Rst, and Sns are cell adhesion molecules from the immunoglobulin superfamily (Klapper et al. 2002). Duf expressed in the FCs attracts clusters of myoblasts for the subsequent fusion (Ruiz-Gómez et al. 2000). Rst is involved in cell sorting before apoptosis, fusion during muscle development, and axonal pathfinding (Apitz et al. 2004). Sns interacts with the Type IIa receptor-like protein tyrosine phosphatases (RPTP) Lar during cell fusion and neuromuscular junction development (Bali et al. 2022; Kocherlakota et al. 2008).

Acknowledgments Our greatest thanks to Prof. Szczepan Biliński and Dr. Sylwia Niżnik for the images.

References

Anderson KV (2000) Toll signaling pathways in the innate immune response. Curr Opin Immunol 12(1):13–19. https://doi.org/10.1016/s0952-7915(99)00045-x

Apitz H, Kambacheld M, Höhne M, Ramos RG, Straube A, Fischbach KF (2004) Identification of regulatory modules mediating specific expression of the roughest gene in *Drosophila melanogaster*. Dev Genes Evol 214(9):453–459. https://doi.org/10.1007/s00427-004-0423-3

Bader M, Arama E, Steller H (2010) A novel F-box protein is required for caspase activation during cellular remodeling in *Drosophila*. Development 137(10):1679–1688. https://doi.org/10.1242/dev.050088

Bailey EC, Dehn AS, Gjelsvik KJ, Besen-McNally R, Losick VP (2020) A *Drosophila* model to study wound-induced polyploidization. J Vis Exp 160. https://doi.org/10.3791/61252

Bali N, Lee HP, Zinn K (2022) Sticks and Stones, a conserved cell surface ligand for the Type IIa RPTP Lar, regulates neural circuit wiring in *Drosophila*. elife 31(11):e71469. https://doi.org/10.7554/eLife.71469

Barletta ABF, Saha B, Trisnadi N, Talyuli OAC, Raddi G, Barillas-Mury C (2022 Sep) Hemocyte differentiation to the megacyte lineage enhances mosquito immunity against *Plasmodium*. elife 2(11):e81116. https://doi.org/10.7554/eLife.81116

Berndt B, Zänker KS, Dittmar T (2013) Cell fusion is a potent inducer of aneuploidy and drug resistance in tumor cell/normal cell hybrids. Crit Rev Oncog 18(1–2):97–113. https://doi.org/10.1615/critrevoncog.v18.i1-2.60

Besen-McNally R, Gjelsvik KJ, Losick VP (2021) Wound-induced polyploidization is dependent on Integrin-Yki signaling. Biol Open 10:bio055996

Biliński SM, Kloc M (2002) Accessory nuclei revisited: the translocation of snRNPs from the germinal vesicle to the periphery of the future embryo. Chromosoma 111(1):62–68. https://doi.org/10.1007/s00412-002-0186-4

Bilinski SM, Kubiak JZ, Kloc M (2017) Asymmetric divisions in oogenesis. Results Probl Cell Differ 61:211–228. https://doi.org/10.1007/978-3-319-53150-2_9

Brubacher JL. Female germline cysts in animals: evolution and function (in this volume)

Brukman NG, Uygur B, Podbilewicz B, Chernomordik LV (2019) How cells fuse. J Cell Biol 218:1436–1451. https://doi.org/10.1083/jcb.201901017

Büning J (1994) The insect ovary: ultrastructure, previtellogenic growth and evolution. Chapman and Hall, London

Chen JV, Megraw TL (2014) Spermitin: a novel mitochondrial protein in *Drosophila* spermatids. PLoS One 9(9):e108802. https://doi.org/10.1371/journal.pone.0108802

Cinege G, Lerner Z, Magyar LB, Soós B, Tóth R, Kristó I, Vilmos P, Juhász G, Kovács AL, Hegedűs Z, Sensen CW, Kurucz É, Andó I (2020) Cellular immune response involving multinucleated giant hemocytes with two-step genome amplification in the Drosophilid *Zaprionus indianus*. J Innate Immun 12(3):257–272. https://doi.org/10.1159/000502646

Cinege G, Magyar LB, Kovács AL, Lerner Z, Juhász G, Lukacsovich D, Winterer J, Lukacsovich T, Hegedűs Z, Kurucz É, Hultmark D, Földy C, Andó I (2022) Broad ultrastructural and transcriptomic changes underlie the multinucleated giant hemocyte mediated innate immune response against parasitoids. J Innate Immun 14(4):335–354. https://doi.org/10.1159/000520110

Cooper D, Eleftherianos I (2017) Memory and specificity in the insect immune system: current perspectives and future challenges. Front Immunol 9(8):539. https://doi.org/10.3389/fimmu.2017.00539

Deng S, Azevedo M, Baylies M (2017) Acting on identity: myoblast fusion and the formation of the syncytial muscle fiber. Semin Cell Dev Biol 72:45–55. https://doi.org/10.1016/j.semcdb.2017.10.033

Donoughe S, Hoffmann J, Nakamura T, Rycroft CH, Extavour CG (2022) Nuclear speed and cycle length co-vary with local density during syncytial blastoderm formation in a cricket. Nat Commun 13(1):3889. https://doi.org/10.1038/s41467-022-31212-8

Fabrizio JJ, Aqeel N, Cote J, Estevez J, Jongoy M, Mangal V, Tema W, Rivera A, Wnukowski J, Bencosme Y (2012) Mulet (mlt) encodes a tubulin-binding cofactor E-like homolog required for spermatid individualization in *Drosophila melanogaster*. Fly (Austin) 6(4):261–272. https://doi.org/10.4161/fly.21533

Fabrizio JJ, Rollins J, Bazinet CW, Wegener S, Koziy I, Daniel R, Lombardo V, Pryce D, Bharrat K, Innabi E, Villanobos M, Mendoza G, Ferrara E, Rodway S, Vicioso M, Siracusa V, Dailey E, Pronovost J, Innabi S, Patel V, DeSouza N, Quaranto D, Niknejad A (2020) Tubulin-binding cofactor E-like (TBCEL), the protein product of the mulet gene, is required in the germline for the regulation of inter-flagellar microtubule dynamics during spermatid individualization. Biol Open 9(2):bio049080. https://doi.org/10.1242/bio.049080

Frappaolo A, Piergentili R, Giansanti MG (2022) Microtubule and actin cytoskeletal dynamics in male meiotic cells of *Drosophila melanogaster*. Cell 11(4):695. https://doi.org/10.3390/cells11040695

Friedländer M (1997) Control of the eupyrene-apyrene sperm dimorphism in Lepidoptera. J Insect Physiol 43(12):1085–1092. https://doi.org/10.1016/s0022-1910(97)00044-9

Grigorian M, Mandal L, Hartenstein V (2011) Hematopoiesis at the onset of metamorphosis: terminal differentiation and dissociation of the *Drosophila* lymph gland. Dev Genes Evol 221(3):121–131. https://doi.org/10.1007/s00427-011-0364-6

Huh JR, Vernooy SY, Yu H, Yan N, Shi Y, Guo M, Hay BA (2004) Multiple apoptotic caspase cascades are required in nonapoptotic roles for *Drosophila* spermatid individualization. PLoS Biol 2(1):E15. https://doi.org/10.1371/journal.pbio.0020015

Jaglarz MK (1992) Peculiarities of the organization of egg chambers in carabid ground beetles and their phylogenetic implications. Tissue Cell 24:397–409. https://doi.org/10.1016/0040-8166(92)90056-d

Jaglarz MK, Kloc M, Bilinski SM (2008) Accessory nuclei in insect oogenesis: in search of the function of enigmatic organelles. Int J Dev Biol 52(2–3):179–185. https://doi.org/10.1387/ijdb.072303mj

Kakanj P, Bhide S, Moussian B, Leptin M (2022) Autophagy-mediated plasma membrane removal promotes the formation of epithelial syncytia. EMBO J 41(12):e109992. https://doi.org/10.15252/embj.2021109992

Kim JH, Chen EH (2019) The fusogenic synapse at a glance. J Cell Sci 132(18):jcs213124. https://doi.org/10.1242/jcs.213124

Kim JH, Jin P, Duan R, Chen EH (2015) Mechanisms of myoblast fusion during muscle development. Curr Opin Genet Dev 32:162–170. https://doi.org/10.1016/j.gde.2015.03.006

Klapper R, Stute C, Schomaker O, Strasser T, Janning W, Renkawitz-Pohl R, Holz A (2002) The formation of syncytia within the visceral musculature of the *Drosophila* midgut is dependent on duf, sns and mbc. Mech Dev 110(1-2):85–96. https://doi.org/10.1016/s0925-4773(01)00567-6

Kloc M (2019) The rove beetle *Creophilus maxillosus* as a model system to study asymmetric division, oocyte specification, and the germ-somatic cell signaling. Results Probl Cell Differ 68:217–230. https://doi.org/10.1007/978-3-030-23459-1_9

Kloc M, Bilinski S, Dougherty MT, Brey EM, Etkin LD (2004) Formation, architecture and polarity of female germline cyst in *Xenopus*. Dev Biol 266(1):43–61. https://doi.org/10.1016/j.ydbio.2003.10.002

Kloc M, Jaglarz M, Dougherty M, Stewart MD, Nel-Themaat L, Bilinski S (2008) Mouse early oocytes are transiently polar: three-dimensional and ultrastructural analysis. Exp Cell Res 314(17):3245–3254. https://doi.org/10.1016/j.yexcr.2008.07.007

Kloc M, Subuddhi A, Uosef A, Kubiak JZ, Ghobrial RM (2022a) Monocyte-macrophage lineage cell fusion. Int J Mol Sci 23(12):6553. https://doi.org/10.3390/ijms23126553

Kloc M, Uosef A, Subuddhi A, Kubiak JZ, Piprek RP, Ghobrial RM (2022b) Giant multinucleated cells in aging and senescence-an abridgement. Biology (Basel) 11(8):1121. https://doi.org/10.3390/biology11081121

Kloc M, Kubiak JZ, Zdanowski R, Ghobrial RM (2023) Memory macrophage. Int J Mol Sci 24(1):38. https://doi.org/10.3390/ijms24010038

Kocherlakota KS, Wu JM, McDermott J, Abmayr SM (2008) Analysis of the cell adhesion molecule sticks-and-stones reveals multiple redundant functional domains, protein-interaction motifs and phosphorylated tyrosines that direct myoblast fusion in *Drosophila melanogaster*. Genetics 178(3):1371–1383. https://doi.org/10.1534/genetics.107.083808

Konagaya T, Idogawa N, Watanabe M (2020) Destination of apyrene sperm following migration from the bursa copulatrix in the monandrous swallowtail butterfly Byasa alcinous. Sci Rep 10(1):20907. https://doi.org/10.1038/s41598-020-77683-x

Kuckwa J, Fritzen K, Buttgereit D, Rothenbusch-Fender S, Renkawitz-Pohl R (2016) A new level of plasticity: *Drosophila* smooth-like testes muscles compensate failure of myoblast fusion. Development 143(2):329–338. https://doi.org/10.1242/dev.126730

López JH, Schuehly W, Crailsheim K, Riessberger-Gallé U (2014) Trans-generational immune priming in honeybees. Proc Biol Sci 281(1785):20140454. https://doi.org/10.1098/rspb.2014.0454

Losick VP, Duhaime LG (2021) The endocycle restores tissue tension in the *Drosophila* abdomen post wound repair. Cell Rep 37(2):109827. https://doi.org/10.1016/j.celrep.2021.109827

Losick VP, Fox DT, Spradling AC (2013) Polyploidization and cell fusion contribute to wound healing in the adult *Drosophila* epithelium. Curr Biol 23(22):2224–2232. https://doi.org/10.1016/j.cub.2013.09.029

Luo Z, Shi J, Pandey P, Ruan ZR, Sevdali M, Bu Y, Lu Y, Du S, Chen EH (2022) The cellular architecture and molecular determinants of the zebrafish fusogenic synapse. Dev Cell 57(13):1582–1597.e6. https://doi.org/10.1016/j.devcel.2022.05.016

Márkus R, Lerner Z, Honti V, Csordás G, Zsámboki J, Cinege G, Párducz Á, Lukacsovich T, Kurucz É, Andó I (2015) Multinucleated giant hemocytes are effector cells in cell-mediated immune responses of *Drosophila*. J Innate Immun 7(4):340–353. https://doi.org/10.1159/000369618

Mathieu J, Michel-Hissier P, Boucherit V, Huynh JR (2022) The deubiquitinase USP8 targets ESCRT-III to promote incomplete cell division. Science 376(6595):818–823. https://doi.org/10.1126/science.abg2653

Matias NR, Mathieu J, Huynh JR (2015) Abscission is regulated by the ESCRT-III protein shrub in *Drosophila* germline stem cells. PLoS Genet 11(2):e1004653. https://doi.org/10.1371/journal.pgen.1004653

McKearin D (1997) The *Drosophila* fusome, organelle biogenesis and germ cell differentiation: if you build it. BioEssays 19(2):147–152. https://doi.org/10.1002/bies.950190209

Moret Y (2006) Trans-generational immune priming: specific enhancement of the antimicrobial immune response in the mealworm beetle, *Tenebrio molitor*. Proc Biol Sci 273(1592):1399–1405. https://doi.org/10.1098/rspb.2006.3465

Nandakumar S, Grushko O, Buttitta LA (2020) Polyploidy in the adult *Drosophila* brain. elife 9:e54385. https://doi.org/10.7554/eLife.54385

Nappi A, Poirié M, Carton Y (2009) The role of melanization and cytotoxic by-products in the cellular immune responses of *Drosophila* against parasitic wasps. Adv Parasitol 70:99–121. https://doi.org/10.1016/S0065-308X(09)70004-1

Niżnik S (2007) Structure and development of hermaphroditic gonad in the scale insect *Icerya purchasi* (Insecta, Hemiptera). Ph.D. Thesis, Jagiellonian University, Kraków, Poland (in Polish)

Ondracka A, Dudin O, Ruiz-Trillo I (2018) Decoupling of nuclear division cycles and cell size during the COENOCYTIC growth of the Ichthyosporean *Sphaeroforma arctica*. Curr Biol 28(12):1964–1969.e2. https://doi.org/10.1016/j.cub.2018.04.074

Oren-Suissa M, Podbilewicz B (2007) Cell fusion during development. Trends Cell Biol 17(11):537–546. https://doi.org/10.1016/j.tcb.2007.09.004

Padilla JR, Ferreira LM, Folker ES (2022) Nuclear movement in multinucleated cells. Development 149(21):dev200749. https://doi.org/10.1242/dev.200749

Petrany MJ, Millay DP (2019) Cell fusion: merging membranes and making muscle. Trends Cell Biol 29(12):964–973. https://doi.org/10.1016/j.tcb.2019.09.002

Phillips DM (1970) Insect sperm: their structure and morphogenesis. J Cell Biol 44(2):243–277. https://doi.org/10.1083/jcb.44.2.243

Robison WG Jr (1966) Microtubules in relation to the motility of a sperm syncytium in an armored scale insect. J Cell Biol 29(2):251–265. https://doi.org/10.1083/jcb.29.2.251

Robison WG Jr (1990) Sperm ultrastructure, behaviour, and evolution. In: Rosen D (ed) World crop pests, volume 4A, Armored scale insects: their biology, natural enemies and control. Elsevier, Amsterdam, pp 205–220

Rosales C (2017) Cellular and molecular mechanisms of insect immunity. In: Shields VD (ed) Insect physiology and ecology. IntechOpen, London. https://doi.org/10.5772/67107

Ross J, Robison WG Jr (1969) Unusual microtubular patterns and three-dimensional movement of mealybug sperm and sperm bundles. J Cell Biol 40(2):426–445. https://doi.org/10.1083/jcb.40.2.426

Rout P, Preußner M, Önel SF (2022) *Drosophila melanogaster*: a model system to study distinct genetic programs in myoblast fusion. Cell 11(3):321. https://doi.org/10.3390/cells11030321

Ruiz-Gómez M, Coutts N, Price A, Taylor MV, Bate M (2000) *Drosophila* dumbfounded: a myoblast attractant essential for fusion. Cell 102(2):189–198. https://doi.org/10.1016/s0092-8674(00)00024-6

Sawyer EM, Brunner EC, Hwang Y, Ivey LE, Brown O, Bannon M, Akrobetu D, Sheaffer KE, Morgan O, Field CO, Suresh N, Gordon MG, Gunnell ET, Regruto LA, Wood CG, Fuller MT, Hales KG (2017) Testis-specific ATP synthase peripheral stalk subunits required for tissue-specific mitochondrial morphogenesis in *Drosophila*. BMC Cell Biol 18(1):16. https://doi.org/10.1186/s12860-017-0132-1

Spradling AC, Niu W, Yin Q, Pathak M, Maurya B (2022) Conservation of oocyte development in germline cysts from *Drosophila* to mouse. elife 11:e83230. https://doi.org/10.7554/eLife.83230

Steinberg F, Gerber SD, Rieckmann T, Trueb B (2010) Rapid fusion and syncytium formation of heterologous cells upon expression of the FGFRL1 receptor. J Biol Chem 285(48):37704–37715. https://doi.org/10.1074/jbc.M110.140517

Storto PD, King RC (1989) The role of polyfusomes in generating branched chains of cystocytes during *Drosophila* oogenesis. Dev Genet 10:70–86. https://doi.org/10.1002/dvg.1020100203

Strom NB, Bushley KE (2016) Two genomes are better than one: history, genetics, and biotechnological applications of fungal heterokaryons. Fungal Biol Biotechnol 3:4. https://doi.org/10.1186/s40694-016-0022-x

Szklarzewicz T (1997) Structure and development of the telotrophic ovariole in ensign scale insects (Hemiptera, Coccomorpha: Ortheziidae). Tiss Cell 29:31–38. https://doi.org/10.1016/s0040-8166(97)80069-9

Szklarzewicz T (1998) The ovaries of scale insects (Hemiptera, Coccinea). Morphology and phylogenetic conclusions. Folia Histochem Cytobiol 36:157–165

Szklarzewicz T, Kalandyk-Kołodziejczyk M, Michalik A (2022) Ovary structure and symbiotic associates of a ground mealybug, *Rhizoecus albidus* (Hemiptera, Coccomorpha: Rhizoecidae) and their phylogenetic implications. J Anat 41:860–872. https://doi.org/10.1111/joa.13712

Tóth KF, Pezic D, Stuwe E, Webster A (2016) The piRNA pathway guards the germline genome against transposable elements. Adv Exp Med Biol 886:51–77. https://doi.org/10.1007/978-94-017-7417-8_4

Tworzydło W, Biliński SM, Kočárek P, Haas F (2010) Ovaries and germline cysts and their evolution in Dermaptera (Insecta). Arthropod Struct Develop 39:360–368. https://doi.org/10.1016/j.asd.2010.05.004

Vedelek V, Laurinyecz B, Kovács AL, Juhász G, Sinka R (2016) Testis-specific Bb8 is essential in the development of spermatid mitochondria. PLoS One 11(8):e0161289. https://doi.org/10.1371/journal.pone.0161289

Xu W, Sato H, Bente H, Santos-González J, Köhler C (2022) Endosperm cellularization failure induces a dehydration stress response leading to embryo arrest. Plant Cell 35:koac337. https://doi.org/10.1093/plcell/koac337

Yang X, Chen D, Zheng S, Yi M, Liu Z, Liu Y, Yang D, Liu Y, Tang L, Zhu C, Huang Y (2022a) BmHen1 is essential for eupyrene sperm development in *Bombyx mori* but PIWI proteins are not. Insect Biochem Mol Biol 151:103874. https://doi.org/10.1016/j.ibmb.2022.103874

Yang D, Xu J, Chen K, Liu Y, Yang X, Tang L, Luo X, Liu Z, Li M, Walters JR, Huang Y (2022b) BmPMFBP1 regulates the development of eupyrene sperm in the silkworm, *Bombyx mori*. PLoS Genet 18(3):e1010131. https://doi.org/10.1371/journal.pgen.1010131

Yip EC, Mikó I, Ulmer JM, Cherim NA, Townley MA, Poltak S, Helms AM, De Moraes CM, Mescher MC, Tooker JF (2021) Giant polyploid epidermal cells and male pheromone production in the tephritid fruit fly *Eurosta solidaginis* (Diptera: Tephritidae). J Insect Physiol 130:104210. https://doi.org/10.1016/j.jinsphys.2021.104210

Zohar-Fux M, Ben-Hamo-Arad A, Arad T, Volin M, Shklyar B, Hakim-Mishnaevski K, Porat-Kuperstein L, Kurant E, Toledano H (2022) The phagocytic cyst cells in *Drosophila* testis eliminate germ cell progenitors via phagoptosis. Sci Adv 8(24):eabm4937. https://doi.org/10.1126/sciadv.abm4937

Zoller R, Schulz C (2012) The *Drosophila* cyst stem cell lineage: partners behind the scenes? Spermatogenesis 2(3):145–157. https://doi.org/10.4161/spmg.21380

Part II
Syncytia in Embryogenesis and Development

Chapter 4
Reshaping the Syncytial Drosophila Embryo with Cortical Actin Networks: Four Main Steps of Early Development

Rebecca Tam and Tony J. C. Harris

Abstract Drosophila development begins as a syncytium. The large size of the one-cell embryo makes it ideal for studying the structure, regulation, and effects of the cortical actin cytoskeleton. We review four main steps of early development that depend on the actin cortex. At each step, dynamic remodelling of the cortex has specific effects on nuclei within the syncytium. During axial expansion, a cortical actomyosin network assembles and disassembles with the cell cycle, generating cytoplasmic flows that evenly distribute nuclei along the ovoid cell. When nuclei move to the cell periphery, they seed Arp2/3-based actin caps which grow into an array of dome-like compartments that house the nuclei as they divide at the cell cortex. To separate germline nuclei from the soma, posterior germ plasm induces full cleavage of mono-nucleated primordial germ cells from the syncytium. Finally, zygotic gene expression triggers formation of the blastoderm epithelium via cellularization and simultaneous division of ~6000 mono-nucleated cells from a single internal yolk cell. During these steps, the cortex is regulated in space and time, gains domain and sub-domain structure, and undergoes mesoscale interactions that lay a structural foundation of animal development.

4.1 Introduction

Beneath the plasma membrane, a thin protein network forms the cell cortex. The cortex is mostly made of actin cytoskeletal networks. These networks can focus at specific structures, e.g. cell adhesion complexes (Harris and Tepass 2010), or form broad cortical domains based on polymerization of filamentous actin (F-actin) and F-actin linkage to the plasma membrane (Svitkina 2020; Chugh and Paluch 2018). Non-muscle myosin II ("myosin", hereafter) pulls on F-actin to contract such

R. Tam · T. J. C. Harris (✉)
Department of Cell & Systems Biology, University of Toronto, Toronto, ON, Canada
e-mail: tony.harris@utoronto.ca

M. Kloc, A. Uosef (eds.), *Syncytia: Origin, Structure, and Functions*, Results and Problems in Cell Differentiation 71,
https://doi.org/10.1007/978-3-031-37936-9_4

domains. In contrast, the Arp2/3 protein complex binds F-actin and nucleates branched actin networks to expand cortical domains. These myosin- and Arp2/3-based actin networks are induced by the signalling of distinct Rho family small G proteins (Banerjee et al. 2020). Various components of the cell can activate or inhibit these signalling pathways, and such changes are often coupled with membrane trafficking. For example, internal centrosomes can induce new cortical domains by promoting F-actin growth and/or membrane exocytosis, as occurs at the immune synapse or the leading edge of migratory cells (Kopf and Kiermaier 2021). Once formed, cytoskeletal networks pull and push on a variety of subcellular components. For example, a major effect of cortical contractility is on the bulk cytoplasm: uniform cortical contraction and cytoplasmic counter pressure produces a spherical cell shape during mitotic cell rounding (Kelkar et al. 2020), whereas non-uniform cortical contraction can constrict a portion of the cell inward and displace cytoplasm to other regions which bulge outward, e.g. during cytokinesis (Sedzinski et al. 2011) and bleb-based cell migration (Bodor et al. 2020). During health and disease, dynamic cortical changes control cell division (Sugioka 2022), cell movement (Bodor et al. 2020), cell–cell fusion (Lee and Chen 2019), and cell–cell interactions (Lecuit et al. 2011).

Early animal embryos are superb models for studying the control and effects of cortical actin networks. It is important to understand the cell cortex in naturally evolved contexts, and probing the first steps of animal development has special appeal. The large cells of early embryos are also ideal for microscopy and can be manipulated with a variety of approaches. Such studies reveal fundamental behaviours of the cell cortex. For example, studies of the one-cell *C. elegans* embryo have shown how a cell can polarize through centrosome-induced flow of the cortical actomyosin network to one pole and resulting counter flow of cytoplasm to the opposite pole (Gubieda et al. 2020; Munro et al. 2004). Analyses of frog and starfish blastomeres have revealed cortical feedback loops that generate dynamic waves of cytoskeletal activity associated with embryo cleavage (Michaud et al. 2021; Bement et al. 2015). Compaction of the mouse embryo also involves cortical waves of actomyosin (Maître et al. 2015), and cell–cell adhesion is reinforced by microtubule-based redistribution of cortical F-actin from the apical cell surface to cell–cell contacts (Zenker et al. 2018).

Early Drosophila embryogenesis is a special case, since the first 3 hours of development occur within a syncytium (Blake-Hedges and Megraw 2019; Schmidt and Grosshans 2018; Sokac et al. 2023; Lv et al. 2021; Sullivan and Theurkauf 1995; Mazumdar and Mazumdar 2002; Schejter and Wieschaus 1993b). The one-cell embryo develops outside the mother and is encased within an egg shell. For a single cell, its ~0.5 mm by ~0.2 mm size (Markow et al. 2009) is huge! Its ovoid shape exposes a broad plane of embryo surface, allowing detailed imaging of the cell cortex. Much of syncytial development is directed by maternally supplied gene products (Vastenhouw et al. 2019), which can be manipulated by Drosophila genetic approaches, optogenetics, and injections of chemical inhibitors. The syncytium is also pre-patterned with molecular distributions that direct gastrulation and later stages of embryogenesis (Stathopoulos and Newcomb 2020).

Here, we review four main steps of syncytial development, starting with the first nuclear divisions and ending with formation of the blastoderm epithelium. Two of these steps, nuclear redistribution along the embryo long axis and nuclear division at the embryo periphery, involve networks of conserved molecular players that dynamically reshape the syncytium. Two additional steps, pole cell formation and cellularization, involve specialized factors that coopt cytoskeletal machinery to form mono-nucleated cells at specific locations and times. For all four steps, we review the induction and regulation of cortical actin networks, the structural organization of the networks, and how the networks function with each other and additional cell components to reshape the syncytial embryo.

4.2 Axial Expansion: Spreading Nuclei Along the Anterior–posterior Axis of the Embryo

4.2.1 General Context

During the early stages of syncytial Drosophila development, nuclei and their associated centrosomes undergo nine rounds of synchronous division away from the cell cortex (Blake-Hedges and Megraw 2019; Lv et al. 2021). Nuclei within the embryo are individually associated with centrosomes which generate microtubule-based mitotic spindles for nuclear division. Otherwise, the inner depths of the early embryo are mainly populated by yolk. However, a three-dimensional mesh of F-actin has been reported throughout the embryo (Von Dassow and Schubiger 1994), and F-actin accumulates at mitotic spindles of the dividing nuclei (Telley et al. 2012; Von Dassow and Schubiger 1994). Additionally, the posterior pole contains the germ plasm, a heterogeneous composite of phase-separated molecular condensates that ultimately distinguishes the germline from the soma (Dodson and Kennedy 2020).

As early embryogenesis begins, the periphery of the one-cell embryo is relatively homogeneous (Schmidt and Grosshans 2018), with the possible exception of the posterior pole which has a distinct actomyosin network during oogenesis (Doerflinger et al. 2022). Microtubules form a thick network close to the plasma membrane and organize an interconnected system of endoplasmic reticulum (ER) and Golgi membranes (Frescas et al. 2006). Actin and myosin form a thinner fibrous network closely associated with the plasma membrane (Von Dassow and Schubiger 1994; Royou et al. 2002).

During nuclear division cycles 4–6, a process called axial expansion spreads nuclei from an initial anterior cluster to become evenly distributed along the anterior-posterior axis of the embryo (Blake-Hedges and Megraw 2019; Lv et al. 2021). Axial expansion is ultimately required for an even distribution of mono-nucleated cells across the embryo (Royou et al. 2002). The process is dependent on contraction of the actomyosin cortex and involves mutual repulsion of mitotic spindles. Elegant embryo extract studies have shown the control of nuclear distancing and arrangement by astral microtubule repulsion (Telley et al. 2012; De-Carvalho et al. 2022), but this mechanism is insufficient to redistribute nuclei from an initial semi-spherical cluster to a distribution that spans the full length of the ovoid embryo. This section focuses on the actomyosin cortex and its role in redistributing internal nuclei.

4.2.2 The Role and Regulation of an Embryo-Wide Actomyosin Network

For axial expansion, cortical myosin accumulation is key. The myosin network is first induced as a band that encircles the short axis of the embryo above the anterior cluster of nuclei (Fig. 4.1). This band assembles and disassembles with the cell cycle, with a maximum accumulation at interphase (Royou et al. 2002). The level of recruited myosin and width of the band both increase from cycle 4 to cycle 6, as axial expansion of the nuclei occurs (Royou et al. 2002; Deneke et al. 2019). Indicating the requirement of the induced myosin networks, embryos lacking myosin activity fail to disperse the nuclei (Wheatley et al. 1995). As the myosin network forms, its contraction is evident from the cortex indenting (Royou et al. 2002) and from cortical flow (Deneke et al. 2019). Myosin turnover and redeployment are also important for axial expansion to occur (Von Dassow and Schubiger 1994). Further studies have shown how cortical myosin moves the internal nuclei and how the internal nuclei pattern cortical myosin.

Deneke et al. (2019) found that the localized cortical contraction induces cytoplasmic streaming that disperses the nuclei. Cytoplasmic streaming, detected by the movement of yolk granules, tightly correlates with myosin network contraction and nuclear dispersal. As a band of myosin contracts above the group of nuclei, the cytoplasm and nuclei move away from the site in a common flow. As the myosin network grows from cycles 4–6, so does the intensity of the cytoplasmic flows. And when the myosin network becomes uniformly distributed over the embryo cortex after cycle 6, the cytoplasmic flows subside. Optogenetic induction of a uniform distribution of myosin before cycle 6 also blocks the cytoplasmic flows (Deneke et al. 2019), indicating that localized myosin accumulation is needed to displace the cytoplasm. Thus, contraction of the band of cortical myosin around the nuclei leads to an internal flow of cytoplasm and nuclei towards the anterior and posterior of the one-cell embryo. From cycles 4–6, this repeated effect evenly distributes nuclei along the anterior–posterior axis, and when cortical myosin levels subside during mitosis, inter-spindle repulsion would additionally contribute to the nuclear spread (Telley et al. 2012; De-Carvalho et al. 2022) (Fig. 4.1). During cycles 6–9, when myosin repeatedly accumulates with a uniform distribution and when cytoplasmic flows are minimal (Deneke et al. 2019), the nuclear distribution progressively expands toward the cortex (Foe and Alberts 1983), presumably due to mitotic

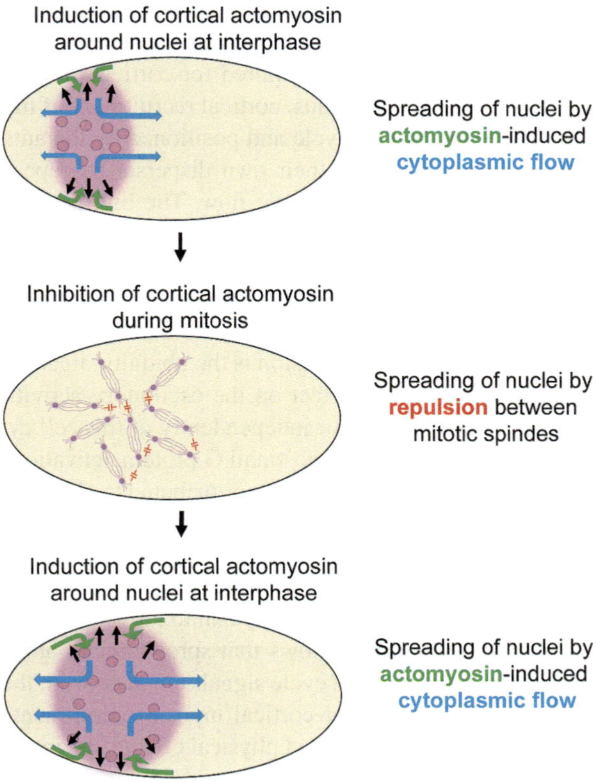

Fig. 4.1 Axial expansion of the early nuclear distribution. During nuclear division cycles 4–6, actomyosin network contractions promote cytoplasmic flows leading to even spreading of nuclei along the embryo. During interphase, signals from the inner cluster of nuclei promote actomyosin activity at the cortex. Cortical flow and inward contraction of the actomyosin network displace cytoplasm and nuclei. During mitosis, the actomyosin network disassembles, and nuclei are distanced by repulsions among mitotic spindles. These steps repeat, with broadening of the nuclear distribution coupled with broadening of the induced actomyosin network. As more nuclei accumulate, they induce stronger actomyosin activity and stronger cytoplasmic flow results in more nuclear spread. See main text for details

spindles repelling each other throughout the inner volume of the syncytium (Telley et al. 2012, De-Carvalho et al. 2022).

The cycles of cortical myosin assembly and disassembly are entrained with the cell cycle. This role of the cell cycle was shown by cell cycle inhibitors blocking the cyclic accumulations of cortical myosin (Royou et al. 2002). A biosensor of phosphorylation by the kinase Cdk1 and de-phosphorylation by the phosphatase PP1 revealed oscillations in the activities of these cell cycle regulators during the syncytial divisions. During mitosis, there is high Cdk1 activity and low PP1 activity, and as nuclei exit mitosis, Cdk1 activity decreases, and PP1 activity increases (Deneke et al. 2019, 2016). In addition to providing temporal information, this phosphoregulatory mechanism also provides spatial information, as PP1 activity emanates from where the nuclei are found (Deneke et al. 2019). Higher PP1 activity precedes Rho small G protein activation which is followed by myosin assembly (Deneke et al. 2019), and Rho kinase activity is required for cortical myosin accumulation and contraction (Royou et al. 2002). Thus, cortical recruitment of myosin is dependent on signalling timed by the cell cycle and positioned by the anterior cluster of nuclei. In this way, the nuclei direct their own dispersal by repeatedly inducing cortical myosin contraction and cytoplasmic flow. The link between PP1 activity and Rho small G protein activation remains unknown.

Additional factors regulate the cycles of cortical myosin assembly. For example, the centrosomal protein CP190 is required for recruitment of myosin to the cortex and for axial expansion (Chodagam et al. 2005). Another factor needed for the cortical myosin accumulation and axial expansion is the ubiquitin ligase Cul-5. Loss-of-function mutants of Cul-5 have no effect on the oscillatory activities of PP1 and Cdk1, suggesting it acts downstream or independently of this cell cycle regulation. The mutants do display reductions to Rho small G protein activation, cortical myosin accumulation, and cytoplasmic flow, defects attributed to abnormally high activity of the non-receptor tyrosine kinase Src (Hayden et al. 2022). How Src affects Rho and myosin activation is unclear, and whether the regulation is linked to the cell cycle or is constitutive remains to be determined.

In summary, the repeated recruitment of myosin to the one-cell embryo cortex is important for generating cytoplasmic flows that spread nuclei along the anterior-posterior axis for axial expansion. Cell cycle signals emanate from the internal cluster of nuclei to induce each period of cortical myosin recruitment. Through this coordination of biochemical signalling and physical effects, the internal nuclei control the cortical contractions that displace them. Additional factors promote the myosin accumulation but remain incompletely understood, and the nuclear dispersion also involves physical repulsion between mitotic spindles.

4.3 Pole Cell Budding: Separating Mono-nucleated Germline Cells from the Syncytial Soma

4.3.1 General Context

After axial expansion, nuclei that arrive at the posterior pole come into proximity to germ plasm and bud from the syncytium as mono-nucleated primordial germ cells, also known as pole cells (Foe and Alberts 1983; Schejter and Wieschaus 1993b) (Fig. 4.2). During earlier oogenesis, cytoplasmic determinants of the germline are recruited below the posterior cortex of the egg to form the germ plasm (Mahowald 2001). After nuclei of the one-cell embryo are carried by cytoplasmic flows to the posterior cortex, they use their centrosomal microtubule arrays and associated dynein motor activity to integrate with the germ plasm (Lerit and Gavis 2011). At nuclear division cycle 10, the resulting pole cell nuclei then cellularize individually,

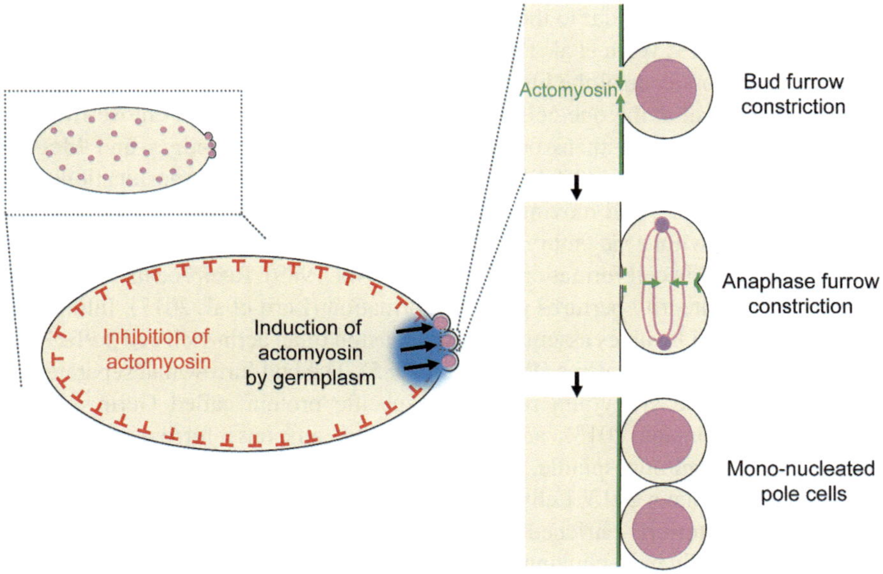

Fig. 4.2 Budding and cleavage of mono-nucleated primordial germ cells (pole cells). At nuclear division cycle 10, a posterior actomyosin network promotes the separation of pole cells from the syncytial soma. Throughout most of the embryo, actomyosin contraction is restrained. At the posterior end of the embryo, actomyosin activity is promoted by the germ plasm. Myosin accumulates at the base of the budding pole cell and the bud furrow constricts. As the pole cell separates from the syncytium, an anaphase furrow divides it into two mono-nucleated cells. See main text for details

thereby separating the germline from the syncytial soma (Strome and Lehmann 2007). This cellularization process produces a cluster of round pole cells at the posterior tip of the much larger syncytium (Blake-Hedges and Megraw 2019). The pole cell cluster remains at the posterior surface until it is internalized with the posterior mid-gut invagination during gastrulation (Dansereau and Lasko 2008). Many factors contribute to pole cell formation (Dansereau and Lasko 2008; Wilson and Macdonald 1993). Here, we will focus on the regulation and roles of cortical actomyosin networks.

4.3.2 *The Roles and Regulation of Local Actomyosin Networks*

The posterior cluster of pole cells forms through two types of myosin-based cleavage: one that asymmetrically divides the pole cell from the syncytial soma and another that symmetrically divides the pole cell (Cinalli and Lehmann 2013) (Fig. 4.2). After an initial budding of the cortex envelopes a pole cell nucleus, the first cleavage furrow, called the bud furrow, separates the pole cell from the rest of the syncytial embryo. At anaphase, a second cleavage furrow, called the anaphase furrow, forms perpendicular to the cell base and divides the pole cell in two (Cinalli and Lehmann 2013; Warn et al. 1985).

Specific factors control the budding and cleavage of pole cells. A key trigger is centrosome arrival at the one-cell embryo cortex. Normally, nuclei move from the interior to the cortex with associated centrosomes (Blake-Hedges and Megraw 2019). However, inhibition of DNA synthesis de-couples this translocation, with centrosomes dividing and moving to the cortex on their own, and disrupted nuclei remaining deep within the embryo. The arrival of these centrosomes alone is sufficient to trigger pole cell formation (Raff and Glover 1989). Reciprocally, disruption of centrosome integrity perturbs pole cell formation (Lerit et al. 2017). Initial budding of a pole cell involves assembly and expansion of an actin-rich cell surface cap through unknown mechanisms (Warn et al. 1985). The bud furrow that separates the pole cell from the syncytium requires a specific protein called Germ cell-less (Cinalli and Lehmann 2013), and the subsequent anaphase furrow seems to be directed by the mitotic spindle, similar to the division of mono-nucleated cells (Cinalli and Lehmann 2013; Lehner 1992; Mishima 2016). Germ cell-less was discovered as a posteriorly enriched protein required for pole cell formation (Jongens et al. 1992). Cinalli and Lehmann (2013) not only discovered that Germ cell-less is required for the bud furrow but also that the level of Germ cell-less expression directly correlates with the speed of bud furrow constriction. Additionally, the myosin-enriched bud furrow requires Rho signalling and downstream effectors. Specifically, RhoGEF2 is localized to the bud furrow and is needed for its constriction (Padash Barmchi et al. 2005), as are Rho1 and Rho kinase (Cinalli and Lehmann 2013), as well as the formin Diaphanous (Dia) (Afshar et al. 2000) and Anillin (Field et al. 2005). Overall, the combination of a nucleus-associated centrosome and germ plasm-associated Germ cell-less protein promotes pole cell cleavage from the syncytium via Rho signalling of actomyosin activity. How Germ cell-less protein promotes the bud furrow remains unclear, but crosstalk with centrosome regulation is evident (Lerit et al. 2017).

As germline nuclei form pole cells at the posterior, somatic nuclei have moved to the periphery of the rest of the embryo but only induce transient cortical compartments (discussed in the next section). There is the potential for these transient compartments to fully separate from the syncytium, but this is normally prevented. Restricting Germ cell-less to the germ plasm is important since ectopic expression of Germ cell-less throughout the syncytium increases the contraction of the transient compartments (Cinalli and Lehmann 2013). Depletion of Steppke, an Arf-guanine nucleotide exchange factor (GEF), also results in elevated myosin activity that excessively constricts the bases of the transient syncytial compartments (Lee and Harris 2013). At the anterior pole, depletion of Steppke results in the full cleavage of mono-nucleated somatic cells from the syncytium, a pronounced effect that might involve the unique geometry and/or specific protein localization of the pole. Reciprocally, over-expression of Steppke inhibits the normal cleavage of pole cells from the posterior pole (Lee et al. 2015). Normally, global expression of Steppke seems to prevent mono-nucleated cells from budding at ectopic sites, and local Germ cell-less activity allows pole cells to overcome this inhibition at the posterior pole, an example of pattern formation through the coupling of global inhibition and local activation (Fig. 4.2).

In summary, centrosome-associated nuclei arrive at the posterior pole, join the germ plasm, reshape the cell cortex, and separate from the syncytium as a cluster of mono-nucleated cells that become the germ line. The arrival of somatic nuclei to the rest of the syncytial cortex results in distinct cortical remodelling.

4.4 The Syncytial Blastoderm: Repeated Cortical Compartmentalization for Dividing Somatic Nuclei

4.4.1 General Context

From nuclear division cycles 10–13, somatic nuclei divide at the one-cell embryo periphery in close association with the cell cortex (Fig. 4.3). After axial expansion completes, nuclei progressively move to the embryo periphery by microtubule-based propulsion (Baker et al. 1993; Blake-Hedges and Megraw 2019; Lv et al.

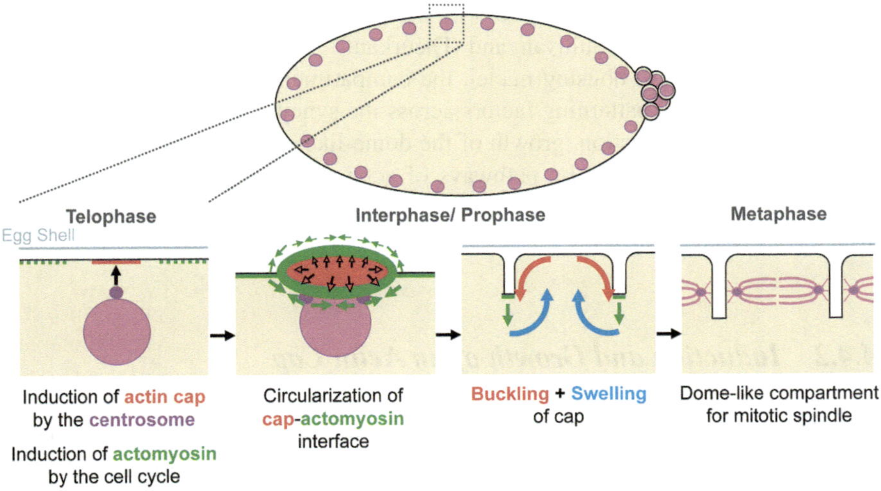

Fig. 4.3 Forming dome-like compartments for dividing nuclei of the syncytial blastoderm. At nuclear division cycle 10, somatic nuclei begin dividing in close association with the one-cell embryo cortex. Assembly of a dome-like compartment for a dividing nucleus begins with centrosome induction of an expanding actin cap and cell cycle induction of a surrounding actomyosin network. As the cap grows against the actomyosin network, it induces a smooth and circular actomyosin boundary which restrains cap growth in the embryo surface plane. With continued growth, the cap buckles to form a dome-like compartment. The actomyosin network is fenestrated with many actin caps and appears to promote swelling of the caps by constricting toward the embryo interior and generating cytoplasmic counterforces. Each dome-like compartment houses a mitotic spindle to attach it to the cortex and to prevent collisions with neighbouring spindles. See main text for details

2021). At cycle 10, an unknown trigger induces close association of most nuclei with the cortex. As the nuclei gain cortical association, they use their centrosomal microtubule networks to divide the previously continuous endoplasmic reticulum among them. In this way, each nucleus organizes its own endomembrane system within the syncytium (Frescas et al. 2006). The cell cortex also gains a domain structure linked to nuclear positioning. An actin cap forms above each nucleus, and an embryo-wide actomyosin network assembles between and around all of the caps (Foe et al. 2000). Plasma membrane compartmentalization is evident from minimal lateral exchange of membrane components between neighbouring caps (Mavrakis et al. 2009). Each actin cap grows into a dome-like compartment that remains basally open. These compartments form synchronously as embryo-wide arrays with structural analogy to cardboard egg trays. Neighbouring compartments ingress in close association. The close apposition of their sides forms pseudocleavage furrows. As these furrows ingress, actomyosin localizes at their base and maintains an embryo-wide network that integrates across the compartments (Foe et al. 2000; Zhang et al. 2018). The compartments form temporarily to house dividing nuclei. By providing attachment sites for spindles and preventing the collision of neighbouring spindles, the compartments prevent aberrant nuclear fusions and maintain nuclei at the embryo periphery, where they will ultimately form the first embryonic epithelium of mono-nucleated cells. The compartments partially regress after each nuclear division cycle and then reform with a doubling of number for the next nuclear division cycle (Sullivan and Theurkauf 1995; Schejter and Wieschaus 1993b). In addition to housing nuclei, the compartmentalization also restricts the diffusion embryonic patterning factors across the syncytium (Daniels et al. 2012). As discussed in this section, growth of the dome-like compartments is directed by the centrosome and driven by pathways of actin network growth and membrane delivery. Biophysical interactions further coordinate the caps, spindles, and embryo-wide actomyosin network.

4.4.2 Induction and Growth of an Actin Cap

Signals from the centrosome are key for inducing the actin cap of each nucleus (Fig. 4.3). Actin cap growth and disassembly are synchronous with the cell cycle (Warn et al. 1984). At anaphase, actin accumulation is initiated above the centrosome at each pole of the mitotic apparatus (Karr and Alberts 1986). Showing the sufficiency of centrosomes to induce the caps, induction of actin accumulations occurs above centrosomes even when they migrate to the embryo periphery without nuclei (Raff and Glover 1989). The requirement of centrosomes is evident from analyses of mutants with compromised centrosome integrity (Megraw et al. 1999; Vaizel-Ohayon and Schejter 1999). With loss of the core centrosome component Centrosomin, actin accumulation is diminished at nascent caps, and later compartments are fragmented (Vaizel-Ohayon and Schejter 1999). A mutant allele affecting a specific domain of Centrosomin separates its role in building the mitotic spindle from its role in inducing actin caps, with only the latter disrupted. This domain was found to bind a protein called Centrocortin, which localizes to both the centrosome and the actin cortex and is needed for inducing the actin caps. Thus, Centrocortin seems to be specifically involved in signalling from the centrosome to the actin cap (Kao and Megraw 2009). Another candidate for mediating signalling from the centrosome to the actin cortex is the protein Scrambled, which also associates with both locations, and is needed for actin accumulation at the caps (Stevenson et al. 2001).

A major contributor to the growth of an actin cap is the Arp2/3 complex, activated downstream of a conserved small G protein signalling pathway. A dedicator of cytokinesis (DOCK) family GEF, Sponge, and its adaptor protein, Engulfment and Cell Motility (ELMO), both localize to the surface cap and sides of the dome-like compartment, and are both required for cap growth into the compartment (Postner et al. 1992; Zhang et al. 2018; Schmidt et al. 2018). Rac-GTP also accumulates at the cap and compartment sides, and this accumulation requires Sponge (Zhang et al. 2018), consistent with the small G protein specificity of DOCK3 and DOCK4 (Kukimoto-Niino et al. 2021), the predicted human orthologs of Sponge (Flybase). An effector of Rac-GTP is the Scar/WAVE complex, which binds and activates the Arp2/3 network for F-actin assembly (Rottner et al. 2021). Both Scar and components of the Arp2/3 complex localize to the surface cap and sides of the dome-like compartment, and are needed for growth of the cap into the compartment (Zallen et al. 2002; Stevenson et al. 2002; Zhang et al. 2018). Rac-GTP promotes the cap localization of Scar, and Sponge is also required for the cap localization of Scar and Arp2/3 (Henry et al. 2022). Arp2/3 seems to induce puncta of F-actin across the cap (Jiang and Harris 2019), suggestive of a growing cortical domain composed of interconnected actin asters (Svitkina 2020). It is tempting to speculate that these cortical domains grow in coordination with new membrane addition: new portions of plasma membrane and new actin asters might be added in concert to expand the domain of a cap. However, F-actin turnover during cap expansion must also be taken into account (Cao et al. 2010).

Additional regulators of Arp2/3 networks affect growth of the cap. Specifically, the proteins Coronin and Cortactin localize across the early cap, and with cap growth, Cortactin becomes uniquely localized to the cap periphery (Xie et al. 2021). Sponge is needed for the accumulation of both proteins (Henry et al. 2022), and Coronin is needed for the peripheral localization of Cortactin (Xie et al. 2021). Accumulation of F-actin in the cap relies more on Coronin at earlier stages and more on Cortactin at later stages, and both proteins are needed for caps to attain normal areas (Xie et al. 2021). As the caps bend inward, the sides of the resulting dome-like compartments also differentiate into sub-domains with additional components (Schmidt and Grosshans 2018). How elaborations to the cortex affect the transformation of a 2D actin cap into a 3D dome-like compartment will be important to investigate further.

The expanding cap is also a composite of Arp2/3-based actin asters and Dia-based actin bundles. The formin Dia is needed for full cap growth and proper formation of a dome-like compartment (Afshar et al. 2000; Cao et al. 2010; Jiang and Harris 2019). Although most strongly localized to cortical actomyosin networks by

Rho small G protein signalling (Afshar et al. 2000; Grosshans et al. 2005), Dia also localizes to the cortex of the expanding actin cap, a domain with relatively low Rho-GTP (Jiang and Harris 2019). Dia localizes to actin bundles at the base of inward folds that criss-cross the cap surface, and growth of Arp2/3 networks in between the bundles appears to push them apart as the cap grows. Thus, the bundles seem to provide cap sub-structures that may coordinate local Arp2/3 network expansions for overall cap growth (Jiang and Harris 2019). Notably, the Arp2/3-dependent actin asters form independently of Dia, and the Dia-based actin bundles form independently of Arp2/3 (Jiang and Harris 2019), and the separate networks appear to complete for actin monomers (Xie et al. 2021). The kinase Par-1 is specifically required for the Dia-based actin bundles, Dia localization to the bundles, and overall cap growth. Providing another potential link between the centrosome and cap, Par-1 localizes to both the centrosome and the Dia-based bundles of the cap (Jiang and Harris 2019).

Membrane trafficking is also an important contributor to growth of the dome-like compartment. From cycles 10–13, pseudocleavage furrows ingress more and more deeply, and during each cycle they ingress rapidly (Holly et al. 2015). This ingression requires expansion of the cortical actin cytoskeleton coupled with delivery of membrane from internal stores. The small G proteins RalA and Rab8 are required for transport of internal membrane to supply the growing furrows (Holly et al. 2015; Mavor et al. 2016). Each protein has been linked to the exocyst complex that tethers exocytic vesicles to the plasma membrane (Wu and Guo 2015). RalA localizes to the plasma membrane and is required for recruitment of the exocyst and Rab8 (Holly et al. 2015). Rab8 additionally localizes to a vesicle population partially overlapping with recycling endosomes and Golgi mini-stacks and requires the exocyst for plasma membrane localization (Mavor et al. 2016). The small G protein Arf1, a Golgi regulator, is also needed for furrow growth as well as the proper structure of Golgi mini-stacks. Arf1 localizes most strongly to Golgi mini-stacks and at lower levels to other internal membranes and the plasma membrane. ASAP, an Arf1 GTPase activating protein (GAP), is similarly required for furrow growth and for Arf1 localization at the Golgi. Intriguingly, ASAP localizes strongly to the plasma membrane, suggesting it may promote the recycling of Arf1-GDP back to the Golgi for activation and function, although ASAP also undergoes a dramatic redistribution to the nuclear matrix during mitosis (Rodrigues et al. 2016). Another key regulator of intramembrane trafficking is Nuclear-fallout, an ortholog of arfophilins which bind and function with both Arf5 and Rab11 (Hickson et al. 2003). Nuclear-fallout cycles between the cytoplasm and a pericentrosomal localization associated with recycling endosomes (Rothwell et al. 1998; Riggs et al. 2003). It is needed for forming the dome-like compartments (Rothwell et al. 1998) and for trafficking to the pseudo-cleavage furrow from internal membranes (Rothwell et al. 1999). To do so, Nuclear-fallout cooperates with Rab11 which it binds through a conserved motif (Riggs et al. 2003). The centrosomal localization of Nuclear-fallout relies on microtubules and dynein and occurs specifically when pseudocleavage furrows form during the cell cycle (Riggs et al. 2007). This cell cycle-dependent localization of Nuclear-fallout is controlled by phospho-regulation downstream of Cdk1 and Polo kinase (Brose et al. 2017). Thus, Nuclear-fallout is spatially and temporally linked with centrosome induction of the dome-like compartments. It will be interesting to learn how the effects of Nuclear-fallout integrate with those of other trafficking regulators, such as RalA, Rab8, and Arf1, and with the induction of cortical Arp2/3 actin networks. Additionally, endocytosis is associated with formation of the compartments (Sokac and Wieschaus 2008; Rikhy et al. 2015) and is substantially increased as the compartments regress (Sokac and Wieschaus 2008).

4.4.3 Coupling Cap Growth with the Mitotic Spindle

Growth of the dome-like compartment is closely coupled with assembly and lengthening of the mitotic spindle, both locally and across the embryo. Nascent actin caps are roughly circular as they form over each centrosome at anaphase and telophase (Sullivan and Theurkauf 1995; Schejter and Wieschaus 1993b). After centrosome duplication at interphase, continued surface expansion of the cap is needed for full separation of the two centrosomes (Cao et al. 2010), and excessive expansion of a cap leads to an abnormal increase in centrosome separation (Sommi et al. 2011). Indicating that these effects are physical, centrosomal microtubules associate closely with the cortex of the actin cap (Cao et al. 2010; Sommi et al. 2011). By interphase, the caps have transformed into shallow and circular dome-like compartments. As the compartments continue to ingress, DE-cadherin-based adhesion between neighbouring compartments transforms the circular compartments into hexagonal compartments (Dey and Rikhy 2020). From metaphase to anaphase, each compartment then elongates in concert with lengthening of the mitotic spindle (Sullivan and Theurkauf 1995). These elongations present a potential challenge: there might not be enough embryo surface area for all compartments to expand simultaneously. This crowding problem is avoided since the formation, expansion, and regression of the dome-like compartments occurs in waves across the embryo in response to a dynamic cell cycle signalling network (Deneke et al. 2016). Because of these waves, mechanical properties differ across the interconnected cortex of the one-cell embryo at any one time. Over time, neighbouring regions expand and contract in coordination, and the embryo cortex displays so-called yo-yo motions, with nuclei moving from their original positions and then returning. Analyses of Kinesin-5 and Map 60 mutants indicated that spindle elongation is required for regions to expand, and actin networks are needed for the return movement, possibly due to elasticity of the network, to cell cycle-dependent induction of actomyosin networks, or to both effects (Lv et al. 2020). It remains unclear how these movements might affect diffusing signals of embryo pre-patterning, but mixing effects may be minimized by both the nucleus-associated endomembrane networks (Frescas et al. 2006) and the pseudocleavage furrows (Daniels et al. 2012).

4.4.4 Coupling Cap Growth with the Surrounding Actomyosin Network

An interconnected actomyosin network surrounds and separates the actin caps and then localizes to the basal rims of the dome-like compartments as an embryo-wide network. Like pre-cycle 10, the broad actomyosin network assembles and disassembles with the cell cycle. Cortical myosin appears at the start of telophase (Royou et al. 2002), just after actin caps begin to grow (Sullivan and Theurkauf 1995, Schejter and Wieschaus 1993b). It accumulates maximally at interphase and prophase, as dome-like compartments arise, and it disassembles to a minimum by metaphase (Royou et al. 2002), when the compartments are at their deepest (Sullivan and Theurkauf 1995, Schejter and Wieschaus 1993b).

Strikingly, myosin perturbations that disrupt either earlier axial expansion or later cellularization do not prevent formation of dome-like compartments of the syncytial blastoderm (Royou et al. 2004). With diminished myosin activity, a compartment seems to form by an expanding cap growing against a non-contractile composite of remaining components of the actomyosin network, including Anillin and the septin Peanut. As individually expanding caps grow against these domains, and each other, they appear to transform into dome-like compartments by buckling (Zhang et al. 2018).

When both networks are active, they affect each other. Normally, intervening actomyosin networks thin as the actin caps expand, and actomyosin is restricted to the tips of the ingressing pseudocleavage furrows (Royou et al. 2002; Zhang et al. 2018). Weakening of the caps leads to expanded distributions of the actomyosin domains (Zhang et al. 2018; Schmidt et al. 2021), and tracking of individual elements of the actomyosin networks suggests they are physically displaced by the cap (Zhang et al. 2018). Reciprocally, weakening of the actomyosin domain leads to excessive lateral expansions of caps into irregular shapes (Zhang et al. 2018; Sommi et al. 2011; Sherlekar et al. 2020). The restriction of cap growth by the surrounding actomyosin network seems to unify pseudocleavage furrow length, since myosin perturbation leads to irregular centrifugal growth of individual caps and non-uniform depths of pseudocleavage furrows around dome-like compartments (Zhang et al. 2018; Sherlekar and Rikhy 2016). Additionally, myosin is responsible for tensile stress detected across cap surface (Zhang et al. 2018). Since myosin activity restricts cap expansion in the embryo surface plane (Zhang et al. 2018, Sommi et al. 2011, Sherlekar et al. 2020), the actomyosin network does not seem to pull cap circumferences centrifugally in this plane. Instead, the tensile stress across the cap surface may arise from swelling of the cap by myosin-dependent displacement of cytoplasm. As discussed, earlier axial expansion of nuclei involves displacement of internal cytoplasm by inward contraction of an embryo-encircling myosin network. When the myosin network becomes fenestrated with actin caps, its inward squeezing may displace cytoplasm into the caps, swelling them and generating tensile stress across their surfaces. Thus, formation of the dome-like compartment seems to involve myosin-corralled buckling and myosin-promoted swelling of the growing cap (Fig. 4.3), although an irregularly shaped compartment can still form when myosin is depleted.

The actomyosin network surrounding the caps appears to be regulated both biochemically and mechanically. One level of biochemical regulation is the assembly and disassembly under the control of cell cycle regulators (Royou et al. 2002). Induction involves Rho small G protein signalling, and full assembly involves multiple proteins (including Anillin, Peanut, and Dia) that recruit each other to the actomyosin domains and promote integrity of the pseudocleavage furrow (Schmidt and Grosshans 2018). Also, actomyosin activity is held in check to restrain contractility around the base of each dome-like compartment. Myosin phosphatase localizes to the actomyosin domain and prevents over-constriction of the actomyosin ring that encircles the base of the dome-like compartment (Zhang et al. 2018). Steppke also localizes to the domain and uses its Arf-GEF activity to prevent actomyosin ring hyper-constriction that can abnormally dislodge nuclei from the embryo periphery (Lee and Harris 2013). Involving an apparent physical effect, actin cap expansions are also required for forming full-sized clearances across the embryo-wide actomyosin network (Zhang et al. 2018). Moreover, Arp2/3 activity is needed for enrichment of myosin in a smooth circle around the cap (Sharma et al. 2021), a circle that becomes the base of the dome-like compartment (Zhang et al. 2018). In silico reconstitution studies of interactions between the two networks indicate that the local induction of a circular actomyosin network requires both centrifugal pushing of the Arp2/3 network and a local mechanical response of surrounding myosin mini-filaments that increases their activity and results in a contractile ring around the cap (Sharma et al. 2021). In this way, an actomyosin ring can be locally induced in a mainly non-contractile network to restrain cap growth and properly shape the dome-like compartment for mitosis (Fig. 4.3). This lateral interaction between the expanding cap and actomyosin border is most relevant at the start of the syncytial blastoderm division cycles, a stage when both the caps and surrounding actomyosin networks assemble near the embryo surface plane (Sharma et al. 2021; Zhang et al. 2018; Foe et al. 2000). At later cycles, when pseudocleavage furrows of the dome-like compartments regress to a lesser degree (Holly et al. 2015), lateral interactions between Arp2/3 and actomyosin networks may only occur at the flatter cortex between the sister nuclei of a division, and it is less clear how actin caps induced above the centrosome spread down the sides of non-regressed pseudocleavage furrows to meet newly assembled actomyosin networks at their base. Another unknown is the nature of the interface between the Arp2/3 and actomyosin cortical domains that prevents them from mixing as they laterally interact to reshape the cell.

The syncytial blastoderm is a model for how the cortex can be rapidly and stereotypically reshaped by coordinated activities of the centrosome, membrane trafficking, expansive cortical actin networks, and contractile cortical actin networks. The factors involved are conserved across animals and are deployed uniquely to form arrays of cortical compartments that house nuclei as they divide at the periphery of a syncytium.

4.5 Cellularization: Making the First Embryonic Epithelium

4.5.1 General Context

As the nuclear division cycles of the syncytial blastoderm occur, the embryo undergoes a transition from maternal to zygotic gene expression (Vastenhouw et al. 2019). Some of the newly expressed proteins direct dramatic cortical changes responsible for transforming the syncytial blastoderm into the first embryonic epithelium of mono-nucleated cells, the cellular blastoderm (Fig. 4.4). This process of cellularization occurs after nuclear division cycle 13, during an extended interphase of cycle 14. At the onset of cellularization, centrosomes duplicate atop each nucleus and nucleate microtubule networks that extend downward to form inverted baskets that encase each nucleus and provide a scaffold to build each columnar epithelial cell (Fig. 4.4). Like earlier pole cell formation, the cellularization of somatic nuclei involves a coincidence detector, but instead of spatially localized germ plasm acting with centrosomes, cellularization involves temporally induced zygotic gene

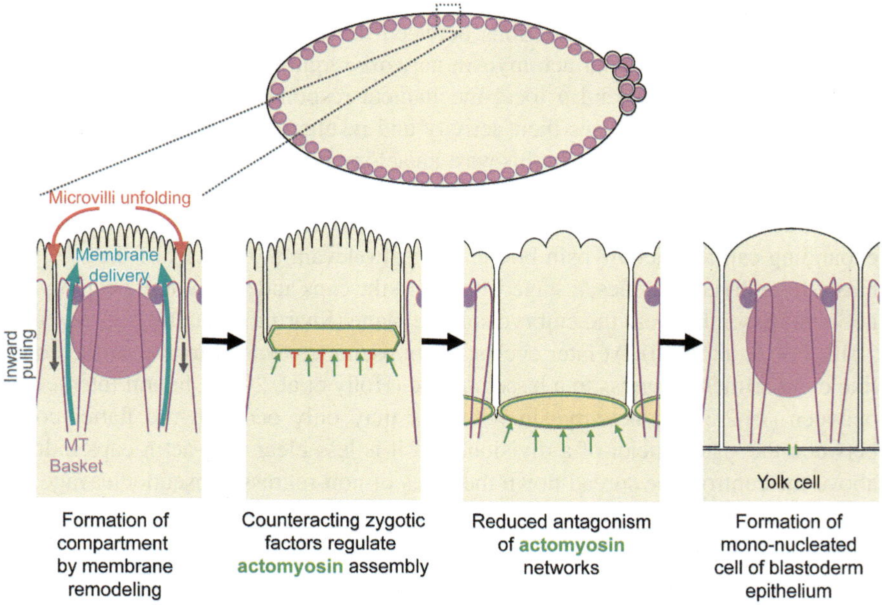

Fig. 4.4 Forming mono-nucleated epithelial cells of the cellular blastoderm. During interphase of nuclear division cycle 14, the soma undergoes cellularization to form mono-nucleated cells. Membrane delivery along polarized microtubules supplies apical microvilli which unfold to form lateral membranes. The lateral furrows are also pulled inward. Around the base of each compartment, a hexagonal actomyosin network assembles but is inhibited. After lateral membranes grow below the nucleus, inhibition of actomyosin contraction is reduced. The base of each somatic cell is then closed off by the contraction of an actomyosin ring, simultaneously separating ~6000 epithelial cells from the large yolk cell within. See main text for details

products acting in conjunction with centrosome-directed activities. For an overview of cellularization, we will compare and contrast key aspects of the process with the earlier transient compartmentalization of the syncytial blastoderm. For detailed explanations of cellularization, comprehensive reviews are available (Mazumdar and Mazumdar 2002; Sokac et al. 2023).

4.5.2 Coupling Membrane Trafficking and Actin Networks

Similar to actin cap induction, nucleus-associated centrosomes are key to the initiation of cellularization. At cellularization, however, the major role of the centrosomal microtubule network is to direct membrane trafficking for growth of lateral furrows. Initially, membrane exocytosis is directed to the apical surface of the forming compartments (Lecuit and Wieschaus 2000) (Fig. 4.4). This exocytosis contributes to growth of numerous apical microvilli, which assemble and disassemble through the dynamic remodelling of F-actin cores (Figard et al. 2016). The membrane delivery involves both recycling and biosynthetic pathways (Sokac et al. 2023). For example, Golgi components are trafficked to the apical surface of the cell via the polarized microtubule network and minus-end-directed dynein motor activity (Papoulas et al. 2005), and the exocyst complex localizes to the apical surface (Murthy et al. 2010). Although organized by a centrosomal microtubule network, the cellularization compartment does not appear to form from an actin cap. Upstream cap inducers, Sponge and ELMO, no longer enrich at the apical surface and instead accumulate in an apicolateral ring (Schmidt et al. 2018). Compared to actin caps of the syncytial blastoderm, the apical surfaces of cellularization compartments display a high density of microvilli (Turner and Mahowald 1976). During cellularization, the dense network of apical microvilli continually unfolds and joins the ingressing furrows (Figard et al. 2013) and is continually reformed by polarized exocytosis (Figard et al. 2016). The shift from forming a compartment from a centrosome-organized cap to a centrosome-organized microvillar network involves the maternal-to-zygotic transition of gene expression (Schmidt et al. 2022). The structural order and stereotyped timing of cellularization (Mazumdar and Mazumdar 2002, Sokac et al. 2023) suggests additional drivers of ingression, and plus-end-directed microtubule motor activity has been implicated (Sommi et al. 2010). Embryo-wide actomyosin networks at the base of cellularization furrows also unify their ingression (Sokac et al. 2023).

Similar to compartment formation of the syncytial blastoderm, actomyosin networks are displaced from the apical domain above the centrosome at the beginning of cellularization and then accumulate at the basal tips of ingressing furrows. However, cellularization-specific mechanisms are involved in the redistribution. Instead of an actin cap displacing actomyosin from above the centrosome, initial assembly of an apical actomyosin network is followed by a cortical flow to early furrows (He et al. 2016). As they flow from the apical domain, the actomyosin networks require a zygotically expressed protein called Dunk for their even furrow distribution. Dunk colocalizes with actomyosin at the furrows, and its requirement for even actomyosin localization can be explained by a role in maintaining uniform tension gradients around the furrow base. Uneven actomyosin localization that arises in the absence of Dunk is rescued by the activity of another zygotically expressed protein called Slam (He et al. 2016). Slam localization to the furrow base involves centrosome-directed membrane trafficking (Acharya et al. 2014), and once there, Slam protein promotes the localization of its own mRNA for local translation and self-reinforced Slam accumulation at the furrow tip (Yan et al. 2017). A major effector of Slam is RhoGEF2 (Wenzl et al. 2010), which promotes actomyosin network assembly that unifies furrow growth across the embryo together with a number of other players (Sokac et al. 2023). Slam (an acronym for slow-as-molasses) is required for the rapid ingression of cellularization furrows (Lecuit et al. 2002), but normal ingression speed does not require myosin-based contractility (Royou et al. 2004), implicating additional effects downstream of Slam, including other actin network activities (Sokac et al. 2023) and membrane trafficking (Lecuit et al. 2002).

As cellularization furrows ingress, specific factors antagonize actomyosin activity to prevent the base of the cell from closing prematurely (Fig. 4.4). A key regulator of contractility around the compartment base is the zygotically expressed protein Bottleneck. Without Bottleneck, premature constriction of the compartment base can pinch the nucleus and displace it to the embryo interior (Schejter and Wieschaus 1993a). Two Rho-GAPs also regulate the actomyosin ring constriction (Mason et al. 2016; Sharma and Rikhy 2021), as does the Arf-GEF Steppke which downregulates Rho1 protein levels at the furrow tip independently of RhoGEF2 (Lee and Harris 2013). After the furrows pass the base of the nuclei, the regulatory balance shifts, and the compartment base constricts through both myosin activity and F-actin disassembly (Sokac et al. 2023). The trigger for this shift remains unknown, but it results in the synchronous division of ~6000 mono-nucleated cells of the blastoderm epithelium from a single large yolk cell that fills the embryo interior (Mazumdar and Mazumdar 2002, Sokac et al. 2023) (Fig. 4.4). By the end of cellularization, each cell has gained initial adherens junctions and apicobasal polarity (Schmidt and Grosshans 2018; Harris 2012), and the embryo embarks on the epithelial rearrangements of gastrulation under the direction of gene expression pre-patterns established during syncytial development (Stathopoulos and Newcomb 2020).

4.6 Concluding Remarks

The four main steps of syncytial Drosophila development provide multiple examples of cell shape change based on cortical actin networks. During these changes, the cortex is controlled by the cell cycle, the centrosome, and developmental factors that accumulate at specific locations and times. Overall, the system exemplifies how an early embryo takes form through the combined effects of genetics, biochemistry, and biophysics. However, many unknowns remain. Inductive pathways and regulatory loops are incompletely defined, as are the molecular mechanisms that drive transitions from one step of syncytial development to another. It is also important to dissect how molecular assemblies interact at the mesoscale, as well as the mechanical properties, molecular complexities, and dynamics of the subcellular materials involved. Approaches combining genetic manipulations with cutting-edge microscopy will continue to bear fruit. Comprehensive understanding of how the early Drosophila embryo takes shape will be gained by augmenting these approaches with targeted proteomics (e.g. Müller et al. 2010), rheology (e.g. D'angelo et al. 2019; Doubrovinski et al. 2017; Wessel et al. 2015), optogenetics (e.g. Deneke et al. 2019), and mathematical modelling (e.g. Deneke et al. 2019; Sharma et al. 2021; He et al. 2016; Lv et al. 2020). Lessons learned are relevant to cell shape changes across animals and in disease states. They may also inspire strategies to construct devices from cytoskeletal smart materials (Banerjee et al. 2020; Galland et al. 2013) and shape-shifting metamaterials (Pishvar and Harne 2020; Holmes 2019).

Acknowledgements Our studies of the syncytial Drosophila embryo are supported by a Natural Sciences and Engineering Research Council of Canada Discovery grant (RGPIN-2016-05617).

References

Acharya S, Laupsien P, Wenzl C, Yan S, Großhans J (2014) Function and dynamics of slam in furrow formation in early Drosophila embryo. Dev Biol 386:371–384

Afshar K, Stuart B, Wasserman SA (2000) Functional analysis of the Drosophila diaphanous FH protein in early embryonic development. Development 127:1887–1897

Baker J, Theurkauf WE, Schubiger G (1993) Dynamic changes in microtubule configuration correlate with nuclear migration in the preblastoderm Drosophila embryo. J Cell Biol 122:113–121

Banerjee S, Gardel M, Schwarz U (2020) The actin cytoskeleton as an active adaptive material. Ann Rev Condens Matter Phys 11:421–439

Bement WM, Leda M, Moe AM, Kita AM, Larson ME, Golding AE, Pfeuti C, Su KC, Miller AL, Goryachev AB, Von Dassow G (2015) Activator-inhibitor coupling between Rho signalling and actin assembly makes the cell cortex an excitable medium. Nat Cell Biol 17:1471–1483

Blake-Hedges C, Megraw TL (2019) Coordination of embryogenesis by the centrosome in Drosophila melanogaster. Results Probl Cell Differ 67:277–321

Bodor DL, Pönisch W, Endres RG, Paluch EK (2020) Of cell shapes and motion: the physical basis of animal cell migration. Dev Cell 52:550–562

Brose L, Crest J, Tao L, Sullivan W (2017) Polo kinase mediates the phosphorylation and cellular localization of Nuf/FIP3, a Rab11 effector. Mol Biol Cell 28:1435–1443

Cao J, Crest J, Fasulo B, Sullivan W (2010) Cortical actin dynamics facilitate early-stage centrosome separation. Curr Biol 20:770–776

Chodagam S, Royou A, Whitfield W, Karess R, Raff JW (2005) The centrosomal protein CP190 regulates myosin function during early Drosophila development. Curr Biol 15:1308–1313

Chugh P, Paluch EK (2018) The actin cortex at a glance. J Cell Sci 131:jcs186254

Cinalli RM, Lehmann R (2013) A spindle-independent cleavage pathway controls germ cell formation in Drosophila. Nat Cell Biol 15:839–845

D'angelo A, Dierkes K, Carolis C, Salbreux G, Solon J (2019) In vivo force application reveals a fast tissue softening and external friction increase during early embryogenesis. Curr Biol 29:1564–1571.e6

Daniels BR, Rikhy R, Renz M, Dobrowsky TM, Lippincott-Schwartz J (2012) Multiscale diffusion in the mitotic Drosophila melanogaster syncytial blastoderm. Proc Natl Acad Sci USA 109:8588–8593

Dansereau DA, Lasko P (2008) The development of germline stem cells in Drosophila. Methods Mol Biol 450:3–26

De-Carvalho J, Tlili S, Hufnagel L, Saunders TE, Telley IA (2022) Aster repulsion drives short-ranged ordering in the Drosophila syncytial blastoderm. Development 149:dev199997

Deneke VE, Melbinger A, Vergassola M, Di Talia S (2016) Waves of Cdk1 activity in S phase synchronize the cell cycle in Drosophila embryos. Dev Cell 38:399–412

Deneke VE, Puliafito A, Krueger D, Narla AV, De Simone A, Primo L, Vergassola M, De Renzis S, Di Talia S (2019) Self-organized nuclear positioning synchronizes the cell cycle in Drosophila embryos. Cell 177:925–941.e17

Dey B, Rikhy R (2020) DE-cadherin and Myosin II balance regulates furrow length for onset of polygon shape in syncytial Drosophila embryos. J Cell Sci 133:jcs240168

Dodson AE, Kennedy S (2020) Phase separation in germ cells and development. Dev Cell 55:4–17

Doerflinger H, Zimyanin V, St Johnston D (2022) The Drosophila anterior-posterior axis is polarized by asymmetric myosin activation. Curr Biol 32:374–385.e4

Doubrovinski K, Swan M, Polyakov O, Wieschaus EF (2017) Measurement of cortical elasticity in Drosophila melanogaster embryos using ferrofluids. Proc Natl Acad Sci USA 114:1051–1056

Field CM, Coughlin M, Doberstein S, Marty T, Sullivan W (2005) Characterization of anillin mutants reveals essential roles in septin localization and plasma membrane integrity. Development 132:2849–2860

Figard L, Xu H, Garcia HG, Golding I, Sokac AM (2013) The plasma membrane flattens out to fuel cell-surface growth during Drosophila cellularization. Dev Cell 27:648–655

Figard L, Wang M, Zheng L, Golding I, Sokac AM (2016) Membrane supply and demand regulates F-actin in a cell surface reservoir. Dev Cell 37:267–278

Foe VE, Alberts BM (1983) Studies of nuclear and cytoplasmic behaviour during the five mitotic cycles that precede gastrulation in Drosophila embryogenesis. J Cell Sci 61:31–70

Foe VE, Field CM, Odell GM (2000) Microtubules and mitotic cycle phase modulate spatiotemporal distributions of F-actin and myosin II in Drosophila syncytial blastoderm embryos. Development 127:1767–1787

Frescas D, Mavrakis M, Lorenz H, Delotto R, Lippincott-Schwartz J (2006) The secretory membrane system in the Drosophila syncytial blastoderm embryo exists as functionally compartmentalized units around individual nuclei. J Cell Biol 173:219–230

Galland R, Leduc P, Guérin C, Peyrade D, Blanchoin L, Théry M (2013) Fabrication of three-dimensional electrical connections by means of directed actin self-organization. Nat Mater 12:416–421

Grosshans J, Wenzl C, Herz HM, Bartoszewski S, Schnorrer F, Vogt N, Schwarz H, Müller HA (2005) RhoGEF2 and the formin Dia control the formation of the furrow canal by directed actin assembly during Drosophila cellularisation. Development 132:1009–1020

Gubieda AG, Packer JR, Squires I, Martin J, Rodriguez J (2020) Going with the flow: insights from Caenorhabditis elegans zygote polarization. Philos Trans R Soc Lond Ser B Biol Sci 375:20190555

Harris TJ (2012) Adherens junction assembly and function in the Drosophila embryo. Int Rev Cell Mol Biol 293:45–83

Harris TJ, Tepass U (2010) Adherens junctions: from molecules to morphogenesis. Nat Rev Mol Cell Biol 11:502–514

Hayden L, Chao A, Deneke VE, Vergassola M, Puliafito A, Di Talia S (2022) Cullin-5 mutants reveal collective sensing of the nucleocytoplasmic ratio in Drosophila embryogenesis. Curr Biol 32:2084–2092.e4

He B, Martin A, Wieschaus E (2016) Flow-dependent myosin recruitment during Drosophila cellularization requires zygotic dunk activity. Development 143:2417–2430

Henry SM, Xie Y, Rollins KR, Blankenship JT (2022) Sponge/DOCK-dependent regulation of F-actin networks directing cortical cap behaviors and syncytial furrow ingression. Dev Biol 491:82–93

Hickson GR, Matheson J, Riggs B, Maier VH, Fielding AB, Prekeris R, Sullivan W, Barr FA, Gould GW (2003) Arfophilins are dual Arf/Rab 11 binding proteins that regulate recycling endosome distribution and are related to Drosophila nuclear fallout. Mol Biol Cell 14:2908–2920

Holly RM, Mavor LM, Zuo Z, Blankenship JT (2015) A rapid, membrane-dependent pathway directs furrow formation through RalA in the early Drosophila embryo. Development 142:2316–2328

Holmes D (2019) Elasticity and stability of shape-shifting structures. Curr Opin Colloid Interface Sci 40:118–137

Jiang T, Harris TJC (2019) Par-1 controls the composition and growth of cortical actin caps during Drosophila embryo cleavage. J Cell Biol 218:4195–4214

Jongens TA, Hay B, Jan LY, Jan YN (1992) The germ cell-less gene product: a posteriorly localized component necessary for germ cell development in Drosophila. Cell 70:569–584

Kao LR, Megraw TL (2009) Centrocortin cooperates with centrosomin to organize Drosophila embryonic cleavage furrows. Curr Biol 19:937–942

Karr TL, Alberts BM (1986) Organization of the cytoskeleton in early Drosophila embryos. J Cell Biol 102:1494–1509

Kelkar M, Bohec P, Charras G (2020) Mechanics of the cellular actin cortex: from signalling to shape change. Curr Opin Cell Biol 66:69–78

Kopf A, Kiermaier E (2021) Dynamic microtubule arrays in leukocytes and their role in cell migration and immune synapse formation. Front Cell Dev Biol 9:635511

Kukimoto-Niino M, Ihara K, Murayama K, Shirouzu M (2021) Structural insights into the small GTPase specificity of the DOCK guanine nucleotide exchange factors. Curr Opin Struct Biol 71:249–258

Lecuit T, Wieschaus E (2000) Polarized insertion of new membrane from a cytoplasmic reservoir during cleavage of the Drosophila embryo. J Cell Biol 150:849–860

Lecuit T, Samanta R, Wieschaus E (2002) Slam encodes a developmental regulator of polarized membrane growth during cleavage of the Drosophila embryo. Dev Cell 2:425–436

Lecuit T, Lenne PF, Munro E (2011) Force generation, transmission, and integration during cell and tissue morphogenesis. Annu Rev Cell Dev Biol 27:157–184

Lee DM, Chen EH (2019) Drosophila myoblast fusion: invasion and resistance for the ultimate union. Annu Rev Genet 53:67–91

Lee DM, Harris TJ (2013) An Arf-GEF regulates antagonism between endocytosis and the cytoskeleton for Drosophila blastoderm development. Curr Biol 23:2110–2120

Lee DM, Wilk R, Hu J, Krause HM, Harris TJ (2015) Germ cell segregation from the Drosophila soma is controlled by an inhibitory threshold set by the Arf-GEF Steppke. Genetics 200:863–872

Lehner CF (1992) The pebble gene is required for cytokinesis in Drosophila. J Cell Sci 103(Pt 4):1021–1030

Lerit DA, Gavis ER (2011) Transport of germ plasm on astral microtubules directs germ cell development in Drosophila. Curr Biol 21:439–448

Lerit DA, Shebelut CW, Lawlor KJ, Rusan NM, Gavis ER, Schedl P, Deshpande G (2017) Germ cell-less promotes centrosome segregation to induce germ cell formation. Cell Rep 18:831–839

Lv Z, Rosenbaum J, Mohr S, Zhang X, Kong D, Preiß H, Kruss S, Alim K, Aspelmeier T, Großhans J (2020) The emergent yo-yo movement of nuclei driven by cytoskeletal remodeling in pseudo-synchronous mitotic cycles. Curr Biol 30:2564–2573.e5

Lv Z, De-Carvalho J, Telley IA, Großhans J (2021) Cytoskeletal mechanics and dynamics in the Drosophila syncytial embryo. J Cell Sci 134:jcs246496

Mahowald AP (2001) Assembly of the Drosophila germ plasm. Int Rev Cytol 203:187–213

Maître JL, Niwayama R, Turlier H, Nédélec F, Hiiragi T (2015) Pulsatile cell-autonomous contractility drives compaction in the mouse embryo. Nat Cell Biol 17:849–855

Markow TA, Beall S, Matzkin LM (2009) Egg size, embryonic development time and ovoviviparity in Drosophila species. J Evol Biol 22:430–434

Mason FM, Xie S, Vasquez CG, Tworoger M, Martin AC (2016) RhoA GTPase inhibition organizes contraction during epithelial morphogenesis. J Cell Biol 214:603–617

Mavor LM, Miao H, Zuo Z, Holly RM, Xie Y, Loerke D, Blankenship JT (2016) Rab8 directs furrow ingression and membrane addition during epithelial formation in Drosophila melanogaster. Development 143:892–903

Mavrakis M, Rikhy R, Lippincott-Schwartz J (2009) Plasma membrane polarity and compartmentalization are established before cellularization in the fly embryo. Dev Cell 16:93–104

Mazumdar A, Mazumdar M (2002) How one becomes many: blastoderm cellularization in Drosophila melanogaster. BioEssays 24:1012–1022

Megraw TL, Li K, Kao LR, Kaufman TC (1999) The centrosomin protein is required for centrosome assembly and function during cleavage in Drosophila. Development 126:2829–2839

Michaud A, Swider ZT, Landino J, Leda M, Miller AL, Von Dassow G, Goryachev AB, Bement WM (2021) Cortical excitability and cell division. Curr Biol 31:R553–r559

Mishima M (2016) Centralspindlin in Rappaport's cleavage signaling. Semin Cell Dev Biol 53:45–56

Müller H, Schmidt D, Steinbrink S, Mirgorodskaya E, Lehmann V, Habermann K, Dreher F, Gustavsson N, Kessler T, Lehrach H, Herwig R, Gobom J, Ploubidou A, Boutros M, Lange BM (2010) Proteomic and functional analysis of the mitotic Drosophila centrosome. EMBO J 29:3344–3357

Munro E, Nance J, Priess JR (2004) Cortical flows powered by asymmetrical contraction transport PAR proteins to establish and maintain anterior-posterior polarity in the early C. elegans embryo. Dev Cell 7:413–424

Murthy M, Teodoro RO, Miller TP, Schwarz TL (2010) Sec5, a member of the exocyst complex, mediates Drosophila embryo cellularization. Development 137:2773–2783

Padash Barmchi M, Rogers S, Häcker U (2005) DRhoGEF2 regulates actin organization and contractility in the Drosophila blastoderm embryo. J Cell Biol 168:575–585

Papoulas O, Hays TS, Sisson JC (2005) The golgin Lava lamp mediates dynein-based Golgi movements during Drosophila cellularization. Nat Cell Biol 7:612–618

Pishvar M, Harne RL (2020) Foundations for soft, smart matter by active mechanical metamaterials. Adv Sci (Weinh) 7:2001384

Postner MA, Miller KG, Wieschaus EF (1992) Maternal effect mutations of the sponge locus affect actin cytoskeletal rearrangements in Drosophila melanogaster embryos. J Cell Biol 119:1205–1218

Raff JW, Glover DM (1989) Centrosomes, and not nuclei, initiate pole cell formation in Drosophila embryos. Cell 57:611–619

Riggs B, Rothwell W, Mische S, Hickson GR, Matheson J, Hays TS, Gould GW, Sullivan W (2003) Actin cytoskeleton remodeling during early Drosophila furrow formation requires recycling endosomal components Nuclear-fallout and Rab11. J Cell Biol 163:143–154

Riggs B, Fasulo B, Royou A, Mische S, Cao J, Hays TS, Sullivan W (2007) The concentration of Nuf, a Rab11 effector, at the microtubule-organizing center is cell cycle regulated, dynein-dependent, and coincides with furrow formation. Mol Biol Cell 18:3313–3322

Rikhy R, Mavrakis M, Lippincott-Schwartz J (2015) Dynamin regulates metaphase furrow formation and plasma membrane compartmentalization in the syncytial Drosophila embryo. Biol Open 4:301–311

Rodrigues FF, Shao W, Harris TJ (2016) The Arf GAP Asap promotes Arf1 function at the Golgi for cleavage furrow biosynthesis in Drosophila. Mol Biol Cell 27:3143–3155

Rothwell WF, Fogarty P, Field CM, Sullivan W (1998) Nuclear-fallout, a Drosophila protein that cycles from the cytoplasm to the centrosomes, regulates cortical microfilament organization. Development 125:1295–1303

Rothwell WF, Zhang CX, Zelano C, Hsieh TS, Sullivan W (1999) The Drosophila centrosomal protein Nuf is required for recruiting Dah, a membrane associated protein, to furrows in the early embryo. J Cell Sci 112(Pt 17):2885–2893

Rottner K, Stradal TEB, Chen B (2021) WAVE regulatory complex. Curr Biol 31:R512–r517

Royou A, Sullivan W, Karess R (2002) Cortical recruitment of nonmuscle myosin II in early syncytial Drosophila embryos: its role in nuclear axial expansion and its regulation by Cdc2 activity. J Cell Biol 158:127–137

Royou A, Field C, Sisson JC, Sullivan W, Karess R (2004) Reassessing the role and dynamics of nonmuscle myosin II during furrow formation in early Drosophila embryos. Mol Biol Cell 15:838–850

Schejter ED, Wieschaus E (1993a) bottleneck acts as a regulator of the microfilament network governing cellularization of the Drosophila embryo. Cell 75:373–385

Schejter ED, Wieschaus E (1993b) Functional elements of the cytoskeleton in the early Drosophila embryo. Annu Rev Cell Biol 9:67–99

Schmidt A, Grosshans J (2018) Dynamics of cortical domains in early Drosophila development. J Cell Sci 131:jcs212795

Schmidt A, Lv Z, Großhans J (2018) ELMO and Sponge specify subapical restriction of Canoe and formation of the subapical domain in early Drosophila embryos. Development 145:dev157909

Schmidt A, Li L, Lv Z, Yan S, Großhans J (2021) Dia- and Rok-dependent enrichment of capping proteins in a cortical region. J Cell Sci 134:jcs258973

Schmidt A, Li L, Yan S, Großhans J (2022) Restriction of subapical proteins during cellularization depends on the onset of zygotic transcription and the formin Dia. Dev Biol 487:110–121

Sedzinski J, Biro M, Oswald A, Tinevez JY, Salbreux G, Paluch E (2011) Polar actomyosin contractility destabilizes the position of the cytokinetic furrow. Nature 476:462–466

Sharma S, Rikhy R (2021) Spatiotemporal recruitment of RhoGTPase protein GRAF inhibits actomyosin ring constriction in Drosophila cellularization. elife 10:e63535

Sharma M, Jiang T, Jiang ZC, Moguel-Lehmer CE, Harris TJ (2021) Emergence of a smooth interface from growth of a dendritic network against a mechanosensitive contractile material. elife 10:e66929

Sherlekar A, Rikhy R (2016) Syndapin promotes pseudocleavage furrow formation by actin organization in the syncytial Drosophila embryo. Mol Biol Cell 27:2064–2079

Sherlekar A, Mundhe G, Richa P, Dey B, Sharma S, Rikhy R (2020) F-BAR domain protein Syndapin regulates actomyosin dynamics during apical cap remodeling in syncytial Drosophila embryos. J Cell Sci 133:jcs235846

Sokac AM, Wieschaus E (2008) Local actin-dependent endocytosis is zygotically controlled to initiate Drosophila cellularization. Dev Cell 14:775–786

Sokac AM, Biel N, De Renzis S (2023) Membrane-actin interactions in morphogenesis: Lessons learned from Drosophila cellularization. Semin Cell Dev Biol 133:107–122

Sommi P, Ananthakrishnan R, Cheerambathur DK, Kwon M, Morales-Mulia S, Brust-Mascher I, Mogilner A (2010) A mitotic kinesin-6, Pav-KLP, mediates interdependent cortical reorganization and spindle dynamics in Drosophila embryos. J Cell Sci 123:1862–1872

Sommi P, Cheerambathur D, Brust-Mascher I, Mogilner A (2011) Actomyosin-dependent cortical dynamics contributes to the prophase force-balance in the early Drosophila embryo. PLoS One 6:e18366

Stathopoulos A, Newcomb S (2020) Setting up for gastrulation: D. melanogaster. Curr Top Dev Biol 136:3–32

Stevenson VA, Kramer J, Kuhn J, Theurkauf WE (2001) Centrosomes and the Scrambled protein coordinate microtubule-independent actin reorganization. Nat Cell Biol 3:68–75

Stevenson V, Hudson A, Cooley L, Theurkauf WE (2002) Arp2/3-dependent pseudocleavage [correction of psuedocleavage] furrow assembly in syncytial Drosophila embryos. Curr Biol 12:705–711

Strome S, Lehmann R (2007) Germ versus soma decisions: lessons from flies and worms. Science 316:392–393

Sugioka K (2022) Symmetry-breaking of animal cytokinesis. Semin Cell Dev Biol 127:100–109

Sullivan W, Theurkauf WE (1995) The cytoskeleton and morphogenesis of the early Drosophila embryo. Curr Opin Cell Biol 7:18–22

Svitkina TM (2020) Actin cell cortex: structure and molecular organization. Trends Cell Biol 30:556–565

Telley IA, Gáspár I, Ephrussi A, Surrey T (2012) Aster migration determines the length scale of nuclear separation in the Drosophila syncytial embryo. J Cell Biol 197:887–895

Turner FR, Mahowald AP (1976) Scanning electron microscopy of Drosophila embryogenesis. 1. The structure of the egg envelopes and the formation of the cellular blastoderm. Dev Biol 50:95–108

Vaizel-Ohayon D, Schejter ED (1999) Mutations in centrosomin reveal requirements for centrosomal function during early Drosophila embryogenesis. Curr Biol 9:889–898

Vastenhouw NL, Cao WX, Lipshitz HD (2019) The maternal-to-zygotic transition revisited. Development 146:dev161471

Von Dassow G, Schubiger G (1994) How an actin network might cause fountain streaming and nuclear migration in the syncytial Drosophila embryo. J Cell Biol 127:1637–1653

Warn RM, Magrath R, Webb S (1984) Distribution of F-actin during cleavage of the Drosophila syncytial blastoderm. J Cell Biol 98:156–162

Warn RM, Smith L, Warn A (1985) Three distinct distributions of F-actin occur during the divisions of polar surface caps to produce pole cells in Drosophila embryos. J Cell Biol 100:1010–1015

Wenzl C, Yan S, Laupsien P, Grosshans J (2010) Localization of RhoGEF2 during Drosophila cellularization is developmentally controlled by Slam. Mech Dev 127:371–384

Wessel AD, Gumalla M, Grosshans J, Schmidt CF (2015) The mechanical properties of early Drosophila embryos measured by high-speed video microrheology. Biophys J 108:1899–1907

Wheatley S, Kulkarni S, Karess R (1995) Drosophila nonmuscle myosin II is required for rapid cytoplasmic transport during oogenesis and for axial nuclear migration in early embryos. Development 121:1937–1946

Wilson JE, Macdonald PM (1993) Formation of germ cells in Drosophila. Curr Opin Genet Dev 3:562–565

Wu B, Guo W (2015) The exocyst at a glance. J Cell Sci 128:2957–2964

Xie Y, Budhathoki R, Blankenship JT (2021) Combinatorial deployment of F-actin regulators to build complex 3D actin structures in vivo. elife 10:e63046

Yan S, Acharya S, Gröning S, Großhans J (2017) Slam protein dictates subcellular localization and translation of its own mRNA. PLoS Biol 15:e2003315

Zallen JA, Cohen Y, Hudson AM, Cooley L, Wieschaus E, Schejter ED (2002) SCAR is a primary regulator of Arp2/3-dependent morphological events in Drosophila. J Cell Biol 156:689–701

Zenker J, White MD, Gasnier M, Alvarez YD, Lim HYG, Bissiere S, Biro M, Plachta N (2018) Expanding actin rings zipper the mouse embryo for blastocyst formation. Cell 173:776–791.e17

Zhang Y, Yu JC, Jiang T, Fernandez-Gonzalez R, Harris TJC (2018) Collision of expanding actin caps with actomyosin borders for cortical bending and mitotic rounding in a syncytium. Dev Cell 45:551–564.e4

Chapter 5
Cell-Mediated Branch Fusion in the *Drosophila* Trachea

Lan Jiang

Abstract The *Drosophila* trachea is an interconnected network of epithelial tubes, which delivers gases throughout the entire organism. It is the premier model to study the development of tubular organs, such as the human lung, kidney, and blood vessels. The *Drosophila* embryonic trachea derives from a series of segmentally repeated clusters. The tracheal precursor cells in each cluster migrate out in a stereotyped pattern to form primary branches. Thereafter, the neighboring branches need to fuse to form an interconnected tubular network. The connection between neighboring branches is orchestrated by specialized cells, called fusion cells. These cells fuse with their counterparts to form a tube with a contiguous lumen. Branch fusion is a multi-step process that includes cell migration, cell adhesion, cytoskeleton track formation, vesicle trafficking, membrane fusion, and lumen formation. This review summarizes the current knowledge on fusion process in the *Drosophila* trachea. These mechanisms will greatly contribute to our understanding of branch fusion in mammalian systems.

5.1 Overview of Tracheal Development and Branch Fusion

Tubular organs, such as the vertebrate vasculature, kidney, and the *Drosophila* trachea, are composed of tubes that arise through budding from the epithelium and then are connected to form tubular networks to deliver gases, cells, and wastes within the organisms. The *Drosophila* trachea is a ramifying network of epithelial tubes with apical surfaces facing the lumen and basal surfaces facing the surrounding tissues. It is the premier model to study branch morphogenesis. The *Drosophila* embryonic trachea derives from a series of segmentally repeated clusters. The tracheal precursor cells in each cluster respond to fibroblast growth factor (FGF)

L. Jiang (✉)
Department of Biological Sciences, Oakland University, Rochester, MI, USA
e-mail: jiang23@oakland.edu

M. Kloc, A. Uosef (eds.), *Syncytia: Origin, Structure, and Functions*, Results and Problems in Cell Differentiation 71,
https://doi.org/10.1007/978-3-031-37936-9_5

signaling (Sutherland et al. 1996) as well as branch-specific signaling (Llimargas and Casanova 1999; Chen et al. 1998a; Chihara and Hayashi 2000a) and migrate out in a stereotyped pattern to form primary branches (Fig. 5.1a). These branches include the dorsal trunk (DT) and transverse connective (TC), which are multicellular tubes; the lateral trunk (LT), dorsal branch (DB), and ganglionic branch (GB), which are unicellular tubes; and the lateral ganglionic branch (LG), which is an intracellular tube (Caussinus et al. 2008; Ribeiro et al. 2004; Ghabrial et al. 2003). Thereafter, the neighboring branches fuse to form an interconnected tubular network with a continuous lumen for gas exchange throughout the entire organism. The formation of connections between neighboring branches is orchestrated by specialized cells, called fusion cells (yellow circles outlined by the red line in Fig. 5.1a, b), which are at or close to the end of most tracheal branches. Fusion cells fuse with their counterparts in either anterior or posterior hemisegments (DT and LT) or with their counterparts in the contralateral hemisegment (DB and three anterior-most GBs) (Manning and Krasnow 1993). Once fusion partners make contact with each other (Fig. 5.2a), morphological changes occur, and fusion cells become toroidal or

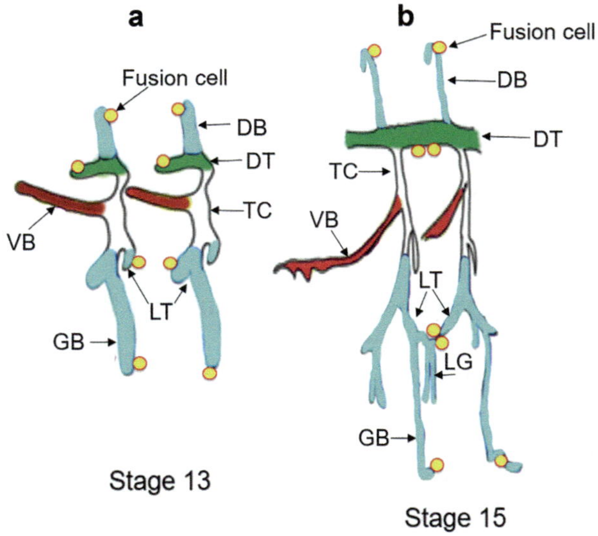

Fig. 5.1 The cells within each tracheal metamere migrate out in certain directions to form stereotyped branches. (**a**) Tracheal segments at stage 13. Cells remain at the site of invagination to form TC (white). TGFβ (DPP) signaling pathway specifies branches at the dorsal and ventral sides to form the DB, LT, and GB branches (blue). Wingless signaling (green) specifies the DT (green), whereas cells that do not receive Wingless signal form VB (red) at the central region. (**b**) Tracheal segments at stage 15. As the trachea develops, DT branches fuse with their neighboring branches to form the fully connected DT at stage 14; LT branches fuse with their neighboring branches to form LT at stage 15; and DB and GB fuse at the dorsal and ventral midlines, respectively, with their counterparts that emanate from the other side of the embryo at stage 16. *DB* dorsal branch, *DT* dorsal trunk, *LT* lateral trunk, *GB* ganglionic branch, *VB* visceral branch, *LG* lateral ganglionic branch, Fusion cells are yellow circles outlined by red lines

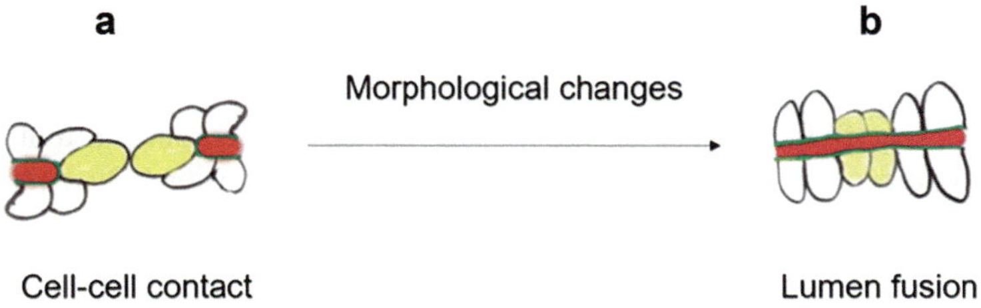

Fig. 5.2 Fusion cell-mediated branch fusion in an optical section along the plane of contact. (**a**) Branch fusion starts with two fusion cells in neighboring branches migrating toward each other and making contact. (**b**) The fusion process completes by the formation of the apical lumen (red outlined by the green apical membrane) between two donut-shaped fusion cells through a series of morphological changes. Fusion cells: yellow cells outlined by black lines. Tracheal stock cells: white cells outlined by black lines

"donut-shaped" (Yellow cells in Fig. 5.2b). A contiguous lumen (red lumen outlined by the green apical membrane in Fig. 5.2b) extends through the holes of the donut and connects with the lumen of the stalk cells (white cells outlined by the black line). The fusion cell lumen ultimately expands to acquire the same diameter as the corresponding tracheal branch (Samakovlis et al. 1996; Tanaka et al. 1996).

5.2 Fusion Cell Specification

Fusion cells are derived from tip cells. During the migration of the tracheal precursor cells, the tip cells with the highest FGF receptor activity take the lead position and guide the movement of the rest of the stalk cells of the branch (Ghabrial and Krasnow 2006; Lebreton and Casanova 2014). When the branch tip reaches its final destination, tip cells further differentiate into two different cell types: the fusion cell, which leads branch fusion to form an interconnected network, and the terminal cell, which forms cytoplasmic extensions with the intracellular lumen for gas exchange.

The specification of fusion cell fate depends on the transforming growth factor ß (TGFß, Dpp in *Drosophila*), Notch, and Wingless-related integration site (Wnt, Wingless in *Drosophila*) signaling pathways that select one fusion cell at the tip of each branch and guide adjacent cells to take non-fusion cell fates (Steneberg et al. 1999; Ikeya and Hayashi 1999; Llimargas 1999, 2000). Loss of components of the TGFß (Dpp) signaling pathway leads to a loss of fusion cells, whereas overactivation of the pathway results in additional fusion cells in DBs (Steneberg et al. 1999). To limit the number of fusion cells to one per DB, fusion cells express Delta as a signal for neighboring cells, which express the Delta receptor, Notch, to remain stalk cells (Steneberg et al. 1999; Ikeya and Hayashi 1999; Llimargas 1999). In addition, the loss of components of the Wnt signaling pathway leads to failed branch fusion and the absence of fusion cell markers. Instead, the fusion cells seem to acquire other tracheal identities, such as terminal cells (Llimargas 2000). This regulation is through the expression of Delta as Wnt signaling is required for both the localized and ectopic Delta expression (Chihara and Hayashi 2000b). Signaling pathways specify fusion cells to express early fusion markers such as transcription factors Escargot (Esg) (Steneberg et al. 1999) and Dysfusion (Dys) (Jiang and Crews 2003, 2006). Both Esg and Dys are required for branch fusion in LT, DB, and GB but not in DT (Jiang and Crews 2003, 2006). This difference suggests that additional transcription factors might play redundant roles in DT fusion cells to ensure DT branch fusion. Besides fusion cell-specific transcription factors, anterior open (AOP), an ETS-domain transcriptional repressor prevents stock cells to acquire either fusion cell or tip cell fate (Caviglia and Luschnig 2013). Taken together, transcription factors regulate the dynamic expression of downstream targets during branch fusion.

A fluorescence in situ hybridization screen revealed 14 genes with high expression in fusion cells (Chandran et al. 2014). Among these genes, 9 genes that encode zinc-finger transcription factors Esg (Steneberg et al. 1999), bHLH-PAS transcription factor Dys (Jiang and Crews 2003, 2006), Arf-like 3 small GTPase Dead end (Arl-3) (Kakihara et al. 2008), potential cytoskeleton protein CG15252, zona pellucida domain (ZP) structure protein CG13196, cell adhesion molecule E-cadherin (Shotgun in *Drosophila*), protein kinase Center divider (Cdi) (Matthews and Crews 1999), nucleocytoplasmic transporter Members only (Mbo) (Uv et al. 2000), and *CG9743* are only expressed in fusion cells. One gene, *rebuff* (Moussian et al. 2015; Chandran et al. 2018), is only expressed in DT fusion cells at stage 14 followed by similar levels of expression in non-fusion DT cells. Five genes that encode cell adhesion molecule E-cadherin (Tanaka et al. 1996), chitin-based cuticle protein Cpr66D (Stahl et al. 2017), chitin-binding protein CG15786, ZP domain protein Quasimodo (Chen et al. 2011), and membrane protein Apnoia (Scholl et al. 2019; Skouloudaki et al. 2019) have relative high abundance in fusion cells compared to other non-fusion tracheal cells. These genes are likely to be involved in morphological changes during branch fusion. For example, the zinc-finger protein Esg suppresses FGF signaling in fusion cells and prevents these cells from responding to FGF signaling to form long cytoplasmic extensions, which occur in terminal cells (Miao and Hayashi 2016). Dys activates genes involved in cell adhesion and cytoskeletal changes (Jiang and Crews 2003, 2006). Arl-3 is required for the intracellular fusion of the plasma membranes in fusion cells (Kakihara et al. 2008). E-cadherin is involved in forming the adhesive tip between two fusion cells when they contact each other (Tanaka et al. 1996). Nevertheless, the functions of other genes in the fusion process remain to be investigated.

5.3 Branch Fusion Process

Branch fusion is a multi-step process that includes cell migration, cell adhesion, cytoskeleton track formation, vesicle trafficking, membrane fusion, and lumen formation.

5.3.1 Cell–Cell Contact Between Two Fusion Cells

During migration, fusion cells (yellow cells in Fig. 5.3a) from neighboring branches form filopodia-like extensions (black arrowheads in Fig. 5.3a). The fusion cells and stock cells are connected by intercellular adherens junction (AJ, blue arrowheads in Fig. 5.3a). The bridge cell (purple cell in Fig. 5.3a), which is located between two fusion cells, secretes guidance cues to bring fusion cells close to each other and then make contact (Wolf and Schuh 2000; Wolf et al. 2002). To facilitate cell–cell contact, filopodia-like cell extensions in fusion cells attach to the bridge-cell surface. Once two fusion cells make contact, they initiate the formation of an adhesive tip, which is characterized by the accumulation of the cell adhesion protein E-cadherin. This tip expands further into a ring-shaped intercellular AJ (blue arrowhead, small blue ring between two fusion cells in Fig. 5.3b) (Tanaka et al. 1996; Gervais et al. 2012). This AJ ring is smaller than the AJ rings between the fusion cell and the stock cell (larger blue rings in Fig. 5.3b). Thereafter, E-cadherin potentially recruits additional apical determinants, apical polarity complex (Par3, aPKC, Crumbs), and apical membrane components (Stranded at second and Discs Lost), to form an apical domain (red arrowhead, red circle in Fig. 5.3b) at the contact site between the two fusion cells (Gervais et al. 2012; Lee and Kolodziej 2002). Thus, each fusion cell possesses two apical domains, one at the junction with its partner fusion cell (red arrowhead in Fig. 5.3b) and another at the interface between the fusion cell and the stock cell (red arrows in Fig. 5.3b).

5.3.2 Formation of the Cytoskeleton Track and Vesicular Trafficking

After fusion cells make contact, a cytoskeleton track that contains actin, microtubule, and the plakin short stop spans the fusion cells and connects the apical domains (green lines in Fig. 5.3b). Short stop and AJ component E-cadherin are both required for the assembly of the cytoskeleton track by recruiting actin and microtubule to the track (Lee and Kolodziej 2002; Lee et al. 2003). Then, the cytoskeleton track contracts, and the adjacent stock cells further invade the fusion cells, resembling a "finger poking into a balloon" (Gervais et al. 2012; Uv et al. 2003). The contractile forces could be generated by the sliding of actin filaments along the myosin

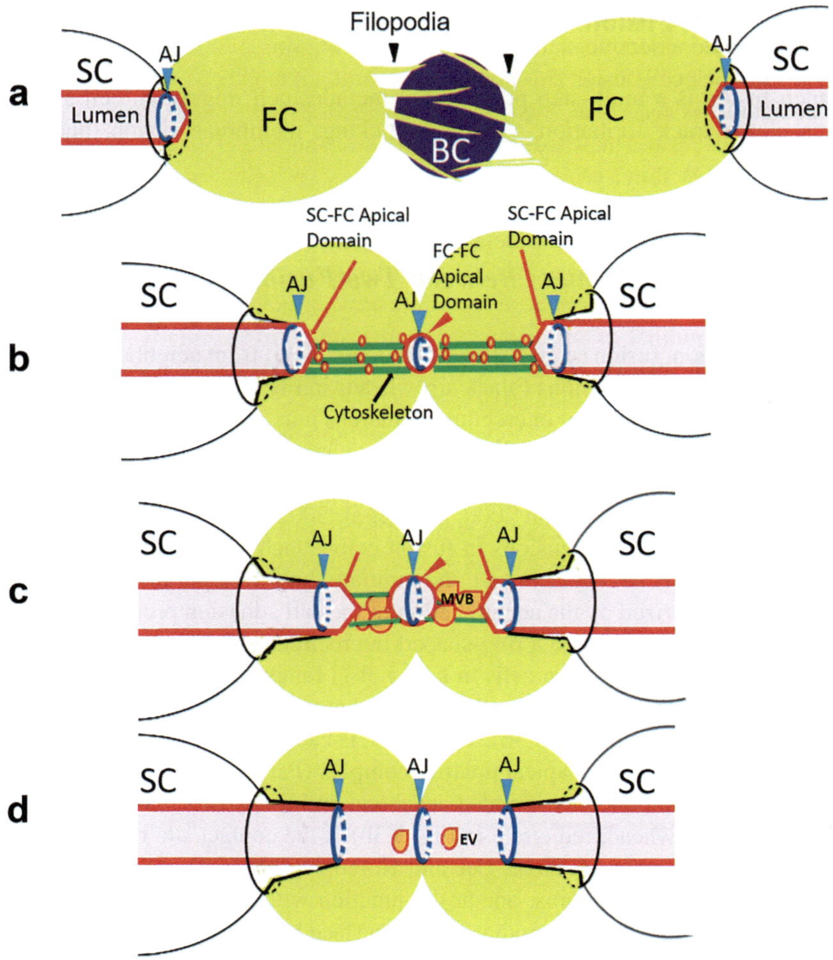

Fig. 5.3 The current model of branch fusion. (**a**) Fusion cells (FC, yellow), which form filopodia-like extensions (black arrowheads), migrate toward each other. Bridge cell (BC, purple) between two fusion cells secrets guidance cues to pull fusion cells closer to each other. The filopodia-like cell extensions attach to the bridge-cell surface to facilitate the cell-cell contact between fusion cells. AJs (blue arrowheads) between fusion cells (FC) and stock cells (SC) are formed. (**b**) Once establishing cell-cell contact, an E-cadherin containing AJ ring (the blue arrowhead, the small blue ring in the middle) is formed at the interface between two fusion cells. E-cadherin recruits additional apical proteins to form an apical domain between two fusion cells (red arrowhead, red circle between two fusion cells). Meanwhile, a cytoskeleton track (green lines) that contains actin, microtubule, and the plakin short stop spans the fusion cells and connects the apical domains. Vesicles (small orange circles outlined by red lines) are deposited along the track and the apical membrane to expand the apical domains (red arrows and arrowhead). (**c**). The Stac-positive MVBs (large orange vesicles) localize between the invading apical domain (red arrows) of the stock cell and the central apical domain (red arrowhead) at the fusion cell-fusion cell interface. (**d**) The Stac-positive MVBs fuse with the two apical domains in fusion cells. Then Stac-containing EVs (orange vesicles) are released into the tracheal lumen upon the fusion of Stac-MVBs with the luminal plasma membrane (the red line outlines the pink lumen). A continuous lumen is formed between two fusion cells

filaments and microtubules (Gervais et al. 2012; Dollar et al. 2016). The invading stock cell could assemble stable apical microtubules parallel to the luminal axis to propel the stock cell-fusion cell junction forward, while the actin/microtubule track inside the fusion cell could contract and thereby pull the fusion cell-stock cell and fusion cell-fusion cell junctions closer (compare blue arrowheads in Fig. 5.3b and in Fig. 5.3c) (Gervais et al. 2012).

The cytoskeleton track serves as a transport route for motor proteins, which carry vesicles (small orange circles outlined by the red line in Fig. 5.3b) that contain membrane and lumen proteins to the track. These vesicles coalesce to generate expanding membrane compartments by inserting apical membrane and luminal components. Not surprisingly, exocyst component Sec5 and Rab11-containing recycling endosomes are accumulated at the fusion cell apical membrane as well as vesicles along the track (Kakihara et al. 2008; Jiang et al. 2007). Exocyst is a protein complex that tethers secretory vesicles at the plasma membrane and controls vesicle fusion (Heider and Munson 2012; Wu and Guo 2015). Rab11 is involved in the trafficking of internalized proteins back to the plasma membrane (Horgan et al. 2010) and from the trans-Golgi network to the plasma membrane (Horgan et al. 2010; Chen et al. 1998b; Li et al. 2007). Thus, both exocyst-mediated vesicle fusion and recycling endosome-mediated trafficking are required for the expansion of the two apical domains in fusion cells.

5.3.3 *Lumen Fusion*

Finally, the two individual apical domains (red arrow and arrowhead in Fig. 5.3c) in each fusion cell are fused to form a continuous lumen. Recently, Stefan Luschnig's lab showed that a subset of extracellular vesicles (EVs) that contains Staccato (Stac), a homolog of the vesicle priming factor Munc13-4, is an essential component of the membrane fusion machinery in fusion cells (Caviglia et al. 2016; Camelo et al. 2022). In fusion cells, Arl3 GTPase activates two downstream targets, Rab27 and Rab35 (Kakihara et al. 2008; Jiang et al. 2007). Rab27 and Rab35 act in a partially redundant fashion to promote the formation of Stac-positive multivesicular bodies (Stac-MVBs) (large orange vesicles in Fig. 5.3c). The local calcium increase that is released from ER exit sites triggers the fusion of Stac-MVBs with the luminal membranes of fusion cells, resulting in the release of Stac-containing EV vesicles in the lumen (large orange vesicles in Fig. 5.3d) (Caviglia et al. 2016). This process is required to fuse the two individual apical domains in fusion cells, resulting in the formation of a continuous lumen (Fig. 5.3d).

5.4 Summary

This review summarizes the current knowledge on branch fusion in the *Drosophila* trachea. Despite significant progress in fusion cell specification, cell migration, cell–cell contact, cytoskeleton track formation, vesicular trafficking, and membrane fusion, there are still questions that remain unsolved. For example, what transcriptional changes take place in fusion cells during branch fusion? What signals are involved in the dynamic regulation of various aspects of the branch fusion process? How do fusion cells generate diversity in different branches? Which membrane trafficking pathways are employed to expand the apical lumen? To reveal the underlying mechanisms of branch fusion, the identification and characterization of genes that play important roles in different steps of the branch fusion process will improve our knowledge of branch fusion in the *Drosophila* trachea. The mechanisms of branch fusion in the *Drosophila* trachea will contribute to our understanding of branch fusion in other systems, such as the vertebrate vascular, renal, and respiratory organs.

References

Camelo C, Körte A, Jacobs T, Luschnig S (2022) Tracheal tube fusion in Drosophila involves release of extracellular vesicles from multivesicular bodies. J Cell Sci 135(3)

Caussinus E, Colombelli J, Affolter M (2008) Tip-cell migration controls stalk-cell intercalation during Drosophila tracheal tube elongation. Curr Biol 18(22):1727–1734

Caviglia S, Luschnig S (2013) The ETS domain transcriptional repressor Anterior open inhibits MAP kinase and Wingless signaling to couple tracheal cell fate with branch identity. Development 140(6):1240–1249

Caviglia S, Brankatschk M, Fischer EJ, Eaton S, Luschnig S (2016) Staccato/Unc-13-4 controls secretory lysosome-mediated lumen fusion during epithelial tube anastomosis. Nat Cell Biol 18(7):727–739

Chandran RR, Iordanou E, Ajja C, Wille M, Jiang L (2014) Gene expression profiling of Drosophila tracheal fusion cells. Gene Expr Patterns 15(2):112–123

Chandran RR, Scholl A, Yang Y, Jiang L (2018) Rebuff regulates apical luminal matrix to control tube size in Drosophila trachea. Biol Open 7(9). https://doi.org/10.1242/bio.036848

Chen CK, Kuhnlein RP, Eulenberg KG, Vincent S, Affolter M, Schuh R (1998a) The transcription factors KNIRPS and KNIRPS RELATED control cell migration and branch morphogenesis during Drosophila tracheal development. Development 125(24):4959–4968

Chen W, Feng Y, Chen D, Wandinger-Ness A (1998b) Rab11 Is required for trans-Golgi network–to–plasma membrane transport and a preferential target for GDP dissociation inhibitor. Mol Biol Cell 9(11):3241–3257

Chen K, Peschel N, Zavodska R, Sehadova H, Stanewsky R (2011) QUASIMODO, a novel GPI-anchored zona pellucida protein involved in light input to the Drosophila circadian clock. Curr Biol 21(9):719–729

Chihara T, Hayashi S (2000a) Control of tracheal tubulogenesis by Wingless signaling. Development 127(20):4433–4442

Chihara T, Hayashi S (2000b) Control of tracheal tubulogenesis by Wingless signaling. Development (Cambridge) 127(20):4433–4442

Dollar G, Gombos R, Barnett AA, Sanchez Hernandez D, Maung SM, Mihaly J et al (2016) Unique and overlapping functions of formins Frl and DAAM during ommatidial rotation and neuronal development in Drosophila. Genetics 202(3):1135–1151

Gervais L, Lebreton G, Casanova J (2012) The making of a fusion branch in the Drosophila trachea. Dev Biol 362(2):187–193

Ghabrial AS, Krasnow MA (2006) Social interactions among epithelial cells during tracheal branching morphogenesis. Nature 441(7094):746–749

Ghabrial A, Luschnig S, Metzstein MM, Krasnow MA (2003) Branching morphogenesis of the Drosophila tracheal system. Annu Rev Cell Dev Biol 19:623–647

Heider MR, Munson M (2012) Exorcising the exocyst complex. Traffic (Copenhagen, Denmark) 13(7):898–907

Horgan CP, Hanscom SR, Jolly RS, Futter CE, McCaffrey MW (2010) Rab11-FIP3 links the Rab11 GTPase and cytoplasmic dynein to mediate transport to the endosomal-recycling compartment. J Cell Sci 123(2):181–191

Ikeya T, Hayashi S (1999) Interplay of Notch and FGF signaling restricts cell fate and MAPK activation in the *Drosophila* trachea. Development 126(20):4455–4463

Jiang L, Crews ST (2003) The Drosophila dysfusion basic Helix-Loop-Helix (bHLH)-PAS gene controls tracheal fusion and levels of the Trachealess bHLH-PAS protein. Mol Cell Biol 23(16):5625–5637

Jiang L, Crews ST (2006) Dysfusion transcriptional control of Drosophila tracheal migration, adhesion, and fusion. Mol Cell Biol 26(17):6547–6556

Jiang L, Rogers SL, Crews ST (2007) The Drosophila dead end Arf-like3 GTPase controls vesicle trafficking during tracheal fusion cell morphogenesis. Dev Biol 311(2):487–499

Kakihara K, Shinmyozu K, Kato K, Wada H, Hayashi S (2008) Conversion of plasma membrane topology during epithelial tube connection requires Arf-like 3 small GTPase in Drosophila. Mech Dev 125(3–4):325–336

Lebreton G, Casanova J (2014) Specification of leading and trailing cell features during collective migration in the Drosophila trachea. J Cell Sci 127(Pt 2):465–474

Lee S, Kolodziej PA (2002) The plakin Short Stop and the RhoA GTPase are required for E-cadherin-dependent apical surface remodeling during tracheal tube fusion. Development 129(6):1509–1520

Lee M, Lee S, Zadeh AD, Kolodziej PA (2003) Distinct sites in E-cadherin regulate different steps in Drosophila tracheal tube fusion. Development 130(24):5989–5999

Li BX, Satoh AK, Ready DF (2007) Myosin V, Rab11, and dRip11 direct apical secretion and cellular morphogenesis in developing Drosophila photoreceptors. J Cell Biol 177(4):659–669

Llimargas M (1999) The *Notch* pathway helps to pattern the tips of the *Drosophila* tracheal branches by selecting cell fates. Development 126(11):2355–2364

Llimargas M (2000) Wingless and its signalling pathway have common and separable functions during tracheal development. Development 127(20):4407–4417

Llimargas M, Casanova J (1999) EGF signalling regulates cell invagination as well as cell migration during formation of tracheal system in Drosophila. Dev Genes Evol 209(3):174–179

Manning G, Krasnow MA (1993) Development of the *Drosophila* tracheal system. In: Bate M, Martinez Arias A (eds) The Development of *Drosophila melanogaster*. Cold Spring Harbor Laboratory Press, Cold Spring Harbor, NY, pp 609–685

Matthews BB, Crews ST (1999) Drosophila center divider gene is expressed in CNS midline cells and encodes a developmentally regulated protein kinase orthologous to human TESK1. DNA Cell Biol 18(6):435–448

Miao G, Hayashi S (2016) Escargot controls the sequential specification of two tracheal tip cell types by suppressing FGF signaling in Drosophila. Development (Cambridge) 143(22):4261–4271

Moussian B, Letizia A, Martinez-Corrales G, Rotstein B, Casali A, Llimargas M (2015) Deciphering the genetic programme triggering timely and spatially-regulated chitin deposition. PLoS Genet 11(1):e1004939

Ribeiro C, Neumann M, Affolter M (2004) Genetic control of cell intercalation during tracheal morphogenesis in Drosophila. Curr Biol 14(24):2197–2207

Samakovlis C, Hacohen N, Manning G, Sutherland D, Guillemin K, Krasnow MA (1996) Development of the *Drosophila* tracheal system occurs by a series of morphologically distinct but genetically coupled branching events. Development 122:1395–1407

Scholl A, O'Brien MJ, Chandran RR, Jiang L (2019) The novel gene apnoia regulates Drosophila tracheal tube size. Dev Dyn 248(6):477–487

Skouloudaki K, Papadopoulos DK, Tomancak P, Knust E (2019) The apical protein Apnoia interacts with Crumbs to regulate tracheal growth and inflation. PLoS Genet 15(1):e1007852

Stahl AL, Charlton-Perkins M, Buschbeck EK, Cook TA (2017) The cuticular nature of corneal lenses in Drosophila melanogaster. Dev Genes Evol 227(4):271–278

Steneberg P, Hemphala J, Samakovlis C (1999) Dpp and Notch specify the fusion cell fate in the dorsal branches of the *Drosophila* trachea. MechDev 87(1–2):153–163

Sutherland D, Samakovlis C, Krasnow MA (1996) Branchless encodes a Drosophila FGF homolog that controls tracheal cell migration and the pattern of branching. Cell 87(6):1091–1101

Tanaka-Matakatsu M, Uemura T, Oda H, Takeichi M, Hayashi S (1996) Cadherin-mediated cell adhesion and cell motility in Drosophila trachea regulated by the transcription factor Escargot. Development 122(12):3697–3705

Uv AE, Roth P, Xylourgidis N, Wickberg A, Cantera R, Samakovlis C (2000) Members only encodes a Drosophila nucleoporin required for Rel protein import and immune response activation. Genes Dev 14(15):1945–1957

Uv A, Cantera R, Samakovlis C (2003) Drosophila tracheal morphogenesis: intricate cellular solutions to basic plumbing problems. Trends Cell Biol 13(6):301–309

Wolf C, Schuh R (2000) Single mesodermal cells guide outgrowth of ectodermal tubular structures in Drosophila. Genes Dev 14(17):2140–2145

Wolf C, Gerlach N, Schuh R (2002) Drosophila tracheal system formation involves FGF-dependent cell extensions contacting bridge-cells. EMBO Rep 3(6):563–568

Wu B, Guo W (2015) The exocyst at a glance. J Cell Sci 128(16):2957–2964

Chapter 6
Trophoblast Syncytialization: A Metabolic Crossroads

Tina Podinić, Andie MacAndrew, and Sandeep Raha

Abstract During placentation, villous cytotrophoblast (CTB) stem cells proliferate and fuse, giving rise to the multinucleated syncytiotrophoblast (STB), which represents the terminally differentiated villous layer as well as the maternal-fetal interface. The syncytiotrophoblast is at the forefront of nutrient, gas, and waste exchange while also harboring essential endocrine functions to support pregnancy and fetal development. Considering that mitochondrial dynamics and respiration have been implicated in stem cell fate decisions of several cell types and that the placenta is a mitochondria-rich organ, we will highlight the role of mitochondria in facilitating trophoblast differentiation and maintaining trophoblast function. We discuss both the process of syncytialization and the distinct metabolic characteristics associated with CTB and STB sub-lineages prior to and during syncytialization. As mitochondrial respiration is tightly coupled to redox homeostasis, we emphasize the adaptations of mitochondrial respiration to the hypoxic placental environment. Furthermore, we highlight the critical role of mitochondria in conferring the steroidogenic potential of the STB following differentiation. Ultimately, mitochondrial function and morphological changes centrally regulate respiration and influence trophoblast fate decisions through the production of reactive oxygen species (ROS), whose levels modulate the transcriptional activation or suppression of pluripotency or commitment genes.

Tina Podinić and Andie MacAndrew contributed equally with all other contributors.

T. Podinić · A. MacAndrew · S. Raha (✉)
Department of Pediatrics and Graduate Program in Medical Sciences, McMaster University, Hamilton, ON, Canada
e-mail: rahas@mcmaster.ca

M. Kloc, A. Uosef (eds.), *Syncytia: Origin, Structure, and Functions*, Results and Problems in Cell Differentiation 71,
https://doi.org/10.1007/978-3-031-37936-9_6

6.1 Overview of Syncytialization in Villous Trophoblasts

The continuous fusion of mononucleated cells to assemble multinucleated structures is a process referred to as syncytialization. This involves the mixing of plasma membrane components, cytoplasmic contents (lipids, proteins, and RNAs), and cellular organelles (mitochondria, endoplasmic reticulum, and the nucleus) resulting in the formation of multinucleated structures that are now termed a syncytium. Typical examples of this cell fusion in humans are observed during skeletal muscle development, osteoclast formation, and placental syncytiotrophoblast development (Gerbaud and Pidoux 2015; Huppertz and Gauster 2011). Although syncytialization occurs in various biological contexts, the steps comprising cellular fusion are relatively conserved across physiological systems. Competent cells first commit to differentiation and express fusogenic proteins. This is followed by cells communicating and associating with adjacent cells via membrane contacts and gap junctions in order to exchange fusogenic signals. Finally, appropriate protein complexes form to facilitate the acquisition of both cellular outer membranes and subsequently promote the opening of a fusion pore and dissolution of cell membranes ensuing cytoplasmic continuity, thus completing the fusion process (Chernomordik and Kozlov 2005; Pidoux et al. 2014).

Human placental cytotrophoblasts are progenitor cells that biochemically and morphologically differentiate early in gestation to become villous trophoblasts or invasive extravillous trophoblasts: two general pathways that are strictly regulated. The villous pathway involves the fusion of mononucleated cytotrophoblasts (CTBs) into the multinucleated syncytiotrophoblast (STB) (Ji et al. 2013). The STB is a unique surface that is surrounded by maternal blood and forms the primary interface between mother and fetus (Zeldovich et al. 2013). The STB and CTBs are jointly located on a basement membrane and comprise an epithelial-like layer that covers the placental villous tree referred to as the villous trophoblast. The villous trophoblast forms the outermost fetal layer of the placenta that is in direct contact with maternal circulation. As the primary maternal-fetal interface, oxygen and important nutrients from maternal blood are able to pass through the STB by simple diffusion, facilitated diffusion, and active transport in order to reach the developing fetus (Lager and Powell 2012). The maternal-fetal interface is crucial for the maintenance of pregnancy. Any impairment in the formation of this structure could lead to complications, such as defective or decreased implantation, resulting in outcomes such as infertility, recurrent miscarriage, or intrauterine growth retardation (Ochoa-Bernal and Fazleabas 2020; Yang et al. 2019). CTBs, located adjacent to the basement membrane, constitute an assemblage of progenitor cells that proliferate, differentiate, and undergo fusogenic events with the overlying, multinucleated layer designated as the STB (Huppertz and Gauster 2011). The STB lacks the proliferative capacity of CTBs and has a finite lifespan, undergoing highly regulated turnover facilitated by the differentiation and fusion of CTBs (Renaud and Jeyarajah 2022).

The differentiation of CTBs to the STB is considered a downstream fusogenic event that is initiated by human chorionic gonadotropin (hCG) binding to and activating a 7-transmembrane receptor, LH-CG, and initiating an increase in intracellular cyclic adenosine monophosphate (cAMP) via the activation of adenylate cyclase (AC) and subsequent activation of protein kinase A (PKA). hCG is secreted early in embryonic development by embryonic trophoblast cells and is continuously produced by the STB promoting syncytialization in an autocrine-paracrine manner (Pidoux et al. 2014; Shi et al. 1993; Yang et al. 2003). PKA activates cAMP Response element binding protein which associates with CREB binding protein (CBP) and P300 to facilitate fusogenic gene transcription. cAMP/PKA signaling further enhances transcription of fusogenic genes, such as connexin (Hussain et al. 2003), cadherin (Coutifaris et al. 1991), and syncytins, by promoting the association of transcription factors, glial cells missing-1 (GCM-1), and CBP (Gerbaud and Pidoux 2015).

6.1.1 The Role of Adhesive and Junctional Proteins in Trophoblast Fusion

The commitment of primary cells is initiated by adherens junctions, tight junctions, and gap junctions, which permit cell-cell adhesion and cellular communication, ultimately triggering fusion (Gerbaud and Pidoux 2015). Cadherins are a family of Ca^{2+} dependent, transmembrane glycoproteins involved in cell-cell adhesion (Oda and Takeichi 2011; Takeichi 1995). This adhesion is achieved through clustering of homotypic and heterotypic cadherins present on adjacent cells, thereby stabilizing and anchoring neighboring cells to each other allowing for the membrane apposition step of cellular fusion. In the context of STB formation, the most studied cadherin is E-cadherin, which is most densely populated on boundaries in-between neighboring CTBs. E-cadherin is also localized on CTBs at the interface between CTBs and the STB, but are reported to be significantly less expressed in fused CTBs forming the STB (Coutifaris et al. 1991). As such, E-cadherin levels and syncytialization seem to demonstrate an inverse relationship. E-cadherins play an important role in trophoblast fusion as they are essential for dynamic membrane alterations (Coutifaris et al. 1991; Renaud and Jeyarajah 2022). Cadherin-11 is another adheren that is implicated in trophoblast syncytialization. In contrast to E-cadherin, cadherin-11 densities tend to increase during trophoblast cell fusion, illustrating the importance of cadherins throughout the entire trophoblast fusion process. Although the direct mechanism by which cadherin-11 mediates syncytialization is poorly understood, it is apparent that cadherin-11 is integral in the morphological and functional differentiation and fusion of trophoblasts (Getsios and MacCalman 2003). Similar to cadherins, tight junctions are proteins that also promote cell-cell anchoring and adhesion required for cell fusion. Specifically, zonula occludens-1 (ZO-1) is implicated in CTB fusion. As seen with E-cadherin, ZO-1 is expressed on CTBs, and its expression decreases with syncytialization. Silencing of ZO-1 blocks cell adhesion and subsequently cell fusion in human trophoblasts and decreases expression of connexin-43 (Cx-43) (Pidoux et al. 2010). Cx-43 is the only gap junction expressed in human CTBs and the STB, localized at the interface between the trophoblast sub-lineages (Cronier et al. 2002). Following activation of the LH-GC receptor by hCG and the successive increase in intracellular cAMP, activated PKA phosphorylates Cx43 which promotes the opening of this gap junction and transfer of fusogenic signals. This instigates the transcription of genes, such as GCM1 and CREB, and initiates the assembly of fusogenic machinery to trigger CTB fusion (Gerbaud and Pidoux 2015; Pidoux et al. 2014).

The final stage of trophoblast syncytialization is the cell-cell fusion process which constitutes the fusing of cellular membranes and combination of cellular components. This fusion is catalyzed by specific proteins that facilitate apposition of CTBs and the opening of a fusion pore promoting a shift in cytoskeletal and plasma membrane structures allowing for cell fusion. The most studied proteins involved in STB formation are called syncytins. Two major syncytins have been identified in humans: syncytin-1 (syn-1) and syncytin-2 (syn-2). These fusogenic proteins are encoded by endogenous retroviral envelope (ERV) genes embedded in human DNA (Gerbaud and Pidoux 2015) and are almost exclusively expressed in the placenta (Pötgens et al. 2004). Syn-1, encoded by ERVW-1, was the first glycoprotein reported to demonstrate fusogenic properties in trophoblasts via retroviral envelope-like properties (Frendo et al. 2003). Syn-1 expression is maintained throughout gestation and is mainly localized in STB (Ji et al. 2013; Mi et al. 2000; Okahara et al. 2004), and it additionally plays an important role in triggering CTB cell fusion (Knerr et al. 2005; Mi et al. 2000) via interactions with human sodium-dependent neutral amino acid transporter type 1 or 2 (hASCT1, hASCT2) mainly localized on CTBs (Lavillette et al. 2002). As opposed to Syn-1, Syn-2 is densely expressed at the apical portion of CTB clusters, at the interface between CTBs and the STB, where it binds to major facilitator domain-containing protein 2 (MFSD2A) receptor which is expressed by the STB (Esnault et al. 2008). Syn-2, encoded by HERV-FRD, is also required for trophoblast fusion (Vargas et al. 2009); however, Syn-2 expression significantly decreases following STB formation (Malassiné et al. 2008). Although syncytins are considered to be the major proteins mediating trophoblast fusion, in vivo knock-out of these proteins does not entirely abolish cell fusion, suggesting other fusogenic proteins may be more involved in syncytialization (Dupressoir et al. 2009; Ji et al. 2013).

6.1.2 Consequences of Improper Syncytialization: The Role of Apoptosis in the Formation of Syncytial Knots

Considering the continuous fusion of CTBs during syncytialization that facilitates the formation of the STB, there must exist a delicate balance between proliferation, differentiation, and trophoblast cell death to avoid hyper fusion or tissue loss. Trophoblast syncytialization comprises the fusion of mononucleated cells to form a multinucleated syncytium. The propensity of the syncytium to continuously accumulate nuclei throughout gestation is supported by the range of differentially aged nuclei which have been detected in the syncytium (Burton and Jones 2009; Ji et al. 2013). Alternatively, evidence supports the shedding of aged nuclei through the formation of syncytial knots. Syncytial knots are aggregates of STB nuclei packaged into membrane-bound vesicles that are released into the maternal circulation (Rajakumar et al. 2012; Yasuda et al. 1995). A two-stage apoptotic model has been proposed to elucidate a potential mechanism of trophoblast turnover. The model consists of an initial pre-apoptotic stage that commits CTBs to STB formation. The next stage encompasses an execution phase where syncytial knots are shed into maternal circulation (Huppertz et al. 1998; Mayhew et al. 1999). Further, studies have associated events in reversible, early-phase apoptosis in proper syncytialization. Activation of caspase-8 (Cas-8), an inhibitor caspase that is an early effector of apoptosis, is required for CTB fusion (Black et al. 2004). Externalization of phosphatidylserine is another characteristic in both syncytialization and apoptosis underlining the overlap between the molecular pathways (Huppertz et al. 2001). Oxygen tension has been shown to impact apoptosis in CTBs and the STB. CTBs isolated from placentas of ranging gestational timelines spontaneously fuse to form the STB under atmospheric (~21%) oxygen (Chang et al. 2018). Lower oxygen levels (>1%) have been found to have a pro-apoptotic effect in human CTBs, upregulating pro-apoptotic proteins p53 and Bax and decreasing anti-apoptotic protein, Bcl-2 (Chen et al. 2010; Humphrey et al. 2008). In the STB, hypoxia has been found to enhance apoptosis through a p53-independent pathway (Chen et al. 2010). Therefore, oxygen tension may play a role in regulating the balance of autophagy and apoptosis in relation to trophoblast shedding. While the tendency for trophoblasts to turn over is still under debate, there is a multitude of consequential evidence supporting some level of STB tissue loss along with the continuous STB cell renewal, and further research may elucidate detailed mechanisms supporting this balance.

6.1.3 Hypoxic Signaling Regulates Trophoblast Differentiation

Microenvironmental factors, like oxygen tension, have also been implicated in the differentiation and cellular fate of trophoblasts. Syncytialization is impaired with reduction of oxygen levels with most CTBs remaining as mononuclear structures and, further, by decreasing normal hormone release characteristic of the ST (Alsat et al. 1996). This inhibition of CTB differentiation has been proposed to be partly related to hypoxia-inducible factor (HIF) complex. HIF is a transcriptional complex that mediates cellular responses and cellular fate under low oxygen conditions (Wakeland et al. 2017). In addition, lower oxygen tension (1%) reduces transcription factor GCM1 and its downstream effector, fusogenic protein, Syn-1, which are major regulators involved in STB formation (Wich et al. 2009). The downregulation of GCM1 has also been found to be regulated by HIF in murine trophoblast stem cells (Maltepe et al. 2005). Reductions in GCM1 are associated with defects in syncytialization along with placental complications such as preeclampsia (PE) illustrating the potentially imperative role of oxygen tension in syncytialization and proper placental development (Bainbridge et al. 2012). Composition and rigidity of the extracellular matrix (ECM) is an additional microenvironment determinant connected with the regulation of CTB fusion. A recent study reported ECM thickness impacts not only mRNA levels of fusogenic markers but also alters trophoblast secreted proteins (Wong et al. 2018). In addition, Choi and colleagues demonstrated changes in ECM composition in which trophoblast stem cells are cultured alter their differentiation propensity toward ST formation rather than trophoblast giant cells in vitro and block oxygen-independent HIF induction (Choi et al. 2013), linking the regulation of oxygen sensors with ECM arrangements. Interestingly, biochemical mechanisms related to placental-related disorders, such as PE, have also been associated with ECM variations (Parameshwar et al. 2021), further implicating the potential importance of ECM matrix formation in proper placental formation and regulation.

6.2 Mitochondrial Dynamics and Signaling in Syncytialization

Despite their infamous role as the powerhouse of the cell, mitochondria contribute much more to cellular function than just energy production. Mitochondria support various cellular functions by mediating programmed cell death, ion homeostasis and calcium signaling, cell growth and differentiation, as well as redox homeostasis. In order to perform these crucial tasks, mitochondria undergo a series of quality control events, involving a dynamic balance between mitochondrial fusion and fission. This process integrates mitochondrial biogenesis and mitophagy and results in the alteration of mitochondrial morphology, numbers, and distribution during periods of cellular growth or stress. In some cases, mitochondrial regulation of these cellular processes is fundamental to cellular function, such as with trophoblast syncytialization.

6.2.1 *Mitochondrial Dynamics: A Balance Between Fission and Fusion*

The dynamic formation and dissociation of mitochondrial networks are largely the product of opposing processes known as mitochondrial fusion and fission, which are tightly regulated by a diverse family of GTPases. As mitochondria are double membrane-bound organelles, mitochondrial elongation therefore requires the coordination of specific GTPases to amalgamate each membrane. Initially, a family of dynamin-related GTPases, mitofusin 1 and 2 (Mfn1/2), associate to coordinate outer mitochondrial membrane (OMM) fusion, whereas another GTPase, optic atrophy protein (Opa1), is responsible for inner mitochondrial membrane (IMM) fusion by remodeling cristae networks (Cipolat et al. 2004; Frezza et al. 2006; Song et al. 2009). At the physiological level, mitochondrial elongation is favored during periods of acute cytotoxic stress, wherein select mitochondria become defective and mitochondrial fusion allows for the allocation and exchange of mitochondrial DNA (mtDNA) (Ono et al. 2001), proteins/lipids, or metabolites to mitigate the mitochondrial stress. Alternatively, when mtDNA becomes damaged, the elongation of mitochondrial networks allows for the imminent consequences of those defects to be minimized as the damaged mtDNA become diluted. Moreover, mitochondrial elongation phenotypes have been shown to predominate in certain metabolic disorders, such as with obese and diabetic individuals. In fact, Parra and colleagues demonstrated that the treatment of cardiomyocytes with insulin in vitro resulted in elongated mitochondrial phenotypes, as evidenced by an increased expression of Opa1 (Parra et al. 2014). Indeed, following Opa1 and Mfn2 knockdowns, insulin-mediated alterations to cardiomyocyte mitochondrial morphology were prevented, suggesting that mitochondria integrate external metabolic signals to mediate mitochondrial fusion capacity (Parra et al. 2014). Furthermore, increased placental mitochondrial fusion has been observed in certain gestational metabolic disorders (Abbade et al. 2020), suggesting that trophoblast mitochondria might possess adaptive features to mitigate metabolic deficits and cellular stress in order to preserve the developing fetus, which will be discussed later in this review.

Ultimately, mitochondrial fission is a quality control mechanism aimed at disconnecting or selectively removing dysfunctional mitochondria and is carried out by a specific GTPase known as dynamin-related protein 1 (Drp1) (Burman et al. 2017). Inactive Drp1 is sequestered in the cytosol, and upon undergoing various posttranslational modifications, it subsequently translocates to the OMM where it associates with mitochondrial fission protein 1 (Fis1) and mitochondrial fission factor (Mff) in response to intracellular cues, including apoptotic stimuli (Kornfeld et al. 2018; Losón et al. 2013). In contrast to mitochondrial fusion proteins, the phosphorylation state of Drp1 serine residues is crucial in regulating its activity. For instance, PKA-dependent Ser637 phosphorylation prevents the translocation of $Drp1^{pSer637}$ from the cytosol to the outer mitochondrial membrane, thus inhibiting mitochondrial fission and easing mitochondrial fusion. Additionally, during periods of sustained mitochondrial membrane depolarization, calcineurin-dependent Ser637 dephosphorylation leads to Drp1[Ser637] activation and localization to the mitochondrial membrane where actin polymerization ultimately facilitates mitochondrial constriction (Cribbs and Strack 2007; Kashatus et al. 2015). At the crux of mitochondrial energetics and mitochondrial structure is the electrochemical proton gradient defined as mitochondrial membrane potential ($\Delta\Psi$m), which largely drives ATP synthesis. In addition to its role in facilitating mitochondrial fragmentation, Duarte et al. demonstrate impaired mitochondrial fusion in MA-10 cells following CCCP-induced mitochondrial uncoupling and subsequent depolarization, effects which were reversed by a 3-h washout. To determine whether mitochondrial fusion was exclusively impaired through $\Delta\Psi$m dissipation, they confirmed that complex V inhibition alone did not hinder mitochondrial fusion (Duarte et al. 2012). In a broader context, $\Delta\Psi$m has been demonstrated to regulate mitophagy (Elmore et al. 2001; Hu et al. 2016b; Narendra et al. 2008), the selective degradation of mitochondria, as well as impart steroidogenic capacity to certain cell types by altering mitochondrial structure (Allen et al. 2006; Duarte et al. 2012).

To emphasize the importance of mitochondrial dynamics in cellular function, several studies have highlighted the critical roles of mitochondrial fission-fusion machinery in the initiation, progression, or inhibition of apoptosis (Frank et al. 2001). In general, early apoptotic cells exhibit predominantly fragmented and round mitochondria, suggesting that mitochondrial fission is involved during these initial phases of apoptosis. Consistent with this observation, Karbowski et al. demonstrated the co-localization of mitochondrial fission proteins, Drp1 and Fis1, with the pro-apoptotic member, Bax, at the OMM (Karbowski et al. 2002). Additionally, the downregulations of Fis1 and Drp1 have both been associated with fewer apoptotic events; however, this resistance to apoptosis was relatively more pronounced following Fis1 downregulation (Lee et al. 2004). Indeed, sensitivity to apoptosis is restored upon Fis1 overexpression, as was assessed by greater cytochrome c release and nuclear fragmentation events, suggesting that the inhibition to apoptosis is tightly linked to Fis1 depletion (Lee et al. 2004). In a similar manner, the loss of Opa1 renders greater sensitivity to apoptotic events, as mitochondrial fragmentation and mitochondrial membrane become unopposed and depleted, respectively. To emphasize the link between cristae structure and apoptosis promotion, Olichon et al. further demonstrate that Opa1 depletion in vitro disrupts IMM cristae organization and exacerbates cytochrome *c* release after 72 h compared to control (Olichon et al. 2003). Cristae remodeling has been well-characterized during apoptosis (Germain et al. 2005), as cristae reorganization allows for the permeabilization of the IMM and subsequent cytochrome *c* release from mitochondria, which potentiates apoptosis through the caspase cascade (Breckenridge et al. 2003; Scorrano et al. 2002). Alternatively, mitochondrial fusion proteins may be regarded as anti-apoptotic proteins, as some studies have established that the overexpression of both Mfn1 and Mfn2 successfully inhibit apoptosis. Using prion disease models, Wu et al. found that Opa1 overexpression rescued against mitochondrial fragmentation, thus limiting cytochrome *c* release and effectively preventing neuronal apoptosis (Wu et al. 2019). In essence, Opa1 stimulates oxidative phosphorylation (OXPHOS) through the promotion of mitochondrial fusion, which may provide an additional basis for its anti-apoptotic activity (Hong et al. 2022a; Patten et al. 2014). Overall, these mitochondrial morphological alterations are intended to mediate cellular functions by participating in intracellular communication and adapting to stressors, through the transient generation of metabolic by-products, such as reactive oxygen species (ROS) and ATP.

6.3 Characteristic Energetic Profiles Associated with Mitochondrial Morphologies

Mitochondrial ultrastructural changes are intimately linked to unique bioenergetic states. This feature is crucial for energy preservation or generation during periods of low cellular activity or high metabolic demand, respectively, and serves to distinguish between cell types. In general, fragmented mitochondrial phenotypes are reminiscent of a primarily glycolytic energy state, whereas elongated mitochondria offer greater surface area for electron transport chain (ETC) complexes and ATP synthase subunits, thus favoring OXPHOS (Khacho and Slack 2017; Yao et al. 2019). To emphasize the link between cristae morphology and mitochondrial respiration, cryotomography images have demonstrated the strategic positioning of ETC complexes within cristae invaginations, and computational models have found that the extent of ATP production is directly proportional to that of inner membrane (IM) folding, suggesting that mitochondria harboring well-defined cristae are more efficient at ATP synthesis (Afzal et al. 2021). To add another layer of complexity, it has been proposed that cristae morphology affects respiration efficiency through the modulation of supercomplex assembly, which refers to the higher-order structures that form between complexes I-III-IV (Azuma et al. 2020; Cogliati et al. 2016; Ikeda et al. 2013). These supercomplexes or respirasomes have been shown to enhance mitochondrial respiration efficiency under conditions of metabolic stress, by decreasing the distance between complexes and limiting the diffusion of cytochrome *c* between complexes (den Brave and Becker 2020; Letts and Sazanov 2017). In addition to certain phospholipids, several mammalian supercomplex assembly-promoting factors have been identified, such as supercomplex assembly factor 1 (Scaf1) and stomatin-like protein 2 (Slp2), which are differentially expressed in human tissues (García-Poyatos et al. 2020; Mitsopoulos et al. 2015). Scaf1 possesses a well-established super complex-promoting role in muscle, where it primarily mediates the structural association between complexes III and IV, thus increasing NADH-dependent respiration coupled with lower ROS production (Calvo et al. 2020). Apart from its localization at the IMM, the roles of Slp2 have been less elucidated, although most studies have identified Slp2 in stabilizing mitochondrial prohibitins, as well as complexes I and IV (Da Cruz et al. 2008). In turn, prohibitin 2 (Phb2) has been implicated in the processing of Opa1 isoforms, wherein Phb2-deficient mouse embryonic fibroblasts (MEFs) exhibited a loss of L-Opa1 isoforms and greater mitochondrial fragmentation (Merkwirth et al. 2008). Taken together, mitochondrial morphology and energy production are interconnected in that respiratory enzyme efficiency is highly influenced by cristae ultrastructure. Under conditions of stress, the dynamic bioenergetic demands of the cell may be accommodated through alterations to the mitochondrial network. One example of mitochondrial dynamics altering the energetic state of the cell to adapt to starvation and prevent cell death is exemplified through the process of macroautophagy, wherein cytoplasmic contents are sequestered and degraded (Wu et al. 2020). Nutrient depletion is a well-characterized autophagic stimulus, which promotes the catabolic degradation of existing intracellular nutrients to maintain cellular viability. During the early phases of autophagy, in vitro studies have characterized mitochondrial elongation after 30 min of starvation, whereas in vivo studies have characterized mitochondrial elongation in both muscle and liver tissues of 12-h fasted mice (Gomes et al. 2011; Rambold et al. 2011). The proposed mechanism underlying these changes is based on starvation leading to increased cAMP levels and the activation of protein kinase A (PKA), which phosphorylates Drp^{pS637} and ensures the cytosolic retention of Drp1. Thus, mitochondrial elongation is unopposed and better supports the formation of the autophagosome, an otherwise energy-consuming process that demands greater ATP production.

6.4 Metabolic Nuances in Stem Cell Fate Decisions

Increasing evidence suggests the crucial interplay between metabolism and stem cell fate decisions. Stem cells are a unique population of cells that possess two defining characteristics: the ability (1) to self-renew, thereby re-populating the stem cell pool, and (2) to commit and differentiate into more specialized cell types, functioning in post-injury tissue repair or during organogenesis (Alison et al. 2002; Bigarella et al. 2014; Shyh-Chang and Ng 2017; Tatapudy et al. 2017). The potency of stem cells describes their ability to give rise to broader or narrower cell lineage pathways. Pluripotent stem cells (PSCs), such as embryonic stem cells (ESCs), are able to generate specialized cells belonging to either of three primary germ layers (Kelly 1977). Whereas following development, adult stem cells (ASCs) are undifferentiated, tissue-resident stem cell populations that give rise to distinct, specialized somatic cells, rendering them multipotent (Li and Xie 2005). To begin, undifferentiated and proliferative stem cells are dependent on glycolytic metabolism to maintain stemness programs due to glycolytic intermediates being consumed for cellular biosynthesis (Lunt and Vander Heiden 2011). Under conditions of hypoxia, anaerobic glycolysis becomes most favorable; however, aerobic glycolysis may still prevail in oxygen-rich environments due to the transformation of glycolytic intermediates to satisfy anabolic demands of proliferative cells (Lunt and Vander Heiden 2011). As opposed to OXPHOS, glycolysis is favorable in that ROS levels are dramatically diminished, thus minimizing the potential for oxidative stress. Though the exact stimuli promoting stem cell commitment remain unclear, evidence suggests that the transition from self-renewal into commitment and differentiation are accompanied by an apparent "metabolic switch" from glycolysis to OXPHOS, which maintains ATP production and relatively elevates ROS production (Gu et al. 2016; Hu et al. 2016a). To emphasize, similar metabolic reprogramming from glycolysis to OXPHOS at the onset of differentiation has been characterized in neural (Zheng et al. 2016), muscle (Hong et al. 2022b), and hair follicle (Tang et al. 2016) stem cells. As mitochondrial structure, bioenergetics, and mtROS production are tightly coupled, several studies have demonstrated how the cellular redox environment impacts transcriptional programs of commitment and differentiation. Using neural stem cells (NSCs), Khacho et al. observed aberrant mitochondrial fragmentation as well as defects in the self-renewal capacity of NSCs following gene knockdowns of Mfn1 and 2 (Khacho et al. 2016). Moreover, they found that ROS signaling stimulates Nrf2-dependent transcriptional activation and suppression of genes related to pluripotency and self-renewal, respectively, whereas the loss of Nrf2 restores self-renewal capacity of NSCs. In vivo studies have found that mouse embryonic stem cells (mESCs) overexpressing Mff failed to produce live-born offspring and determined that mitochondrial dynamics crucially regulate β-catenin stability to maintain pluripotency (Zhong et al. 2019). Likewise, somatic dedifferentiation requires metabolic reprogramming from oxidative to glycolytic metabolism, as has been evidenced by increased glycolytic intermediates and downregulated OXPHOS complexes during this transition (Folmes et al. 2011). Although the role of mitochondrial ROS (mtROS) production has previously been undervalued, recent studies are uncovering the crucial intracellular signaling role of mtROS during stem cell differentiation. In low oxygen environments, studies have uncovered that ROS-mediated signaling likely precedes HIF1α expression and anaerobic glycolysis in hematopoietic stem cells (HSCs) to maintain quiescence (Lacher et al. 2018; Simsek et al. 2010). Furthermore, studies have revealed distinct mitochondrial morphological changes associated with either undifferentiated or differentiated stem cells, which presumably precede and necessitate the metabolic switch. Following the knockdown of Opa1 in murine muscle tissue (Opa1-MKO), Baker et al. observed greater mitochondrial fragmentation, reduced numbers of $Pax7^+$ MuSCs, and increased sensitivity to activation and commitment stimuli (Baker et al. 2022). Their findings coincide with the existing literature suggesting that mitochondrial fragmentation promotes stem cell commitment (Forni et al. 2016; Zhong et al. 2019). As hair follicle stem cell (HFSC) differentiation progresses, mitochondria become elongated with greater cristae definition, which reflects their transition away from glycolytic metabolism (Tang et al. 2016). Increases in mtDNA copy number, ETC complexes and enhanced mitochondrial elongation have been observed during cortical neuron differentiation in vitro (Agostini et al. 2016), suggesting similar trends in metabolic and mitochondrial phenotypes that govern stem cell pluripotency and differentiation. However, less is known about the role of mitochondria during the syncytialization of trophoblasts, despite exhibiting marked differences in mitochondrial and metabolic characteristics in undifferentiated CTBs and the differentiated STB.

6.5 Mitochondria in Trophoblast Syncytialization

The placenta is a semi-permeable organ that provides the developing fetus with structural and endocrine support while simultaneously facilitating nutrient, waste, and gas exchange at the maternal-fetal interface. To ensure successful pregnancy progression and fetal development, the placenta ultimately relies on mitochondria to satisfy the energetic demands and confer steroidogenic properties of the syncytiotrophoblast. Previous studies have characterized the distinct mitochondrial phenotypes associated with either CTB or STB, suggesting that mitochondrial morphological changes may be involved in the process of syncytialization (Bartho et al. 2020). Below the syncytium, proliferative CTBs possess larger, ovoid-shaped mitochondria containing abundant and well-defined cristae (Fig. 6.1) (Fisher et al. 2019a). In addition, they are highly metabolically active as evidenced by their

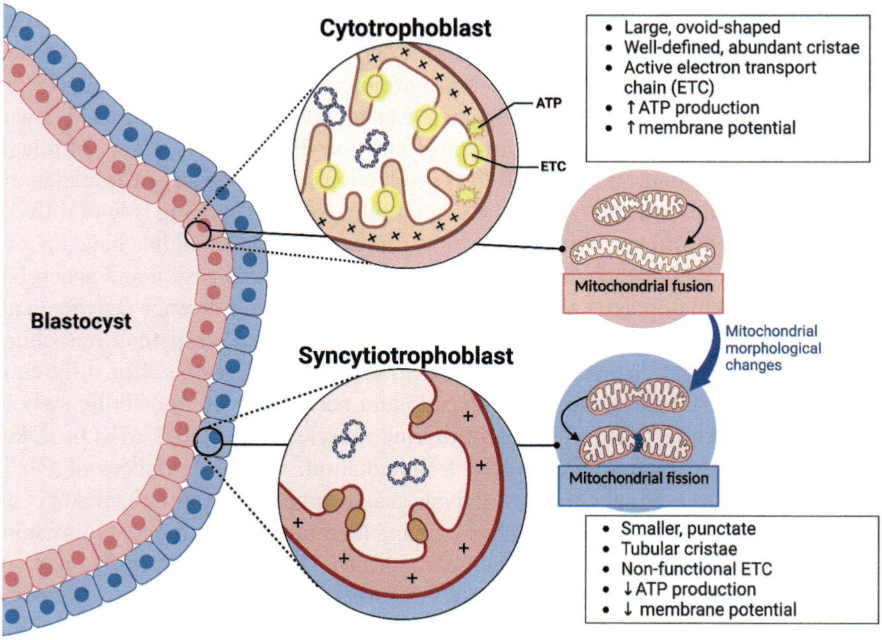

Fig. 6.1 Characteristics of villous trophoblast mitochondria. In the villous pathway, villous cytotrophoblasts (CTBs) give rise to the syncytiotrophoblast (STB) through terminal differentiation. Trophoblast differentiation is accompanied by morphological and functional changes to mitochondria. CTBs possess elongated mitochondrial morphologies relative to STB mitochondria, which are smaller and punctate in nature. CTB mitochondria are hyperpolarized and capable of generating ATP through their active ETC complexes, whereas STB mitochondria possess relatively depolarized mitochondrial membrane potentials coupled with lower ATP production. Mitochondrial polarization is indicated by greater proton (+) accumulation in the intermembrane (IM) space. ATP molecules are represented by stars in the IM space, whereas ETC complexes are situated on the inner mitochondrial membrane (IMM) and are either glowing (active) or dull (inactive). Figure created with BioRender.com

elevated $\Delta\Psi m$ (Bustamante et al. 2014) and ATP production (Fisher et al. 2019a; Kolahi et al. 2017); however, they lack the expression of the P450scc enzyme responsible for converting cholesterol to pregnenolone—a progesterone precursor—indicating that CTBs are impartial to progesterone production (Martínez et al. 1997). It has been reported that mitochondrial content is increased at the onset of cellular differentiation in several stem cell types (Zhang et al. 2013). Indeed, the estrogen-related receptor γ (ERRγ), which has been identified at high levels in the placenta (Takeda et al. 2009) and in human CTBs (Poidatz et al. 2012), has also been implicated in promoting mitochondrial biogenesis in purified human CTB cultures (Poidatz et al. 2015). Silencing of ERRγ leads to a loss of mtDNA content (Poidatz et al. 2015), which ultimately limits their oxidative respiration efficiency considering that mtDNA genes chiefly encode 13 respiratory complex proteins and subunits (Anderson et al. 1981). During syncytialization, ERRγ upregulation not only precedes but regulates aromatase (CYP19) gene transcription, which in turn confers additional steroid-metabolizing and estrogen-generating functions to the syncytium (Kumar and Mendelson 2011; Poidatz et al. 2015). In the context of the metabolic switch, ERRγ has been found to be upregulated during neuronal differentiation along with PGC1α, a master regulator of mitochondrial biogenesis, where they collectively function to sustain energy metabolism through the transcription of mitochondrial genes (Zheng et al. 2016). Thus, it is becoming increasingly apparent that the regulation of mitochondrial respiration, biogenesis, and mtDNA content is not only integral during cellular differentiation, rather a metabolic trend that extends to trophoblast syncytialization.

As compared to CTBS, the smaller, punctate mitochondria with poorly defined cristae of the STB correspond to their mitotically inactive state and lower metabolic activity (Fig. 6.1) (Fisher et al. 2019a). STB mitochondria possess fewer ETC subunits and complex V (De los Rios Castillo et al. 2011; Fisher et al. 2019b), greater mitochondrial fragmentation (relative to their CTB precursors) (Kolahi et al. 2017), and elevated P450scc expression (Martinez et al. 2015) and exhibit a relatively diminished $\Delta\Psi m$ (Bustamante et al. 2014). To summarize, the ETC consists of four major respiratory complexes (I, II, III, IV) situated on inner mitochondrial cristae folds, which shuttle electrons to complex V where ATP is synthesized or hydrolyzed. More specifically, ATP synthase contains two major regions: the F_1 portion, located along the inner surface of the mitochondrial matrix and primarily functioning to hydrolyze ATP, and the F_0 portion, which is embedded in the IMM and couples proton translocation to ATP synthesis (Jonckheere et al. 2012). ATP synthase exists as two identical monomers, whose dimerization determines the extent of cristae folding (Blum et al. 2019). Due to intrinsic mitochondrial homeostatic mechanisms, these complex subunits typically require either posttranslational modifications or stabilizing proteins to ensure their functional regulation. The transient regulation of ATP synthase dimerization and activity is dependent on the phosphorylation status of ATPase inhibitory factor 1 (IF_1), which is determined by the mitochondrial cAMP-PKA signaling cascade upstream. Unphosphorylated IF_1 binds between α- and β-subunits located in the F_1 region and inhibits ATP hydrolysis; however, more recent studies have demonstrated that IF_1 binding may inhibit both ATP hydrolysis and synthesis (Esparza-Moltó et al. 2017). In CTBs, levels of IF_1 were elevated compared to STB, suggesting enhanced dimerization of ATP synthase and more efficient ATP synthesis coupled with well-defined cristae and orthodox IM folding (De los Rios Castillo et al. 2011).

Isolated human CTBs and STB grown on a two-sided culture model were initially reliant anaerobic glycolysis as determined by elevated lactate and low glycogen levels; however, this metabolic profile was attributed to the recovery phase post-isolation and perhaps reminiscent of the hypoxic placental environment (Bax and Bloxam 1997). Following 24 and 48–64 h of culture, the two-sided trophoblast culture experienced a burst of glycogenolysis/glycolysis followed by an increase in oxygen consumption, with no changes to glycolytic rate, respectively. According to the literature, hCG secretion first appears around 48–64 h period of STB differentiation (Kolahi et al. 2017), which presumably corresponds with greater reliance on OXPHOS and clarifies the surge in oxygen consumption (Bax and Bloxam 1997). Overall, it can be concluded that although pre- and post-differentiated trophoblasts rely primarily on aerobic glycolysis, functional differentiation necessitates both glycolytic and oxidative energy production in CTBs, whereas STB depends on aerobic respiration. More recent investigations aimed at distinguishing CTB and STB metabolic profiles revealed that CTB are highly glycolytic and metabolically active compared to the STB lineage and that syncytialization is accompanied by progressive mitochondrial fragmentation (Kolahi et al. 2017). In fact, fragmented mitochondrial morphologies in the STB have been observed in several studies (Fisher et al. 2019a; Martínez et al. 1997; Wasilewski et al. 2012). Similar to the metabolic reprogramming observed in other stem cells, a shift from glycolysis to OXPHOS is crucial during syncytialization. Increases in OXPHOS-related genes concomitant with decreases in lactate production, a key metabolite produced from glycolysis, are observed during syncytialization (Liu et al. 2017). Although the intracellular mechanism governing this shift remains largely unknown, Liu et al. identified the downregulation of a mitochondrial matrix protein, PDK4, during syncytialization to be necessary for the suppression of glycolysis and promotion of OXPHOS. In addition, they determined that hCG-mediated cAMP stimulation suppresses PDK4 expression, highlighting an indirect role for the hCG/cAMP/PKA pathway in STB metabolism (Liu et al. 2017). Overall, it has been thought that villous cytotrophoblasts favor glycolytic respiration to better oxygenate the developing fetus whilst accommodating their hypoxic niche. However, as pregnancy progresses, the STB becomes the dominant trophoblast lineage, allowing for excess carbohydrates to be shunted to the fetus due to suppressed metabolic activity.

The placenta is responsible for forming and secreting several hormones related to pregnancy maintenance, such as hCG and progesterone (P_4). Earlier we discussed the role of hCG in promoting trophoblast fusion and facilitating the metabolic switch; however, STB mitochondria possess a well-established role in the P_4 synthesis pathway. Placental P_4 production is crucial in maintaining the uterine integrity and preventing uterine contractions may provide an explanation for why P_4 deficiencies have been implicated in miscarriage (Daya 1994) and premature births (Csapo et al. 1974). In brief, cholesterol is transported into the STB mitochondrial matrix and is subsequently converted into pregnenolone by an inner membrane-bound mitochondrial enzyme, CYP450scc (Shikita and Hall 1973). Following this conversion, pregnenolone is further converted into P_4 through the activity of another mitochondrial enzyme, 1,3-β-hydroxysteroid dehydrogenase (3βHSD) (Tuckey 2005). Moreover, mitochondrial dynamics proteins have been implicated in conferring the steroidogenic potential of the STB. The apparent increase in CYP450scc levels of differentiated STB is accompanied by mitochondrial fragmentation, as well as increased Drp1 expression (4-fold) and diminished levels of Opa1 and Mfn2. Interestingly, Wasilewski et al. found that pregnenolone biosynthesis was either increased or decreased following Opa1 knockdown or overexpression in differentiated BeWo cells, respectively (Wasilewski et al. 2012). Though the morphological characteristics of STB mitochondria reflect their steroidogenic role, other steroidogenic tissues possess similarly unique mitochondrial structures. Typically, steroidogenic mitochondria contain tubular, vesicular, or tubulovesicular mitochondria, as in the case of testosterone-producing Leydig cells of the male gonads. In fact, Mfn2 was upregulated following steroidogenic stimulation in MA-10 cells, and the knockdown of Mfn2 ultimately impaired steroidogenesis, suggesting a similar structure-function link between mitochondrial dynamics and steroidogenic capacity in male reproductive tissues (Duarte et al. 2012, 2014). Another group concluded that P_4 synthesis in Leydig cells is $\Delta\Psi m$-dependent, after observing decreases in P_4 levels following the treatment of MA-10 cells with CCCP (Allen et al. 2006). Although $\Delta\Psi m$ is capable of modifying the rate and efficiency of ATP hydrolysis in STB mitochondria, the dissipation of $\Delta\Psi m$ did not result in impaired STB steroidogenesis due to increases in NADPH content and ATP hydrolysis, which outweighed the decreased $\Delta\Psi m$ and sustained progesterone synthesis in isolated human STB mitochondria (Flores-Herrera et al. 2015). Overall, steroidogenesis is supported by mitochondrial function and may be further regulated through cristae ultrastructure in the placenta.

6.6 Clinical Implications of Mitochondrial Morphology and Bioenergetics in Syncytialization

Considering the extensive role of mitochondria in the generation of ROS and mediation of toxic oxidative stress (Fariss et al. 2005), mitochondrial bioenergetics, morphology, and dynamic regulation have been implicated in numerous diseases. Indeed, various placental mitochondrial alterations have been associated with PE, a placental disease related to impaired angiogenesis (Agarwal and Karumanchi 2011) and altered syncytialization (Li et al. 2003; Newhouse et al. 2007). Preeclamptic placentae have been demonstrated to have decreased total mitochondria densities, along with increased constituents of glycolysis, implying increased glycolytic activity (Vangrieken et al. 2021). As previously discussed, glycolytic energy production is mainly utilized by undifferentiated, proliferative stem cells as well as during periods of oxidative stress, whereas differentiated cells are mainly dependent on OXPHOS metabolism (Hu et al. 2016a). This study also reports increased gene and protein expression of mitochondrial fission marker, DRP1, with no alterations to fusion markers (Vangrieken et al. 2021). The pathogenesis of PE is also linked with increased ROS levels (Madazli et al. 2002; Matsubara et al. 2015; Myatt 2010). ROS production directly regulates mitochondrial dynamics, suggesting variations in redox homeostasis related to ROS levels may be associated with mitochondrial dynamic balance and morphological alterations (Willems et al. 2015). Gestational diabetes mellitus (GDM), another pregnancy-related disorder, has also been related to mitochondrial homeostasis. GDM is a metabolic disorder associated with decreased fetal/placental weight ratio (Daskalakis et al. 2008), in addition to increased placental oxidative stress and ER stress (Lappas et al. 2011; Yung et al. 2016; Zhu et al. 2015). In contrast to preeclamptic placentae, enhanced mitochondria fusion is seen in placentae from pregnancies complicated by GDM (Daskalakis et al. 2008). From a cellular bioenergetic standpoint, mitochondrial fusion is associated with increased oxidative phosphorylation demand and as a compensatory mechanism for mitochondrial damage attributable to placental cell stress by means of preserving mtDNA content, mitochondrial membrane potential and respiratory function (Abbade et al. 2020; Mishra and Chan 2016; Youle and van der Bliek 2012). Similar to PE, decreased placental mitochondrial levels are seen with GDM (Daskalakis et al. 2008), potentially linking both disorders to altered metabolic outputs. Ultimately, both PE and GDM pregnancy disorders comprise metabolic alterations related to mitochondrial morphology and bioenergetics directly impacting trophoblast cell fate, and thereby, further investigation into direct processes by which these effects manifest is imperative into the elucidation of potential therapeutic mechanisms.

References

Abbade J, Klemetti MM, Farrell A, Ermini L, Gillmore T, Sallais J, Caniggia I (2020) Increased placental mitochondrial fusion in gestational diabetes mellitus: an adaptive mechanism to optimize feto-placental metabolic homeostasis? BMJ Open Diabetes Res Care 8(1). https://doi.org/10.1136/bmjdrc-2019-000923

Afzal N, Lederer WJ, Jafri MS, Mannella CA (2021) Effect of crista morphology on mitochondrial ATP output: a computational study. Curr Res Physiol 4:163–176. https://doi.org/10.1016/j.crphys.2021.03.005

Agarwal I, Karumanchi SA (2011) Preeclampsia and the Anti-Angiogenic State. Pregnancy Hypertens 1(1):17–21. https://doi.org/10.1016/j.preghy.2010.10.007

Agostini M, Romeo F, Inoue S, Niklison-Chirou MV, Elia AJ, Dinsdale D et al (2016) Metabolic reprogramming during neuronal differentiation. Cell Death Differ 23(9):1502–1514. https://doi.org/10.1038/cdd.2016.36

Alison MR, Poulsom R, Forbes S, Wright NA (2002) An introduction to stem cells. J Pathol 197(4):419–423. https://doi.org/10.1002/path.1187

Allen JA, Shankara T, Janus P, Buck S, Diemer T, Hales KH, Hales DB (2006) Energized, polarized, and actively respiring mitochondria are required for acute Leydig cell steroidogenesis. Endocrinology 147(8):3924–3935. https://doi.org/10.1210/en.2005-1204

Alsat E, Wyplosz P, Malassiné A, Guibourdenche J, Porquet D, Nessmann C, Evain-Brion D (1996) Hypoxia impairs cell fusion and differentiation process in human cytotrophoblast, in vitro. J Cell Physiol 168(2):346–353. https://doi.org/10.1002/(SICI)1097-4652(199608)168:2<346::AID-JCP13>3.0.CO;2-1

Anderson S, Bankier AT, Barrell BG, de Bruijn MH, Coulson AR, Drouin J et al (1981) Sequence and organization of the human mitochondrial genome. Nature 290(5806):457–465. https://doi.org/10.1038/290457a0

Azuma K, Ikeda K, Inoue S (2020) Functional mechanisms of mitochondrial respiratory chain supercomplex assembly factors and their involvement in muscle quality. Int J Mol Sci 21(9). https://doi.org/10.3390/ijms21093182

Bainbridge SA, Minhas A, Whiteley KJ, Qu D, Sled JG, Kingdom JC, Adamson SL (2012) Effects of reduced Gcm1 expression on trophoblast morphology, fetoplacental vascularity, and pregnancy outcomes in mice. Hypertension 59(3):732–739. https://doi.org/10.1161/HYPERTENSIONAHA.111.183939

Baker N, Wade S, Triolo M, Girgis J, Chwastek D, Larrigan S et al (2022) The mitochondrial protein OPA1 regulates the quiescent state of adult muscle stem cells. Cell Stem Cell 29(9):1315–1332.e1319. https://doi.org/10.1016/j.stem.2022.07.010

Bartho LA, Fisher JJ, Cuffe JSM, Perkins AV (2020) Mitochondrial transformations in the aging human placenta. Am J Physiol Endocrinol Metab 319(6):E981–E994. https://doi.org/10.1152/ajpendo.00354.2020

Bax BE, Bloxam DL (1997) Energy metabolism and glycolysis in human placental trophoblast cells during differentiation. Biochim Biophys Acta 1319(2–3):283–292. https://doi.org/10.1016/s0005-2728(96)00169-7

Bigarella CL, Liang R, Ghaffari S (2014) Stem cells and the impact of ROS signaling. Development 141(22):4206–4218. https://doi.org/10.1242/dev.107086

Black S, Kadyrov M, Kaufmann P, Ugele B, Emans N, Huppertz B (2004) Syncytial fusion of human trophoblast depends on caspase 8. Cell Death Differ 11(1):90–98. https://doi.org/10.1038/sj.cdd.4401307

Blum TB, Hahn A, Meier T, Davies KM, Kühlbrandt W (2019) Dimers of mitochondrial ATP synthase induce membrane curvature and self-assemble into rows. Proc Natl Acad Sci U S A 116(10):4250–4255. https://doi.org/10.1073/pnas.1816556116

Breckenridge DG, Stojanovic M, Marcellus RC, Shore GC (2003) Caspase cleavage product of BAP31 induces mitochondrial fission through endoplasmic reticulum calcium signals, enhancing cytochrome c release to the cytosol. J Cell Biol 160(7):1115–1127. https://doi.org/10.1083/jcb.200212059

Burman JL, Pickles S, Wang C, Sekine S, Vargas JNS, Zhang Z et al (2017) Mitochondrial fission facilitates the selective mitophagy of protein aggregates. J Cell Biol 216(10):3231–3247. https://doi.org/10.1083/jcb.201612106

Burton GJ, Jones CJ (2009) Syncytial knots, sprouts, apoptosis, and trophoblast deportation from the human placenta. Taiwan J Obstet Gynecol 48(1):28–37. https://doi.org/10.1016/S1028-4559(09)60032-2

Bustamante J, Ramírez-Vélez R, Czerniczyniec A, Cicerchia D, Aguilar de Plata AC, Lores-Arnaiz S (2014) Oxygen metabolism in human placenta mitochondria. J Bioenerg Biomembr 46(6):459–469. https://doi.org/10.1007/s10863-014-9572-x

Calvo E, Cogliati S, Hernansanz-Agustín P, Loureiro-López M, Guarás A, Casuso RA et al (2020) Functional role of respiratory supercomplexes in mice: SCAF1 relevance and segmentation of the Qpool. Sci Adv 6(26):eaba7509. https://doi.org/10.1126/sciadv.aba7509

Chang CW, Wakeland AK, Parast MM (2018) Trophoblast lineage specification, differentiation and their regulation by oxygen tension. J Endocrinol 236(1):R43–R56. https://doi.org/10.1530/JOE-17-0402

Chen B, Longtine MS, Sadovsky Y, Nelson DM (2010) Hypoxia downregulates p53 but induces apoptosis and enhances expression of BAD in cultures of human syncytiotrophoblasts. Am J Physiol Cell Physiol 299(5):C968–C976. https://doi.org/10.1152/ajpcell.00154.2010
Chernomordik LV, Kozlov MM (2005) Membrane hemifusion: crossing a chasm in two leaps. Cell 123(3):375–382. https://doi.org/10.1016/j.cell.2005.10.015
Choi HJ, Sanders TA, Tormos KV, Ameri K, Tsai JD, Park AM et al (2013) ECM-dependent HIF induction directs trophoblast stem cell fate via LIMK1-mediated cytoskeletal rearrangement. PLoS One 8(2):e56949. https://doi.org/10.1371/journal.pone.0056949
Cipolat S, Martins de Brito O, Dal Zilio B, Scorrano L (2004) OPA1 requires mitofusin 1 to promote mitochondrial fusion. Proc Natl Acad Sci U S A 101(45):15927–15932. https://doi.org/10.1073/pnas.0407043101
Cogliati S, Calvo E, Loureiro M, Guaras AM, Nieto-Arellano R, Garcia-Poyatos C et al (2016) Mechanism of super-assembly of respiratory complexes III and IV. Nature 539(7630):579–582. https://doi.org/10.1038/nature20157
Coutifaris C, Kao LC, Sehdev HM, Chin U, Babalola GO, Blaschuk OW, Strauss JF (1991) E-cadherin expression during the differentiation of human trophoblasts. Development 113(3):767–777. https://doi.org/10.1242/dev.113.3.767
Cribbs JT, Strack S (2007) Reversible phosphorylation of Drp1 by cyclic AMP-dependent protein kinase and calcineurin regulates mitochondrial fission and cell death. EMBO Rep 8(10):939–944. https://doi.org/10.1038/sj.embor.7401062
Cronier L, Defamie N, Dupays L, Théveniau-Ruissy M, Goffin F, Pointis G, Malassiné A (2002) Connexin expression and gap junctional intercellular communication in human first trimester trophoblast. Mol Hum Reprod 8(11):1005–1013. https://doi.org/10.1093/molehr/8.11.1005
Csapo AI, Pohanka O, Kaihola HL (1974) Progesterone deficiency and premature labour. Br Med J 1(5899):137–140. https://doi.org/10.1136/bmj.1.5899.137
Da Cruz S, Parone PA, Gonzalo P, Bienvenut WV, Tondera D, Jourdain A et al (2008) SLP-2 interacts with prohibitins in the mitochondrial inner membrane and contributes to their stability. Biochim Biophys Acta 1783(5):904–911. https://doi.org/10.1016/j.bbamcr.2008.02.006
Daskalakis G, Marinopoulos S, Krielesi V, Papapanagiotou A, Papantoniou N, Mesogitis S, Antsaklis A (2008) Placental pathology in women with gestational diabetes. Acta Obstet Gynecol Scand 87(4):403–407. https://doi.org/10.1080/00016340801908783
Daya S (1994) Issues in the etiology of recurrent spontaneous abortion. Curr Opin Obstet Gynecol 6(2):153–159
De los Rios Castillo D, Zarco-Zavala M, Olvera-Sanchez S, Pardo JP, Juarez O, Martinez F et al (2011) Atypical cristae morphology of human syncytiotrophoblast mitochondria: role for complex V. J Biol Chem 286(27):23911–23919. https://doi.org/10.1074/jbc.M111.252056
den Brave F, Becker T (2020) Supercomplex formation boosts respiration. EMBO Rep 21(12):e51830. https://doi.org/10.15252/embr.202051830
Duarte A, Poderoso C, Cooke M, Soria G, Cornejo Maciel F, Gottifredi V, Podestá EJ (2012) Mitochondrial fusion is essential for steroid biosynthesis. PLoS One 7(9):e45829. https://doi.org/10.1371/journal.pone.0045829
Duarte A, Castillo AF, Podestá EJ, Poderoso C (2014) Mitochondrial fusion and ERK activity regulate steroidogenic acute regulatory protein localization in mitochondria. PLoS One 9(6):e100387. https://doi.org/10.1371/journal.pone.0100387
Dupressoir A, Vernochet C, Bawa O, Harper F, Pierron G, Opolon P, Heidmann T (2009) Syncytin-A knockout mice demonstrate the critical role in placentation of a fusogenic, endogenous retrovirus-derived, envelope gene. Proc Natl Acad Sci U S A 106(29):12127–12132. https://doi.org/10.1073/pnas.0902925106
Elmore SP, Qian T, Grissom SF, Lemasters JJ (2001) The mitochondrial permeability transition initiates autophagy in rat hepatocytes. FASEB J 15(12):2286–2287. https://doi.org/10.1096/fj.01-0206fje

Esnault C, Priet S, Ribet D, Vernochet C, Bruls T, Lavialle C et al (2008) A placenta-specific receptor for the fusogenic, endogenous retrovirus-derived, human syncytin-2. Proc Natl Acad Sci U S A 105(45):17532–17537. https://doi.org/10.1073/pnas.0807413105

Esparza-Moltó PB, Nuevo-Tapioles C, Cuezva JM (2017) Regulation of the H + -ATP synthase by IF1: a role in mitohormesis. Cell Mol Life Sci 74(12):2151–2166. https://doi.org/10.1007/s00018-017-2462-8

Fariss MW, Chan CB, Patel M, Van Houten B, Orrenius S (2005) Role of mitochondria in toxic oxidative stress. Mol Interv 5(2):94–111. https://doi.org/10.1124/mi.5.2.7

Fisher J, McKeating D, Pennell E, Cuffe J, Holland O, Perkins A (2019a) Mitochondrial isolation, cryopreservation and preliminary biochemical characterisation from placental cytotrophoblast and syncytiotrophoblast. Placenta 82:1–4. https://doi.org/10.1016/j.placenta.2019.05.004

Fisher JJ, McKeating DR, Cuffe JS, Bianco-Miotto T, Holland OJ, Perkins AV (2019b) Proteomic analysis of placental mitochondria following trophoblast differentiation. Front Physiol 10:1536. https://doi.org/10.3389/fphys.2019.01536

Flores-Herrera O, Olvera-Sánchez S, Esparza-Perusquía M, Pardo JP, Rendón JL, Mendoza-Hernández G, Martínez F (2015) Membrane potential regulates mitochondrial ATP-diphosphohydrolase activity but is not involved in progesterone biosynthesis in human syncytiotrophoblast cells. Biochim Biophys Acta 1847(2):143–152. https://doi.org/10.1016/j.bbabio.2014.10.002

Folmes CD, Nelson TJ, Martinez-Fernandez A, Arrell DK, Lindor JZ, Dzeja PP et al (2011) Somatic oxidative bioenergetics transitions into pluripotency-dependent glycolysis to facilitate nuclear reprogramming. Cell Metab 14(2):264–271. https://doi.org/10.1016/j.cmet.2011.06.011

Forni MF, Peloggia J, Trudeau K, Shirihai O, Kowaltowski AJ (2016) Murine mesenchymal stem cell commitment to differentiation is regulated by mitochondrial dynamics. Stem Cells 34(3):743–755. https://doi.org/10.1002/stem.2248

Frank S, Gaume B, Bergmann-Leitner ES, Leitner WW, Robert EG, Catez F et al (2001) The role of dynamin-related protein 1, a mediator of mitochondrial fission, in apoptosis. Dev Cell 1(4):515–525. https://doi.org/10.1016/s1534-5807(01)00055-7

Frendo JL, Olivier D, Cheynet V, Blond JL, Bouton O, Vidaud M et al (2003) Direct involvement of HERV-W Env glycoprotein in human trophoblast cell fusion and differentiation. Mol Cell Biol 23(10):3566–3574. https://doi.org/10.1128/MCB.23.10.3566-3574.2003

Frezza C, Cipolat S, Martins de Brito O, Micaroni M, Beznoussenko GV, Rudka T et al (2006) OPA1 controls apoptotic cristae remodeling independently from mitochondrial fusion. Cell 126(1):177–189. https://doi.org/10.1016/j.cell.2006.06.025

García-Poyatos C, Cogliati S, Calvo E, Hernansanz-Agustín P, Lagarrigue S, Magni R et al (2020) Scaf1 promotes respiratory supercomplexes and metabolic efficiency in zebrafish. EMBO Rep 21(7):e50287. https://doi.org/10.15252/embr.202050287

Gerbaud P, Pidoux G (2015) Review: an overview of molecular events occurring in human trophoblast fusion. Placenta 36(Suppl 1):S35–S42. https://doi.org/10.1016/j.placenta.2014.12.015

Germain M, Mathai JP, McBride HM, Shore GC (2005) Endoplasmic reticulum BIK initiates DRP1-regulated remodelling of mitochondrial cristae during apoptosis. EMBO J 24(8):1546–1556. https://doi.org/10.1038/sj.emboj.7600592

Getsios S, MacCalman CD (2003) Cadherin-11 modulates the terminal differentiation and fusion of human trophoblastic cells in vitro. Dev Biol 257(1):41–54. https://doi.org/10.1016/s0012-1606(03)00041-1

Gomes LC, Di Benedetto G, Scorrano L (2011) During autophagy mitochondria elongate, are spared from degradation and sustain cell viability. Nat Cell Biol 13(5):589–598. https://doi.org/10.1038/ncb2220

Gu W, Gaeta X, Sahakyan A, Chan AB, Hong CS, Kim R et al (2016) Glycolytic metabolism plays a functional role in regulating human pluripotent stem cell state. Cell Stem Cell 19(4):476–490. https://doi.org/10.1016/j.stem.2016.08.008

Hong HS, Mbah NE, Shan M, Loesel K, Lin L, Sajjakulnukit P et al (2022a) OXPHOS promotes apoptotic resistance and cellular persistence in T. Sci Immunol 7(77):eabm8182. https://doi.org/10.1126/sciimmunol.abm8182

Hong X, Isern J, Campanario S, Perdiguero E, Ramírez-Pardo I, Segalés J et al (2022b) Mitochondrial dynamics maintain muscle stem cell regenerative competence throughout adult life by regulating metabolism and mitophagy. Cell Stem Cell 29(9):1298–1314.e1210. https://doi.org/10.1016/j.stem.2022.07.009

Hu C, Fan L, Cen P, Chen E, Jiang Z, Li L (2016a) Energy metabolism plays a critical role in stem cell maintenance and differentiation. Int J Mol Sci 17(2):253. https://doi.org/10.3390/ijms17020253

Hu Y, Li XC, Wang ZH, Luo Y, Zhang X, Liu XP et al (2016b) Tau accumulation impairs mitophagy via increasing mitochondrial membrane potential and reducing mitochondrial Parkin. Oncotarget 7(14):17356–17368. https://doi.org/10.18632/oncotarget.7861

Humphrey RG, Sonnenberg-Hirche C, Smith SD, Hu C, Barton A, Sadovsky Y, Nelson DM (2008) Epidermal growth factor abrogates hypoxia-induced apoptosis in cultured human trophoblasts through phosphorylation of BAD Serine 112. Endocrinology 149(5):2131–2137. https://doi.org/10.1210/en.2007-1253

Huppertz B, Gauster M (2011) Trophoblast fusion. Adv Exp Med Biol 713:81–95. https://doi.org/10.1007/978-94-007-0763-4_6

Huppertz B, Frank HG, Kingdom JC, Reister F, Kaufmann P (1998) Villous cytotrophoblast regulation of the syncytial apoptotic cascade in the human placenta. Histochem Cell Biol 110(5):495–508. https://doi.org/10.1007/s004180050311

Huppertz B, Tews DS, Kaufmann P (2001) Apoptosis and syncytial fusion in human placental trophoblast and skeletal muscle. Int Rev Cytol 205:215–253. https://doi.org/10.1016/s0074-7696(01)05005-7

Hussain MU, Kremer M, Zoidl G, Dermietzel R (2003) Transcriptional and translational regulation of zebrafish connexin 55.5 (zf.Cx.55.5) and connexin 52.6 (zf.Cx52.6). Cell Commun Adhes 10(4–6):227–231

Ikeda K, Shiba S, Horie-Inoue K, Shimokata K, Inoue S (2013) A stabilizing factor for mitochondrial respiratory supercomplex assembly regulates energy metabolism in muscle. Nat Commun 4:2147. https://doi.org/10.1038/ncomms3147

Ji L, Brkić J, Liu M, Fu G, Peng C, Wang YL (2013) Placental trophoblast cell differentiation: physiological regulation and pathological relevance to preeclampsia. Mol Aspects Med 34(5):981–1023. https://doi.org/10.1016/j.mam.2012.12.008

Jonckheere AI, Smeitink JA, Rodenburg RJ (2012) Mitochondrial ATP synthase: architecture, function and pathology. J Inherit Metab Dis 35(2):211–225. https://doi.org/10.1007/s10545-011-9382-9

Karbowski M, Lee YJ, Gaume B, Jeong SY, Frank S, Nechushtan A et al (2002) Spatial and temporal association of Bax with mitochondrial fission sites, Drp1, and Mfn2 during apoptosis. J Cell Biol 159(6):931–938. https://doi.org/10.1083/jcb.200209124

Kashatus JA, Nascimento A, Myers LJ, Sher A, Byrne FL, Hoehn KL et al (2015) Erk2 phosphorylation of Drp1 promotes mitochondrial fission and MAPK-driven tumor growth. Mol Cell 57(3):537–551. https://doi.org/10.1016/j.molcel.2015.01.002

Kelly SJ (1977) Studies of the developmental potential of 4- and 8-cell stage mouse blastomeres. J Exp Zool 200(3):365–376. https://doi.org/10.1002/jez.1402000307

Khacho M, Slack RS (2017) Mitochondrial and reactive oxygen species signaling coordinate stem cell fate decisions and life long maintenance. Antioxid Redox Signal. https://doi.org/10.1089/ars.2017.7228

Khacho M, Clark A, Svoboda DS, Azzi J, MacLaurin JG, Meghaizel C et al (2016) Mitochondrial dynamics impacts stem cell identity and fate decisions by regulating a nuclear transcriptional program. Cell Stem Cell 19(2):232–247. https://doi.org/10.1016/j.stem.2016.04.015

Knerr I, Schubert SW, Wich C, Amann K, Aigner T, Vogler T et al (2005) Stimulation of GCMa and syncytin via cAMP mediated PKA signaling in human trophoblastic cells under normoxic and hypoxic conditions. FEBS Lett 579(18):3991–3998. https://doi.org/10.1016/j.febslet.2005.06.029

Kolahi KS, Valent AM, Thornburg KL (2017) Cytotrophoblast, not syncytiotrophoblast, dominates glycolysis and oxidative phosphorylation in human term placenta. Sci Rep 7:42941. https://doi.org/10.1038/srep42941

Kornfeld OS, Qvit N, Haileselassie B, Shamloo M, Bernardi P, Mochly-Rosen D (2018) Interaction of mitochondrial fission factor with dynamin related protein 1 governs physiological mitochondrial function in vivo. Sci Rep 8(1):14034. https://doi.org/10.1038/s41598-018-32228-1

Kumar P, Mendelson CR (2011) Estrogen-related receptor gamma (ERRgamma) mediates oxygen-dependent induction of aromatase (CYP19) gene expression during human trophoblast differentiation. Mol Endocrinol 25(9):1513–1526. https://doi.org/10.1210/me.2011-1012

Lacher SE, Levings DC, Freeman S, Slattery M (2018) Identification of a functional antioxidant response element at the HIF1A locus. Redox Biol 19:401–411. https://doi.org/10.1016/j.redox.2018.08.014

Lager S, Powell TL (2012) Regulation of nutrient transport across the placenta. J Pregnancy 2012:179827. https://doi.org/10.1155/2012/179827

Lappas M, Hiden U, Desoye G, Froehlich J, Hauguel-de Mouzon S, Jawerbaum A (2011) The role of oxidative stress in the pathophysiology of gestational diabetes mellitus. Antioxid Redox Signal 15(12):3061–3100. https://doi.org/10.1089/ars.2010.3765

Lavillette D, Marin M, Ruggieri A, Mallet F, Cosset FL, Kabat D (2002) The envelope glycoprotein of human endogenous retrovirus type W uses a divergent family of amino acid transporters/cell surface receptors. J Virol 76(13):6442–6452. https://doi.org/10.1128/jvi.76.13.6442-6452.2002

Lee YJ, Jeong SY, Karbowski M, Smith CL, Youle RJ (2004) Roles of the mammalian mitochondrial fission and fusion mediators Fis1, Drp1, and Opa1 in apoptosis. Mol Biol Cell 15(11):5001–5011. https://doi.org/10.1091/mbc.e04-04-0294

Letts JA, Sazanov LA (2017) Clarifying the supercomplex: the higher-order organization of the mitochondrial electron transport chain. Nat Struct Mol Biol 24(10):800–808. https://doi.org/10.1038/nsmb.3460

Li L, Xie T (2005) Stem cell niche: structure and function. Annu Rev Cell Dev Biol 21:605–631. https://doi.org/10.1146/annurev.cellbio.21.012704.131525

Li H, Dakour J, Kaufman S, Guilbert LJ, Winkler-Lowen B, Morrish DW (2003) Adrenomedullin is decreased in preeclampsia because of failed response to epidermal growth factor and impaired syncytialization. Hypertension 42(5):895–900. https://doi.org/10.1161/01.HYP.0000095613.41961.6E

Liu X, Zuo R, Bao Y, Qu X, Sun K, Ying H (2017) Down-regulation of PDK4 is critical for the switch of carbohydrate catabolism during syncytialization of human placental trophoblasts. Sci Rep 7(1):8474. https://doi.org/10.1038/s41598-017-09163-8

Losón OC, Song Z, Chen H, Chan DC (2013) Fis1, Mff, MiD49, and MiD51 mediate Drp1 recruitment in mitochondrial fission. Mol Biol Cell 24(5):659–667. https://doi.org/10.1091/mbc.E12-10-0721

Lunt SY, Vander Heiden MG (2011) Aerobic glycolysis: meeting the metabolic requirements of cell proliferation. Annu Rev Cell Dev Biol 27:441–464. https://doi.org/10.1146/annurev-cellbio-092910-154237

Madazli R, Benian A, Aydin S, Uzun H, Tolun N (2002) The plasma and placental levels of malondialdehyde, glutathione and superoxide dismutase in pre-eclampsia. J Obstet Gynaecol 22(5):477–480. https://doi.org/10.1080/0144361021000003573

Malassiné A, Frendo JL, Blaise S, Handschuh K, Gerbaud P, Tsatsaris V et al (2008) Human endogenous retrovirus-FRD envelope protein (syncytin 2) expression in normal and trisomy 21-affected placenta. Retrovirology 5:6. https://doi.org/10.1186/1742-4690-5-6

Maltepe E, Krampitz GW, Okazaki KM, Red-Horse K, Mak W, Simon MC, Fisher SJ (2005) Hypoxia-inducible factor-dependent histone deacetylase activity determines stem cell fate in the placenta. Development 132(15):3393–3403. https://doi.org/10.1242/dev.01923

Martínez F, Kiriakidou M, Strauss JF (1997) Structural and functional changes in mitochondria associated with trophoblast differentiation: methods to isolate enriched preparations of syncytiotrophoblast mitochondria. Endocrinology 138(5):2172–2183. https://doi.org/10.1210/endo.138.5.5133

Martinez F, Olvera-Sanchez S, Esparza-Perusquia M, Gomez-Chang E, Flores-Herrera O (2015) Multiple functions of syncytiotrophoblast mitochondria. Steroids 103:11–22. https://doi.org/10.1016/j.steroids.2015.09.006

Matsubara K, Higaki T, Matsubara Y, Nawa A (2015) Nitric oxide and reactive oxygen species in the pathogenesis of preeclampsia. Int J Mol Sci 16(3):4600–4614. https://doi.org/10.3390/ijms16034600

Mayhew TM, Leach L, McGee R, Ismail WW, Myklebust R, Lammiman MJ (1999) Proliferation, differentiation and apoptosis in villous trophoblast at 13-41 weeks of gestation (including observations on annulate lamellae and nuclear pore complexes). Placenta 20(5–6):407–422. https://doi.org/10.1053/plac.1999.0399

Merkwirth C, Dargazanli S, Tatsuta T, Geimer S, Löwer B, Wunderlich FT et al (2008) Prohibitins control cell proliferation and apoptosis by regulating OPA1-dependent cristae morphogenesis in mitochondria. Genes Dev 22(4):476–488. https://doi.org/10.1101/gad.460708

Mi S, Lee X, Li X, Veldman GM, Finnerty H, Racie L et al (2000) Syncytin is a captive retroviral envelope protein involved in human placental morphogenesis. Nature 403(6771):785–789. https://doi.org/10.1038/35001608

Mishra P, Chan DC (2016) Metabolic regulation of mitochondrial dynamics. J Cell Biol 212(4):379–387. https://doi.org/10.1083/jcb.201511036

Mitsopoulos P, Chang YH, Wai T, König T, Dunn SD, Langer T, Madrenas J (2015) Stomatin-like protein 2 is required for in vivo mitochondrial respiratory chain supercomplex formation and optimal cell function. Mol Cell Biol 35(10):1838–1847. https://doi.org/10.1128/MCB.00047-15

Myatt L (2010) Review: reactive oxygen and nitrogen species and functional adaptation of the placenta. Placenta 31 Suppl(Suppl):S66–S69. https://doi.org/10.1016/j.placenta.2009.12.021

Narendra D, Tanaka A, Suen DF, Youle RJ (2008) Parkin is recruited selectively to impaired mitochondria and promotes their autophagy. J Cell Biol 183(5):795–803. https://doi.org/10.1083/jcb.200809125

Newhouse SM, Davidge ST, Winkler-Lowen B, Demianczuk N, Guilbert LJ (2007) In vitro differentiation of villous trophoblasts from pregnancies complicated by intrauterine growth restriction with and without pre-eclampsia. Placenta 28(10):999–1003. https://doi.org/10.1016/j.placenta.2007.04.008

Ochoa-Bernal MA, Fazleabas AT (2020) Physiologic events of embryo implantation and decidualization in human and non-human primates. Int J Mol Sci 21(6). https://doi.org/10.3390/ijms21061973

Oda H, Takeichi M (2011) Evolution: structural and functional diversity of cadherin at the adherens junction. J Cell Biol 193(7):1137–1146. https://doi.org/10.1083/jcb.201008173

Okahara G, Matsubara S, Oda T, Sugimoto J, Jinno Y, Kanaya F (2004) Expression analyses of human endogenous retroviruses (HERVs): tissue-specific and developmental stage-dependent expression of HERVs. Genomics 84(6):982–990. https://doi.org/10.1016/j.ygeno.2004.09.004

Olichon A, Baricault L, Gas N, Guillou E, Valette A, Belenguer P, Lenaers G (2003) Loss of OPA1 perturbates the mitochondrial inner membrane structure and integrity, leading to cytochrome c release and apoptosis. J Biol Chem 278(10):7743–7746. https://doi.org/10.1074/jbc.C200677200

Ono T, Isobe K, Nakada K, Hayashi JI (2001) Human cells are protected from mitochondrial dysfunction by complementation of DNA products in fused mitochondria. Nat Genet 28(3):272–275. https://doi.org/10.1038/90116

Parameshwar PK, Sagrillo-Fagundes L, Fournier C, Girard S, Vaillancourt C, Moraes C (2021) Disease-specific extracellular matrix composition regulates placental trophoblast fusion efficiency. Biomater Sci 9(21):7247–7256. https://doi.org/10.1039/d1bm00799h

Parra V, Verdejo HE, Iglewski M, Del Campo A, Troncoso R, Jones D et al (2014) Insulin stimulates mitochondrial fusion and function in cardiomyocytes via the Akt-mTOR-NFκB-Opa-1 signaling pathway. Diabetes 63(1):75–88. https://doi.org/10.2337/db13-0340

Patten DA, Wong J, Khacho M, Soubannier V, Mailloux RJ, Pilon-Larose K et al (2014) OPA1-dependent cristae modulation is essential for cellular adaptation to metabolic demand. EMBO J 33(22):2676–2691. https://doi.org/10.15252/embj.201488349

Pidoux G, Gerbaud P, Gnidehou S, Grynberg M, Geneau G, Guibourdenche J et al (2010) ZO-1 is involved in trophoblastic cell differentiation in human placenta. Am J Physiol Cell Physiol 298(6):C1517–C1526. https://doi.org/10.1152/ajpcell.00484.2008

Pidoux G, Gerbaud P, Dompierre J, Lygren B, Solstad T, Evain-Brion D, Taskén K (2014) A PKA-ezrin-Cx43 signaling complex controls gap junction communication and thereby trophoblast cell fusion. J Cell Sci 127(Pt 19):4172–4185. https://doi.org/10.1242/jcs.149609

Poidatz D, Dos Santos E, Brulé A, De Mazancourt P, Dieudonné MN (2012) Estrogen-related receptor gamma modulates energy metabolism target genes in human trophoblast. Placenta 33(9):688–695. https://doi.org/10.1016/j.placenta.2012.06.002

Poidatz D, Dos Santos E, Gronier H, Vialard F, Maury B, De Mazancourt P, Dieudonné MN (2015) Trophoblast syncytialisation necessitates mitochondrial function through estrogen-related receptor-γ activation. Mol Hum Reprod 21(2):206–216. https://doi.org/10.1093/molehr/gau102

Pötgens AJ, Drewlo S, Kokozidou M, Kaufmann P (2004) Syncytin: the major regulator of trophoblast fusion? Recent developments and hypotheses on its action. Hum Reprod Update 10(6):487–496. https://doi.org/10.1093/humupd/dmh039

Rajakumar A, Cerdeira AS, Rana S, Zsengeller Z, Edmunds L, Jeyabalan A et al (2012) Transcriptionally active syncytial aggregates in the maternal circulation may contribute to circulating soluble fms-like tyrosine kinase 1 in preeclampsia. Hypertension 59(2):256–264. https://doi.org/10.1161/HYPERTENSIONAHA.111.182170

Rambold AS, Kostelecky B, Elia N, Lippincott-Schwartz J (2011) Tubular network formation protects mitochondria from autophagosomal degradation during nutrient starvation. Proc Natl Acad Sci U S A 108(25):10190–10195. https://doi.org/10.1073/pnas.1107402108

Renaud SJ, Jeyarajah MJ (2022) How trophoblasts fuse: an in-depth look into placental syncytiotrophoblast formation. Cell Mol Life Sci 79(8):433. https://doi.org/10.1007/s00018-022-04475-z

Scorrano L, Ashiya M, Buttle K, Weiler S, Oakes SA, Mannella CA, Korsmeyer SJ (2002) A distinct pathway remodels mitochondrial cristae and mobilizes cytochrome c during apoptosis. Dev Cell 2(1):55–67. https://doi.org/10.1016/s1534-5807(01)00116-2

Shi QJ, Lei ZM, Rao CV, Lin J (1993) Novel role of human chorionic gonadotropin in differentiation of human cytotrophoblasts. Endocrinology 132(3):1387–1395. https://doi.org/10.1210/endo.132.3.7679981

Shikita M, Hall PF (1973) Cytochrome P-450 from bovine adrenocortical mitochondria: an enzyme for the side chain cleavage of cholesterol. I. Purification and properties. J Biol Chem 248(16):5598–5604

Shyh-Chang N, Ng HH (2017) The metabolic programming of stem cells. Genes Dev 31(4):336–346. https://doi.org/10.1101/gad.293167.116

Simsek T, Kocabas F, Zheng J, Deberardinis RJ, Mahmoud AI, Olson EN et al (2010) The distinct metabolic profile of hematopoietic stem cells reflects their location in a hypoxic niche. Cell Stem Cell 7(3):380–390. https://doi.org/10.1016/j.stem.2010.07.011

Song Z, Ghochani M, McCaffery JM, Frey TG, Chan DC (2009) Mitofusins and OPA1 mediate sequential steps in mitochondrial membrane fusion. Mol Biol Cell 20(15):3525–3532. https://doi.org/10.1091/mbc.e09-03-0252

Takeda Y, Liu X, Sumiyoshi M, Matsushima A, Shimohigashi M, Shimohigashi Y (2009) Placenta expressing the greatest quantity of bisphenol A receptor ERR{gamma} among the human reproductive tissues: predominant expression of type-1 ERRgamma isoform. J Biochem 146(1):113–122. https://doi.org/10.1093/jb/mvp049

Takeichi M (1995) Morphogenetic roles of classic cadherins. Curr Opin Cell Biol 7(5):619–627. https://doi.org/10.1016/0955-0674(95)80102-2

Tang Y, Luo B, Deng Z, Wang B, Liu F, Li J et al (2016) Mitochondrial aerobic respiration is activated during hair follicle stem cell differentiation, and its dysfunction retards hair regeneration. PeerJ 4:e1821. https://doi.org/10.7717/peerj.1821

Tatapudy S, Aloisio F, Barber D, Nystul T (2017) Cell fate decisions: emerging roles for metabolic signals and cell morphology. EMBO Rep 18(12):2105–2118. https://doi.org/10.15252/embr.201744816

Tuckey RC (2005) Progesterone synthesis by the human placenta. Placenta 26(4):273–281. https://doi.org/10.1016/j.placenta.2004.06.012

Vangrieken P, Al-Nasiry S, Bast A, Leermakers PA, Tulen CBM, Schiffers PMH et al (2021) Placental mitochondrial abnormalities in preeclampsia. Reprod Sci 28(8):2186–2199. https://doi.org/10.1007/s43032-021-00464-y

Vargas A, Moreau J, Landry S, LeBellego F, Toufaily C, Rassart E et al (2009) Syncytin-2 plays an important role in the fusion of human trophoblast cells. J Mol Biol 392(2):301–318. https://doi.org/10.1016/j.jmb.2009.07.025

Wakeland AK, Soncin F, Moretto-Zita M, Chang CW, Horii M, Pizzo D et al (2017) Hypoxia directs human extravillous trophoblast differentiation in a hypoxia-inducible factor-dependent manner. Am J Pathol 187(4):767–780. https://doi.org/10.1016/j.ajpath.2016.11.018

Wasilewski M, Semenzato M, Rafelski SM, Robbins J, Bakardjiev AI, Scorrano L (2012) Optic atrophy 1-dependent mitochondrial remodeling controls steroidogenesis in trophoblasts. Curr Biol 22(13):1228–1234. https://doi.org/10.1016/j.cub.2012.04.054

Wich C, Kausler S, Dotsch J, Rascher W, Knerr I (2009) Syncytin-1 and glial cells missing a: hypoxia-induced deregulated gene expression along with disordered cell fusion in primary term human trophoblasts. Gynecol Obstet Invest 68(1):9–18. https://doi.org/10.1159/000209396

Willems PH, Rossignol R, Dieteren CE, Murphy MP, Koopman WJ (2015) Redox homeostasis and mitochondrial dynamics. Cell Metab 22(2):207–218. https://doi.org/10.1016/j.cmet.2015.06.006

Wong MK, Shawky SA, Aryasomayajula A, Green MA, Ewart T, Selvaganapathy PR, Raha S (2018) Extracellular matrix surface regulates self-assembly of three-dimensional placental trophoblast spheroids. PLoS One 13(6):e0199632. https://doi.org/10.1371/journal.pone.0199632

Wu W, Zhao D, Shah SZA, Zhang X, Lai M, Yang D et al (2019) OPA1 overexpression ameliorates mitochondrial cristae remodeling, mitochondrial dysfunction, and neuronal apoptosis in prion diseases. Cell Death Dis 10(10):710. https://doi.org/10.1038/s41419-019-1953-y

Wu C, Yao W, Kai W, Liu W, Wang W, Li S et al (2020) Mitochondrial fusion machinery specifically involved in energy deprivation-induced autophagy. Front Cell Dev Biol 8:221. https://doi.org/10.3389/fcell.2020.00221

Yang M, Lei ZM, Rao CV (2003) The central role of human chorionic gonadotropin in the formation of human placental syncytium. Endocrinology 144(3):1108–1120. https://doi.org/10.1210/en.2002-220922

Yang F, Zheng Q, Jin L (2019) Dynamic function and composition changes of immune cells during normal and pathological pregnancy at the maternal-fetal interface. Front Immunol 10:2317. https://doi.org/10.3389/fimmu.2019.02317

Yao CH, Wang R, Wang Y, Kung CP, Weber JD, Patti GJ (2019) Mitochondrial fusion supports increased oxidative phosphorylation during cell proliferation. Elife 8. https://doi.org/10.7554/eLife.41351

Yasuda M, Umemura S, Osamura RY, Kenjo T, Tsutsumi Y (1995) Apoptotic cells in the human endometrium and placental villi: pitfalls in applying the TUNEL method. Arch Histol Cytol 58(2):185–190. https://doi.org/10.1679/aohc.58.185

Youle RJ, van der Bliek AM (2012) Mitochondrial fission, fusion, and stress. Science 337(6098):1062–1065. https://doi.org/10.1126/science.1219855

Yung HW, Alnæs-Katjavivi P, Jones CJ, El-Bacha T, Golic M, Staff AC, Burton GJ (2016) Placental endoplasmic reticulum stress in gestational diabetes: the potential for therapeutic intervention with chemical chaperones and antioxidants. Diabetologia 59(10):2240–2250. https://doi.org/10.1007/s00125-016-4040-2

Zeldovich VB, Clausen CH, Bradford E, Fletcher DA, Maltepe E, Robbins JR, Bakardjiev AI (2013) Placental syncytium forms a biophysical barrier against pathogen invasion. PLoS Pathog 9(12):e1003821. https://doi.org/10.1371/journal.ppat.1003821
Zhang Y, Marsboom G, Toth PT, Rehman J (2013) Mitochondrial respiration regulates adipogenic differentiation of human mesenchymal stem cells. PLoS One 8(10):e77077. https://doi.org/10.1371/journal.pone.0077077
Zheng X, Boyer L, Jin M, Mertens J, Kim Y, Ma L et al (2016) Metabolic reprogramming during neuronal differentiation from aerobic glycolysis to neuronal oxidative phosphorylation. Elife 5. https://doi.org/10.7554/eLife.13374
Zhong X, Cui P, Cai Y, Wang L, He X, Long P et al (2019) Mitochondrial dynamics is critical for the full pluripotency and embryonic developmental potential of pluripotent stem cells. Cell Metab 29(4):979–992.e974. https://doi.org/10.1016/j.cmet.2018.11.007
Zhu C, Yang H, Geng Q, Ma Q, Long Y, Zhou C, Chen M (2015) Association of oxidative stress biomarkers with gestational diabetes mellitus in pregnant women: a case-control study. PLoS One 10(4):e0126490. https://doi.org/10.1371/journal.pone.0126490

Chapter 7
Early Syncytialization of the Ovine Placenta Revisited

Heewon Seo, Fuller W. Bazer, and Gregory A. Johnson

Abstract Placentation is the development of a temporary arrangement between the maternal uterus and blastocyst-derived placental tissues designed to transport nutrients, gases, and other products from the mother to the embryo and fetus. Placentation differs histologically among species, but all types of placentation share the common trait of utilizing highly complex cell-to-cell and tissue-to-tissue morphological and biochemical interactions to remodel the uterine-placental interface. An elegant series of electron microscopy (EM) images supports the classification of ovine placentation as synepitheliochorial, because uterine luminal epithelial (LE) cells are maintained at the uterine-placental interface through incorporation into trophoblast syncytial plaques. In this review, we utilize immunofluorescence microscopy to provide further insights into early syncytialization of the ovine placenta. These observations, based on results using immunofluorescence microscopy, complement and expand, not replace, our understanding of syncytialization in sheep.

7.1 Introduction

Ovine conceptuses (embryo/fetus and associated placental membranes) remain free floating within the uterine lumen as they transform from spherical blastocysts to filamentous conceptuses between Days 6 and 17 of gestation (Johnson et al. 2018; Johnson 2018). The majority of the trophoblast cells within elongating ovine conceptuses are mononucleate, and these cells secrete interferon tau to signal pregnancy recognition and participate in the initial attachment of the trophoblast to the

H. Seo (✉) · G. A. Johnson
Department of Veterinary Integrative Biosciences, Texas A&M University, College Station, TX, USA
e-mail: hseo@cvm.tamu.edu; gjohnson@cvm.tamu.edu

F. W. Bazer
Department of Animal Science, Texas A&M University, College Station, TX, USA

M. Kloc, A. Uosef (eds.), *Syncytia: Origin, Structure, and Functions*, Results and Problems in Cell Differentiation 71,
https://doi.org/10.1007/978-3-031-37936-9_7

uterine luminal epithelium (LE) for implantation (Godkin et al. 1984; Farin et al. 1989; Guillomot et al. 1990). During this peri-implantation period of pregnancy, binucleate trophoblast giant cells (BNCs) begin to differentiate from the mononucleate trophoblast cells. BNCs first appear between Days 14 and 16 of gestation in sheep conceptuses and comprise 15–20% of the trophoblast during the apposition and attachment phases of implantation (Wooding et al. 1986; Wooding 2022). It has been proposed that BNCs arise through consecutive nuclear divisions without cytokinesis, also termed mitotic polyploidy (Wimsatt 1951; Hoffman and Wooding 1993; Klisch et al. 1999). A role for the endogenous beta retroviruses (enJSRV) in this process is likely (Dunlap et al. 2006; Spencer et al. 2007) suggesting the potential for BNCs to also, or in total, form through cell-cell fusion similar to what has been proposed for syncytins in humans and mice (Mi et al. 2000; Blond et al. 2000; Blaise et al. 2003; Frendo et al. 2003; Mallet et al. 2004; Dupressoir et al. 2005). This provides a possible explanation for the trinucleate cells observed in the trophoblast layer of cows that are presently considered to be the result of aberrant mitotic polyploidization (Klisch et al. 1999, 2004) (see Fig. 7.1) and for reports of multinucleated cells in the sheep trophoblast (Seo et al. 2019). Initial BNCs contain small numbers of granules but become fully granulated as they mature. Mature BNCs migrate through the microvillar junctions between the trophoblast and uterine LE to initiate syncytialization within the LE layer of the uterine-placental interface. In addition to being instrumental in syncytia formation, BNCs secrete hormones such as placental lactogen and pregnancy-associated glycoproteins (PAGs). The measurement of PAGs levels in serum can be used reliably to assess BNC development and diagnose pregnancy in sheep and cattle (Ruder et al. 1988; Karen et al. 2003; Wallace et al. 2015).

Placentation is the close juxtaposition of the microcirculatory systems of the mother and placenta for the exchange of nutrients, gases, and wastes, as well as to

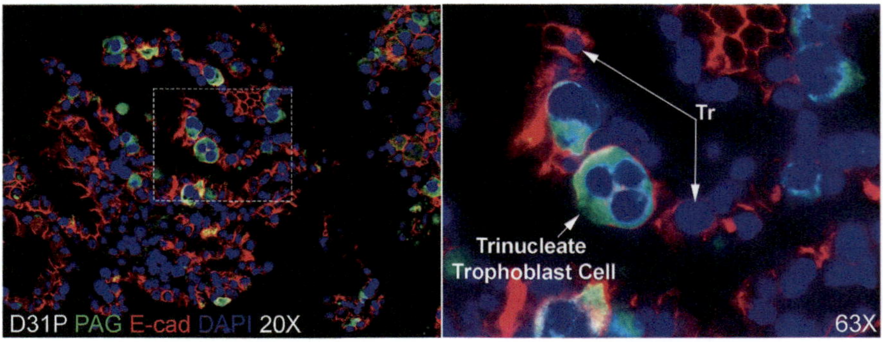

Fig. 7.1 Immunofluorescence staining for pregnancy-associated glycoprotein (PAG, green color) and E-cadherin (E-cad, red color) within the trophoblast layer of a bovine conceptus. Immunostaining for PAG was observed in a small number of trophoblast giant cells containing three nuclei. DAPI (4′,6-diamidino-2-phenylindole) stains cell nuclei blue and is used for histological reference. D, day; P, pregnancy; Tr, trophoblast cells. The width of field for immunofluorescence images captured at 20× and 63× is 142 μm and 220 μm, respectively

provide protection and serve as a source of hormones. Placentation is classified based on histological analyses regarding the number of uterine and placental cell layers/types that are present, as well as the extent of trophoblast invasion into uterine tissue, degree of trophoblast differentiation, and the number of trophoblast layers separating maternal and fetal blood (Wimsatt 1951; Grosser 1909; Amoroso 1952; Wimsatt 1975; Mossman 1991). Variations within this framework provide the basis for the definition of diverse types of placentation ranging from noninvasive epitheliochorial to highly invasive hemochorial. Although differences in histology at the uterine-placental interface of ruminants vary among species, placentation in ruminants, as a group, is a unique intermediate between noninvasive and highly invasive forms of placentation (Green et al. 2021). Trophoblast invasion is limited to the uterine LE, with varying degrees of syncytialization observed. There is no invasion of trophoblast cells into the uterine stroma, and all cell layers within the uterine stroma remain intact, although connective tissue decreases to minimize interhemal distance between the uterine and placental vasculatures. Ovine placentation is classified as synepitheliochorial because uterine LE is considered to be modified through syncytialization with trophoblast cells and not syndesmochorial in which uterine LE is not present at the uterine-placental interface (Wooding 1992).

What has been understood about trophoblast differentiation and the formation of syncytia at the uterine-placental interface of sheep was derived from analyses of an elegant series of electron microscopy (EM) images (Wooding 1984). Conventional wisdom is that BNCs migrate to the uterine LE and fuse with individual LE cells to form trinucleate trophoblast-LE syncytial hybrid cells (Fig. 7.2, Panel a). Through continued migration and fusion of BNCs with individual LE cells, as well as fusion

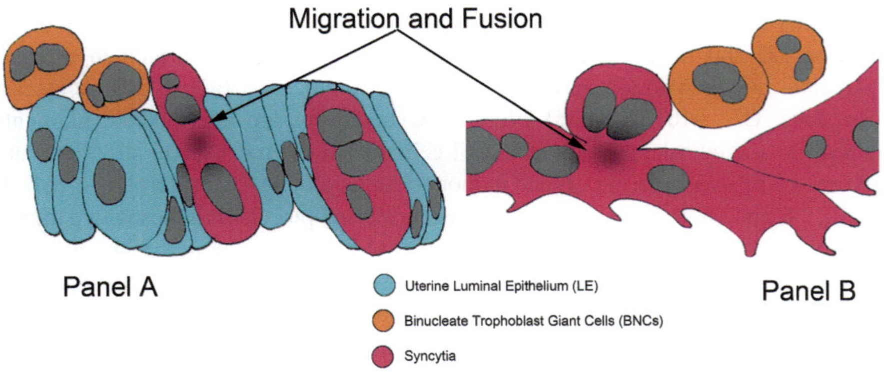

Fig. 7.2 Panel **a**. Binucleate trophoblast giant cells (BNCs) migrate and fuse with individual uterine luminal epithelial (LE) cells to form trinucleate syncytial cells (beginning about Day 16 of pregnancy in sheep), thereby assimilating within the uterine LE. Panel **b**. The syncytia of sheep subsequently enlarge through continued BNC migration and fusion to form syncytial plaques. The syncytial plaques are conceptus-maternal hybrid cells composed of uterine LE and conceptus BNCs, and they eventually form the epithelial interface between uterine and placental tissues within the placentome. In sheep, the syncytial plaques are a consistent feature in placentomes throughout pregnancy

with newly formed syncytia with the uterine LE, the trinucleate syncytial cells subsequently enlarge into syncytial plaques containing large numbers of nuclei (Fig. 7.2, Panel b). In this review, we will revisit the early stages of syncytialization in sheep based on recent observations made through immunofluorescence microscopy.

7.2 Immunofluorescence Microscopy Analyses Identify the Possible Destiny of Specific Cells Engaged in Syncytialization

The uterine and placental tissues were processed by conventional means. Pregnant ewes were hysterectomized, and several 1–1.5 cm sections of the uterine wall and associated placenta from different regions of each uterine horn were (1) snap-frozen in Tissue-Tek Optimal Cutting Temperature compound (OCT; Miles Inc., Oneonta, NY), cooled in liquid nitrogen vapor, and stored at −80 °C, or (2) fixed in fresh 4% paraformaldehyde in PBS (pH 7.2), changed to 70% ethanol after 24 h, and then dehydrated and embedded in Paraplast-Plus (Oxford Labware, St. Louis, MO). Uterine-placental tissues embedded in OCT were cryo-sectioned at 8 μm, whereas tissues embedded in Paraplast-Plus were sectioned at 5 μm and then deparaffinized in xylene and rehydrated to water through a graded alcohol series. All tissue sections were affixed to Superfrost Plus slides (Fisher Scientific, Pittsburgh, PA).

The antibodies utilized in these studies were selected for their immunoreactivity to proteins expressed by uterine LE, mononucleate trophoblast, BNCs, or syncytia. Epithelial cadherin (E-cadherin) is a transmembrane protein unique to adherens junctions that connect mononuclear epithelial cells, and it is expressed by ovine uterine LE and mononuclear trophoblast. E-cadherin was identified in uterine-placental tissues with a mouse anti-E-cadherin monoclonal IgG (BD Biosciences; San Jose, CA; 610182; 1:200 dilution). Cytokeratins are intermediate filaments unique to the cytoskeleton of epithelial cells that resist mechanical stress and are expressed by ovine uterine LE and mononuclear trophoblast. Cytokeratins 1, 5, 6, 7, 8, 10, 11, and 18 were identified in ovine uterine and placental tissues with a mouse anti-cytokeratin monoclonal IgG (Sigma-Aldrich; C-6909; 1:500 dilution). Phosphoserine phosphatase (PSPH) is an enzyme required to convert 3-phosphoserine to serine and is expressed by the ovine uterine LE (Seo et al. 2021; Johnson et al. 2022). PSPH was identified in uterine-placental tissues by a rabbit anti-PSPH polyclonal antibody (Lifespan Biosciences, Seattle, WA; LS-B2935; 1:100). Pregnancy-associated glycoproteins (PAGs) are expressed by ovine placental trophoblast cells, and most members of the PAG family are expressed predominantly in BNCs. PAGs were identified in BNCs, syncytia, and some mononucleate trophoblast in uterine-placental tissues by a rabbit anti-PAG polyclonal IgG (kindly provided by Jonathan A. Green, University of Missouri-Columbia, Columbia, MO; 1:100 dilution). Serine hydroxymethyltransferase 2 (SHMT2) is an enzyme required for the conversion of tetrahydrofolate (THF) to methylene-THF within the one-carbon metabolic pathway and is expressed by ovine trophoblast cells (Sah et al. 2022). SHMT2 was identified in uterine-placental tissues by a rabbit anti-SHMT2 IgG (Sigma-Aldrich; HPA020543; 1:100 dilution).

Slides were incubated overnight at 4 °C with the primary antibody and then were incubated for 1 h at room temperature with a secondary antibody conjugated to Alexa Fluor 488 or Alexa Fluor 594. Slides were then counterstained with Prolong Gold Antifade reagent containing DAPI (4′,6-diamidino-2-phenylindole; Life Technologies) to identify nuclei. Images were taken using an Axioplan 2 microscope (Carl Zeiss, Thornwood, NY) interfaced with an Axioplan HR digital camera. For dual immunofluorescence staining, the two primary antibodies generated in different species (rabbit or mouse) were added to the tissue sections simultaneously on the first day, and the two secondary antibodies (Alexa Fluor 488-conjugated and Alexa Fluor 594-conjugated) were added to the tissue sections simultaneously on the second day.

Immunofluorescence microscopy allows the examination of large tissue sections that include the entire interface of a uterine-placental cross section to identify specific cell types according to the differential expression of a protein(s) unique to each cell type (Fig. 7.3).

7.3 Are BNCs the Only Cells That Migrate into the Uterine Luminal Epithelium?

Although PAGs are predominantly expressed by BNCs, PAGs are also expressed by mononuclear trophoblast cells in sheep (Green et al. 2021). As demonstrated in the uterine/placental tissue section in Fig. 7.3, both PAG-positive BNCs and PAG-positive mononucleate trophoblast cells are present within the conceptus trophoblast. Both cell types are also evident in the uterine-placental tissue section shown in panel a of Fig. 7.4. The PAG-positive mononucleate trophoblast cells in Figs. 7.3 and 7.4 may be the BNCs in which the observed nucleus obscured perception of the second nucleus, but the frequency of the occurrence of these cells suggests that some PAG-positive cells are mononuclear as reported previously (Nagel et al. 1993; Xie et al. 1997; Green et al. 2000). If trophoblast cell entry into the uterine LE is limited to the fusion of BNCs to uterine LE cells or fusion of BNCs to growing syncytial plaques, then PAG-positive mononuclear cells and PAG-positive BNCs should not be present in the uterine LE. However, the uterine LE of uterine-placental tissue sections in Fig. 7.4 include PAG-positive mononucleate trophoblast cells and BNCs. Again, it is possible that the two nuclei observed in these PAG-positive BNCs have obscured a third nucleus and these cells are actually trinucleate syncytia. However, it is unlikely that the single nuclei observed in the PAG-positive mononucleate cells obscure the perception of two other nuclei within the cell, so these cells are likely either mononuclear cells or BNCs. It is unlikely that these

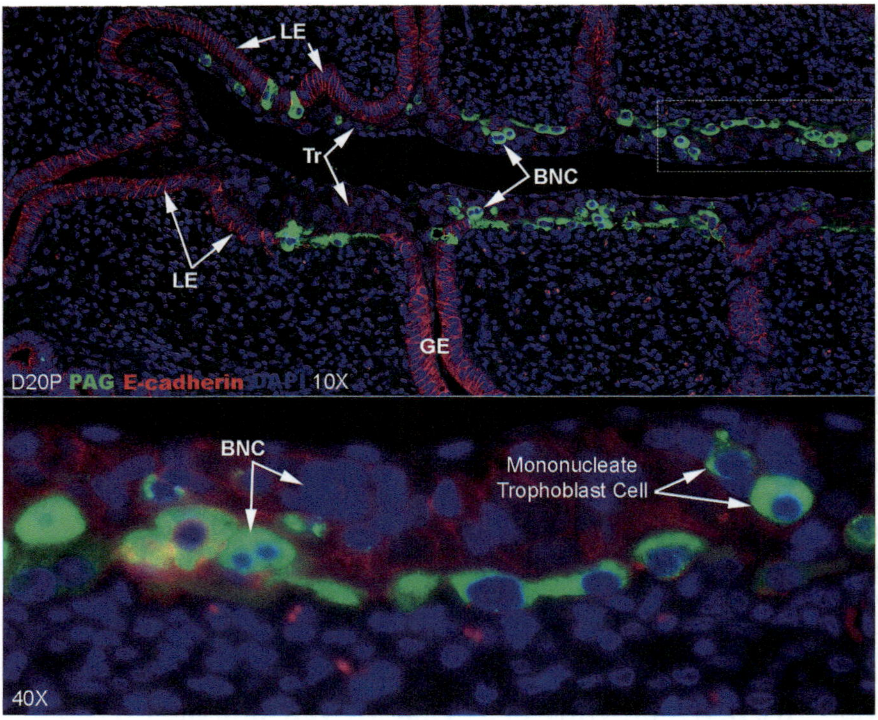

Fig. 7.3 (**Top Panel**) Immunofluorescence staining illustrating global views of the ovine uterine-placental interface in sheep during the early stages of syncytia formation (Day 20). Immunofluorescence staining for pregnancy-associated glycoprotein (PAG, green color) identifies binucleate trophoblast giant cells (BNCs) and immunofluorescence staining for epithelial cadherin (E-cadherin) identifies the uterine luminal epithelium (LE) and mononuclear trophoblast cells. DAPI (4′,6-diamidino-2-phenylindole, blue color) identifies cell nuclei. It is clear that the formation of binucleate cells (green color) is not uniform across the uterine-placental interface on Day 20 of gestation. (**Bottom Panel**) Both BNC and PAG-positive mononucleate cells are present in the trophoblast layer. D, Day; P, pregnancy; GE, glandular epithelium. The width of field for the immunofluorescence image captured at 10× and 40× is 890 μm and 220 μm, respectively

PAG-positive mononuclear trophoblast cells and PAG-positive BNCs penetrated the uterine LE via the fusion of BNCs with the apical ends of individual uterine LE cells; therefore, they likely migrated into the uterine LE through limited dissociation of uterine LE and mechanical intrusion. This possibility was discounted in the rabbit due to a lack of evidence from analyses of EM images, but the authors only "suggested" that fusion is the "normal" method of epithelial penetration by the trophoblast of the rabbit (Enders and Schlafke 1971). Based on results in Fig. 7.4, PAG-positive mononucleate trophoblast cells likely migrate between the lateral surfaces of uterine LE cells to reside within the uterine LE as noted for the mononucleate trophoblast cell apparently inserting itself between cells in the uterine LE. Immune cells routinely traffic through intact layers of epithelial cells (Zemans et al. 2009; Strazielle et al. 2016), and invasion of trophoblast cells into the uterine

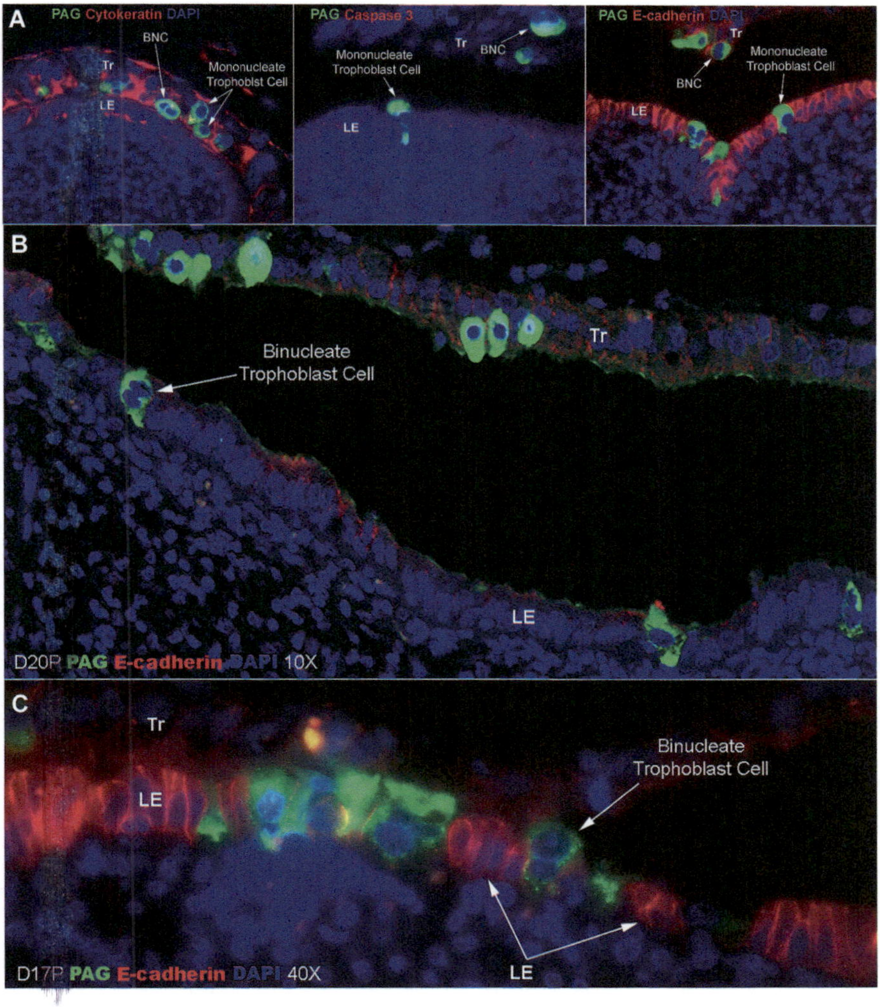

Fig. 7.4 Mononuclear and binuclear trophoblast cells (BNCs) migrate from the trophoblast layer to the uterine luminal epithelium (LE) between Day 17 and Day 20 of gestation in sheep. (**Panel a**) Immunofluorescence staining for pregnancy-associated glycoprotein (PAG, all panels green color), cytokeratin (left panel red color), caspase 3 (middle panel red color), and E-cadherin (right panel red color) detected anti-PAG immunostaining in cells (PAG-positive cells) that have a single nucleus within both the trophoblast layer (Tr) and uterine luminal epithelium (LE). The PAG-positive mononuclear trophoblast cells appear to be inserting themselves between uterine LE. The width of field for immunofluorescence images is 220 μm. (**Panels b, c**) Immunofluorescence staining for PAG (green color) and E-cadherin (red color) detected BNCs within the uterine LE layer. The width of field for immunofluorescence images for Panels **b** and **c** is 593 μm and 220 μm, respectively. DAPI (4′,6-diamidino-2-phenylindole) stains cell nuclei blue and is used for histological reference

LE by human blastocysts is proposed to occur in that way (Aplin and Ruane 2017; Siriwardena and Boroviak 2022). The presence of PAG-positive mononuclear cells in the uterine LE provides the opportunity for BNCs in the trophoblast to fuse with mononuclear trophoblast cells to form trinucleate syncytia entirely of trophoblast origin as is illustrated in Fig. 7.5.

7.4 What Is the Fate of Uterine LE During Syncytialization?

Ovine placentation was initially classified as syndesmochorial based on evidence that the uterine LE cells were lost during placentation (Grosser 1909, 1927). Subsequently, investigators concluded that uterine LE cells are incorporated into syncytial plaques and, therefore, classified sheep as having synepitheliochorial

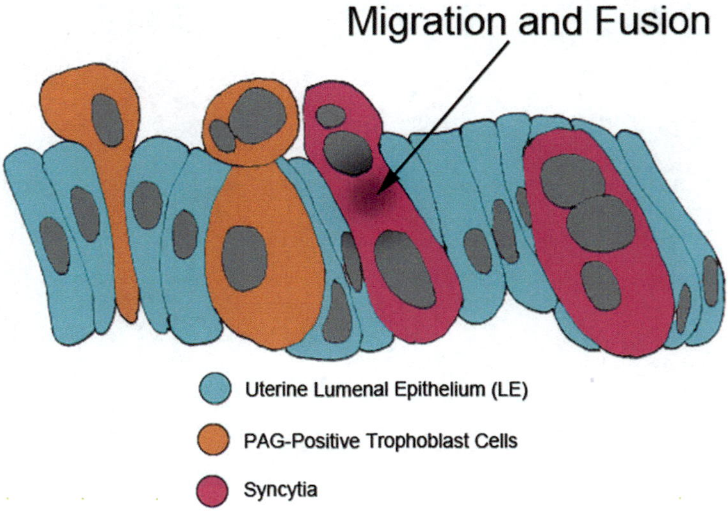

Fig. 7.5 (**Panel a**) The presence of pregnancy-associated glycoprotein (PAG)-positive mononuclear cells within the uterine LE suggests that in some cases the trinucleate cells observed in the uterine LE layer may form through fusion between mononuclear trophoblast cells and BNCs. PAG-positive mononuclear trophoblast cells migrate between uterine LE cells where their unusually large size compared to the uterine LE cells results in the physical stretching of the neighboring LE cells to accommodate their presence. Trophoblast binucleate cells then migrate to the uterine LE and potentially fuse with these PAG-positive mononucleate cells to form trinucleate syncytial cells that are entirely of trophoblast origin. (**Panel b**) Immunofluorescence staining for PAG (green color, top panels), E-cadherin (red color, top and bottom right panels), phosphoserine phosphatase (PSPH; red color, bottom left panel), and serine hydroxymethyltransferase 2 (SHMT2; green color, bottom right panel) on Days 17, 18, and 20 of gestation shows what appear to be trophoblast cells (arrow heads) invading into the uterine luminal epithelial (LE) layer. D, Day; P, pregnancy; Tr, trophoblast. DAPI (4′,6-diamidino-2-phenylindole) stains cell nuclei. The width of field for the immunofluorescence images captured at 40× and 63× is 220 μm and 140 μm, respectively

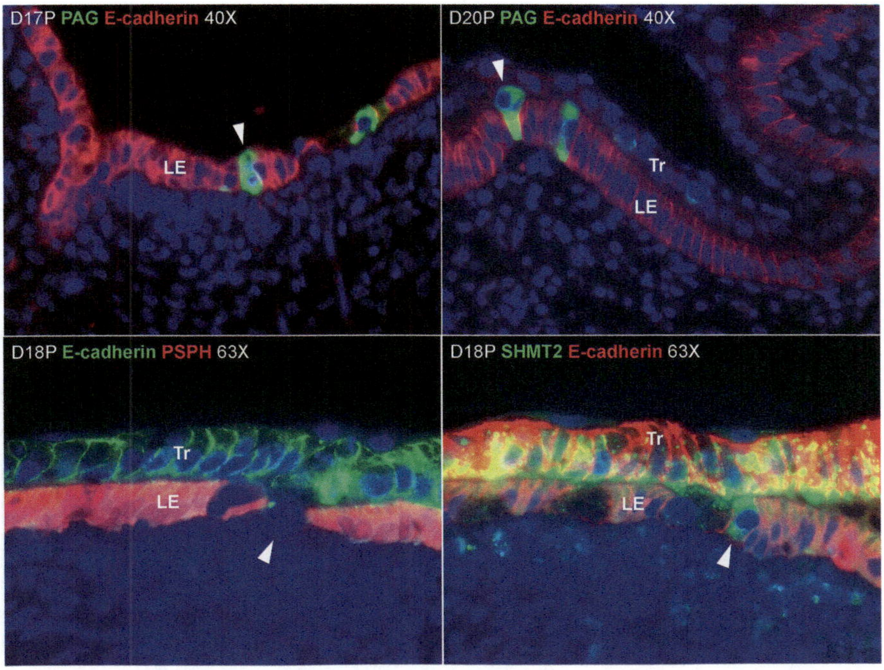

Fig. 7.5 (continued)

placentation (Wooding 1992), as illustrated in Fig. 7.2. As stated earlier in this review, the presence of PAG-positive mononuclear trophoblast cells in the uterine LE suggests the possibility of the formation of some syncytial plaques that are devoid of uterine LE. Further, gaps form in the uterine LE during the process of syncytialization (Seo et al. 2019; Wooding 1984). Those observations suggest that some LE cells are not incorporated into syncytial plaques but that they are eliminated from the uterine-placental interface of sheep.

Two ideas have been put forward to address the elimination of some uterine LE at the uterine-trophoblast interface, both involving trophoblast cells. One idea is that uterine LE cells that do not participate in fusion with BNCs die and the remaining residues are phagocytized by trophoblasts resulting in gaps in the uterine LE (Wooding 2022). The other idea involves trophoblast cell migration into the uterine LE, phagocytosis of apoptotic LE cells by trophoblast cells, further migration of these trophoblasts into the uterine stroma, and elimination of uterine LE within the stroma, resulting in gaps in the uterine LE (Seo et al. 2019).

The possible resolution of the two ideas is addressed based on the results shown in Fig. 7.6. Panel a and Panel b of Fig. 7.6 reveal H&E staining of a region of the uterine-placental interface on Day 20 of gestation indicating that uterine LE has been eliminated and no intact cells remain to separate trophoblast and uterine stroma. Interestingly, neutrophils have accumulated within the uterine stroma adjacent to the region lacking uterine LE. This suggests an innate immune response has

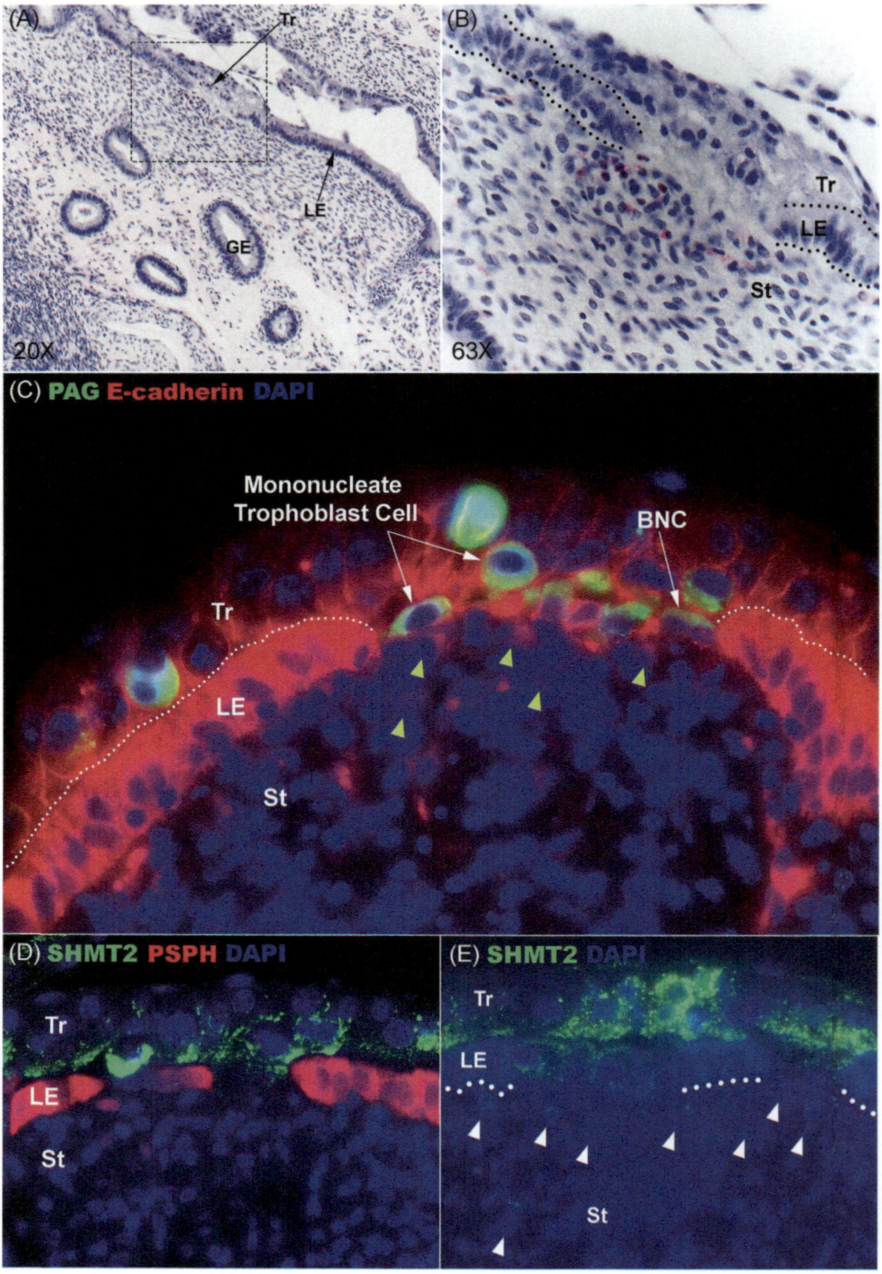

Fig. 7.6 (**Panels a, b**) H&E staining reveals a region of the uterine-placental interface without uterine LE cells separating trophoblast and endometrial stroma on Day 20 of gestation. Neutrophils, some of which are undergoing apoptosis, are present in the uterine stroma adjacent to this region of the uterine-placental interface, suggesting an innate immune response, within this region of the uterine stroma. Dotted lines approximate the uterine LE layer. The width of field for the H&E

(continued)

images captured at 20× and 63× is 738 μm and 234 μm, respectively. (**Panel c**) Immunofluorescence staining for pregnancy-associated glycoprotein (PAG, green color) and E-cadherin (red color) on Day 17 of gestation shows the presence of PAG-positive cells within the uterine LE layer. These cells appear to be mononuclear PAG-positive, presumably trophoblast cells. In this same region of the uterine-placental interface, E-cadherin-positive uterine LE cells are rounded-up and dissociated from the columnar epithelial cells. Yellow arrow heads indicate E-cadherin-positive cells within the uterine stroma. Dotted lines approximate the interface between trophoblast and uterine LE. The width of field for the immunofluorescence image is 220 μm. (**Panel d**) Immunofluorescence staining for serine hydroxymethyltransferase 2 (SHMT2; green color) and phosphoserine phosphatase (PSPH; red color) on Day 18 of gestation shows SHMT2-positive trophoblast cells penetrating the uterine LE that is positive for immunoreactive PSPH. DAPI (4′,6-diamidino-2-phenylindole) stains cell nuclei. The width of field for the immunofluorescence images is 140 μm. (**Panel e**) SHMT2-positive cells are present in the trophoblast (Tr) and uterine LE and within the uterine stroma (St) on Day 18 of gestation. White arrow heads indicate cells positive for immunoreactive SHMT2 within the uterine stroma. Dotted line approximates the basement membrane of the uterine LE. The width of field for the immunofluorescence images is 140 μm. DAPI (4′,6-diamidino-2-phenylindole) stains cell nuclei blue for histological reference

drawn these highly phagocytic cells to this specific region of the uterine stroma, supporting the idea that immune cells traffic to regions of uterine LE loss, perhaps to phagocytize degenerating cells as has been previously proposed (Seo et al. 2019). Panel c of Fig. 7.6 shows a region of uterine LE from Day 17 of pregnancy that contains a mixture of PAG-positive trophoblast cells and E-cadherin-positive cells that appear to have rounded up and dissociated from the columnar epithelial cells. Within the stroma adjacent to this region of remodeling uterine LE, there are cells whose cytoplasm immunostains for E-cadherin. During apoptosis, expression of E-cadherin is not detected at the cell surface, but it accumulates in the cytosol (Steinhusen et al. 2001). Panel d of Fig. 7.6 shows two areas of uterine LE from Day 18 of pregnancy that include SHMT2-positive trophoblast cells penetrating the uterine LE. Although the immunostaining is faint, SHMT2-positive cells are present in the uterine stroma adjacent to the region in which there is remodeling of the uterine LE (see panel e of Fig. 7.6). Collectively, these results demonstrate that (1) uterine LE is eliminated from some regions of syncytialization leaving the gaps in the uterine LE; (2) phagocytic immune cells, including, but perhaps not limited to, neutrophils, are present in the uterine stroma adjacent to these regions of uterine LE remodeling and syncytialization; (3) E-cadherin positive cells are present in the uterine stroma adjacent to these regions of uterine LE remodeling and syncytialization; and (4) SHMT2-positive cells, likely of trophoblast lineage, are present in the uterine stroma adjacent to these regions of uterine LE remodeling and syncytialization. Therefore, conceptus trophoblast cells potentially migrate to the uterine LE, engulf uterine LE, and transport uterine LE into the uterine stroma for phagocytosis by immune cells (Seo et al. 2019).

7.5 Based on This Scenario, Several Factors Remain to Be Clarified

First, the factors that induce the death of uterine LE cells are unknown. The integrity of uterine LE depends on proteins making up junctional complexes, but the insertion of trophoblast cells between uterine LE may result in the dissociation of cell-cell junctional complexes and subsequent apoptosis of the affected uterine LE (Bruner and Derksen 2018). Alternatively, conceptuses may secrete soluble factors that induce apoptosis of uterine LE (Welsh 1993; Ashary et al. 2018). Second, although EM has been used to detect the uptake of residues from dead uterine LE (Wooding 2022), there is no direct evidence for the engulfment of apoptotic LE by trophoblast cells. Third, the basement membrane of uterine LE must be breached. Evidence for E-cadherin-positive cells that have transformed from a columnar to a round morphology suggests that these cells have dissociated from the basement membrane. Further, the removal of the uterine LE suggests that the integrity of the basement membrane is compromised. The basement membrane is not a true membrane but is composed of a basal lamina that includes transmembrane integrins that bind laminins that bind type 4 collagen to anchor the cells to the connective tissue and an underlying reticular lamina composed of type 3 collagen. The integrity of the basal lamina is dependent on the integrity of the epithelial cells, and maintenance of that integrity is questionable during syncytialization. The reticular lamina does not represent a significant barrier to cell migration. Another possible explanation for the delivery of apoptotic LE cells into the stroma is that invading trophoblast cells exert a physical force that displaces uterine LE cells and pushes them through a compromised basement membrane (Chang and Chaudhuri 2019). Fourth, there is ample precedence for trophoblast invasion into the uterine stroma. In humans, neighboring cytotrophoblast cells fuse to generate the syncytiotrophoblast at the time of adhesion of the blastocyst to the uterine LE that initiates implantation. It is the syncytiotrophoblast that penetrates through the uterine LE during the implantation of human blastocysts (Huppertz and Gauster 2011). The mechanism by which syncytiotrophoblasts create a route for penetration is unknown, but the blastocyst may migrate between uterine LE through transepithelial migration which is similar to transendothelial migration by leukocytes (Uchida et al. 2016). However, histological studies suggest that the syncytiotrophoblast eliminates uterine LE cells at sites of implantation in non-human primates and humans (Enders 2000). Therefore, it is likely that syncytialization precedes trophoblast invasion and that it is the syncytiotrophoblasts that actively breach the uterine LE to allow the blastocyst to invade the endometrial stroma (Aplin and Ruane 2017; Siriwardena and Boroviak 2022).

7.6 Conclusions

Placentation in ruminants is intermediate between the noninvasive type, as observed in the epitheliochorial placenta of pigs, and the invasive type, as observed in the hemochorial placentae of mice and humans. In ruminants, trophoblasts invade uterine tissue, but the invasion is limited to the uterine LE wherein there are varying degrees of syncytialization among species. There is no invasion of trophoblast cells into the uterine stroma, and all cell layers within the uterine stroma remain intact, although connective tissue is decreased to minimize the interhemal distance between the uterine and placental vasculatures. Through careful examination of a series of EM images, ovine placentation was classified as synepitheliochorial because uterine LE is considered to be modified through syncytialization with trophoblast cells and not syndesmochorial in which uterine LE is removed in areas of the uterine-placental interface. This review provides evidence based on careful analyses of immunofluorescence microscopy images to complement and extend existing knowledge of placentation in sheep. Interpretation of those analyses is summarized in Fig. 7.7 to expand understanding of mechanisms responsible for syncytialization of cells during placentation in sheep. Two points to consider are the possibility that PAG-positive mononuclear trophoblast cells participate in syncytialization of the uterine-placental interface of sheep and that the majority of uterine LE may be eliminated from the uterine-placental interface through phagocytosis within the

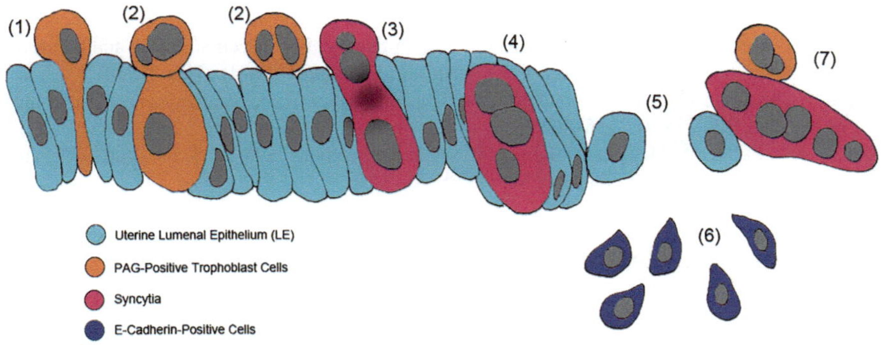

Fig. 7.7 (**1**) Pregnancy-associated glycoprotein (PAG)-positive mononuclear trophoblast cells potentially migrate and insert themselves between uterine luminal epithelial (LE) cells. (**2**) Binucleate trophoblast giant cells (BNCs) physically interact with uterine LE or potentially PAG-positive mononuclear trophoblast cells in preparation for fusion. (**3**) BNCs fuse with mononuclear cells in the uterine LE. (**4**) A trinucleate syncytial cell forms and is composed of trophoblast and LE or only trophoblast. (**5**) In regions of the uterine-placental interface, uterine LE cells round up, dissociate, and begin to die, perhaps through apoptosis, to form gaps in the uterine LE. (**6**) E-cadherin-positive cells are present in the uterine stroma in regions where the uterine LE has developed gaps. (**7**) BNCs continue to fuse with growing syncytia which will eventually compose what was the uterine LE of the uterine-placental interface. In this model, only BNCs were drawn to illustrate fusion with cells in the uterine LE layer. However, PAG-positive mononuclear trophoblast cells may also fuse with mononuclear cells or growing syncytial plaques in the uterine LE during placentation in sheep

uterine stroma. It is important to note that these results, based on histological analyses of cells at the uterine-placental interface, imply physiological aspects of pregnancy that are subject to debate and future alteration with new scientific findings. Interpretations, based on histological analyses, are key to developing a better understanding of syncytialization in the sheep placenta.

Acknowledgments This work was supported by National Research Initiative Competitive Grant no. 2022-67015-36376 from the USDA National Institute of Food and Agriculture.

References

Amoroso EC (1952) Placentation. In: Parkes AS (ed) Marshall's physiology of reproduction. Longmans Green, London, pp 127–311. Chapter 15

Aplin JD, Ruane PT (2017) Embryo-epithelium interactions during implantation at a glance. J Cell Sci 130(1):15–22

Ashary N, Tiwari A, Modi D (2018) Embryo implantation: war in times of love. Endocrinology 159(2):1188–1198

Blaise S, de Parseval N, Bénit L, Heidmann T (2003) Genomewide screening for fusogenic human endogenous retrovirus envelopes identifies syncytin 2, a gene conserved on primate evolution. Proc Natl Acad Sci U S A 100(22):13013–13018

Blond JL, Lavillette D, Cheynet V, Bouton O, Oriol G, Chapel-Fernandes S, Mandrand B, Mallet F, Cosset FL (2000) An envelope glycoprotein of the human endogenous retrovirus HERV-W is expressed in the human placenta and fuses cells expressing the type D mammalian retrovirus receptor. J Virol 74(7):3321–3329

Bruner HC, Derksen PWB (2018) Loss of E-cadherin-dependent cell-cell adhesion and the development and progression of cancer. Cold Spring Harb Perspect Biol 10(3):a029330

Chang J, Chaudhuri O (2019) Beyond proteases: basement membrane mechanics and cancer invasion. J Cell Biol 218(8):2456–2469

Dunlap KA, Palmarini M, Spencer TE (2006) Ovine endogenous betaretroviruses (enJSRVs) and placental morphogenesis. Placenta Suppl A:S135–S140

Dupressoir A, Marceau G, Vernochet C, Bénit L, Kanellopoulos C, Sapin V, Heidmann T (2005) Syncytin-A and syncytin-B, two fusogenic placenta-specific murine envelope genes of retroviral origin conserved in Muridae. Proc Natl Acad Sci U S A 102(3):725–730

Enders AC (2000) Trophoblast-uterine interactions in the first days of implantation: models for the study of implantation events in the human. Semin Reprod Med 18:255–363

Enders AC, Schlafke S (1971) Penetration of the uterine epithelium during implantation in the rabbit. Am J Anat 132(2):219–230

Farin CE, Imakawa K, Roberts RM (1989) In situ localization of mRNA for the interferon, ovine trophoblast protein-1, during early embryonic development of the sheep. Mol Endocrinol 3:1099–1107

Frendo JL, Olivier D, Cheynet V, Blond JL, Bouton O, Vidaud M, Rabreau M, Evain-Brion D, Mallet F (2003) Direct involvement of HERV-W Env glycoprotein in human trophoblast cell fusion and differentiation. Mol Cell Biol 23(10):3566–3574

Godkin JD, Bazer FW, Thatcher WW, Roberts RM (1984) Proteins released by cultured day 15–16 conceptuses prolong luteal maintenance when introduced into the uterine lumen of cyclic ewes. J Reprod Fertil 71:57–64

Green JA, Xie S, Quan X, Bao B, Gan X, Mathialagan N, Beckers JF, Roberts RM (2000) Pregnancy-associated bovine and ovine glycoproteins exhibit spatially and temporally distinct expression patterns during pregnancy. Biol Reprod 62(6):1624–1631

Green JA, Geisert RD, Johnson GA, Spencer TE (2021) Implantation and placentation in ruminants. Adv Anat Embryol Cell Biol 234:129–154
Grosser O (1909) Vergleichende anatomie und entwicklungsgeschichte der eihaute und der placenta. Braumuller W, Vienna and Leipzig
Grosser O (1927) Fruhentwicklung, Eihautbidung und Placentation des Menschen und der Saugetiere. J. F. Bergmann, Munchen
Guillomot M, Michel C, Gaye P, Charlier N, Trojan J, Martal J (1990) Cellular localization of an embryonic interferon, ovine trophoblastin and its mRNA in sheep embryos during early pregnancy. Biol Cell 68:205–211
Hoffman LH, Wooding FB (1993) Giant and binucleate trophoblast cells of mammals. J Exp Zool 266:559–577
Huppertz B, Gauster M (2011) Trophoblast fusion. Adv Exp Med Biol 713:81–95
Johnson GA (2018) Domestic animal placentation. In: Skinner MK (ed) Encyclopedia of reproduction, vol 2. Academic Press, Elsevier, pp 448–454
Johnson GA, Bazer FW, Burghardt RC, Wu G, Seo H, Kramer AC, McLendon BA (2018) Cellular events during ovine implantation and impact for gestation. Anim Reprod 15(Suppl 1):843–855
Johnson GA, Seo H, Bazer FW, Wu G, Kramer AC, McLendon BA, Cain JW (2022) Metabolic pathways utilized by the porcine conceptus, uterus and placenta. Mol Reprod Dev. https://doi.org/10.1002/mrd.23570
Karen A, Beckers JF, Sulon J, de Sousa NM, Szabados K, Reczigel J, Szenci O (2003) Early pregnancy diagnosis in sheep by progesterone and pregnancy-associated glycoprotein tests. Theriogenology 59(9):1941–1948
Klisch K, Hecht W, Pfarrer C, Schuler G, Hoffmann B, Leiser R (1999) DNA content and ploidy level of bovine placentomal trophoblast giant cells. Placenta 20:451–458
Klisch K, Thomsen PD, Dantzer V, Leiser R (2004) Genome multiplication is a generalised phenomenon in placentomal and interplacentomal trophoblast giant cells in cattle. Reprod Fertil Dev 16(3):301–306
Mallet F, Bouton O, Prudhomme S, Cheynet V, Oriol G, Bonnaud B, Lucotte G, Duret L, Mandrand B (2004) The endogenous retroviral locus ERVWE1 is a bona fide gene involved in hominoid placental physiology. Proc Natl Acad Sci U S A 101(6):1731–1736
Mi S, Lee X, Li X, Veldman GM, Finnerty H, Racie L, LaVallie E, Tang XY, Edouard P, Howes S, Keith JC Jr, McCoy JM (2000) Syncytin is a captive retroviral envelope protein involved in human placental morphogenesis. Nature 403(6771):785–789
Mossman HW (1991) Classics revisited: comparative morphogenesis of the fetal membranes and accessory uterine structures. Placenta 12(1):1–5
Nagel RJ, Xie S, Roberts RM (1993) Aspartic proteinases as markers of trophoblast differentiation in sheep. Biol Reprod 48(suppl 1):93. (abstract 139)
Ruder CA, Stellflug JN, Dahmen JJ, Sasser RG (1988) Detection of pregnancy in sheep by radioimmunoassay of sera for pregnancy-specific protein B. Theriogenology 29(4):905–912
Sah N, Stenhouse C, Halloran KM, Moses RM, Seo H, Burghardt RC, Johnson GA, Wu G, Bazer FW (2022) Creatine metabolism at the uterine-conceptus interface during early gestation in sheep. Biol Reprod 107(6):1528–1539
Seo H, Bazer FW, Burghardt RC, Johnson GA (2019) Immunohistochemical examination of trophoblast syncytialization during early placentation in sheep. Int J Mol Sci 20:4530
Seo H, Johnson GA, Bazer FW, Wu G, McLendon BA, Kramer AC (2021) Cell-specific expression of enzymes required for serine biosynthesis and glutaminolysis in farm animals. Adv Exp Med Biol 1285:17–28
Siriwardena D, Boroviak TE (2022) Evolutionary divergence of embryo implantation in primates. Philos Trans R Soc Lond B Biol Sci 377(1865):20210256
Spencer TE, Johnson GA, Bazer FW, Burghardt RC, Palmarini M (2007) Pregnancy recognition and conceptus implantation in domestic ruminants: roles of progesterone, interferons and endogenous retroviruses. Reprod Fertil Dev 19:65–78

Steinhusen U, Weiske J, Badock V, Tauber R, Bommert K, Huber O (2001) Cleavage and shedding of E-cadherin after induction of apoptosis. J Biol Chem 276(7):4972–4980

Strazielle N, Creidy R, Malcus C, Boucraut J, Ghersi-Egea JF (2016) T-lymphocytes traffic into the brain across the blood-CSF barrier: evidence using a reconstituted choroid plexus epithelium. PLoS One 11(3):e0150945

Uchida H, Maruyama T, Masuda H, Uchida S, Miki F, Hihara H, Katakura S, Yoshimasa Y, Tanaka M (2016) How to create an embryo penetration route. Am J Reprod Immunol 75:326–332

Wallace RM, Pohler KG, Smith MF, Green JA (2015) Placental PAGs: gene origins, expression patterns, and use as markers of pregnancy. Reproduction 149(3):R115–R126

Welsh AO (1993) Uterine cell death during implantation and early placentation. Microsc Res Tech 25(3):223–245

Wimsatt WA (1951) Observations on the morphogenesis, cytochemistry, and significance of the binucleate giant cells of the placenta of ruminants. Am J Anat 89(2):233–281

Wimsatt WA (1975) Some comparative aspects of implantation. Biol Reprod 12(1):1–40

Wooding FB (1984) Role of binucleate cells in fetomaternal cell fusion at implantation in the sheep. Am J Anat 170(2):233–250

Wooding FB (1992) Current topic: the synepitheliochorial placenta of ruminants: binucleate cell fusions and hormone production. Placenta 13(2):101–113

Wooding FBP (2022) The ruminant placental trophoblast binucleate cell: an evolutionary breakthrough. Biol Reprod 107(3):705–716

Wooding FB, Flint AP, Heap RB, Morgan G, Buttle HL, Young IR (1986) Control of binucleate cell migration in the placenta of sheep and goats. J Reprod Fertil 76(2):499–512

Xie S, Green J, Bao B, Beckers JF, Valdez KE, Hakami L, Roberts RM (1997) Multiple pregnancy-associated glycoproteins are secreted by day 100 ovine placental tissue. Biol Reprod 57(6):1384–1393

Zemans RL, Colgan SP, Downey GP (2009) Transepithelial migration of neutrophils: mechanisms and implications for acute lung injury. Am J Respir Cell Mol Biol 40(5):519–535

Chapter 8
Syncytia in *Utricularia*: Origin and Structure

Bartosz J. Płachno, Małgorzata Kapusta, and Piotr Świątek

Abstract In animals and plants, multinucleate cells (syncytia and coenocytes) are essential in ontogeny and reproduction. Fuso-morphogenesis is the formation of multinucleated syncytia by cell–cell fusion, but coenocytes are formed as a result of mitosis without cytokinesis. However, in plants, coenocytes are more widespread than true syncytia. Except for articulated laticifers, most plant syncytia have a trophic function. Here, we summarize the results of histological, histochemical, and ultrastructural analyses of syncytia in the *Utricularia* species from the Lentibulariaceae family. *Utricularia* syncytia, known only from a few species, are heterokaryotic because the syncytium possesses nuclei from two different sources: cells of maternal sporophytic nutritive tissue (placenta) and endosperm haustorium. Thus, syncytium contains both maternal and paternal genetic material. In species from section *Utricularia*, syncytia are highly active structures (with hypertrophied nuclei, cell wall ingrowths, and extensive cytoskeleton) that exist only during embryo development. They serve as an example of evolutionary unique trophic structures in the plant kingdom.

B. J. Płachno (✉)
Department of Plant Cytology and Embryology, Institute of Botany, Faculty of Biology, Jagiellonian University in Kraków, Kraków, Poland
e-mail: bartosz.plachno@uj.edu.pl

M. Kapusta
Department of Plant Cytology and Embryology, Faculty of Biology, University of Gdańsk, Gdańsk, Poland

P. Świątek
Institute of Biology, Biotechnology and Environmental Protection, Faculty of Natural Sciences, University of Silesia in Katowice, Katowice, Poland

M. Kloc, A. Uosef (eds.), *Syncytia: Origin, Structure, and Functions*, Results and Problems in Cell Differentiation 71,
https://doi.org/10.1007/978-3-031-37936-9_8

8.1 Introduction

In both plants and animals, there are two main types of multinucleate cells: syncytia (syncytium from Greek: *σύν syn* "together" and *κύτος kytos* "box") and coenocytes. Although these cells may be very similar in both structure and function, they differ in origin (unfortunately, many authors use the terms "syncytium" and "syncytial" to describe coenocytic structures). The syncytium is formed by cell fusion (fuso-morphogenesis). Also, an egg cell could be classified as two nucleate syncytium shortly after fertilization but before karyogamy. However, coenocyte is formed by multiple nuclear divisions without their accompanying cytokinesis. Syncytia and coenocytes are formed during ontogeny and reproduction processes and may have a trophic function (Shemer and Podbilewicz 2000; Kloc et al. 2004; Świątek and Urbisz 2019). For example, in gymnosperms, angiosperms, and some pteridophytes, coenocytic developmental phases (megagametophyte, nuclear endosperm) are essential in sexual reproduction (Rudall and Bateman 2019). Also, in animal development, coenocytic phase is common, because, in the initial stages of gametogenesis, male and female germ cells develop in full synchrony as interconnected cells called germline cysts (Pepling et al. 1999; Greenbaum et al. 2011; Chaigne and Brunet 2022). Syncytial cells have also been utilized in metazoans to form many complex organs, e.g., bones, muscles, and placentae (Shemer and Podbilewicz 2000, 2003). However, syncytia occur more rarely in plants than coenocytes and have mainly the trophic role. They are specialized, terminally differentiated multinucleated structures which are very active but shortly lived: an amoeboid tapetum (Anger and Weber 2006; Pacini and Juniper 1983; Tiwari and Gunning 1986), a nucellar plasmodium of Podostemaceae (Arekal and Nagendran 1975; Mukkada 1962; Jäger-Zürn 1997; Murguía-Sánchez et al. 2002), the endospermal syncytium of *Nothapodytes foetida* (family Icacinaceae) (Swamy and Ganapathy 1957), and syncytia of *Utricularia* (e.g., Khan 1954; Płachno and Świątek 2011). Other examples of plant syncytia are articulated laticifers (Esau 1965) which produce latex and associated metabolites for the defense of plants against insects (Hagel et al. 2008) and plant syncytia caused by parasitic nematodes (Sobczak and Golinowski 2009; Sobczak and Matuszkiewicz 2023, in this book).

Why are these syncytia in *Utricularia* remarkable, unusual, and worth mentioning? Because this type of syncytium possesses nuclei from two different sources: cells of maternal sporophytic nutritive tissue (placental tissue) and endosperm haustorium (paternal + maternal genetic material).

The present chapter summarizes and discusses the results of past and recent ultrastructural and histochemical analyses of *Utricularia* syncytia. This chapter is based mainly on a series of our works (Płachno et al. 2011, 2013; Płachno and Świątek 2011) but also on our recent unpublished results.

8.2 *Utricularia* (Bladderworts)

Utricularia (family Lentibulariaceae, Lamiales, Eudicots) are herbaceous, most often small to medium-sized flowering carnivorous plants (Fig. 8.1). *Utricularia* show great phenotypic plasticity and are aquatic plants (affixed or free-floating or rheophytes) or terrestrials, lithophytes, and epiphytes. Generally, they are associated with wetlands or seasonally wet habitats (Taylor 1989; Miranda et al. 2021). These plants are rootless and known for unusual vegetative organs, which follow structural features differing from typical flowering plants (Rutishauser 2016; Reut et al. 2021; Reut and Płachno 2023), but also known for the fastest closing traps in

Fig. 8.1 *Utricularia intermedia* Hayne in natural habitat at the Jeleniak-Mikuliny Nature Reserve near the town Lubliniec, Poland

the plant world (Poppinga et al. 2017). The genus *Utricularia* includes around 250 species (Miranda et al. 2021), classified into three subgenera, *Polypompholyx*, *Bivalvaria*, and *Utricularia*, and about 35 sections (Müller and Borsch 2005; Jobson et al. 2017).

8.2.1 Syncytium Occurrence

Unfortunately, given the vast number of *Utricularia* species, the embryology of only a few has been studied in detail (see for references Khan 1992; Płachno 2011). Thus, our knowledge about the occurrence of syncytia is insufficient. Some sections of *Utricularia* have not been studied at all. Also, most researchers, who studied *Utricularia* female gametophyte and endosperm development, used the paraffin method and did not give photographic documentation but gave line drawings only (which sometimes are very esthetic and detailed; see, e.g., Wylie and Yocom 1923; Khan 1954). These are already a form of subjective interpretation of the results and make revising challenging. Syncytia were not found in members of subgenus *Polypompholyx* (Lang 1901; Siddiqui 1978); however, only two species from this subgenus were studied in the case of endosperm development. Syncytia have been recorded in terrestrial species only in *U. caerulea* (Kausik 1938), section *Nigrescentes* and *U. reticulata* (Kausik and Raju 1955), section *Oligocista*; both species are classified into subgenus *Bivalvaria*. Syncytia have also been described in species of the section *Utricularia*, subgenus *Utricularia*: *U. macrorhiza* (Wylie and Yocom 1923), *U. aurea* (Khan 1954), *U. inflexa* (Farooq 1964), *U. vulgaris* (Jankun and Płachno 2000; Płachno and Świątek 2011), *U. intermedia* (Płachno and Świątek 2011; Płachno et al. 2013), and *U. minor* (Płachno et al. 2013). Thus, syncytia were found in almost all examined representatives of section *Utricularia*. However, with current limited knowledge, it is difficult to say whether syncytia have appeared multiple times in different *Utricularia* lineages or once in the ancestor of *Bivalvaria* + *Utricularia* lineage.

8.2.2 Placental Nutritive Tissues and "Naked" Embryo Sacs

When discussing syncytia, two characteristic features of *Utricularia* cannot be overlooked: placental nutritive tissue and naked embryo sac, which are essential for syncytia formation. Placental nutritive tissue (also termed "Nährgewebe," "Drüsengewebe," and "receptive tissue") (Merz 1897; Merl 1915; Wylie and Yocom 1923; Khan 1954) develops near the base of the ovule in the placenta in members of subgenera *Bivalvaria* and *Utricularia* (Fig. 8.2a). In members of subgenus *Polypompholyx*, nutritive tissue occurs in the funiculus (Płachno and Świątek 2008). Placental nutritive tissue cells stain more intensely compared to cells of surrounding tissues, and nutritive tissue cells differ in shape and size from other placenta cells

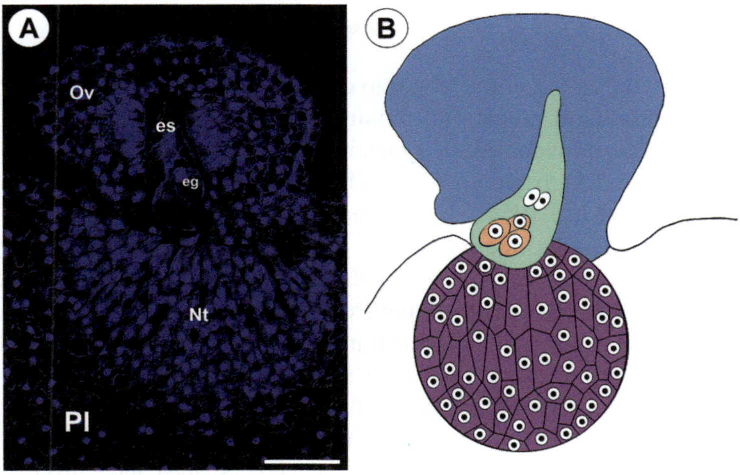

Fig. 8.2 *Utricularia intermedia*. (**a**) Placenta fragment showing a nutritive tissue (Nt), an ovule (Ov), an embryo sac (es), and an egg cell (eg); Differential interference contrast (DIC) optics. Section stained with DAPI (blue fluorescence). Bar is equal to 50 μm. (**b**) Schematic drawing of *Utricularia* ovule (blue color) showing relationships between the embryo sac (green color), nutritive tissue (violet color)

(Wylie and Yocom 1923). Also, nutritive tissue cells differ from placental parenchyma cells in lacking giant vacuoles and large amyloplasts with starch grains. In some species, placental nutritive tissue consists of colenchymatous cells (Płachno and Świątek 2008; Płachno 2011). In *U. nelumbifolia*, nuclei located in placental nutritive tissues have spindle-like tubular projections—chromatubules (Płachno et al. 2017). Depending on the species, placental nutritive tissues may be small group cells (e.g., in *U. sandersonii*, Płachno and Świątek 2008) or consist of several thousand cells (e.g., in *U. macrorhiza*, Wylie and Yocom 1923). Various researchers agree that placental nutritive tissue supplies nutrients to the embryo sac and later to the embryo (e.g., Merz 1897; Wylie and Yocom 1923; Khan 1954; Płachno and Świątek 2008); however, there is no experimental confirmation of this suggestion. But the occurrence of arabinogalactan proteins in the placental nutritive tissue suggests that it may function as an obturator in some *Utricularia* species (Płachno et al. 2021).

In some *Utricularia* (members of sections: *Vesiculina*, *Utricularia*, *Orchidioides*, *Foliosa*, *Calpidisca*, *Nigrescentes*), the embryo sac grows beyond the limit of the integument and has contact with placental nutritive tissue (Płachno et al. 2021, 2022; Płachno 2011 and references therein) (Fig. 8.2a, b). In members of section *Utricularia*, the embryo sac is aggressive and invades the placental nutritive tissue even at the 4-nucleate stage (Khan 1954; Farooq and Siddiqui 1964; Farooq 1965).

8.2.3 Syncytium Development

In *Utricularia*, the endosperm is ab initio cellular, and the mature endosperm consists of the endosperm proper and terminal haustoria (micropylar and chalazal) (Khan 1992). The development of syncytium in *Utricularia* species from section *Utricularia* (Figs. 8.3a, b and 8.4a–c) can be divided into several stages (see Wylie and Yocom 1923; Khan 1954; Płachno and Świątek 2011). Initially, a micropylar haustorium occupies the micropylar part of the former central cell and is two-nucleate, thus having direct contact with placental nutritive tissue. In the next stage, the cell walls separating the haustorium from the nutrient tissue brakes and syncytium formation starts. When the syncytium is formed, the cell walls of placental nutritive tissue cells gradually lose their structure (Fig. 8.4c), firstly in the cells closest to the former haustorium. In a fully formed syncytium, there is a complete absence of cell walls between the haustorium and the syncytium. In the cytoplasm of the mature syncytium, there are two giant, ameboid-shaped nuclei from the endospermal haustorium and many nutritive-tissue cell nuclei, which are much smaller (Fig. 8.4a, b). Khan (1954) reported more than 150 nuclei in the syncytia of *U. aurea*. Khan (1954) and Farooq (1964) suggested that the giant nuclei (from endospermal haustorium) may break into independent nuclei. Płachno and Świątek (2011) observed in *U. vulgaris* that the lobes of these giant nuclei were only connected by thin strands, which suggests a similar mechanism to that of *U. aurea* and *U. inflexa*. These authors proposed that the size of giant nuclei and their nucleoli and their irregular shape may indicate high ploidy. Thus, the structure of these nuclei and their gradual enlargement suggests the occurrence of endoreduplication cycles.

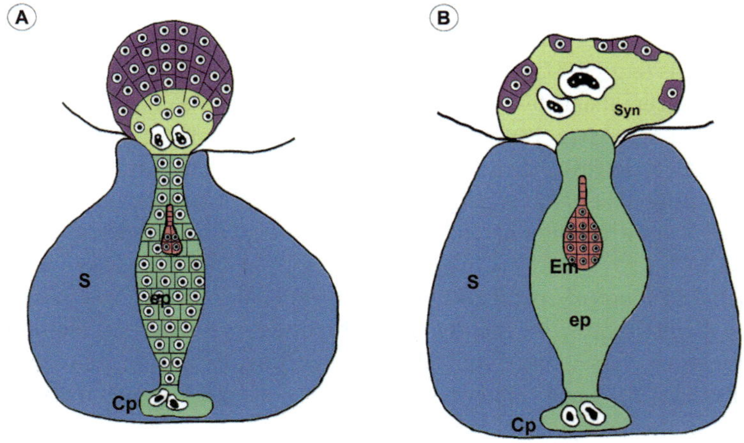

Fig. 8.3 (**a**, **b**) Schematic drawings of two developmental stages of young *Utricularia* seeds (according to Płachno et al. 2013, changed), showing relationships between the endosperm–placenta syncytium, chalazal haustorium, endosperm proper and embryo; placenta (P), seed (S, blue color), nutritive tissue (violet color), endosperm proper (ep, green color), embryo (Em, red color), syncytium (Syn, yellow color), Cp (chalazal endosperm haustorium)

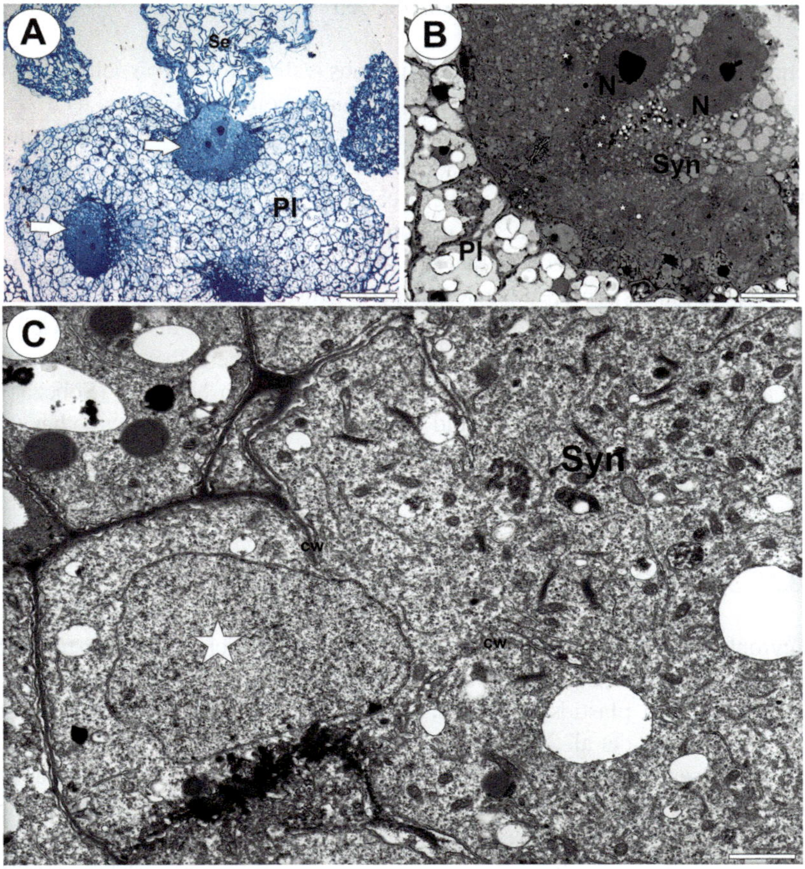

Fig. 8.4 *Utricularia minor*. (**a–c**) Syncytium structure. (**a**) Placenta (Pl) fragment showing two syncytia (arrows), seed (Se); Bar is equal to 100 μm. (**b**) Section through the syncytium; syncytium (Syn), giant endosperm nucleus (N), nucleus from the nutritive tissue cell (star), placenta (Pl); Bar is equal to 20 μm. (**c**) The peripheral part of the syncytium (Syn), where the protoplast of nutritive cell merge with the syncytium; digested nutritive tissue cell walls (cw), nucleus from nutritive tissue cell (star); Bar is equal to 1.2 μm. (**a**, **b**) Semithin sections stained with methylene blue; light microscopy image. (**c**) Transmission electron microscopy image

The syncytium functions during embryo development; once the embryo has accumulated nutrients, the syncytium fades away. Therefore, the syncytium should be regarded as a short-lived trophic structure serving as an organ that draws substances from the placenta to nourish the embryo.

8.2.3.1 Syncytium as "Super" Transfer Cell

Płachno and Świątek (2011) observed numerous wall ingrowths on cell walls of the syncytium in *U. intermedia*. Thus, this syncytium should be regarded as a "super" transfer cell. The transfer cells develop for the intensive short-distance transport between the symplast and apoplast (Gunning and Pate 1974; Offler et al. 2003; Offler and Patrick 2020). Thus, occurring cell wall ingrowths in syncytium supports the idea that this structure is active in nutrient transport between the placenta and endosperm + embryo. Especially that cell wall ingrowths are frequent in the various active trophic plant structures, e.g., endosperm haustoria (Nagl 1992; Świerczyńska et al. 2013a, b) and suspensors (Kozieradzka-Kiszkurno et al. 2012). Cell wall ingrowths also occur in nematode-induced syncytia, where there is an increased water and nutrient transport between syncytia and xylem elements (Sobczak and Golinowski 2009). However, it should be noted that cell wall ingrowths have been found in syncytia in only one *Utricularia* species, so only the future studies will determine whether all *Utricularia* syncytia are transfer cells.

8.2.3.2 Syncytium Ultrastructure and Organization

The cytoplasm of mature trophic plant cells is dense and rich in endoplasmic reticulum, active dictyosomes, microtubules, bundles of microfilaments, microbodies, mitochondria, and plastids (see, e.g., Nagl 1992; Kozieradzka-Kiszkurno et al. 2012; Świerczyńska et al. 2013a). Also, syncytia induced in plants during a nematode infection are characterized by a proliferation of cytoplasm, endoplasmic reticulum, ribosomes, and other organelles but also by a reduction of vacuoles (Sobczak and Golinowski 2009). In terms of these features, *Utricularia* syncytia are like other types of plant trophic cells and syncytia (Fig. 8.5) (Płachno and Świątek 2011; Płachno et al. 2011).

Many multinucleate plant structures are organized into regular nuclear cytoplasmic domains (Brown and Lemmon 2001; Baluška et al. 2004a, b; Nguyen et al. 2002). Using syncytia, Płachno et al. (2013) tried to verify the hypothesis that the size of the nuclear-cytoplasmic domains/cell bodies depends on the nuclei volume (see Baluška et al. 2004a, b, 2006). They found that, in *Utricularia* syncytium, there were different-sized cytoplasmic domains whose architecture depended on the source and size of the nuclei. Two giant nuclei, derived from endosperm haustorium, were surrounded by a three-dimensional microtubule cage, forming a huge cytoplasmic domain. However, at the periphery of the syncytium, where new protoplasts of the nutritive cells joined the syncytium, the microtubules formed a network, which surrounded small, nutritive tissue-derived nuclei. Comparing these results with other heterokaryotic coenocytes and syncytia from other plant species would be interesting.

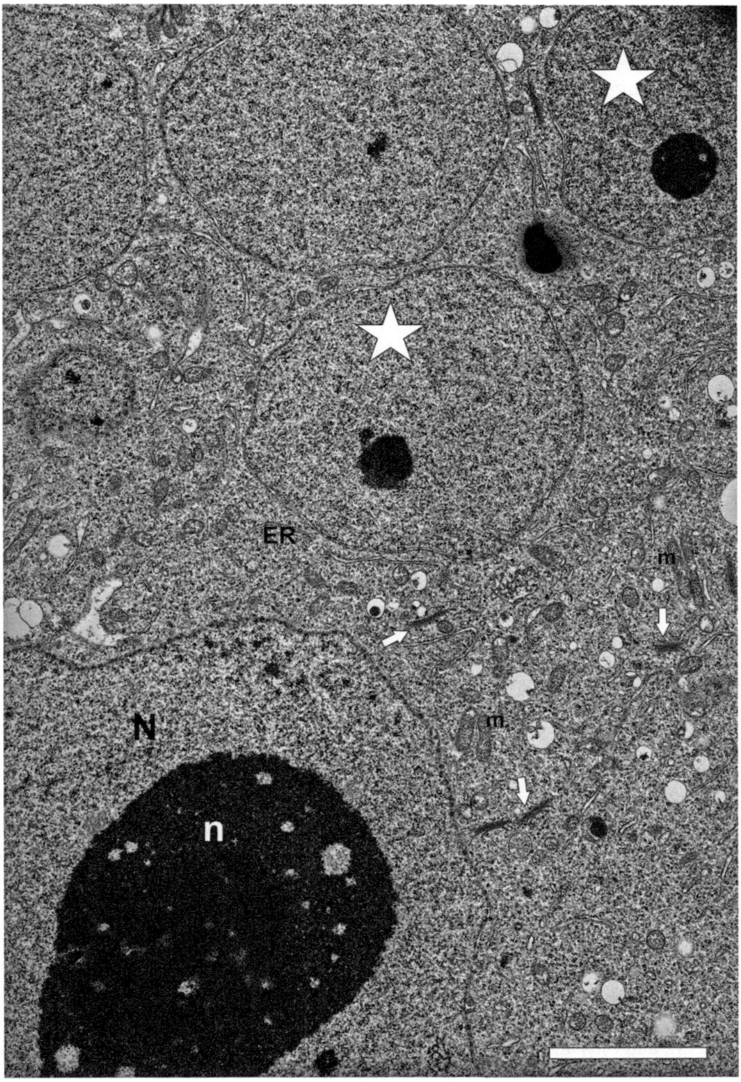

Fig. 8.5 *Utricularia minor*. Mature syncytium ultrastructure; giant endosperm nucleus (N), nucleolus (n), nucleus from the nutritive tissue cell (star), mitochondrion (m), dictyosome (arrow), endoplasmic reticulum (ER). Bar is equal to ~3 μm. Transmission electron microscopy image

8.3 Conclusions

According to Płachno and Świątek (2011) formation of a syncytium in *Utricularia* creates a huge surface for the exchange of nutrients between the placenta and the endosperm. This provides an economical way to redistribute cell components, to release nutrients from digested cell walls and fused protoplasts which can be used for the embryo. Thus, the formation of a syncytium in *Utricularia* is part of the specialized reproductive strategies of these unusual carnivorous plants and seems to be an apomorphic feature. Unfortunately, at the present state of the investigations, the available data do not provide a clear answer about the occurrence and evolution of *Utricularia* syncytia. Only future detailed embryological studies of various representatives of the various *Utricularia* sections will give the answer. *Utricularia* syncytia should be studied using modern techniques of three-dimensional imaging and molecular analyses, especially since genomes of two *Utricularia* species from section *Utricularia* were already sequenced (Ibarra-Laclette et al. 2013; Bárta et al. 2015).

Acknowledgments The chapter is dedicated to the 90th anniversary of Department of Plant Cytology and Embryology (Institute of Botany, Faculty of Biology, Jagiellonian University in Kraków). We would like to cordially thank Professor Małgorzata Kloc for invitation to "Syncytia: Origin, Structure, and Functions" volume. This research was partially supported financially by the Ministry of Science and Higher Education of Poland as part of the statutory activities of the Department of Plant Cytology and Embryology, Institute of Botany, Faculty of Biology, Jagiellonian University in Kraków (N18/DBS/000002), and the Department of Plant Cytology and Embryology, Faculty of Biology, University of Gdańsk (531-D030-D847-22).

References

Anger EM, Weber M (2006) Pollen-wall formation in *Arum alpinum*. Ann Bot (Lond) 97:239–244

Arekal GD, Nagendran CR (1975) Is there a Podostemum type of embryo sac in the genus *Farmeria*? Caryologia 28:229–235

Baluška F, Volkmann D, Barlow PW (2004a) Eukaryotic cells and their cell bodies: cell theory revised. Ann Bot Lond 94:9–32

Baluška F, Volkmann D, Barlow PW (2004b) Cell bodies in a cage. Nature 428:371

Baluška F, Volkmann D, Barlow PW (2006) Cell-cell channels and their implications for cell theory. In: Baluška F, Volkmann D, Barlow PW (eds) Cell-cell channels. Landes Bioscience-Springer, Berlin, pp 1–18

Bárta J, Stone JD, Pech J, Sirová D, Adamec L, Campbell MA, Štorchová H (2015) The transcriptome of *Utricularia vulgaris,* a rootless plant with minimalist genome, reveals extreme alternative splicing and only moderate sequence similarity with *Utricularia gibba*. BMC Plant Biol 15:78. https://doi.org/10.1186/s12870-015-0467-8

Brown RC, Lemmon BE (2001) The cytoskeleton and spatial control of cytokinesis in the plant life cycle. Protoplasma 215:35–49

Chaigne A, Brunet T (2022) Incomplete abscission and cytoplasmic bridges in the evolution of eukaryotic multicellularity. Curr Biol 32:385–R397

Esau K (1965) Plant anatomy. Wiley, London

Farooq M (1964) Studies in the Lentibulariaceae I. The embryology of *Utricularia stellaris* L. var. *inflexa* Clarke. Part II. Microsporangium, male gametophyte, fertilization, endosperm, embryo, and seed. Proc Natl Inst Sci India 30:280–299

Farooq M (1965) Studies in the Lentibulariaceae III. The embryology of *Utricularia uliginosa* Vahl. Phytomorphology 15:123–131

Farooq M, Siddiqui SA (1964) The embryology of *Utricularia stellaris* Linn. f. Sci Cult 30:394–395

Greenbaum MP, Iwamori T, Buchold GM, Matzuk MM (2011) Germ cell intercellular bridges. Cold Spring Harb Perspect Biol 3:a005850

Gunning BES, Pate JS (1974) Transfer cells. In: Robards AW (ed) Dynamic aspects of plant ultrastructure. McGraw-Hill, London, pp 441–476

Hagel JM, Yeung EC, Facchini PJ (2008) Got milk? The secret life of laticifers. Trends Plant Sci 13(12):631–639

Ibarra-Laclette E, Lyons E, Hernández-Guzmán G, Pérez-Torres CA, Carretero-Paulet L, Chang TH, Lan T, Welch AJ, Juárez MJ, Simpson J, Fernández-Cortés A, Arteaga-Vázquez M, Góngora-Castillo E, Acevedo-Hernández G, Schuster SC, Himmelbauer H, Minoche AE, Xu S, Lynch M, Oropeza-Aburto A, Cervantes-Pérez SA, de Jesús Ortega-Estrada M, Cervantes-Luevano JI, Michael TP, Mockler T, Bryant D, Herrera-Estrella A, Albert VA, Herrera-Estrella L (2013) Architecture and evolution of a minute plant genome. Nature 498(7452):94–98. https://doi.org/10.1038/nature12132

Jäger-Zürn I (1997) Embryological and floral studies in *Weddellina squamulosa* Tul. (Podostemaceae, Tristichoideae). Aquat Bot 57:151–182

Jankun A, Płachno B (2000) Embryology of *Utricularia vulgaris* L. Acta Biol Cracov Bot 42(1):23. Abstracts of the XXIV Conference on Embryology Plants, Animals, Humans May 24–26, 2000, Podlesice, Poland

Jobson RW, Baleeiro PC, Guisande C (2017) Systematics and evolution of Lentibulariaceae: III. *Utricularia*. In: Ellison A, Adamec L (eds) Carnivorous plants: physiology, ecology, and evolution (Oxford, 2017; online edn, Oxford Academic, 15 Feb. 2018). https://doi.org/10.1093/oso/9780198779841.003.0008. Accessed 20 Nov 2022

Kausik SB, Raju MVS (1955) A contribution to the floral morphology and embryology of *Utricularia reticulata*. Proc Indian Acad Sci 41:155–166

Kausik SD (1938) Pollen development and seed formation in *Utricularia caerulea*. Beih Bot Zbl 58A:365–378

Khan R (1954) A contribution to the embryology of *Utricularia flexuosa* Vahl. Phytomorphology 4:80–117

Khan R (1992) Lentibulariaceae. In: Johri BM, Ambegaokar KB, Srivastava PS (eds) Comparative embryology of angiosperms. Springer Verlag, Berlin, pp 755–762

Kloc M, Bilinski S, Dougherty MT, Brey EM, Etkin LD (2004) Formation, architecture and polarity of female germline cyst in *Xenopus*. Dev Biol 266(1):43–61

Kozieradzka-Kiszkurno M, Płachno BJ, Bohdanowicz J (2012) New data about the suspensor of succulent angiosperms: ultrastructure and cytochemical study of the embryo-suspensor of *Sempervivum arachnoideum* L. and *Jovibarba sobolifera* (Sims) Opiz. Protoplasma 249:613–624

Lang FX (1901) Untersuchungen über Morphologie, Anatomie und Samenentwicklung von *Polypompholyx* und *Byblis gigantea*. Flora 88:149–206

Merl EH (1915) Beiträge zur Kenntnis der Utricularien und Genlisen. Flora 108:127–200

Merz M (1897) Untersuchungen über die Samenentwicklung der Utricularien. Flora 84:69–87

Miranda VFO, Silva SR, Reut MS, Dolsan H, Stolarczyk P, Rutishauser R, Płachno BJ (2021) A historical perspective of bladderworts (*Utricularia*): traps, carnivory and body architecture. Plants 10(12):2656. https://doi.org/10.3390/plants10122656

Mukkada AJ (1962) Some observations on the embryology of *Dicraeea stylosa* Wight. In: Plant embryology—a symposium. CSIR, New Delhi, pp 139–145

Müller KF, Borsch T (2005) Phylogenetics of *Utricularia* (Lentibulariaceae) and molecular evolution of the trnK intron in a lineage with high substitutional rates. Plant Syst Evol 250:39–67

Murguía-Sánchez G, Alejandro Novelo R, Thomas Philbrick C, Márquez Guzmán GJ (2002) Embryo sac development in *Vanroyenella plumosa*, Podostemaceae. Aquat Bot 73:201–210

Nagl W (1992) The polytenic endosperm haustorium of *Rhinanthus minor* (Scrophulariaceae): functional ultrastructure. Can J Bot 70:1997–2004

Nguyen H, Brown RC, Lemmon BE (2002) Cytoskeletal organization of the micropylar endosperm in *Coronopus didymus* L. (Brassicaceae). Protoplasma 219:210–220

Offler CE, Patrick JW (2020) Transfer cells: what regulates the development of their intricate wall labyrinths? New Phytol 228:427–444

Offler CE, McCurdy DW, Patrick JW, Talbot MJ (2003) Transfer cells: cells specialized for a special purpose. Annu Rev Plant Biol 54:431–454

Pacini E, Juniper BE (1983) The ultrastructure of the formation and development of the amoeboid tapetum in *Arum italicum* Miller. Protoplasma 117:116–129

Pepling ME, De Cuevas M, Spradling AC (1999) Germline cysts: a conserved phase of germ cell development? Trends Cell Biol 9:257–262. https://doi.org/10.1016/S0962-8924(99)01594-9

Płachno BJ (2011) Female germ unit in *Genlisea* and *Utricularia,* with remarks about the evolution of the extraovular female gametophyte in members of Lentibulariaceae. Protoplasma 248:391–404. https://doi.org/10.1007/s00709-010-0185-x

Płachno BJ, Świątek P (2008) Cytoarchitecture of *Utricularia* nutritive tissue. Protoplasma 234:25–32. https://doi.org/10.1007/s00709-008-0020-9

Płachno BJ, Świątek P (2011) Syncytia in plants: cell fusion in endosperm-placental syncytium formation in *Utricularia* (Lentibulariaceae). Protoplasma 248(2):425–435. https://doi.org/10.1007/s00709-010-0173-1

Płachno BJ, Świątek P, Kozieradzka-Kiszkurno M (2011) The F-actin cytoskeleton in syncytia from non-clonal progenitor cells. Protoplasma 248:623–629

Płachno BJ, Świątek P, Sas-Nowosielska H, Kozieradzka-Kiszkurno M (2013) Organisation of the endosperm and endosperm-placenta syncytia in bladderworts (*Utricularia*, Lentibulariaceae) with emphasis on the microtubule arrangement. Protoplasma 250(4):863–873. https://doi.org/10.1007/s00709-012-0468-5

Płachno BJ, Świątek P, Jobson RW, Małota K, Brutkowski W (2017) Serial block face SEM visualization of unusual plant nuclear tubular extensions in a carnivorous plant (*Utricularia*, Lentibulariaceae). Ann Bot 120(5):673–680. https://doi.org/10.1093/aob/mcx042

Płachno BJ, Kapusta M, Świątek P, Banaś K, Miranda VFO, Bogucka-Kocka A (2021) Spatio-temporal distribution of cell wall components in the placentas, ovules and female gametophytes of *Utricularia* during pollination. Int J Mol Sci 22(11):5622. https://doi.org/10.3390/ijms22115622

Płachno BJ, Kapusta M, Stolarczyk P, Bogucka-Kocka A (2022) Spatiotemporal distribution of homogalacturonans and hemicelluloses in the placentas, ovules and female gametophytes of *Utricularia nelumbifolia* during pollination. Cells 11(3):475. https://doi.org/10.3390/cells11030475

Poppinga S, Daber LE, Westermeier AS et al (2017) Biomechanical analysis of prey capture in the carnivorous Southern bladderwort (*Utricularia australis*). Sci Rep 7:1776. https://doi.org/10.1038/s41598-017-01954-3

Reut M, Świątek P, Miranda VFO, Płachno BJ (2021) Living between land and water – structural and functional adaptations in bladderworts. Plant Soil 464:237–255. https://doi.org/10.1007/s11104-021-04929-6

Reut MS, Płachno BJ (2023) Development, diversity and dynamics of plant architecture in *Utricularia* subgenus *Polypompholyx* –towards understanding evolutionary processes in the Lentibulariaceae. Bot Rev 89:201–236. https://doi.org/10.1007/s12229-022-09283-5

Rudall PJ, Bateman RM (2019) Coenocytic growth phases in land plant development: a paleo-evo-devo perspective. Int J Plant Sci 180(6):607–622

Rutishauser R (2016) Evolution of unusual morphologies in Lentibulariaceae (bladderworts and allies) and Podostemaceae (river-weeds): a pictorial report at the interface of developmental biology and morphological diversification. Ann Bot 117:811–832. https://doi.org/10.1093/aob/mcv172

Shemer G, Podbilewicz B (2000) Fusomorphogenesis: cell fusion in organ formation. Dev Dyn 218:30–51

Shemer G, Podbilewicz B (2003) The story of cell fusion: big lessons from little worms. Bioessays 25(7):672–682

Siddiqui SA (1978) Studies in the Lentibulariaceae 9. Pollination, fertilisation, endosperm, embryo and seed in *U. dichotoma* Labill. Bot Jahrb Syst 100:237–245

Sobczak M, Golinowski W (2009) Structure of cyst nematode feeding sites. In: Berg RH, Taylor CG (eds) Cell biology of plant nematode parasitism. Plant Cell Monographs 15

Sobczak M, Matuszkiewicz M (2023) Syncytium induced by plant parasitic nematodes. Chapter 22

Swamy BGL, Ganapathy PM (1957) A new type of endosperm haustorium in *Nothapodytes foetida*. Phytomorphology 7:331–336

Świątek P, Urbisz AZ (2019) Architecture and life history of female germ-line cysts in clitellate annelids. In: Results and problems in cell differentiation, pp 515–551. https://doi.org/10.1007/978-3-030-23459-1_21

Świerczyńska J, Kozieradzka-Kiszkurno M, Bohdanowicz J (2013a) *Rhinanthus serotinus* (Schönheit) Oborny (Scrophulariaceae): immunohistochemical and ultrastructural studies of endosperm chalazal haustorium development. Protoplasma 250:1369–1380. https://doi.org/10.1007/s00709-013-0520-0

Świerczyńska J, Kozieradzka-Kiszkurno M, Bohdanowicz J (2013b) Developmental and cytochemical studies of the endosperm chalazal haustorium of *Rhinanthus serotinus* (Scrophulariaceae). Acta Biol Crac Ser Bot 55(1):99–106

Taylor P (1989) The genus *Utricularia*—a taxonomic monograph; Kew Bulletin Additional Series XIV; Royal Botanic Gardens, Kew: London, UK

Tiwari SC, Gunning BES (1986) Colchicine inhibits plasmodium formation and disrupts pathways of sporopollenin secretion in the anther tapetum of *Tradescantia virginiana* L. Protoplasma 133:115–128

Wylie R, Yocom AE (1923) The endosperm of *Utricularia*. U Iowa Stud Nat Hist 10:3–18

Part III
Fungal and Somatic Cell Syncytia and Genomic View of Extremophiles as the Ancestral Precursors of Eukaryotic Syncytia

Chapter 9
Syncytial Assembly Lines: Consequences of Multinucleate Cellular Compartments for Fungal Protein Synthesis

Alex Mayer, Grace McLaughlin, Amy Gladfelter, N. Louise Glass, Alexander Mela, and Marcus Roper

Abstract Fast growth and prodigious cellular outputs make fungi powerful tools in biotechnology. Recent modeling work has exposed efficiency gains associated with dividing the labor of transcription over multiple nuclei, and experimental innovations are opening new windows on the capacities and adaptations that allow nuclei to behave autonomously or in coordination while sharing a single, common cytoplasm. Although the motivation of our review is to motivate and connect recent work toward a greater understanding of fungal factories, we use the analogy of the assembly line as an organizing idea for studying coordinated gene expression, generally.

A. Mayer
Department of Mathematics, University of California Los Angeles, Los Angeles, CA, USA

G. McLaughlin
Department of Cell Biology, Duke University, Durham, NC, USA

Department of Biology, University of North Carolina Chapel Hill, Chapel Hill, NC, USA

A. Gladfelter
Department of Cell Biology, Duke University, Durham, NC, USA

N. L. Glass · A. Mela
Department of Plant and Microbial Biology, University of California Berkeley, Berkeley, CA, USA

M. Roper (✉)
Department of Mathematics, University of California Los Angeles, Los Angeles, CA, USA

Department of Computational Medicine, University of California Los Angeles, Los Angeles, CA, USA
e-mail: mroper@math.ucla.edu

M. Kloc, A. Uosef (eds.), *Syncytia: Origin, Structure, and Functions*, Results and Problems in Cell Differentiation 71,
https://doi.org/10.1007/978-3-031-37936-9_9

9.1 Introduction

Among the most diverse of all of the kingdoms of life, fungi can be found in every biosphere, from deep bedrock to atmospheric particulates (Kendrick 2017). Fungal syncytism—the ability of a single fungal mycelium to harbor tens, or even millions, of nuclei, bathed in a single, common cytoplasm—has shaped fungal life histories and created distinctive routes to phenotypic plasticity, reproduction, and dispersal (Roper et al. 2011). In particular, the tolerance of the fungal syncytia for internal genetic diversity, acquired via mutation or, more rarely, through a transfer of nuclei or genetic materials between genetically dissimilar mycelia, allows for levels of genetic experimentation that make pathogenic fungi formidable adversaries to their hosts and allows fungi to adapt in real-time to new niches and habitats. Their potential to rapidly adapt to new hosts inspired (Buxton 1960) to call fungi the mutable and treacherous tribe" (though Caten traces the phrase back even earlier (Caten 1996)). Internal diversity and fungal nuclear totipotence challenge definitions of what counts as a fungal individual and at what level in the hierarchy from gene to mycelium selection occurs (Buss 2014). However, almost 80 years ago, Pontecorvo, one of the pioneers of microbial genetics, predicted that fungal heterokaryosis would be both a challenge and an opportunity for the then nascent science of population genetics (Pontecorvo 1946). He argued that the fluid nuclear populations existing in a fungal syncytium had proportions that are set by the dynamics of competition and cooperation, intrinsic fitness, and drift (Pontecorvo 1946). Recent experiments promise to turn Pontecorvo's vision into a quantitative scientific model (Roper et al. 2011). In particular, live cell imaging of histone-labeled nuclei allows quantification of nuclear dispersal (Roper et al. 2013), while quantitative PCR allows measurement of the number of nuclei present from different populations and has been used to infer the presence of selection upon individual nuclear populations (Samils et al. 2014; Meunier et al. 2018) and to demonstrate that selfish nuclear populations can invade a syncytium (Bastiaans et al. 2014; Grum-Grzhimaylo et al. 2021). These results on the genetic dynamics of nuclei within fungal syncytia may provide a conceptual template for explaining the widespread appearance of syncytia across the Tree of Life (Skejo et al. 2021).

Although syncytism has profound consequences for fungal evolution, the subject of this review is the less scrutinized question of how syncytism has allowed fungi to evolve into remarkable factories for the production of secreted enzymes. Although secreted enzymes have direct importance in biotechnology, we will expand our focus to study nuclear coordination for protein production generally. Recent modeling has highlighted the importance of the distinctive form of syncytism adopted by fungi in enabling their prodigious cellular outputs (Heaton et al. 2020). The absence of cell walls between nuclei permits rapid translocation of resources and organelles across substrates and allows the labor involved in multistep synthesis pathways to be divided among many nuclei—creating syncytial assembly lines—with corresponding efficiency gains. Here we will discuss evidence and ideas from biological experiments and mathematical modeling, focusing on the adaptations which in a single mycelium allow nuclei to coordinate across an entire organism or to maintain autonomy. We will discuss the implications of sharing mRNAs or proteins between nuclei upon the amounts of proteins that can be expressed, the stability and predictability of expression, and of the amount of labor that each nucleus must perform.

Our work ties into an ancient story of microbial domestication: humans have utilized enzymes from domesticated filamentous fungi in food production since the beginning of written records. Reports of the use of kōji or qū molds (predominantly *Aspergillus oryzae* and other species) to produce soy sauce, miso, and fermented beverages date back at least as far as 300 BCE (Chen et al. 2022). Even further back, insects have practiced farming of filamentous fungi for some 50 million years, with the best known example of insect fungal-domestication being the farming of *Lepiotaceae* fungi by leaf-cutter ants (Chapela et al. 1994). The common motif in these systems is that filamentous fungi produce suites of secreted enzymes whose complexity and quantity are unmatched within the other kingdoms of life. The story of fungal domestication is largely one of humans and other animals discovering and exploiting these remarkable abilities.

Despite the likely importance of syncytial cell organization to fungal cellular factories and the growing utility these factories have found in biotechnology (Punt et al. 2002; Xiao and Zhong 2016; Lübeck and Lübeck 2022; Meyer et al. 2021), how nuclei sharing a common cytoplasm manifest different identities while dwelling in the same cell remains a mystery. The study of adaptations that enable nuclei to function autonomously or collectively and switch between such behaviors remains in its infancy. The purpose of this review is to describe the features of fungal syncytia that permit nuclei to marshal coordinated responses to global cues received by the mycelium, such as the fungivory, while simultaneously allowing individualized expression profiles, including the capacity of nuclei to divide autonomously of each other. A particular focus of our review is on how the sharing of mRNAs and proteins between nuclei increases efficiency by stabilizing protein abundances against noise and how the division of labor drastically reduces the burden of mRNA expression upon individual nuclei.

Our focus is on fungal syncytia; however, many of the adaptations identified for filamentous fungi, such as RNA-protein condensates, are also known to be present in syncytia from the other kingdoms of life and also in uninucleate cells. The ability of fungi to knit together syncytia over immense spatial scales, to build immortal networks that can stretch for meters and even kilometers, and to do so despite a radically fluid cellular environment (which includes nuclear and organelle translocation at mm/second and growth rates that can reach 10 μm/s) makes them a powerful tool for studying robust models of syncytial organization.

9.2 Heterogeneous Responses, Nuclear Autonomy, and Collective Behavior

Filamentous fungi can have hundreds to millions of nuclei residing within a continuous cytosol (Fig. 9.1a). These nuclei are highly dynamic, moving from a combination of cytoskeletal forces and fluid flow. As they are all bathed in a common cytosol, diffusible mRNAs and proteins can, theoretically, be widely exchanged between nuclei. Thus, both the cues that the nuclei receive from the cytosol and the macromolecules that each nucleus produces can be shared among many nuclei.

Despite a fungus' continuous cytosol, fungal syncytia are capable of heterogeneous responses, such that individual hyphae or parts of the mycelium may have very different expression profiles. In *Aspergillus niger*, different hyphal tips show widely separated levels of expression of the glucoamylase gene *glaA* (Vinck et al. 2005) and different RNA profiles generally (de Bekker et al. 2011). Closing of septal pores is one method that *Aspergillus* uses to maintain expression heterogeneity between hyphae (Bleichrodt et al. 2012). On a larger scale, microarray experiments in *N. crassa* mycelia show functional differentiation of different parts of the mycelium, through stark contrasts in mRNA abundance, that reflect functional specializations of different parts of the colony, such as commitment to growth or sporulation (Kasuga and Glass 2008). The Live Canvas project of Luis Larrondo and collaborators provides a complementary and visually arresting demonstration of the potential of different parts of an *N. crassa* mycelium to function autonomously of each other (Luis Larrondo, pers. comm.). In these experiments, luciferase reporting (Morgan et al. 2003) is used to monitor the circadian time of an *N. crassa* mycelium. Masks

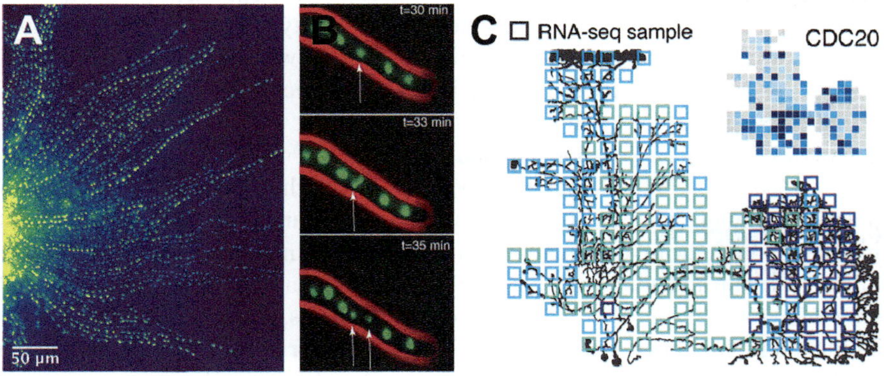

Fig. 9.1 Multinucleated fungal mycelia and slime molds exhibit heterogeneity in expression levels and behavior. (**a**) The filamentous fungi *Ashbya gossypii* contains hundreds of nuclei in a common cytosol. (**b**) *Ashbya* nuclei divide asynchronously from immediate neighbors despite residing in a common cytosol (Gladfelter et al. 2006). (**c**) Although slime molds display coordinated nuclear division, a significant degree of heterogeneity can be observed across the plasmodium for various genes, according to spatially resolved RNA-seq analyses. For instance, the clustering of CDC20 expression level reveals distinct expression patterns even among neighboring nuclei (top right) (Gerber et al. 2022)

applied to the mycelium allow only the targeted regions of the mycelium to be exposed to light. After this exposure, regions of the mycelium inside and outside of the mask maintain out of phase rhythms of luciferase expression, with some parts bright and others dark, and the relative roles of the two parts alternating every 12 h. Using this method, Larrondo uses the fungus as an artistic canvas on which to recreate images.

Where the distribution of nuclear products has been examined in fungal-like organisms, it appears that syncytial cells can divide labor for some processes while simultaneously sharing the labor for others. Thus, heterogeneity or independent nuclear behaviors can exist for some tasks, while simultaneously other processes are performed collectively by all nuclei. Although not fungi, slime molds have morphological similarities to filamentous fungi with large numbers of nuclei flowing through complex networks of interconnected tubes. The slime mold *Physarum polycephalum* was an early workhorse of cell cycle research because of its dramatically synchronized nuclear division cycles, which might suggest a uniform sharing of goods and products (Sudbery and Grant 1976). However, single nuclear sequencing has revealed a much more complex story (Gerber et al. 2022): in fact, even nearby nuclei may be transcribing distinct clusters of genes, indicating that in addition to being shared among many nuclei, tasks can be subdivided among individual nuclei within the same syncytium (Fig. 9.1c). Taken together, these data indicate that free exchange and compartmentalization of molecules must be selective, with some pathways segregated into different nuclei while others freely mix.

The cell cycle provides a window on the level of coordination of gene expression among nuclei across different syncytium-forming species and systems. In early cell fusion experiments, Rao and Johnson (1970) exposed HeLa cells to Sendai virus, inducing them to fuse. By fusing cells in different cell cycle stages, they showed that sharing cytoplasm quickly synchronized nuclear division among the individual nuclei present within these synthetic syncytia. Synchronized nuclear phases also occur in binucleate *Saccharomyces cerevisiae* cells carrying mutations that prevent pairs of nuclei originating from a single mitosis from separating into mother and daughter cell compartments (see Fig. 7 of Hoepfner et al. 2002). The synchronous nuclear division also occurs in natural syncytia, including in network-forming slime molds as noted above (Nygaard et al. 1960). In the slime mold *Physarum polycephalum*, a fusion of two plasmodiums started at different cell cycle stages rapidly synchronizes when protoplasmic continuity has been established (Loidl and Sachsenmaier 1982). A variant of complete synchronization, sometimes called parasynchronization, occurs in the filamentous fungus *Aspergillus nidulans*, in which waves of mitosis are initiated at the hyphal tips and then propagate subapically (Clutterbuck 1970). Synchrony of nuclear division also appears to dominate in insect embryos such as *Drosophila* which have rapid waves of coordinated mitosis in early development (Foe et al. 1993; Hayden et al. 2022). In fact, the appearance of nuclear-division synchrony in syncytia was foundational to understanding that cell division was controlled by diffusible factors in the cytosol. The factors were eventually discovered to be small cyclin proteins, which push cells toward mitosis by activating cyclin-dependent kinases (CDKs). Cell cycle states are controlled by families of small proteins that should be diffusible enough to pass between and synchronize adjacent nuclei in the same cytosol.

In the model ascomycetes *Neurospora crassa* and *Ashbya gossypii*, nuclear division is asynchronous, showing that nuclei may be isolated from each other in the absence of septa (Fig. 9.1b). Specifically, in *A. gossypii*, careful reconstruction of nuclear division sequences, both spatially and across nuclear lineages, has shown that mother-daughter nuclei retain some correlation in time of division, but more distantly related nuclei have uncorrelated division times, even when they neighbor each other within the fungal cytoplasm (Anderson et al. 2013). The syncytium's capacity for nuclear asynchronicity is even more remarkable, because *Ashbya gossypii* is genetically highly similar to brewer's yeast, a cell cycle model, and regulates its cell cycle in a similar manner, through a cyclin/CDK biochemical oscillator (Gladfelter et al. 2006).

Intriguingly, the asynchronicity of nuclear division is not necessarily absolute, but may vary depending on the mycelium's environment. For example, although the nuclei of *N. crassa* divide asynchronously early in mycelial development (Roca et al. 2010), mitosis in mature mycelia is entrained to the fungus's circadian clock, like much of *Neurospora*'s cellular machinery (Vitalini et al. 2006; Sancar et al. 2015). Hong et al. showed that in mycelia grown under conditions of variable light or temperature, more than half of nuclei entered mitosis during the late evening hours, a significant increase from the rest of the day (Hong et al. 2014). This pattern of nighttime division likely ensures that the nuclear abundances anticipate the demands of the fungus's characteristic growth rhythm: expanding the mycelium through hyphal tip extension during the daytime and packing nuclei into profuse conidia (asexual spores) during nighttime (Hong et al. 2014). Hence, while nuclei in *Neurospora* do not completely synchronize their phases, the circadian clock can at least partially coordinate their divisions over the entire network. Conversely, in *Aspergillus nidulans*, which exhibits parasynchronous or wave-synchronized mitoses when grown under optimal conditions, synchrony is lost when the fungus is grown on nutrient-poor media (Rosenberger and Kessel 1967). Since the fungus' growth and conidiation rates are slowed on nutrient-poor substrates, it is possible that *A. nidulans* transitions to an asynchronous state to avoid drastic variability in the nuclear-to-cytoplasmic ratio, which occurs when the mycelium rapidly transitions from n nuclei to $2n$ nuclei.

Although we are not aware of experimental measurements of fitness impacts of synchronous or asynchronous mitosis, it is likely that the ability of fungal syncytia to toggle between asynchronous and synchronous divisions confers adaptive advantages to fungal syncytia. Synchronous division requires the syncytium to tolerate large variations and rapid changes in nuclear-cytoplasmic ratio and may be inefficient, since all nuclei must commit resources to cell cycle initiation and nuclear-synthesis at the same time. But, synchronous cell cycles may be beneficial by allowing all nuclei to respond to homogeneous external stimuli in unison. Conversely, autonomous nuclei enable large syncytia to respond locally to external stimuli, such as spatially heterogeneous soil or environmental conditions that may modulate rates of cell expansion on fine scales. But, for synchronous and asynchronous division to be modifiable behaviors, fungi much have a mechanism for remodeling cytoplasm to allow or limit the diffusion of proteins or mRNAs that control cell cycle.

9.3 Partitioning of the Cytoplasm

Individualized nuclear behaviors require that resources and nuclear products remain controllably localized near the sites of their production. In the higher fungi (*dikarya*), internal cell walls, or septa, divide the protoplasmic space into smaller compartments (Fig. 9.2a) (Kendrick 2017). Septa are often porous, however, allowing adjacent compartments to exchange signals and materials (Pieuchot et al. 2015). Specialized structures act to rapidly close the pores, allowing them to act as bulkheads that limit outflow of protoplasm if the cell wall is breached (Dhavale and Jedd 2007), or as part of the normal aging and divestment of resources from subapical compartments (Bleichrodt et al. 2015), or as a precursor to cell death in a compartment in which the vegetative incompatibility response has been triggered (Glass and Dementhon 2006).

Although septa perform in the same way as cell membranes by creating physical barriers between populations of nuclei, the ability of nuclei in the same compartment and only a few microns apart to undergo cell cycle asynchronously implies

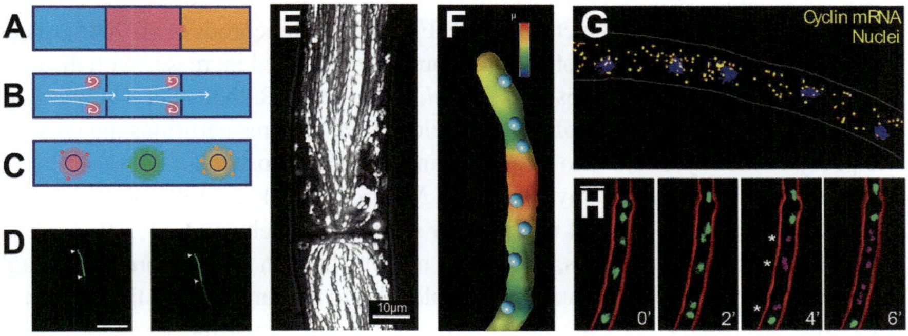

Fig. 9.2 Mechanisms of compartmentalizing fungal cells. (**a**) Septal pores can create distinct compartments within hyphae. Here, the left septum is closed, restricting the flow of cytoplasm and diffusion of macromolecules between the left and middle compartments. (**b**) Mechanical forces, such as eddies formed by cytoplasmic flow, can create locally distinct pockets of cytosol (pink). (**c**) Liquid-liquid phase separation creates membraneless compartments near nuclei. (**d**) The closure of septal pores completely blocks the diffusion of cytoplasmic GFP into upstream compartments (left). When the pore is open, its diffusion is considerably restricted (right). Bar = 100 μm (Bleichrodt et al. 2015). (**e**) Fluid flow within *Neurospora* hyphae creates eddies near the septa, which function to create locally distinct regions of cytosol (Pieuchot et al. 2015). Image provided courtesy of Laurent Pieuchot. (**f**) Particle tracking of genetically expressed nanoparticles revealed heterogeneity in cytoplasmic crowding along *Ashbya* hyphae (McLaughlin et al. 2020). (**g**) Within *Ashbya*, CLN3 mRNA (a G1 cyclin) is clustered non-randomly around nuclei by the RNA binding protein Whi3 (Lee et al. 2013). (**h**) Deleting Whi3 leads to more synchronous mitosis, as illustrated by the actively dividing nuclei in magenta (scale: 5 μm) (Lee et al. 2013)

that fungal syncytia also have the ability to localize cell cycle control molecules, and likely other mRNAs and proteins, within the protoplasm (Fig. 9.2c). Localization of these molecules means overcoming their tendency to diffuse due to random, thermal fluctuations or active stirring of the cytoplasm due to motor proteins (Lin et al. 2016) and due to bulk protoplasmic flows associated with growth (Abadeh and Lew 2013).

The relative importance of the two mechanisms (random mixing and bulk flow) in preventing localization is mathematically encapsulated into a quantity called the Pèclet number, *Pe*, which is made up of the bulk cytoplasmic velocity, *U*; typical length scale (e.g., hyphal diameter), *a*; and the diffusivity, *D*, of the species of molecule under consideration: $Pe = Ua/D$. If $Pe > 1$ (large distances or flow rates), then bulk flow is the dominant challenge to keeping molecules localized, whereas if $Pe < 1$ (large diffusivities), then diffusion is the greater challenge (Roper et al. 2015). The dependence of molecular diffusivity upon size means that diffusion may be the dominant force of delocalization for small molecules (large *D*) such as metabolites or ions, but flows may dominate the delocalization of mRNAs or large proteins (small *D*).

In fungi with moderate growth speeds, such as *Aspergillus* spp., proteins may diffuse along the length of cytoplasmic compartments. Septal pores, even open, provide a significant barrier to free diffusion of molecules: Bleichrodt et al. used FRAP experiments to measure the diffusion of cytoplasmic GFP in *Aspergillus nidulans* mycelia, and pore closure effectively eliminated all diffusion across a pore, and even open pores severely constrained protein diffusion, reducing the effective diffusivity of the GFP by a factor of 500 (Fig. 9.2d) (Bleichrodt et al. 2015). They also found that the likelihood of a septal pore being plugged increases with distance from the extending hyphal tips of *A. nidulans* mycelia. Rather than a single connected cytoplasm, the system of open and closed septal pores partitions the growing periphery of the mycelium into a set of island-like neighborhoods.

In the fast-growing ascomycete fungi *N. crassa*, bulk cytoplasmic flows can range from microns to 100s of microns per second (Abadeh and Lew 2013). These flows readily carry organelles, including nuclei, through septal pores. Although pores do not limit advective transport of molecules to the same degree that they limit diffusion, macromolecules present in a downstream compartment cannot diffuse upstream through a pore (Roper et al. 2015), explaining how signals can be passed so that apical compartments receive cues originating upstream, but not conversely (Roper et al. 2015).

The pore itself carves out a neighborhood within the syncytium: the rush of cytoplasm through the small septal pore pushes on the cytoplasm trapped in the corners of the compartment between hyphal wall and septum and sets it wheeling into a slowly turning eddy (Pieuchot et al. 2015) (Fig. 9.2b, e). Nuclei and other organelles may become trapped within the eddy, accumulating there for minutes before being swept through the pore. Eddy-trapped nuclei accumulate SPA-19, a protein involved in maintaining the septum (Pieuchot et al. 2015). Flow-maintained compartments offer a solution to the problem of functional localization in syncytia with high Peclet numbers (e.g., fast growing mycelia), because the hydrodynamic forces keeping nuclei trapped increase in proportion to the flow speeds. Fungal mycelia have complex and anisotropic geometries, with branching points, tips, and bends all acting as other potential locations of fluid-flow driven eddies that could trap nuclei, though we are not aware of any systematic effort to identify trapping regions outside of septal pore-eddies.

Septa create insulated functional territories containing handfuls to hundreds of nuclei, but evidence of nuclear asynchrony indicates that insulation can be achieved within a continuous cytoplasm. Through tracking of genetically expressed nanoparticles, the cytosol along *Ashbya* hyphae was shown to be highly heterogeneous, with higher levels of crowding observed locally around nuclei in a cell cycle-dependent manner (Fig. 9.2f) (McLaughlin et al. 2020). Lee et al. showed that *Ashbya* G1 cyclin mRNAs are held close to nuclei through phase separation occurring due to RNAs binding with a partner protein (Fig. 9.2g) (Lee et al. 2015; Zhang et al. 2015). Mutants lacking the RNA-binding protein have a significantly more synchronous nuclear division (Fig. 9.2h). The RNA-protein ("RNP") condensates are attached to the ER, helping anchor them in the vicinity of nuclei (Snead et al. 2022). Unlike compartments produced via cell membranes, RNP-condensates are fluid, able to change both their sizes and their locations. Moreover, two condensates can coalesce if brought into contact (Langdon et al. 2018). The physicochemical foundations of how compartments stably form and how compartment specificity is maintained (i.e., how non-target mRNAs are excluded from condensates) remain subjects of active investigation, but RNA secondary structure has been shown to exclude cyclin-*CLN3* mRNAs from droplets containing septin-*BNI1* mRNAs, even though the two mRNA-species condense with the same protein partner (Whi3) (Langdon et al. 2018).

9.4 Coordinating Across the Mycelium

In the last section, we focused on methods for limiting communications between nuclei, identifying adaptations that allow individual nuclei or groups of nuclei to behave autonomously within a continuous protoplasm. However, response to stimuli from the environment and efficient utilization of resources often require that fungi marshal responses of many nuclei require that cues, metabolites, and resources be swiftly translocated over long distances within the mycelium.

As with uninucleate cells, the cytosol surrounding each nucleus carries cues from the environment outside the mycelium. However, unlike uninucleate cells, the local cytosol also contains dispersible molecules produced by the other nuclei present within the syncytium. The extent to which nuclei readout these signals is a topic of ongoing research. Recent work has highlighted the diversity of nuclear transport proteins present in the membranes of *Aspergillus nidulans* nuclei, enabling the nuclei to sample their surrounding cytosol (Markina-Iñarrairaegui et al. 2011). In addition to identifying specific functions for these proteins in this and other fungal species, an open question is whether and how a given nucleus can distinguish the cues or signals it receives from its neighbors from the signals it produces itself.

Although we are not aware of experiments seeking to answer this question, the problem is held in common with social unicellular organisms. For example, the social amoeba *Dictyostelium* spp. uses self-generated cAMP gradients to migrate toward common focii (Gerisch and Wick 1975). Oscillatory expression of cAMP produces traveling waves of the chemoattractant, allowing individual cells to "listen" for a collectively generated signal with a distinctive time character (Schaap and Wang 1986). Prior to the formation of a single connected mycelium, clonal *N. crassa* germlings also use time varying signals to coordinate their homing growth (Fleissner et al. 2009): each pair of attracting cells expresses synchronized, but out of phase, levels of the proteins *MAK-2* and *SO*, which has been likened to making one chemoattractant while listening for the other one (Fleissner et al. 2009). Advances in single nucleus RNA-profiling may produce evidence that syncytial nuclei are using one or both mechanisms to solve the highly similar problems of communication.

The establishment and maintenance of polar-growth provide a useful paradigm for understanding both cytoplasmic partitioning and nuclear coordination. In addition to requiring nuclear division, hyphal growth requires a constant traffic of vesicles containing hyphal wall proteins to each growing tip (Steinberg 2007); these vesicles are trafficked via motor proteins, along microtubules, which is a mechanism for relatively short-distance translocation. Additionally, the phenomenal rate of growth of some fungi (e.g., *N. crassa*, growing at 1 μm/s) seems too large for a single sub-apical nucleus to supply all of the cell wall needed for growth. We infer that multiple nuclei close to the hyphal tip are likely expressing cell wall proteins. How are these nuclei marshalled? Ca^{2+} gradients are known to be primary players in maintaining polarized growth (Brand et al. 2007), and recent innovations in genetically encoded Ca^{2+} indicators allow for visualization of the polarity gradients (Kim et al. 2021). It would be very interesting to explore the allometry between these gradients and the rate of growth of the tip, as a first step toward understanding whether they play a role, also, in nuclear coordination.

Many fungal mycelia can extend themselves essentially indefinitely, so long as they continue to encounter new resources to maintain their growth (Stajich et al. 2009). In these species, mycelial expansion is a form of foraging in which new nutrient sources or hosts are continuously integrated into the syncytium, and resources for growth, ingested in one location, must be transported to others. Communication of changing conditions across the mycelium may allow the fungus to react more effectively to emerging threats, such as fungivory, or to encounters with new hosts or resource patches.

For both resource movement and communications, the mycelium itself serves as a highly efficient transportation network, in which either motors or bulk protoplasmic flow carry resources through the mycelium (Roper et al. 2015), along with nuclei themselves (Fig. 9.3a).

Work on foraging basidiomycete fungi has revealed some of the physical optimization principles that shape the growth of mycelial transportation networks. Foraging basidiomycete fungi travel through leaf litter and soil and can link carbon sources or hosts for meters or kilometers (Dowson et al. 1989). Since the fungus has no prior information about the locations of carbon sources, network organization

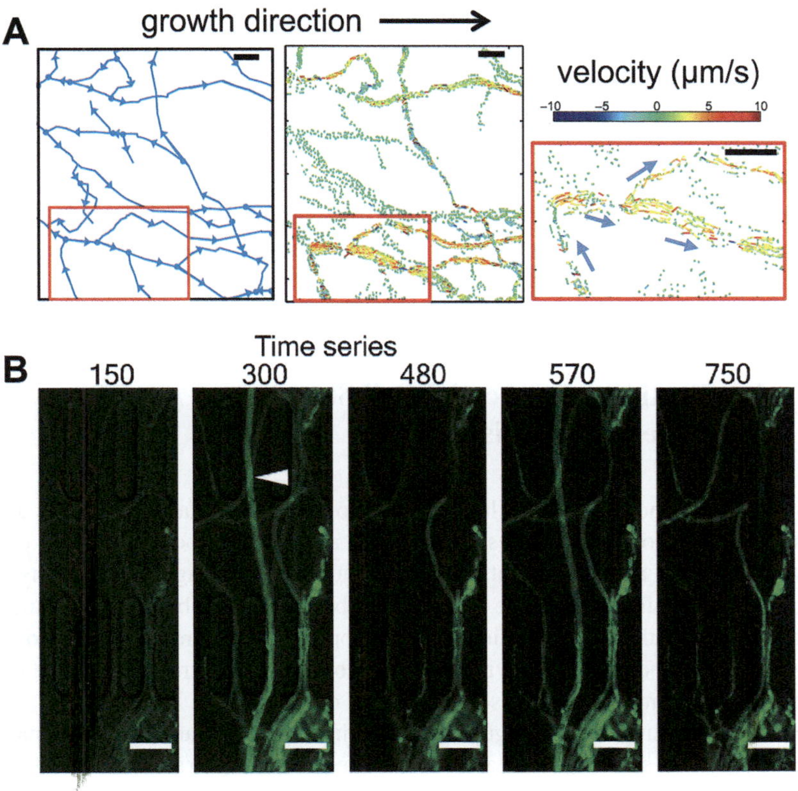

Fig. 9.3 Transport networks in filamentous fungi. (**a**) Cytoplasmic flow in *N. crassa* hyphae. On the left, arrows indicate the direction of nuclear flows as individual nuclei stream toward hyphal tips. In the center, the nuclear trajectories from the same region over a 4-second period are displayed with line segments indicating their displacements over 0.2-second intervals and color-coded based on their velocity in the direction of growth. On the right, a subsample of nuclear displacements is shown in a magnified region of the image, accompanied by the mean flow direction in each hypha indicated by blue arrows (Roper et al. 2013). (**b**) A *C. cinerea* mycelium embedded in a microfluidic device designed by Schmeider et al. (2019). The signal of a glucose analog, 2-NBDG, exhibits periodic fluorescence in specific hyphae, which are labeled by the white arrow. Here, time is in seconds, and the scale bar is 20 μm

must be adaptable and indeterminate. Once sources are found, the network rapidly integrates them into an extended network that is efficient in its use of materials, energetically thrifty in the sense that the amount of work that it does in transporting resources is kept as low as possible and resilient to damage (Heaton et al. 2012).

Although there has been extensive study of fungal network architectures, much less data exists on how cytoplasmic traffic is managed on a network scale. Schmieder et al. (2019) used microfluidic devices to tease apart the hyphae of *Coprinopsis cinerea*. The fungi grew through microfluidic chambers that served as ports in which a fluorescently labeled glucose analog, or fungivorous nematode worms, could be introduced. The microfluidic setup enabled the observation of the spread of the glucose analog and the activation of defense-response genes induced by fungivory at the scale of individual hyphae (Fig. 9.3b). The study showed that the mycelial network operates like a human-built highway system, with most transport occurring within trunk hyphae. Intriguing cytoplasmic transport within trunk hyphae oscillates between phases of flow toward the mycelial periphery and flow toward the interior. The physical mechanism by which flow can switch direction remains undiscovered; protoplasmic flows typically follow growth (Abadeh and Lew 2013), bidirectional protoplasmic flows may require that the fungus alternate growth in two directions (Roper and Dressaire 2019).

9.5 Implications of Syncytial Cell Structure for Protein Synthesis

It is clear that syncytia exploit global integration and coordinated responses as well as regional and localized processes, with the ability to harbor both kinds of processes in the cell and switch between these modes depending on conditions. How can we begin to understand the mechanistic benefits of global or local modes of coordination? In these next sections we develop several theoretical frameworks for considering the benefits of a syncytial state for conserving resources, sustainable production, and minimizing noise in synthesis or signaling.

Filamentous fungi are distinguished for their use as cellular factories, capable of synthesizing and secreting a wealth of enzymes at an industrial scale (Lübeck and Lübeck 2022). A central question in the study of global gene expression in syncytia is how this vast output of enzymes is achieved while still maintaining consistent expression of housekeeping proteins, which are essential for basic cell function. One potential evolutionary advantage of syncytism is through the economical division of transcriptional labor. In Hausser et al. (2019), analysis of high-throughput sequencing data in several model organisms reveals a strong evolutionary bias toward combining low transcription and high translation to achieve a given protein abundance, which they attribute to cells being adapted to minimize the energetic costs of mRNA transcription. Another possible advantage toward infrequent transcription may come from reducing transcriptional interference, which occurs when the transcription of one gene interferes with the transcription of an independently regulated gene (Shearwin et al. 2005). However, transcription rates are generally prevented from being driven too low, because small mRNA copy numbers are more strongly affected by expression noise, leading to larger fluctuations in protein abundance (Raser and O'shea 2005). In our recent modeling work, we have studied whether syncytism allows for even lower floors on mRNA copy numbers and more stochastic, less frequent transcription because the dispersion of proteins between nuclear neighborhoods smooths out fluctuations in single nucleus expression. In this section, we present a simple mathematical model of gene expression in multinucleated cells and illustrate how nuclear cooperation for common goods may emerge spontaneously through stochastic transcription.

Mathematical models for gene expression vary extensively in scope and methodology. In this review, we focus on simplified *mass-action kinetics-based* models, which involve modeling the synthesis, decay, and binding of biochemical components, to form a dynamical picture of the gene network. Typically, mass-action models incorporate the transcription of RNAs, translation of proteins, and their corresponding degradation as single-step reactions that occur with certain propensities. In uninucleate gene expression models, the cytoplasm is generally assumed to be well-mixed, which allows for gene networks to be studied purely temporally. The well-mixed assumption necessitates that gene products are highly diffusive in their cytoplasmic environment, allowing for spatial heterogeneities to be smoothed out. The nucleus may additionally be modeled as a well-mixed compartment within the cell, in which case each participating molecule will be prescribed nuclear import/export rates. This added layer of complexity is necessary when modeling feedback mechanisms such as the binding of transcription factors that regulate circadian clocks (Forger and Peskin 2003; Gonze et al. 2003).

When extending gene expression models from uninucleate cells to multinucleate syncytia, the partitioning of mRNAs and proteins becomes an important model element. To model this partitioning, we must measure or make assumptions about the spatial organization of nuclei in the cytosol, as well as the extent to which they enjoy transcriptional autonomy. Indeed, nuclear spacing is highly regulated in fungi; in *A. gossypii*, microtubule and motor protein-assisted repulsion between neighboring nuclei space them regularly through the syncytium (Anderson et al. 2013). Even in syncytia in which nuclei are rapidly moved by protoplasmic flows, eddy currents can lead to formation of stable nuclear aggregates. For these reasons, we treat the cytosol as being divided into distinct *nuclear neighborhoods*, which may function as "cells-within-cells" (Gladfelter et al. 2006).

In syncytia, nuclei can exhibit different degrees of nuclear cooperation and nuclear autonomy. Nuclear cooperation refers to the situation where nuclei within the syncytium express the same set of genes, potentially in a synchronous fashion, and work together to control the overall function of the syncytium. Conversely, in the case of transcriptional autonomy, nuclei are shielded from the influence of one another and exhibit distinct transcriptional programs (Youn et al. 2010).

Nuclear cooperation may emerge spontaneously, without the need for explicit programming or direct communication between nuclei. Even in genotypically identical nuclei, transcription of RNAs will generally not be precisely coordinated, due to stochastic effects, and resulting in heterogeneity in gene product levels between nuclear compartments, which is exacerbated in the case of transcriptional bursting, where mRNAs are synthesized in infrequent pulses rather than individually. However, due to the homogenizing ability of molecular diffusion, the sharing of proteins between nuclear compartments may suppress noise in compartmental protein abundance, providing the time scale of diffusion is sufficiently fast. To illustrate this idea, we consider an elementary model for syncytial gene expression, which consists of a linear arrangement of N nuclear compartments, which we assume to be uniform in size (Fig. 9.4). This model resembles the typical morphology of hyphal compartments in *A. gossypii*, which consists of $N \approx 8$–10 nuclei spaced $L \approx 2$–5 μm apart (Anderson et al. 2013; Kaufmann and Philippsen 2009). A protein, modeled as a Brownian particle with diffusivity D, will diffuse half a cell length $NL/2$ in time $\Delta\tau_p = N^2L^2/4\pi^2D$, at which point there is roughly an even probability that it occupies any of the N compartments. Taking the protein GFP for example, which has diffusivity $D \approx 30\ \mu m^2/s$ in eukaryotic cytoplasm (Mika et al. 2014), the homogenization time in a hyphal compartment of *A. gossypii* is $\Delta\tau_p \approx .75s$. As the kinetics of gene expression often operates on the time scale of minutes (Hausser et al. 2019), we may make the limiting assumption that at any given time t, the total protein content $p(t)$ is disseminated with uniform probability between nuclear compartments. In Mayer et al. (preprint), the magnitude of relative fluctuations of protein abundance in a compartment in this model is found to have the approximate scaling

$$CV^2\left(p_{comp}\right) = \frac{CV^2\left(p_{mono}\right)}{N}$$

(see Fig. 9.5), where p_{mono} is the distribution of protein in a mononucleated cell with identical expression kinetics and

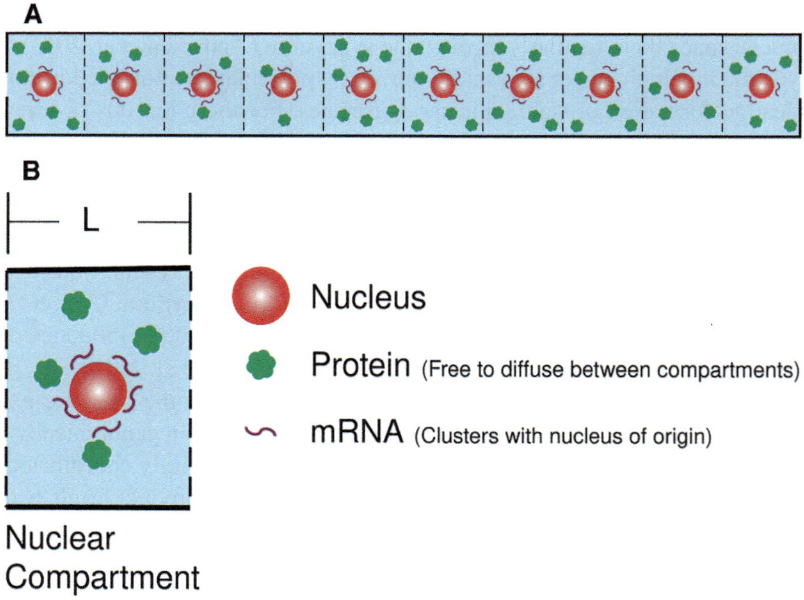

Fig. 9.4 Modeling mRNA and protein distributions in a syncytial cell. (**a**) A linear arrangement of N identical nuclear compartments. Proteins can freely diffuse between compartments, whereas mRNAs are confined to their nucleus of origin. (**b**) A single nuclear compartment with characteristic length L

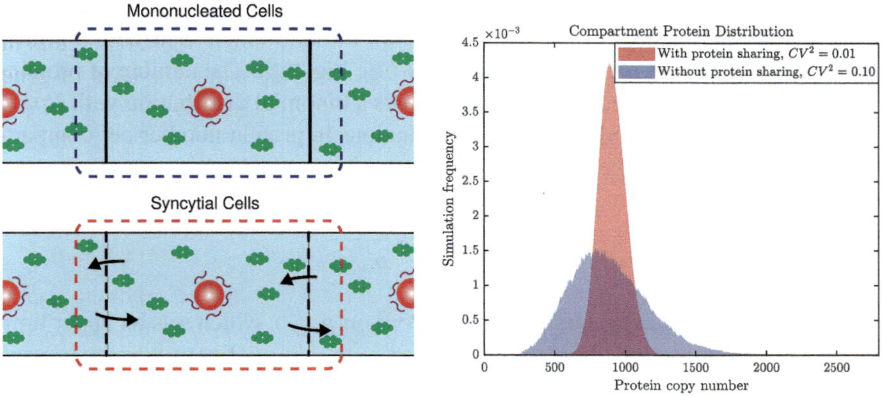

Fig. 9.5 The sharing of proteins in syncytial cells permits consistent protein expression under identical transcriptional kinetics. Left panel: in mononucleated cells (top), global protein abundance strictly depends on the synthesis rates of mRNA in the nucleus and translation rates in the cytoplasm. In syncytial cells (bottom), proteins are free to diffuse into other nuclear neighborhoods, and typically protein diffusive time scales are substantially faster than the gene expression time scales of transcription, translation, and decay. Right panel: Stochastic simulations derived from standard and syncytial models of gene expression reveal that spatial averaging of diffusive proteins results in a tighter distribution of local proteins in nuclear neighborhoods of syncytial cells. In the syncytial model, we simulate the sharing of proteins among ten nuclei, all of which have identical transcription kinetics. In the limit of rapid diffusion, CV^2 is reduced by a factor of 1/10 compared to the scenario where protein sharing is prohibited

$$CV^2\left(p_{mono}\right) \approx \frac{\tau_B}{\tau_m + \tau_p}$$

Here τ_m is the average mRNA lifetime, τ_p is the average protein lifetime, and τ_B is the average waiting time between transcriptional bursts. As transcription is often an energetic bottleneck in gene circuits (Hausser et al. 2019), this result suggests that the syncytia permit more stochastic transcription (i.e., larger and more infrequent bursts) before reaching a compartmental noise ceiling driven by selection. Preliminary analysis of smFISH data in *A. gossypii* on the cell cycle-associated genes encoding cln2 and clb2 supports this hypothesis revealing potentially divergent modes of transcription to analogous genes in the genetically similar *S. cerevisiae* (Mayer et al.).

While the sharing of protein between nuclei can act to suppress fluctuations due to noisy transcription, it introduces a degree of fluctuations via the random diffusion of proteins away from their nuclei of origin. This effect is most pronounced when nuclei are each independently expressing identical numbers of proteins, which may occur if the protein's abundance is tightly regulated by nondiffusive molecules such as RNAs (Singh 2011). In our syncytial model, we assume each nucleus expresses a target protein copy number of p_{target} which provides maximum fitness benefit per nuclear neighborhood. Given that the N compartments are uniform, the total number of protein in the syncytium is given by $p_{tot} = Np_{target}$. At steady state, the probability that a single protein occupies a particular compartment is simply $1/N$. Therefore, the distribution of proteins in a syncytium follows a multinomial distribution with a uniform compartment occupancy probability (see Fig. 9.6). The number of proteins a particular nucleus has access to, p_{comp}, follows a binomial distribution with Np_{target} trials and success probability $1/N$. The fluctuations in protein number per compartment is then given by

$$CV^2\left(p_{comp}\right) = \frac{1 - 1/N}{p_{target}}$$

It can be shown that for large N, $p_{comp} \sim \text{Poisson}(p_{target})$, which sets an upper limit on the noise contribution of diffusion to $CV^2(p_{comp}) = 1/p_{target}$. For proteins expressed at high levels, i.e., on the order of 10^2 molecules/nucleus, this degree of noise is negligible. Although little data exists on whole proteome abundances in *A. gossypii*, if it follows its near relative, the yeast *S. cerevisiae*, for which 90% of proteins are expressed with median levels over 822 per cell, then we expect protein partitioning noise to be negligible.

What about low abundance proteins, such as transcription factors? For such proteins, random diffusions can lead to very uneven distributions through the syncytium. However, even when instant-to-instant distributions of proteins may be very uneven, function may only be impaired if the protein level in a given nuclear neighborhood is persistently too low or too high (Wang and Zhang 2011). If the

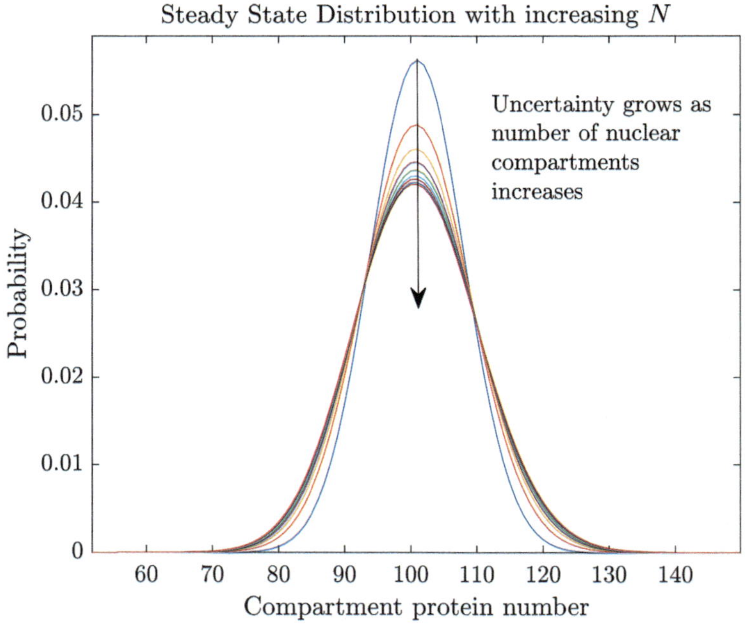

Fig. 9.6 Compartmental protein abundance is subject to Poissonian diffusive noise in the limit where protein synthesis is deterministic (i.e., cellular protein abundance fixed)

time-varying protein in each nuclear neighborhood is $p(t)$, then the phenotype associated with the protein likely reflects the time averaged abundance of the protein. Supposing this time averaging occurs over a characteristic time scale T (e.g., the time taken for protein recruitment to a particular organelle, which is often on the order of seconds or 10s of seconds), then

$$p_T = \frac{1}{T}\int_{t_0}^{t_0+T} p(t)\,dt$$

Although $p(t)$ fluctuates on a time scale τ_P, if these fluctuations are rapid, meaning that $\tau_P \ll T$, then deviations from the mean are suppressed (see Fig. 9.7).

9.6 Circadian Rhythm Through the Lens of Nuclear Coordination

Although much biochemical interest in fungal syncytia is associated with secreted proteins, due to their many-fold uses in biotechnologies, specific pathways for secreted protein production are usually not understood at the level of individual nuclei. By contrast, the circadian clock in *N. crassa* is likely to provide an essential case study for understanding space and time-resolved patterns of nuclear

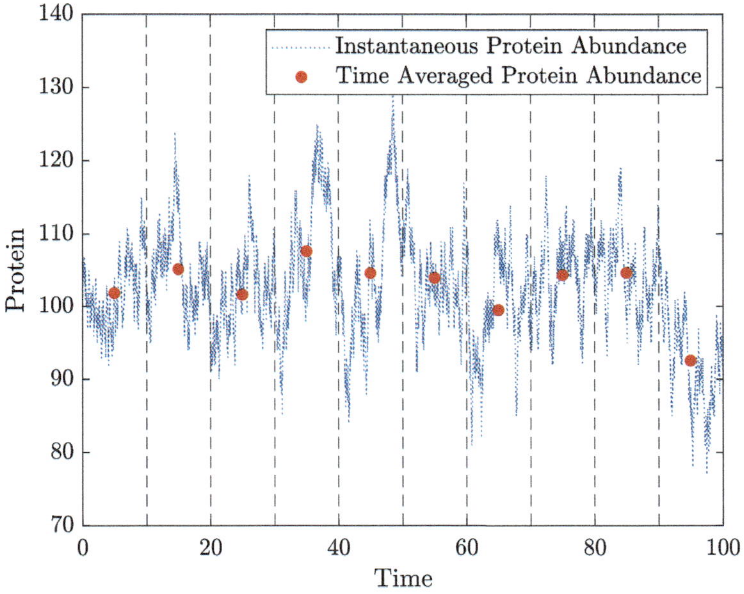

Fig. 9.7 A stochastic time series of compartmental protein abundance in syncytial gene expression model with finite diffusion rate. Averaging over characteristic time periods reduces the perceived magnitude of protein copy number fluctuations

coordination, because of recent innovations both in developing carefully parameterized models for the expression pathways that drive the clock and innovative new cell manipulation tools allowing the contributions of individual nuclei to be systematically examined. *N. crassa* is a common model organism for studying circadian rhythm, as many of the molecules and regulatory motifs involved in its circadian rhythm are conserved in other organisms, including mammals (Dunlap and Loros 2017; Ko and Takahashi 2006). The fundamental mechanism of the circadian clock in *N. crassa* involves two interdependent feedback loops—one positive and one negative—that pivot on the phosphorylation-mediated interactions between the White Collar Complex (WCC) and the protein *frq*. In its active state, the WCC is a transcriptional activator for *frq*, promoting the synthesis of *frq* mRNAs. Within the cytoplasm, the *frq* protein associates with the FRQ-interacting RNA helicase (FRH) and the casein kinase-1a (CK1), culminating in the formation of the FRQ-FRH complex (FFC), which drives the incremental phosphorylation of *frq* and ultimately directs its degradation. During this process, the FCC may be imported into the nucleus, where it interacts with and phosphorylates the WCC, eventually leading to its inactivation (see Fig. 9.8a) (Dunlap et al. 2007). Due to the tractability of its principal molecular components, data-calibrated mathematical models have been created to analyze the self-sustaining and entrainable characteristics of the circadian clock (Upadhyay et al. 2019; Bellman et al. 2018), as well as to model the circadian-linked oscillations in metabolism (Dovzhenok et al. 2015) and cell cycle (Hong et al. 2014).

The synchronization of circadian clocks is a fundamental characteristic of circadian rhythms in populations of cells. Circadian clocks in uninucleate cells distributed through a tissue typically exhibit phase synchronization (Sgro et al. 2015), but stochastic variation in gene expression hinders this synchronization (Deng et al. 2016).

Cheong et al. (2022) used fluorescent imaging of an mCherry labeled clock associated promoter to study the spatial scales over which synchronization among *N. crassa* nuclei that were not cytoplasmically connected: they packed unfused *N. crassa* conidial cells into two specialized microfluidic devices, a large chamber that simulated an artificial tissue (Fig. 9.8b) and a microwell device which inhibited cell-to-cell contact. Cells that were packed into microwells of increasing density showed higher levels of synchronization, evidencing the presence of a quorum-sensing molecule that can diffuse between individual cells. Based on the time scales over which cells became synchronized, they inferred this molecule must be protein-sized. Although only minimally syncytial (since each conidium may contain one, two, or even up to 10 nuclei), the ability to track the emergence of synchrony between nuclear neighborhoods in syncytia is a tantalizing prospect for understanding how circadian clocks are assembled.

Microfluidic tools are likely to play a key role in allowing individual nuclear activities to be monitored. A recent study by Lee et al. (2016) introduced a novel method for tracking single nuclei over circadian periods in *N. crassa*. To overcome the challenge posed by the fungus's rapid growth rate, which could cause nuclei to migrate outside of the imaging region, the researchers utilized a microfluidic device

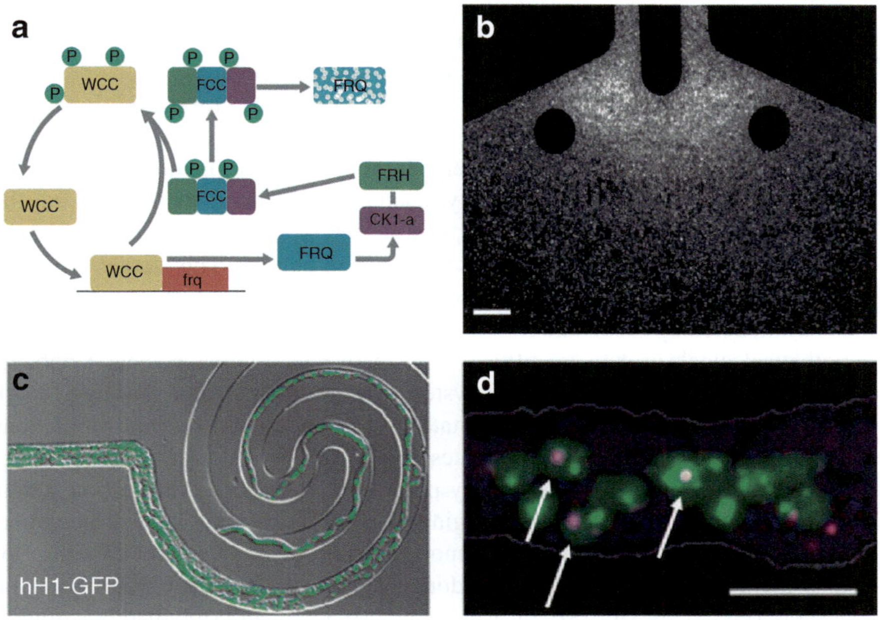

Fig. 9.8 The circadian clock of *N. crassa* offers valuable context for understanding nuclear coordination in syncytial cells. (**a**) A simplified diagram of the Neurospora circadian clock. When active, the White Collar Complex (WCC) upregulates FRQ expression. FRQ forms a component of the FRQ-FRH complex (FFC), which may inactivate WCC through a series of phosphorylations. (**b**) Cheong et al. (2022) employed a large chamber microfluidic device to investigate the synchronization of circadian clocks in densely packed conidial cells of *N. crassa*. In the study, fluorescence intensity was used to represent the "hand" of the circadian clock. Scale bar: 50 μm. (**c**) A microfluidic design proposed by Lee et al. (2016) to perform single nucleus tracking over long time periods in hyphal *N. crassa* cells. (**d**) Bartholomai et al. (2022) performed smFISH on *N. crassa* to localize the circadian clock associated mRNA frq at various time points. Maximum-intensity Z projections of composite images of frq (depicted in magenta) and nuclei (depicted in green) are presented, with arrows indicating nuclei that contain colocalized frq transcripts (scale bar, 5 μm)

with a spiral design that effectively held nuclei within specified regions of interest (see Fig. 9.8c). smFISH (single-molecule fluorescence in situ hybridization), already exploited in yeast, and to study fungal cell cycle, is a powerful and flexible tool for creating static snap shots of the coronae of mRNAs surrounding each nucleus. Bartholomai et al. (2022) used it to quantify and locate *frq* mRNA at various points in the circadian cycle. They discovered regular clustering of *frq* mRNA near nuclei (Fig. 9.8d), mediated by interactions with the RNA-binding protein PRD-2. Surprisingly, *frq* mRNA levels were found to not exceed single digits per nucleus during a cycle, and only a small fraction of nuclei was found to be actively transcribing at any given time, despite being synchronized to the same oscillator. These low copy numbers are quantitatively discordant with existing mathematical models that tend to require much higher transcription rates. They also suggest that circadian rhythm may be one of the first direct test beds for the general result, given in the previous section, that syncytia can stably and economically operate with more stochastic and lower transcription rates than uninucleate cells.

To probe whether syncytism allows for robust circadian clocks with low, noisy transcription rates, Zinn-Brooks and Roper (2021) employed a syncytial gene expression model for linked nuclear neighborhoods, akin to the model outlined in Fig. 9.4. This simplified model only encompassed the negative arm of the transcription-translation feedback loop, simplifying the circuit to the level of a solitary protein controlling its own transcriptional inhibition. This single feedback model closely mimics the circadian oscillations of *per* proteins observed in mammals, as modeled by Wang and Peskin (2018), but with the added capability of the hypothetical clock protein to diffuse freely between different nuclear neighborhoods. Stochastic simulations on this system demonstrated that the sharing of clock proteins between neighborhoods facilitated precise and synchronized oscillations with a period of 24 h at transcription rates nearly 1000-fold slower than a uninucleate system with comparable oscillatory precision. As discussed in Section 5, this noise insensitivity is due to the averaging of proteins between nuclear neighborhoods. Importantly, the mathematical models exposed how the labor of transcription is divided between nuclei; a readout that is hard to obtain from the static snapshots offered by smFISH measurements. At low transcription rates, often a single nucleus was responsible for synthesizing the majority of mRNAs for the entire syncytial compartment, and this energy-saving division of labor occurred spontaneously and without additional mechanisms for coordination existing outside of the negative feedback loop.

9.7 Perspectives

In this review, we have highlighted specific adaptive characteristics that are present in fungal syncytia, allowing nuclei to coexist in a shared cytoplasm. The diverse mechanisms employed by filamentous fungi to compartmentalize their cytosol and transport resources allow for independent nuclear behavior when required, such as with asynchronous cell division, while also facilitating coordinated responses to environmental stimuli. An essential trait found in syncytial cells is the ability to allocate the workload of RNA synthesis among multiple nuclei. We have termed this a "syncytial assembly line," based on our presented mathematical model of gene expression in syncytia that indicates that nuclei can achieve the necessary copy number precision for a specific gene, even with increased transcriptional stochasticity, by sharing proteins across a large, continuous cytosol. The efficiency gains resulting from this division of transcriptional labor could explain the remarkable enzymatic output of filamentous fungi, enabling them to thrive in challenging environments and serve as a valuable resource for industrial protein production.

The themes of nuclear cooperation in syncytia explored in this review are not limited to filamentous fungi. Windner et al. (2019) outline a top-down system of nuclear regulation in *Drosophila* muscle cells, beginning with the global scaling of muscle cell size and number of nuclei conserved across cell types. As in some filamentous fungi, nuclei are distributed uniformly throughout the cell via microtubule repulsion (Deshpande et al. 2021), demarcating cytoplasmic territories which contain nuclei which may vary considerably in size and DNA content. In spite of this heterogeneity, the authors suggest that the diffusion of the nucleolar protein such as Fibrillarin across neighborhood boundaries may perform a critical role in homogenizing transcriptional activity, by regulating nucleolar-nuclear volume ratios. This mechanism suggests that the sharing of proteins between cytoplasmic territories can coordinate nuclear activity, which can help to smooth out expression heterogeneity at the cell level (Windner et al. 2019). This global coordination of nuclei is important in tissues such as mammalian myofibers, where the nuclei work in concert to control muscle contraction (Dos Santos et al. 2020).

Although we have highlighted circadian rhythm as a natural first place for bringing together mathematical modeling with experimental observations, we hypothesize that the sharing of gene products between compartments in syncytial fungi may contribute to robust gene expression in a variety of cellular processes. Further research and mathematical modeling in this area is necessary to fully understand the complex mechanisms underlying fungal syncytial behavior and to identify novel strategies for enhancing protein production and optimizing biotechnological processes.

References

Abadeh A, Lew RR (2013) Mass flow and velocity profiles in Neurospora hyphae: partial plug flow dominates intra-hyphal transport. Microbiology 159(Pt 11):2386–2394

Anderson CA, Eser U, Korndorf T, Borsuk ME, Skotheim JM, Gladfelter AS (2013) Nuclear repulsion enables division autonomy in a single cytoplasm. Curr Biol 23(20):1999–2010

Bartholomai BM, Gladfelter AS, Loros JJ, Dunlap JC (2022) Prd-2 mediates clock-regulated perinuclear localization of clock gene RNAs within the circadian cycle of Neurospora. Proc Natl Acad Sci 119(31):e2203078119

Bastiaans E, Aanen D, Debets A, Hoekstra R, Lestrade B, Maas M (2014) Regular bottlenecks and restrictions to somatic fusion prevent the accumulation of mitochondrial defects in Neurospora. Philos Trans R Soc B: Biol Sci 369(1646):20130448

Bellman J, Kim JK, Lim S, Hong CI (2018) Modeling reveals a key mechanism for light-dependent phase shifts of Neurospora circadian rhythms. Biophys J 115(6):1093–1102

Bleichrodt R-J, van Veluw GJ, Recter B, Maruyama J-I, Kitamoto K, Wösten HA (2012) Hyphal heterogeneity in Aspergillus oryzae is the result of dynamic closure of septa by woronin bodies. Mol Microbiol 86(6):1334–1344

Bleichrodt R-J, Hulsman M, Wösten HA, Reinders MJ (2015) Switching from a unicellular to multicellular organization in an Aspergillus niger hypha. MBio 6(2):e00111–e00115

Brand A, Shanks S, Duncan VM, Yang M, Mackenzie K, Gow NA (2007) Hyphal orientation of Candida albicans is regulated by a calcium-dependent mechanism. Curr Biol 17(4):347–352

Buss LW (2014) The evolution of individuality, vol 796. Princeton University Press

Buxton E (1960) Heterokaryosis, saltation and adaptation. Plant Pathol 2(S 359):405

Caten C (1996) The mutable and treacherous tribe revisited. Plant Pathol 45(1):1–12

Chapela IH, Rehner SA, Schultz TR, Mueller UG (1994) Evolutionary history of the symbiosis between fungus-growing ants and their fungi. Science 266(5191):1691–1694

Chen W, Lv X, Tran V-T, Maruyama J-I, Han K-H, Yu J-H (2022) From traditional to modern: progress of molds and yeasts in fermented-food production. Front Microbiol 13

Cheong JH, Qiu X, Liu Y, Al-Omari A, Griffith J, Schüttler H-B, Mao L, Arnold J (2022) The macroscopic limit to synchronization of cellular clocks in single cells of Neurospora crassa. Sci Rep 12(1):6750

Clutterbuck A (1970) Synchronous nuclear division and septation in Aspergillus nidulans. Microbiology 60(1):133–135

de Bekker C, Bruning O, Jonker MJ, Breit TM, Wösten HA (2011) Single cell transcriptomics of neighboring hyphae of Aspergillus niger. Genome Biol 12:1–12

Deng Z, Arsenault S, Caranica C, Griffith J, Zhu T, Al-Omari A, Schüttler HB, Arnold J, Mao L (2016) Synchronizing stochastic circadian oscillators in single cells of Neurospora crassa. Sci Rep 6(1):1–18

Deshpande O, De-Carvalho J, Vieira DV, Telley IA (2021) Astral microtubule cross-linking safeguards uniform nuclear distribution in the Drosophila syncytium. J Cell Biol 221(1):e202007209

Dhavale T, Jedd G (2007) The fungal woronin body. Biol Fungal Cell:87–96

Dos Santos M, Backer S, Saintpierre B, Izac B, Andrieu M, Letourneur F, Relaix F, Sotiropoulos A, Maire P (2020) Single-nucleus RNA-seq and fish identify coordinated transcriptional activity in mammalian myofibers. Nat Commun 11(1):5102

Dovzhenok AA, Baek M, Lim S, Hong CI (2015) Mathematical modeling and validation of glucose compensation of the Neurospora circadian clock. Biophys J 108(7):1830–1839

Dowson C, Springham P, Rayner A, Boddy L (1989) Resource relationships of foraging mycelial systems of Phanerochaete velutina and Hypholoma fasciculare in soil. New Phytol 111(3):501–509

Dunlap JC, Loros JJ (2017) Making time: conservation of biological clocks from fungi to animals. Microbiol Spectr 5(3):5–3

Dunlap J, Loros J, Colot H, Mehra A, Belden W, Shi M, Hong C, Larrondo L, Baker C, Chen C-H et al (2007) A circadian clock in Neurospora: how genes and proteins cooperate to produce a sustained, entrainable, and compensated biological oscillator with a period of about a day. In: Cold Spring Harbor Symposia on Quantitative Biology, vol 72. Cold Spring Harbor Laboratory Press, pp 57–68

Foe VE, Odell GM, Edgar BA (1993) Mitosis and morphogenesis in the Drosophila embryo: point and counterpoint. In: Bate M, Martinez Arias A (eds) The development of Drosophila Melanogaster. Cold Spring Harbor Laboratory Press, pp 149–300

Fleissner A, Leeder AC, Roca MG, Read ND, Glass NL (2009) Oscillatory recruitment of signaling proteins to cell tips promotes coordinated behavior during cell fusion. Proc Natl Acad Sci 106(46):19387–19392

Forger DB, Peskin CS (2003) A detailed predictive model of the mammalian circadian clock. Proc Natl Acad Sci 100(25):14806–14811

Gerber T, Loureiro C, Schramma N, Chen S, Jain A, Weber A, Weigert A, Santel M, Alim K, Treutlein B et al (2022) Spatial transcriptomic and single-nucleus analysis reveals heterogeneity in a gigantic single-celled syncytium. Elife 11:e69745

Gerisch G, Wick U (1975) Intracellular oscillations and release of cyclic amp from Dictyostelium cells. Biochem Biophys Res Commun 65(1):364–370

Gladfelter AS, Hungerbuehler AK, Philippsen P (2006) Asynchronous nuclear division cycles in multinucleated cells. J Cell Biol 172(3):347–362

Glass NL, Dementhon K (2006) Non-self recognition and programmed cell death in filamentous fungi. Curr Opin Microbiol 9(6):553–558

Gonze D, Halloy J, Leloup J-C, Goldbeter A (2003) Stochastic models for circadian rhythms: effect of molecular noise on periodic and chaotic behaviour. Comptes rendus biologies 326(2):189–203

Grum-Grzhimaylo AA, Bastiaans E, van den Heuvel J, Berenguer Millanes C, Debets AJ, Aanen DK (2021) Somatic deficiency causes reproductive parasitism in a fungus. Nat Commun 12(1):783
Hausser J, Mayo A, Keren L, Alon U (2019) Central dogma rates and the trade-off between precision and economy in gene expression. Nat Commun 10(1):68
Hayden L, Hur W, Vergassola M, Di Talia S (2022) Manipulating the nature of embryonic mitotic waves. Curr Biol 32(22):4989–4996
Heaton L, Obara B, Grau V, Jones N, Nakagaki T, Boddy L, Fricker MD (2012) Analysis of fungal networks. Fungal Biol Rev 26(1):12–29
Heaton LL, Jones NS, Fricker MD (2020) A mechanistic explanation of the transition to simple multicellularity in fungi. Nat Commun 11(1):2594
Hoepfner D, Schaerer F, Brachat A, Wach A, Philippsen P (2002) Reorientation of mispositioned spindles in short astral microtubule mutant spc72 δ is dependent on spindle pole body outer plaque and kar3 motor protein. Mol Biol Cell 13(4):1366–1380
Hong CI, Zámborszky J, Baek M, Labiscsak L, Ju K, Lee H, Larrondo LF, Goity A, Chong HS, Belden WJ et al (2014) Circadian rhythms synchronize mitosis in Neurospora crassa. Proc Natl Acad Sci 111(4):1397–1402
Kasuga T, Glass NL (2008) Dissecting colony development of Neurospora crassa using mRNA profiling and comparative genomics approaches. Eukaryot Cell 7(9):1549–1564
Kaufmann A, Philippsen P (2009) Of bars and rings: Hof1-dependent cytokinesis in multiseptated hyphae of Ashbya gossypii. Mol Cell Biol 29(3):771–783
Kendrick B (2017) The fifth kingdom. Hackett Publishing
Kim H-S, Kim J-E, Hwangbo A, Akerboom J, Looger LL, Duncan R, Son H, Czymmek KJ, Kang S (2021) Evaluation of multicolor genetically encoded Ca2+ indicators in filamentous fungi. Fungal Genet Biol 149:103540
Ko CH, Takahashi JS (2006) Molecular components of the mammalian circadian clock. Hum Mol Genet 15(suppl 2):R271–R277
Langdon EM, Qiu Y, Ghanbari Niaki A, McLaughlin GA, Weidmann CA, Gerbich TM, Smith JA, Crutchley JM, Termini CM, Weeks KM et al (2018) mRNA structure determines specificity of a polyQ-driven phase separation. Science 360(6391):922–927
Lee C, Zhang H, Baker AE, Occhipinti P, Borsuk ME, Gladfelter AS (2013) Protein aggregation behavior regulates cyclin transcript localization and cell-cycle control. Dev Cell 25(6):572–584
Lee C, Occhipinti P, Gladfelter AS (2015) PolyQ-dependent RNA–protein assemblies control symmetry breaking. J Cell Biol 208(5):533–544
Lee KK, Labiscsak L, Ahn CH, Hong CI (2016) Spiral-based microfluidic device for long-term time course imaging of Neurospora crassa with single nucleus resolution. Fungal Genet Biol 94:11–14
Lin C, Schuster M, Guimaraes SC, Ashwin P, Schrader M, Metz J, Hacker C, Gurr SJ, Steinberg G (2016) Active diffusion and microtubule-based transport oppose myosin forces to position organelles in cells. Nat Commun 7(1):11814
Loidl P, Sachsenmaier W (1982) Control of mitotic synchrony in Physarum polycephalum. Phase shifting by fusion of Peterophasic plasmodia contradicts a limit cycle oscillator model. Eur J Cell Biol 28(2):175–179
Lübeck M, Lübeck PS (2022) Fungal cell factories for efficient and sustainable production of proteins and peptides. Microorganisms 10(4):753
Markina-Iñarrairaegui A, Etxebeste O, Herrero-García E, Araújo-Bazán L, Fernández-Martínez J, Flores JA, Osmani SA, Espeso EA (2011) Nuclear transporters in a multinucleated organism: functional and localization analyses in Aspergillus nidulans. Mol Biol Cell 22(20):3874–3886
McLaughlin GA, Langdon EM, Crutchley JM, Holt LJ, Forest MG, Newby JM, Gladfelter AS (2020) Spatial heterogeneity of the cytosol revealed by machine learning-based 3d particle tracking. Mol Biol Cell 31(14):1498–1511
Meunier C, Hosseini S, Heidari N, Maryush Z, Johannesson H (2018) Multilevel selection in the filamentous ascomycete Neurospora tetrasperma. Am Nat 191(3):290–305

Meyer V, Cairns T, Barthel L, King R, Kunz P, Schmideder S, Müller H, Briesen H, Dinius A, Krull R (2021) Understanding and controlling filamentous growth of fungal cell factories: novel tools and opportunities for targeted morphology engineering. Fungal Biol Biotechnol 8(1):8

Mika JT, Schavemaker PE, Krasnikov V, Poolman B (2014) Impact of osmotic stress on protein diffusion in Lactococcus lactis. Mol Microbiol 94(4):857–870

Morgan LW, Greene AV, Bell-Pedersen D (2003) Circadian and light-induced expression of luciferase in Neurospora crassa. Fungal Genet Biol 38(3):327–332

Nygaard OF, Güttes S, Rusch HP (1960) Nucleic acid metabolism in a slime mold with synchronous mitosis. Biochim Biophys Acta 38:298–306

Pieuchot L, Lai J, Loh RA, Leong FY, Chiam K-H, Stajich J, Jedd G (2015) Cellular subcompartments through cytoplasmic streaming. Dev Cell 34(4):410–420

Pontecorvo G (1946) Genetic systems based on hetero caryosis. In: Cold Spring Harbor Symposia on Quantitative Biology, vol 11. Cold Spring Harbor Laboratory Press, pp 193–201

Punt PJ, van Biezen N, Conesa A, Albers A, Mangnus J, van den Hondel C (2002) Filamentous fungi as cell factories for heterologous protein production. Trends Biotechnol 20(5):200–206

Rao PN, Johnson RT (1970) Mammalian cell fusion: studies on the regulation of DNA synthesis and mitosis. Nature 225:159–164

Raser JM, O'shea EK (2005) Noise in gene expression: origins, consequences, and control. Science 309(5743):2010–2013

Roca MG, Kuo H-C, Lichius A, Freitag M, Read ND (2010) Nuclear dynamics, mitosis, and the cytoskeleton during the early stages of colony initiation in Neurospora crassa. Eukaryot Cell 9(8):1171–1183

Roper M, Dressaire E (2019) Fungal biology: bidirectional communication across fungal networks. Curr Biol 29(4):R130–R132

Roper M, Ellison C, Taylor JW, Glass NL (2011) Nuclear and genome dynamics in multinucleate ascomycete fungi. Curr Biol 21(18):R786–R793

Roper M, Simonin A, Hickey PC, Leeder A, Glass NL (2013) Nuclear dynamics in a fungal chimera. Proc Natl Acad Sci 110(32):12875–12880

Roper M, Lee C, Hickey PC, Gladfelter AS (2015) Life as a moving fluid: fate of cytoplasmic macromolecules in dynamic fungal syncytia. Curr Opin Microbiol 26:116–122

Rosenberger R, Kessel M (1967) Synchrony of nuclear replication in individual hyphae of Aspergillus nidulans. J Bacteriol 94(5):1464–1469

Samils N, Oliva J, Johannesson H (2014) Nuclear interactions in a heterokaryon: insight from the model Neurospora tetrasperma. Proc R Soc B Biol Sci 281(1786):20140084

Sancar C, Sancar G, Ha N, Cesbron F, Brunner M (2015) Dawn-and dusk-phased circadian transcription rhythms coordinate anabolic and catabolic functions in Neurospora. BMC Biol 13:1–16

Schaap P, Wang M (1986) Interactions between adenosine and oscillatory camp signaling regulate size and pattern in Dictyostelium. Cell 45(1):137–144

Schmieder SS, Stanley CE, Rzepiela A, van Swaay D, Saboti˘c J, Nørrelykke SF, deMello AJ, Aebi M, Künzler M (2019) Bidirectional propagation of signals and nutrients in fungal networks via specialized hyphae. Curr Biol 29(2):217–228

Sgro AE, Schwab DJ, Noorbakhsh J, Mestler T, Mehta P, Gregor T (2015) From intracellular signaling to population oscillations: bridging size-and time-scales in collective behavior. Mol Syst Biol 11(1):779

Shearwin KE, Callen BP, Egan JB (2005) Transcriptional interference–a crash course. Trends Genet 21(6):339–345

Singh A (2011) Negative feedback through mRNA provides the best control of gene-expression noise. IEEE Trans Nanobiosci 10(3):194–200

Skejo J, Garg SG, Gould SB, Hendriksen M, Tria FD, Bremer N, Franjevic D, Blackstone NW, Martin WF (2021) Evidence for a syncytial origin of eukaryotes from ancestral state reconstruction. Genome Biol Evol 13(7):evab096

Snead WT, Jalihal AP, Gerbich TM, Seim I, Hu Z, Gladfelter AS (2022) Membrane surfaces regulate assembly of ribonucleoprotein condensates. Nat Cell Biol 24(4):461–470
Stajich JE, Berbee ML, Blackwell M, Hibbett DS, James TY, Spatafora JW, Taylor JW (2009) Primer–the fungi. Curr Biol: CB 19(18):R840
Steinberg G (2007) Hyphal growth: a tale of motors, lipids, and the spitzenkorper. Eukaryot Cell 6(3):351–360
Sudbery P, Grant W (1976) The control of mitosis in Physarum polycephalum: the effect of delaying mitosis and evidence for the operation of the control mechanism in the absence of growth. J Cell Sci 22(1):59–65
Upadhyay A, Brunner M, Herzel H (2019) An inactivation switch enables rhythms in a Neurospora clock model. Int J Mol Sci 20(12):2985
Vinck A, Terlou M, Pestman WR, Martens EP, Ram AF, van den Hondel CA, Wösten HA (2005) Hyphal differentiation in the exploring mycelium of Aspergillus niger. Mol Microbiol 58(3):693–699
Vitalini MW, de Paula RM, Park WD, Bell-Pedersen D (2006) The rhythms of life: circadian output pathways in Neurospora. J Biol Rhythm 21(6):432–444
Wang G, Peskin CS (2018) Entrainment of a cellular circadian oscillator by light in the presence of molecular noise. Phys Rev E 97(6):062416
Wang Z, Zhang J (2011) Impact of gene expression noise on organismal fitness and the efficacy of natural selection. Proc Natl Acad Sci 108(16):E67–E76
Windner SE, Manhart A, Brown A, Mogilner A, Baylies MK (2019) Nuclear scaling is coordinated among individual nuclei in multinucleated muscle fibers. Dev Cell 49(1):48–62
Xiao H, Zhong J-J (2016) Production of useful terpenoids by higher-fungus cell factory and synthetic biology approaches. Trends Biotechnol 34(3):242–255
Youn M-Y, Takada I, Imai Y, Yasuda H, Kato S (2010) Transcriptionally active nuclei are selective in mature multinucleated osteoclasts. Genes Cells 15(10):1025–1035
Zhang H, Elbaum-Garfinkle S, Langdon EM, Taylor N, Occhipinti P, Bridges AA, Brangwynne CP, Gladfelter AS (2015) RNA controls polyQ protein phase transitions. Mol Cell 60(2):220–230
Zinn-Brooks L, Roper ML (2021) Circadian rhythm shows potential for mrna efficiency and self-organized division of labor in multinucleate cells. PLOS Comput Biol 17(8):e1008828

Chapter 10
Ancestors in the Extreme: A Genomics View of Microbial Diversity in Hypersaline Aquatic Environments

Lulit Tilahun, Asfawossen Asrat, Gary M. Wessel, and Addis Simachew

Abstract The origin of eukaryotic cells, and especially naturally occurring syncytial cells, remains debatable. While a majority of our biomedical research focuses on the eukaryotic result of evolution, our data remain limiting on the prokaryotic precursors of these cells. This is particularly evident when considering extremophile biology, especially in how the genomes of organisms in extreme environments must have evolved and adapted to unique habitats. Might these rapidly diversifying organisms have created new genetic tools eventually used to enhance the evolution of the eukaryotic single nuclear or syncytial cells? Many organisms are capable of surviving, or even thriving, in conditions of extreme temperature, acidity, organic composition, and then rapidly adapt to yet new conditions. This study identified organisms found in extremes of salinity. A lake and a nearby pond in the Ethiopian Rift Valley were interrogated for life by sequencing the DNA of populations of organism collected from the water in these sites. Remarkably, a vast diversity of microbes were identified, and even though the two sites were nearby each other, the populations of organisms were distinctly different. Since these microbes are capable

Supplementary Information The online version contains supplementary material available at https://doi.org/10.1007/978-3-031-37936-9_10.

L. Tilahun · A. Simachew
Institute of Biotechnology, Addis Ababa University, Addis Ababa, Ethiopia
e-mail: lulit.tilahun@aau.edu.et; addis.simachew@aau.edu.et

A. Asrat
Department of Mining and Geological Engineering, Botswana International University of Science and Technology, Palapye, Botswana

School of Earth Sciences, Addis Ababa University, Addis Ababa, Ethiopia
e-mail: kassayea@biust.ac.bw; asfawossen.asrat@aau.edu.et

G. M. Wessel (✉)
Department of Molecular and Cell Biology and Biochemistry, Brown University, Providence, RI, USA
e-mail: rhet@brown.edu

M. Kloc, A. Uosef (eds.), *Syncytia: Origin, Structure, and Functions*, Results and Problems in Cell Differentiation 71,
https://doi.org/10.1007/978-3-031-37936-9_10

of living in what for humans would be inhospitable conditions, the DNA sequences identified should inform the next step in these investigations; what new gene families, or modifications to common genes, do these organisms employ to survive in these extreme conditions. The relationship between organisms and their environment can be revealed by decoding genomes of organisms living in extreme environments. These genomes disclose new biological mechanisms that enable life outside moderate environmental conditions, new gene functions for application in biotechnology, and may even result in identification of new species. In this study, we have collected samples from two hypersaline sites in the Danakil depression, the shorelines of Lake As'ale and an actively mixing salt pond called Muda'ara (MUP), to identify the microbial community by metagenomics. Shotgun sequencing was applied to high density sampling, and the relative abundance of Operational Taxonomic Units (OTUs) was calculated. Despite the broad taxonomic similarities among the salt-saturated metagenomes analyzed, MUP stood out from Lake As'ale samples. In each sample site, Archaea accounted for 95% of the total OTUs, largely to the class Halobacteria. The remaining 5% of organisms were eubacteria, with an unclassified strain of *Salinibacter ruber* as the dominant OTU in both the Lake and the Pond. More than 40 different genes coding for stress proteins were identified in the three sample sites of Lake As'ale, and more than 50% of the predicted stress-related genes were associated with oxidative stress response proteins. Chaperone proteins (DnaK, DnaJ, GrpE, and ClpB) were predicted, with percentage of query coverage and similarities ranging between 9.5% and 99.2%. Long reads for ClpB homologous protein from Lake As'ale metagenome datasets were modeled, and compact 3D structures were generated. Considering the extreme environmental conditions of the Danakil depression, this metagenomics dataset can add and complement other studies on unique gene functions on stress response mechanisms of thriving bio-communities that could have contributed to cellular changes leading to single and/or multinucleated eukaryotic cells.

10.1 Introduction

Microorganisms are extraordinary in their abilities to evolve, adapt, maintain, and change under vast extremes of terrestrial, aquatic, or potentially extraterrestrial environments (Tait et al. 2017; Viswadeepika and Bramhachari 2022; Coleine and Delgado-Baquerizo 2022). Many unique and extreme environments exist globally that harbor diverse populations of microbiota, many containing diverse gene sets that enhance future evolvability (Viswadeepika and Bramhachari 2022; Berg et al. 2020; Rodriguez and Durán 2020; Berg et al. 2021; Oren 2006). Among the multiple extreme conditions known, salinity is widespread as 96% of the global water (corresponding to 70% of Earth's surface) is either ocean or sea water (https://www.usgs.gov/special-topics/water-science-school/science/how-much-water-there-earth). In addition, salinization of groundwater, caused by encroachment of sea level, is increasing the concentration of soluble salts in soils (Srivastava et al. 2019).

Hence, organisms dwelling in the vast saline Earth surface, both aquatic and terrestrial, are exposed to osmotic and oxidative stresses caused by high salt concentrations of their environment. Most high-stress organisms are prokaryotic, especially in the extremophile environments, but a continuum of such environmental conditions suggests that transitions in rapid adaptability could lead to changes in organismal biology that we might not consider when thinking of our 37 °C CO_2 incubators (Oren 2006; Adem et al. 2014; Shrivastava and Kumar 2015). We believe that dynamic changes in moderate to extreme environments, and back, may have selected for prokaryotic cells able to find new ways to survive and thrive. Perhaps those prokaryotic cells on their own did not give rise to nuclear, or even syncytial cell types, but we consider favorable a cell type undergoing significant adaptations to environmental conditions as a template for a most extreme change—acquiring endosymbionts, and coopting them for adaptations to more moderate environmental conditions.

Microbes which habitat high salt conditions have devised mechanisms at least for salt-tolerance but likely as well for other extreme conditions often associated with hypersaline environments. Often the hypersaline niches include environmental stresses like osmotic stress, temperature changes, diverse light conditions and intensities, exposure to heavy metals and organic pollutants, etc. Strategies and the adaptations acquired by hypersaline-dwelling microbes have made them critical for consideration of agricultural, industrial, and pharmaceutical applications (Dindhoria et al. 2023 and references therein).

Lake As'ale (also known as Lake Karum or Assale) is a paradigm for hypersaline lakes (Oren 2010; Makhdoumi-Kakhki et al. 2012; Lanzén et al. 2013; Mernagh et al. 2014; Viver et al. 2018; Asrat 2016) and is located in the northern part of the Danakil depression, south of the Dallol volcano (Asrat 2016; Cavalazzi et al. 2019; Abbate et al. 2015). The Danakil depression is part of the East African Rift System (EARS), which is a segment of proto-oceanic crust and characterized by low annual rainfall, high temperature, and intense sunlight (Asrat 2016; Cavalazzi et al. 2019; Abbate et al. 2015; Ahrens 2011; Beyth 1978; Fazzini et al. 2015; Hughes and Hughes 1992). Various extreme geologic features, such as the acidic hypersaline hydrothermal systems on and around mount Dallol, are present in the depression (Cavalazzi et al. 2019; Belilla et al. 2019; Tilahun et al. 2021a, b). A previous study indicated that Lake As'ale used to be part of the Red Sea until a volcanic barrier flow separated it less than 32,000 years ago (Beyth 1978). Thus, Lake As'ale is considered to be formed after the auxiliary sinking of the Danakil depression resulting in periodic flooding of the Red Sea (Beyth 1978; Hughes and Hughes 1992). Moreover, the Lake is dotted by actively mixing salt ponds in many places with colorful and bubbling "cold" springs (Asrat 2016). One such pond, called Mud'ara (MUP), is located at the northwestern margin of the lake (Asrat 2016).

The general poly-extreme environmental conditions of Danakil depression (high temperature, high salinity, and high UV radiation) are also characteristics of Lake As'ale (Asrat 2016; Cavalazzi et al. 2019). Prior investigations using metagenomics revealed the dominance of prokaryotes in the open water of Lake As'ale and in the other hyper-acidic and saline hydrothermal systems of the depression (Cavalazzi et al. 2019; Belilla et al. 2019; Tilahun et al. 2021a, b). This poly-extreme environmental condition of Lake As'ale can inflict osmotic and oxidative stresses on the inhabiting microbes in addition to the hypersaline conditions (Xie et al. 2019; Pedone et al. 2020). Oxidative stress can be a source for elevation of mutation rates or modification of solvent accessibility of amino acids that leads to misfolding of proteins (Vidovic et al. 2014). Naturally, prokaryotes are well equipped at a molecular level to overcome oxidative stress and protect themselves from disruptions of protein stability and misfolding (Vidovic et al. 2014; Susin et al. 2006; Voth and Jakob 2017; Verghese et al. 2012; Tittelmeier et al. 2020). Resistance to misfolding of proteins due to oxidative damage is intrinsic and can be assisted by molecular chaperones (Vidovic et al. 2014; Susin et al. 2006; Voth and Jakob 2017; Verghese et al. 2012; Tittelmeier et al. 2020). Heat shock proteins (HSP) such as DnaK, DnaJ, GrpE, and ClpB are molecular chaperones involved in the proper folding and refolding of newly synthesized and damaged (misfolded) proteins, respectively (Vidovic et al. 2014; Susin et al. 2006; Voth and Jakob 2017; Verghese et al. 2012; Tittelmeier et al. 2020). Yet, synthesis of HSPs is not only activated as a response to exposure to heat shock stress but also to diverse environmental stresses such as freezing (hypothermia), chemicals such as ethanol, free radicals, and ultraviolet radiation (Verghese et al. 2012; Tittelmeier et al. 2020). Accordingly, the main function of HSPs is to stabilize and assist proteins proper folding of nascent polypeptides and refolding of misfolded proteins that otherwise may accumulate and become deadly to any organism (Vidovic et al. 2014; Susin et al. 2006; Voth and Jakob 2017; Verghese et al. 2012; Tittelmeier et al. 2020).

In this study, our focus of investigation is the shorelines of Lake As'ale and MUP in the Danakil depression (Fig. 10.1a and b). In contrast to open waters, shorelines of any lake can be more stressful to its inhabitants because of the constant fluctuations of water level and hydrochemistry (Mortsch 1998; Maihemuti et al. 2020). Therefore, through shotgun metagenome analysis, extremophiles in the brine samples from the shoreline of Lake As'ale and MUP were studied to check their similarity to the Lake's open water diversity. Further in silico queries were accomplished to identify the genome composition and proteins predicted to overcome the environmental stresses. In addition, from the generated assembled metagenome data, we predicted the open reading frames (ORFs) for chaperone proteins and 3D protein configurations to determine if their structure is acclimatizing, and distinct from respective reference proteins, and how viral and fusogenic proteins may have evolved.

10.2 Materials and Methods

10.2.1 The Study Site

The brine samples from Lake As'ale were collected randomly from accessible sites in the shoreline but with mindfulness of minimal human and animal contact in order to reduce the chances of sample contamination (Fig. 10.2a). The selected three different locations of Lake As'ale were at −115 m and −120 m below sea level and labeled as LA (UTM 0646610E, 1558846N), LADP5 (UTM 0648365E, 1558709N), and LADP6 (UTM 0645195E, 1558770N). Brine samples from Muda'ara pond (MUP) were collected at UTM location 0645190E, 1558764N, −109 m below sea level. MUP is a small, greenish, and highly mixing pond located at the northwestern shore of Lake As'ale (Figs. 10.1b and 10.2b). The distances between the three sample sites of Lake As'ale and MUP were calculated using an online Latitude/Longitude Distance Calculator (https://www.nhc.noaa.gov/gccalc.shtml). Hence, LA and LADP5 are approximately 1.8 km apart; LADP5 and LADP6 are approximately 1.4 km apart, and MUP is located adjacent to LADP6.

10.2.2 Sample Collection and Processing

The shorelines of Lake As'ale were very shallow (cms), while MUP was very small but strongly mixing. The distances between the shoreline and the sample collection sites LADP5, LA, and LADP6/MUP are 176 m, 125 m, 767 m, respectively. All brine samples were collected from the surface water, in triplicates and indiscriminately using sterile High Density Polyethylene (HDPE) bottles on February 3 and 4, 2015.

A total of 2000 ml surface brine samples were collected from each sample site, where 200 ml per filter was filtered without dilution using 0.22 μm GE® polycarbonate filter membranes. The residual microbial cells on the membranes were stored in 2 ml Lasany® internal threaded Cryovials with sucrose lysis buffer (0.75 M sucrose, 20 mM EDTA, 200 mM NaCl, and 50 mM Tris-HCl, pH 9.0). When analyzing the physico-chemical properties of the samples, 200 ml of water filtrate was placed in sterile containers. All the prepared samples for molecular and physico-chemical analysis were transported to the Microbial Biotechnology Laboratory, Addis Ababa University, under controlled cooling conditions subsequently stored at −20 °C and 4 °C, respectively, until further analysis was conducted.

10.2.3 Physico-chemical Analysis and Water Isotope Measurement

In situ measurements of pH, conductivity, and temperature were performed using the 430 Enterprise Portable pH and Conductivity Meter 430271 (Jenway). Salinity was measured using a refractometer (DIGIT-0120 ATC, VWR) after diluting the brine samples 1 to 10 times with deionized water. Selective ex situ hydrochemical analyses and water isotope measurement were carried out as described in Tilahun et al. (2021a).

Fig. 10.1 Satellite map of Lake As'ale (**a**) and image of sampling locations (**b**) on the shore of Lake As'ale and the Mud'ara pond (Source: Google Map)

10.2.4 DNA Extraction and Metagenome Sequence Processing

Community genomic DNA was extracted at the Microbial Biotechnology Laboratory (Addis Ababa University, Ethiopia) and at the PrIMO Laboratory (Brown University, Providence RI, USA) using a modified CTAB method adapted from Zhou et al. (1996). DNA was cleaned and concentrated using DNA Clean & Concentrator™-5 from Zymo Research. The quantity and quality of DNA was checked using PicoGreen Assay and Thermo Scientific NanoDrop 3300 Fluorospectrometer. All

Fig. 10.2 Sample sites photo. Lake As'ale (**a**); Mud'ara pond (**b**) (Photo by Tilahun, 2015)

filtered brine samples yielded genomic DNA quantities exceeding the required 1 ng for Illumina 2500 library preparation (Table 10.3). DNA library preparation and post sequencing quality control (quality filtering and trimming) was performed as described in Tilahun et al. (2021a).

10.2.5 Taxonomy Profiling and Statistical Analysis

10.2.5.1 Mapping of Direct Assembly-Free Sequence Reads

The sequence reads of the samples from the three runs were merged together and put into MetaPhlAn 2 bioinformatics tool which maps direct sequence reads against a reduced set of clade-specific marker sequences (Segata et al. 2012; Truong et al.

2015). Microbial relative abundance table was generated using default parameters and bowtie2 alignment. Distribution and heatmaps were generated with "metaphlan_hclust_heatmap.py" script using default options and the "-d braycurtis," "-minv 0.01" flags.

R Studio with Vegan and Mass packages (RStudio Team 2020; Venables and Ripley 2002; Oksanen et al. 2017) was used for calculating similarity indices (Bray-Curtis index and Jaccard's index), diversity indices (Shannon's index and Simpson's index), and nonparametric estimation of number of species in a community. The functions "vegdist" and "diversity" in the Vegan package were applied to calculate the dissimilarity indices (Bray-Curtis index and Jaccard's index) and diversity indices (Shannon's index and Simpson's index), respectively. For calculating Bray-Curtis dissimilarity, the default for the function "vegdist" was used. Hierarchical clustering was plotted based on Bray-Curtis dissimilarity distance using function "spantree," which finds a minimum spanning tree for dissimilarities connecting all points.

Principal Component Analysis (PCA) of the microbial datasets was generated using ClustVis (Metsalu and Vilo 2015), a web tool for visualizing clustering of multivariate data (BETA). For calculating PCA, default method SVD with imputation was used, which performed imputation and singular value decomposition (SVD) iteratively until estimates of missing values converge (Yata and Aoshima 2010).

10.2.5.2 Taxonomy Binning of Metagenome-Assembled Reads

Quality checked reads were assembled using metaSPADes with a flag "meta" and kmers 21, 33, and 55 (Nurk et al. 2017). The resulted metagenome contigs were aligned against NCBI non-redundant protein database using double index alignment of next generation data (DIAMOND) v0.9.24; BLASTx with the sensitive mode, frameshift alignment for assembled long sequences, and a default e-value cut-off of 0.001 (Buchfink et al. 2015). MEtaGenome analyzer 6 Community Edition (MEGAN6 CE) was used to perform taxonomic assignment of assembled contigs (Huson et al. 2016, 2018). The parameters used to retrieve metagenome-assembled genomes and compute estimation of the taxonomical content from the samples using MEGAN6 Long-read (MEGAN-LR) are described in Tilahun et al. (2021a).

Comparison of the multiple samples in this study was accomplished by using the "Compare" dialog on MEGAN6 CE and by selecting the "Use Normalized Counts" to normalize all counts to the smallest number of reads of the studied samples. Using "Cluster analysis viewer" dialog on MEGAN6 CE, Bray-Curtis dissimilarity index was used to calculate distances based on the leaves of the corresponding Phylogeny and Principle Coordinates Analysis (PcoA) as well as hierarchical clustering (UPGMA tree) were plotted.

10.2.5.3 Functional Binning of Metagenome-Assembled Genomes (MAGs) and Prediction of Chaperone Proteins

Each assembled long read was mapped onto a gene that has a known functional role and then compared to one or more different subsystems by computing SEED RefSeq id and using "Compare" dialog on MEGAN6 CE, respectively. The comparison was calculated by selecting "use Absolute count" to compare the original counts of reads for each sample and by ignoring all unassigned reads. Functional profile of stress response genes for all sample sites was studied by generating "Stress_Response.rma" file using SEED-MEGAN and compared using "Compare" dialog on MEGAN6 CE. Long reads of consensus amino acid sequences, associated with chaperone proteins from SEED's protein metabolism class, were retrieved using MEGAN's-Alignment tool and exported in fasta format. The consensus translated ORFs for chaperone proteins were mapped to respective reference protein sequences using NCBI. NCBI's Blastp was used (with cut-off value 1e-10) to analyze sequence similarity of translated ORFs grouped specifically in chaperone protein DnaJ and DnaK as well as ClpB protein families. Furthermore, BLAST's "Distance tree of results" was applied to compute pairwise alignment of query (that have greater than 94% query coverage) against non-redundant protein sequences (nr) database using default parameters. The selection of query translated sequences for homology as well as further structural analysis was performed using InterProScan version: 5.56–89.0 and SWISS-MODEL server as described in Tilahun et al. (2021a).

10.3 Results

10.3.1 Geochemical and Physico-chemical Properties

The studied shoreline sites of Lake As'ale and MUP are categorized as hypersaline environments with salinity exceeding or equal to 32% (Gomez 2014). Lake As'ale and the nearby groundwater systems, including MUP, are generally fed by the periodic runoff from the highlands located west of the Danakil depression (Abbate et al. 2015; Hughes and Hughes 1992). However, due to the phreatic eruption on mount Dallol on January 2015, the hydrochemistry of the nearby groundwater systems was likely affected by the released volcanic gases, which can readily mix and alter the pH (Tilahun et al. 2021a; Master 2016; Jasim et al. 2018). Hence, contrary to the report on the open water (Belilla et al. 2019), the pH of the brine samples collected from the shoreline of Lake As'ale and MUP was acidic (pH 5.6 and 4.25, respectively). The results of in situ geochemical measurements are listed in Table 10.1.

The two selected sampling sites for further hydrochemistry analysis (LA and MUP) were approximately 3.2 km apart from each other. MUP is characterized by depleted values of $\delta^{18}O$ and $\delta^{2}H$ (−1.38/mil and 3.42/mil, respectively) while brine of LA is characterized by enriched $\delta^{18}O$ and $\delta^{2}H$ (11.51/mil and 46.36/mil,

Table 10.1 Geochemical data of the sample sites

Lake/ pond labeling code	GPS location		Altitude (m)	Description of sampling sites	pH	Average salinity	Average T°	Average EC (mS/cm)
	UTM. EW	UTM. NS						
LA	646610	1558846	−120	Large shallow lake on white salt plain	5.6	0.32	32 °C	237 at 32 °C
LADP5	648365	1558709	−115	Large shallow lake on white salt plain	5.61	0.32	32 °C	236 at 32 °C
LADP6	645195	1558770	−115	Large shallow lake on white salt plain	5.61	0.32	31 °C	237 at 31 °C
MUP	645190	1558764	−111	Small mixing greenish color pond adjacent to Lake As\ale	4.25	0.36	30 °C	235 at 30 °C

Table 10.2 Selected hydrochemical and stable water isotope measurements

Sample site	Mg^{2+} (g/l)	NO_2^- (mg/l)	NO_3^- (g/l)	TP (g/l)	COD	SO_4^{2-} (g/l)	Cl^- (g/l)	$\delta^{18}O$ (/mil)	δ^2H (/mil)
LA	59.4	1.34	0.2	Nil	2450	0.4	217.3	11.51	46.36
MUD	7.8	2.4	0.13	Nil	1925	0.1	219.1	−1.38	3.42

respectively). Evapo-transpiration affects more shallow surface water bodies than groundwater (Yeh and Lee 2018). As a result, the values of oxygen and hydrogen isotopes indicated intense evaporation with selective heavier isotopes enrichment in the Lake's waters. Hence, the cold spring in the MUP is less evaporated, is mixed, and is continuously fed by groundwater from the highlands whose source is predominantly rainfall depleted in $\delta^{18}O$ and δ^2H. In general, the observed variations of amounts of Mg^{2+}, NO_2^-, SO_4^{2-}, and isotopes of oxygen and hydrogen in MUP and LA (Table 10.2) indicate extreme evaporation and insignificant precipitation in the Danakil depression (Ahrens 2011; Fazzini et al. 2015).

10.3.2 Microbial Richness in the Shorelines of Lake As'ale and MUP

Species richness and diversity of microbial communities are inversely affected by salinity in hypersaline aquatic environments (Oren 2006; Lanzén et al. 2013; Simachew et al. 2016; Podell et al. 2014; Yang et al. 2016; Plominsky et al. 2018; Banda et al. 2020). In our study, the OTU abundance calculated by MetaPhlAn 2 identified less than 50 OTUs from the surface brine samples of Lake As'ale's

Table 10.3 Summary of metagenomics and statistical measures of diversities based on non-assembled sequence reads

Sample site/ sequence ID	Estimated DNA concentration for sequencing	Total number of reads	Observed OTU	Fisher's alpha index	Shannon's index	Simpson's index
LADP5 (LTW0004)	79 ng/μl	13103002	38	2.45	1.34	0.4
LADP6 (LTW0005)	52 ng/μl	14021795	47	3.06	1.01	0.59
LA (LTW0006)	137 ng/μl	10440347	35	2.28	0.91	0.36
MUP (LTW0007)	58 ng/μl	11016361	49	3.26	1.56	0.68

shoreline and MUP (Table 10.3). Likewise, less than 161 OTUs were identified from studied sample sites by MEGAN6 CE (Table 10.4). The consequences of extreme salinity, pH, and some ions such as $Mg^{2+}$are known to impact microbial community compositions and richness in extreme environments such as the ones inhabiting the Dead Sea (Oren 2006; Oren 2010; Belilla et al. 2019; Yang et al. 2016; Oren 2005; Nissenbaum 1975; Hallsworth et al. 2007; Logares et al. 2013).

Results from investigating the assembled and non-assembled metagenomic DNA sequence reads from studied sample sites revealed more diverse microbial communities in MUP than the shoreline sites of Lake As'ale (Tables 10.3 and 10.4; Supplementary Table 1, Figs. S1–S3). Detailed analysis of the taxonomic content of sample sites was obtained from MEGAN6 CE after analyzing the metagenome-assembled reads (Table 10.4; Supplementary Table 2). More than 35% of the total ~9.6 million aligned bases were assigned to 205 OTUs at species level (Table 10.4). Similar to MetaPhlAn 2 results, MEGAN6 CE outputs showed higher number OTUs (154 OTUs) identified at species level for MUP compared to the sample sites of Lake As'ale (111–132 OTUs) (Supplementary Table 2). The multivariate analyses performed to visualize sample sites similarities based on the results of MetaPhlan2 and MEGAN6 CE clearly indicated the difference between Lake As'ale and MUP species wise (Fig. 10.3, Supplementary Figs. S1 and S2) (Iscen et al. 2008; Oksanen 2015).

10.3.3 Diversity of Extremophilic Prokaryotes

All the sampled sites in this study were highly dominated by prokaryotes (Figs. 10.4 and 10.5, Supplementary Table 1, Supplementary Fig. S3). The archaeal OTUs were abundant in the shoreline sites of Lake As'ale and MUP as in the case of many other hypersaline or soda or slightly acidic or neutral lakes (Oren 2006; Oren 2010; Makhdoumi-Kakhki et al. 2012; Lanzén et al. 2013; Mernagh et al. 2014; Belilla et al. 2019; Simachew et al. 2016; Plominsky et al. 2018; Banda et al. 2020). Though

Table 10.4 Overview of metagenome assemblies with MetaSPAde and taxonomic binning of assembled reads using DIAMOND and MEGAN6 CE

	LADP5	LADP6	LA	MUP
No. of contigs	83343	95819	98941	59342
Max. contig length (bp)	24552	20601	27707	48514
N50 (bp)	468	596	483	383
Total length (bp)	32250632	46005719	40523474	22881057
Metagenome size (bp/read)	32250632/13103002	46005719/14021795	40523474/10440347	22881057/11016361
# of bases that must be assigned to taxon/taxa[a]	3486	6562	4520	2836
Total # of contigs aligned by DIAMOND	42305	62841	56019	42549
Average # of normalized counts of aligned bases per contig aligned by DIAMOND	335.2 bases/contig	225.7 bases/contig	253.2 bases/contig	333.3 bases/contig
Total # of normalized counts of aligned bases assigned to NCBI taxonomy	9632187	9804509	9751492	10068334
Total # of normalized count of aligned bases assigned to NCBI taxonomy at species level	3597502	3747214	3684591	3591818
Total normalized count with no hit	665	627	637	715
Total normalized count not assigned	4548374	4376078	4429083	4112165
No. of OTUs profiled at species level	119	111	132	154
Archaea OTU	83 (70%)	73 (66%)	99 (75%)	129 (84%)
Bacteria OTU	36 (30%)	37 (33.3%)	32 (24.2%)	23 (15%)
Virus OTU	0	1 (0.01%)	1 (0.01%)	2 (0.01%)
# of OTUs profiled at higher taxonomic level	26	18	22	6

[a] Number of bases that must be assigned to taxon/taxa to appear in the result at Min support percent = 0.02

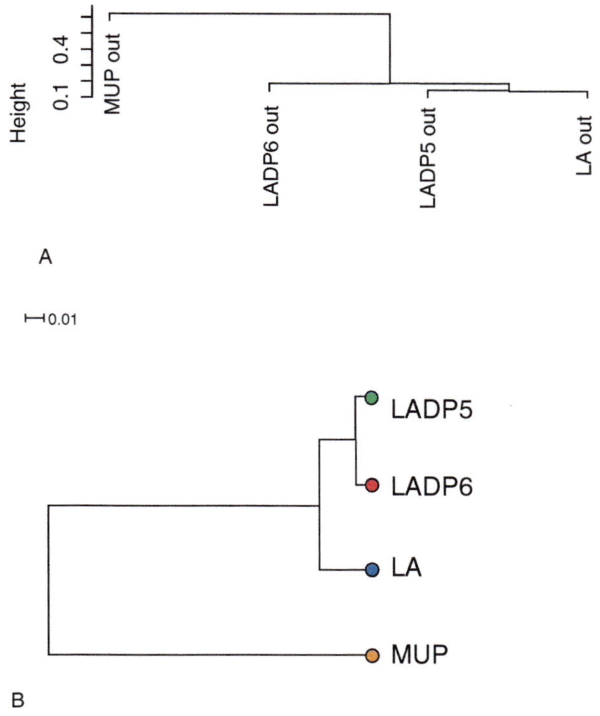

Fig. 10.3 Minimum spanning tree of the Bray-Curtis distances between the samples based on MetaPhlAn2 taxonomic profiling from direct metagenomic shotgun sequencing data using R studio [**a**]. Hierarchical clustering using Unweighted Pair Group Method with Arithmetic Mean (UPGMA) was also plotted by calculating Bray-Curtis distances of taxonomic profiles of assembled contigs using MEGAN 6 CE [**b**]

bacterial and viral OTUs were also encountered, their representation was at a minimum percentage in all studied sites (Table 10.4 and Figs. 10.4 and 10.5). Presence of eukaryotic OTUs was reported in a previous study on the open water of Lake As'al by Belilla et al. (2019). However, eukaryotic OTU were absent in the current study on the shoreline sites of Lake As'ale and MUP (Figs. 10.4 and 10.5, Supplementary Table 1, Supplementary Fig. S3). The reason for the absence of eukaryotes might be due to the molecular fitness of prokaryotes to adjust to different stressful environmental conditions of shallow parts of any lake such as new pattern of water level fluctuations and hydrochemistry, intense UV radiation, and low rainfall (Mortsch 1998; Maihemuti et al. 2020).

The class Halobacteria was dominant in all studied sites (Fig. 10.5, Supplementary Fig. S3). Members of archaeal class Halobacteria are well known and common inhabitants of hypersaline environments (Oren 2010; Burns et al. 2004; Ventosa et al. 2004; Yang et al. 2007; Sorokin et al. 2014). Five families of Halobacteria (Halobacteriaceae, Haloarculaceae, Halococcaceae, Natriabaceae, and Halorubraceaea) were identified after Metaphlan2 and MEGAN 6 CE mapping (Supplementary Tables 1–3). MEGAN 6 CE performed well in assigning reads at species level because long assembled reads were analyzed (Tran and Phan 2020). As a result, more than 125 OTUs from each sampled site in this study were profiled at species and higher taxonomic level from all domains of life using MEGAN 6 CE

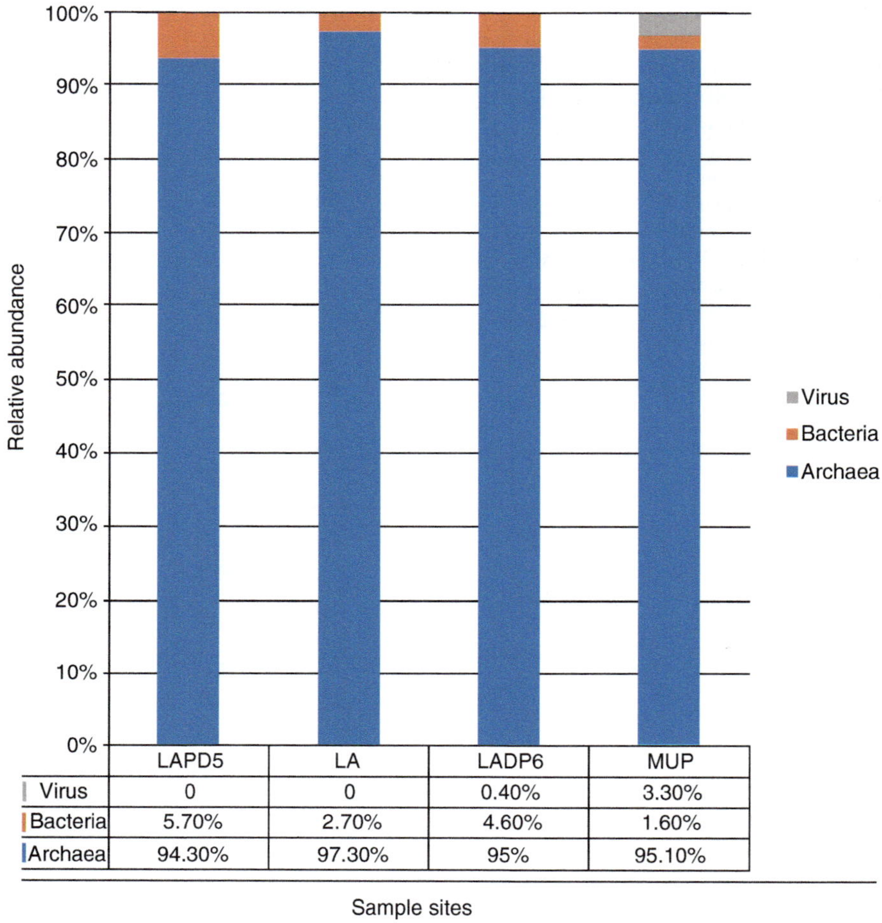

	LAPD5	LA	LADP6	MUP
Virus	0	0	0.40%	3.30%
Bacteria	5.70%	2.70%	4.60%	1.60%
Archaea	94.30%	97.30%	95%	95.10%

Fig. 10.4 Distribution of domains of organisms and viruses along the subsampling sites of Lake As'ale and Mud'ara pond based on MetaPhlAn2

(Table 10.4; Supplementary Table 2). Moreover, additional groups of archaea (Candidatus Nanohaloarchaeota, Methanonatronarchaeaceae, Thermoplasmata, Unclassified Archaea, and Euryarchaeota) were identified by MEGAN6 CE (Supplementary Fig. S3). Bacterial and viral OTUs were also identified in the studied sample sites. Surprisingly, diverse groups of bacterial OTUs were identified even if the majority of the sequence reads were binned to archaeal domain (Fig. 10.7; Supplementary Tables 1 and 2). *Salinibacter ruber* was the most represented bacterial OTU in the current study (Figs. 10.6 and 10.7; Supplementary Tables 1 and 2). *S. ruber* is an intraspecifically diverse, strictly halophile, and highly distributed organism in different hypersaline environments (Oren 2006; Lanzén et al. 2013; Viver et al. 2018; Simachew et al. 2016; Antón et al. 2002; Mormile et al. 2009; Ventosa and Arahal 2009; González-Torres and Gabaldón 2018). Archaeal dsDNA

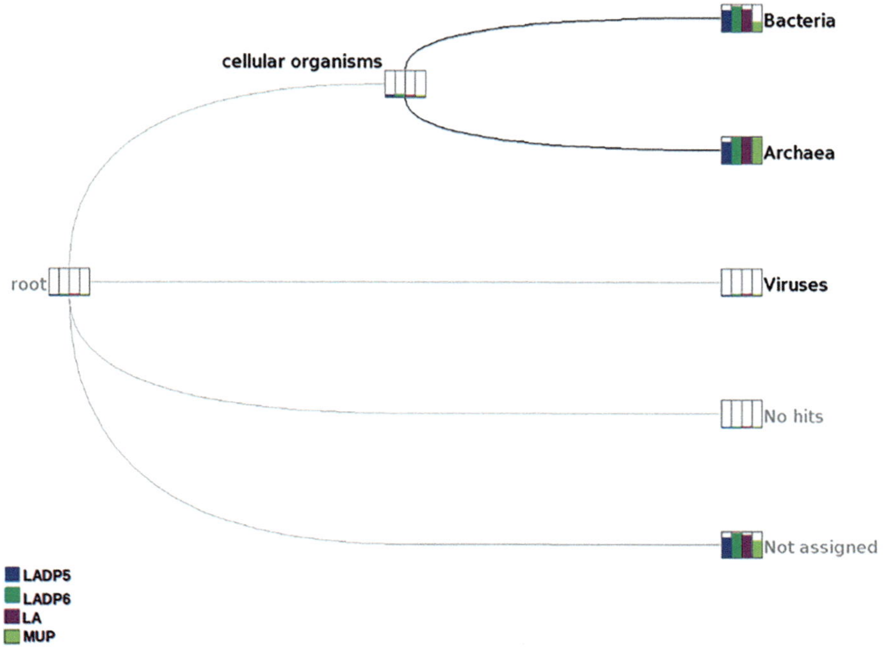

Fig. 10.5 Domain level classification of assembled reads based on normalized counts of assigned bases using DIAMOND-MEGAN6 CE

viruses were also identified by MEGAN6 CE, in either MUP or one or more shoreline sample sites of Lake As'ale. Though extreme environments are known troves of tremendous diversity of unknown viruses, the limitations in finding universally conserved markers in viral genomes makes them challenging to study (Edwards and Rohwer 2005; Paez-Espino et al. 2016). Moreover, as viral quasispecies, they are distributed across very different habitats, attributing viral OTUs from the metagenome data to particular hosts was carried out in caution (Kazlauskas et al. 2019). Hence, MUP and LADP6 were the only sample sites with non-assembled NGS reads mapped to viral OTUs (Supplementary Table 1) while MEGAN6 CE was able to bin numerous aligned bases to archaeal Halovirus in LADP6, LA, and MUP (Table 10.4; Supplementary Table 2; Supplementary Fig. S3).

10.3.4 Prediction of Proteins Involved in Stress Response in Lake As'ale and MUP

More than a thousand different genes were predicted in samples from Lake As'ale compared to the total 668 different genes predicted from MUP samples (Supplementary Table 4). While various stress response protein families were

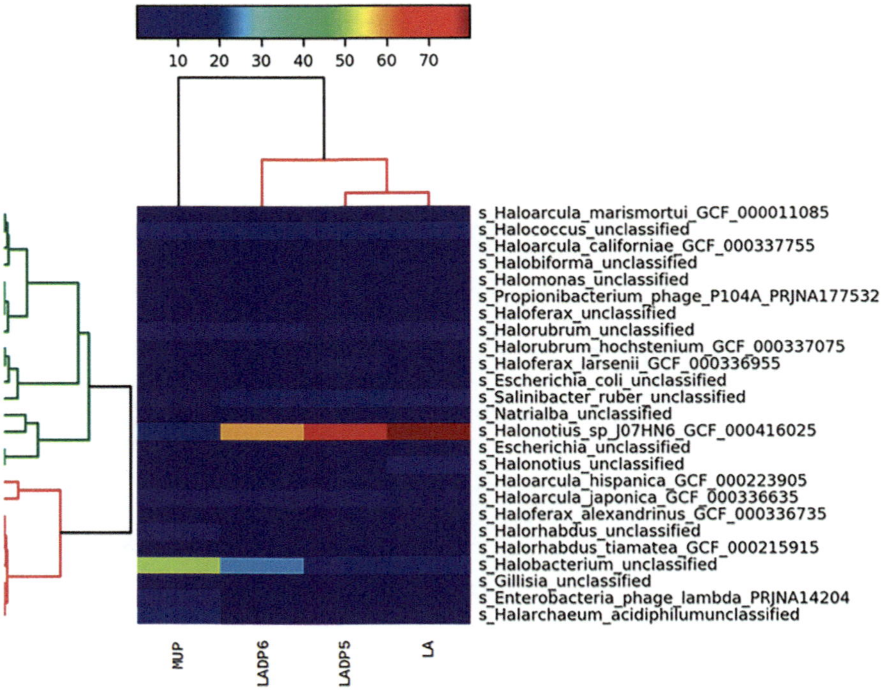

Fig. 10.6 Heat map of top 25 abundant OTUs in sampling sites of Lake As'ale and Mud'ara pond generated by MetPhlAn2

predicted in Lake As'ale and MUP, only 27 stress proteins were identified in MUP, and 40 different stress proteins were identified in the three sample sites of Lake As'ale (Supplementary Table 5). Constant fluctuations of environmental conditions at the shorelines of Lake As'ale, as in the case of any lake, can intensify stresses that cause the observable high number of stress-related proteins and also differences in survival strategies (Mortsch 1998; Maihemuti et al. 2020; Matarredona et al. 2020; Fulda et al. 2010).

Oxidative stress inside cells contributes to the disproportional production and accumulation of reactive oxygen species (ROS), decrease ATP levels that will inactivate molecular chaperones, and increase of aggregated misfolded proteins (Xie et al. 2019; Pedone et al. 2020; Matarredona et al. 2020; Fulda et al. 2010; Zhang et al. 2016; Winter et al. 2005). Thus, from the total of 86 predicted protein families, more than 50% are either scavenging enzymes for ROS, or proteases or regulatory proteins involved in response to oxidative stress (Supplementary Table 5). In connection with the differences in the geochemistry and hydrochemistry results (Tables 10.1 and 10.2), only twelve of the predicted stress protein families were commonly shared in all studied sites (Supplementary Table 5). These shared protein families have known functions in response to acid stress, to ROS, hyperosmotic and heat-shock, regulation/reprogramming of cellular metabolism through cell signaling, and anaerobic

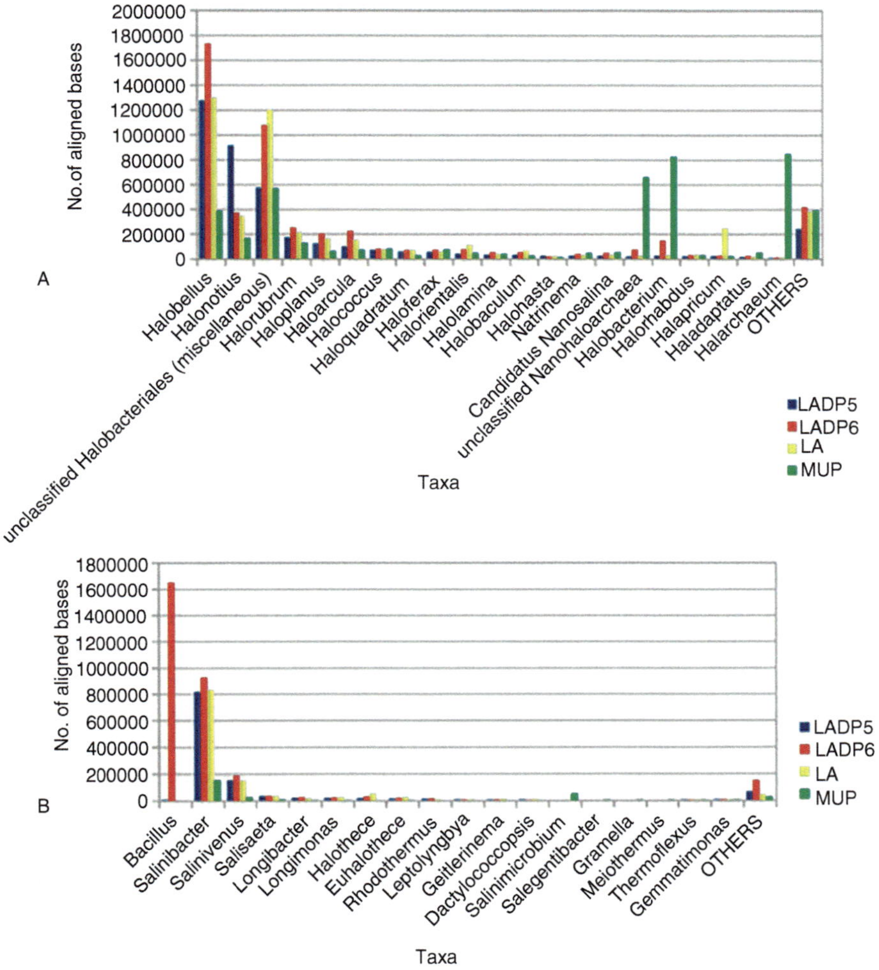

Fig. 10.7 Distribution of Archaea (**a**) and Bacteria (**b**) OTUs at Genus Taxonomy level along the sample sites based on MEGAN-DIAMOND alignment. More than 91% of the aligned reads are assigned to the top 21 archaeal genera in both Lake As'ale and MUP. In addition, more than 94% and 90% of the aligned reads are assigned to the top 18 bacterial genera in both Lake As'ale and MUP, respectively

respiration (Zhang et al. 2016; Broadbent et al. 2010; Rajput et al. 2021; Rawat and Maupin-Furlow 2020; Price et al. 2008; Audrito et al. 2020; Jamali et al. 2020; Valette et al. 2017; Zhang et al. 2015). Furthermore, four protein families (gamma-glutamyl transpeptidase (GGT) (EC 2.3.2.2), glutaredoxin-related protein, glutathione S-transferase omega (EC 2.5.1.18), and peroxiredoxin (EC 1.11.1.15) were uniquely predicted from Halobacteria class in MUP. Under extreme oxidative conditions, GGT and glutaredoxin (alternative to thioredoxin) proteins can aid scavenging glutathione as a source of amino acid and/or sulfur (Rawat and Maupin-Furlow 2020;

Sánchez-Riego et al. 2013; Saini et al. 2021). Peroxiredoxin (EC 1.11.1.15) and glutathione S-transferase omega (EC 2.5.1.18) are powerful antioxidants that can regulate and detoxify chemicals such as aliphatic/aromatic hydroperoxides and peroxynitrite from cell by conjugating the toxins with glutathione, respectively (Rajput et al. 2021; Poole et al. 2011).

In general, posttranslational modifications of proteins such as carbonylation, side chain oxidations, misfolding, and accumulation are triggered by severe oxidative stress and attended by the expression of proteases and molecular chaperones (Vidovic et al. 2014; Zhang et al. 2016; Winter et al. 2005; Zhang et al. 2015). The molecular chaperones, or HSPs, are intrinsically upregulated with ROS-scavenging enzymes, mainly to resist protein misfolding and maintain normal intracellular ATP concentration, respectively (Vidovic et al. 2014; Fulda et al. 2010; Zhang et al. 2016; Kim et al. 2021; Chatterjee et al. 2020; Yusof et al. 2022; Kim et al. 2008; Zolkiewski et al. 2012; Wen et al. 2020). Though ROS scavenging enzymes were annotated in all sample sites of the current study, only one chaperone protein, DnaK, was commonly predicted in all sites (Supplementary Table 5). In addition to the heat-shock stress response, DnaK/DnaJ chaperone system has been found to protect protein against carbonylation under oxidative stress (Zhang et al. 2016; Kim et al. 2021).

Further investigation on amino acid sequences of the chaperone proteins was done to check the novelty of the proteins. Partial or complete consensuses ORFs for HSP were obtained from all sample sites (Table 10.5). DnaJ (Hsp60) and ClpB (Hsp100s family) were primarily obtained from bacteria; heat-shock protein GrpE was from archaea, and DnaK (Hsp70) was predicted from bacteria as well as archaea in all sampling sites of Lake As'ale (Table 10.5). In the case of MUP, short but important ORFs for chaperone proteins such as DnaK, ClpB, and high-temperature protein G/HtpG (Hsp90) instead of DnaJ and GrpE were annotated (Table 10.5).

We used BLAST to visualize the distance tree for the long sequences with query coverage greater than 94% to the reference sequences (Supplementary Fig. S4a–c). The distance trees indicate the predicted ClpB proteins are ATP-dependent chaperone ClpB proteins potentially transferred from *Salinibacter* to *Bacteroidetes* (Supplementary Fig. S4 a[I], b[I], c[II]). In addition, the predicted chaperone DnaJ protein (Supplementary Fig. S4 a[II]) is potentially considered to be transferred from *Salinibacter* while the predicted chaperone DnaK proteins from LADP6 and LA (Supplementary Fig. S4 b[II], c[I]) are considered to be transferred from Bacillaceae and euryarchaeotes, respectively. These predicted proteins are conserved among bacteria and class Halobacteria (particularly DnaK) of eukaryarchaea, and the protein-protein comparison result can implicate the possibility of new taxonomic branches of strains of the respective reference microbes (Chatterjee et al. 2020; Mogk et al. 2003; Gabaldón 2007; Khaledian et al. 2020).

The extended ORFs of ClpB obtained from Lake As'ale's metagenome sequence data were associated to the genus *Salinibacter*. The heavy Hsp families (HtpG [Hsp90] and ClpB [Hsp100s]) are generally absent in any archaeal species (Matarredona et al. 2020; Zolkiewski et al. 2012; Large et al. 2009). The predicted ClpB proteins were as compact as their respective reference homologous proteins

Table 10.5 List of predicted heat-shock/chaperone proteins

Sample sites	Node(s) number/NCBI query ID	Predicted protein	Organism (Genus)	Query (a.a) length	Reference seq. ID	Query coverage %	%identity similarity of query to reference
LADP5	2613	DnaJ (J domain containing protein)	*Salinibacter*	178	WP_011404352.1	97.3	63
	35755_36305	Chaperone protein DnaJ	*Salinibacter*	166	WP_105013661.1	43.5	76
	812_2669_38494	Chaperone protein DnaK	*Salinibacter*	562	WP_103021287.1	84.4	95
	36495_58067	Chaperone protein DnaK	*Halothece*	163	WP_015226668.1	23.7	78
	410_23518	ClpB protein	*Salinibacter*	889	WP_011405536.1	99.2	91
LADP6	17008	Chaperone protein DnaJ	*Salinibacter*	170	WP_103027025.1	43.8	87
	41686_73815	Chaperone protein DnaJ	*Bacillus*	130	WP_113927497.1	34.7	98
	1499	Chaperone protein DnaK	*Bacillus*	572	WP_113927496.1	94.1	99
	1916_17978_67169	Chaperone protein DnaK	*Salinibacter*	570	WP_103021287.1	85.6	94
	7647	Chaperone protein DnaK	*Haloquadratum*	219	WP_011571867.1	34.2	88
	36752_49702	Chaperone protein DnaK	*Halomicrobium*	217	WP_208626431.1	34.1	96
	632	ClpB protein	*Salinibacter*	872	WP_118831603.1	98	91

(continued)

Table 10.5 (continued)

Sample sites	Node(s) number/NCBI query ID	Predicted protein	Organism (Genus)	Query (a.a) length	Reference seq. ID	Query coverage %	%identity similarity of query to reference
LA	26569	Chaperone protein DnaJ	*Salinibacter*	78	WP_118841212.1	42.6	65
	51586	Chaperone protein DnaJ	*Salinibacter*	79	WP_103027025.1	20.4	85
	1292_7788	Chaperone protein DnaK	*Salinibacter*	562	WP_103021287.1	84.4	96
	8636_40604	Chaperone protein DnaK (Partial)	*Halobacterium*	148	WP_235168930.1	96.1	60
	46972_72870	Chaperone protein DnaK	*Haloarcula*	152	WP_165896661.1	24.2	94
	61410_72230	Chaperone protein DnaK	*Halothece*	126	WP_015226668.1	18.3	85
	546	ClpB protein	*Salinibacter*	871	WP_105011162.1	97.2	90
	8389	ClpB protein	*Acetothermia bacterium*	257	MBI1730593.1	31.4	71
MUP	49187_19000_20800	Chaperone protein DnaK	*Halomicrobium*	189	WP_232570114.1	29.9	87
	49187_48077	Chaperone protein DnaK	*Halorhabdus*	154	WP_136687911.1	23.8	78
	50526	Chaperone protein HtpG	*Salinimicrobium*	60	WP_097056766.1	9.5	93
	46507_11372_17801	ClpB protein	*Salinibacter*	317	WP_118840038.1	35.5	79

but were super coiled (Supplementary Table 6; Supplementary Figs. S5–S7). Unlike ClpB identified in *E. coli*, the compactness of the predicted homologous protein entails the adaptation strategies of the extremophiles in extreme environments such as Lake As'ale (Zolkiewski et al. 2012). A single amino acid replacement (N-T) on ATP binding site of ClpB was also observed (Supplementary Table 6). Though it is difficult to be certain on the functionality of the specific predicted protein, the substitution of a non-charged polar amino acid by another non-charged polar residue may be tolerated to keep the functionality of the protein (Betts and Russell 2003). ClpB protein was also predicted in MUP; however the sequence was partial, and only the ClpA/ClpB_AAA_lid domain was identified (Table 10.5).

Long consensus translated sequences for DnaK (Hsp70) were retrieved from all metagenome data (Table 10.5). DnaK is known to be induced as response to environmental stressors in both eukaryotes and prokaryotes, and their synthesis can usually be increased or decreased during stressful conditions (Susin et al. 2006; Yusof et al. 2022). Generally, DnaK carries out its chaperone functions (stabilize and assist proteins refolding) in collaboration with Hsp40 (DnaJ) and GrpE as Hsp70 system (Voth and Jakob 2017). Furthermore, together with ClpB, the Hsp70 system extracts and refolds aggregated proteins once non-stress environmental conditions have been restored (Susin et al. 2006; Ungelenk et al. 2016). Identification of numerous partial sequences of Hsp40 (DnaJ) showed the DnaK/J chaperone machinery together with ClpB in *Salinibacter* and *Bacillus* are needed for survival of cells after exposure of multiple environmental stresses at the shorelines of Lake As'ale. Partial ORFs for DnaK were binned to a number of genera in class Halobacteria, and no genes for DnaJ from archaea were predicted (Table 10.5). Though DnaK is one of the two most common families of heat-shock proteins in haloarchaea (Matarredona et al. 2020), the gene hsp70 (dnaK) may be absent in several archaeal species of different phylogenetic branches (Macario and de Macario 1999). Further in-depth investigation on the gene hsp70 system in the haloachaea inhabitants of Lake As'ale and MUP will help unveil evolutionary puzzle that raises the important question of what replaces the product of this gene, Hsp70 (DnaK), in protein biogenesis and refolding and for stress resistance.

10.4 Discussion

In this investigation, we learned that Lake As'ale and the MUP are two different extreme habitats found in the Danakil depression. We anticipated, based on the nature of their hypersalinity, that we would find consensus of their microbial contents. Instead, the results of the physico-chemical and the metagenomics analysis showed discord between these two aquatic environments, even though they are in proximity of each other. Differences in ionic contents such as Mg^{2+} and SO_4^{2-} and H^+ appear to drive the differences in diversity and abundance of prokaryotes inhabiting the studied environments. Furthermore, different stress response proteins were also predicted, indicating different strategies to overcome their respective environmental strains. The diversity in microbial niches is rich in this environment leading to marked diversity in microbial populations. Further, the microbes inhabiting the shoreline brine of Lake As'ale differ from the previously studied open water inhabitants of the Lake. So, even in the same lake, a diversity in microbial populations reflect diverse physical environments within the lake. In accordance with the high concentration of divalent ions and lower pH measured in the current study, various genes from the inhabiting prokaryotes were annotated as proteins known for their involvement in responding to oxidative stress. The prediction of different chaperone Hsp genes was indicative of how the bacterial and archaeal communities in the two studied places stabilize and assist proteins refolding under stressful conditions. More compact and coiled chaperone ClpB homologous proteins were predicted in Lake As'ale. These data can be used to model stress-response mechanisms for oxidation of extremophiles of the Danakil depression and in organisms which may experience such stresses outside of this specific hypersaline environment. Additional investigation is recommended to understand the individual extremophile's stress adaptation mechanisms and their role in the ecosystem.

10.5 Conclusion

We conclude that the niches within these hypersaline aquatic environments are richly diverse for the microbes discovered. While physiochemical measurements in such studies are of high sensitivity and accuracy, they appear to be less sensitive, and with lower acuity, than the microbial populations present in identifying unique niches in the wild. Future analyses will benefit from micro-sampling of these extreme environments both for physiochemical conditions, but also for localized microbial populations. Just as a forest is rich in four-dimensional niches, we conclude that these hypersaline aquatic environments are similarly dynamic and diverse, especially when considering the extreme stresses seen and predicted in these microenvironments. With protocols used and developed herein, we conclude that future studies of extreme environments will do better at identifying the physiochemical niche environments by analysis of the microbial populations inhabiting them, than that of the gross physiochemical analysis predicting microbial inhabitants. We are particularly interested in testing the changes in the cells along gradients of extreme environments with the hypothesis that those cells subjected to dynamically high stress to extreme conditions are ripe for new biological inventions such as membranous organelles and membrane-bound genomes.

Acknowledgments Yara Dallol potash mining company for providing lodging.

Funding

This research was funded by the National Institutes of Health USA grant number 1R35GM140897.

Author Contributions
Each author has approved the submitted version and agrees to be accountable for the author's own contributions and for ensuring that questions related to the accuracy or integrity of any part of the work, even ones in which the author was not personally involved, are appropriately investigated, resolved, and documented in the literature. Author contributions are as follows: Conceptualization, L.T., A.A., and A.S.; Methodology, L.T., A.A., and A.S.; Formal Analysis, L.T., A.A., and A.S.; Investigation, L.T., A.A., and A.S.; Resources, L.T., A.A., G.W., and A.S.; Data Curation, L.T., A.A., G.W., and A.S.; Writing—Original Draft Preparation, L.T.; Writing—Review and Editing, L.T., A.A., G.W., and A.S.; Supervision, A.A., G.W., and A.S.; Project Administration, L.T., A.A., G.W., and A.S.; Funding Acquisition, L.T., A.A., G.W., and A.S.

Institutional Review Board Statement
Not applicable.

Informed Consent Statement
Not applicable.

Data Availability Statement
All data are available directly upon request. Consensus assembled sequences of Chaperone proteins used in this study are available as additional supplementary file. The Metagenome sequence data is available upon request.

Conflicts of Interest
The authors declare no conflict of interest.

References

Abbate E, Bruni P, Sagri M (2015) Geology of Ethiopia: a review and geomorphological perspectives. Landscapes and landforms of Ethiopia. Springer, pp 33–64

Adem GD, Roy SJ, Zhou M, Bowman JP, Shabala S (2014) Evaluating contribution of ionic, osmotic and oxidative stress components towards salinity tolerance in barley. BMC Plant Biol 14:113. https://doi.org/10.1186/1471-2229-14-113

Ahrens CD (2011) Essentials of meteorology: an invitation to the atmosphere. Cengage Learning, p 60

Antón J, Oren A, Benlloch S, Rodríguez-Valera F, Amann R, Rosselló-Mora R (2002) *Salinibacter ruber* gen. nov., sp. nov., a novel, extremely halophilic member of the Bacteria from saltern crystallizer ponds. Int J System Evolution Microbiol 52(2):485–491. https://doi.org/10.1099/00207713-52-2-485

Asrat AK (2016) The Danakil Depression: an exceptional place where different types of extreme environments coexist. Africa's Top Geological Sites. Struik Nature, pp 189–196

Audrito V, Messana VG, Deaglio S (2020) NAMPT and NAPRT: two metabolic enzymes with key roles in inflammation. Front Oncol 10:358. https://doi.org/10.3389/fonc.2020.00358

Banda JF, Lu Y, Hao C, Pei L, Du Z, Zhang Y, Wei P, Dong H (2020) The effects of salinity and pH on microbial community diversity and distribution pattern in the brines of soda lakes in Badain Jaran Desert, China. Geomicrobiol J 37(1):1–12. https://doi.org/10.1080/01490451.2019.1654568

Belilla J, Moreira D, Jardillier L, Reboul G, Benzerara K, López-García JM, Bertolino P, López-Archilla AI, López-García P (2019) Hyperdiverse archaea near life limits at the polyextreme geothermal Dallol area. Nat Ecol Evol 3:1552–1561. https://doi.org/10.1038/s41559-019-1005-0

Berg G, Rybakova D, Fischer D, Cernava T, Vergès MC, Charles T, Chen X, Cocolin L, Eversole K, Corral GH, Kazou M, Kinkel L, Lange L, Lima N, Loy A, Macklin JA, Maguin E, Mauchline T, McClure R, Mitter B, Ryan M, Sarand I, Smidt H, Schelkle B, Roume H, Kiran GS, Selvin J, de Souza RSC, Overbeek L, Singh BK, Wagner M, Walsh A, Sessitsch A, Schloter M (2020) Microbiome definition re-visited: old concepts and new challenges. Microbiome 8(1):103. https://doi.org/10.1186/s40168-020-00875-0

Berg G, Kusstatscher P, Abdelfattah A, Cernava T, Smalla K (2021) Microbiome modulation—toward a better understanding of plant microbiome response to microbial inoculants. Front Microbiol 12:650610. https://doi.org/10.3389/fmicb.2021.650610

Betts MJ, Russell RB (2003) Amino acid properties and consequences of substitutions. Bioinform Geneticists 317:289. https://doi.org/10.1002/9780470059180.ch13

Beyth M (1978) A comparative study of the sedimentary fills of the Danakil Depression (Ethiopia) and Dead-Sea Rift (Israel). Tectonophysics 46(3–4):357–367

Broadbent JR, Larsen RL, Deibel V, Steele JL (2010) Physiological and transcriptional response of *Lactobacillus casei ATCC334* to acid stress. J Bacteriol 192(9):2445–2458. https://doi.org/10.1128/JB.01618-09

Buchfink B, Xie C, Huson DH (2015) Fast and sensitive protein alignment using DIAMOND. Nat Methods 12:59–60. https://doi.org/10.1038/nmeth.3176

Burns D, Camakaris H, Janssen P, Dyall-Smith M (2004) Combined use of cultivation-dependent and cultivation-independent methods indicates that members of most haloarchaeal groups in an Australian crystallizer pond are cultivable. Appl Environ Microbiol 70(9):5258–5265

Cavalazzi B, Barbieri R, Gómez F, Capaccioni B, Olsson-Francis K, Pondrelli M, Rossi AP, Hickman-Lewis K, Agangi A, Gasparotto G, Glamoclija M, Ori GG, Rodriguez N, Hagos M (2019) The Dallol Geothermal Area, Northern Afar (Ethiopia)—an exceptional planetary field analog on Earth. Astrobiology 19(4):553–578. https://doi.org/10.1089/ast.2018.1926

Chatterjee A, Rajarshi K, Ghosh H, Singh MK, Roy OP, Ray S (2020) Molecular chaperones in protein folding and stress management in cyanobacteria. In: Advances in cyanobacterial biology. Academic Press, pp 119–128

Coleine C, Delgado-Baquerizo M (2022) Unearthing terrestrial extreme microbiomes for searching terrestrial-like life in the Solar System. Trends Microbiol 30(11). https://doi.org/10.1016/j.tim.2022.04.002

Dindhoria K, Manyapu V, Ali A, Kumar R (2023) Unveiling the role of emerging metagenomics for the examination of hypersaline environments. Biotechnol Genet Eng Rev 5:1–39. https://doi.org/10.1080/02648725.2023.2197717. Epub ahead of print. PMID: 37017219

Edwards R, Rohwer F (2005) Viral metagenomics. Nat Rev Microbiol 3:504–510. https://doi.org/10.1038/nrmicro1163

Fazzini M, Bisci C, Billi P (2015) The climate of Ethiopia. Landscapes and landforms of Ethiopia. Springer, pp 65–87

Fulda S, Gorman AM, Hori O, Samali A (2010) Cellular stress responses: cell survival and cell death. Int J Cell Biol 214074. https://doi.org/10.1155/2010/214074

Gabaldón T (2007) Evolution of proteins and proteomes: a phylogenetics approach. Evol Bioinform Online 1:51–61

Gomez F (2014) Extreme environment. In: Amils R et al (eds) Encyclopedia of astrobiology. Springer, Berlin, Heidelberg. https://doi.org/10.1007/978-3-642-27833-4_566-2

González-Torres P, Gabaldón T (2018) Genome variation in the model halophilic bacterium *Salinibacter ruber*. Front Microbiol 9:1499. https://doi.org/10.3389/fmicb.2018.01499

Hallsworth JE, Yakimov MM, Golyshin PN, Gillion JLM, D'Auria G, Alves F, La Cono V, Genovese M, McKew BA, Hayes SL, Harris G, Giuliano L, Timmis KN, McGenity TG (2007) Limits of life in $MgCl_2$-containing environments: chaotropicity defines the window. Environ Microbiol 9(3):801–813. https://doi.org/10.1111/j.1462-2920.2006.01212.x

Hughes RH, Hughes JS (1992) A directory of African Wetlands. The World Conservation Union (IUCN), Cambridge UK, The United Nations Environmental Programme (UNEP), Nairobi Kenya, The World Conservation Monitoring Center (WCMC), Cambridge, UK, pp 167–168

Huson D, Beier S, Flade I, Górska A, El-Hadidi M, Mitra S, Ruscheweyh H-J, Tappu R (2016) MEGAN Community Edition - interactive exploration and analysis of large-scale microbiome sequencing data. PLoS Comput Biol 12(6):e1004957. https://doi.org/10.1371/journal.pcbi.1004957

Huson DH, Albrecht B, Bağcı C, Bessarab I, Górska A, Jolic D, Williams RBH (2018) MEGAN-LR: new algorithms allow accurate binning and easy interactive exploration of metagenomic long reads and contigs. Biol Direct 13(1):6. https://doi.org/10.1186/s13062-018-0208-7

Iscen CF, Emiroglu Ö, Ilhan S, Arslan N, Yilmaz V, Ahiska S (2008) Application of multivariate statistical techniques in the assessment of surface water quality in Uluabat Lake, Turkey. Environ Monitor Assess 144(1–3):269–276

Jamali M.A.M., Gopalasingam C.C., Johnson R.M., Tosha T., Muramoto K., Muench S.P, Antonyuk S.V., Shiroa Y. And Hasnainb S.S. (2020). The active form of quinol-dependent nitric oxide reductase from *Neisseria meningitidis* is a dimer. IUCrJ 7(3):404-415. https://doi.org/10.1107/S2052252520003656

Jasim A, Hemmings B, Mayer K, Scheu B (2018) Groundwater flow and volcanic unrest. Adv Volcanol. https://doi.org/10.1007/11157_2018_33

Kazlauskas D, Varsani A, Koonin EV, Krupovic M (2019) Multiple origins of prokaryotic and eukaryotic single-stranded DNA viruses from bacterial and archaeal plasmids. Nat Commun 10:3425. https://doi.org/10.1038/s41467-019-11433-0

Khaledian E, Brayton KA, Broschat SL (2020) A systematic approach to bacterial phylogeny using order level sampling and identification of HGT using network science. Microorganisms 8(2):312. https://doi.org/10.3390/microorganisms8020312

Kim YJ, Moon MH, Song JY, Smith CP, Hong SK, Chang YK (2008) Acidic pH shock induces the expressions of a wide range of stress-response genes. BMC Genomics 9(1):1–10. https://doi.org/10.1186/1471-2164-9-604

Kim JS, Liu L, Vázquez-Torres A (2021) The DnaK/DnaJ chaperone system enables RNA Polymerase-DksA complex formation in salmonella experiencing oxidative stress. Mbio 12(3):e03443–e03420. https://doi.org/10.1128/mBio.03443-20

Lanzén A, Simachew A, Gessesse A, Chmolowska D, Jonassen I, Øvreås L (2013) Surprising prokaryotic and eukaryotic diversity, community structure and biogeography of Ethiopian Soda Lakes. PLoS ONE 8(8):e72577. https://doi.org/10.1371/journal.pone.0072577

Large AT, Goldberg MD, Lund PA (2009) Chaperones and protein folding in the archaea. Biochem Soc Trans 37:46–51. https://doi.org/10.1042/BST0370046

Logares R, Lindström E, Langenheder S, Logue JB, Paterson H, Laybourn-Parry J, Rengefors K, Tranvik L, Bertilsson S (2013) Biogeography of bacterial communities exposed to progressive long-term environmental change. ISME J 7:937–948. https://doi.org/10.1038/ismej.2012.168

Macario AJ, de Macario EC (1999) The archaeal molecular chaperone machine: peculiarities and paradoxes. Genetics 152(4):1277–1283

Maihemuti B, Aishan T, Simayi Z, Alifujiang Y, Yang S (2020) Temporal scaling of water level fluctuations in shallow lakes and its impacts on the lake eco-environments. Sustainability 12:3541. https://doi.org/10.3390/su12093541

Makhdoumi-Kakhki A, Amoozegar MA, Kazemi B, PaiC L, Ventosa A (2012) Prokaryotic diversity in Aran-Bidgol Salt Lake, the largest hypersaline playa in Iran. Microb Environ 27(1):87–93

Master S (2016) Gaet'ale- a reactivated thermal spring and potential tourist hazard in the Asale salt flats, Danakil Depression, Ethiopia. J Appl Volcanol 5:1. https://doi.org/10.1186/s13617-015-0042-x

Matarredona L, Camacho M, Zafrilla B, Bonete M-J, Esclapez J (2020) The role of stress proteins in haloarchaea and their adaptive response to environmental shifts. Biomolecules 10(10):1390. https://doi.org/10.3390/biom10101390

Mernagh TP, Bastrakov EN, Clarke JDA, de Caritat P, Dulfer H, English PM, Jaireth S, Thomas M, Trafford J (2014) Lake Tyrrell, Australia, and its potential for strategic resources. Promotional brochure. Geoscience Australia, Canberra, 6 p

Metsalu T, Vilo J (2015) ClustVis: a web tool for visualizing clustering of multivariate data using Principal Component Analysis and heatmap. Nucleic Acids Res 43(1):566–570. https://doi.org/10.1093/nar/gkv468

Mogk A, Deuerling E, Vorderwülbecke S, Vierling E, Bukau B (2003) Small heat shock proteins, ClpB and the DnaK system form a functional triade in reversing protein aggregation. Mol Microbiol 50:585–595. https://doi.org/10.1046/j.1365-2958.2003.03710.x

Mormile MR, Hong BY, Benison KC (2009) Molecular analysis of the microbial communities of Mars analog lakes in Western Australia. Astrobiology 9(10):919–930

Mortsch LD (1998) Assessing the impact of climate change on the Great Lakes Shoreline Wetlands. Climatic Change 40:391–416. https://doi.org/10.1023/A:1005445709728

Nissenbaum A (1975) The microbiology and biogeochemistry of the Dead Sea. Microb Ecol 2(2):139–161

Nurk S, Meleshko D, Korobeynikov A, Pevzner PA (2017) metaSPAdes: a new versatile metagenomic assembler. Genome Res 27(5):824–834. https://doi.org/10.1101/gr.213959.116

Oksanen J (2015) Multivariate Analysis of Ecological Communities in R: vegan tutorial

Oksanen FJ et al (2017) Vegan: Community Ecology Package. R package Version 2.4-3. https://CRAN.R-project.org/package=vegan

Oren A (2005) A hundred years of Dunaliella research: 1905–2005. Saline Syst 1(2). https://doi.org/10.1186/1746-1448-1-2

Oren A (2006) Life at high salt concentrations. Prokaryotes 2:263–282. https://doi.org/10.1007/0-387-30742-7_9

Oren A (2010) The dying Dead Sea: the microbiology of an increasingly extreme environment. Lakes Reserv: Res Manag 15:215–222. https://doi.org/10.1111/j.1440-1770.2010.00435.x

Paez-Espino D, Eloe-Fadrosh E, Pavlopoulos G, Thomas AD, Huntemann M, Mikhailova N, Rubin E, Ivanova NN, Kyrpides NC (2016) Uncovering Earth's virome. Nature 536:425–430. https://doi.org/10.1038/nature19094

Pedone E, Fiorentino G, Bartolucci S, Limauro D (2020) Enzymatic antioxidant signatures in Hyperthermophilic Archaea. Antioxidants (Basel) 9(8):703. https://doi.org/10.3390/antiox9080703

Plominsky AM, Henríquez-Castillo C, Delherbe N, Podell S, Ramirez-Flandes S, Ugalde JA, Santibañez JF, van den Engh G, Hanselmann K, Ulloa O, De la Iglesia R, Allen EE, Trefault N (2018) Distinctive archaeal composition of an artisanal crystallizer pond and functional insights into salt-saturated hypersaline environment adaptation. Front Microbiol 9:1800. https://doi.org/10.3389/fmicb.2018.01800

Podell S, Emerson JB, Jones CM, Ugalde JA, Welch S, Heidelberg KB, Allen EE et al (2014) Seasonal fluctuations in ionic concentrations drive microbial succession in a hypersaline lake community. ISME J 8(5):979–990

Poole LB, Hall A, Nelson KJ (2011) Overview of peroxiredoxins in oxidant defense and redox regulation. Curr Protoc Toxicol 2011;Chapter 7:Unit7.9. https://doi.org/10.1002/0471140856.tx0709s49

Price CT, Bukka A, Cynamon M, Graham JE (2008) Glycine betaine uptake by the ProXVWZ ABC transporter contributes to the ability of Mycobacterium tuberculosis to initiate growth in human macrophages. J Bacteriol 190(11):3955–3961. https://doi.org/10.1128/JB.01476-07

Rajput VD, Harish SRK, Verma KK, Sharma L, Quiroz-Figueroa FR, Meena M, Gour VS, Minkina T, Sushkova S, Mandzhieva S (2021) Recent developments in enzymatic antioxidant defence mechanism in plants with special reference to abiotic stress. Biology 10(4):267. https://doi.org/10.3390/biology10040267

Rawat M, Maupin-Furlow JA (2020) Redox and thiols in Archaea. Antioxidants 9(5):381. https://doi.org/10.3390/antiox9050381

Rodriguez R, Durán P (2020) Natural holobiome engineering by using native extreme microbiome to counteract the climate change effects. Front Bioeng/Biotechnol 8:568. https://doi.org/10.3389/fbioe.2020.00568

RStudio Team (2020) RStudio: Integrated Development Environment for R, Boston, MA. Available at: http://www.*rstudio*.com/

Saini M, Kashyap A, Bindal S, Saini K, Gupta R (2021) Bacterial gamma-glutamyl transpeptidase, an emerging biocatalyst: insights into structure-function relationship and its biotechnological applications. Front Microbiol 12:641251. https://doi.org/10.3389/fmicb.2021.641251

Sánchez-Riego AM, López-Maury L, Florencio FJ (2013) Glutaredoxins are essential for stress adaptation in the cyanobacterium *Synechocystis sp. PCC6803*. Front Plant Sci 4:428. https://doi.org/10.3389/fpls.2013.00428

Segata N, Waldron L, Ballarini A, Narasimhan V, Jousson O, Huttenhower C (2012) Metagenomic microbial community profiling using unique clade-specific marker genes. Nat Methods 9(8):811–814

Shrivastava P, Kumar R (2015) Soil salinity: a serious environmental issue and plant growth promoting bacteria as one of the tools for its alleviation. Saudi J Biol Sci 22(2):123–131. https://doi.org/10.1016/j.sjbs.2014.12.001

Simachew A, Lanzén A, Gessesse A, Øvreås L (2016) Prokaryotic community diversity along an increasing salt gradient in a soda ash concentration pond. Microb Ecol 71(2):326–338. https://doi.org/10.1007/s00248-015-0675-7

Sorokin DY, Berben T, Melton ED, Overmars L, Vavourakis CD, Muyzer G (2014) Microbial diversity and biogeochemical cycling in soda lakes. Extremophiles 18(5):791–809. https://doi.org/10.1007/s00792-014-0670-9

Srivastava P, Wu QS, Giri B (2019) Salinity: an overview. In: Giri B, Varma A (eds) Microorganisms in saline environments: strategies and functions. Soil biology, vol 56. Springer, Cham. https://doi.org/10.1007/978-3-030-18975-4_1

Susin MF, Baldini RL, Gueiros-Filho F, Gomes SL (2006) GroES/GroEL and DnaK/DnaJ have distinct roles in stress responses and during cell cycle progression in Caulobacter crescentus. J Bacteriol 188(23):8044–8053. https://doi.org/10.1128/JB.00824-06

Tait AW, Gagen EJ, Wilson SA, Tomkins AG, Southam G (2017) Microbial populations of stony meteorites: substrate controls on first colonizers. Front Microbiol 8. https://doi.org/10.3389/fmicb.2017.01227

Tilahun L, Asrat A, Wessel GM, Simachew A (2021a) In silico determination of nitrogen metabolism in microbes from extreme conditions using metagenomics. Arch Microbiol 203:2521–2540. https://doi.org/10.1007/s00203-021-02227-4

Tilahun L, Asrat A, Wessel GM, Simachew A (2021b) Prediction of genes that function in methanogenesis and CO_2 pathways in extremophiles. Microorganisms 9:2211. https://doi.org/10.3390/microorganisms9112211

Tittelmeier J, Nachman E, Nussbaum-Krammer C (2020) Molecular chaperones: a double-edged sword in neurodegenerative diseases. Front Aging Neurosci 12. https://doi.org/10.3389/fnagi.2020.581374

Tran Q, Phan V (2020) Assembling reads improves taxonomic classification of species. Genes 11:946. https://doi.org/10.3390/genes11080946

Truong DT, Franzosa EA, Tickle TL, Scholz M, Weingart G, Pasolli E, Segata N et al (2015) MetaPhlAn2 for enhanced metagenomic taxonomic profiling. Nat Methods 12(10):902

Ungelenk S, Moayed F, Ho CT, Grousl T, Scharf A, Mashaghi A, Tans S, Mayer MP, Mogk A, Bukau B (2016) Small heat shock proteins sequester misfolding proteins in near-native conformation for cellular protection and efficient refolding. Nat Commun 7:13673. https://doi.org/10.1038/ncomms13673

Valette O, Tran TTT, Cavazza C, Caudeville E, Brasseur G, Dolla A, Talla E, Pieulle L (2017) Biochemical function, molecular structure and evolution of an atypical thioredoxin reductase from *Desulfovibrio vulgaris*. Front Microbiol 8:1855. https://doi.org/10.3389/fmicb.2017.01855

Venables WN, Ripley BD (2002) Modern applied statistics with S, 4th edn. Springer, New York. ISBN 0-387-95457-0, https://www.stats.ox.ac.uk/pub/MASS4/

Ventosa A, Arahal D (2009) Physico-chemical characteristics of hypersaline environments and their biodiversity. Extremophiles 2:1–6

Ventosa A, Gutiérrez MC, Kamekura M, Zvyagintseva IS, Oren A (2004) Taxonomic study of *Halorubrum distributum* and proposal of *Halorubrum terrestre* sp. nov. Int J System Evolution Microbiol 54(2):389–392. https://doi.org/10.1099/ijs.0.02621-0

Verghese J, Abrams J, Wang Y, Morano KA (2012) Biology of the heat shock response and protein chaperones: budding yeast (*Saccharomyces cerevisiae*) as a model system. Microbiol Mol Biol Rev 76(2):115–158

Vidovic A, Supek F, Nikolic A, Krisko A (2014) Signatures of conformational stability and oxidation resistance in proteomes of pathogenic bacteria. Cell Rep 7:1393–1400. https://doi.org/10.1016/j.celrep.2014.04.057

Viswadeepika K, Bramhachari PV (2022) Mechanistic adaptation of microbiomes in extreme environments. In: Veera Bramhachari P (ed) Understanding the microbiome interactions in agriculture and the environment. Springer, Singapore. https://doi.org/10.1007/978-981-19-3696-8_3

Viver T, Orellana L, González-Torres P, Díaz S, Urdiain M, Farías ME, Benes V, Kaempfer P, Shahinpei A, Amoozegar MA, Amann R, Josefa AJ, Konstantinidis KT, Rosselló-Móra R (2018) Genomic comparison between members of the Salinibacteraceae family, and description of a new species of *Salinibacter (Salinibacter altiplanensis sp. nov.)* isolated from high altitude hypersaline environments of the Argentinian Altiplano. System Appl Microbiol 41(3):198–212. https://doi.org/10.1016/j.syapm.2017.12.004

Voth W, Jakob U (2017) Stress-activated chaperones: a first line of defense. Trends Biochem Sci 42(11):899–913. https://doi.org/10.1016/j.tibs.2017.08.006

Wen J, Wang W, Xu K, Ji D, Xu Y, Chen C, Xie C (2020) Comparative analysis of proteins involved in energy metabolism and protein processing in *Pyropia haitanensis* at different salinity levels. Front Mar Sci 7. https://doi.org/10.3389/fmars.2020.00415

Winter J, Linke K, Jatzek A, Jakob U (2005) Severe oxidative stress causes inactivation of DnaK and activation of the redox-regulated chaperone Hsp33. Mol Cell 17(3):381–392. https://doi.org/10.1016/j.molcel.2004.12.027

Xie X, He Z, Chen N, Tang Z, Wang Q, Cai Y (2019) The roles of environmental factors in regulation of oxidative stress in plant. Biomed Res Int 2019:9732325. https://doi.org/10.1155/2019/9732325

Yang Y, Cui H-L, Zhou P-J, Liu S-J (2007) *Haloarcula amylolytica* sp. nov., an extremely halophilic archaeon isolated from Aibi Salt Lake in Xin-Jiang, China. Int J System Evolution Microbiol 57(1):103–106. https://doi.org/10.1099/ijs.0.64647-0

Yang J, Ma L, Jiang H, Wu G, Dong H (2016) Salinity shapes microbial diversity and community structure in surface sediments of the Qinghai-Tibetan Lakes. Sci Rep 6:25078. https://doi.org/10.1038/srep25078

Yata K, Aoshima M (2010) Effective PCA for high-dimension, low-sample-size data with singular value decomposition of cross data matrix. J Multivar Anal 101(9):2060–2077. https://doi.org/10.1016/j.jmva.2010.04.006

Yeh H, Lee J (2018) Stable hydrogen and oxygen isotopes for groundwater sources of Penghu Islands, Taiwan. Geosciences 8:84. https://doi.org/10.3390/geosciences8030084

Yusof NA, Masnoddin M, Charles J, Thien YQ, Nasib FN, Wong CMVL, Abdul Murad AM, Mahadi NM, Bharudin I (2022) Can heat shock protein 70 (HSP70) serve as biomarkers in Antarctica for future ocean acidification, warming and salinity stress? Polar Biol:1–24

Zhang Y, Mandava CS, Cao W, Li X, Zhang D, Li N, Zhang Y, Zhang X, Qin Y, Mi K, Lei J, Sanyal S, Gao N (2015) HflX is a ribosome-splitting factor rescuing stalled ribosomes under stress conditions. Nat Struct Mol Biol 22(11):906–913. https://doi.org/10.1038/nsmb.3103

Zhang H, Yang J, Wu S, Gong W, Chen C, Perrett S (2016) Glutathionylation of the bacterial Hsp70 chaperone DnaK provides a link between oxidative stress and the heat shock response. J Biol Chem 291(13):6967–6981. https://doi.org/10.1074/jbc.M115.673608

Zhou J, Bruns MA, Tiedje JM (1996) DNA recovery from soils of diverse composition. Appl Environ Microbiol 62:316–322

Zolkiewski M, Zhang T, Nagy M (2012) Aggregate reactivation mediated by the Hsp100 chaperones. Arch Biochem Biophys 520(1):1–6. https://doi.org/10.1016/j.abb.2012.01.012

Chapter 11
Somatic Cell Fusion in Host Defense and Adaptation

Jeffrey L. Platt and Marilia Cascalho

Abstract Evidence of fusion of somatic cells has been noted in health and in disease for more than a century. The most obvious but uncertain hallmark has been the presence of multiple nuclei in cells. Although multinucleated cells are found in normal and diseased tissues, the benefit or harm of such cells can be difficult to elucidate. Still more difficult however is the identification of mononuclear cells previously formed by fusion of somatic cells with one or more nuclei disposed. The later process can introduce mutations that promote viral diversification, cancer, and tissue senescence. Less obvious the potential benefits of cell fusion. Recent work in cell biology, immunology, and genomic analysis however makes it possible to postulate benefits and potentially arrive at novel therapeutic agents and approaches that replicate or enhance these benefits.

11.1 Introduction

The division of one cell into two occurs roughly 10^{16} times in a human lifetime (Cairns 1975) and few processes have been investigated more thoroughly and are better understood. The reverse process, cell fusion, has been recognized and investigated for nearly two centuries (Barski et al. 1961; Schwann 1847), yet the frequency, mechanisms, and consequences of fusion of somatic cells in various conditions remain largely inscrutable, sometimes sparking divergent theories and controversy. Using our initial observations on cell fusion as an illustration, we shall discuss the hurdles to investigation of the process and a potential solution and then

J. L. Platt (✉) · M. Cascalho
Departments of Surgery and Microbiology & Immunology University of Michigan, Ann Arbor, MI, USA
e-mail: plattjl@med.umich.edu

M. Kloc, A. Uosef (eds.), *Syncytia: Origin, Structure, and Functions*, Results and Problems in Cell Differentiation 71,
https://doi.org/10.1007/978-3-031-37936-9_11

consider how cell fusion might benefit or impose risks on higher eukaryotic organisms.

11.2 Identification of Somatic Cell Fusion

We were drawn to the subject of cell fusion by unanticipated observations made in the course of research aimed at testing whether human thymus tissue and T cells could be generated de novo by introduction of human hematopoietic stem cells (HSC) into fetal pigs (Ogle et al. 2009). The experiments had both practical and theoretical rationales, as de novo formation and selection of T cells from stem cells of a given individual might be used to repair immunodeficiencies, and since T cell development requires presence of thymus cells. Besides practical uses, the experimental approach could provided an opportunity to examine trans-differentiation of adult stem cells, then a matter of controversy. After administration of human HSC at mid gestation, we detected human cells in the porcine thymus and in the blood at birth and at various times up to 2 years of age. The human cells included lymphocytes expressing markers and exhibiting functional properties characteristic of human T cells. The human T cells comprised a diverse antigen receptor repertoire and proliferated in response to allogeneic human cells but not in response to porcine cells or cells from the human source of HSC. The T cells also proliferated in response to antigenic components of vaccines that had been administered to the young pigs. A large fraction of the human T cells contained "T cell receptor excision circles," indicating the T cells had acquired antigen receptors by recent recombination of T cell antigen receptor genes in the chimeric thymus of the porcine host (and did not reflect undetected contamination of HSC by mature T cells).

In addition to "normal" human lymphocytes, we also found a population of mononuclear cells with both human proteins and porcine proteins on the surface. The presence of porcine proteins on the surface of human cells might be explained if the human cells had adsorbed proteins released from porcine cells and circulating in the plasma. The human cells might also acquire proteins with lipid anchors released from porcine cells into the lipid fraction of blood, as previously observed in two other systems (Kooyman et al. 1995; McCurry et al. 1995).

To explore how human cells (or porcine cells) acquired heterologous surface markers, we used Epstein-Barr virus (EBV) to transform and expand human B cells present in the blood of chimeric pigs. Since EBV transforms human but not porcine B cells, the human B cells can be cultivated, and the population expanded in the absence of porcine plasma and lipid. After serial passage, the expanded population of human cells continued to express markers of human B cells, as expected, and a population of the cells also continued to express porcine markers, such as the saccharide Galα1-3Gal, a product of α1,3-glactosyltransferase expressed by pigs but not humans. Consistent with ongoing synthesis and expression of Galα1-3Gal on the expanded human B cells, the cells had mRNA encoding α1,3-glactosyltransferase and certain other porcine markers.

The human B cells thus contained coding and regulatory sequences needed to enable ongoing expression of one or more porcine proteins. The mechanism of gene transfer or cell hybridization was not obvious given occurrence in piglets in multiple experiments with human HSC from multiple sources and the high level and enduring expression (cells obtained up to 2 years after and cells subjected to multiple passages). The mechanism became apparent however when nuclei were probed for human and porcine DNA and karyotyping was performed. Chromosomal DNA contained abundant ALU sequences (as expected, since the cells could be transformed with EBV) but also hybridized porcine repeat sequences. The chromosome number varied, usually between numbers typical of human and pig; the banding patterns varied, some chromosomes hybridizing both human and porcine probes. In situ hybridization in tissue sections revealed hybrid cells in various tissues, including blood, spleen, skin, and thymus. These findings were consistent with changes in karyotype, gene expression, and protein products of inter-species hybrid cells generated by spontaneous or virus-induced fusion in vitro and explored for many decades (Ephrussi and Weiss 1965; Harris et al. 1965). Our observations provided the first report of spontaneous fusion of somatic cells of disparate species in vivo.

11.3 Adaptive Evolution of Hybrid Cells and Viruses

Hybrid B cells in the human-pig chimera appeared to have been subject to selection in utero or after birth or both. One type of selection observed in hematopoietic cell transplantation is exerted by natural killer cells which eliminate "foreign cells" and abnormal autogenous cells lacking discernable major histocompatibility complex class 1 (h*MHCI*)-encoded polypeptides. The expression or lack of expression of h*MHCI*-encoded polypeptides on the surface of hybrid B cells in a given piglet might differ owing to generation of hybrid cells in multiple fusion events. However, since the same source of human hematopoietic stem cells and hence the same *MHCI* was introduced into fetuses of a given pregnancy, expressed sequences should manifest the same allotypic properties. On the other hand, cells lacking expression of *MHCI* might be subject to selection by natural killer cells in hybrid animals leading to uniformity of expression in a given individual. Consistent with the impact of selection, hybrid B cells in a given piglet always expressed or always lacked hMHCI suggesting selection by natural killer cells might have determined survival of the hybrid cells.

Selection and evolution also potentially influenced viral activation and transmission in chimeric animals. Pigs harbor an endogenous retrovirus, porcine endogenous retrovirus, that potentially infect human cells under certain conditions (Patience et al. 1997). However, the virus has not been reported to infect human recipients of porcine xenotransplants (Irgang et al. 2003; Paradis et al. 1999). It was striking therefore to observe that porcine endogenous retrovirus sequences were detected in hybrid cells in chimeric pigs, and perhaps more importantly that the virus acquired capacity for transmission to normal human cells (Ogle et al. 2004b). This observation was consistent with the concept that cell fusion potentially fuels viral evolution and potentially recombination in hybrid cells enables selection of multiple disparate genomic sequences that could adaptively confer resistance (Ogle et al. 2005).

11.4 Hurdles to Detecting and Investigating Somatic Cell Fusion

The discovery of spontaneously generated human-porcine hybrid cells in chimeric pigs (Ogle et al. 2004b) highlights some vexing obstacles to advancing knowledge of similar phenomena. Fusion of one cell with another in vivo cannot be observed directly but must be deduced by exclusion of other explanations for detection of bi-nucleated or multi-nucleated cells and a hybrid genome or from experimental observations on cells in culture. Likewise the mechanisms promoting somatic cell fusion must be surmised indirectly, e.g., by detecting changes in numbers of hybrid or multinucleated cells after blockade or induced expression of putative fusogen and are not amenable to direct observation in vivo (Brukman et al. 2019; Dittmar and Hass 2022; Leikina et al. 2018). That is not to suggest conclusions about cell fusion have been erroneous but rather that the conditions including tissue compositions and microenvironments conducive to cell fusion are incompletely understood.

Much is known about the cellular and molecular processes by which cell fusion generates certain multinucleated cells and how the physiology and pathology of such cells depends on cell fusion. Thus, trophoblast, osteoclast, and skeletal muscle, in mature mammals, are known to be formed by fusion of precursor cells and the function of the multinucleated cells to depart significantly from the functions of the precursor cells (Lucas and Cooper 2023; Renaud and Jeyarajah 2022; Whitlock et al. 2023). Recent investigation of regional physiology of syncytiotrophoblast provides examples of investigation aligning location and function that could be undertaken (Arutyunyan et al. 2023). The locations and sequential fusion events that generate syncytiotrophoblast over time are not subject to observation, and therefore the origin of any one of the multiple nuclei in these multinucleated cells must be assumed. However, these limitations do not appear to detract from achieving a thorough understanding of the process of fusion and function of the multinucleated cells.

Far more difficult has been the investigation of cell fusion in settings in which binuclear or multinucleated cells transition by one of several pathways to mononuclear cells. Thus, binuclear and multi-nuclear cells formed by fusion can undergo ongoing change in the number and structure of chromosomes, "ploidy reduction, polyploid reversal or "depolyploidation" (Hoehn et al. 1975; Ruddle and Creagan 1975; Wang et al. 2021; Wu et al. 2022) to form more or less stable mononuclear cells with varying karyotype and somatic genome. Indeed, the resolution of polyploidy in cultured cells provides a powerful tool for investigation of dynamic processes and for gene mapping (Hoehn et al. 1987; Ruddle 1984). The most thoroughly investigated example of such resolution occurs in hybridoma cells formed by fusion of activated B cells with multiple myeloma cells (Kohler and Milstein 1975). However, observations are potentially influenced by the cell type used, conditions favoring cell fusion (where deliberately induced), selection for metaphase or resistance to cytotoxins, and dependence on other variables (e.g., survival or expansion in culture).

As might be expected, mitosis of tetraploid, polyploid, and aneuploid cells and other processes associated with conversion of multi-nucleated cells to mononuclear cells have been connected with chromosomal instability and mutagenesis culminating in malignant transformation (Kloc et al. 2022; Matsumoto et al. 2021; Wang et al. 2021; Zhou et al. 2015). However, similar changes are observed in cells thought to have adapted to injury, ischemia, or other untoward processes. Therefore, whether cell fusion and ensuing changes in ploidy, chromosomal structure or sequence, including those accompanying transition of ploidy are pathogenic or adaptive or merely incidental has been impossible to establish in "normal cells" or normal organisms.

The presumed occurrence and consequences of cell fusion in homologous species has been studied in chimeras. Human bone marrow and stem cell transplant recipients have been found to harbor cells with hybrid DNA reflecting co-integration of donor and the recipient DNA (Dittmar 2022; D. Duelli and Lazebnik 2007; Gast et al. 2018; Ogle et al. 2005; Pawelek and Chakraborty 2008). Co-integration of genomic DNA from a transplant and recipient was evident as described above in some cells of the human-pig hematopoietic cell chimera we reported (Ogle et al. 2004b; Ogle et al. 2009) and described above. Although co-integration of donor and recipient DNA was detected in human subjects with the context of malignancy, we never observed malignancy in human-pig chimeras. However, our observations led us to postulate that cell fusion might nonetheless facilitate evolution of viruses and of species, development, adaptation, and development of malignancy in chimeras (Ogle et al. 2005; Platt and Cascalho 2019). Whether and how cell fusion fuels evolution, adaptation, and disease in non-chimeric individuals remains a matter of speculation.

Recent advances enabling rapid single cell genomic sequencing and analytic methodologies encourage us to think that "footprints" of prior cell fusion in non-chimeric individuals might be found. We, like some others, currently use high throughput sequencing to trace clonal evolution of antigen-specific B cells and viruses (de Mattos Barbosa et al. 2021a, b; Platt et al. 2019). Although this work focuses on limited and rapidly mutating regions of the genome, similar approaches could be applied to broader regions or to whole genome sequences. Indeed, evidence of broad genome diversification in individual cells in normal tissues and cancers already suggest several distinguishable mechanisms (as examples see (Acha-Sagredo et al. 2022; Balachandran et al. 2022; Dolle et al. 2000; Martincorena and Campbell 2015)). We postulate that cells generated by fusion would exhibit appreciable variation in chromosome number and structure (recombination), consistent with our observations on in vitro cell fusion (Zhou et al. 2015) and with detailed observations on the properties of tumor cell-macrophage hybrids (Gast et al. 2018). On the other hand, both point mutation and structural variation would be prominent in tissues with frequent mitosis and/or endoreplication but rare or absent cell fusion. As later discussed, these mechanisms might be exploited for adaptation and control of disease, including malignancy.

11.5 Consideration of Cell Fusion in Host Defense and Adaptation to Environmental Challenge

The observation of cell fusion in various conditions suggests cell fusion could induce or amplify those conditions. Thus, multi-nucleated cells and evidence of cell fusion in infection and cancer was taken to suggestion cell fusion could have pathogenic import and led to experimental work elucidating mechanisms (see (Berndt et al. 2013; Duelli et al. 2005, 2007; Fortuna et al. 1990; Gast et al. 2018; Harkness et al. 2013; Houghton et al. 2004; Lazova et al. 2013; Lu and Kang 2009; Ogle et al. 2004a; Platt et al. 2016; Powell et al. 2011; Yoshizaki et al. 1994; Zhou et al. 2015) as examples). We do not doubt that fusion of non-malignant cells can sometimes induce malignant transformation and that fusion of macrophages or other cells with transformed cells can induce progression of malignancy. However, an observation connecting cell fusion with pathology might also suggest that cell fusion confers a biological benefit in that condition. Given the complexity of disease and the difficulties identifying the progeny of and circumstances inciting somatic cell fusion, quantitative weighing of harm and benefit in individuals or in populations cannot be proved. We shall list certain conditions in which cell fusion could have mediated resistance to or recovery from disease or injury and offer suggestions about inquiry that could eventually determine the balance.

11.5.1 Immunity

Cell fusion potentially improves ability to mount adaptive immune responses by providing a source of antigen and agonists that prompt leukocytes to engulf and process antigen for presentation of T cells (see (Platt and Cascalho 2019) for detailed discussion). Since the repertoire of T cells is selected to allow recognition of "foreign" or novel peptides and to limit recognition of "self" peptides, mutations in cancer genomes potentially enable the development of protective immune responses (Ward et al. 2016) and by causing mutation and recombination at multiple loci (Zhou et al. 2015), cell fusion might enable the immune system to recognize at least some hybrid cells. Consistent with that concept are clinical observations correlating protective tumor immunity with extent of mutation in human cancers.

While appealing, this concept linking mutagenesis and tumor immunity has at least two potential flaws. One flaw reflects concern that oncogenesis is integrally connected with mutation and tumors subject to investigation escaped control by tumor immunity (Schreiber et al. 2011). Similar concerns could be posed regarding immunity to mutable viruses such as HIV and hepatitis c. Ultimately, these concerns do not argue against the import of anti-tumor or anti-viral immunity, as the incidence and severity malignancy and viral disease are profoundly worse in immunodeficient individuals. Put in another way, the ability of adaptive immunity, particularly cell-mediated immunity to recognize and respond to products of mutant genes, likely averts development of some tumors and infections and converts otherwise acute and rapidly fatal disease to chronic conditions. Indeed, the presence of oncogenic mutations in the absence of discernable cancer (Kennedy et al. 2019) might reflect effective host defense, potentially involving cell fusion.

The second concern about the concept linking mutation in tumor and viral genomes to cell fusion and protective immunity concerns diversification of the tumor or infecting virus and the possibility and perhaps likelihood that escape from immunity will ensue. Mutagenesis of a nidus of malignancy or infection generates antigenic variants expressed by some cells but not others and cells lacking a given mutation will not express the novel antigens that spark protective immunity. This concern underlies the evolution of viruses and tumors in a given host, but it does not necessarily preclude immune control. A tumor or nidus of infection is surrounded by and in part depends on neighboring capillaries. Both the capillaries and antigen presenting cells that migrate through the capillaries can engulf cellular debris and antigens and present novel peptides for recognition of responding T cells. The protective immunity that ensues is not limited to cytotoxicity directed at tumor cells, although that is part of the response, but includes generation of agonists that modify the physiology of capillaries to suppress blood flow and inflict ischemia on cells dependent on the local capillary network (Saadi et al. 2002). Indeed, these regional changes are countered by tumor-induced immunosuppression.

Cell fusion could have a greater impact than commonly appreciated on protective immunity. An important limitation on the putative connection between mutagenesis and protective immunity may stem from dilution of trace amounts novel polypeptide by abundant amounts of "self" polypeptide produced in tumors and infected cells. Recognition of antigen by T cells (and B cells) requires ligation of a threshold number of antigen receptors. Although the threshold for activation of memory T cells (and B cells) is markedly lower than the threshold for naïve cells, activation of memory cells nonetheless requires appreciable stimulation of antigen receptors. Therefore, presence of only trace amounts of antigen in the vicinity of an antigen presenting cell or dilution of this neoantigen by self-antigen potentially hinders activation of T cells. Put in another way, insufficient neoantigen or excess competition can prevent a protective cell-mediated immune response. Because availability of antigen is often limited in these ways, vaccines are typically designed not only to provide abundant antigen but also to provoke local antigen-presenting cells to take up, process, and present antigen coordinately. For example, vaccines are often designed for injection at sites where tissue damage together with adjuvant substances activates regional antigen-presenting cells and the antigen of vaccines is often aggregated in ways that provide multiple copies for the presenting cell. Thus, heparan sulfate and possibly other substances released from damaged cells can activate antigen presenting cells (Johnson et al. 2002), causing selective processing and presenting of antigen by newly expressed major histocompatibility antigens (Wrenshall et al. 1991) and by fixing and prolonging such expression (Kodaira et al. 2000). Cellular damage also generates activation of the complement system thereby generating a powerful agonist, C3d, that lowers the threshold for activation of protective cell-mediated immunity (Platt et al. 2017). Where fusion of cells engenders damage and death of cells expressing novel proteins, the conditions underlying cell fusion potentially facilitate generation and action of protective immunity.

The focus we and others have place on the cells that thrive after fusion potentially obscurs an important adaptive impact of the process. Many cells that survive and proliferate after fusion ultimately die (Hoehn et al. 1975; Zhou et al. 2015). Cell death could reflect genomic instability or crowding by more rapidly proliferating neighboring cells. Transient survival and proliferation generates agonists and potentially liberates sufficient neoantigen to elicit an immune response. Our observation that a complement end-product promotes immunity to rapidly mutating cells and cell lines (Platt et al. 2017) is consistent with this concept. The concept might also explain how cell fusion promotes the functions of antigen-presenting cells (Gong et al. 1997; Koido 2016) and the ready detection of oncogenic mutations in tissues in which cancer is not found (Kennedy et al. 2019).

11.5.2 Regeneration

Cell fusion potentially confers adaptive benefits beyond protective immunity. One benefit long postulated but difficult to prove concerns healing of injury and regeneration of tissues. Although cell fusion has been found to underlie healing and regeneration in some invertebrate organisms (Hernández and Podbilewicz 2017; Kasprzycka et al. 2019), it can be difficult to determine the extent to which cell fusion observed during the course of healing and regeneration in higher animals is essential for or a consequence of healing and regeneration.

Cell fusion would seem most likely to promote healing of tissues the function of which depends on anatomical alignment, for example muscle and nerve. In these tissues, cell division of parenchymal cells or stem cells potentially could impair anatomic relationships. A full understanding of both may depend not only on resolving the formation and properties of multinucleated cells, but also on determining fate of nuclei that were shed or disposed as described above. Structural diversification of chromosomal DNA might prove useful in tracing prior cell fusion events in such cells.

The occurrence and consequences of cell fusion during regeneration of liver represent a different and illuminating system. Regeneration and maintenance of liver is characterized at least in part by proliferation of hepatocytes; consistent with that concept is a broad distribution of point mutations and diversification of hepatocyte

DNA (Blokzijl et al. 2016). However, cell fusion also clearly occurs, and hybrid tetraploid cells can be detected (Myerson and Parkin 2019).

11.6 Concluding Remarks

The involvement of cell fusion in various aspects of health and disease has been postulated for more than a century but investigation has been hindered by technical and conceptual challenges. While the potential contribution of cell fusion to development of polyploidy was obvious, the possibility that mononuclear cells might represent the progeny of fused somatic cells was infrequently considered. We think increasingly available methodologies for single cell genomic sequencing and analysis soon might overcome this limitation and in doing so reveal how cell fusion contributes to host defense and adaptation.

Evidence that tumor cells commonly exhibit a propensity to fuse with other cells has led to the idea that cell fusion might fuel cell transformation and tumor progression. That concept potentially explains the high incidence of malignancy in tissues and in cell types like breast, prostate, and lung with little or no appreciable baseline proliferation. However, until recently the concept eclipsed consideration of the possibility that cell fusion could also provide an adaptive benefit in cancer, especially the promoting of protective immunity. We hope that recent reports and the present discussion will spark consideration of how the products of fused cancer cells, including dead and dying hybrid cells, potentially benefit the host.

Certain viruses, such as HIV and hepatitis c, are immunogenic yet eminently capable of evading clearance by immunity. These viruses typically promote cell fusion which together with intrinsic lability of the viral genome enables mutation and selection of viral genomes and over time enables the virus to escape control by the immune system of the host. While adaptive immunity cannot control mutable viruses, the processes that eventuate in diversification, including cell fusion, potentially shed light on therapeutic and biological solutions. If mutable viruses use cell fusion to diversify, therapeutic or prophylactic solutions might exploit the products of fusion to hasten protective immunity and slow viral evolution.

We offered only a brief comment on the broad subject of cell fusion in repair and regeneration of tissues. Work on this subject is advancing too quickly to offer enduring insights. However, a full understanding of cell fusion in this setting may face the hurdles like those slowing appreciation of cell fusion in immunity. One hurdle reflects still limited insight into the fate of fused cells that have undergone depolyploidization. A tangentially related subject we did not address concerns aging and the apparent senescence of cells and organs. Cell fusion and endoreplication have been associated with senescence (Kloc et al. 2022) and with characteristic genomic changes (Dolle et al. 2000). It may be reasonable to postulate that both processes underlie accelerating incidence of cancer with aging. As efforts are made to counter aging by reversal of these processes, there may be value in elucidating more fully by-products of cell fusion that confer protection.

Acknowledgment Work by the authors discussed in this communication was supported by grants from the National Institutes of Health (USA), the Department of Defense (USA), and Michigan Translational Research and Commercialization hub.

References

Acha-Sagredo A, Ganguli P, Ciccarelli FD (2022) Somatic variation in normal tissues: friend or foe of cancer early detection? Ann Oncol 33(12):1239–1249. https://doi.org/10.1016/j.annonc.2022.09.156

Arutyunyan A, Roberts K, Troulé K, Wong FCK, Sheridan MA, Kats I, Garcia-Alonso L, Velten B, Hoo R, Ruiz-Morales ER, Sancho-Serra C, Shilts J, Handfield LF, Marconato L, Tuck E, Gardner L, Mazzeo CI, Li Q, Kelava I, Wright GJ, Prigmore E, Teichmann SA, Bayraktar OA, Moffett A, Stegle O, Turco MY, Vento-Tormo R (2023) Spatial multiomics map of trophoblast development in early pregnancy. Nature 616(7955):143–151. https://doi.org/10.1038/s41586-023-05869-0

Balachandran S, Pozojevic J, Sreenivasan VKA, Spielmann M (2022) Comparative single-cell analysis of the adult heart and coronary vasculature. Mamm Genome. (online ahead of print):1–9. https://doi.org/10.1007/s00335-022-09968-7

Barski G, Sorieul S, Cornefert F (1961) "Hybrid" type cells in combined cultures of two different mammalian cell strains. J Natl Cancer Inst 26:1269–1291. http://www.ncbi.nlm.nih.gov/entrez/query.fcgi?cmd=Retrieve&db=PubMed&dopt=Citation&list_uids=13687369

Berndt B, Zanker KS, Dittmar T (2013) Cell fusion is a potent inducer of aneuploidy and drug resistance in tumor cell/ normal cell hybrids. Crit Rev Oncog 18(1–2):97–113. http://www.ncbi.nlm.nih.gov/pubmed/23237554

Blokzijl F, de Ligt J, Jager M, Sasselli V, Roerink S, Sasaki N, Huch M, Boymans S, Kuijk E, Prins P, Nijman IJ, Martincorena I, Mokry M, Wiegerinck CL, Middendorp S, Sato T, Schwank G, Nieuwenhuis EE, Verstegen MM, van der Laan LJ, de Jonge JJN, Vries RG, van de Wetering M, Stratton MR, Clevers H, Cuppen E, van Boxtel R (2016) Tissue-specific mutation accumulation in human adult stem cells during life. Nature 538(7624):260–264. https://doi.org/10.1038/nature19768

Brukman NG, Uygur B, Podbilewicz B, Chernomordik LV (2019) How cells fuse. J Cell Biol 218(5):1436–1451. https://doi.org/10.1083/jcb.201901017

Cairns J (1975) Mutation selection and the natural history of cancer. Nature 255(5505):197–200. http://www.ncbi.nlm.nih.gov/pubmed/1143315

de Mattos Barbosa MG, Lefferts AR, Huynh D, Liu H, Zhang Y, Fu B, Barnes J, Samaniego M, Bram RJ, Geha RS, Shikanov A, Prak ETL, Farkash EA, Platt JL, Cascalho M (2021a) TNFRSF13B genotypes control immune-mediated pathology by regulating the functions of innate B cells. JCI Insight 6(17):e150483. https://doi.org/10.1172/jci.insight.150483

de Mattos Barbosa MG, Liu H, Huynh D, Shelley G, Keller ET, Emmer BT, Sherman E, Ginsberg D, Kennedy AA, Tai AW, Wobus C, Mirabeli C, Lanigan TM, Samaniego M, Meng W, Rosenfeld AM, Prak ETL, Platt JL, Cascalho M (2021b) IgV somatic mutation of human anti-SARS-CoV-2 monoclonal antibodies governs neutralization and breadth of reactivity. JCI Insight 6(9):e147386. https://doi.org/10.1172/jci.insight.147386

Dittmar T (2022) Generation of cancer stem/initiating cells by cell-cell fusion. Int J Mol Sci 23(9):e4514. https://doi.org/10.3390/ijms23094514

Dittmar T, Hass R (2022) Extracellular events involved in cancer cell-cell fusion. Int J Mol Sci 23(24):e16071. https://doi.org/10.3390/ijms232416071

Dolle ME, Snyder WK, Gossen JA, Lohman PH, Vijg J (2000) Distinct spectra of somatic mutations accumulated with age in mouse heart and small intestine. Proc Natl Acad Sci U S A 97(15):8403–8408. http://www.ncbi.nlm.nih.gov/pubmed/10900004

Duelli D, Lazebnik Y (2007) Cell-to-cell fusion as a link between viruses and cancer. Nat Rev Cancer 7(12):968–976. https://doi.org/10.1038/nrc2272

Duelli DM, Hearn S, Myers MP, Lazebnik Y (2005) A primate virus generates transformed human cells by fusion. J Cell Biol 171(3):493–503. https://doi.org/10.1083/jcb.200507069

Duelli DM, Padilla-Nash HM, Berman D, Murphy KM, Ried T, Lazebnik Y (2007) A virus causes cancer by inducing massive chromosomal instability through cell fusion. Curr Biol 17(5):431–437. https://doi.org/10.1016/j.cub.2007.01.049

Ephrussi B, Weiss MC (1965) Interspecific hybridization of somatic cells. Proc Natl Acad Sci U S A 53(5):1040–1042. http://www.ncbi.nlm.nih.gov/entrez/query.fcgi?cmd=Retrieve&db=PubMed&dopt=Citation&list_uids=5222546

Fortuna MB, Dewey MJ, Furmanski P (1990) Enhanced lung colonization and tumorigenicity of fused cells isolated from primary MCA tumors. Cancer Lett 55(2):109–114. http://www.ncbi.nlm.nih.gov/entrez/query.fcgi?cmd=Retrieve&db=PubMed&dopt=Citation&list_uids=2265408

Gast CE, Silk AD, Zarour L, Riegler L, Burkhart JG, Gustafson KT, Parappilly MS, Roh-Johnson M, Goodman JR, Olson B, Schmidt M, Swain JR, Davies PS, Shasthri V, Iizuka S, Flynn P, Watson S, Korkola J, Courtneidge SA, Fischer JM, Jaboin J, Billingsley KG, Lopez CD, Burchard J, Gray J, Coussens LM, Sheppard BC, Wong MH (2018) Cell fusion potentiates tumor heterogeneity and reveals circulating hybrid cells that correlate with stage and survival. Sci Adv 4(9):7828. https://doi.org/10.1126/sciadv.aat7828

Gong J, Chen D, Kashiwaba M, Kufe D (1997) Induction of antitumor activity by immunization with fusions of dendritic and carcinoma cells. Nat Med 3(5):558–561. https://www.ncbi.nlm.nih.gov/pubmed/9142127

Harkness T, Weaver BA, Alexander CM, Ogle BM (2013) Cell fusion in tumor development: accelerated genetic evolution. Crit Rev Oncog 18(1–2):19–42. http://www.ncbi.nlm.nih.gov/pubmed/23237551

Harris H, Watkins JF, Campbell GL, Evans EP, Ford CE (1965) Mitosis in hybrid cells derived from mouse and man. Nature 207(997):606–608

Hernández JM, Podbilewicz B (2017) The hallmarks of cell-cell fusion. Development 144(24):4481–4495. https://doi.org/10.1242/dev.155523

Hoehn H, Bryant EM, Johnston P, Norwood TH, Martin GM (1975) Non-selective isolation, stability and longevity of hybrids between normal human somatic cells. Nature 258(5536):608–610. https://doi.org/10.1038/258608a0

Hoehn H, Koch H, Köhler J, Bettecken T, Kubbies M, Heckl W, Salk D, Rabinovitch PS (1987) Interphase cell flow cytometry as a means of monitoring genomic size in normal and neoplastoid cell cultures. Cancer Genet Cytogenet 24(2):191–204. https://doi.org/10.1016/0165-4608(87)90099-9

Houghton J, Stoicov C, Nomura S, Rogers AB, Carlson J, Li H, Cai X, Fox JG, Goldenring JR, Wang TC (2004) Gastric cancer originating from bone marrow-derived cells. Science 306(5701):1568–1571. http://www.ncbi.nlm.nih.gov/entrez/query.fcgi? cmd=Retrieve&db=PubMed&dopt=Citation&list_uids=15567866

Irgang M, Sauer IM, Karlas A, Zeilinger K, Gerlach JC, Kurth R, Neuhaus P, Denner J (2003) Porcine endogenous retroviruses: no infection in patients treated with a bioreactor based on porcine liver cells. J Clin Virol 28(2):141–154

Johnson GB, Brunn GJ, Kodaira Y, Platt JL (2002) Receptor-mediated monitoring of tissue Well-being via detection of soluble heparan sulfate by toll-like receptor 4. J Immunol 168(10):5233–5239

Kasprzycka P, Archacka K, Kowalski K, Mierzejewski B, Zimowska M, Grabowska I, Piotrowski M, Rafałko M, Ryzko A, Irhashava A, Senderowski K, Gołąbek M, Stremińska W, Jańczyk-Ilach K, Koblowska M, Iwanicka-Nowicka R, Fogtman A, Janowski M, Walczak P, Ciemerych MA, Brzoska E (2019) The factors present in regenerating muscles impact bone marrowderived mesenchymal stromal/stem cell fusion with myoblasts. Stem Cell Res Ther 10(1):e343. https://doi.org/10.1186/s13287-019-1444-1

Kennedy SR, Zhang Y, Risques RA (2019) Cancer-associated mutations but no cancer: insights into the early steps of carcinogenesis and implications for early cancer detection. Trends Cancer 5(9):531–540. https://doi.org/10.1016/j.trecan.2019.07.007

Kloc M, Uosef A, Subuddhi A, Kubiak JZ, Piprek RP, Ghobrial RM (2022) Giant multinucleated cells in aging and senescence-an abridgement. Biology (Basel) 11(8):e1121. https://doi.org/10.3390/biology11081121

Kodaira Y, Nair SK, Wrenshall LE, Gilboa E, Platt JL (2000) Phenotypic and functional maturation of dendritic cells modulated by heparan sulfate. J Immunol 165:1599–1604

Kohler G, Milstein C (1975) Continuous cultures of fused cells secreting antibody of predefined specificity. Nature 256(5517):495–497

Koido S (2016) Dendritic-tumor fusion cell-based cancer vaccines. Int J Mol Sci 17(6):e828. https://doi.org/10.3390/ijms17060828

Kooyman DL, Byrne GW, McClellan S, Nielsen D, Tone M, McCurry KR, Coffman TM, Waldmann H, Platt JL, Logan JS (1995) In vivo transfer of GPI-linked complement restriction factors from erythrocytes to the endothelium. Science 269:89–92

Lazova R, Laberge GS, Duvall E, Spoelstra N, Klump V, Sznol M, Cooper D, Spritz RA, Chang JT, Pawelek JM (2013) A melanoma brain metastasis with a donor-patient hybrid genome following bone marrow transplantation: first evidence for fusion in human cancer. PLoS One 8(6):e66731. https://doi.org/10.1371/journal.pone.0066731

Leikina E, Gamage DG, Prasad V, Goykhberg J, Crowe M, Diao J, Kozlov MM, Chernomordik LV, Millay DP (2018) Myomaker and Myomerger work independently to control distinct steps of membrane remodeling during myoblast fusion. Dev Cell 46(6):767–780.e767. https://doi.org/10.1016/j.devcel.2018.08.006

Lu X, Kang Y (2009) Cell fusion as a hidden force in tumor progression. Cancer Res 69(22):8536–8539. https://doi.org/10.1158/0008-5472.CAN-09-2159

Lucas L, Cooper TA (2023) Insights into cell-specific functions of microtubules in skeletal muscle development and homeostasis. Int J Mol Sci 24(3):e2903. https://doi.org/10.3390/ijms24032903

Martincorena I, Campbell PJ (2015) Somatic mutation in cancer and normal cells. Science 349(6255):1483–1489. https://doi.org/10.1126/science.aab4082

Matsumoto T, Wakefield L, Peters A, Peto M, Spellman P, Grompe M (2021) Proliferative polyploid cells give rise to tumors via ploidy reduction. Nat Commun 12(1):e646. https://doi.org/10.1038/s41467-021-20916-y

McCurry KR, Kooyman DL, Alvarado CG, Cotterell AH, Martin MJ, Logan JS, Platt JL (1995) Human complement regulatory proteins protect swine-to-primate cardiac xenografts from humoral injury. Nat Med 1:423–427

Myerson D, Parkin RK (2019) Donor-derived hepatocytes in human hematopoietic cell transplant recipients: evidence of fusion. Virchows Arch 474(3):365–374. https://doi.org/10.1007/s00428-018-2497-8

Ogle BM, Butters KA, Plummer TB, Ring KR, Knudsen BE, Litzow MR, Cascalho M, Platt JL (2004a) Spontaneous fusion of cells between species yields transdifferentiation and retroviral transfer in vivo. FASEB J 18(3):548–550. https://doi.org/10.1096/fj.03-0962fje.03-0962fje [pii]

Ogle BM, Butters KB, Plummer TB, Ring KR, Knudsen B, Litzow MR, Cascalho M, Platt JL (2004b) Spontaneous fusion of cells between species yields transdifferentiation and retroviral in vivo. FASEB J 18:548–550

Ogle BM, Cascalho M, Platt JL (2005) Biological implications of cell fusion. Nat Rev Mol Cell Biol 6:567–575

Ogle BM, Knudsen BE, Nishitai R, Ogata K, Platt JL (2009) Toward the development of human T cells in swine for potential use in adoptive T cell immunotherapy. Tissue Eng 15(5):1031–1040

Paradis K, Langford G, Long Z, Heneine W, Sandstrom P, Switzer WM, Chapman LE, Lockey C, Onions D (1999) Search for cross-species transmission of porcine endogenous retrovirus in patients treated with living pig tissue. Science 285:1236–1241

Patience C, Takeuchi Y, Weiss RA (1997) Infection of human cells by an endogenous retrovirus of pigs. Nat Med 3:282–286
Pawelek JM, Chakraborty AK (2008) Fusion of tumour cells with bone marrow-derived cells: a unifying explanation for metastasis. Nat Rev Cancer 8(5):377–386. https://doi.org/10.1038/nrc2371
Platt JL, Cascalho M (2019) Cell fusion in malignancy: a cause or consequence? A provocateur or cure? Cell 8(6):e587. https://doi.org/10.3390/cells8060587
Platt JL, de Mattos G, Barbosa M, Cascalho M (2019) The five dimensions of B cell tolerance. Immunol Rev 292(1):180–193. https://doi.org/10.1111/imr.12813
Platt JL, Silva I, Balin SJ, Lefferts AR, Farkash E, Ross TM, Carroll MC, Cascalho M (2017) C3d regulates immune checkpoint blockade and enhances antitumor immunity. JCI Insight 2(9):e90201. https://doi.org/10.1172/jci.insight.90201
Platt JL, Zhou X, Lefferts AR, Cascalho M (2016) Cell fusion in the war on cancer: a perspective on the inception of malignancy. Int J Mol Sci 17(7):e1118. https://doi.org/10.3390/ijms17071118
Powell AE, Anderson EC, Davies PS, Silk AD, Pelz C, Impey S, Wong MH (2011) Fusion between intestinal epithelial cells and macrophages in a cancer context results in nuclear reprogramming. Cancer Res 71(4):1497–1505. https://doi.org/10.1158/0008-5472.CAN-10-3223
Renaud SJ, Jeyarajah MJ (2022) How trophoblasts fuse: an in-depth look into placental syncytiotrophoblast formation. Cell Mol Life Sci 79(8):e433. https://doi.org/10.1007/s00018-022-04475-z
Ruddle FH (1984) The William Allan memorial award address: reverse genetics and beyond. Am J Hum Genet 36(5):944–953
Ruddle FH, Creagan RP (1975) Parasexual approaches to the genetics of man. Annu Rev Genet 9:407–486. https://doi.org/10.1146/annurev.ge.09.120175.002203
Saadi S, Wrenshall LE, Platt JL (2002) Regional manifestations and control of the immune system. FASEB J 16(8):849–856
Schreiber RD, Old LJ, Smyth MJ (2011) Cancer immunoediting: integrating immunity's roles in cancer suppression and promotion. Science 331(6024):1565–1570. https://doi.org/10.1126/science.1203486
Schwann, T. (1847). Microscopical researches into the accordance in the structure and growth of animals and plants (trans: Smith H). The Sydenham Society, London
Wang XF, Yang SA, Gong S, Chang CH, Portilla JM, Chatterjee D, Irianto J, Bao H, Huang YC, Deng WM (2021) Polyploid mitosis and depolyploidization promote chromosomal instability and tumor progression in a notch-induced tumor model. Dev Cell 56(13):1976–1988.e1974. https://doi.org/10.1016/j.devcel.2021.05.017
Ward JP, Gubin MM, Schreiber RD (2016) The role of Neoantigens in naturally occurring and therapeutically induced immune responses to cancer. Adv Immunol 130:25–74. https://doi.org/10.1016/bs.ai.2016.01.001
Whitlock JM, Leikina E, Melikov K, De Castro LF, Mattijssen S, Maraia RJ, Collins MT, Chernomordik LV (2023) Cell surfacebound La protein regulates the cell fusion stage of osteoclastogenesis. Nat Commun 14(1):e616. https://doi.org/10.1038/s41467-023-36168-x
Wrenshall LE, Cerra FB, Carlson A, Bach FH, Platt JL (1991) Regulation of murine splenocyte responses by heparan sulfate. J Immunol 147:455–459
Wu CS, Lu WH, Hung MC, Huang YS, Chao HW (2022) From polyploidy to polyploidy reversal: its role in normal and disease states. Trends Genet 38(10):991–995. https://doi.org/10.1016/j.tig.2022.05.007
Yoshizaki T, Takimoto T, Takeshita H, Tanaka S, Furukawa M, Seiki M, Sato H (1994) Epstein-Barr virus lytic cycle spreads via cell fusion in a nasopharyngeal carcinoma hybrid cell line. Laryngoscope 104(1 Pt 1):91–94
Zhou X, Merchak K, Lee W, Grande JP, Cascalho M, Platt JL (2015) Cell fusion connects oncogenesis with tumor evolution. Am J Pathol 185(7):2049–2060. https://doi.org/10.1016/j.ajpath.2015.03.014

Chapter 12
Osteoclasts at Bone Remodeling: Order from Order

Jiro Takito and Naoko Nonaka

Abstract Osteoclasts are multinucleated bone-resorbing cells derived from the monocyte/macrophage lineage. The macrophage colony-stimulating factor/receptor activator of nuclear factor κB ligand (M-CSF/RANKL) signaling network governs the differentiation of precursor cells into fusion-competent mononucleated cells. Repetitive fusion of fusion-competent cells produces multinucleated osteoclasts. Osteoclasts are believed to die via apoptosis after bone resorption. However, recent studies have found that osteoclastogenesis in vivo proceeds by replacing the old nucleus of existing osteoclasts with a single newly differentiated mononucleated cell. Thus, the formation of new osteoclasts is minimal. Furthermore, the sizes of osteoclasts can change via cell fusion and fission in response to external conditions. On the other hand, osteoclastogenesis in vitro involves various levels of heterogeneity, including osteoclast precursors, mode of fusion, and properties of the differentiated osteoclasts. To better understand the origin of these heterogeneities and the plasticity of osteoclasts, we examine several processes of osteoclastogenesis in this review. Candidate mechanisms that create heterogeneity involve asymmetric cell division, osteoclast niche, self-organization, and mode of fusion and fission. Elucidation of the plasticity or fluctuation of the M-CSF/RANKL network should be an important topic for future researches.

J. Takito (✉) · N. Nonaka
Department of Oral Anatomy and Developmental Biology, School of Dentistry, Showa University, Tokyo, Japan
e-mail: takito@dent.showa-u.ac.jp; choku@dent.showa-u.ac.jp

M. Kloc, A. Uosef (eds.), *Syncytia: Origin, Structure, and Functions*, Results and Problems in Cell Differentiation 71,
https://doi.org/10.1007/978-3-031-37936-9_12

12.1 Hierarchical Regulation of Osteoclastogenesis: Interaction Among Homeostatic Systems

Bones in the adults maintain specific morphologies and mechanical properties in the process known as bone homeostasis. Bone homeostasis is controlled by distant organs including the brain, kidney, muscle, and immune cells (Zaidi et al. 2018). The life of osteoclasts, which are bone-resorbing multinucleated cells, is involved in bone homeostasis. The main component of bone is hydroxyapatite, which is a composite of calcium and inorganic phosphate. Bone degradation and formation correlate with the rise and fall in serum calcium ion (Ca^{2+}) and inorganic phosphates concentrations, respectively. Thus, bone homeostasis is coupled with systemic Ca^{2+} and phosphate homeostasis (Fig. 12.1). These two homeostatic systems dispatch distal signals via vascular networks, such as parathyroid hormone, active form of vitamin D3 ($1\alpha,25(OH)_2D_3$), and calcitonin, to regulate osteoclastogenesis. Hormone delivery to target cells invokes an intracellular signaling cascade, resulting in the induction of proximal signals that act on osteoclast precursor cells. Osteoclast precursors, which are mononucleated cells located in the blood or bone marrow, change their gene expression and begin to migrate towards bone surfaces. During migration, precursors encounter different environments and receive distinct signals, resulting in a gradual change in gene expression (Baron et al. 1986). The precursors reaching at the bone surface fuse with a partner mononucleated cell or an existing osteoclast. Therefore, osteoclastogenesis is regulated hierarchically in a spatiotemporal manner.

This grand scheme of information processing during osteoclastogenesis has been established by intensive studies over the past 40 years. Our knowledge of osteoclastogenesis is mostly based on in vitro biochemical studies and in vivo phenotypic analyses of gene modified animals. Although osteoclast precursors are highly responsive to environmental cues, most biochemical studies have been carried out on two-dimensional (2D) plastic dishes, which lack in vivo three-dimensional (3D) structural settings. Therefore, it is difficult to directly extrapolate in vitro experimental results to in vivo osteoclastogenesis. In particular, this issue is critical for the development of therapeutics against bone diseases. Although biochemical studies assume a single species of osteoclasts, osteoclast-like cells can be induced from various types of precursor cells in vitro. At a given time of differentiation, cells in the intermediate stage of differentiation often show variations in the level of gene

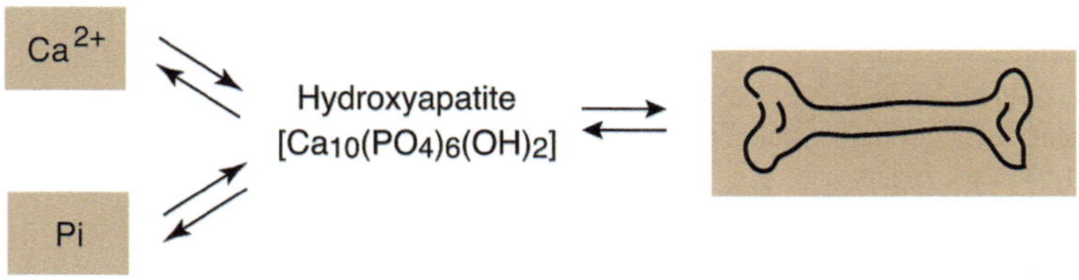

Fig. 12.1 Coupling of bone homeostasis with Ca^{2+} and phosphate homeostasis

expression. The most prominent process of osteoclastogenesis, cell fusion, occurs between heterotypic cells in vitro. Osteoclasts isolated from distinct bone sites show differences in gene expression, enzymatic activity, and bone-resorbing activity. To address these inconsistencies, this review aims to delineate the whole processes of in vivo osteoclastogenesis at the bone remodeling state from our fragmentary knowledge. Therefore, we will not provide a comprehensive consideration of specific issues. Rather we focus on separate research fields to gain insight into the role of the environment in osteoclastogenesis. Osteoclastogenesis occurs in the isolated space of bone, into which precursors are supplied from blood vessels and bone marrow. During the migration to bone surfaces, precursors react to mixed cues provided from molecules embedded in the matrix and secreted from distinct cells. In addition, cell-to-cell interaction plays an important role in differentiation. The formation and maintenance of osteoclasts on bone surfaces also depends on these environmental signals. Because these environments vary from bone-to-bone and site-to-site, the interaction between precursors and the specific environment may determine the mode of osteoclastogenesis. The interaction may generate heterogeneity on the cells of interest.

12.2 Bone, Bone Marrow, and Vascular Network

The importance of the environment for osteoclastogenesis is evidenced by the observations that the number and size of osteoclasts generated in vitro are highly sensitive to the experimental conditions. In the physiological state, osteoclastogenesis occurs in a compartment that is predetermined by bone homeostasis. This section describes the general features of bones. Bone tissue is a connective tissue that consists of bone, bone marrow, and well-developed vascularity. Bone is formed by the hierarchical organization principle (Reznikov et al. 2013: Reznikov et al. 2018). Active osteoblasts produce and secrete type I collagen fibrils (Fig. 12.2). The building unit of bone is the aggregate of hydroxyapatite [$Ca_{10}(PO_4)_6(OH)_2$] precipitated on the collagen fibrils. The stacks of collagen-hydroxyapatite bundles form an aggregate called a tesselle (Buss et al. 2022). Depending on the orientation of the collagen bundles, the tesselles are divided into two types: ordered and disordered. The disordered tesselle is softer than the ordered tesselle. Ordered and disordered tesselles are layered to form lamellae. The cylindrical arrangement of lamellae forms osteons in large animals and humans. Packets of lamellar bone form trabecular bone. An individual bone is usually a composite of trabecular and cortical bone.

The body contains more than 60 distinct bones with unique shapes and properties. Osteoblasts and osteoclasts live on the surface of bone (Fig. 12.3). Some osteoblasts that stopped secreting collagen are embedded in the bone and become osteocytes. Osteocytes are the most abundant cells in the bone, comprising about 90% of all cells. They live in the soft disordered region of bone and extend many long projections that connect with the projections of their neighbors (Reznikov et al. 2013). All osteocytic connections, called the lacunocanaliculi network, run in

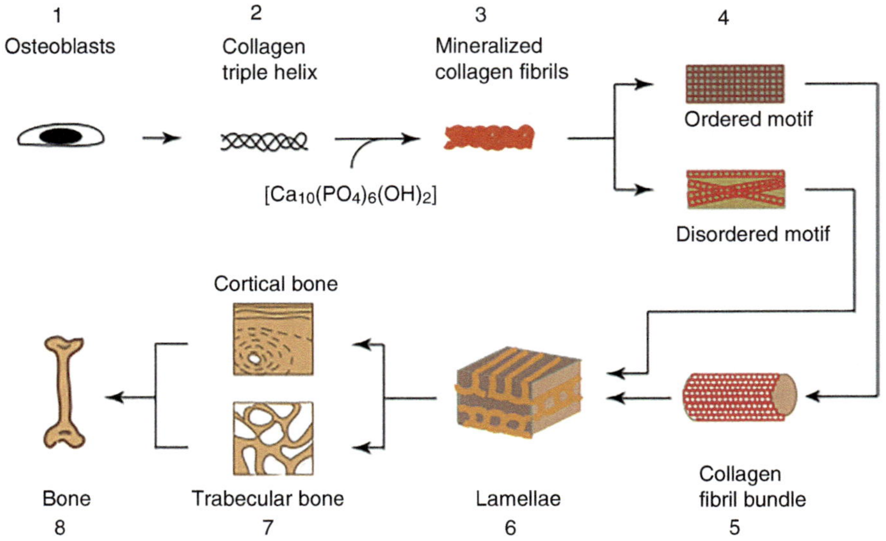

Fig. 12.2 Hierarchical organization of bone. In bone, the smaller structural unit forms larger structural units. The hierarchical organization confers stiffness and toughness on bones. (1) Osteoblasts produce collagens and secrete collagen fibers. (2) The secreted fibers form the triple helix. (3) Hydroxyapatite precipitates on the collagen fibers. (4) Mineralized collagen fibers form aggregate in ordered or disordered arrays. (5) Ordered motif forms a collagen fibril bundle. (6) The alternate stacking of ordered (light brown) and disordered motif (yellow) forms lamellae. (7) The differential stacking of lamellae forms the trabecular and cortical bone. (8) Bone is the composite of trabecular and cortical bones. Adopted and modified from Reznikov et al. (2018)

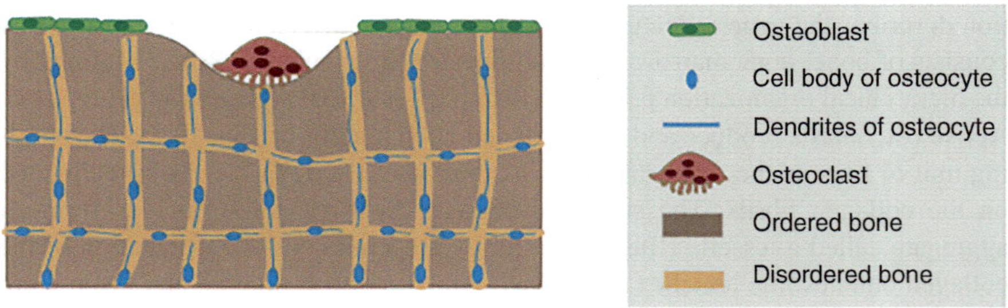

Fig. 12.3 Three types of cells in bone. Osteoblasts and osteoclasts live on the bone surface, whereas osteocytes reside in the bone. Osteocytes form the lacunocanaliculi network in the disordered region of bone

disordered bone with a total length of 175,000 km. This giant informational network senses the change in local mechanical force in the bone and transmits signals to osteoblasts and osteoclasts to regulate bone mass (Burger and Klein-Nulend 1999; Buenzli and Sims 2015). Recent studies have revealed new functions of osteocytes that include bone formation, bone resorption, and presentation of receptor activator of nuclear factor κB ligand (RANKL) to osteoclast precursors (Robling and Bonewald 2020).

Blood vessels form a vast and complex network and provide nutrients and information to bone cells. The bone marrow is a site for hematopoiesis and immune cells. Hematopoietic stem cells (HSCs), the most distant ancestor of osteoclasts, reside in the bone marrow. Because osteoclastogenesis begins with the mobilization of proximal precursors from the blood and bone marrow, the microenvironment of the blood and bone marrow is critically important for osteoclastogenesis. However, osteoclast niche in the vasculature and bone marrow remains unclear and is beyond the scope of this review. Readers are recommended to refer to previous reviews for bone marrow as an HSC niche (Pinho and Frenette 2019) and the role of blood vessels for osteoclastogenesis (Brandi and Collin-Osdoby 2006; Sivaraj and Adams 2016).

12.3 Bone Remodeling

The mechanism of osteoclastogenesis in vivo varies in the modeling, remodeling, and pathological state (Boyce and Xing 2008; Feng et al. 2019; Sharma et al. 2022). In addition, there is no guarantee that the same mechanism works in different types of bones in the remodeling state (Sims and Vrahnas 2014; Takito and Nakamura 2020). This review focuses on osteoclastogenesis during the bone remodeling in the cortical (also called compact) and trabecular (also called spongy) bones. The mechanisms of remodeling are best known in these two bones (Kenkre and Bassett 2018; Sims and Martin 2020). To maintain bone homeostasis, old or damaged bone is constantly renewed by the coordinated activities of osteoclasts and osteoblasts in a confined space called the basic multicellular unit (BMU). The BMU is 1 to 2 mm long and 0.2 to 0.4 mm wide (Parfitt 1994). The BMU consists of osteoclasts, osteoblasts, and a blood vessel. Approximately, one million BMUs operate at any moment, and their lifespan is 6 to 9 months in human adults (Manolagas 2000). The concept of BMU has been extended to the bone remodeling compartment (BRC), in which canopy cells cover the BMU to create an isolated space on the trabecular bone (Hauge et al. 2001; Eriksen 2010). Bone remodeling is divided into four stages: resorption, reversal, formation, and quiescence. The resorption stage involves the initiation of osteoclastogenesis and bone resorption by differentiated osteoclasts. The reversal phase is the period between bone resorption and bone formation, in which osteoclasts stop bone resorption and transmit signals for osteoblastogenesis. During this period, osteoblast-lineage cells remove unmineralized collagen from the resorbed bone surface and deposit a non-collagenous mineralized matrix for osteoblast adhesion. In the formation stage, differentiated osteoblasts actively secrete collagen type I and form osteoids, which slowly turn into mineralized bone. After bone repair, some osteoblasts disappear due to apoptosis. Others become flat bone-lining cells on the bone surface or are encased in the bone and differentiate into osteocytes during the quiescence stage. Although bone remodeling proceeds with these sequential events in all bones, the reality of remodeling differs depending on the bone site. For example, in trabecular bone, the BRC is located on the bone surface. Bone remodeling starts with the induction of osteoclasts in the BRC. The active osteoclasts form the resorption pits or trenches, called Howship's lacunae (Fig. 12.4a). After the reversal stage, active osteoblasts refill the Howship's lacunae to their original shape. Thus, osteoclasts and osteoblasts appear in the same location at the different times in BRC. In contrast, remodeling in the cortical bone proceeds within a pyramidal-shaped BRC in an osteon (Fig. 12.4b). The osteoclast precursors are provided from the central blood vessel in the BRC. Osteoclasts appear at the tip of the cutting cone and tunnel in the cortex. Osteoblast-lineage cells line the bone surface at the back of osteoclast-rich zone. This zone follows active osteoblasts that deposit osteoids into the central canal. The osteoid becomes mineralized in the backward direction, and the central vessel forms the Haversian canal of the new osteon. It is possible to observe all stages of bone remodeling in cortical BRC. Thus, cellular events of bone remodeling within the trabecular BRC are segregated in time, while those in the cortical BRC are separated in space. In summary, osteoclastogenesis during the bone remodeling is integrated into a determined sequence of multicellular events within a confined space. This suggests that the form of osteoclastogenesis that deviates from bone remodeling may be associated with bone diseases.

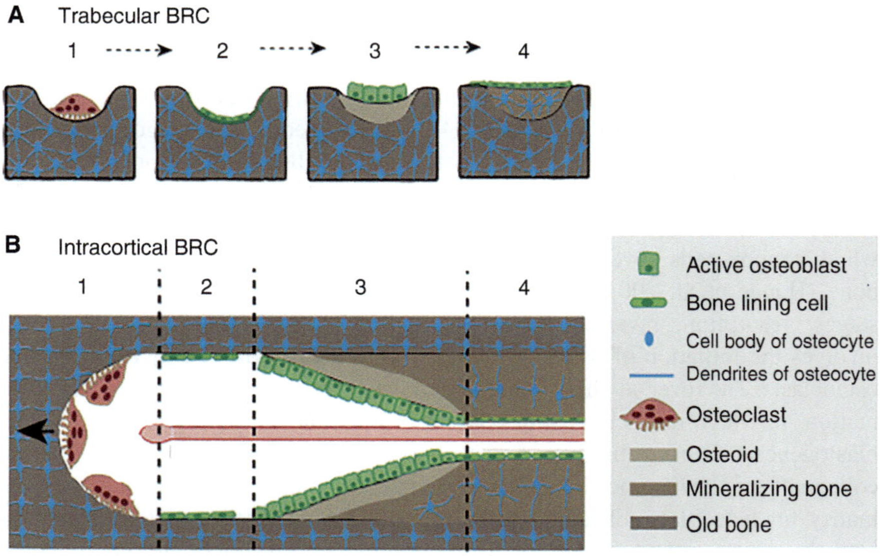

Fig. 12.4 Model of bone remodeling in the trabecular (**a**) and cortical bone (**b**). Bone remodeling occurs in a confined space. Bone remodeling is divided into resorption (1), reversal (2), formation (3), and quiescence stages (4). Each stage is characterized by the appearance of specific cells. See text for details. Adopted and modified from Sims and Martin (2020)

12.4 Precursor Migration

Osteoclastogenesis starts with the mobilization of mononucleated precursors from the blood or bone marrow to the bone surface (Walker 1975). During the migration, precursors encounter distinct environments and undergo cell proliferation and differentiation (Baron et al. 1986). The translocation of precursors among the three compartments, blood, bone marrow, and bone surface, is regulated by diverse factors. These factors may promote or inhibit the intra- and trans-compartment migrations. These include stromal-cell-derived factor 1 (SDF-1; Yu et al. 2003), CCL2 (Binder et al. 2009), CCL9 (Yang et al. 2006), CXC motif chemokine ligand 1 (CXCL1; Onan et al. 2009), and CXCL12 (Gronthos and Zannettino 2007). Here we refer to the directional migration of precursors by sphingosine-1-phosphate (S1P)/S1P receptors (S1PRs) (Ishii et al. 2009; Ishii et al. 2010) and Epstein-Barr virus-induced G-protein coupled receptor 2 (EBI2)/EBI2 ligand signaling (Nevius et al. 2015).

S1P is a lipid mediator. The levels are high in the blood and low in tissues. The S1P gradient induces chemotaxis of precursors from the blood to the bone compartment. Precursors express two types of S1P receptors: S1PR1 and S1PR2 (Ishii et al. 2009). Precursors displayed biphasic migratory behavior in an in vitro migration assay (Ishii et al. 2010). The low concentration of S1P ($<10^{-9}$ M) acts as a chemoattractant, while a high concentration of S1P ($>10^{-7}$ M) causes chemorepulsion of precursors. This biphasic behavior is explained by the rapid internalization of S1PR1 bound to S1P. In vitro analyses revealed that S1P signaling mediated by S1PR1 in precursors induces positive chemotaxis. In contrast, the signaling mediated by S1PR2 causes negative chemotaxis. Consistent with these in vitro observations, intravital two-photon imaging revealed increased and decreased ratio of osteoclasts attached to the calvarial bone of conditional S1PR1- and SIPR2-deficient mice, respectively, compared to the controls (Ishii and Kikuta 2013). The former mice exhibited osteoporosis in the femur trabecular bone. The latter mice exhibited an osteopetrotic phenotype. Because S1P itself has no effect on osteoclastogenesis in vitro, the authors concluded that S1P/SIPR1 signaling facilitates the directional migration of precursors from the bone compartment to the blood circulation via positive chemotaxis, whereas S1P/S1PR2 signaling contributes to reverse migration via negative chemotaxis (Ishii and Kikuta 2013).

Translocation of precursors between bone marrow and bone surfaces is regulated by the Gαi protein coupled receptor, EBI2, and its ligand, 7α, 25-dihyodroxycholesterol (Nevius et al. 2015). EBI2 is highly expressed in precursors and osteoclasts. Osteoblasts synthesize and secrete 7α, 25-dihyodroxycholesterol by the enzymatic activities of cholesterol 25-hydroxylase (CH25H) and 25-hydroxycholesterol 7-alpha-hydroxylase (CYP7B1). The combination of adoptive bone marrow transfer and intravital two-photon imaging was used to assess the migration of precursors from bone marrow to bone surfaces. When EBI2-deficient or wild-type (WT) cFms+ osteoclast precursor cells were transferred into irradiated recipient mice, there was no difference in bone marrow homing between the two precursors. However, the number of osteoclasts derived from EBI2-deficient precursors was lower than that in controls. When precursors overexpressing EBI2 were transferred into WT recipients, there was an increase in the number of precursors located near bone surfaces and osteoclasts at bone surfaces compared to the controls. In contrast, CH25H-deficient precursors transferred into WT recipients showed normal localization near the bone surfaces. WT precursors transferred to CH25H-deficient recipients showed less localization near bone surfaces and formed fewer osteoclasts. Finally, both EBI2- and CH25H-dificient mice displayed increased trabecular bone mass in the femur, but not in cortical bone thickness. These results suggest that EBI2/CH25H signaling promotes the directional migration of precursors from bone marrow to bone surfaces. In summary, the translocation of precursors to the blood, bone marrow, and bone surfaces is tightly regulated by external factors and the intrinsic cell mechanisms. Future studies using advanced imaging combined with the identification of markers of distinct stages of differentiation (Tsukasaki et al. 2020) will clarify the relationship between the migration and differentiation of precursors in distinct bones.

Do precursors migrate to the matrix without any guidance? Søe et al. (2019) proposed that collagen fibers act as "a track" for the migration. The proposal seems reasonable, because type I collagen is the most abundant protein, and bone tesselles are made by relatively regular orientation of collagen fibers. A dense mesh of collagen fibers surrounded the capillaries. In the future, it will be important to determine whether precursors recognize the polarity of collagen fibers.

12.5 Cell Proliferation

During osteoclastogenesis, precursors exit the cell cycle after several rounds of proliferation and become cell cycle-arrested quiescent cells prior to fusion (Mizoguchi et al. 2009). As described below, osteoclast fusion occurs between the heterotypic cells. However, the macrophage colony-stimulating factor (M-CSF)/RANKL signaling network for osteoclastogenesis does not explicitly provide a mechanism that creates heterogeneity. Here we consider whether cell proliferation is involved in the creation of heterogeneous precursors.

Cell proliferation involves the doubling of DNA, cell growth, and division of equal amounts of DNA into daughter cells. Because the distribution of cytoplasmic contents is not uniform, daughter cells do not have the same number of cytoplasmic components after cell division (Fig. 12.5). Accordingly, cell division inevitably produces the heterogeneity among daughter cells. An increase in the number of rounds of cell division may expand the heterogeneity. M-CSF is the main factor involved in cell proliferation during osteoclastogenesis (Yoshida et al. 1990; Tanaka et al. 1993). The cytokine induces two rounds of cell division at the early stage of differentiation in vitro (Motiur Rahman et al. 2015). Interestingly, a low dose of RANKL enhances the proliferative activity of M-CSF. Experimental evidence suggests the heterogeneity of precursors after cell proliferation. Incubation of the homogeneous

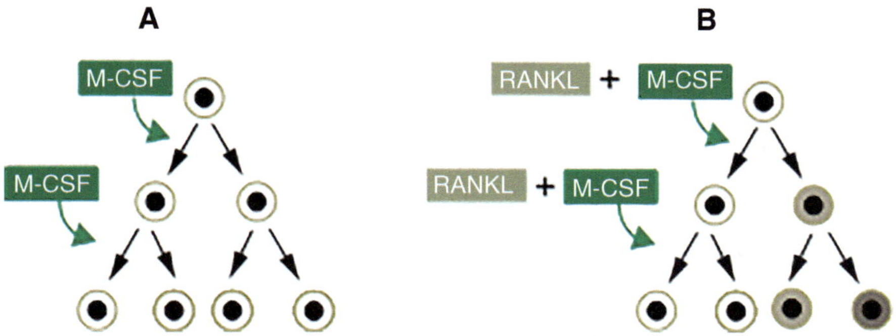

Fig. 12.5 Model of cell proliferation that creates heterogeneity. (**a**) M-CSF increases the number of cells with producing little heterogeneity. (**b**) M-CSF in the presence of RANKL signaling increases the heterogeneity in daughters. The rounds of cell cycle expand the heterogeneity

population of precursors sorted by fluorescence-activated cell sorting (FACS) with M-CSF and RANKL splits them into the dendritic cell-specific transmembrane protein (DC-STAMP)^{high} and DC-STAMP^{low} subpopulations (Mensah et al. 2010). Incubation of the clonal cells derived from a single RAW 264.7 cell with M-CSF and RANKL produces a difference in the expression level and cytoplasmic localization of nuclear factor of activated T cells 1 (NFATc1) in descendants (Levaot et al. 2015). In rat osteoclast precursors cultured with $1\alpha,25(OH)_2D_3$, RANKL, and tumor necrosis factor-alpha (TNFα), the dividing cells in the late mitotic phase exhibit asymmetric expression of adrenomedullin receptors (Kukita et al. 2021). These results support the hypothesis that cell proliferation induced by M-CSF creates heterogeneity in the precursors. In particular, other signaling inputs, such as RANKL, appear to increase the heterogeneity. In this regard, it is notable that $\alpha_v\beta3$ integrin is indispensable for adhesion-dependent proliferation of osteoclast precursors (Miyamoto et al. 2000).

12.6 Osteoclast Niche

During in vivo osteoclastogenesis, heterogeneous precursors produced by cell proliferation migrate towards bone surfaces and receive input signals from relevant cells at a specific location for differentiation. Precursors also sense and respond to physical cues, such as force by fluid flow, loading force, gravitation, and stiffness of the matrix. The environment varies from bone-to-bone and site-to-site. We term this environment setting as an osteoclast niche. Below, we discuss the unsolved issues of signaling input of M-CSF and RANKL in vivo with special reference to heterogeneity. Søe et al. (2021) have already attempted to review in vivo osteoclastogenesis in this context.

The secreted cytokine, M-CSF, acts on precursors in a paracrine manner. The M-CSF producing cells in bone include vascular endothelial cells, bone lining cells, osteoblasts, and osteocytes. The same type of cells can also present RANKL to the precursors. These include vascular endothelial cells (Collin-Osdoby et al. 2001), stromal cells (Udagawa et al. 1999; Kartsogiannis et al.1999), hypertrophic chondrocytes (Lacey et al. 1998), bone lining cells (Streicher et al. 2017), osteoblasts (Kartsogiannis et al. 1999), and osteocytes (Silvestrini et al. 2005). Which cells mainly provide M-CSF or RANKL to the precursors? This question may be inappropriate to answer. Bone marrow cells are not present in the periosteum, and osteocytes are not present in the growth plate (O'Brien 2010). Because cell fusion only occurs on bone surfaces in BRU, it is reasonable to assume that RANKL presented by osteocytes contributes to the fusion reaction. This idea is consistent with the observations that the number of osteoclasts significantly decreases in the conditional RANKL-deficient mice in the osteocyte lineage (Nakashima et al. 2011; Xiong et al. 2011; Xiong et al. 2015). Conditional RANKL-deficient mice exhibit high bone mass in vertebra and trabecular bones in femur, but not in the cortical thickness of the femur (Xiong et al. 2015). These findings may suggest that RANKL presentation does not depend on the cell type. On the other hand, RANKL-producing cells express RANKL via different mechanisms (O'Brien 2010). The RANKL-producing cells should be arranged along the migration path of precursors to bone surfaces. If we assume that one stimulus of M-CSF and RANKL is sufficient for osteoclastogenesis, the signal must be in the "ON" state over 3 days. However, the duration of effective signaling may be shorter. Within a few hours after ligand binding, cytokine receptors on the plasma membranes are internalized via endocytosis and sorted towards degradation or recycling pathways (Platta and Stenmark 2011; Cendrowski et al. 2017). Although some receptors are still active in the endocytic compartment (signaling endosome), internalized receptors become refractory to ligand binding. This is consistent with the in vitro experimental results that a continuous supply of M-CSF and RANKL is needed for optimal osteoclastogenesis (Tanaka et al. 1993; Jimi et al. 1999). M-CSF and RANKL play multiple roles, including proliferation, differentiation, survival, and cell fusion, suggesting a need for cytokines at different stages of osteoclastogenesis (Takahashi et al. 1999; Jimi et al. 1999). To meet these requirements, we posit that the osteoclast niche provides multiple M-CSF and RNAKL stimuli to the precursors (Fig. 12.6). Because each osteoclast niche differs from bone-to-bone and site-to-site, this unique niche may determine the unique pattern of osteoclastogenesis. The osteoclast niche remains unchanged in bone homeostasis, which reproduces the same type of osteoclasts, leading to the generation of site-specific osteoclasts.

What can we learn from in vitro osteoclastogenesis? The precursors receive continuous M-CSF and soluble RANKL (sRANKL) stimuli during the whole period of osteoclastogenesis. They are cultured with a limited number of cell species and grown on a plastic or glass surface. Under these conditions, the intrinsic robust integrin/M-CSF/RANKL signaling network may exhibit all repertoires of osteoclastogenesis in vivo and more.

Fig. 12.6 Model of osteoclast niche for osteoclast differentiation. Heterogeneous precursors enter the process of differentiation. During migration to bone surfaces, precursors receive the M-CSF and RANKL signaling inputs multiple times at different locations depending on the osteoclast niche. They also respond to environmental cues. The osteoclast niche at the specific site maintains its characteristics in the remodeling state. Osteoclastogenesis at a given site proceeds in a unique pattern determined by the niche. The model follows the multiple stimuli model (see text for details)

12.7 Regulation of RANKL Binding with RANK

The mouse RANKL gene produces three RANKL isoforms: RANKL 1, RANKL 2, and RANKL 3 (Ikeda et al. 2001). RANKL 1 and 2 form a functional homo- or heterotrimer, respectively (Lam et al. 2001; Ito et al. 2002). The RANKL trimer exists as a type II membrane protein or soluble protein. Although both forms can induce osteoclasts from precursors, the membrane form is believed to be more potent than the soluble form (Nakashima et al. 2000; Honma et al. 2013). The receptor for RANKL, RANK, is a type I membrane protein that may exist as a monomer on the plasma membrane. Because a RANK trimer elicits downstream signals without RANKL binding (Kanazawa and Kudo 2005), binding of a RANKL trimer to a RANK monomer probably induces the trimerization of RANK (Liu et al. 2010). The conventional view is that distal factors, such as PTH or $1\alpha,25(OH_2)D_3$, induces the expression of RANKL in the RANKL-producing cells. The membrane form of RANKL directly binds to RANK expressed on precursors, resulting in the activation of downstream signaling cascades for osteoclastogenesis (Udagawa et al. 1999; Asagiri and Takayanagi 2007; Park et al. 2017). Based on this concept, transcriptional regulation of RANKL gene has been a central theme in elucidating the mechanism of RANKL presentation (O'Brien 2010).

Unexpectedly, the results from studies on intracellular trafficking of RANKL have shaken the conventional view. Newly synthesized green fluorescent protein (GFP)-RANKL is predominantly concentrated in lysosomes in RANKL-producing cells, such as ST2 cells and primary mouse calvarial osteoblasts (Kariya et al. 2009). The distribution of GFP-RANKL in the plasma membrane is negligible. In contrast, in HeLa and HEK 293 cells, which do not produce RANKL, most transfected GFP-RANKL was reportedly localized at the plasma membranes. Interestingly, stimulation with RANK-Fc-beads induces the translocation of GFP-RANKL from the lysosomes to the plasma membranes. These results suggest that RANKL is normally stored in secretory lysosomes, and that the stimulation of RANK triggers the translocation of the ligand in RANKL-producing cells. How do RANKL-expressing cells sense the RANK stimulus? The authors speculated that a small fraction of RANKL, which is directly transported from the Golgi to the plasma membrane via the minor pathway, senses RANK. The same group also reported a novel role of osteoprotegerin (OPG) as a regulator of RANKL trafficking (Aoki et al. 2010). They found that the binding of OPG to RANKL in the Golgi is necessary for the translocation of RANKL to lysosomes. The OPG/RANKL complex transported to the lysosomes becomes RANKL alone because of the degradation of OPG by proteases. In OPG-deficient osteoblasts, RANKL without OPG cannot be sorted to lysosomes and is directly destined to plasma membranes via the minor pathway, leading to the excess generation of osteoclasts. Thus, OPG acts as a regulator of RANKL trafficking, in addition to its conventional role as a soluble decoy receptor for RANKL in osteoclastogenesis (Sheikh and Fornace 2000; Shin et al. 2008). These results suggest that the regulation of RANKL localization at the plasma membranes, in addition to transcriptional gene regulation, is important for RANKL presentation to RANK.

Although the downstream signaling cascade of activated RANK has attracted much attention, the fate of the RANKL/RANK complex remains elusive. According to the current model, the extracellular domain of the RANK trimer binds to the extracellular domain of the RANKL trimer between the juxtaposed membranes of the two cells (Liu et al. 2010). Indeed, RANK-Fc-beads can capture RANKL overexpressed in ST2 cells (Kariya et al. 2009). In an in vivo setting, precursors that receive RANKL signals migrate towards bone surfaces or move looking for fusion partners on bone. By what mechanism do precursors dissociate from RANKL expressed in another cell? Because the affinity of RANKL/RANK binding is strong ($K_d = 3 \times 10^{-9}$ M; Hsu et al. 1999), it is not reasonable to assume the spontaneous dissociation. Possible scenarios for the dissociation of the complex are presented in Fig. 12.7. In one scenario, the extracellular domain of the RANKL trimer is cleaved by sheddases, such as matrix metalloprotease (MMP)9 and MMP13, and a disintegrin and metalloproteinase (ADAMS). The cleaved RANKL bound to RANK is digested in the lysosomes after internalization. Another scenario is that an unknown mechanism releases RANKL from RANK, and two proteins enter the recycling or degradation pathway in the respective cells. Understanding the fate of the RANKL/RANK complex will be useful in devising RANKL- and RANK-targeted therapeutics.

12.8 Lifespan of Osteoclasts

Although new osteoclasts are generated in the modeling and pathological states, the main mode of osteoclastogenesis in the remodeling state is the renovation of existing osteoclasts (Fig. 12.8). Jacome-Galarza et al. (2019) estimated the dynamics and lifespan of osteoclasts in mouse femurs in a remodeling state. At the modeling

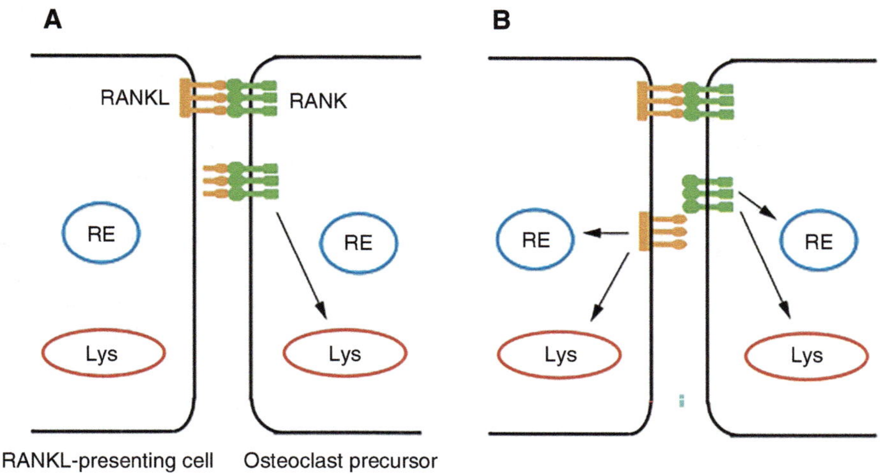

Fig. 12.7 Scenarios for the RANKL/RANK complex degradation. (**a**) RANKL in the RANKL/RANK complex is cleaved by protease. RANK with the RANKL fragment is routed to lysosomal degradation. (**b**) RANKL is removed from RANK via unknown mechanism. The intact RANKL and RANK is destined to recycling endosomes or lysosomes. RE, recycling endosomes. Lys, lysosomes

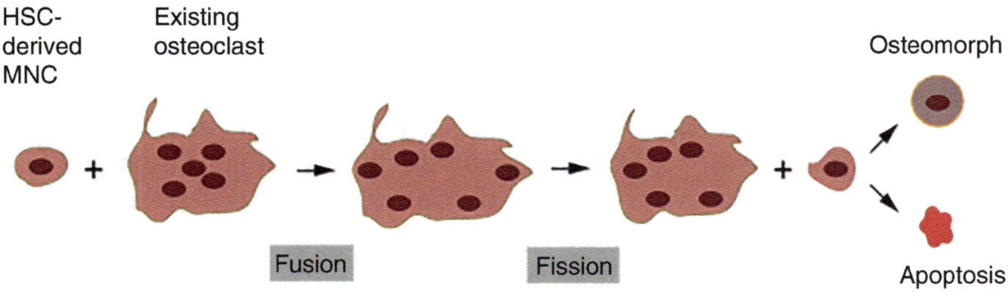

Fig. 12.8 Dynamics of osteoclast in the remodeling state. The most probable mode of fusion at the remodeling stage is the addition of the differentiated mononucleated cell by M-CSF and RANKL to an existing osteoclast. The fused osteoclast removes old nuclei by fission. The released mononucleated cell may die via apoptosis or remain as the pool of osteomorphs that have the ability to fuse with osteoclasts again. HSC, hematopoietic stem cell; MNC, mononucleated cell

stage (within 2 months of birth), embryonic erythro-myeloid progenitor cells generate osteoclasts. Osteoclasts formed in the modeling stage are slowly renewed by fusion with an HSC-derived precursor cell during the remodeling stage. A mononucleated cell derived from HSC fuses with an existing osteoclast once every 1 to 2 months. Each osteoclast has about five nuclei. This number does not change for 6 months, suggesting that the lifespan of osteoclasts is over 6 months. These estimates are contrary to the conventional view that osteoclasts have a short life ranging from a few days to 2 weeks, with cell death occurring via apoptosis after resorption (Parfitt 1994; Manolagas 2000; Tanaka et al. 2006). On the other hand, classical studies have supported the longevity of osteoclasts. In the evolving secondary

Harvesian systems of dog ribs, the lifespan of osteoclasts was estimated to be equal to that of the entire evolving osteon (Jaworski et al. 1981). In the extreme, osteoclasts with nuclear renewal via fusion can survive for the lifetime of animals (Hall 2005). How long can osteoclasts survive without fusion? The lifespan of osteoclasts was estimated in the proximal tibial and distal femoral metaphyses of irradiated mice rescued by spleen cells from beige mice (Marks and Seifert 1985). Approximately, half of the osteoclasts without fusion died within 4 weeks, and no osteoclasts without fusion could survive for more than 6 weeks. These figures agree with the iterative fusion of existing osteoclasts (Jacome-Galarza et al. 2019). The collective findings indicate that cell fusion in the remodeling state is the addition of a new mononuclear cell to an existing osteoclast. Iterative fusion is required to prolong the lifespan of osteoclasts. Bone resorption at the remodeling stage appears to be executed by a relatively constant number of osteoclasts without the formation of new osteoclasts. This conclusion is based on animal experiments. The lifespan of human osteoclasts remains an open question.

12.9 Fusion and Fission of Osteoclasts

Cell fusion can be divided into cell contact, hemifusion, fusion pore formation, and pore expansion. Previous reviews provide excellent overviews of the mechanisms and energetics of these processes (Cohen and Melikyan 2004; Chernomordik et al. 2006). Cell fusion has been discussed based on reactions at equilibrium. However, recent research has shed light on the role of self-organization in osteoclast fusion. Self-organized actin-based structures appear specifically during fusion and link fusion partner cells. Below, we discuss the contribution of self-organization to osteoclast fusion.

12.9.1 Actin-Based Linking Structures During Fusion

The reorganization of the actin cytoskeleton plays various roles at different stages of cell fusion (Martin 2016; Takito and Nakamura 2020). This section focuses on the structures formed by actin waves. Osteoclast fusion in vitro involves the formation of actin-based contacts or linkages between partner cells (Fig. 12.9). These include lamellipodia (Søe et al. 2015; Fiorino and Harrison 2016), filopodia (Zambonin Zallone et al. 1984; Jansen et al. 2012; Takito and Nakamura 2012; Takahashi et al. 2013; Shin et al. 2014; Song et al. 2014; Pennanen et al. 2017), invasive protrusions (Oikawa et al. 2012; Shin et al. 2014; Søe et al. 2015), zipper-like structure (Takito et al. 2012; Pauksch et al. 2014; Wang et al. 2018), and phagocytic cups (Søe et al. 2015). Surprisingly, cells linked via filopodia or tunneling nanotubes can exchange soluble proteins and membrane lipids (Takahashi et al. 2013; Verma et al. 2014; Pennanen et al. 2017), suggesting the establishment of

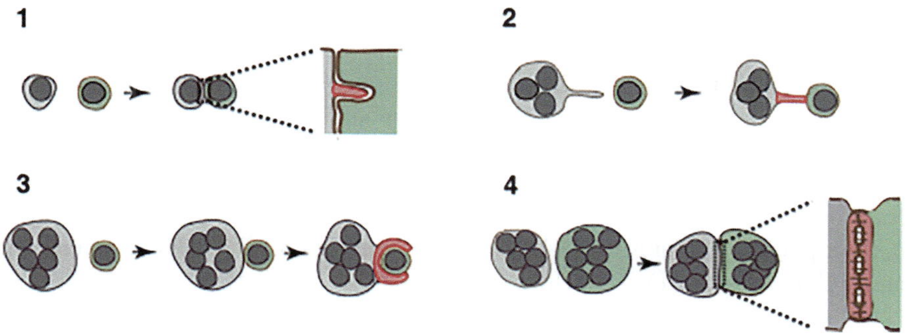

Fig. 12.9 Variations of actin-linking structures observed during fusion in vitro. (1) Invasive protrusion. (2) Filopodia-like tube. (3) Phagocytic cup. (4) Zipper-like structure. Structures colored in pink indicate the actin-based linkage that specifically appears at the pre-fusion stage. Adopted and modified from Takito and Nakamura (2020)

fusion pores. The classical view of membrane fusion between lipid bilayers restricts cytoplasmic mixing prior to pore formation. On the other hand, content exchange before cell fusion, termed "kiss-and-run," is observed in the fusion between biological membranes. Content exchange can be explained by the fact that pore enlargement is the most energy-demanding process in the fusion of biomembranes (Chernomordik and Kozlov. 2003; Cohen and Melikyan 2004). In this context, partner cells in osteoclast fusion choose between closing a fusion pore or complete fusion (Chernomordik et al. 2006). The formation of various actin-based contacts or linkage structures indicates that cell fusion during osteoclastogenesis involves multiple pathways. Because the number of nuclei of precursors varies from one to over 100, the mode of fusion may differ depending on the size of the fusion partners (Hobolt-Pedersen et al. 2014; Levaot et al. 2015; Takito and Nakamura 2012). In general, fusion between multinucleated cells occurs at a broader contact area and takes longer than fusion between mononucleated cells. Fusion between mononucleated cells does not require formation of a macroscopic actin-based linkage. Mononucleated cells tend to fuse with a multinucleated cells using filopodia. Fusion between multinucleated cells requires a large linking structure, such as a zipper-like structure.

12.9.2 Actin Wave at the Nonequilibrium State

The assembly of molecules forms the structure or order by two physical principles: self-assembly and self-organization. The former occurs at or near thermodynamic equilibrium, whereas the latter requires energy input and occurs at nonequilibrium (McCusker 2020). The formation of a self-organized structure is nondeterministic and depends on subtle changes in inherent signaling networks and environments. One of the self-organized structures, actin waves, represents the vectorial propagation of waves at the ventral surface of a cell (Weiner et al. 2007; Bretschneider et al. 2009). The driving force of actin waves is the elongation of actin filaments by actin related protein 2/3 (Arp2/3) complex. Actin waves transform into another actin-based structure, such as a ring, lamellipodium (Gerhart et al. 2014), and phagocytic cups (Gerisch et al. 2009). The theoretical basis of this transformation has been elusive. Recent progress has provided the theoretical framework for this transformation. The clue came from the discovery that actin waves are coupled with distinct phosphatidylinositol (3,4,5)-trisphosphate (PtdIns (3,4,5)P3, also abbreviated PIP3) waves in Dictyostelium (Gerhardt et al. 2014) (Fig. 12.10a). PIP3, a component of the plasma membrane, stimulates local actin polymerization at the membranes (Insall and Weiner 2001). Khamviwath et al. (2013) presented a model in which actin waves are formed by a signaling network that independently regulates actin polymerization. Miao et al. (2019) dissected the molecular components of waves using biosensors and divided them into Signal Transduction Excitable Network (STEN) waves and Cytoskeleton Excitable Network (CEN) waves (Fig. 12.10b). STEN contains activated Rap and Ras, PIP3, and PKB. CEN consists of F-actin, activated Rac, Rac GEF1, and coronin. In the STEN-CEN model, STEN orchestrates CEN and dictates the wave speed and range, while CEN controls the triggering of STEN through positive and negative feedback. This model has been validated experimentally. Perturbations in the STEN or CEN network in Dictyostelium resulted in a change in cell size and morphology, with the formation of protrusions resembling filopodia, lamellipodia, and ruffles (Fig. 12.10c).

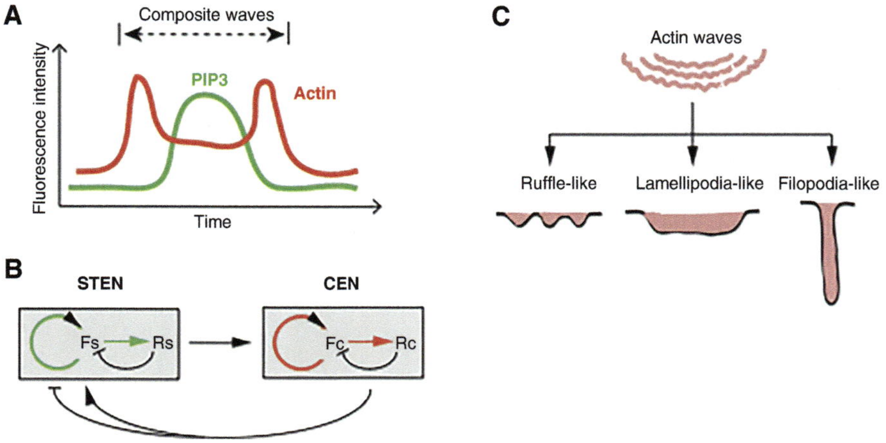

Fig. 12.10 Transformation of actin waves into cellular protrusions. (**a**) Coupling of actin waves with PIP3 waves. Actin waves (orange) and PIP3 waves (green) form composite waves. Adopted and modified from Gerhardt et al. (2014). (**b**) Architecture of STEN-CEN model. Fs and Rs (Fc and Rc) indicate the activator and inhibitor of STEN (CEN), respectively. →, positive feedback. ⊥, negative feedback. Adopted and modified from Miao et al. (2019). (**c**) Waves with different properties form the different types of cell protrusions. High RacGEF1, low PIP2 or high Rap/Ras, and high PKBs produce the ruffle-like, lamellipodia-like, and filopodia-like protrusions, respectively. Adopted and modified from Miao et al. (2019)

Therefore, the STEN-CEN model successfully explains the mechanism of actin wave transformation into various cell protrusions. This is the first manifestation of the propagation of informational waves transforming actin waves into macroscopic cellular structures. The findings by Miao et al. (2019) contrast with the conventional explanation that a linear cascade of specific regulatory molecules determines the formation of a particular cellular structure (Rottner et al. 2017).

Podosomes are dynamic actin puncta that are the characteristics of monocyte/macrophage lineages, including osteoclasts. The structure of podosome in osteoclast-like cells is subdivided into actin core, adhesion ring, and actin cloud (Chabadel et al. 2007; Saltel et al. 2008). These substructures are washed with vectorial actin waves in osteoclast-like cells (Takito et al. 2017b). The propagation of actin waves is restricted to an area in which podosomes appear. The intensity and propagation patterns discriminated the actin wave from a general ventral actin wave. Based on these observations, we propose that self-organized actin structures are produced from actin waves in a specified area, termed podosome field (Takito et al. 2017a, Takito et al. 2017b; Takito et al. 2018). In this scenario, the area should contain the regulatory wave, because self-organized actin waves (CEN) are controlled by regulatory waves (STEN), as described above (Miao et al. 2019). During podosome formation in fibroblasts, PIP3 is concentrated in the prospective podosome core region (Yu et al. 2013). The phosphoinositide is localized in podosome rings in NIH-Src cells (Oikawa et al. 2008) and the plasma membranes of fusion site in RAW 264.7 cells (Oikawa et al. 2012). Thus, the podosome field appears to contain a regulatory wave involving PIP3. Self-organization produces podosomes, podosome rings, and podosome belts during osteoclastogenesis in vitro (Destaing et al. 2003). The same principle may also apply to the formation of actin-based contacts or linkage structures during fusion, including filopodia, invasive protrusions, and zipper-like structures. These arguments lead to the idea that at least some processes of osteoclast fusion occur at nonequilibrium.

12.9.3 Fission of Osteoclasts

By what mechanism do osteoclasts maintain a relatively constant number of nuclei for a long period? Although cell fusion has been the main theme in osteoclastogenesis, fission is not a rare event in cultured osteoclast-like cells. Solari et al. (1995) reported the generation of mononucleated and binucleated cells from oversized multinucleated giant cells derived from chicken macrophages in vitro. Isolated rabbit osteoclasts and osteoclast-like cells differentiated from mouse bone marrow were split into two or three in vitro (Jansen et al. 2012). Nevius et al. (2015) reported that both fusion and fission occur spontaneously during in vitro osteoclastogenesis.

McDonald et al. (2021a) first reported in vivo fission of large osteoclasts in mice administered a pharmacological dose of sRANKL. Administration of sRANKL activated existing osteoclasts and promoted fusion between osteoclasts, resulting in giant osteoclasts in the mouse tibia. Large osteoclasts often undergo fission and produce two smaller osteoclasts, called osteomorphs. Osteomorphs are motile, transcriptionally distinct from authentic osteoclasts, and have the ability to resorb bone. Osteomorphs can re-fuse with the existing osteoclasts, which leads to a new concept of recycling osteoclasts. Because osteomorphs reside in bone marrow and blood, they may function as a pool of fusion-competent cells. The fact that osteomorphs represent the distinct species of osteoclasts may explain the heterogeneity in the bone-resorbing osteoclasts; pit-forming and trench-forming osteoclasts (Sharma et al. 2022). Database searches revealed that 17 out of 40 mouse lines deleted with one or both copies of the osteomorph upregulated gene exhibit abnormal bone phenotypes (McDonald et al. (2021a). Twenty-two out of 520 human orthologs of osteomorph genes are associated with monogenic human skeletal dysplasia, suggesting the linkage between osteomorphs and human bone diseases. Clinically, the "rebound effect," a loss in bone mass after discontinuation of denosumab becomes a problem during osteoporosis treatment (Anastasilakis et al. 2017; Bone et al. 2017). This rebound effect may be explained by the rapid re-fusion of osteomorphs to form osteoclasts that actively resorb bone (McDonald et al. (2021a). Readers can find the intensive discussions on these issues in the previous reviews (McDonald et al. 2021b; Elson et al. 2022; Sharma et al. 2022). The finding of osteomorphs has impacted on both basic and clinical researches and highlighted "fission" in osteoclast biology.

The dynamics of osteoclasts in vivo (Jacome-Galarza et al. 2019) combined with the results of McDonald et al. (2021a) led to the idea that osteoclasts in the remodeling state maintain their size by the addition of a new nucleus and removal of an old nucleus (Fig. 12.8). Thus, differentiated osteoclasts are in dynamic equilibrium both in vitro and in vivo (Nevius et al. 2015; Jacome-Galarza et al. 2019). Environmental parameters, such as sRANKL and OPG, determine the equilibrium point, the size of osteoclasts (McDonald et al. 2021a). In summary, fusion and fission are the mechanisms that increase the heterogeneity of differentiated osteoclasts. Furthermore, the concept of osteoclasts in dynamic equilibrium challenges the conventional notion of the individuality of a cell in cell biology.

12.9.4 Fission at Nonequilibrium

Actin waves are involved in cell fission. Cell division during cytokinesis relies on formation of an actomyosin contractile ring (Alberts et al. 2002). There are two forms of cell division that occur without a contractile ring: traction-mediated and wave-mediated cytofission. The former depends on the traction force between the cell and substrate (Knecht and Loomis 1987; De Lozanne and Spudich 1987). The latter utilizes self-organized cortical actin waves (Flemming et al. 2020). In electrofused oversized multinucleated Dictyostelium cells, traveling waves collide with the cell border and become unstable. This leads to their splitting into two independent waves that move in opposite directions. The force generated by antidirectional

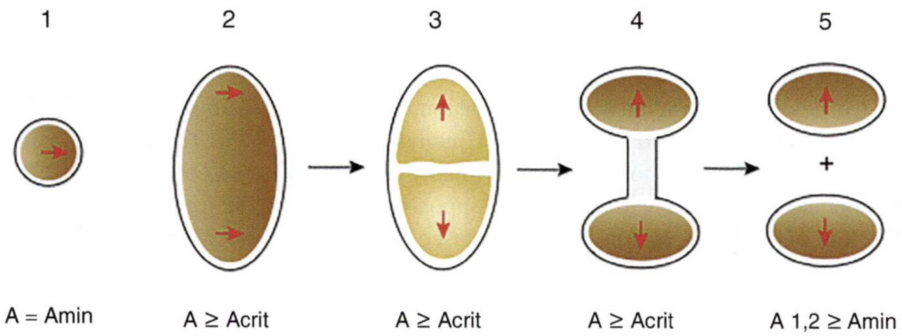

Fig. 12.11 Model of wave-mediated binary cytofission. (1) Ventral actin waves occur in a giant cell. (2) When the cell size exceeds the critical size, waves becomes unstable. (3) Unstable waves split into two waves. (4) The antidirectional waves push the membrane forward, and the connection to the main cell body gradually narrows. (5) The rupture of the cytoplasmic connection produces two daughter cells. The released cell exhibits a stable elongated wave segment and move in highly persistent fashion. A_{min} and A_{crit} indicate the minimal and critical cell size. Arrows show the direction of wave propagation. Adopted and modified from Flemming et al. (2020)

waves induces binary cytofission in a pinch-off manner (Fig. 12.11). Computer simulation predicts that wave-mediated cytofission depends on cell size and occurs at almost a 100% probability when the size is more than five times the size of a single cell. The simulation results have been experimentally validated. Interestingly, 70% of the cytofission fragments (daughter cells) contained one nucleus. These figures are remarkable coincidental with those of mouse osteoclasts in the remodeling state (Jacome-Galarza et al. 2019), suggesting that wave-mediated cytofission may be a mechanism for maintaining the constant size of osteoclasts. Interestingly, cell size is an important factor in the formation of actin waves and wave-mediated cytofission (Flemming et al. 2020). Size dependence may be characteristics of self-organized structures in biology.

12.10 Cortical Actin

Cortical actin is a dense meshwork (approximately 100 nm thick) of actin filaments underneath the plasma membrane (Chugh and Paluch 2018). Cortical actin consists of various actin binding proteins, such as actin nucleators, regulators of actin filament turnover, actin crosslinkers, and myosins. The actin cortex is linked to the plasma membrane by ezrin-radixin-moesin (ERM) proteins (Bretscher et al. 2002). The polymerization of actin filaments in the cortex depends on the activity of Arp2/3 and formin mDia1 (Bovellan et al. 2014). The actin filaments formed by Arp2/3 and mDia1 are approximately 120 and 1200 nm long, respectively (Fritzsche et al. 2013). The short filaments account for 90% of the total actin filaments in the cortex. Although the turnover kinetics of the two filaments differ, the half-life of actin filaments in the cortex is approximately 15s (Guha et al. 2005; Mukhina et al. 2007). Thus, cortical actin is a dynamic factor contributing to the mechanical properties of cell membranes.

Cortical actin generates cortical tension, which directly linked to cell shape deformation (Köster and Mayor 2016; Chugh and Paluch 2018). The formation and enlargement of fusion pores must accompany reorganization of cortical actin at the fusion site. Cortical tension acts as an opposing factor for fusion pore formation in the low pH-induced fusion of Sf9 insect cells transfected with gp64 (Chen et al. 2008). In the electropulse-induced fusion of Dictyostelium cells, cortical actin accumulates at the fusion zone, bridging the gap between the disrupted membranes and delaying pore expansion (Gerisch et al. 2013). The reorganization of plasma membranes at the fusion site takes a minute to an hour for fusion of osteoclast-like cells (Oikawa et al. 2012; Takito and Nakamura 2012; Levaot et al. 2015; McDonald et al. 2021a). In contrast, the fusion pore expands at a rate of 5 cm/sec during the fusion of lipid vesicles (Haluska et al. 2006). These observations underscore the role of cortical actin in osteoclast fusion. As previously described, fusion pore expansion is the most energy-demanding process in biomembrane fusion (Chernomordik and Kozlov 2003; Cohen and Melikyan 2004). Future studies are needed to elucidate the regulatory mechanism of the reorganization of cortical actin at the fusion site, especially to determine whether it is integrated in the integrin/M-CSF/RANKL network.

12.11 Role of Fusion in Osteoclastogenesis

The degree of multinucleation in osteoclasts correlates with an increase in the in vitro bone-resorbing activity (Piper et al. 1992; Yagi et al. 2005). Thus, osteoclasts acquire the higher activity via cell fusion. Osteoclasts dynamically construct and degrade a variety of new intracellular structures in the context of bone resorption. They form the sealing zone to create the isolated area between the plasma membrane and bone (Marchisio et al. 1984) and the ruffled border membrane to secrete protons and digestive enzymes into the isolated area (Baron et al. 1985). Osteoclasts take up the massive amount of digested materials and calcium through the uptake zone (Mulari et al. 2003). The digested materials are transcytosed via transport vesicles and discharged through the functional secretory domain at the opposite side (Nesbitt and Horton 1997; Salo et al. 1997). Actin rings are used for the saltatory movement on bone (Hu et al. 2011). Reorganization of actin cytoskeleton generates the diverse linkage structures during cell fusion as described above. Podosomes at the ventral membrane senses the physical property of the matrix (Collin et al. 2008). All of these structures are made up from phospholipids or actin or both. However, the content of phospholipids and actin per differentiating cell does not significantly increase due to cell division. Stimulus of RANKL does not change the content of phospholipids except for phosphatidylethanolamine (Irie et al. 2017). Also, the transcription of β-actin, one of housekeeping genes, does not appreciably change during osteoclastogenesis (Stephens et al. 2011). Osteoclasts may adopt fusion to circumvent the difficulties (Fig. 12.12). In this context, there may be another way for osteoclasts. A given cell doubles its cellular contents through growth during the cell cycle. Incomplete cytokinesis results in the increase in the phospholipids and actin levels. Indeed, approximately 10% of osteoclast-like cells induced by RANKL were experienced incomplete cytokinesis (Takegahara et al. 2016). Thus, fusion and incomplete cytokinesis are alternative strategies used by osteoclasts to attain a large pool of actin and phospholipids, which enables osteoclasts to create new domains in plasma membranes, new endomembrane vesicles, and various forms of actin-based intracellular organelles.

12.12 Perspectives

The two seminal in vivo studies (Jacome-Galarza et al. 2019; McDonald et al. 2021a) have changed our view of osteoclasts. We must update our understanding of the lifespan of osteoclasts, the mode of fusion, and the role of fission in maintaining osteoclasts. Although osteoclast-like cells generated in vitro die in approximately a week, osteoclasts at the remodeling bone can live for longer than 6 months. Moreover, the size of osteoclasts in vivo is in a dynamic equilibrium. Factors that may dominate, change, or maintain the equilibrium point include the calcium and phosphate homeostatic systems and the osteoclast niche. These factors treat osteoclasts as a bone-resorbing machinery. To this end, osteoclasts are generated and maintained by various processes, including cell proliferation, migration, differentiation, cell fusion, incomplete cytokinesis, and fission (Fig. 12.13). There are several paths for multinucleation and fusion. The released osteomorphs can re-fuse with

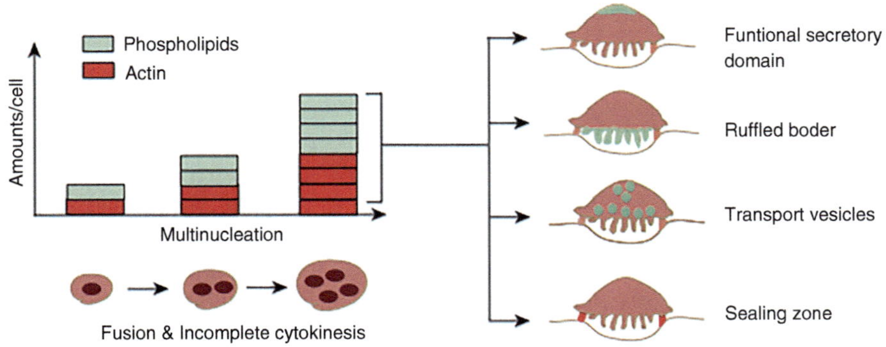

Fig. 12.12 Osteoclasts acquire a large pool of actin and phospholipids via fusion and incomplete cytokinesis. Fusion increases the cell contents in proportion to the number of nuclei. Incomplete cytokinesis during cell cycle doubles the cell contents. Osteoclasts with a large pool of phospholipids and actin can create new domains in the plasma membranes, new endomembrane vesicles, and various forms of actin-based intracellular organelles

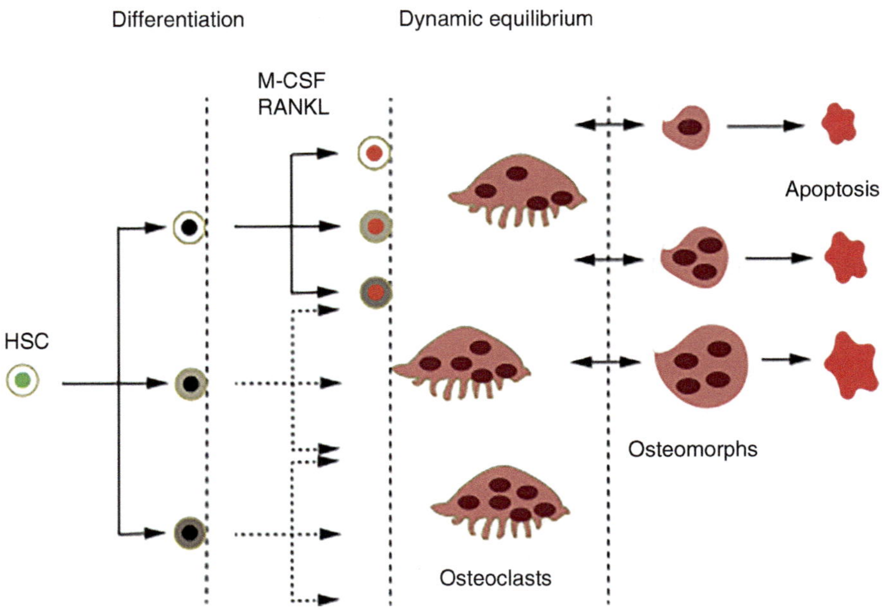

Fig. 12.13 Divergence in osteoclast differentiation and plasticity of osteoclasts. The differentiation of fusion-competent mononucleated cells by M-CSF and RANKL involves heterogeneity. The fusion-competent cells fuse with osteoclasts in multiple paths. Osteoclasts can release osteomorphs via fission and re-fuse with them. The size of osteoclasts in the BRC is regulated through multiple routes. HSC, hematopoietic stem cell

osteoclasts. Thus, osteoclasts in the BRC are generated and maintained through multiple routes, but not a fixed mechanism.

Osteoclastogenesis is characterized by heterogeneity and plasticity. The heterogeneity may be created by asymmetric cell division, osteoclast niche, self-organization, fusion, and fission. We introduce nondeterministic mechanisms in cell fusion and cytofission in this review. The maintenance of osteoclasts in bone remodeling depends on both the deterministic and nondeterministic mechanisms. Although we do not consider the significance of the deterministic integrin/M-CSF/RANKL signaling network, it is evident that the network has primary importance in osteoclastogenesis. Exploring the mechanism of fluctuation or plasticity of the network will deepen our understanding of osteoclasts. In this regard, we emphasize that monitoring the events of a single cell leads to the discovery of heterogeneity during osteoclastogenesis. The endpoint population assay has not revealed the mechanism of time-dependent processes. Cutting-edge imaging technology combined with monitoring the expression of multiple genes in a single cell will provide novel insights into the mechanism of osteoclast differentiation.

Finally, we refer to the impact of changes in our view of osteoclasts in clinical medicine. Anti-RANKL antibodies and bisphosphonates are widely used therapeutically. They are designed to alter the number of osteoclasts or lower the osteoclastic bone resorption. Both are pan-osteoclast drugs, which theoretically act on all osteoclasts in the body and have the potential to cause unexpected secondary effects. Updated knowledge indicates that osteoclasts are in dynamic equilibrium, determined by systemic and local environments, including the osteoclast niche. To develop new drugs that target a given bone disease, more information on the site- and disease-specific bone niche is needed. This information may lead to the development of novel osteoclasts-targeted therapeutics.

References

Alberts B, Johnson A, Lewis J, Raff M, Roberts K, Walter P (2002) Molecular biology of the cell, 4th edn. Garland Science, p 983

Anastasilakis AD, Polyzos SA, Makras P, Aubry-Rozier B, Kaouri S, Lamy O (2017) Clinical features of 24 patients with rebound-associated vertebral fractures after Denosumab discontinuation: systematic review and additional cases. J Bone Miner Res 32:1291–1296

Aoki S, Honma M, Kariya Y, Nakamichi Y, Ninomiya T, Takahashi N, Udagawa N, Suzuki H (2010) Function of OPG as a traffic regulator for RANKL is crucial for controlled osteoclastogenesis. J Bone Miner Res 25:1907–1921

Asagiri M, Takayanagi H (2007) The molecular understanding of osteoclast differentiation. Bone 40:251–264

Baron R, Neff L, Louvard D, Courtoy PJ (1985) Cell-mediated extracellular acidification and bone resorption: evidence for a low pH in resorbing lacunae and localization of a 100-kD lysosomal membrane protein at the osteoclast ruffled border. J Cell Biol 101:2210–2222

Baron R, Neff L, Van PT, Nefussi JR, Vignery A (1986) Kinetic and cytochemical identification of osteoclast precursors and their differentiation into multinucleated osteoclasts. Am J Pathol 122:363–378

Binder NB, Niederreiter B, Hoffmann O, Stange R, Pap T, Stulnig TM, Mack M, Erben RG, Smolen JS, Redlich K (2009) Estrogen-dependent and C-C chemokine receptor-2-dependent pathways determine osteoclast behavior in osteoporosis. Nat Med 15:417–424

Bone HG, Wagman RB, Brandi ML, Brown JP, Chapurlat R, Cummings SR, Czerwinski E, Fahrleitner-Pammer A, Kendler DL, Lippuner K, Reginster J-Y, Roux C, Malouf J, Bradley MN, Daizadeh NS, Wang A, Dakin P, Pannacciulli N, Dempster DW, Papapoulos S (2017) 10 Years of denosumab treatment in postmenopausal women with osteoporosis: results from the phase 3 randomised FREEDOM trial and open-label extension. Lancet Diabetes Endocrinol 5:513–523

Bovellan M, Romeo Y, Biro M, Boden A, Chugh P, Yonis A, Vaghela M, Fritzsche M, Moulding D, Thorogate R, Jégou A, Thrasher AJ, Romet-Lemonne G, Roux PP, Paluch EK, Charras G (2014) Cellular control of cortical actin nucleation. Curr Biol 24:1628–1635

Boyce BF, Xing L (2008) Functions of RANKL/RANK/OPG in bone modeling and remodeling. Arch Biochem Biophys 473:139–146

Brandi ML, Collin-Osdoby P (2006) Vascular biology and the skeleton. J Bone Miner Res 21:183–192

Bretscher A, Edwards K, Fehon RG (2002) ERM proteins and merlin: integrators at the cell cortex. Nat Rev Mol Cell Biol 3:586–599

Bretschneider T, Anderson K, Ecke M, Müller-Taubenberger A, Schroth-Diez B, Ishikawa-Ankerhold HC, Gerisch G (2009) The three-dimensional dynamics of actin waves, a model of cytoskeletal self-organization. Biophys J 96:2888–2900

Buenzli PR, Sims NA (2015) Quantifying the osteocyte network in the human skeleton. Bone 75:144–150

Burger EH, Klein-Nulend J (1999) Mechanotransduction in bone–role of the lacuno-canalicular network. FASEB J 13:S101–S112
Buss DJ, Kröger R, McKee MD, Reznikov N (2022) Hierarchical organization of bone in three dimensions: A twist of twists. J Struct Biol X 6:10057
Cendrowski J, Mamińska A, Miaczynska M (2017) Endocytic regulation of cytokine receptor signaling. Cytokine Growth Factor Rev 32:63–73
Chabadel A, Bañon-Rodríguez I, Cluet D, Rudkin BB, Wehrle-Haller B, Genot E, Jurdic P, Anton IM, Saltel F (2007) CD44 and beta3 integrin organize two functionally distinct actin-based domains in osteoclasts. Mol Biol Cell 18:4899–4910
Chen A, Leikina E, Melikov K, Podbilewicz B, Kozlov MM, Chernomordik LV (2008) Fusion-pore expansion during syncytium formation is restricted by an actin network. J Cell Sci 121:3619–3628
Chernomordik LV, Kozlov MM (2003) Protein-lipid interplay in fusion and fission of biological membranes. Annu Rev Biochem 72:175–207
Chernomordik LV, Zimmerberg J, Kozlov MM (2006) Membranes of the world unite! J Cell Biol 175:201–207
Chugh P, Paluch EK (2018) The actin cortex at a glance. J Cell Sci 131:jcs186254
Cohen FS, Melikyan GB (2004) The energetics of membrane fusion from binding, through hemifusion, pore formation, and pore enlargement. J Membr Biol 199:1–14
Collin O, Na S, Chowdhury F, Hong M, Shin ME, Wang F, Wang N (2008) Self-organized podosomes are dynamic mechanosensors. Curr Biol 18:1288–1294
Collin-Osdoby P, Rothe L, Anderson F, Nelson M, Maloney W, Osdoby P (2001) Receptor activator of NF-kappa B and osteoprotegerin expression by human microvascular endothelial cells, regulation by inflammatory cytokines, and role in human osteoclastogenesis. J Biol Chem 276:20659–20672
De Lozanne A, Spudich JA (1987) Disruption of the Dictyostelium myosin heavy chain gene by homologous recombination. Science 236:1086–1091
Destaing O, Saltel F, Géminard JC, Jurdic P, Bard F (2003) Podosomes display actin turnover and dynamic self-organization in osteoclasts expressing actin-green fluorescent protein. Mol Biol Cell 14:407–416
Elson A, Anuj A, Barnea-Zohar M, Reuven N (2022) The origins and formation of bone-resorbing osteoclasts. Bone 164:116538
Eriksen EF (2010) Cellular mechanisms of bone remodeling. Rev Endocr Metab Disord 11:219–227
Feng W, Guo J, Li M (2019) RANKL-independent modulation of osteoclastogenesis. J Oral Biosci 61:16–21
Fiorino C, Harrison RE (2016) E-cadherin is important for cell differentiation during osteoclastogenesis. Bone 86:106–118
Flemming S, Font F, Alonso S, Beta C (2020) How cortical waves drive fission of motile cells. Proc Natl Acad Sci USA 117:6330–6338
Fritzsche M, Lewalle A, Duke T, Kruse K, Charras G (2013) Analysis of turnover dynamics of the submembranous actin cortex. Mol Biol Cell 24:757–767
Gerhardt M, Ecke M, Walz M, Stengl A, Beta C, Gerisch G (2014) Actin and PIP3 waves in giant cells reveal the inherent length scale of an excited state. J Cell Sci 127:4507–4517
Gerisch G, Ecke M, Schroth-Diez B, Gerwig S, Engel U, Maddera L, Clarke M (2009) Self-organizing actin waves as planar phagocytic cup structures. Cell Adhes Migr 3:373–382
Gerisch G, Ecke M, Neujahr R, Prassler J, Stengl A, Hoffmann MS, Neumann US, E. (2013) Membrane and actin reorganization in electropulse-induced cell fusion. J Cell Sci 126:2069–2078
Gronthos S, Zannettino AC (2007) The role of the chemokine CXCL12 in osteoclastogenesis. Trends Endocrinol Metab 18:108–113
Guha M, Zhou MW, YL. (2005) Cortical actin turnover during cytokinesis requires myosin II. Curr Biol 15:732–736
Hall BK (2005) Bones and cartilage: developmental and evolutionary skeletal biology. Elsevier, p 207

Haluska CK, Riske KA, Marchi-Artzner V, Lehn JM, Lipowsky R, Dimova R (2006) Time scales of membrane fusion revealed by direct imaging of vesicle fusion with high temporal resolution. Proc Natl Acad Sci USA 103:15841–15846
Hauge EM, Qvesel D, Eriksen EF, Mosekilde L, Melsen F (2001) Cancellous bone remodeling occurs in specialized compartments lined by cells expressing osteoblastic markers. J Bone Miner Res 16:1575–1582
Hobolt-Pedersen AS, Delaissé JM, Søe K (2014) Osteoclast fusion is based on heterogeneity between fusion partners. Calcif Tissue Int 95:73–82
Honma M, Ikebuchi Y, Kariya Y, Hayashi M, Hayashi N, Aoki S, Suzuki H (2013) RANKL subcellular trafficking and regulatory mechanisms in osteocytes. J Bone Miner Res 28:1936–1949
Hsu H, Lacey DL, Dunstan CR, Solovyev I, Colombero A, Timms E, Tan H-L, Elliott G, Kelly MJ, Sarosi I, Wang L, Xia X-Z, Elliott R, Chiu L, Black T, Scully S, Capparelli C, Morony S, Shimamoto G, Bass MB, Boyle WL (1999) Tumor necrosis factor receptor family member RANK mediates osteoclast differentiation and activation induced by osteoprotegerin ligand. Proc Natl Acad Sci USA 96:3540–3545
Hu S, Planus E, Georgess D, Place C, Wang X, Albiges-Rizo C, Jurdic P, Géminard JC (2011) Podosome rings generate forces that drive saltatory osteoclast migration. Mol Biol Cell 22:3120–3126
Ikeda T, Kasai M, Utsuyama M, Hirokawa K (2001) Determination of three isoforms of the receptor activator of nuclear factor-kappaB ligand and their differential expression in bone and thymus. Endocrinology 142:1419–1426
Insall RH, Weiner OD (2001) PIP3, PIP2, and cell movement–similar messages, different meanings? Dev Cell 1:743–747
Irie A, Yamamoto K, Miki Y, Murakami M (2017) Phosphatidylethanolamine dynamics are required for osteoclast fusion. Sci Rep 7:46715
Ishii M, Kikuta J (2013) Sphingosine-1-phosphate signaling controlling osteoclasts and bone homeostasis. Biochim Biophys Acta 1831:223–227
Ishii M, Egen JG, Klauschen F, Meier-Schellersheim M, Saeki Y, Vacher J, Proia RL, Germain RN (2009) Sphingosine-1-phosphate mobilizes osteoclast precursors and regulates bone homeostasis. Nature 458:524–528
Ishii M, Kikuta J, Shimazu Y, Meier-Schellersheim M, Germain RN (2010) Chemorepulsion by blood S1P regulates osteoclast precursor mobilization and bone remodeling in vivo. J Exp Med 207:2793–2798
Ito S, Wakabayashi K, Ubukata O, Hayashi S, Okada F, Hata T (2002) Crystal structure of the extracellular domain of mouse RANK ligand at 2.2-A resolution. J Biol Chem 277:6631–6636
Jacome-Galarza CE, Percin GI, Muller JT, Mass E, Lazarov T, Eitler J, Rauner M, Yadav VK, Crozet L, Bohm M, Loyher PL, Karsenty G, Waskow C, Geissmann F (2019) Developmental origin, functional maintenance and genetic rescue of osteoclasts. Nature 568:541–545
Jansen ID, Vermeer JA, Bloemen V, Stap J, Everts V (2012) Osteoclast fusion and fission. Calcif Tissue Int 90:515–522
Jaworski ZF, Duck B, Sekaly G (1981) Kinetics of osteoclasts and their nuclei in evolving secondary Haversian systems. J Anat 133:397–405
Jimi E, Akiyama S, Tsurukai T, Okahashi N, Kobayashi K, Udagawa N, Nishihara T, Takahashi N, Suda T (1999) Osteoclast differentiation factor acts as a multifunctional regulator in murine osteoclast differentiation and function. J Immunol 163:434–442
Kanazawa K, Kudo A (2005) Self-assembled RANK induces osteoclastogenesis ligand-independently. J Bone Miner Res 20:2053–2060
Kariya Y, Honma M, Aoki S, Chiba A, Suzuki H (2009) Vps33a mediates RANKL storage in secretory lysosomes in osteoblastic cells. J Bone Miner Res 24:1741–1752
Kartsogiannis V, Zhou H, Horwood NJ, Thomas RJ, Hards DK, Quinn JM, Niforas P, Ng KW, Martin TJ, Gillespie MT (1999) Localization of RANKL (receptor activator of NF kappa B ligand) mRNA and protein in skeletal and extraskeletal tissues. Bone 25:525–534
Kenkre JS, Bassett J (2018) The bone remodelling cycle. Ann Clin Biochem 55:308–327

Khamviwath V, Hu J, Othmer HG (2013) A continuum model of actin waves in Dictyostelium discoideum. PLoS One 8:e64272

Knecht DA, Loomis WF (1987) Antisense RNA inactivation of myosin heavy chain gene expression in Dictyostelium discoideum. Science 236:1081–1086

Köster DV, Mayor S (2016) Cortical actin and the plasma membrane: inextricably intertwined. Curr Opin Cell Biol 38:81–89

Kukita T, Hiura H, Gu JY, Zhang JQ, Kyumoto-Nakamura Y, Uehara N, Murata S, Sonoda S, Yamaza T, Takahashi I, Kukita A (2021) Modulation of osteoclastogenesis through adrenomedullin receptors on osteoclast precursors: initiation of differentiation by asymmetric cell division. Lab Invest 101:1449–1457

Lacey DL, Timms E, Tan HL, Kelley MJ, Dunstan CR, Burgess T, Elliott R, Colombero A, Elliott G, Scully S, Hsu H, Sullivan J, Hawkins N, Davy E, Capparelli C, Eli A, Qian YX, Kaufman S, Sarosi I, Shalhoub V, Senaldi G, Guo J, Delaney J, Boyle WJ (1998) Osteoprotegerin ligand is a cytokine that regulates osteoclast differentiation and activation. Cell 93:165–176

Lam J, Nelson CA, Ross FP, Teitelbaum SL, Fremont DH (2001) Crystal structure of the TRANCE/RANKL cytokine reveals determinants of receptor-ligand specificity. J Clin Invest 108:971–979

Levaot N, Ottolenghi A, Mann M, Guterman-Ram G, Kam Z, Geiger B (2015) Osteoclast fusion is initiated by a small subset of RANKL-stimulated monocyte progenitors which can fuse to RANKL-unstimulated progenitors. Bone 79:21–28. https://doi.org/10.1016/j.bone.2015.05.021

Liu C, Walter TS, Huang P, Zhang S, Zhu X, Wu Y, Wedderburn LR, Tang P, Owens RJ, Stuart DI, Ren J, Gao B (2010) Structural and functional insights of RANKL-RANK interaction and signaling. J Immunol 184:6910–6919

Manolagas SC (2000) Birth and death of bone cells: basic regulatory mechanisms and implications for the pathogenesis and treatment of osteoporosis. Endocr Rev 21:115–137

Marchisio PC, Cirillo D, Naldini L, Primavera MV, Teti A, Zambonin-Zallone A (1984) Cell-substratum interaction of cultured avian osteoclasts is mediated by specific adhesion structures. J Cell Biol 99:1696–1705

Marks SC, Seifert MF (1985) The lifespan of osteoclasts: experimental studies using the giant granule cytoplasmic marker characteristic of beige mice. Bone 6:451–455

Martin SG (2016) Role and organization of the actin cytoskeleton during cell-cell fusion. Semin Cell Dev Biol 60:121–126

McCusker D (2020) Cellular self-organization: generating order from the abyss. Mol Biol Cell 31:143–148

McDonald MM, Khoo WH, Ng PY, Xiao Y, Zamerli J, Thatcher P, Kyaw W, Pathmanandavel K, Grootveld AK, Moran I, Butt D, Nguyen A, Corr A, Warren S, Biro M, Butterfield NC, Guilfoyle SE, Komla-Ebri D, Dack MRG, Dewhurst HF, Logan JG, Li Y, Mohanty ST, Byrne N, Terry RL, Simic MK, Chai R, Quinn JMW, Youlten SE, Pettitt JA, Abi-Hanna D, Jain R, Weninger W, Lundberg M, Sun S, Ebetino FH, Timpson P, Lee WM, Baldock PA, Rogers MJ, Brink R, Williams GR, Bassett JHD, Kemp JP, Pavlos NJ, Croucher PI, Phan TG (2021a) Osteoclasts recycle via osteomorphs during RANKL-stimulated bone resorption. Cell 184:1330–1347

McDonald MM, Kim AS, Mulholland BS, Rauner M (2021b) New insights into osteoclast biology. JBMR Plus 5:e10539

Mensah KA, Ritchlin CT, Schwarz EM (2010) RANKL induces heterogeneous DC-STAMP(lo) and DC-STAMP(hi) osteoclast precursors of which the DC-STAMP(lo) precursors are the master fusogens. J Cell Physiol 223:76–83

Miao Y, Bhattacharya S, Banerjee T, Abubaker-Sharif B, Long Y, Inoue T, Iglesias PA, Devreotes PN (2019) Wave patterns organize cellular protrusions and control cortical dynamics. Mol Syst Biol 15:e8585

Miyamoto T, Arai F, Ohneda O, Takagi K, Anderson DM, Suda T (2000) An adherent condition is required for formation of multinuclear osteoclasts in the presence of macrophage colony-stimulating factor and receptor activator of nuclear factor kappa B ligand. Blood 96:4335–4343

Mizoguchi T, Muto A, Udagawa N, Arai A, Yamashita T, Hosoya A, Ninomiya T, Nakamura H, Yamamoto Y, Kinugawa S, Nakamura M, Nakamichi Y, Kobayashi Y, Nagasawa S, Oda K,

Tanaka H, Tagaya M, Penninger JM, Ito M, Takahashi N (2009) Identification of cell cycle-arrested quiescent osteoclast precursors in vivo. J Cell Biol 184:541–554
Motiur Rahman M, Takeshita S, Matsuoka K, Kaneko K, Naoe Y, Sakaue-Sawano A, Miyawaki A, Ikeda K (2015) Proliferation-coupled osteoclast differentiation by RANKL: Cell density as a determinant of osteoclast formation. Bone 81:392–399
Mukhina S, Wang YL, Murata-Hori M (2007) Alpha-actinin is required for tightly regulated remodeling of the actin cortical network during cytokinesis. Dev Cell 13:554–565
Mulari MT, Zhao H, Lakkakorpi PT, Väänänen HK (2003) Osteoclast ruffled border has distinct subdomains for secretion and degraded matrix uptake. Traffic 4:113–125
Nakashima T, Kobayashi Y, Yamasaki S, Kawakami A, Eguchi K, Sasaki H, Sakai H (2000) Protein expression and functional difference of membrane-bound and soluble receptor activator of NF-kappaB ligand: modulation of the expression by osteotropic factors and cytokines. Biochem Biophys Res Commun 275:768–775
Nakashima T, Hayashi M, Fukunaga T, Kurata K, Oh-Hora M, Feng JQ, Bonewald LF, Kodama T, Wutz A, Wagner EF, Penninger JM, Takayanagi H (2011) Evidence for osteocyte regulation of bone homeostasis through RANKL expression. Nat Med 17:1231–1234
Nesbitt SA, Horton MA (1997) Trafficking of matrix collagens through bone-resorbing osteoclasts. Science 276:266–269
Nevius E, Pinho F, Dhodapkar M, Jin H, Nadrah K, Horowitz MC, Kikuta J, Ishii M, Pereira JP (2015) Oxysterols and EBI2 promote osteoclast precursor migration to bone surfaces and regulate bone mass homeostasis. J Exp Med 212:1931–1946
O'Brien CA (2010) Control of RANKL gene expression. Bone 46:911–919
Oikawa T, Itoh T, Takenawa T (2008) Sequential signals toward podosome formation in NIH-src cells. J Cell Biol 182:157–169
Oikawa T, Oyama M, Kozuka-Hata H, Uehara S, Udagawa N, Saya H, Matsuo K (2012) Tks5-dependent formation of circumferential podosomes/invadopodia mediates cell-cell fusion. J Cell Biol 197:553–568
Onan D, Allan EH, Quinn JM, Gooi JH, Pompolo S, Sims NA, Gillespie MT, Martin TJ (2009) The chemokine Cxcl1 is a novel target gene of parathyroid hormone (PTH)/PTH-related protein in committed osteoblasts. Endocrinology 150:2244–2253
Parfitt AM (1994) Osteonal and hemi-osteonal remodeling: the spatial and temporal framework for signal traffic in adult human bone. J Cell Biochem 55:273–286
Park JH, Lee NK, Lee SY (2017) Current Understanding of RANK Signaling in Osteoclast Differentiation and Maturation. Mol Cells 40:706–713
Pauksch L, Rohnke M, Schnettler R, Lips KS (2014) Silver nanoparticles do not alter human osteoclastogenesis but induce cellular uptake. Toxicol Rep 1:900–908
Pennanen P, Alanne MH, Fazeli E, Deguchi T, Näreoja T, Peltonen S, Peltonen J (2017) Diversity of actin architecture in human osteoclasts: network of curved and branched actin supporting cell shape and intercellular micrometer-level tubes. Mol Cell Biochem 432:131–139
Pinho S, Frenette PS (2019) Haematopoietic stem cell activity and interactions with the niche. Nat Rev Mol Cell Biol 20:303–320
Piper K, Boyde A, Jones SJ (1992) The relationship between the number of nuclei of an osteoclast and its resorptive capability in vitro. Anat Embryol (Berl) 186:291–299
Platta HW, Stenmark H (2011) Endocytosis and signaling. Curr Opin Cell Biol 23:393–403
Reznikov N, Shahar R, Weiner S (2013) Three-dimensional structure of human lamellar bone: The presence of two different materials and new insights into the hierarchical organization. Bone 59:93–104
Reznikov N, Bilton M, Lari L, Stevens MM, Kroger R (2018) Fractal-like hierarchical organization of bone begins at the nanoscale. Science 360:eaao2189
Robling AG, Bonewald LF (2020) The osteocyte: new insights. Annu Rev Physiol 82:485–506
Rottner K, Faix J, Bogdan S, Linder S, Kerkhoff E (2017) Actin assembly mechanisms at a glance. J Cell Sci 130:3427–3435

Salo J, Lehenkari P, Mulari M, Metsikkö K, Väänänen HK (1997) Removal of osteoclast bone resorption products by transcytosis. Science 276:270–273
Saltel F, Chabadel A, Bonnelye E, Jurdic P (2008) Actin cytoskeletal organisation in osteoclasts: a model to decipher transmigration and matrix degradation. Eur J Cell Biol 87:459–468
Sharma N, Weivoda MM, Søe K (2022) Functional heterogeneity within osteoclast populations-a critical review of four key publications that may change the paradigm of osteoclasts. Curr Osteoporos Rep 20:344–355
Sheikh MS, Fornace AJ (2000) Death and decoy receptors and p53-mediated apoptosis. Leukemia 14:1509–1513
Shin J, Kim YM, Li SZ, Lim SK, Lee W (2008) Structure-function of the TNF receptor-like cysteine-rich domain of osteoprotegerin. Mol Cells 25:352–357
Shin NY, Choi H, Neff L, Wu Y, Saito H, Ferguson SM, De Camilli P, Baron R (2014) Dynamin and endocytosis are required for the fusion of osteoclasts and myoblasts. J Cell Biol 207:73–89
Silvestrini G, Ballanti P, Patacchioli F, Leopizzi M, Gualtieri N, Monnazzi P, Tremante E, Sardella D, Bonucci E (2005) Detection of osteoprotegerin (OPG) and its ligand (RANKL) mRNA and protein in femur and tibia of the rat. J Mol Histol 36:59–67
Sims NA, Martin TJ (2020) Osteoclasts provide coupling signals to osteoblast lineage cells through multiple mechanisms. Annu Rev Physiol 82:507–529
Sims NA, Vrahnas C (2014) Regulation of cortical and trabecular bone mass by communication between osteoblasts, osteocytes and osteoclasts. Arch Biochem Biophys 561:22–28
Sivaraj KK, Adams RH (2016) Blood vessel formation and function in bone. Development 143:2706–2715
Søe K, Hobolt-Pedersen AS, Delaisse JM (2015) The elementary fusion modalities of osteoclasts. Bone 73:181–189
Søe K, Andersen TL, Hinge M, Rolighed L, Marcussen N, Delaisse JM (2019) Coordination of Fusion and Trafficking of Pre-osteoclasts at the Marrow-Bone Interface. Calcif Tissue Int 105:430–445
Søe K, Delaisse JM, Borggaard XG (2021) Osteoclast formation at the bone marrow/bone surface interface: importance of structural elements matrix and intercellular communication. Semin Cell Dev Biol 112:8–15. https://doi.org/10.1016/j.semcdb.2020.05.016
Solari F, Domenget C, Gire V, Woods C, Lazarides E, Rousset B, Jurdic P (1995) Multinucleated cells can continuously generate mononucleated cells in the absence of mitosis: a study of cells of the avian osteoclast lineage. J Cell Sci 108:3233–3241
Song RL, Liu XZ, Zhu JQ, Zhang JM, Gao Q, Zhao HY, Sheng AZ, Yuan Y, Gu JH, Zou H, Wang QC, Liu ZP (2014) New roles of filopodia and podosomes in the differentiation and fusion process of osteoclasts. Genet Mol Res 13:4776–4787
Stephens AS, Stephens SR, Morrison NA (2011) Internal control genes for quantitative RT-PCR expression analysis in mouse osteoblasts, osteoclasts and macrophages. BMC Res Notes 4:410
Streicher C, Heyny A, Andrukhova O, Haigl B, Slavic S, Schüler C, Kollmann K, Kantner I, Sexl V, Kleiter M, Hofbauer LC, Kostenuik PJ, Erben RG (2017) Estrogen regulates bone turnover by targeting RANKL expression in bone lining cells. Sci Rep 7:6460
Takahashi N, Udagawa N, Suda T (1999) A new member of tumor necrosis factor ligand family, ODF/OPGL/TRANCE/RANKL, regulates osteoclast differentiation and function. Biochem Biophys Res Commun 256:449–455
Takahashi A, Kukita A, Li YJ, Zhang JQ, Nomiyama H, Yamaza T, Ayukawa Y, Koyano K, Kukita T (2013) Tunneling nanotube formation is essential for the regulation of osteoclastogenesis. J Cell Biochem 114:1238–1247
Takegahara N, Kim H, Mizuno H, Sakaue-Sawano A, Miyawaki A, Tomura M, Kanagawa O, Ishii M, Choi Y (2016) Involvement of receptor activator of nuclear factor-κB ligand (RANKL)-induced incomplete cytokinesis in the polyploidization of osteoclasts. J Biol Chem 291:3439–3454
Takito J, Nakamura M (2012) Precursors linked via the zipper-like structure or the filopodium during the secondary fusion of osteoclasts. Commun Integr Biol 5:453–457

Takito J, Nakamura M (2020) Heterogeneity and actin cytoskeleton in osteoclast and macrophage multinucleation. Int J Mol Sci 21:6629

Takito J, Nakamura M, Yoda M, Tohmonda T, Uchikawa S, Horiuchi K, Toyama Y, Chiba K (2012) The transient appearance of zipper-like actin superstructures during the fusion of osteoclasts. J Cell Sci 125:662–672

Takito J, Inoue S, Nakamura M (2017a) Emerging role of actin flow in the organization of podosomes in osteoclasts. Macrophage 4:e1614

Takito J, Otsuka H, Inoue S, Kawashima T, Nakamura M (2017b) Symmetrical retrograde actin flow in the actin fusion structure is involved in osteoclast fusion. Biol Open 6:1104–1114

Takito J, Inoue S, Nakamura M (2018) The sealing zone in osteoclasts: a self-organized structure on the bone. Int J Mol Sci 19:984

Tanaka S, Takahashi N, Udagawa N, Tamura T, Akatsu T, Stanley ER, Kurokawa T, Suda T (1993) Macrophage colony-stimulating factor is indispensable for both proliferation and differentiation of osteoclast progenitors. J Clin. Invest 91:257–263

Tanaka S, Miyazaki T, Fukuda A, Akiyama T, Kadono Y, Wakeyama H, Kono S, Hoshikawa S, Nakamura M, Ohshima Y, Hikita A, Nakamura I, Nakamura K (2006) Molecular mechanism of the life and death of the osteoclast. Ann N Y Acad Sci 1068:180–186

Tsukasaki M, Huynh NC, Okamoto K, Muro R, Terashima A, Kurikawa Y, Komatsu N, Pluemsakunthai W, Nitta T, Abe T, Kiyonari H, Okamura T, Sakai M, Matsukawa T, Matsumoto M, Kobayashi Y, Penninger JM, Takayanagi H (2020) Stepwise cell fate decision pathways during osteoclastogenesis at single-cell resolution. Nat Metab 12:1382–1390

Udagawa N, Takahashi N, Jimi E, Matsuzaki K, Tsurukai T, Itoh K, Nakagawa N, Yasuda H, Goto M, Tsuda E, Higashio K, Gillespie MT, Martin TJ, Suda T (1999) Osteoblasts/stromal cells stimulate osteoclast activation through expression of osteoclast differentiation factor/RANKL but not macrophage colony-stimulating factor. Bone 25:517–523

Verma SK, Leikina E, Melikov K, Chernomordik LV (2014) Late stages of the synchronized macrophage fusion in osteoclast formation depend on dynamin. Biochem J 464:293–300

Walker DG (1975) Control of bone resorption by hematopoietic tissue. The induction and reversal of congenital osteopetrosis in mice through use of bone marrow and splenic transplants. J Exp Med 142:651–663

Wang D, Gu JH, Feng LL, Tong XS, Song RL, Zhao HY, Bian JC, Liu XZ, Yuan Y, Liu ZP (2018) 1-alpha,25-dihydroxyvitamin D3 potentiates avian osteoclast activation by increasing the formation of zipper-like structure via Src/Rac1 signaling. Biochem Biophys Res Commun 501:576–583

Weiner OD, Marganski WA, Wu LF, Altschuler SJ, Kirschner MW (2007) An actin-based wave generator organizes cell motility. PLoS Biol 5:e221

Xiong J, Onal M, Jilka RL, Weinstein RS, Manolagas SC, O'Brien CA (2011) Matrix-embedded cells control osteoclast formation. Nat Med 17:1235–1241

Xiong J, Piemontese M, Onal M, Campbell J, Goellner JJ, Dusevich V, Bonewald L, Manolagas SC, O'Brien CA (2015) Osteocytes, not osteoblasts or lining cells, are the main source of the RANKL required for osteoclast formation in remodeling bone. PLoS One 10:e0138189

Yagi M, Miyamoto T, Sawatani Y, Iwamoto K, Hosogane N, Fujita N, Morita K, Ninomiya K, Suzuki T, Miyamoto K, Oike Y, Takeya M, Toyama Y, Suda T (2005) DC-STAMP is essential for cell-cell fusion in osteoclasts and foreign body giant cells. J Exp Med 202:345–351

Yang M, Mailhot G, MacKay CA, Mason-Savas A, Aubin J, Odgren PR (2006) Chemokine and chemokine receptor expression during colony stimulating factor-1-induced osteoclast differentiation in the toothless osteopetrotic rat: a key role for CCL9 (MIP-1gamma) in osteoclastogenesis in vivo and in vitro. Blood 107:2262–2270

Yoshida H, Hayashi S, Kunisada T, Ogawa M, Nishikawa S, Okamura H, Sudo T, Shultz LD, Nishikawa S (1990) The murine mutation osteopetorosis is in the coding region of macrophage colony stimulating factor gene. Nature 345:442–444

Yu X, Huang Y, Collin-Osdoby P, Osdoby P (2003) Stromal cell-derived factor-1 (SDF-1) recruits osteoclast precursors by inducing chemotaxis, matrix metalloproteinase-9 (MMP-9) activity, and collagen transmigration. J Bone Miner Res 18:1404–1418
Yu CH, Rafiq NB, Krishnasamy A, Hartman KL, Jones GE, Bershadsky AD, Sheetz MP (2013) Integrin-matrix clusters form podosome-like adhesions in the absence of traction forces. Cell Rep 5:1456–1468
Zaidi M, Yuen T, Sun L, Rosen CJ (2018) Regulation of skeletal homeostasis. Endocr Rev 39:701–718
Zambonin Zallone A, Teti A, Primavera MV (1984) Monocytes from circulating blood fuse in vitro with purified osteoclasts in primary culture. J Cell Sci 66:335–342

Chapter 13
Muscle Progenitor Cell Fusion in the Maintenance of Skeletal Muscle

Jarred M. Whitlock

Abstract Skeletal muscle possesses a resident, multipotent stem cell population that is essential for its repair and maintenance throughout life. Here I highlight the role of this stem cell population in muscle repair and regeneration and review the genetic control of the process; the mechanistic steps of activation, migration, recognition, adhesion, and fusion of these cells; and discuss the novel recognition of the membrane signaling that coordinates myogenic cell-cell fusion, as well as the identification of a two-part fusogen system that facilitates it.

13.1 Introduction

Muscle fibers are the individual, cellular units comprising skeletal muscle tissues and represent a highly specialized cell type. Skeletal muscle tissue represents ~40% of overall body mass and is essential for indispensable tasks such as the consumption of food and breathing (Proctor et al. 1999). Each muscle fiber is created by the fusion of hundreds to thousands of muscle progenitor cells during development to produce a single, continuous cell that can run up to 600 mm in length in humans (Yang et al. 1998; Abmayr and Pavlath 2012). These fibers consist of a specialized plasma membrane, the sarcolemma (SL); a specialized cytoplasm, the sarcoplasm (SP); and many sarcomeres, bundles of actin/myosin contractile units organized into long chains termed myofibrils, which underpin skeletal muscle's contractile force. Moreover, skeletal muscle fibers also contain many specialized organelles, including many large specialized mitochondria, sarcosomes; a modified endoplasmic reticulum, the sarcoplasmic reticulum (SR); and hundreds to thousands of

J. M. Whitlock (✉)
Section on Membrane Biology, *Eunice Kennedy Shrive* National Institute of Child Health and Human Development, National Institutes of Health, Bethesda, MD, USA
e-mail: jarred.whitlock@nih.gov

M. Kloc, A. Uosef (eds.), *Syncytia: Origin, Structure, and Functions*, Results and Problems in Cell Differentiation 71,
https://doi.org/10.1007/978-3-031-37936-9_13

myonuclei, whose numbers are tightly maintained following pre- and postnatal development (Fig. 13.1a) (White et al. 2010; Spalding et al. 2013).

Much of the specialization surrounding muscle fiber anatomy revolves around their contractile role and the requirement that each muscle fiber must transduce electrical signals, which initiate from innervations at the neuromuscular junction (NMJ), along their entire length to tightly coordinate Ca^{2+} release from the SR and subsequent sarcomere contraction. The tight regulation of this Ca^{2+} signaling is essential for the coordination of sarcomere contraction along the length of the fiber in synchronization with neighboring fibers to produce a concerted contraction of the tissue. To coordinate this tightly regulated Ca^{2+} cascade, the transverse tubules (invaginated SL) rapidly transmit action potentials to activate resident L-type Ca^{2+} channels. The activity of these L-type Ca^{2+} channels stimulates dihydropyridine receptors that elicit Ca^{2+} release from the SR. The specialized SR network, which

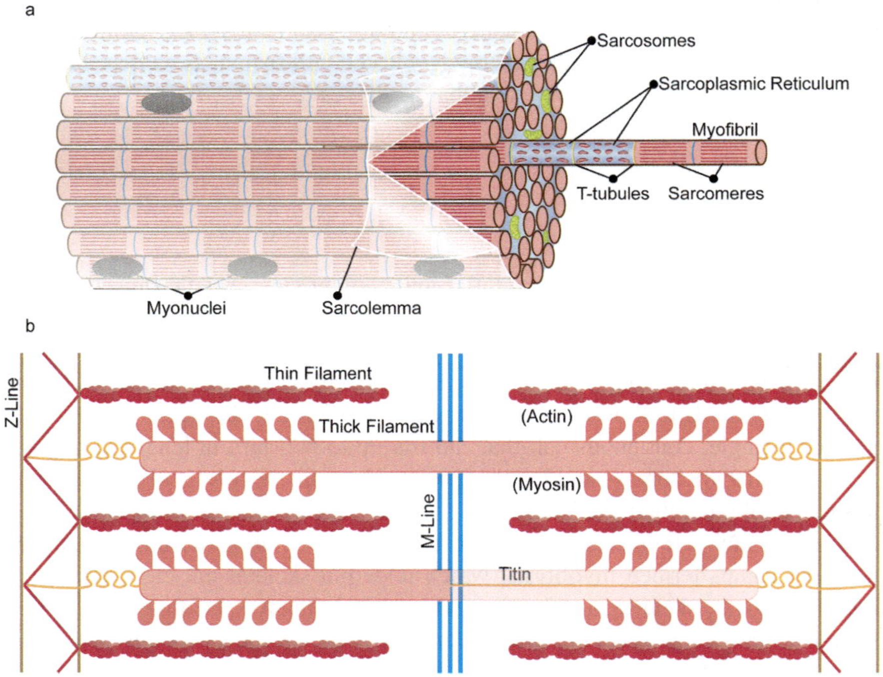

Fig. 13.1 An illustrative representation of skeletal muscle cell ultrastructure. (**a**) Ultrastructure of a single skeletal muscle fiber along its longitudinal axis. (**b**) The general organization of a single sarcomere. Skeletal muscle relies on at least two essential repair processes that work to maintain its integrity and function. These processes include a membrane patch repair process, present in all mammalian cell types, and a specialized repair/regeneration mechanism that relies on a resident, mononucleated stem cell population. These processes work in parallel to maintain muscle function throughout life. Perturbations in either of these repair processes have been shown to lead to a variety of myopathies, primarily disorders that result in muscle weakness and wasting [muscular dystrophies (MDs)]. Here we will review these repair processes and what is known concerning their mechanisms

winds through the entire fiber in a regulated pattern, quickly releases Ca^{2+} into the SP that is filled largely with contractile units referred to as myofibrils [iterative collections of sarcomeres (Fig. 13.1b)]. This intracellular signaling cascade elicits the coordinated contraction of the myofibrils and converts a chemical signal elicited by the original electrical impulse at the NMJ into a physical action within skeletal muscle tissue (reviewed in greater detail (Calderon et al. 2014)). Skeletal muscle is akin to a marvelous biological machine; however like any machine, this evolutionary marvel requires a system of repair processes that dynamically maintain its function.

13.2 Skeletal Muscle Requires Rapid Repair/Regeneration Mechanisms for Lifelong Maintenance

13.2.1 Plasma Membrane Lesions Undergo Patching via Ca^{2+} Regulated Exocytic Repair

For more than half a century, it has been well recognized that extracellular Ca^{2+} flux through plasma membrane lesions triggers the rapid (seconds) resealing of the membrane to avoid Ca^{2+} toxicity (reviewed (Andrews et al. 2014)). If these injuries are small enough (a few nanometers), the lesions can be rapidly resealed by virtue of membrane lipids rearranging to avoid exposure of the hydrophobic membrane core to the outside, polar environment (Gozen and Dommersnes 2014). However, if lesions exceed several nanometers, active repair processes are required to reinstate membrane integrity (Cooper and McNeil 2015). Active repair of membrane lesions often occurs through the rapid exocytosis of intracellular vesicles (Miyake and McNeil 1995; Bi et al. 1995). The resealing capability of this pathway is highlighted in classical experiments using microneedle damage of echinoderm egg membrane, where these cells can employ exocytic patch repair to reseal lesions of $>2000\ \mu m^2$ in <5 s (Terasaki et al. 1997). Why cells go to such great lengths to maintain the mechanism of membrane patch repair lies in the propensity of this process to protect irreplaceable cell types (neurons) or those that are of significant energetic value (skeletal muscle). The mechanistic intricacy of membrane patch repair combined with its diversity of players in different cell types strongly suggests that it has been governed by immense selective pressure during cellular evolution.

13.2.1.1 General Membrane Patch Repair Mechanism

The general mechanism of membrane lesion patch repair is consistent throughout many diverse cell types; however, the exact proteins regulating patch repair and the membranes contributing to the patch continue to be hotly debated topics and likely differ greatly between cell types and organisms. Much of the controversy surrounding this field likely stems from the many associated processes that are stimulated by intracellular Ca^{2+} flux but may not be essential for membrane patch repair to restore plasma membrane (PM) barrier competence. For instance, the fusion of exocytic vesicles during conventional exocytosis associated with the secretory pathway is regulated by increases in Ca^{2+}; therefore, damaging the PM might cause exocytic vesicles to fuse with the PM due to incidental Ca^{2+} flux. This does not mean that these vesicles are necessarily important for membrane repair or that their fusion machinery should also be considered patch repair machinery. Moreover, because of the importance of closing membrane lesions to the survival of the cell, the probability of built-in redundancy is high.

In short, upon elevated Ca^{2+} flux from the extracellular solution through the lesion site, vesicular membranes are recruited to the site of membrane breach, and rapid exocytic processes elicit the fusion of intracellular vesicles with the lesion. Vesicle-vesicle fusion produces a growing membrane patch, while fusion of these patch vesicles with the lesion membrane restores membrane barrier competence (Cooper and McNeil 2015). This process is likely regulated by exocytic machinery, of which synaptotagmin-7 and a variety of other soluble N-ethylmaleimide sensitivity factor attachment protein receptor SNARE proteins have been highlighted for their roles in Ca^{2+} sensing and fusogenic repair (Togo et al. 1999; Reddy et al. 2001; Detrait et al. 2000; Chakrabarti et al. 2003). Because diverse intracellular vesicles are fusion competent and vesicular fusion is regulated in part by elevated cytosolic Ca^{2+}, most vesicular membranes are probably capable of contributing to membrane patch repair, although some may be preferred. There are also a variety of non-vesicular repair pathways that likely complement this process and may even be sufficient for lesion closure in some special cases (reviewed elsewhere (Andrews et al. 2014)).

Skeletal muscle fibers have a particular dependence on membrane patch repair, as up to 30% of skeletal muscle fibers exhibit membrane wounding in vivo (McNeil and Steinhardt 1997). Muscle fibers possess orders of magnitude more delimiting membrane than any other cell type, and this membrane is continuously subjected to immense mechanical stress that leads to SL lesions (McNeil and Khakee 1992). Although considerable controversy exists over the precise mechanism of skeletal muscle patch repair, a general consensus has been resolved following decades of investigation. Following SL insult, dysferlin is thought to function as a Ca^{2+} sensor and/or a fusogenic protein regulating membrane patch formation by virtue of its C2 domains' Ca^{2+}-dependent affinity for phospholipids. The C2 domains of dysferlin and many other regulators of intracellular membrane fusion are essential for the Ca^{2+}-dependent association of these proteins with membranes (Corbalan-Garcia and Gomez-Fernandez 2014). Dysferlin's muscle-specific role in SL resealing is similar to related ferlin proteins' roles in resealing PMs in other cell (Glover and Brown Jr. 2007; Johnson and Chapman 2010; Bansal et al. 2003). Recent investigation suggests that when SL damage occurs, dysferlin-laden SL is endocytosed and fuses along with endosomes and other endo-lysosomal organelles to form a membrane patch (Glover and Brown Jr. 2007). In addition, dysferlin may remain at the membrane lesion and work to coordinate the fusion of membrane patch vesicles in

SL resealing. Strong evidence suggests that the Ca^{2+}-dependent association of annexins with phosphatidylserine (PS) at the lesion membrane coordinates the action of dysferlin in SL resealing (Lennon et al. 2003). In parallel, mitsugumin 53 oligomerizes on PS containing intracellular vesicles upon Ca^{2+} flux and works to recruit repair vesicles and patch repair machinery to the lesion site (Cai et al. 2009; Cai et al. 2015). Mutations and/or loss of any of these components leads to perturbations of muscle repair in vivo, and some of the genes encoding these proteins are linked to a variety of human MDs (Bansal et al. 2003; Cai et al. 2009; Leikina et al. 2015; Defour et al. 2017).

13.2.2 Skeletal Muscle Employs a Multipotent Stem Cell Population in Fiber Repair/Regeneration

Because of the essential nature of maintaining proper musculature, muscle possesses a second major repair process that relies on a native, multipotent stem cell population. These stem cells are termed satellite cells because of their proximity to the muscle fiber, occupying a niche in the surrounding basal lamina (Mauro 1961). Satellite cell-dependent muscle repair/regeneration is essential for the lifelong maintenance and function of skeletal muscle (Sambasivan et al. 2011; Lepper et al. 2011). Upon fiber injury, satellite cells are activated and begin rampant proliferation. The resulting daughter cells contribute to the repair of damaged skeletal muscle (Snow 1977; Reznik 1969). Like developmental muscle precursors, their multipotent daughter cells proceed through a muscle developmental program (myogenesis) and undergo homotypic (progenitor cell-progenitor cell) and heterotypic fusion with the damaged fiber to restore the fiber (Allen and Boxhorn 1989). Additionally, this muscle precursor population retains the ability to fuse and regenerate lost fibers de novo to restore damaged muscle lost to injury/disease (Gurevich et al. 2016; Collins et al. 2005; Rosenblatt et al. 1995; Corona et al. 2013).

13.3 Satellite Cell-Dependent Muscle Repair: A Trip Back to Development?

13.3.1 Myogenic Progression of Progenitor Cells in Skeletal Muscle

Our understanding of satellite cell-dependent muscle repair is largely colored by the characterization of myogenic progression during development. Skeletal muscle develops from the dorsal portion of the somites, with the exception of many head muscles (Gros et al. 2005). Embryonic myogenic specification is controlled primarily by the transcription factor paired box protein 3 (PAX3), and to a lesser extent

PAX7, which plays a sort of master regulatory role in the process of myogenic specification (Tajbakhsh et al. 1997; Bober et al. 1994). Further commitment to differentiated muscle involves sequential activation of four myogenic regulatory factor (MRF) family transcription factors Myf5, MyoD, myogenin, and MRF4 (reviewed extensively (Dumont et al. 2015)). Some controversy remains involving the hierarchy of these regulators, but the consensus is that Myf5–followed shortly by MyoD–act as determination factors, committing cells to a myogenic fate. Next, myogenin and MRF4 expression increase and fulfill essential roles for terminal muscle differentiation. Largely, myogenic progression of resident muscle stem cells in adult tissue proceeds through a similar myogenic progression (i.e., myogenesis); however some differences exist in the mechanistic action of these general processes.

13.3.2 Satellite Cells Become Activated and Migrate to Tissue Damage upon Muscle Injury

Following muscle injury, skeletal muscle stem cells must become activated and migrate to sites of fiber injury to participate in tissue repair, as illustrated in Fig. 13.2. Adult skeletal muscle maintains a satellite stem cell population in the basal lamina surrounding the muscle fiber that plays essential roles in myogenic repair/regeneration. These stem cells reside in a quiescent state with a high nuclear/cytoplasm ratio and very low transcriptional and metabolic activity at rest (Cheung and Rando 2013). Resting satellite cells are very similar to undifferentiated muscle progenitors in development; however this stem cell population displays an essential requirement for PAX7, which they express at very high levels in contrast to the high reliance on PAX3 developmentally (Sambasivan et al. 2011; Lepper et al. 2011; Bober et al. 1994; Mansouri et al. 1996). Upon muscle injury, nitric oxide is near instantly generated and plays an essential role in the injury-induced activation of satellite cells (Anderson 2000). The extracellular matrix (ECM) of the basal lamina embeds satellite cells and insulates them from the influence of circulating growth factors. Nitric oxide strongly upregulates the production of matrix metalloproteinase that degrades the ECM environment of the satellite cell niche (Tatsumi 2010). Upon ECM damage, satellite cells are exposed to and activated by a variety of circulating factors. One factor, fibroblast growth factor (FGF) 2, induces a rapid intracellular Ca^{2+} spike through transient receptor potential cation channels, which activates nuclear factor activated T-cells, cytoplasmic, and leads to cell cycle reentrance (Liu and Schneider 2014). A similar activating role has also been ascribed to hepatocyte growth factor (Tatsumi et al. 1998; Allen et al. 1995). Cell cycle reentrance marks the initiation of satellite cell-dependent muscle repair.

Following activation, satellite cells become highly mobile and migrate to the site of muscle fiber injury where they proliferate, giving rise to a pool of highly fusogenic muscle precursor cells (Bischoff 1997). Initial migration to the site of injury is essential, as activated satellite cells become less mobile following proliferation

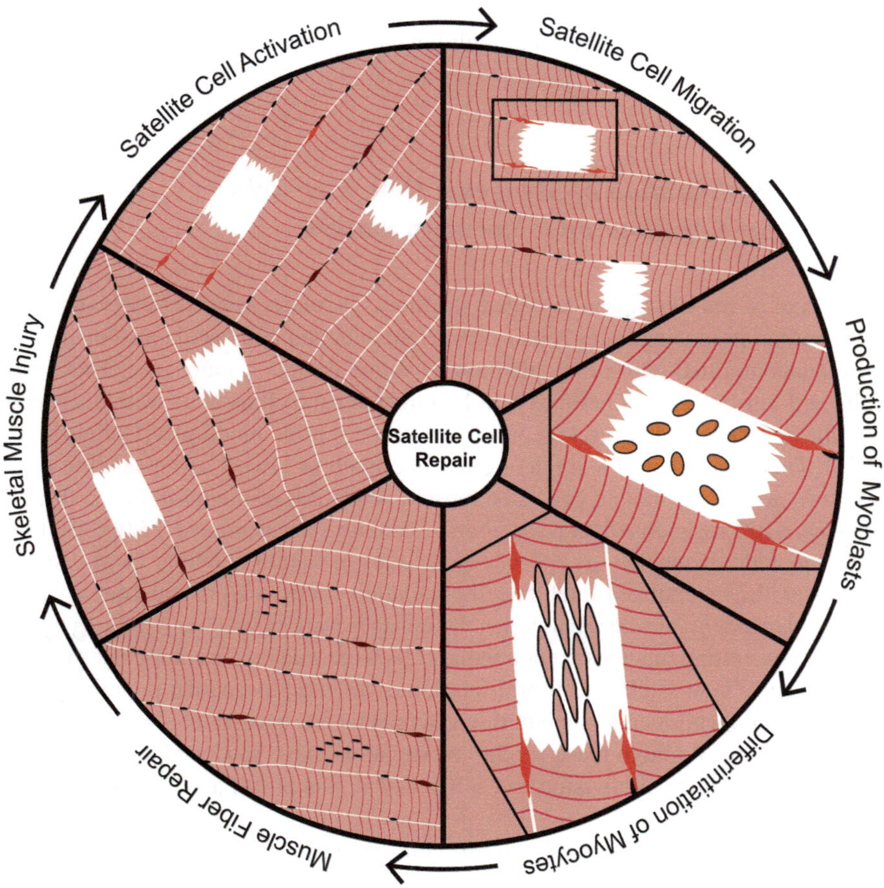

Fig. 13.2 Satellite cell-dependent skeletal muscle repair. A graphical illustration of the progression of satellite cell-dependent skeletal muscle repair following injury. Sequential steps highlight the roles of the myogenic cell types involved at each step of repair: Burgundy, quiescent satellite cells; red, activated satellite cells, orange, proliferative myoblasts; pink, quiescent myocytes; and pink with magenta striations. Multinucleated myofibers. Within damaged fiber windows examples of the many singling factors associated with the progression of satellite cell-dependent repair between steps are highlighted as colored spheres: green, nitric oxide (NO_2), yellow, fibroblast growth factor and hepatocyte growth factor (FGF and HGF); and purple, a general representation of interleukin factors (ILs)

(Dumont et al. 2015). Injured muscle releases a variety of soluble chemoattractant molecules, illustrated by the ability of extract from crushed muscle to promote directional migration of isolated muscle progenitor cells (Bischoff 1997). The chemoattractant abilities of some soluble molecules within crush extracts have been recognized (e.g., transforming growth factor-β, hepatocyte growth factor, FGF2, and FGF6) (Siegel et al. 2009; Neuhaus et al. 2003); however, the exact contributions of various factors to muscle progenitor chemotaxis remain an area of open investigation. Upon reaching the site of damage, activated satellite cells begin proliferating to produce a population of multipotent daughter cells capable of repairing the musculature.

13.3.3 Proliferation of Myogenic Daughter Cells for Contribution to the Musculature

At the site of muscle injury, activated satellite cells undergo successive rounds of proliferation producing many repair-competent daughter cells while also maintaining an uncommitted stem cell population for future muscle maintenance. These satellite cell attributes were perhaps best highlighted by Collins et al. who evaluated satellite cell-dependent muscle repair in irradiated murine hindlimbs (Collins et al. 2005). Transplantation of a single muscle fiber with as few as seven satellite cells was sufficient to repopulate irradiated muscle with >100 resident satellite cells and produce ~100 regenerated muscle fibers containing an estimated 25,000–30,000 differentiated myonuclei in 3 weeks post transplantation (Collins et al. 2005). Moreover, the repopulated satellite cell population contributed to subsequent satellite cell-dependent muscle repair when the muscles were injured with notexin injection, demonstrating that these repopulated cells were true repair competent satellite cells. Exactly how satellite cell proliferation gives rise to separate populations of committed progenitors and uncommitted stem cells has been extensively evaluated in recent years. Although these recent investigations have produced a basic understanding of satellite cell division and self-renewal, this is still a very active area of investigation.

For some time, whether $Pax7^+$ satellite cells were true stem cells or whether they were dedifferentiated myoblasts was an open question. In 2007, the Rudnicki group demonstrated that ~10% of the total satellite cell pool were $Pax7^+$/ myogenic factor 5 $(Myf5)^-$ and had never expressed the commitment factor Myf5 throughout the history of the cell (Kuang et al. 2007). This never committed $Pax7^+$/ $Myf5^-$ population expands to ~30% of the satellite cell pool following injury, suggesting a role for the expansion of satellite stem cells in muscle repair. These $Pax7^+$/ $Myf5^-$ cells are thought to represent the true, uncommitted stem cell population in skeletal muscle; however the precise origin of individual muscle stem cells is complicated, as there is strong evidence that committed muscle progenitor cells can undergo dedifferentiation and revert to quiescent satellite cells in vitro (Zammit et al. 2004). Moreover, the physiological relevance of this dedifferentiation process in maintaining repair competent satellite cells was recently bolstered with the observation of myogenic regression of committed muscle progenitors in vivo (Shea et al. 2010).

Regardless of their previous differentiation states, satellite cells undergo successive rounds of proliferation at the site of muscle injury. This proliferative process is partially regulated by many of the same factors that activate these cells (e.g., FGF2); however recent investigation has highlighted the role of a cascade of proinflammatory cytokines (e.g., interleukin 6) released in injured muscle that coincides with muscle progenitor proliferation and play major roles in regulating this process (Wang et al. 2008; Cantini et al. 1995). Once proliferation is initiated, satellite cells can undergo two separate division processes giving rise primarily to committed satellite cells ($Pax7^{+}/Myf5^{+}$)–and to a lesser extent–maintaining an uncommitted satellite cell population ($Pax7^{+}/Myf5^{-}$) for stem cell self-renewal. $Pax7^{+}/Myf5^{-}$ cells can either undergo asymmetric (apical/basal) or symmetric (planar) division, with respect to the SL (see initial satellite cell division in Fig. 13.3) (Kuang et al. 2007). Asymmetric division gives rise to a daughter cell that maintains $Pax7^{+}/Myf5^{-}$ attributes of the mother satellite cell and a second daughter cell (typically the cell that has divided toward the basal lamella and is no longer in direct contact with the SL) is a committed $Pax7^{+}/Myf5^{+}$ cell. Symmetric division expands the $Pax7^{+}/Myf5^{-}$ satellite cell population by giving rise to two $Pax7^{+}/Myf5^{-}$ daughter cells and is likely important for the expansion of this cellular population

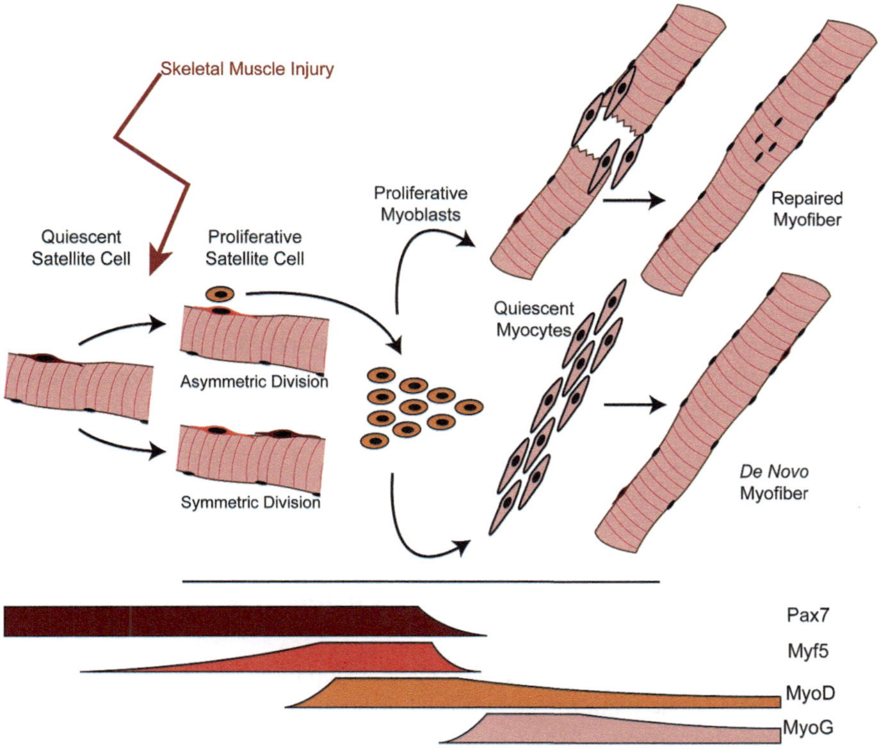

Fig. 13.3 Myogenic progression in satellite cell-dependent skeletal muscle repair. An illustrative model of myogenic progression associated with the activation of satellite cells and the subsequent repair of a damaged muscle fiber or regeneration of a lost skeletal muscle fiber de novo. At the bottom of the figure is a graphical representation of the expression of the four primary myogenic factors that regulate this myogenic process and their temporal relationship to satellite cell-dependent muscle regeneration

following injury and the overall maintenance of the $Pax7^+/Myf5^-$ quiescent cell population following repair. $Pax7^+/Myf5^+$ satellite cells, while having initiated their myogenic program, are actually more proliferative than their uncommitted counterparts but only undergo symmetric cell division in order to expand this committed muscle progenitor population. These processes produce a mixed muscle precursor population that must further differentiate in order to contribute to the musculature and repair damage.

13.4 Satellite Cell Differentiation

Satellite cells are non-fusogenic and must undergo further differentiation to contribute to muscle tissue. Myogenic differentiation of satellite cells first begins when most satellite cells start to express Myf5. Differentiation proceeds when $Pax7^+/Myf5^+$ satellite cells begin expressing myoblast determination protein (MyoD) and are then termed proliferative myoblasts. Along myogenic progression, MyoD initiates the transition from proliferation to quiescence accompanied by further myogenic differentiation. MyoD accomplishes these further steps in myogenic progression by upregulating cell cycle inhibitors and by initiating the expression of myogenin (Fig. 13.3) (Hollenberg et al. 1993; Halevy et al. 1995). As in development, myogenin and myogenic regulatory transcription factor 4 both work to differentiate muscle progenitor cells by activating genes essential for muscle function (e.g., myosin heavy chain and L-type Ca^{2+} channels), as well as activate the expression of muscle proteins required for fusion (e.g., myomaker, myomerger) (Millay et al. 2013; Davie et al. 2007). From this point, quiescent, differentiated muscle progenitors, termed myocytes, proceed through homotypic fusion to produce nascent myofibers or through heterotypic fusion with the ends of damaged fibers to restore the muscle fiber.

13.5 Muscle Fusion in Fiber Repair and Regeneration

13.5.1 Adhesion Proteins in Muscle Fusion

Muscle progenitor cell fusion is the fundamental step by which muscle progenitor cells contribute to the repair/regeneration of skeletal muscle. Quiescent myocytes precede through myogenic fusion by recognizing a fusion partner, forming tight cell-cell adhesion with that partner, and then undergoing a complicated fusion process to form a syncytium.

Myogenic fusion is preceded by the mutual recognition of fusion partners. Fusion partner recognition is perhaps best understood in *Drosophila melanogaster*; however muscle cell fusion is fundamentally different in fly compared to mammals (e.g., the role of founder cells). In *D. melanogaster*, fusion competent myoblasts (FCMs) recognize their fusion partners (founder cells/myotubes) membranes recognize cognate immunoglobulin superfamily proteins Kin-of-IrreC and the functionally redundant Roughest on fusion partner membranes (reviewed extensively (Rochlin et al. 2010)). The association of these factors with their cognate partners on fusion partner membranes plays an essential role in these cells distinguishing one another from the many other cell types in the skeletal muscle tissue, in the adhesive association of fusion partners, and in activating downstream signaling required for fusion initiation (e.g., activation of cytoskeletal rearrangements discussed below).

Nephrin (vertebrate homolog of sticks-and-stones) is the only common recognition/adhesion molecule common to both mammal and *D. melanogaster*. In mice, all myocytes are all considered equivalent, so the segregation of roles for different adhesion proteins between partners is not obvious (i.e., there are no founder cells in mammals; all daughter cells appear equal). However, integrins α3, β1, and α9β1 have been recognized for their roles in muscle progenitor cell adhesion and fusion (Schwander et al. 2003; Lafuste et al. 2005). Moreover, the muscle-specific adhesion protein M-cadherin is enriched at sites of myocyte cell-cell contact and is required for the efficient fusion of immortalized rat myoblasts (Zeschnigk et al. 1995). However, loss of M-cadherin expression in mice leads to no obvious perturbation in skeletal muscle development or regeneration (Hollnagel et al. 2002). The latter finding is hard to interpret in light of the many putative roles of M-cadherin in skeletal muscle; however this may suggest significant redundancy in adhesion/fusion machinery.

13.5.2 Membrane Signaling in Muscle Recognition/Fusion

In addition to specific adhesion proteins, the exofacial exposure of phosphatidylserine (PS) is a major membrane signaling process regulating muscle cell recognition, adhesion, and fusion. Its exposure is well recognized as an important signaling process regulating physiological cell fusion events (e.g., sperm-egg, macrophage, syncytiotrophoblast, osteoclast precursor, and virus-cell fusion) (Zaitseva et al. 2017; Verma et al. 2018; Riddell et al. 2013; Helming and Gordon 2009; Adler et al. 1995; Whitlock and Chernomordik 2021). Van den Eijnde et al. first identified non-apoptotic PS exposure on the exofacial leaflet of developing murine musculature (Van den Eijnde et al. 1997). Shortly after, they followed this observation with the description of PS exposure at sites of contact in fusing immortalized myoblasts and found that masking this PS signal inhibited fusion (van den Eijnde et al. 2001). This transient PS exposure was not associated with the appearance of apoptotic markers and could not be blocked via inhibition of caspase activity. These findings have been confirmed in primary murine myoblasts, and exogenous PS liposomes were found to directly promote primary muscle fusion (Jeong and Conboy 2011). Although exposed PS regulates myocyte fusion, the exact mechanism by which this process regulates fusion is only beginning to be appreciated.

Recently, two receptors have been identified for their role in recognizing PS during myocyte fusion. Brain-specific angiogenesis inhibitor 1 and stablin 2 have both been identified as muscle PS receptors required for efficient fusion in vitro, and loss of either significantly perturbs satellite cell-dependent muscle repair/regeneration following cardiotoxin challenge in vivo (Park et al. 2016; Hochreiter-Hufford et al. 2013). BAI1 is a G protein-coupled receptor previously recognized for its role in mediating PS recognition on apoptotic cells through the engulfment and cell motility (ELMO)/dedicator of cytokinesis (DOCK)180/ Ras-related C3 botulinum toxin substrate (Rac)1 pathway (Park et al. 2007). Because the ELMO/DOCK180/Rac1 pathway plays an essential role in regulating muscle fusion in fly, fish, and mouse (Park et al. 2007; Vasyutina et al. 2009; Moore et al. 2007; Laurin et al. 2008; Geisbrecht et al. 2008) and PS is also implicated in muscle fusion, the role of BAI1 in muscle fusion was investigated. Although loss of BAI1 does perturb muscle repair in vivo, the protein's role in fusion was primarily investigated in immortalized C2C12 cells. The STAB2 receptor is also thought to mediate PS recognition through the ELMO/DOCK180/Rac1 pathway. Despite the previously identified role for BAI1 in muscle fusion, investigators found that *Bai1* mRNA was barely detectible in murine skeletal muscles or in C2C12 myoblasts during differentiation. In contrast, high levels of *Stab2* mRNA were detected in multiple murine limb muscles, and *Stab2* is upregulated during the differentiation of both C2C12 and primary myoblasts (Park et al. 2016). *Stab2* steady-state levels are regulated during myogenic differentiation by increases in cytosol Ca^{2+} concentration under the regulation of NFATc, which as previously discussed, also regulates the activation of satellite cells through FGF2 signaling upon damage. Moreover, STAB2 can confer fusion capability to non-fusogenic fibroblasts, suggesting that PS signaling may play a role in direct cell-cell fusion and not be restricted to only regulating cell recognition and adhesion. In direct contrast to Hochreiter-Hufford et al. (Park et al. 2007), Park et al. found apoptotic inhibition in C2C12 cells had no effect on fusion, in control or STAB2 overexpressing conditions (Park et al. 2016). This finding is in line with previous descriptions of the inhibition of caspase activity having no effect on PS exposure or fusion in myoblasts (van den Eijnde et al. 2001). Some of these discrepancies may be due to when and for how long the cells were treated with caspase inhibitors. Interestingly, the loss of either BAI1 or STAB2 has a very modest effect on skeletal muscle development despite their obvious effects on muscle repair and in vitro muscle progenitor fusion.

Although consensus suggests myoblast PS exposure and fusion are not affected by apoptotic inhibition, sub-apoptotic activity of caspase-2, −3, and −9 play essential roles in myogenesis (Murray et al. 2008; Fernando et al. 2002; Boonstra et al. 2018). Recent work has convincingly demonstrated that sub-apoptotic activity of caspase-9 is initiated in non-apoptotic C2C12 cells during myogenic progression following differentiation. This rise in the initiator caspase-9 coincides with the cleavage/activation of the effector caspase-3 during differentiation at levels below those observed following apoptotic induction of C2C12 cells via staurosporine treatment (Murray et al. 2008). This caspase-3 activity elicited in C2C12 cells was found to activate MST1, a component of the p38 MAPK signaling pathway

(Fernando et al. 2002). The p38 MAPK signaling pathway has been highlighted previously for its role in increasing the endogenous activity of skeletal muscle transcription factors associated with myogenic progression (Ornatsky et al. 1999; Lechner et al. 1996). Interestingly, the inhibition of C2C12 fusion associated with caspase-3 knock-down can be rescued via MST1 overexpression, suggesting this kinase plays an integral function in the progression of myogenic progression/fusion in C2C12 cells downstream of caspase-3, which is downstream of the initiator caspase 9 (Fernando et al. 2002). Moreover, sub-apoptotic caspase-2 activity has also been highlighted for a potential role in regulating cell cycle arrest and myogenic progression upstream of C2C12 cell fusion (Fernando et al. 2002). Thus, the non-apoptotic role of caspase activity in skeletal muscle development is likely epistatic to muscle fusion, and manipulating caspase activity likely perturbs these upstream signaling processes, making the inhibitory effect of caspase on muscle cell fusion difficult to interpret regardless of the effect observed.

Non-apoptotic phospholipid scrambling is known to elicit the rapid exposure of exofacil PS via Ca^{2+}-dependent phospholipid scrambling (Ca^{2+}-PLS). Previous reports have demonstrated that some transmembrane protein (TMEM)16 proteins are the Ca^{2+}-PLSases that produce this exofacil PS signal in diverse cell types (Brunner et al. 2014; Malvezzi et al. 2013; Yu et al. 2015). Mutations in TMEM16E–a close relative of the PLSase TMEM16F–were found to cause limb-gridle muscular dystrophy type 2 L (Bolduc et al. 2010; Hicks et al. 2011). We characterized a murine model of this MD and found that the absence of TMEM16E results in significantly perturbed myoblast fusion (Griffin et al. 2016). Further investigation by the Hartzell group has demonstrated a non-PLSase role for TMEM16E in SL patch repair (Foltz et al. 2021). Interestingly, like loss of *Bai1* and *Stab2,* loss of *Ano5* significantly perturbs muscle repair/regeneration but has a very modest effect on skeletal muscle development (Griffin et al. 2016).

13.5.3 The Skeletal Muscle Bipartite Fusion Machine

Work over the last decade has significantly improved our mechanistic understanding of how muscle precursor cells fuse and form syncytial fibers; however many questions concerning how the machinery that facilitates these cell-cell fusions is regulated remain to be elucidated. Myoblast membrane fusion is facilitated by a bipartite fusion machine created by the cooperative activity of the muscle-specific proteins myomaker and myomerger (Millay et al. 2013; Zhang et al. 2017; Bi et al. 2017; Quinn et al. 2017). All biological membrane fusion events proceed through a shared fusion through hemifusion mechanism (Whitlock and Chernomordik 2021). In skeletal muscle, myomaker promotes the union of exofacial plasma membrane leaflets to form a hemifusion diaphragm between tightly adhered fusion partners (Millay et al. 2013; Leikina et al. 2018). Following myomaker activity, myomerger promotes the formation of fusion pores at the perimeter of the hemifusion diaphragm, resulting in a shared cytoplasm between fusion partners (Leikina et al.

2018; Golani et al. 2021). Interestingly, recent work has shown that the exposure of exofacial PS was found to be a critical trigger for the conformational reengagement of myomerger that promotes fusion pore formation, demonstrating that PS exposure is indeed a fusion trigger in skeletal muscle (Whitlock and Chernomordik 2021; Gamage et al. 2022).

Although myomaker and myomerger are both necessary and, together, sufficient for membrane fusion, skeletal muscle relies on many additional proteins to coordinate cell-cell fusion and syncytial formation in a regulated fashion. Recent work from the Chernomordik group has helped to resolve the mechanistic roles of several fusion regulators, including the role of PS binding proteins, annexins A1 and A5, for their roles in promoting initial hemifusion between fusing myoblasts and the roles of dynamin, phosphatidylinositol(4,5)bisphosphate, and cell metabolism in the expansion of the fusion pore to form organized syncytia (Leikina et al. 2013). In contrast, the mechanosensitive Ca^{2+} channel PIEZO1 is activated by the removal of PS from the plasma membrane of fusion partners and acts as a molecular break, stopping fusion following myotube formation (Tsuchiya et al. 2018). In addition to specific regulatory proteins, cytoskeletal dynamics play an integral role in the formation of skeletal fiber syncytia.

13.5.4 Other Players in Muscle Cell Fusion

Cytoskeletal regulators and rearrangement of the cytoskeleton play essential roles in myogenic fusion, although the precise mechanism is still ambiguous. The essential role of actin rearrangement for muscle fusion in *D. melanogaster* has been discussed extensively elsewhere (Abmayr and Pavlath 2012). Briefly, in *D. melanogaster* actin related protein 2/3 complex generates actin foci that are essential for the fusion of FCMs and FCs. Arp2/3 cytoskeletal rearrangement is regulated by the Scar complex, which in turn is controlled by the Rac1/2 complex. Overarching regulators of these cytoskeletal rearrangements are ELMO/Myoblast city in FCMs. All of these processes are thought to occur downstream of cell-cell adhesion under the regulation of the Sticks-and-stone/Hibris and Kin-of-IrreC/Roughest discussed earlier. Actin regulation through the ELMO/DOCK180/Rac1 pathway is also thought to be a major regulator of muscle fusion in vertebrates (Hochreiter-Hufford et al. 2013; Park et al. 2007; Vasyutina et al. 2009; Moore et al. 2007; Laurin et al. 2008); however the exact mechanism underlying how these changes in intracellular actin occur is unclear. Much confusion surrounding the precise role of the cytoskeleton in muscle fusion is probably created because this system is essential for the migration required for cells to fuse. With this in mind, it is difficult to determine whether loss of these actin signaling processes perturb actual fusion or prevent fusion by perturbing cell contact by altering cell migration.

In vitro evaluation of muscle cell fusion partner recognition from murine models commonly highlights the production of many lamellopodia and filopodia that contact potential fusion partners and are thought to sample the membranes of other cells

(Stadler et al. 2010; Segal et al. 2016). These cellular extensions are enriched in both adhesion and signaling proteins and may provide force driving the union of fusion partners during this match-making process (Mukai et al. 2009; Abramovici and Gee 2007). Similarly, filopodial extensions are described in *D. melanogaster* and are essential for myogenic precursor fusion during development (Yoon et al. 2007; Murphy and Courtneidge 2011). Lamellopodial and filopodial structures appear to carry out vital mechanistic roles in muscle fusion.

Finally, much recent interest has been created around the role of extracellular vesicle (EV) signaling in the regulation of muscle cell fusion. Many tissues throughout the body release EVs that play diverse cell-cell signaling roles in health and disease (extensively reviewed (van Niel et al. 2018)). Skeletal muscle releases EVs that are thought to function in crosstalk with other tissues and may play roles in musculoskeletal disease (Whitham et al. 2018; Murphy et al. 2018). Recently, it has been shown that proliferating and differentiated myoblasts release EVs that differ in their cargo, and which are endocytosed by other muscle cells (Forterre et al. 2014). Exchange of both protein and RNA cargo has been observed, and this exchange has been found to alter myogenic progression (Forterre et al. 2014; De Gasperi et al. 2017). Moreover, EVs from muscle appear to promote neuronal survival and neurite outgrowth, suggesting these signaling units may play an important role in the maintenance of the NMJ (Madison et al. 2014). Much concerning the production, uptake, and regulatory role of EVs has been discovered in the past decade (reviewed extensively (van Niel et al. 2018)). Interestingly, Ca^{2+}-PLS is commonly associated with the production of EVs and may play a role in the release and signaling of EVs in muscle as we have discussed previously (Whitlock and Hartzell 2017).

13.6 Conclusion

Recent work investigating the mechanisms of skeletal muscle repair has highlighted the parallel roles of SL patch repair and satellite cell-dependent repair in the maintenance of skeletal muscle tissue throughout life. Significant progress has been made in elucidating the mechanisms of these repair processes over the last two decades; however the many rapid steps that take place in each process successively have created a significant challenge in characterizing the precise contributions of molecular players to these processes. For satellite cell-dependent repair demonstrate precisely what step in myogenesis/fusion a protein of interest is involved has proven to be a challenging task. Many proteins have been shown to perturb muscle progenitor cell fusion; however, it is conceivable that proteins involved in myogenic differentiation, fusion partner recognition, cell-cell adhesion, or in cell-cell fusion could all lead to this outcome. Although challenging, careful evaluation of each of the processes mentioned above is important for fully characterizing the role of a particular regulatory factor in satellite cell-dependent skeletal muscle repair.

In addition to the challenges of characterizing a particular molecular player's role in muscle repair, current models commonly used to evaluate muscle progenitor cell repair competence (e.g., cultured muscle progenitor cells) likely recapitulate aspects of both embryonic muscle development as well as adult myogenic repair. Although these two processes are similar, many factors appear to have very different requirements during these two myogenic processes (e.g., the different requirements for the PS signaling machinery discussed above). In the future, it may be important to employ new models where in vivo observation of satellite cell-dependent muscle repair is more readily observable (e.g., zebra fish).

References

Abmayr SM, Pavlath GK (2012) Myoblast fusion: lessons from flies and mice. Development 139(4):641–656. https://doi.org/10.1242/dev.068353. PubMed PMID: 22274696; PMCID: PMC3265056

Abramovici H, Gee SH (2007) Morphological changes and spatial regulation of diacylglycerol kinase-zeta, syntrophins, and Rac1 during myoblast fusion. Cell Motil Cytoskeleton 64(7):549–567. https://doi.org/10.1002/cm.20204. PubMed PMID: 17410543

Adler RR, Ng AK, Rote NS (1995) Monoclonal antiphosphatidylserine antibody inhibits intercellular fusion of the choriocarcinoma line, JAR. Biol Reprod 53(4):905–910. PubMed PMID: 8547487

Allen RE, Boxhorn LK (1989) Regulation of skeletal muscle satellite cell proliferation and differentiation by transforming growth factor-beta, insulin-like growth factor I, and fibroblast growth factor. J Cell Physiol 138(2):311–315. https://doi.org/10.1002/jcp.1041380213. PubMed PMID: 2918032

Allen RE, Sheehan SM, Taylor RG, Kendall TL, Rice GM (1995) Hepatocyte growth factor activates quiescent skeletal muscle satellite cells in vitro. J Cell Physiol 165(2):307–312. https://doi.org/10.1002/jcp.1041650211. PubMed PMID: 7593208

Anderson JE (2000) A role for nitric oxide in muscle repair: nitric oxide-mediated activation of muscle satellite cells. Mol Biol Cell 11(5):1859–1874. PubMed PMID: 10793157; PMCID: PMC14889

Andrews NW, Almeida PE, Corrotte M (2014) Damage control: cellular mechanisms of plasma membrane repair. Trends Cell Biol 24(12):734–742. https://doi.org/10.1016/j.tcb.2014.07.008. PubMed PMID: 25150593; PMCID: PMC4252702

Bansal D, Miyake K, Vogel SS, Groh S, Chen CC, Williamson R, McNeil PL, Campbell KP (2003) Defective membrane repair in dysferlin-deficient muscular dystrophy. Nature 423(6936):168–172. https://doi.org/10.1038/nature01573. PubMed PMID: 12736685

Bi GQ, Alderton JM, Steinhardt RA (1995) Calcium-regulated exocytosis is required for cell membrane resealing. J Cell Biol 131(6 Pt 2):1747–1758. PubMed PMID: 8557742; PMCID: PMC2120667

Bi P, Ramirez-Martinez A, Li H, Cannavino J, McAnally JR, Shelton JM, Sanchez-Ortiz E, Bassel-Duby R, Olson EN (2017) Control of muscle formation by the fusogenic micropeptide myomixer. Science 356(6335):323–327. https://doi.org/10.1126/science.aam9361. Epub 2017/04/08. PubMed PMID: 28386024; PMCID: PMC5502127

Bischoff R (1997) Chemotaxis of skeletal muscle satellite cells. Dev Dyn 208(4):505–515. https://doi.org/10.1002/(SICI)1097-0177(199704)208:4<505::AID-AJA6>3.0.CO;2-M. Epub 1997/04/01. PubMed PMID: 9097022

Bober E, Franz T, Arnold HH, Gruss P, Tremblay P (1994) Pax-3 is required for the development of limb muscles: a possible role for the migration of dermomyotomal muscle progenitor cells. Development 120(3):603–612. PubMed PMID: 8162858

Bolduc V, Marlow G, Boycott KM, Saleki K, Inoue H, Kroon J, Itakura M, Robitaille Y, Parent L, Baas F, Mizuta K, Kamata N, Richard I, Linssen WH, Mahjneh I, de Visser M, Bashir R, Brais B (2010) Recessive mutations in the putative calcium-activated chloride channel Anoctamin 5 cause proximal LGMD2L and distal MMD3 muscular dystrophies. Am J Hum Genet 86(2):213–221. https://doi.org/10.1016/j.ajhg.2009.12.013. PubMed PMID: 20096397; PMCID: 2820170

Boonstra K, Bloemberg D, Quadrilatero J (2018) Caspase-2 is required for skeletal muscle differentiation and myogenesis. Biochim Biophys Acta 1865(1):95–104. https://doi.org/10.1016/j.bbamcr.2017.07.016. PubMed PMID: 28765049

Brunner JD, Lim NK, Schenck S, Duerst A, Dutzler R (2014) X-ray structure of a calcium-activated TMEM16 lipid scramblase. Nature 516(7530):207–212. https://doi.org/10.1038/nature13984. PubMed PMID: 25383531

Cai C, Masumiya H, Weisleder N, Matsuda N, Nishi M, Hwang M, Ko JK, Lin P, Thornton A, Zhao X, Pan Z, Komazaki S, Brotto M, Takeshima H, Ma J (2009) MG53 nucleates assembly of cell membrane repair machinery. Nat Cell Biol 11(1):56–64. https://doi.org/10.1038/ncb1812. PubMed PMID: 19043407; PMCID: PMC2990407

Cai C, Lin P, Zhu H, Ko JK, Hwang M, Tan T, Pan Z, Korichneva I, Ma J (2015) Zinc binding to MG53 protein facilitates repair of injury to cell membranes. J Biol Chem 290(22):13830–13839. https://doi.org/10.1074/jbc.M114.620690. PubMed PMID: 25869134; PMCID: PMC4447959

Calderon JC, Bolanos P, Caputo C (2014) The excitation-contraction coupling mechanism in skeletal muscle. Biophys Rev 6(1):133–160. https://doi.org/10.1007/s12551-013-0135-x. PubMed PMID: 28509964; PMCID: PMC5425715

Cantini M, Massimino ML, Rapizzi E, Rossini K, Catani C, Dalla Libera L, Carraro U (1995) Human satellite cell proliferation in vitro is regulated by autocrine secretion of IL-6 stimulated by a soluble factor(s) released by activated monocytes. Biochem Biophys Res Commun 216(1):49–53. PubMed PMID: 7488123

Chakrabarti S, Kobayashi KS, Flavell RA, Marks CB, Miyake K, Liston DR, Fowler KT, Gorelick FS, Andrews NW (2003) Impaired membrane resealing and autoimmune myositis in synaptotagmin VII-deficient mice. J Cell Biol 162(4):543–549. https://doi.org/10.1083/jcb.200305131. PubMed PMID: 12925704; PMCID: PMC2173791

Cheung TH, Rando TA (2013) Molecular regulation of stem cell quiescence. Nat Rev Mol Cell Biol 14(6):329–340. https://doi.org/10.1038/nrm3591. PubMed PMID: 23698583; PMCID: PMC3808888

Collins CA, Olsen I, Zammit PS, Heslop L, Petrie A, Partridge TA, Morgan JE (2005) Stem cell function, self-renewal, and behavioral heterogeneity of cells from the adult muscle satellite cell niche. Cell 122(2):289–301. https://doi.org/10.1016/j.cell.2005.05.010. PubMed PMID: 16051152

Cooper ST, McNeil PL (2015) Membrane repair: mechanisms and pathophysiology. Physiol Rev 95(4):1205–1240. https://doi.org/10.1152/physrev.00037.2014. PubMed PMID: 26336031; PMCID: PMC4600952

Corbalan-Garcia S, Gomez-Fernandez JC (2014) Signaling through C2 domains: more than one lipid target. Biochim Biophys Acta 1838(6):1536–1547. https://doi.org/10.1016/j.bbamem.2014.01.008. Epub 2014/01/21. PubMed PMID: 24440424

Corona BT, Garg K, Ward CL, McDaniel JS, Walters TJ, Rathbone CR (2013) Autologous minced muscle grafts: a tissue engineering therapy for the volumetric loss of skeletal muscle. Am J Physiol Cell Physiol 305(7):C761–C775. https://doi.org/10.1152/ajpcell.00189.2013. PubMed PMID: 23885064

Davie JK, Cho JH, Meadows E, Flynn JM, Knapp JR, Klein WH (2007) Target gene selectivity of the myogenic basic helix-loop-helix transcription factor myogenin in embryonic muscle. Dev Biol 311(2):650–664. https://doi.org/10.1016/j.ydbio.2007.08.014. PubMed PMID: 17904117

De Gasperi R, Hamidi S, Harlow LM, Ksiezak-Reding H, Bauman WA, Cardozo CP (2017) Denervation-related alterations and biological activity of miRNAs contained in exosomes released by skeletal muscle fibers. Sci Rep 7(1):12888. https://doi.org/10.1038/s41598-017-13105-9. PubMed PMID: 29038428; PMCID: PMC5643439

Defour A, Medikayala S, Van der Meulen JH, Hogarth MW, Holdreith N, Malatras A, Duddy W, Boehler J, Nagaraju K, Jaiswal JK (2017) Annexin A2 links poor myofiber repair with inflammation and adipogenic replacement of the injured muscle. Hum Mol Genet 26(11):1979–1991. https://doi.org/10.1093/hmg/ddx065. PubMed PMID: 28334824

Detrait E, Eddleman CS, Yoo S, Fukuda M, Nguyen MP, Bittner GD, Fishman HM (2000) Axolemmal repair requires proteins that mediate synaptic vesicle fusion. J Neurobiol 44(4):382–391. PubMed PMID: 10945894

Dumont NA, Bentzinger CF, Sincennes MC, Rudnicki MA (2015) Satellite cells and skeletal muscle regeneration. Compr Physiol 5(3):1027–1059. https://doi.org/10.1002/cphy.c140068. Epub 2015/07/04. PubMed PMID: 26140708

Fernando P, Kelly JF, Balazsi K, Slack RS, Megeney LA (2002) Caspase 3 activity is required for skeletal muscle differentiation. Proc Natl Acad Sci U S A 99(17):11025–11030. https://doi.org/10.1073/pnas.162172899. PubMed PMID: 12177420; PMCID: PMC123204

Foltz SJ, Cui YY, Choo HJ, Hartzell HC (2021) ANO5 ensures trafficking of annexins in wounded myofibers. J Cell Biol 220(3). https://doi.org/10.1083/jcb.202007059. Epub 2021/01/27. PubMed PMID: 33496727; PMCID: PMC7844426

Forterre A, Jalabert A, Berger E, Baudet M, Chikh K, Errazuriz E, De Larichaudy J, Chanon S, Weiss-Gayet M, Hesse AM, Record M, Geloen A, Lefai E, Vidal H, Coute Y, Rome S (2014) Proteomic analysis of C2C12 myoblast and myotube exosome-like vesicles: a new paradigm for myoblast-myotube cross talk? PLoS One 9(1):e84153. https://doi.org/10.1371/journal.pone.0084153. PubMed PMID: 24392111; PMCID: PMC3879278

Gamage DG, Melikov K, Munoz-Tello P, Wherley TJ, Focke LC, Leikina E, Huffman E, Diao J, Kojetin DJ, Prasad V, Chernomordik LV, Millay DP (2022) Phosphatidylserine orchestrates Myomerger membrane insertions to drive myoblast fusion. Proc Natl Acad Sci U S A 119(38):e2202490119. https://doi.org/10.1073/pnas.2202490119. Epub 2022/09/13. PubMed PMID: 36095199; PMCID: PMC9499509

Geisbrecht ER, Haralalka S, Swanson SK, Florens L, Washburn MP, Abmayr SM (2008) Drosophila ELMO/CED-12 interacts with myoblast city to direct myoblast fusion and ommatidial organization. Dev Biol 314(1):137–149. https://doi.org/10.1016/j.ydbio.2007.11.022. PubMed PMID: 18163987; PMCID: PMC2697615

Glover L, Brown RH Jr (2007) Dysferlin in membrane trafficking and patch repair. Traffic 8(7):785–794. https://doi.org/10.1111/j.1600-0854.2007.00573.x. PubMed PMID: 17547707

Golani G, Leikina E, Melikov K, Whitlock JM, Gamage DG, Luoma-Overstreet G, Millay DP, Kozlov MM, Chernomordik LV (2021) Myomerger promotes fusion pore by elastic coupling between proximal membrane leaflets and hemifusion diaphragm. Nat Commun 12(1):495. https://doi.org/10.1038/s41467-020-20804-x. Epub 2021/01/23. PubMed PMID: 33479215; PMCID: PMC7820291

Gozen I, Dommersnes P (2014) Pore dynamics in lipid membranes. Eur Phys J Spec Top 223(9):1813–1829

Griffin DA, Johnson RW, Whitlock JM, Pozsgai ER, Heller KN, Grose WE, Arnold WD, Sahenk Z, Hartzell HC, Rodino-Klapac LR (2016) Defective membrane fusion and repair in Anoctamin5-deficient muscular dystrophy. Hum Mol Genet 25:1900. https://doi.org/10.1093/hmg/ddw063. Epub 2016/02/26. PubMed PMID: 26911675

Gros J, Manceau M, Thome V, Marcelle C (2005) A common somitic origin for embryonic muscle progenitors and satellite cells. Nature 435(7044):954–958. https://doi.org/10.1038/nature03572. PubMed PMID: 15843802

Gurevich DB, Nguyen PD, Siegel AL, Ehrlich OV, Sonntag C, Phan JM, Berger S, Ratnayake D, Hersey L, Berger J, Verkade H, Hall TE, Currie PD (2016) Asymmetric division of clonal muscle stem cells coordinates muscle regeneration in vivo. Science 353(6295):aad9969. https://doi.org/10.1126/science.aad9969. PubMed PMID: 27198673

Halevy O, Novitch BG, Spicer DB, Skapek SX, Rhee J, Hannon GJ, Beach D, Lassar AB (1995) Correlation of terminal cell cycle arrest of skeletal muscle with induction of p21 by MyoD. Science 267(5200):1018–1021. PubMed PMID: 7863327

Helming L, Gordon S (2009) Molecular mediators of macrophage fusion. Trends Cell Biol 19(10):514–522. https://doi.org/10.1016/j.tcb.2009.07.005. PubMed PMID: 19733078

Hicks D, Sarkozy A, Muelas N, Koehler K, Huebner A, Hudson G, Chinnery PF, Barresi R, Eagle M, Polvikoski T, Bailey G, Miller J, Radunovic A, Hughes PJ, Roberts R, Krause S, Walter MC, Laval SH, Straub V, Lochmuller H, Bushby K (2011) A founder mutation in Anoctamin 5 is a major cause of limb-girdle muscular dystrophy. Brain 134(Pt 1):171–182. https://doi.org/10.1093/brain/awq294. PubMed PMID: 21186264; PMCID: 4038512

Hochreiter-Hufford AE, Lee CS, Kinchen JM, Sokolowski JD, Arandjelovic S, Call JA, Klibanov AL, Yan Z, Mandell JW, Ravichandran KS (2013) Phosphatidylserine receptor BAI1 and apoptotic cells as new promoters of myoblast fusion. Nature 497(7448):263–267. https://doi.org/10.1038/nature12135. PubMed PMID: 23615608; PMCID: PMC3773542

Hollenberg SM, Cheng PF, Weintraub H (1993) Use of a conditional MyoD transcription factor in studies of MyoD trans-activation and muscle determination. Proc Natl Acad Sci U S A. 90(17):8028–8032. PubMed PMID: 8396258; PMCID: PMC47281

Hollnagel A, Grund C, Franke WW, Arnold HH (2002) The cell adhesion molecule M-cadherin is not essential for muscle development and regeneration. Mol Cell Biol 22(13):4760–4770. PubMed PMID: 12052883; PMCID: PMC133893

Jeong J, Conboy IM (2011) Phosphatidylserine directly and positively regulates fusion of myoblasts into myotubes. Biochem Biophys Res Commun 414(1):9–13. https://doi.org/10.1016/j.bbrc.2011.08.128. PubMed PMID: 21910971; PMCID: PMC3195849

Johnson CP, Chapman ER (2010) Otoferlin is a calcium sensor that directly regulates SNARE-mediated membrane fusion. J Cell Biol 191(1):187–197. https://doi.org/10.1083/jcb.201002089. PubMed PMID: 20921140; PMCID: PMC2953437

Kuang S, Kuroda K, Le Grand F, Rudnicki MA (2007) Asymmetric self-renewal and commitment of satellite stem cells in muscle. Cell 129(5):999–1010. https://doi.org/10.1016/j.cell.2007.03.044. PubMed PMID: 17540178; PMCID: PMC2718740

Lafuste P, Sonnet C, Chazaud B, Dreyfus PA, Gherardi RK, Wewer UM, Authier FJ (2005) ADAM12 and alpha9beta1 integrin are instrumental in human myogenic cell differentiation. Mol Biol Cell 16(2):861–870. https://doi.org/10.1091/mbc.E04-03-0226. PubMed PMID: 15574885; PMCID: PMC545917

Laurin M, Fradet N, Blangy A, Hall A, Vuori K, Cote JF (2008) The atypical Rac activator Dock180 (Dock1) regulates myoblast fusion in vivo. Proc Natl Acad Sci U S A 105(40):15446–15451. https://doi.org/10.1073/pnas.0805546105. PubMed PMID: 18820033; PMCID: PMC2563090

Lechner C, Zahalka MA, Giot JF, Moller NP, Ullrich A (1996) ERK6, a mitogen-activated protein kinase involved in C2C12 myoblast differentiation. Proc Natl Acad Sci U S A 93(9):4355–4359. Epub 1996/04/30. PubMed PMID: 8633070; PMCID: PMC39541

Leikina E, Melikov K, Sanyal S, Verma SK, Eun B, Gebert C, Pfeifer K, Lizunov VA, Kozlov MM, Chernomordik LV (2013) Extracellular annexins and dynamin are important for sequential steps in myoblast fusion. J Cell Biol 200(1):109–123. https://doi.org/10.1083/jcb.201207012. Epub 2013/01/02. PubMed PMID: 23277424; PMCID: PMC3542790

Leikina E, Defour A, Melikov K, Van der Meulen JH, Nagaraju K, Bhuvanendran S, Gebert C, Pfeifer K, Chernomordik LV, Jaiswal JK (2015) Annexin A1 deficiency does not affect myofiber repair but delays regeneration of injured muscles. Sci Rep 5:18246. https://doi.org/10.1038/srep18246. PubMed PMID: 26667898; PMCID: PMC4678367

Leikina E, Gamage DG, Prasad V, Goykhberg J, Crowe M, Diao J, Kozlov MM, Chernomordik LV, Millay DP (2018) Myomaker and myomerger work independently to control distinct steps of membrane remodeling during myoblast fusion. Dev Cell 46(6):767–780 e7. https://doi.org/10.1016/j.devcel.2018.08.006. Epub 2018/09/11. PubMed PMID: 30197239; PMCID: PMC6203449

Lennon NJ, Kho A, Bacskai BJ, Perlmutter SL, Hyman BT, Brown RH Jr (2003) Dysferlin interacts with annexins A1 and A2 and mediates sarcolemmal wound-healing. J Biol Chem 278(50):50466–50473. https://doi.org/10.1074/jbc.M307247200. PubMed PMID: 14506282

Lepper C, Partridge TA, Fan CM (2011) An absolute requirement for Pax7-positive satellite cells in acute injury-induced skeletal muscle regeneration. Development 138(17):3639–3646. https://doi.org/10.1242/dev.067595. PubMed PMID: 21828092; PMCID: PMC3152922

Liu Y, Schneider MF (2014) FGF2 activates TRPC and ca(2+) signaling leading to satellite cell activation. Front Physiol 5:38. https://doi.org/10.3389/fphys.2014.00038. PubMed PMID: 24575047; PMCID: PMC3920331

Madison RD, McGee C, Rawson R, Robinson GA (2014) Extracellular vesicles from a muscle cell line (C2C12) enhance cell survival and neurite outgrowth of a motor neuron cell line (NSC-34). J Extracell Vesicles 3:3. https://doi.org/10.3402/jev.v3.22865. PubMed PMID: 24563732; PMCID: PMC3930942

Malvezzi M, Chalat M, Janjusevic R, Picollo A, Terashima H, Menon AK, Accardi A (2013) Ca2+−dependent phospholipid scrambling by a reconstituted TMEM16 ion channel. Nat Commun 4:2367. https://doi.org/10.1038/ncomms3367. PubMed PMID: 23996062; PMCID: PMC3970400

Mansouri A, Stoykova A, Torres M, Gruss P (1996) Dysgenesis of cephalic neural crest derivatives in Pax7−/− mutant mice. Development 122(3):831–838. PubMed PMID: 8631261

Mauro A (1961) Satellite cell of skeletal muscle fibers. J Biophys Biochem Cytol 9:493–495. PubMed PMID: 13768451; PMCID: PMC2225012

McNeil PL, Khakee R (1992) Disruptions of muscle fiber plasma membranes. Role in exercise-induced damage. Am J Pathol 140(5):1097–1109. PubMed PMID: 1374591; PMCID: PMC1886518

McNeil PL, Steinhardt RA (1997) Loss, restoration, and maintenance of plasma membrane integrity. J Cell Biol 137(1):1–4. PubMed PMID: 9105031; PMCID: PMC2139853

Millay DP, O'Rourke JR, Sutherland LB, Bezprozvannaya S, Shelton JM, Bassel-Duby R, Olson EN (2013) Myomaker is a membrane activator of myoblast fusion and muscle formation. Nature 499(7458):301–305. https://doi.org/10.1038/nature12343. Epub 2013/07/23. PubMed PMID: 23868259; PMCID: PMC3739301

Miyake K, McNeil PL (1995) Vesicle accumulation and exocytosis at sites of plasma membrane disruption. J Cell Biol 131(6 Pt 2):1737–1745. PubMed PMID: 8557741; PMCID: PMC2120668

Moore CA, Parkin CA, Bidet Y, Ingham PW (2007) A role for the myoblast city homologues Dock1 and Dock5 and the adaptor proteins Crk and Crk-like in zebrafish myoblast fusion. Development 134(17):3145–3153. https://doi.org/10.1242/dev.001214. PubMed PMID: 17670792

Mukai A, Kurisaki T, Sato SB, Kobayashi T, Kondoh G, Hashimoto N (2009) Dynamic clustering and dispersion of lipid rafts contribute to fusion competence of myogenic cells. Exp Cell Res 315(17):3052–3063. https://doi.org/10.1016/j.yexcr.2009.07.010. PubMed PMID: 19615358

Murphy DA, Courtneidge SA (2011) The 'ins' and 'outs' of podosomes and invadopodia: characteristics, formation and function. Nat Rev Mol Cell Biol 12(7):413–426. https://doi.org/10.1038/nrm3141. PubMed PMID: 21697900; PMCID: PMC3423958

Murphy C, Withrow J, Hunter M, Liu Y, Tang YL, Fulzele S, Hamrick MW (2018) Emerging role of extracellular vesicles in musculoskeletal diseases. Mol Asp Med 60:123–128. https://doi.org/10.1016/j.mam.2017.09.006. PubMed PMID: 28965750; PMCID: PMC5856577

Murray TV, McMahon JM, Howley BA, Stanley A, Ritter T, Mohr A, Zwacka R, Fearnhead HO (2008) A non-apoptotic role for caspase-9 in muscle differentiation. J Cell Sci 121(Pt 22):3786–3793. https://doi.org/10.1242/jcs.024547. PubMed PMID: 18957517

Neuhaus P, Oustanina S, Loch T, Kruger M, Bober E, Dono R, Zeller R, Braun T (2003) Reduced mobility of fibroblast growth factor (FGF)-deficient myoblasts might contribute to dystrophic changes in the musculature of FGF2/FGF6/mdx triple-mutant mice. Mol Cell Biol 23(17):6037–6048. PubMed PMID: 12917328; PMCID: PMC180975

Ornatsky OI, Cox DM, Tangirala P, Andreucci JJ, Quinn ZA, Wrana JL, Prywes R, Yu YT, McDermott JC (1999) Post-translational control of the MEF2A transcriptional regulatory protein. Nucleic Acids Res 27(13):2646–2654. Epub 1999/06/22. PubMed PMID: 10373581; PMCID: PMC148473

Park D, Tosello-Trampont AC, Elliott MR, Lu M, Haney LB, Ma Z, Klibanov AL, Mandell JW, Ravichandran KS (2007) BAI1 is an engulfment receptor for apoptotic cells upstream of the ELMO/Dock180/Rac module. Nature 450(7168):430–434. https://doi.org/10.1038/nature06329. PubMed PMID: 17960134

Park SY, Yun Y, Lim JS, Kim MJ, Kim SY, Kim JE, Kim IS (2016) Stabilin-2 modulates the efficiency of myoblast fusion during myogenic differentiation and muscle regeneration. Nat Commun 7:10871. https://doi.org/10.1038/ncomms10871. PubMed PMID: 26972991; PMCID: PMC4793076

Proctor DN, O'Brien PC, Atkinson EJ, Nair KS (1999) Comparison of techniques to estimate total body skeletal muscle mass in people of different age groups. Am J Phys 277(3 Pt 1):E489–E495. PubMed PMID: 10484361

Quinn ME, Goh Q, Kurosaka M, Gamage DG, Petrany MJ, Prasad V, Millay DP (2017) Myomerger induces fusion of non-fusogenic cells and is required for skeletal muscle development. Nat Commun 8:15665. https://doi.org/10.1038/ncomms15665. Epub 2017/06/02. PubMed PMID: 28569755; PMCID: PMC5461499

Reddy A, Caler EV, Andrews NW (2001) Plasma membrane repair is mediated by ca(2+)-regulated exocytosis of lysosomes. Cell 106(2):157–169. PubMed PMID: 11511344

Reznik M (1969) Thymidine-3H uptake by satellite cells of regenerating skeletal muscle. J Cell Biol 40(2):568–571. PubMed PMID: 5812478; PMCID: PMC2107616

Riddell MR, Winkler-Lowen B, Jiang Y, Davidge ST, Guilbert LJ (2013) Pleiotropic actions of forskolin result in phosphatidylserine exposure in primary trophoblasts. PLoS One 8(12):e81273. https://doi.org/10.1371/journal.pone.0081273. PubMed PMID: 24339915; PMCID: PMC3855289

Rochlin K, Yu S, Roy S, Baylies MK (2010) Myoblast fusion: when it takes more to make one. Dev Biol 341(1):66–83. https://doi.org/10.1016/j.ydbio.2009.10.024. PubMed PMID: 19932206; PMCID: PMC2854170

Rosenblatt JD, Lunt AI, Parry DJ, Partridge TA (1995) Culturing satellite cells from living single muscle fiber explants. In Vitro Cell Dev Biol Anim 31(10):773–779. https://doi.org/10.1007/BF02634119. PubMed PMID: 8564066

Sambasivan R, Yao R, Kissenpfennig A, Van Wittenberghe L, Paldi A, Gayraud-Morel B, Guenou H, Malissen B, Tajbakhsh S, Galy A (2011) Pax7-expressing satellite cells are indispensable for adult skeletal muscle regeneration. Development 138(17):3647–3656. https://doi.org/10.1242/dev.067587. PubMed PMID: 21828093

Schwander M, Leu M, Stumm M, Dorchies OM, Ruegg UT, Schittny J, Muller U (2003) Beta1 integrins regulate myoblast fusion and sarcomere assembly. Dev Cell 4(5):673–685. PubMed PMID: 12737803

Segal D, Dhanyasi N, Schejter ED, Shilo BZ (2016) Adhesion and fusion of muscle cells are promoted by Filopodia. Dev Cell 38(3):291–304. https://doi.org/10.1016/j.devcel.2016.07.010. PubMed PMID: 27505416

Shea KL, Xiang W, LaPorta VS, Licht JD, Keller C, Basson MA, Brack AS (2010) Sprouty1 regulates reversible quiescence of a self-renewing adult muscle stem cell pool during regeneration. Cell Stem Cell 6(2):117–129. https://doi.org/10.1016/j.stem.2009.12.015. PubMed PMID: 20144785; PMCID: PMC2846417

Siegel AL, Atchison K, Fisher KE, Davis GE, Cornelison DD (2009) 3D timelapse analysis of muscle satellite cell motility. Stem Cells 27(10):2527–2538. https://doi.org/10.1002/stem.178. PubMed PMID: 19609936; PMCID: PMC2798070

Snow MH (1977) Myogenic cell formation in regenerating rat skeletal muscle injured by mincing. II. An autoradiographic study. Anat Rec 188(2):201–217. https://doi.org/10.1002/ar.1091880206. PubMed PMID: 869238

Spalding KL, Bergmann O, Alkass K, Bernard S, Salehpour M, Huttner HB, Bostrom E, Westerlund I, Vial C, Buchholz BA, Possnert G, Mash DC, Druid H, Frisen J (2013) Dynamics of hippocampal neurogenesis in adult humans. Cell 153(6):1219–1227. https://doi.org/10.1016/j.cell.2013.05.002. PubMed PMID: 23746839; PMCID: PMC4394608

Stadler B, Blattler TM, Franco-Obregon A (2010) Time-lapse imaging of in vitro myogenesis using atomic force microscopy. J Microsc 237(1):63–69. https://doi.org/10.1111/j.1365-2818.2009.03302.x. PubMed PMID: 20055919

Tajbakhsh S, Rocancourt D, Cossu G, Buckingham M (1997) Redefining the genetic hierarchies controlling skeletal myogenesis: Pax-3 and Myf-5 act upstream of MyoD. Cell 89(1):127–138. PubMed PMID: 9094721

Tatsumi R (2010) Mechano-biology of skeletal muscle hypertrophy and regeneration: possible mechanism of stretch-induced activation of resident myogenic stem cells. Anim Sci J 81(1):11–20. https://doi.org/10.1111/j.1740-0929.2009.00712.x. PubMed PMID: 20163667

Tatsumi R, Anderson JE, Nevoret CJ, Halevy O, Allen RE (1998) HGF/SF is present in normal adult skeletal muscle and is capable of activating satellite cells. Dev Biol 194(1):114–128. https://doi.org/10.1006/dbio.1997.8803. PubMed PMID: 9473336

Terasaki M, Miyake K, McNeil PL (1997) Large plasma membrane disruptions are rapidly resealed by Ca2+−dependent vesicle-vesicle fusion events. J Cell Biol 139(1):63–74. PubMed PMID: 9314529; PMCID: PMC2139822

Togo T, Alderton JM, Bi GQ, Steinhardt RA (1999) The mechanism of facilitated cell membrane resealing. J Cell Sci 112(Pt 5):719–731. PubMed PMID: 9973606

Tsuchiya M, Hara Y, Okuda M, Itoh K, Nishioka R, Shiomi A, Nagao K, Mori M, Mori Y, Ikenouchi J, Suzuki R, Tanaka M, Ohwada T, Aoki J, Kanagawa M, Toda T, Nagata Y, Matsuda R, Takayama Y, Tominaga M, Umeda M (2018) Cell surface flip-flop of phosphatidylserine is critical for PIEZO1-mediated myotube formation. Nat Commun 9(1):2049. https://doi.org/10.1038/s41467-018-04436-w. Epub 2018/05/26. PubMed PMID: 29799007; PMCID: PMC5967302

Van den Eijnde SM, Boshart L, Reutelingsperger CP, De Zeeuw CI, Vermeij-Keers C (1997) Phosphatidylserine plasma membrane asymmetry in vivo: a pancellular phenomenon which alters during apoptosis. Cell Death Differ 4(4):311–316. https://doi.org/10.1038/sj.cdd.4400241. PubMed PMID: 16465246

van den Eijnde SM, van den Hoff MJ, Reutelingsperger CP, van Heerde WL, Henfling ME, Vermeij-Keers C, Schutte B, Borgers M, Ramaekers FC (2001) Transient expression of phosphatidylserine at cell-cell contact areas is required for myotube formation. J Cell Sci 114(Pt 20):3631–3642. PubMed PMID: 11707515

van Niel G, D'Angelo G, Raposo G (2018) Shedding light on the cell biology of extracellular vesicles. Nat Rev Mol Cell Biol 19(4):213–228. https://doi.org/10.1038/nrm.2017.125. PubMed PMID: 29339798

Vasyutina E, Martarelli B, Brakebusch C, Wende H, Birchmeier C (2009) The small G-proteins Rac1 and Cdc42 are essential for myoblast fusion in the mouse. Proc Natl Acad Sci U S A 106(22):8935–8940. https://doi.org/10.1073/pnas.0902501106. PubMed PMID: 19443691; PMCID: PMC2682539

Verma SK, Leikina E, Melikov K, Gebert C, Kram V, Young MF, Uygur B, Chernomordik LV (2018) Cell-surface phosphatidylserine regulates osteoclast precursor fusion. J Biol Chem 293(1):254–270. https://doi.org/10.1074/jbc.M117.809681. Epub 2017/11/05. PubMed PMID: 29101233; PMCID: PMC5766907

Wang X, Wu H, Zhang Z, Liu S, Yang J, Chen X, Fan M, Wang X (2008) Effects of interleukin-6, leukemia inhibitory factor, and ciliary neurotrophic factor on the proliferation and differentiation of adult human myoblasts. Cell Mol Neurobiol 28(1):113–124. https://doi.org/10.1007/s10571-007-9247-9. PubMed PMID: 18240017

White RB, Bierinx AS, Gnocchi VF, Zammit PS (2010) Dynamics of muscle fibre growth during postnatal mouse development. BMC Dev Biol 10:21. https://doi.org/10.1186/1471-213X-10-21. PubMed PMID: 20175910; PMCID: PMC2836990

Whitham M, Parker BL, Friedrichsen M, Hingst JR, Hjorth M, Hughes WE, Egan CL, Cron L, Watt KI, Kuchel RP, Jayasooriah N, Estevez E, Petzold T, Suter CM, Gregorevic P, Kiens B, Richter EA, James DE, Wojtaszewski JFP, Febbraio MA (2018) Extracellular vesicles provide a means for tissue crosstalk during exercise. Cell Metab 27(1):237–251 e4. https://doi.org/10.1016/j.cmet.2017.12.001. PubMed PMID: 29320704

Whitlock JM, Chernomordik LV (2021) Flagging fusion: Phosphatidylserine signaling in cell-cell fusion. J Biol Chem 296:100411. https://doi.org/10.1016/j.jbc.2021.100411. Epub 2021/02/14. PubMed PMID: 33581114; PMCID: PMC8005811

Whitlock JH, Hartzell HC (2017) Anoctamins/TMEM16 proteins: chloride channels flirting with lipids and extracellular vesicles. Annu Rev Physiol 79:79. https://doi.org/10.1146/annurev-physiol-022516-034031

Yang D, Morris SF, Sigurdson L (1998) The sartorius muscle: anatomic considerations for reconstructive surgeons. Surg Radiol Anat 20(5):307–310. PubMed PMID: 9894308

Yoon S, Molloy MJ, Wu MP, Cowan DB, Gussoni E (2007) C6ORF32 is upregulated during muscle cell differentiation and induces the formation of cellular filopodia. Dev Biol 301(1):70–81. https://doi.org/10.1016/j.ydbio.2006.11.002. PubMed PMID: 17150207; PMCID: PMC1779902

Yu K, Whitlock JM, Lee K, Ortlund EA, Cui YY, Hartzell HC (2015) Identification of a lipid scrambling domain in ANO6/TMEM16F. Elife 4:e06901. https://doi.org/10.7554/eLife.06901. Epub 2015/06/10. PubMed PMID: 26057829; PMCID: 4477620

Zaitseva E, Zaitsev E, Melikov K, Arakelyan A, Marin M, Villasmil R, Margolis LB, Melikyan GB, Chernomordik LV (2017) Fusion stage of HIV-1 entry depends on virus-induced cell surface exposure of phosphatidylserine. Cell Host Microbe 22(1):99–110 e7. https://doi.org/10.1016/j.chom.2017.06.012. PubMed PMID: 28704658; PMCID: PMC5558241

Zammit PS, Golding JP, Nagata Y, Hudon V, Partridge TA, Beauchamp JR (2004) Muscle satellite cells adopt divergent fates: a mechanism for self-renewal? J Cell Biol 166(3):347–357. https://doi.org/10.1083/jcb.200312007. PubMed PMID: 15277541; PMCID: PMC2172269

Zeschnigk M, Kozian D, Kuch C, Schmoll M, Starzinski-Powitz A (1995) Involvement of M-cadherin in terminal differentiation of skeletal muscle cells. J Cell Sci 108(Pt 9):2973–2981. PubMed PMID: 8537437

Zhang Q, Vashisht AA, O'Rourke J, Corbel SY, Moran R, Romero A, Miraglia L, Zhang J, Durrant E, Schmedt C, Sampath SC, Sampath SC (2017) The microprotein Minion controls cell fusion and muscle formation. Nat Commun 8:15664. https://doi.org/10.1038/ncomms15664. Epub 2017/06/02. PubMed PMID: 28569745; PMCID: PMC5461507

Part IV
Virus- and Parasite-Induced Syncytia

Chapter 14
Virus-Induced Cell Fusion and Syncytia Formation

Maorong Xie

Abstract Most enveloped viruses encode viral fusion proteins to penetrate host cell by membrane fusion. Interestingly, many enveloped viruses can also use viral fusion proteins to induce cell-cell fusion, both in vitro and in vivo, leading to the formation of syncytia or multinucleated giant cells (MGCs). In addition, some non-enveloped viruses encode specialized viral proteins that induce cell-cell fusion to facilitate viral spread. Overall, viruses that can induce cell-cell fusion are nearly ubiquitous in mammals. Virus cell-to-cell spread by inducing cell-cell fusion may overcome entry and post-entry blocks in target cells and allow evasion of neutralizing antibodies. However, molecular mechanisms of virus-induced cell-cell fusion remain largely unknown. Here, I summarize the current understanding of virus-induced cell fusion and syncytia formation.

14.1 Introduction

The term syncytium or symplasm is used to describe a multinucleated giant cell (MGC) formed by cell fusions of several individual cells, in contrast to coenocyte, which describes a MGC arising from multiple nuclear divisions without cytokinesis. Both result in the formation of MGC containing two or more nuclei in the same cell; however they are not mutually exclusive. For example, abnormal nuclear division could occur in small syncytia followed by virus infection, such as measles, Sendai, and myxoma virus (Poste 1970). Interestingly, nuclear bridges have been described in MGCs in patients with AIDS, suggesting the role of nuclear division in MGC formation (Mizusawa et al. 1987). However, little evidence support that abnormal nuclear division may contribute to the formation of virus-induced MGC

M. Xie (✉)
Division of Infection and Immunity, UCL, London, UK
e-mail: maorong.xie@ucl.ac.uk

M. Kloc, A. Uosef (eds.), *Syncytia: Origin, Structure, and Functions*, Results and Problems in Cell Differentiation 71,
https://doi.org/10.1007/978-3-031-37936-9_14

(Tsukamoto et al. 1999). In general, it is well established that cell-fusion is responsible for syncytia formation in most scenarios (Leroy et al. 2020).

Cell-cell fusion and syncytia formation are universal, active, and frequent phenomena in mammals. This process is evolutionarily conserved among eukaryotes (Ogle et al. 2005). Under certain homeostatic and developmental conditions, mammalian cells can undergo cell fusion and give rise to multinucleated giant cells with specialized functions. At the cellular level, cell fusion of sperm and egg is essential for fertilization in mammals. At the organ level, cell fusion of myoblasts is important for the formation of skeletal muscle with increased myofiber size to improve contractile strength, and it can also take place during the formation of placenta (i.e., the fusion of trophoblast cells)(Buchrieser et al. 2019; Ogle et al. 2005). In addition, fusion of macrophages and formation of MGCs is part of the inflammatory responses; the formation of MGCs by cell fusion is associated with cancer pathogenesis (Duelli and Lazebnik 2007).

All enveloped viruses encode viral fusion proteins for virus entry. These viral fusion proteins can also be used by viruses to mediate cell-to-cell fusion for efficient viral spread. Indeed, many families of enveloped viruses have been shown to induce cell-cell fusion and syncytia formation under appropriate conditions (Duelli and Lazebnik 2007; Leroy et al. 2020; Poste 1972). To date, four classes of viral fusion proteins have been characterized based on their key structural features (Harrison 2008; Podbilewicz 2014).

Class I viral fusion proteins, typified by influenza HA and HIV Env, have a trimeric α-helical coiled coil at N-terminal and three C-terminal helices. They require a proteolytic cleavage of a trimeric and single-chain precursor to make them fusogenic. The cleavage eliminates a single peptide bond to generate two fragments of the fusion proteins. The N-terminal fragment, such as the gp120 fragment of HIV-1 Env or the HA1 fragment of influenza HA, is usually a receptor-binding domain which constrains the C-terminal, fusogenic fragment, such as HA2 or gp41, until triggered by receptor binding to release it. The fusogenic fragment consists of a hydrophobic fusion peptide at or near its N-terminus and a C-terminal transmembrane anchor to hold on to the viral membrane.

Most members of the class II viral fusion proteins, such as the flavivirus envelope protein E, are transmembrane glycoproteins that form an icosahedral scaffold on the mature virion. Class II fusion proteins consist of β-sheet structure forming an elongated ectodomain, with internal fusion loops located at the tips of β-strands. Cleavage of a second membrane protein (i.e., membrane protein M in flaviviruses), which acts as a chaperone, primes viral membrane fusion. This cleavage induces the rearrangement of surface lattice and causes trimerization of the viral fusion protein. The hydrophobic fusion loop engages target cellular membrane during the conformational change (Kielian 2006).

The class III viral fusion proteins, such as rhabdovirus G protein and herpesvirus gB protein, share structural features with Class I and Class II fusion proteins. They have a trimeric α-helical coiled coil, and they are trimeric in both pre- and post-fusion conformations. However, their fusion loops resemble to those of Class II fusion proteins and are located at the tips of β-strands (Backovic and Jardetzky 2009).

The Class IV membrane proteins are fusion-associated small transmembrane (FAST) proteins encoded by some non-enveloped reoviruses (Shmulevitz and Duncan 2000). FAST proteins are the smallest known viral fusogens and are unlikely to form trimeric hairpins. Interestingly, FAST proteins are not involved in virus entry but are dedicated to mediate cell-cell fusion and syncytia formation (Ciechonska and Duncan 2014).

To spread within an infected host, virus has evolved different strategies for insuring viral transmission and persistence. First, cell-free virus particles released from infected cells could infect neighboring or distant cells. Second, virus particles or viral genetic material could be transferred directly from infected cells to neighboring uninfected cells through cell-to-cell contacts. However, virus spread via cell-free virus particles is rather inefficient and rate limited by fluid uptake and virion attachment. Moreover, freely diffusing virions are vulnerable to antibody neutralization and antibody-mediated opsonization (Abela et al. 2012; Marechal et al. 2001; Sigal et al. 2011). In addition, cell-free virus particles must survive the innate and adaptive immune defenses to successfully establish infection in a new host. To circumvent these challenges, many viruses have adapted to benefit from cell-to-cell interactions for efficient virus dissemination without appearing as cell-free virus particles, a mechanism described as virus cell-to-cell transmission (Bracq et al. 2018; Cifuentes-Munoz et al. 2018; Phillips 1994). For example, HIV-1 cell-to-cell transmission between CD4+ T cells mainly involves the formation of virological synapses (Alvarez et al. 2014; Chen et al. 2007; Hubner et al. 2009; Jolly et al. 2004; Jolly et al. 2007; Rudnicka et al. 2009), but also through other membrane structures, such as nanotubes or filopodia, and has been estimated to be 100–1000 times more efficient than the cell-free virus infection process (Dimitrov et al. 1993; Sato et al. 1992). Different modes of virus cell-to-cell infection through distinct intercellular structures enabling close contacts between virus-infected cells and non-infected target cells have been described over the past years. Intercellular transfer of viral material has been described mainly through establishment of the virological or infectious synapses, and also using membrane protrusions such as tunnelling nanotubes (TNTs), filopodia, phagocytosis, and cell-to-cell fusion processes (Bracq et al. 2018; Chen 2012; Jolly and Sattentau 2004; Mothes et al. 2010; Pedro et al. 2019; Sattentau 2008; Zhong et al. 2013).

Several mechanisms for virus cell-to-cell transmission have been described; many viruses could also utilize the fusogenic capacity of viral fusion proteins to induce cell-cell fusion between virus-infected cells and neighboring cells, resulting in the formation of enlarged multinucleated syncytia (Leroy et al. 2020). The formation of syncytia, the fusion between infected and uninfected cells, is the simplest mechanism of virus cell-to-cell spread which can be reproduced in vitro for paramyxoviruses, herpesviruses, and some retroviruses. For example, paramyxoviruses, including respiratory syncytial virus (RSV) and measles virus (MV), can induce the formation of MGCs in vitro and in vivo for efficient viral cell-to-cell spread (Fugier-Vivier et al. 1997; McChesney et al. 1997; Takeuchi et al. 2003). Similarly, alphaherpesviruses express a number of fusogenic glycoproteins and are able to induce syncytia formation (Cole and Grose 2003). Finally, HIV-1 and HTLV-1 can also form syncytia in tissue culture (Benovic et al. 1998; Ceccaldi et al. 2006). Syncytia formation also facilitates many viruses, such as HIV-1 and measles viruses, to spread from cell-to-cell without encountering the extracellular environment (Duprex et al. 1999; Sato et al. 1992). Thus, virus-induced cell-to-cell fusion represents a unique mode of virus cell-to-cell transmission by protecting virus from neutralizing antibodies and may overcome the virus entry block due to limited expression of receptors on target cells and also the inhibitory effects of several antiviral restriction factors, such as SAMHD1 and tetherin/BST2 (Bracq et al. 2017, 2018; Cifuentes-Munoz et al. 2018; Han et al. 2022; Phillips 1994). Indeed, many respiratory viruses, such as respiratory syncytia virus, measles, influenza, and parainfluenza virus, exploit this mechanism for virus cell-to-cell spread (Cifuentes-Munoz et al. 2018). Virus-induced multinucleated syncytia maintain metabolic functions and essential gene expression, which provide an expanded and prolonged localized environment for viral propagation. When multinucleated syncytia succumb to apoptosis and rupture, they release an enormous burst of infectious virus particles, facilitating systemic spread of the infection (Castedo et al. 2002; Salsman et al. 2005). Therefore, the formation of virus-induced multinucleated syncytia causes cytopathic effects and potentially facilitates virus propagation, virus dissemination, and immune evasion (Leroy et al. 2020). The goal of this chapter is to provide an overview of the diverse virus families that can trigger cell-to-cell fusion and syncytia formation for efficient virus cell-to-cell transmission.

14.2 Human Endogenous Retroviruses

The placenta is an autonomous, transient organ that is essential for normal fetal development in mammals. Placentation is a fundamental cell-to-cell fusion process during embryonic development. The cell fusion of progenitor cytotrophoblasts leads to the formation of multinucleated syncytium, namely, syncytiotrophoblast (STB). Syncytins, the fusogens that mediate cytotrophoblast fusion, are derived from the viral fusogens of human endogenous retroviruses (HERVs). Interestingly, HERVs activation and expression are associated with the development of cancer; upregulation of syncytins in cancer cells results in increased cell-cell fusion and increased invasiveness in endometrial carcinoma (Gimenez et al. 2010; Kitsou et al. 2022).

An essential step in the retrovirus replicative cycle is the integration of proviral DNA into the host genome. During evolution, some exogenous retroviruses successfully infected the germline cells of their natural hosts and permanently integrated into the germline chromosomes, a process called retroviral endogenization (Greenwood et al. 2018; Oliveira et al. 2007). These stably integrated endogenous retroviruses (ERVs) are vertically transmitted to subsequent generations in a Mendelian fashion. The continuous accumulation of newly established retroviral integrations in germline cells over millions of years suggests the co-evolution of ERVs with their natural hosts. Indeed, HERVs constitute approximately 8% of the human genome (Griffiths 2001). HERVs have been classified into three classes based on their sequence similarity to different genera of the exogenous counterparts of retroviruses: class I HERVs constitute about 2.3% of the genome and are related to gamma-retroviruses such as murine leukemia virus (MLV); class II HERVs make up to 0.7% of the human genome and are related to beta-retroviruses such as mouse mammary tumor virus (MMTV); class III HERVs consist of 4% of the human genome and are distantly related to spumaretroviruses such as equine foamy virus (EFV) (Vargiu et al. 2016).

Full-length HERVs consist of three main proviral genes, group-specific antigen (gag), polymerase (pol) and envelope (env), flanked by two long-terminal repeats (LTRs). Most HERVs have accumulated genetic defects during evolution and are unable to produce infectious virions or viral proteins. However, some HERVs still possess open reading frames (ORFs) and reserve their potential transcriptional capacity within host genome. Interestingly, these protein-coding sequences of HERVs have been co-opted by their hosts to perform crucial biological roles. For example, the lineage-specific host enhancers and promoters are derived from the LTRs of viral multifunctional regulatory elements (Chuong et al. 2017). Specifically, multiple MER41 elements of ERVs have been co-opted to regulate the adjacent IFN-induced genes (ISGs)(Chuong et al. 2016). The env genes of HERVs encode class I fusion proteins (Syncytins). The fusogenic activity of ERV-derived envelopes has also been co-opted to mediate cell fusion and multinucleated syncytia (the syncytiotrophoblast) formation during human placentation.

In 2003, de Parseval et al. performed a systematic screening of the human genome for retroviral coding *env* genes and identified 16 *env* genes that encode full-length ORFs (de Parseval et al. 2003). Among them, *syncytin 1* (HERV-W) and *syncytin 2* (HERV-FRD) encode envelope glycoproteins with fusogenic activity. Syncytin-1 interacts with its receptor, SLC1A5/ASCT2 (the type D mammalian retrovirus receptor), leading to the cell-to-cell fusion of cytotrophoblast cells and the formation of multinucleated syncytiotrophoblast layer (Blond et al. 2000; Malassine et al. 2005; Mi et al. 2000). Syncytin-2 (HERV-FRD) are specifically expressed in the placenta and are sufficient to induce cell-cell fusion and syncytia formation upon interaction with its receptor MFSD2 (major facilitator superfamily domain containing 2) (Blaise et al. 2003; de Parseval et al. 2005; Esnault et al. 2008). In mice, two retroviral envelope genes, designated as *syncytin-A* and *-B*, have also been co-opted for placenta formation and can trigger cell-to-cell fusion in transfection assays (Dupressoir et al. 2005; Dupressoir et al. 2009; Peng et al. 2007). Interestingly, syncytins have also been implicated in cell-cell fusion during multinucleated osteoclasts formation (Moller et al. 2017; Soe et al. 2011) and in cancer pathogenesis (Bolze et al. 2016; Larsson et al. 2007). More recently, Frank et al. computationally scanned the human genome to identify over 1500 env-derived ORFs, which are much higher than previously appreciated. Many of these putative retroviral-derived *env* genes have been shown to exhibit tissue-specific transcription. Frank et al. characterized a retroviral-derived *env* gene, designated as *Suppressyn*, which encodes a protein lacking a transmembrane domain. Suppressyn, and its hominoid orthologs, has been shown to interact with ASCT2 and block infection by mammalian-type D retroviruses (Frank et al. 2022). Suppressyn also inhibits syncytin-1 mediated cell-cell fusion and syncytia formation by occupying the ASCT2 receptor expressed on trophoblast cells (Sugimoto et al. 2013).

14.3 Human Immunodeficiency Viruses

Virus entry of cell-free HIV-1 particles into a specific host cell is mediated by the sequential interactions of the viral surface protein, known as the viral Envelope (Env) protein, the CD4 receptor, and a seven transmembrane chemokine co-receptor CCR5 or CXCR4. Both CCR5 and CXCR4 are members of seven-transmembrane G-protein-coupled receptor (GPCR) superfamily and are structurally formed of a cytoplasmic C-terminal tail, three intracellular loops, three extracellular loops (ECLs), and an extracellular N-terminal segment. The viral envelope glycoprotein (Env) spikes the exterior of virus particles and comprises both the outer surface glycoprotein gp120 (SU) and the transmembrane glycoprotein gp41 (TM). The surface gp120 SU and the transmembrane gp41 TM are non-covalently associated and form a trimer of 3 molecules of gp120 and 3 molecules of gp41 on each spike of viral particles. Binding of the viral gp120 to the cell surface receptor CD4 triggers a conformation change in gp120 and subsequently exposes the binding sites for the co-receptor, either CCR5 or CXCR4. The third variable loop region (V3) of gp120 is a key determinant of co-receptor binding (Cardozo et al. 2007; Hoffman and Doms 1999), and the overall positively charged residues of the V3 loop significantly affect co-receptor binding (Hartley et al. 2005). Following interaction of CD4-bound gp120 to the co-receptor, the fusion peptide of the transmembrane gp41 is exposed and then inserts into the host cell plasma membrane, leading to the formation of fusion pore that subsequently triggers the fusion between viral and cellular membranes, and release of the viral contents into the host cell cytoplasm (Clapham and McKnight 2001; Wilen et al. 2012).

HIV-1 induced cell-fusion and syncytia formation is also mediated by engagement of CD4 on target cell with Env expressed by infected donor cells during HIV-1 infection (Bracq et al. 2018; Lifson et al. 1986). In fact, HIV-1 strains were initially classified as syncytia-inducing (SI) and non-syncytia inducing (NSI) viral strains, referring to their capacity to induce syncytia formation in vitro (Schuitemaker et al. 1992). Theoretically, cell fusion and syncytium formation could occur when HIV-1 infected mononuclear cells enter the blood and migrate to tissues, such as the spleen, the lungs, and the central nervous system, and then fuse with CD4+ target cells. Earlier studies have shown that HIV-1 infected multinucleated macrophages and multinucleated DCs could be detected in vivo in different tissues of HIV-1-infected patients (Frankel et al. 1996; Granelli-Piperno et al. 1995). The presence of HIV-1-infected MGCs expressing specific DC markers has been found at the surface of nasopharyngeal tonsils of HIV-1-infected individuals (Frankel et al. 1996). Similarly, HIV-1-infected multinucleated giant macrophages were found in vivo in different tissues during HIV-1 infection, including spleen, genital, lymph nodes, lungs, digestive tracts, as well as the central nervous system (CNS) (Bracq et al. 2018; Frankel et al. 1996; Harbison et al. 2014; Koenig et al. 1986; Symeonides et al. 2015; Vicandi et al. 1999). However, the relevance of syncytia formation for HIV-1 transmission in vivo is less clear. Recent studies using humanized mice model and 3D cultures found that HIV-1-infected CD4+ T cells are migratory vehicles for virus dissemination in vivo. HIV-infected T cells can form multinucleated syncytia through Env-mediated cell-cell fusion and migrate to distant tissues to spread infection (Murooka et al. 2012). Indeed, HIV-1-induced small syncytia have been found in the lymph nodes of HIV-1-infected patients and have been recapitulated in vitro in 3D cultures (Bracq et al. 2017; Compton and Schwartz 2017; Law et al. 2016; Murooka et al. 2012; Symeonides et al. 2015).

In 2017, Bracq et al. revealed a very efficient mechanism involved in cell-to-cell transfer from HIV-1-infected T cells to macrophages and subsequent virus spread between macrophages by a two-step cell fusion process. HIV-1-infected T cells first establish contacts and fuse with macrophage targets. The newly formed lymphocyte-macrophage fused cells then acquire the ability to fuse with surrounding uninfected macrophages, leading to the formation of infected multinucleated giant cells that can survive for a long time and can produce large amount of infectious viral particles, as evidenced in vivo in lymphoid organs and the central nervous system (Bracq et al. 2017, 2018).

We also demonstrated that HIV-1 uses a common and specific cell-to-cell fusion mechanism for massive virus transfer from infected T lymphocytes and dissemination in myeloid cell targets, including immature dendritic cells, macrophages, as well as osteoclasts, but not monocytes and mature dendritic cells (Xie et al. 2019). We showed that the establishment of contacts with HIV-1-infected T cells leads to heterotypic cell fusion for the fast and massive transfer of viral material in myeloid cells, which subsequently triggers homotypic fusion with noninfected neighboring osteoclasts or immature dendritic cells for virus dissemination. Both cell fusion steps are mediated by viral envelope-receptor interactions and are highly efficient for macrophage-tropic CCR5- and CXCR4-using viruses, dual-tropic R5X4 viruses, and to a lesser extent for non-macrophage-tropic R5 viruses, including transmitted/founder viruses (Xie et al. 2019). Interestingly, this virus-induced cell-cell fusion bypasses the restriction imposed by the SAMHD1 host cell restriction factor for replication of HIV-1. Furthermore, this cell-to-cell infection process by cell-cell fusion also overcome virus entry block of the non-macrophage-tropic (R5 non-M-tropic) strains in macrophages (Han et al. 2022). These important results revealed the cellular mechanisms involved in the formation of the infected multinucleated giant cells observed in vivo in lymphoid and non-lymphoid tissues of HIV-1-infected patients (Bracq et al. 2018).

The envelope glycoproteins (Env) of HIV and other lentiviruses have unusually long cytoplasmic domains (typically 150–200 amino acids in length) (Postler and Desrosiers 2013). Mutations in the cytoplasmic domain of HIV can influence the incorporation of Env into virus particles (Dubay et al. 1992b), virus infectivity (Chen et al. 1998), as well as syncytia formation (Dubay et al. 1992a). Similarly, truncations of the gp41 cytoplasmic tail of SIV can result in enhanced Env incorporation into virions and increased cell-to-cell fusion (Johnston et al. 1993; Zingler and Littman 1993). The endocytosis motif (YXXΦ) within the gp41 cytoplasmic tails of HIV-1 and SIV has been shown to regulate intracellular trafficking of Env and cell surface expression of Env (LaBranche et al. 1995; Rowell et al. 1995). Mutation of the YXXΦ motif resulted in increased Env surface expression on SIV infected cells, leading to enhanced cell-to-cell fusion and accelerated infection kinetics (Sauter et al. 1996).

14.4 Severe Acute Respiratory Syndrome Coronavirus 2 (SARS-CoV-2)

The severe acute respiratory syndrome coronavirus 2 (SARS-CoV-2) is the causative agent for coronavirus disease 2019 (COVID-19) (Huang et al. 2020). To date, SARS-CoV-2 has caused more than 651 million confirmed infections and resulted in more than 6 million deaths worldwide (https://covid19.who.int). SARS-CoV-2, an enveloped single-strand RNA virus, belongs to the genus Betacoronavirus (β-CoVs), which also includes SARS-CoV-1 and Middle East respiratory syndrome coronavirus (MERS-CoV) (Lu et al. 2020; Zhou et al. 2020). SARS-CoV-2 entry is mediated by the interactions between coronavirus spike protein, which is a class I viral fusion protein, and its receptor, angiotensin-converting enzyme 2 (ACE2)(Lan et al. 2020). ACE2 was initially identified as a functional receptor for SARS-CoV (Li et al. 2003). The spike protein forms a homotrimer and is inserted into the virion membrane. S protein on the virion is comprised of the S1 subunit, which binds to ACE2 receptor, and the S2 subunit, which anchors the spike protein to the membrane. The S1 subunit is comprised of an N-terminal domain (NTD) and a receptor binding domain (RBD). The S2 subunit includes a fusion peptide (FP), two heptapeptide repeat sequences (HR1 and HR2), the transmembrane anchor (TA), and the C-terminal domain (CTD). RBD binding to the ACE2 receptor allows the exposure of the S2 cleavage site upstream of the FP, then the unstructured HR1 becomes helical, and the hydrophobic FP inserts into the target membrane (Jackson et al. 2022).

Beside spike-mediated virus entry, spike protein expressed on plasma membrane of infected cells also induces receptor-dependent cell-fusion and syncytia formation. Many coronaviruses (CoVs), including coronavirus mouse hepatitis virus (MHV), MERS-CoV, SARS-CoV, and SARS-CoV-2, have been shown to induce cell-cell fusion and syncytia formation during infection (Lavi et al. 1996; Leroy et al. 2020; Rajah et al. 2022). Multinucleated syncytia are associated with the lung tissue damage in SARS-CoV and MERS-CoV infections. Autopsy reports from MERS-CoV-infected patients and SARS patients revealed the presence of infected giant multinucleated syncytia cells in the lungs, which are mostly of epithelial origin (Alsaad et al. 2018; Franks et al. 2003; Ng et al. 2016; Nicholls et al. 2003). Recent studies have also identified the presence of infected multinucleated syncytial in the lung tissues of severe COVID-19 patients, and immunohistochemistry analyses have shown that these syncytia mainly originate from cell-to-cell fusion of SARS-CoV-2-infected pneumocytes (Braga et al. 2021; Bussani et al. 2020). The histopathological features described in the lung tissues of severe COVID-19 patients are very similar to those seen in SARS-CoV and MERS-CoV infections (Liu et al. 2020).

SARS-CoV-2 infected cells express S protein at the cell surface and induce large multinucleated syncytia formation upon its engagement of ACE2 receptor on target cells (Fig. 14.1). SARS-CoV-2 spike-mediated cell fusion was further enhanced by the expression of the cellular protease TMPRSS2, which primes viral membrane fusion by mediating the cleavage of S2 subunit (Beucher et al. 2022; Buchrieser et al. 2020). During the pandemic, the major variant of concerns (VOCs) of SARS-CoV-2 has emerged and has accumulated numerous mutations within spike proteins, which greatly impact virus characteristics, including infectivity, antigenicity, and transmissibility (Harvey et al. 2021; Peng et al. 2022). For example, SARS-CoV-2 D614G variant, one of the earlier variants, containing the D614G mutation close to the S1/S2 cleavage site, exhibited much more efficient virus replication and transmission in primary human airway epithelial cells and animal models (Hou et al. 2020). The D614G variant increased the efficiency of virus entry and induced more multinucleated syncytia than the ancestral Wuhan strain (Jiang et al. 2020; Ozono et al. 2021). Moreover, several other variants, including SARS-CoV-2 Alpha (B.1.1.7), Beta (B.1.351), and Delta (B.1.617.2), also displayed enhanced spike mediated cell-cell fusion and syncytia formation (Mlcochova et al. 2021; Rajah et al. 2021; Saito et al. 2022). In addition, mutations within the spike NTD also regulate spike-mediated cell-fusion activity (Rajah et al. 2022). Specifically, spike deletion H69/V70 in the Alpha variant increased cleaved S2 and spike infectivity. This deletion H69/V70 is required for B.1.1.7 spike-induced syncytia formation (Meng et al. 2021).

Studies with the MHV spike protein showed that the two HR regions of S2 subunit are important for spike-mediated cell-cell fusion. Substitutions of bulky hydrophobic amino acids with charged residues within the HR1 region severely affected

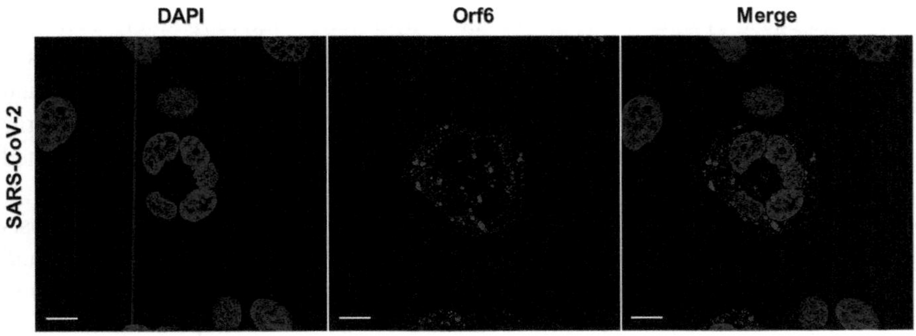

Fig. 14.1 SARS-CoV-2-induced syncytia formation in HeLa-ACE2 cells. HeLa-ACE2 cells were infected with SARS-CoV-2 Alpha (B.1.1.7) variant (500 E copies/cell); cells were then fixed at 16 hpi and stained with anti-SARS-CoV-2 Orf6 and DAPI (scale bar, 10 μm)

cell-cell fusion activity without affecting the processing and surface expression of spike protein (Luo and Weiss 1998). Similarly, codon mutations within the HR2 region also significantly reduced MHV spike-mediated cell-cell fusion (Luo et al. 1999). Interestingly, it has been shown that the HR2 peptide can strongly inhibit spike-mediated virus entry and syncytium formation (Bosch et al. 2003). The HR1 and HR2 regions within the S2 subunit are highly conserved among various CoVs and play important role in spiked-mediated viral entry by forming the six-helix bundle (6-HB) core structure. Indeed, HR-derived peptides have been shown to strongly inhibit cell-cell fusion induced by spike proteins of multiple human CoVs (Xia et al. 2019). More recently, a pan-coronavirus fusion inhibitor, EK1 peptide, has been shown to strongly inhibit infection of multiple CoVs, including MERS-CoV, SARS-CoV, and SARS-CoV-2, by targeting highly conserved HR2 region of spike proteins (Xia et al. 2020; Xia et al. 2019). The CTD of S2 subunit harbors several trafficking signals that involved in COPI/II binding as well as intracellular transport of coronavirus spike proteins (Cattin-Ortola et al. 2021; McBride et al. 2007; Sadasivan et al. 2017). Expression of S protein alone results in its accumulation on the cell surface, leading to formation of multinucleated syncytia. Cytoplasmic tail truncation of SARS-CoV-2 S protein resulted in increased surface expression and enhanced cell-fusion activity (Chen et al. 2021; Havranek et al. 2020). Interestingly, palmitoylation of the cysteine rich CTD of coronavirus spike protein is also critical for spike-mediated virus entry and syncytia formation (Li et al. 2022; Petit et al. 2007; Thorp et al. 2006).

14.5 Herpesviridae

Herpesviridae is a large and diverse family of enveloped dsDNA viruses which have a very broad host range and could establish life-long persistent infections. Herpesviruses have been classified into three subfamilies: *Alphaherpesvirinae*, composed of herpes simplex virus types 1 (HSV-1) and herpes simplex virus 2 (HSV-2), as well as varicella-zoster virus (VZV or HHV-3), with a short replicative cycle and cause cytopathology in the monolayer cell cultures; *Betaherpesvirinae*, including human cytomegalovirus (HCMV, or HHV-5), human herpesviruses 6 (HHV-6), and human herpesviruses 7 (HHV-7), with a restricted host range and long replicative cycle; and *Gammaherpesvirinae*, composed of Epstein-Barr virus (EBV, or HHV-4) and human herpesvirus 8 (HHV-8), with a very restricted host range.

The formation of syncytia following herpesvirus infection in their natural hosts has been well documented (Hoggan and Roizman 1959; Kim et al. 2013; Lee and Spear 1980; Sawitzky 1997). The presence of multinucleate giant cells (MGCs) in skin lesions has long been recognized as the hallmark of herpesvirus infection (Blank et al. 1951), and can be used as diagnostic for herpetic simplex keratitis (Farhatullah et al. 2004). Syncytia formation is a cytopathologic effect (CPE) of herpesvirus infection in lower respiratory tract of the lung (Pritt and Aubry 2017).

The extent of Herpesvirus-mediated cell-to-cell fusion is dependent of the identity of herpesvirus: VZV infection induces extensive syncytial formation in skin lesions (Cole and Grose 2003), while HSV infection results in the formation of small polykaryocytes in lesions of HSV infected patients (Muggeridge et al. 2004). However, the significance of herpesvirus-mediated cell-to-cell fusion for viral replication and spread in vivo remains unclear. The degree of cell-cell fusion induced by different clinical isolates and laboratory-adapted strains varies significantly in tissue culture (Galitska et al. 2018; Wheeler 1960). For example, HSV-1 primary isolates induce limited cell fusion (Ejercito et al. 1968), whereas laboratory strains cause extensive syncytial formation in tissue culture (Cassai et al. 1975; Read et al. 1980).

14.5.1 The Core Fusion Machinery for Herpesvirus

Herpesviruses enter cells by enabling membrane fusion of viral envelopes with cellular membranes, which occurs at plasma membrane or in the endosomal compartment. This viral entry process depends on the identity of herpesvirus and is also cell-type dependent. The core membrane fusion machinery for herpesviruses consist of the glycoprotein gB and the heterodimer gH/gL, which are conserved envelope proteins among all herpesviruses (Connolly et al. 2011; Hutt-Fletcher 2007; Nguyen and Kamil 2018; Nishimura and Mori 2019; Sathiyamoorthy et al. 2017; Spear and Longnecker 2003; Vanarsdall and Johnson 2012). Glycoprotein B (gB) is the most conserved herpesvirus viral fusion protein and has been classified as a type III viral fusogen because of its structural similarities with the vesicular stomatitis virus glycoprotein (VSV-G) (Backovic et al. 2009; Baquero et al. 2015; Connolly et al. 2011; Heldwein et al. 2006). Furthermore, gB is also a major determinant of herpesvirus infectivity in vitro and in vivo (Beitia Ortiz de Zarate et al. 2007; Gerdts et al. 2000). The gH/gL heterodimer interacts with gB and regulates its activity (Atanasiu et al. 2007). Herpesviruses minimally require gH/gL and gB for both virus entry and virus-induced cell-cell fusion. In herpesvirus, this membrane fusion process also requires additional, non-conserved membrane glycoproteins specific to individual herpesvirus, which can bind to cell-specific host receptors. HSV-1 and HSV-2 express a viral glycoprotein gD that can interact with several host cell receptors depending of the target cells (Fan et al. 2014, 2015): nectin-1, a cell adhesion molecule, and the main receptor in epithelial cells and neurons (Di Giovine et al. 2011); HveA or TNFRSF14, a member of the tumor necrosis factor receptor superfamily, expressed on activated lymphocytes (Carfi et al. 2001; Montgomery et al. 1996); and 3-O-sulfated heparan sulfate, a non-protein soluble receptor (Spear 2004; Tiwari et al. 2007). HCMV expresses a glycoprotein gO as well as three small glycoproteins, UL128, UL130, and UL131A, to mediate cellular receptor binding, but the respective host cell receptors remain unclear (Ciferri et al. 2015; Ryckman et al. 2006; Vanarsdall et al. 2011). HHV-6A and HHV-6B express a glycoprotein gO and gQ1/gQ2 proteins to engage human CD46 (Jasirwan et al. 2014; Mori et al.

2004; Santoro et al. 1999); several studies also indicate that HHV-6B uses human CD134 (hCD134) as an alternative receptor for viral entry (Tang and Mori 2015; Tang et al. 2013; Tang et al. 2014). EBV expresses a soluble viral protein gp42 to engage the MHC class II receptor HLA-DR1 on B cells but can also use the gH/gL heterodimer to directly bind cell surface intergrins, including αvβ5, αvβ6, and αvβ8, for viral entry (Chesnokova and Hutt-Fletcher 2011; Chesnokova et al. 2009; Kirschner et al. 2006; Li et al. 1997; Mullen et al. 2002). Some herpesviruses, such as KSHV, only use gH/gL heterodimer to bind cellular receptors (Hahn et al. 2012), while the requirement of gH/gL heterodimer for viral entry indicates a highly conserved function among all herpesviruses (Stampfer and Heldwein 2013). The gH/gL heterodimer receives an activating signal upon binding of cellular receptors by viral glycoproteins and transmits it to gB, subsequently inducing a conformational change of gB (Heldwein 2016; Sathiyamoorthy et al. 2017; Stampfer and Heldwein 2013). Once the glycoprotein gB is triggered, its fusion loops insert into the host cell membrane, followed by the refolding of gB to drive the membrane merger and onset of infection (Connolly et al. 2011). The core fusion machinery is required for both herpesvirus entry and herpesvirus-induced cell-to-cell fusion (Cole and Grose 2003; Spear and Longnecker 2003; Weed and Nicola 2017). However, the mechanisms of herpesvirus-induced cell-to-cell fusion are poorly understood. Cell-to-cell fusion by herpesvirus is highly cell type-dependent, consistent with an important role for cellular factors. For example, VZV induce extensive syncytial formation on primary keratinocytes, but poorly cause cell-to-cell fusion on primary fibroblasts (Cole and Grose 2003). The cellular factors for herpesvirus-induced cell-to-cell fusion are unknown except for viral receptors that herpesvirus use to enter cells (Weed and Nicola 2017).

14.5.2 Herpes Simplex Virus

Multinucleated giant cells (also known as poly-karyocytes) were first described in skin lesions of HSV infected patients many years ago, and have been used for diagnostic purposes for HSV infection (Blank et al. 1951). For herpes simplex virus (HSV), transfer of virus particles from infected cells to uninfected neighboring cells can induce cell-to-cell fusion of plasma membranes of HSV infected cells and uninfected cells, and can even occur in the presence of neutralizing antibody; this virus-induced cell-cell fusion has been initially proposed as another mechanism by which the virus infects host cells and spreads infection (Dingwell et al. 1994; Sarfo et al. 2017). Early studies described the formation of poly-karyocytes when human epithelial cells were infected with HSV wild-type strain isolated from HSV-infected patient (Scherer and Syverton 1954). Moreover, primary strains of HSV isolated from the eye and oral of infected patients also induced poly-karyocytes formation in cultured cells (Doane et al. 1955; Gray et al. 1958; Scott and Mc 1959). These poly-karyocytes were detected in tissue culture and in skin lesions, containing no more than ten nuclei, have been referred to as the "small multinucleated giant cells" (Gray et al. 1958). Many HSV variants that induce extensive cell-cell fusion and syncytia formation in tissue culture were easily isolated from laboratory viral stocks (Cassai et al. 1975; Hoggan and Roizman 1959; Read et al. 1980). The resulting multinucleated syncytia contained hundreds or even thousands of nuclei and resulted from mutations in one or several viral genes (Tognon et al. 1984; Weed and Nicola 2017). Cell-to-cell fusion by HSV syncytial variants requires the activity of gD and the core fusion machinery, consisting of gB and the heterodimer gH/gL, as well as the non-glycosylated membrane protein UL45 (Haanes et al. 1994) and the glycoproteins gM, gE, and gI (Davis-Poynter et al. 1994). It has been reported that the core fusion machinery, including gB and gH/L, and gD are both necessary and sufficient to mediate HSV-1 induced cell-cell fusion and syncytia formation (Turner et al. 1998). Truncations or single amino acid mutations in the cytoplasmic tail of gB can result in extensive syncytia formation in tissue culture (Bzik et al. 1984; Cai et al. 1988a; Cai et al. 1988b; Engel et al. 1993; Gage et al. 1993). HSV-induced cell-cell fusion is also modulated by other viral proteins that regulate the fusion activity of the core fusion machinery. The glycoprotein K (gK) regulates the core fusion machinery as it inhibits cell-cell fusion when co-expressed with gB, gH/gL, and gD(Avitabile et al. 2003). The glycoprotein K (gK) and UL20 protein (UL20p) form a functional protein complex and physically interact with gB, thereby modulating the fusogenic properties of gH and gB(Chouljenko et al. 2012; Chouljenko et al. 2010; Foster et al. 2003). Similarly, the tegument proteins UL11, UL16, and UL21 is tightly associated with gE, and the function of gE requires the assembly of UL11–UL16–UL21 complex on its cytoplasmic tail(Han et al. 2012). The deletion mutants of UL11, UL16, and UL21 fail to induce cell-to-cell fusion in vero cells(Han et al. 2012). Moreover, mutations in UL20 (Foster et al. 2004), UL45 (Haanes et al. 1994) or the cytoplasmic tail of gH negatively regulate HSV-1 mediated cell-cell fusion and syncytia formation (Browne et al. 1996; Wilson et al. 1994). Along with the core fusion machinery, the glycoproteins gE/gI heterodimer is also involved in HSV-mediated cell-cell fusion and multinucleated syncytia formation (Balan et al. 1994; Chatterjee et al. 1989; Davis-Poynter et al. 1994). The glycoprotein gM inhibited HSV-1 mediated cell-cell fusion by downregulating the cell surface expression of gD and the gH/gL heterodimer (Crump et al. 2004; Ren et al. 2012). Moreover, the glycoprotein N (gN) has been shown to physically interact with gM and modulates its membrane fusion activity (El Kasmi and Lippe 2015). Overexpression of gN induced extensive syncytia formation in wild-type HSV-1 infected cells(El Kasmi and Lippe 2015). The gD receptors, including HVEM nectin-1 and nectin-2, are also required for HSV induced cell-cell fusion in CHO cells (Terry-Allison et al. 2001; Terry-Allison et al. 1998). Heparan sulfate (HS) appears to be less important for HSV mediated cell-cell fusion than for virus entry (Shieh and Spear 1994; Terry-Allison et al. 2001). It remains unclear how the interaction of gD with its receptor result in cell-to-cell fusion. One hypothesis is that the binding of gD with its receptor triggers the conformational changes of gD, allowing its interaction with the core fusion machinery, triggering the activation of its fusogenic activity. Receptors for gB and/or gH/gL may also exist, and the binding of either to these additional receptors could also trigger fusion activity and bypass the requirement for gD. Recently, the paired immunoglobulin-like type 2 receptor (PILR) has been identified as the entry coreceptor that physically associate with the gB, and interactions of PILR coreceptor and gB are involved in HSV-1 mediated cell-cell fusion events during HSV-1 infection (Satoh et al. 2008). Interestingly, the protein tyrosine phosphatase 1B (PTP1B), a protein tyrosine phosphatase, has been reported to be specifically required for the HSV-1 mediated cell-cell fusion, and inhibitors of PTP1B dramatically reduced cell-cell fusion (Carmichael et al. 2018).

14.5.3 Human Cytomegalovirus

Human cytomegalovirus (HCMV) infects a broad range of cell types, including leukocytes, monocytes, macrophages, fibroblasts, endothelial cells, and epithelial cells (Gerna et al. 2004). HCMV entry into different cell types requires the interactions of viral glycoproteins with various host receptors (Compton 2004; Nguyen and Kamil 2018; Nishimura and Mori 2019). It has been proposed that HCMV can make use of different forms of gH/gL heterodimer, a trimeric complex of gH/gL/gO and a pentameric complex consisting of gH/gL/ UL128-131 to engage distinct host receptors, and these interactions trigger gB-mediated membrane fusion (Nishimura and Mori 2019; Vanarsdall and Johnson 2012). HCMV clinical isolates have been shown to accumulate mutations when passaged many times on fibroblasts, resulting in a lack of pentameric complex on the mature virion and no longer infect endothelial and epithelial cells (Dargan et al. 2010; Murrell et al. 2013; Nishimura and Mori 2019). These laboratory-adapted strains generally exhibit restricted tropism and can produce high levels of cell-free virus particles. HCMV transmission in vivo is largely cell-associated, and clinical isolates have been shown to induce enlarged flower-shaped syncytial foci in vitro (Galitska et al. 2018; Sinzger et al. 1999; Waldman et al. 1989). However, HCMV laboratory-adapted strains spread via diffusion of cell-free virus and fail to recapitulate the phenotypic characteristics of clinical isolates due to acquisition of genetic mutations (Murrell et al. 2017; Wilkinson et al. 2015). The ability of wild-type HCMV to spread via syncytia formation is attributed to the high-level expression of pentameric complex gH/gL/ UL128-131, and this cell-cell spread is resistant to neutralizing antibodies and can also overcome IFN-induced antiviral factors (Gerna et al. 2016; Murrell et al. 2017). Importantly, the pentameric complex has been shown to be required for wild-type HCMV-induced cell-cell fusion and syncytia formation in epithelial cells (Gerna et al. 2016). Similar to HSV-mediated cell-cell fusion, HCMV gB and gH/gL are sufficient to induce cell-cell fusion and syncytia formation, and this cell-cell fusion process is most efficient at neutral pH (Vanarsdall et al. 2008). Overexpression of gB promotes cell-to-cell fusion in U373 glioblastoma cells (Navarro et al. 1993; Tugizov et al. 1994). Interestingly, gB882stop mutant lacking the 25 C-terminal residues displayed enhanced cell-cell fusion (Wille et al. 2013). A single amino acid mutation within AD169 gB (275Y) induces cell-cell fusion and syncytia formation, resulting in caspase-2 activation and thereby compromising genome stability (Tang et al. 2019). However, mutations in gH/gL heterodimer do not affect gB-mediated cell-cell fusion (Schultz et al. 2015). More recently, it has been reported that gB-mediated cell fusion and multinucleated cell syncytia formation can be potently blocked by neutralizing mAbs specifically targeting the antigenic domain 5 (AD-5) within the ectodomain of gB (Reuter et al. 2020). Glycoprotein O (gO) is also involved in HCMV-mediated cell-cell fusion, and the anti-gO antibodies can block syncytia formation by HCMV (Paterson et al. 2002).

14.5.4 Human Herpesvirus 6

Human herpesvirus 6 (HHV-6), belonging to β-herpesvirus family, is a T lymphotropic virus that productively replicates in T cells and monocytes. HHV-6 isolates can be classified into two groups, HHV-6A and HHV-6B, although the overall identity of genome sequence between HHV-6A and HHV-6B is almost 90% (Dominguez et al. 1999; Salahuddin et al. 1986). HHV-6A can induce cell-to-cell fusion of diverse human cells expressing human CD46, including primary T lymphocytes, primary astrocytes (He et al. 1996), as well as endothelial cells (Rotola et al. 2000; Wu and Shanley 1998), whereas HHV-6B does not replicate and induce syncytium formation in these cells. Syncytia formation induced by HHV-6A can be observed as early as 2 h after infection; this cell-to-cell fusion process could occur even in the absence of viral protein synthesis, also known as fusion from without (FFWO) (Mori et al. 2002). Mechanistically, syncytia formation induced by HHV-6A is dependent on the expression of CD46 on target cells and requires glycoproteins B and H (Mori et al. 2002). Indeed, it has been reported that HHV-6 gH/gL heterodimer forms a complex with glycoproteins gQ1 and gQ2, and that this complex mediate CD46 receptor binding (Mori et al. 2004; Mori et al. 2003; Santoro et al. 1999). The syncytium formation induced by HHV-6A can be inhibited by both anti-gB and anti-CD46 MAbs (Mori et al. 2002). More recently, it has been shown that HHV-6A and HHV-6B both require *cis* expression of gB, gH/ gL, gQ1, and gQ2 on the same cell for cell-to-cell fusion (Tanaka et al. 2013), unlike that of HCMV and HSV, in which the envelope glycoproteins gB, gD, and gH/gL can induce cell-to-cell fusion when expressed either *in trans* or *in cis* (Atanasiu et al. 2010; Vanarsdall et al. 2008).

Human herpesvirus 6 (HHV-6) expresses two homologs of human G protein-coupled receptors (GPCRs), U12 and U51 (Gompels et al. 1995). Stable expression of an siRNA specific for HHV-6 U51 in human T cells prior to infection reduces viral DNA replication and HHV-6-induced syncytia formation. Furthermore, expression of HHV-6 U51 resulted in enhanced cell-to-cell fusion mediated by the vesicular stomatitis virus glycoprotein (VSV-G), indicating U51 positively regulates cell-to-cell fusion in vitro (Zhen et al. 2005). HHV-6A U51 has been shown to bind to several CC chemokines including CCL5 (also known as RANTES), an inflammatory chemokine which can bind to chemokine receptor CCR5. RANTES has been reported to increase the efficiency of HIV-1 Env-mediated cell-to-cell fusion (Brooks et al. 2019; Trkola et al. 1999). Interestingly, HHV-6 dramatically suppressed HIV-1 replication and syncytium formation in the co-infected DC cultures (Asada et al. 1999; Csoma et al. 2006; Grivel et al. 2001), suggesting UL51 may play different roles in virus-induced cell-to-cell fusion.

14.6 Reoviridae

The family Reoviridae includes 15 genera which can be divided into two subfamilies, the Sedoreovirinae and Spinareovirinae (Lefkowitz et al. 2018). Reoviridae consists of diverse human and animal pathogens, such as mammalian orthoreovirus (MRV), rotavirus (RV), and Nelson Bay orthoreovirus (NBV). These viruses share several common features, a nonenveloped capsid and a double-stranded RNA (dsRNA) genome of 9–12 segments. Orthoreovirus and quareovirus, two genera of the family Reoviridae, have been shown to induce cell-cell fusion and form multinucleated cells (MGCs) after infection, which facilitate viral spread and pathogenicity (Duncan 2019; Kanai et al. 2019). Both aquareovirus and orthoreovirus induced cell-cell fusion and syncytial formation are caused by a family of nonstructural viral membrane fusion proteins, the fusion-associated small transmembrane (FAST) proteins (Boutilier and Duncan 2011). FAST proteins are the smallest known viral fusogens that can induce cell-cell fusion between virus-infected cells and neighboring cells. Interestingly, FAST proteins are not required for virus entry, and the expression of FAST proteins alone on host cells is sufficient to induce cell-cell fusion, leading to formation of large multinucleated syncytial (Dawe and Duncan 2002; Duncan 2019; Shmulevitz and Duncan 2000). Although the biological functions of FAST proteins remain unclear, it has been reported that FAST proteins are required for efficient viral replication in vitro and play an important role in viral pathogenesis in vivo (Kanai et al. 2019). FAST proteins consist of three functional domains: an acylated N-terminal ectodomain, a transmembrane (TM) domain, and the C-terminal cytoplasmic endodomain (Dawe et al. 2005; Duncan 2019). The acylation (palmitoylation or myristoylation) at N-terminal ectodomain has been shown to be essential for the cell fusion activity (Corcoran and Duncan 2004; Corcoran et al. 2004; Shmulevitz et al. 2003). The transmembrane domain directs plasma membrane insertion of FAST protein and a bitopic Nexoplasmic/Cendoplasmic type I topology by serving as a reverse signal-anchor sequence (Dawe et al. 2005). The cytoplasmic endodomain of FAST protein contains a membrane-proximal polybasic (PB) motif and an amphipathic α-helix, which functions to promote fusion pore formation (Read et al. 2015). Overexpression of the endodomain has been shown to promote fusion pore expansion and enhance cell-cell fusion (Top et al. 2009). Unlike the fusion peptides (FPs) of most enveloped viruses, FAST proteins of non-enveloped reoviruses lack receptor-binding capacity and are capable of inducing cell-cell fusion and syncytia formation of a wide range of cell types without a specific trigger, such as low pH and/or receptor binding (Salsman et al. 2008). In fact, FAST proteins rely on surrogate cadherins to mediate close membrane apposition and require active actin remodelling to drive cell-cell fusion and syncytia formation (Chan et al. 2021; Salsman et al. 2008).

14.7 Conclusions

Many viruses can induce cell-cell fusion and syncytia formation. For most enveloped viruses, such as HIV-1 and SARS-COV-2, this cell-cell fusion process is mediated by the interactions between viral fusion proteins and their cognate receptors. Some non-enveloped Reoviruses even encode specialized viral proteins, the fusion-associated small transmembrane (FAST) proteins, to mediate cell-cell fusion and syncytia formation. Interestingly, FAST proteins are not involved in virus entry but are dedicated to mediate cell fusion and syncytia formation (Ciechonska and Duncan 2014). Virus-induced cell-cell fusion represents a unique mode of virus cell-to-cell spread by protecting viruses from extracellular environment, such as neutralizing antibodies and antiviral drugs, and may also overcome the inhibitory effects of several host restriction factors, such as SAMHD1 and tetherin/BST2 (Bracq et al. 2017, 2018; Cifuentes-Munoz et al. 2018; Han et al. 2022; Phillips 1994). Strikingly, virus-induced cell-cell fusion can also overcome the restriction to entry of cell-free virus in non-permissive cells (Han et al. 2022; Rauch et al. 2000; Simmons et al. 1995). Virus-induced syncytia maintain metabolic functions and essential gene expression, which provide an expanded and prolonged localized environment for viral propagation by protecting viruses against neutralizing antibodies and antiviral drugs (Fig. 14.2). When multinucleated syncytia succumb to apoptosis and rupture, resulting in a burst of enormous new infectious virus particles release for systemic spread of the infection (Castedo et al. 2002; Salsman et al. 2005). Therefore, the formation of virus-induced multinucleated syncytia causes cytopathic effects and potentially facilitates virus propagation, virus dissemination, and immune evasion (Leroy et al. 2020). However, the significance of virus-mediated cell-cell fusion and syncytia formation for viral replication and spread in vivo remains unclear. The impact of syncytia formation in viral pathogenesis remains an outstanding question.

Cell-cell fusion induced by some viral fusion proteins involves reorganization of the cytoskeleton (Chan et al. 2021; Chan et al. 2020; Chen et al. 2008; Sylwester et al. 1993). It has been shown that actin cytoskeletal reorganizations are involved in HIV-1-induced cell-cell fusion (Pontow et al. 2004). HIV-1 envelope-coreceptor interactions can activate Rac-1 GTPase and induce actin cytoskeletal reorganizations that are required for the cell-cell fusion process (Pontow et al. 2004). Interestingly, some FAST proteins encoded by reoviruses can hijack the actin cytoskeleton to facilitate cell-cell fusion (Chan et al. 2021; Chan et al. 2020). These studies indicate that the actin cytoskeleton could be a general requirement for virus-induced cell-cell fusion (Chen and Olson 2005; Podbilewicz 2014).

The cytoplasmic tails of most enveloped viral fusogens, such as HIV-1 Env and herpesvirus gB, generally inhibit their fusion activity. The cytoplasmic tails of Env proteins of retroviruses have been shown to regulate cell-to-cell fusion efficiency.

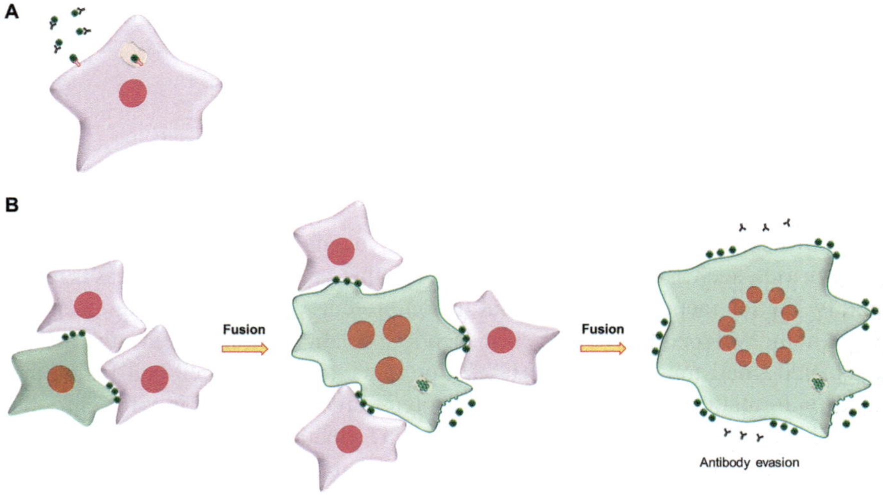

Fig. 14.2 Virus entry and virus-induced cell-cell fusion. (**a**) Virus entry. Cell-free viruses enter cells by enabling membrane fusion of viral envelopes with cellular membranes upon receptors binding, which occurs at plasma membrane or in the endosomal compartment. Cell-free infection of target cell (in red) is rather inefficient, and rate limited by fluid uptake and virion attachment. Moreover, freely diffusing virions are vulnerable to antibody neutralization. (**b**) Virus-induced cell-cell fusion and syncytia formation. Virus-infected donor cells (in green) express viral fusion proteins at cell surface and fuse with surrounding target cells (in red). The newly fused cells can initiate several rounds of cell fusion with neighboring target cells (in red), resulting in the formation of enlarged multinucleated syncytia. Syncytia also contain high amounts of virus particles accumulated in the vesicular compartment, and virus budding can occur at the plasma membrane, indicating active progeny production. For some virus, such as Reovirus, receptors are not required for cell-cell fusion and syncytia formation. Virus-induced syncytia may shield virus from neutralizing antibodies

Deletion of a 16-amino-acid R peptide within the cytoplasmic tail of Moloney Murine Leukemia Virus (MMLV) Env protein greatly enhanced Env-mediated cell-cell fusion (Januszeski et al. 1997; Melikyan et al. 2000; Yang and Compans 1997). Moreover, the expression of cytoplasmic tail-truncated SIV Env proteins in CD4 positive cells resulted in enhanced Env-mediated cell-cell fusion and syncytia formation (Ritter et al. 1993; Zingler and Littman 1993). Indeed, mutations in the cytoplasmic domain of HIV can influence the incorporation of Env into virus particles (Dubay et al. 1992b), virus infectivity (Chen et al. 1998), as well as syncytia formation (Dubay et al. 1992a). Similarly, the cytoplasmic tails of the measles virus (MV) hemagglutinin (H) and fusion protein (F) have been shown to be involved in virus envelope assembly and regulate MV mediated cell-cell fusion efficiency (Cathomen et al. 1998; Moll et al. 2002). Interestingly, the cytoplasmic tail of coronavirus spike protein harbors several trafficking signals that involved in COPI/II binding as well as intracellular transport of coronavirus spike proteins (Cattin-Ortola et al. 2021; McBride et al. 2007; Sadasivan et al. 2017). Indeed, cytoplasmic tail truncation of SARS-CoV-2 S protein resulted in increased surface expression and enhanced cell-fusion activity (Chen et al. 2021; Havranek et al. 2020). These observations suggest that regulation of membrane fusion activity via the cytoplasmic tails of viral fusion proteins is a widespread mechanism.

Recent studies shed light on the mechanism of syncytia formation by several viruses, such as HIV-1 and SARS-CoV-2. However, the contribution of virus-induced syncytia on viral pathogenesis is less clear. It seems that syncytia induced by several viruses, such as HIV-1 and measles virus, have a surprising long lifespan and are highly motile (Bracq et al. 2017; Herschke et al. 2007; Van Goethem et al. 2010; Xie et al. 2019). However, the clearance or turnover of virus-induced syncytia in vivo is poorly understood. Do virus-induced syncytia contribute to long-lived viral reservoir in vivo? One could speculate that virus-induced syncytia may shield virus from neutralizing antibodies and antiviral drugs, which could be the major obstacle to viral clearance in vivo.

Recently, it has been reported that influenza virus infection triggers phosphorylation and deSUMOylation of TRIM28, which is a key repressor of human endogenous retroviruses (HERVs), leading to a de-repression of HERVs and an amplification of host innate immune response (Feng et al. 2022; Schmidt et al. 2019). Interestingly, syncytia formation induced by some viruses, such as SARS-CoV-2 and measles virus, have been shown to amplify the interferon response (Herschke et al. 2007; Liu et al. 2022). Moreover, cell-cell fusion mediated by measles virus or ERVWE1 has been reported to cause cellular senescence in cancer cells and normal cells (Chuprin et al. 2013). Indeed, de-repression of HERVK induces cellular senescence and triggers the innate immune response. The resurrection of human endogenous retroviruses has been reported to reinforce senescence during aging (Liu et al. 2023). However, the mechanism of innate immune activation by virus-induced syncytia in viral infection remains an outstanding question. Thus, it would be of great interest to investigate if other viruses, such as HIV-1 and SARS-CoV-2, can also trigger phosphorylation and deSUMOylation of TRIM28, leading to the de-repression of ERVs and expression of ERV-derived envelopes, which may amplify interferon response and enhance cell-cell fusion.

Acknowledgments This work was supported by a Wellcome Investigator Award (220863, G.J.T.).

References

Abela IA, Berlinger L, Schanz M, Reynell L, Gunthard HF, Rusert P, Trkola A (2012) Cell-cell transmission enables HIV-1 to evade inhibition by potent CD4bs directed antibodies. PLoS Pathog *8*:e1002634. https://doi.org/10.1371/journal.ppat.1002634

Alsaad KO, Hajeer AH, Al Balwi M, Al Moaiqel M, Al Oudah N, Al Ajlan A, AlJohani S, Alsolamy S, Gmati GE, Balkhy H et al (2018) Histopathology of Middle East respiratory syndrome coronovirus (MERS-CoV) infection - clinicopathological and ultrastructural study. Histopathology *72*:516–524. https://doi.org/10.1111/his.13379

Alvarez RA, Barria MI, Chen BK (2014) Unique features of HIV-1 spread through T cell virological synapses. PLoS Pathog *10*:e1004513. https://doi.org/10.1371/journal.ppat.1004513

Asada H, Klaus-Kovtun V, Golding H, Katz SI, Blauvelt A (1999) Human herpesvirus 6 infects dendritic cells and suppresses human immunodeficiency virus type 1 replication in coinfected cultures. J Virol *73*:4019–4028. https://doi.org/10.1128/JVI.73.5.4019-4028.1999

Atanasiu D, Whitbeck JC, Cairns TM, Reilly B, Cohen GH, Eisenberg RJ (2007) Bimolecular complementation reveals that glycoproteins gB and gH/gL of herpes simplex virus interact with each other during cell fusion. Proc Natl Acad Sci U S A *104*:18718–18723. https://doi.org/10.1073/pnas.0707452104

Atanasiu D, Saw WT, Cohen GH, Eisenberg RJ (2010) Cascade of events governing cell-cell fusion induced by herpes simplex virus glycoproteins gD, gH/gL, and gB. J Virol *84*:12292–12299. https://doi.org/10.1128/JVI.01700-10

Avitabile E, Lombardi G, Campadelli-Fiume G (2003) Herpes simplex virus glycoprotein K, but not its syncytial allele, inhibits cell-cell fusion mediated by the four fusogenic glycoproteins, gD, gB, gH, and gL. J Virol *77*:6836–6844. https://doi.org/10.1128/jvi.77.12.6836-6844.2003

Backovic M, Jardetzky TS (2009) Class III viral membrane fusion proteins. Curr Opin Struct Biol *19*:189–196. https://doi.org/10.1016/j.sbi.2009.02.012

Backovic M, Longnecker R, Jardetzky TS (2009) Structure of a trimeric variant of the Epstein-Barr virus glycoprotein B. Proc Natl Acad Sci U S A *106*:2880–2885. https://doi.org/10.1073/pnas.0810530106

Balan P, Davis-Poynter N, Bell S, Atkinson H, Browne H, Minson T (1994) An analysis of the in vitro and in vivo phenotypes of mutants of herpes simplex virus type 1 lacking glycoproteins gG, gE, gI or the putative gJ. J Gen Virol *75(Pt 6)*:1245–1258. https://doi.org/10.1099/0022-1317-75-6-1245

Baquero E, Albertini AA, Gaudin Y (2015) Recent mechanistic and structural insights on class III viral fusion glycoproteins. Curr Opin Struct Biol *33*:52–60. https://doi.org/10.1016/j.sbi.2015.07.011

Beitia Ortiz de Zarate I, Cantero-Aguilar L, Longo M, Berlioz-Torrent C, Rozenberg F (2007) Contribution of endocytic motifs in the cytoplasmic tail of herpes simplex virus type 1 glycoprotein B to virus replication and cell-cell fusion. J Virol *81*:13889–13903. https://doi.org/10.1128/JVI.01231-07

Benovic S, Kok T, Stephenson A, McInnes J, Burrell C, Li P (1998) De novo reverse transcription of HTLV-1 following cell-to-cell transmission of infection. Virology *244*:294–301. https://doi.org/10.1006/viro.1998.9111

Beucher G, Blondot ML, Celle A, Pied N, Recordon-Pinson P, Esteves P, Faure M, Metifiot M, Lacomme S, Dacheux D et al (2022) Bronchial epithelia from adults and children: SARS-CoV-2 spread via syncytia formation and type III interferon infectivity restriction. Proc Natl Acad Sci U S A *119*:e2202370119. https://doi.org/10.1073/pnas.2202370119

Blaise S, de Parseval N, Benit L, Heidmann T (2003) Genomewide screening for fusogenic human endogenous retrovirus envelopes identifies syncytin 2, a gene conserved on primate evolution. Proc Natl Acad Sci U S A *100*:13013–13018. https://doi.org/10.1073/pnas.2132646100

Blank H, Burgoon CF, Baldridge GD, Mc CP, Urbach F (1951) Cytologic smears in diagnosis of herpes simplex, herpes zoster, and varicella. J Am Med Assoc *146*:1410–1412. https://doi.org/10.1001/jama.1951.63670150005012b

Blond JL, Lavillette D, Cheynet V, Bouton O, Oriol G, Chapel-Fernandes S, Mandrand B, Mallet F, Cosset FL (2000) An envelope glycoprotein of the human endogenous retrovirus HERV-W is expressed in the human placenta and fuses cells expressing the type D mammalian retrovirus receptor. J Virol *74*:3321–3329. https://doi.org/10.1128/jvi.74.7.3321-3329.2000

Bolze PA, Patrier S, Cheynet V, Oriol G, Massardier J, Hajri T, Guillotte M, Bossus M, Sanlaville D, Golfier F, Mallet F (2016) Expression patterns of ERVWE1/Syncytin-1 and other placentally expressed human endogenous retroviruses along the malignant transformation process of hydatidiform moles. Placenta *39*:116–124. https://doi.org/10.1016/j.placenta.2016.01.011

Bosch BJ, van der Zee R, de Haan CA, Rottier PJ (2003) The coronavirus spike protein is a class I virus fusion protein: structural and functional characterization of the fusion core complex. J Virol *77*:8801–8811. https://doi.org/10.1128/jvi.77.16.8801-8811.2003

Boutilier J, Duncan R (2011) The reovirus fusion-associated small transmembrane (FAST) proteins: virus-encoded cellular fusogens. Curr Top Membr *68*:107–140. https://doi.org/10.1016/B978-0-12-385891-7.00005-2

Bracq L, Xie M, Lambele M, Vu LT, Matz J, Schmitt A, Delon J, Zhou P, Randriamampita C, Bouchet J, Benichou S (2017) T cell-macrophage fusion triggers multinucleated giant cell formation for HIV-1 spreading. J Virol 91(94):e01237-17. https://doi.org/10.1128/JVI.01237-17

Bracq L, Xie M, Benichou S, Bouchet J (2018) Mechanisms for cell-to-cell transmission of HIV-1. Front Immunol *9*:260. https://doi.org/10.3389/fimmu.2018.00260

Braga L, Ali H, Secco I, Chiavacci E, Neves G, Goldhill D, Penn R, Jimenez-Guardeno JM, Ortega-Prieto AM, Bussani R et al (2021) Drugs that inhibit TMEM16 proteins block SARS-CoV-2 spike-induced syncytia. Nature *594*:88–93. https://doi.org/10.1038/s41586-021-03491-6

Brooks PJ, Glogauer M, McCulloch CA (2019) An overview of the derivation and function of multinucleated giant cells and their role in pathologic processes. Am J Pathol *189*:1145–1158. https://doi.org/10.1016/j.ajpath.2019.02.006

Browne HM, Bruun BC, Minson AC (1996) Characterization of herpes simplex virus type 1 recombinants with mutations in the cytoplasmic tail of glycoprotein H. J Gen Virol *77(Pt 10)*:2569–2573. https://doi.org/10.1099/0022-1317-77-10-2569

Buchrieser J, Degrelle SA, Couderc T, Nevers Q, Disson O, Manet C, Donahue DA, Porrot F, Hillion KH, Perthame E et al (2019) IFITM proteins inhibit placental syncytiotrophoblast formation and promote fetal demise. Science *365*:176–180. https://doi.org/10.1126/science.aaw7733

Buchrieser J, Dufloo J, Hubert M, Monel B, Planas D, Rajah MM, Planchais C, Porrot F, Guivel-Benhassine F, Van der Werf S et al (2020) Syncytia formation by SARS-CoV-2-infected cells. EMBO J *39*:e106267. https://doi.org/10.15252/embj.2020106267

Bussani R, Schneider E, Zentilin L, Collesi C, Ali H, Braga L, Volpe MC, Colliva A, Zanconati F, Berlot G et al (2020) Persistence of viral RNA, pneumocyte syncytia and thrombosis are hallmarks of advanced COVID-19 pathology. EBioMedicine *61*:103104. https://doi.org/10.1016/j.ebiom.2020.103104

Bzik DJ, Fox BA, DeLuca NA, Person S (1984) Nucleotide sequence of a region of the herpes simplex virus type 1 gB glycoprotein gene: mutations affecting rate of virus entry and cell fusion. Virology *137*:185–190. https://doi.org/10.1016/0042-6822(84)90022-9

Cai WH, Gu B, Person S (1988a) Role of glycoprotein B of herpes simplex virus type 1 in viral entry and cell fusion. J Virol *62*:2596–2604. https://doi.org/10.1128/JVI.62.8.2596-2604.1988

Cai WZ, Person S, DebRoy C, Gu BH (1988b) Functional regions and structural features of the gB glycoprotein of herpes simplex virus type 1. An analysis of linker insertion mutants. J Mol Biol *201*:575–588. https://doi.org/10.1016/0022-2836(88)90639-0

Cardozo T, Kimura T, Philpott S, Weiser B, Burger H, Zolla-Pazner S (2007) Structural basis for coreceptor selectivity by the HIV type 1 V3 loop. AIDS Res Hum Retrovir *23*:415–426. https://doi.org/10.1089/aid.2006.0130

Carfi A, Willis SH, Whitbeck JC, Krummenacher C, Cohen GH, Eisenberg RJ, Wiley DC (2001) Herpes simplex virus glycoprotein D bound to the human receptor HveA. Mol Cell *8*:169–179. https://doi.org/10.1016/s1097-2765(01)00298-2

Carmichael JC, Yokota H, Craven RC, Schmitt A, Wills JW (2018) The HSV-1 mechanisms of cell-to-cell spread and fusion are critically dependent on host PTP1B. PLoS Pathog *14*:e1007054. https://doi.org/10.1371/journal.ppat.1007054

Cassai E, Manservigi R, Corallini A, Terni M (1975) Plaque dissociation of herpes simplex viruses: biochemical and biological characters of the viral variants. Intervirology *6*:212–223. https://doi.org/10.1159/000149476

Castedo M, Roumier T, Blanco J, Ferri KF, Barretina J, Tintignac LA, Andreau K, Perfettini JL, Amendola A, Nardacci R et al (2002) Sequential involvement of Cdk1, mTOR and p53 in apoptosis induced by the HIV-1 envelope. EMBO J *21*:4070–4080. https://doi.org/10.1093/emboj/cdf391

Cathomen T, Naim HY, Cattaneo R (1998) Measles viruses with altered envelope protein cytoplasmic tails gain cell fusion competence. J Virol *72*:1224–1234. https://doi.org/10.1128/JVI.72.2.1224-1234.1998

Cattin-Ortola J, Welch LG, Maslen SL, Papa G, James LC, Munro S (2021) Sequences in the cytoplasmic tail of SARS-CoV-2 Spike facilitate expression at the cell surface and syncytia formation. Nat Commun *12*:5333. https://doi.org/10.1038/s41467-021-25589-1

Ceccaldi PE, Delebecque F, Prevost MC, Moris A, Abastado JP, Gessain A, Schwartz O, Ozden S (2006) DC-SIGN facilitates fusion of dendritic cells with human T-cell leukemia virus type 1-infected cells. J Virol *80*:4771–4780. https://doi.org/10.1128/JVI.80.10.4771-4780.2006

Chan KMC, Son S, Schmid EM, Fletcher DA (2020) A viral fusogen hijacks the actin cytoskeleton to drive cell-cell fusion. elife *9*:e51358. https://doi.org/10.7554/eLife.51358

Chan KMC, Arthur AL, Morstein J, Jin M, Bhat A, Schlesinger D, Son S, Stevens DA, Drubin DG, Fletcher DA (2021) Evolutionarily related small viral fusogens hijack distinct but modular actin nucleation pathways to drive cell-cell fusion. Proc Natl Acad Sci U S A *118*:e2007526118. https://doi.org/10.1073/pnas.2007526118

Chatterjee S, Koga J, Whitley RJ (1989) A role for herpes simplex virus type 1 glycoprotein E in induction of cell fusion. J Gen Virol *70*(*Pt 8*):2157–2162. https://doi.org/10.1099/0022-1317-70-8-2157

Chen BK (2012) T cell virological synapses and HIV-1 pathogenesis. Immunol Res *54*:133–139. https://doi.org/10.1007/s12026-012-8320-8

Chen EH, Olson EN (2005) Unveiling the mechanisms of cell-cell fusion. Science *308*:369–373. https://doi.org/10.1126/science.1104799

Chen SS, Lee SF, Hao HJ, Chuang CK (1998) Mutations in the leucine zipper-like heptad repeat sequence of human immunodeficiency virus type 1 gp41 dominantly interfere with wild-type virus infectivity. J Virol *72*:4765–4774. https://doi.org/10.1128/JVI.72.6.4765-4774.1998

Chen P, Hubner W, Spinelli MA, Chen BK (2007) Predominant mode of human immunodeficiency virus transfer between T cells is mediated by sustained Env-dependent neutralization-resistant virological synapses. J Virol *81*:12582–12595. https://doi.org/10.1128/JVI.00381-07

Chen A, Leikina E, Melikov K, Podbilewicz B, Kozlov MM, Chernomordik LV (2008) Fusion-pore expansion during syncytium formation is restricted by an actin network. J Cell Sci *121*:3619–3628. https://doi.org/10.1242/jcs.032169

Chen HY, Huang C, Tian L, Huang X, Zhang C, Llewellyn GN, Rogers GL, Andresen K, O'Gorman MRG, Chen YW, Cannon PM (2021) Cytoplasmic tail truncation of SARS-CoV-2 spike protein enhances titer of pseudotyped vectors but masks the effect of the D614G mutation. J Virol *95*:e0096621. https://doi.org/10.1128/JVI.00966-21

Chesnokova LS, Hutt-Fletcher LM (2011) Fusion of Epstein-Barr virus with epithelial cells can be triggered by alphavbeta5 in addition to alphavbeta6 and alphavbeta8, and integrin binding triggers a conformational change in glycoproteins gHgL. J Virol *85*:13214–13223. https://doi.org/10.1128/JVI.05580-11

Chesnokova LS, Nishimura SL, Hutt-Fletcher LM (2009) Fusion of epithelial cells by Epstein-Barr virus proteins is triggered by binding of viral glycoproteins gHgL to integrins alphavbeta6 or alphavbeta8. Proc Natl Acad Sci U S A *106*:20464–20469. https://doi.org/10.1073/pnas.0907508106

Chouljenko VN, Iyer AV, Chowdhury S, Kim J, Kousoulas KG (2010) The herpes simplex virus type 1 UL20 protein and the amino terminus of glycoprotein K (gK) physically interact with gB. J Virol *84*:8596–8606. https://doi.org/10.1128/JVI.00298-10

Chouljenko DV, Kim IJ, Chouljenko VN, Subramanian R, Walker JD, Kousoulas KG (2012) Functional hierarchy of herpes simplex virus 1 viral glycoproteins in cytoplasmic virion envelopment and egress. J Virol *86*:4262–4270. https://doi.org/10.1128/JVI.06766-11

Chuong EB, Elde NC, Feschotte C (2016) Regulatory evolution of innate immunity through co-option of endogenous retroviruses. Science *351*:1083–1087. https://doi.org/10.1126/science.aad5497

Chuong EB, Elde NC, Feschotte C (2017) Regulatory activities of transposable elements: from conflicts to benefits. Nat Rev Genet *18*:71–86. https://doi.org/10.1038/nrg.2016.139
Chuprin A, Gal H, Biron-Shental T, Biran A, Amiel A, Rozenblatt S, Krizhanovsky V (2013) Cell fusion induced by ERVWE1 or measles virus causes cellular senescence. Genes Dev *27*:2356–2366. https://doi.org/10.1101/gad.227512.113
Ciechonska M, Duncan R (2014) Reovirus FAST proteins: virus-encoded cellular fusogens. Trends Microbiol *22*:715–724. https://doi.org/10.1016/j.tim.2014.08.005
Ciferri C, Chandramouli S, Donnarumma D, Nikitin PA, Cianfrocco MA, Gerrein R, Feire AL, Barnett SW, Lilja AE, Rappuoli R et al (2015) Structural and biochemical studies of HCMV gH/gL/gO and Pentamer reveal mutually exclusive cell entry complexes. Proc Natl Acad Sci U S A *112*:1767–1772. https://doi.org/10.1073/pnas.1424818112
Cifuentes-Munoz N, Dutch RE, Cattaneo R (2018) Direct cell-to-cell transmission of respiratory viruses: The fast lanes. PLoS Pathog *14*:e1007015. https://doi.org/10.1371/journal.ppat.1007015
Clapham PR, McKnight A (2001) HIV-1 receptors and cell tropism. Br Med Bull *58*:43–59. https://doi.org/10.1093/bmb/58.1.43
Cole NL, Grose C (2003) Membrane fusion mediated by herpesvirus glycoproteins: the paradigm of varicella-zoster virus. Rev Med Virol *13*:207–222. https://doi.org/10.1002/rmv.377
Compton T (2004) Receptors and immune sensors: the complex entry path of human cytomegalovirus. Trends Cell Biol *14*:5–8. https://doi.org/10.1016/j.tcb.2003.10.009
Compton AA, Schwartz O (2017) They might be giants: does syncytium formation sink or spread HIV infection? PLoS Pathog *13*:e1006099. https://doi.org/10.1371/journal.ppat.1006099
Connolly SA, Jackson JO, Jardetzky TS, Longnecker R (2011) Fusing structure and function: a structural view of the herpesvirus entry machinery. Nat Rev Microbiol *9*:369–381. https://doi.org/10.1038/nrmicro2548
Corcoran JA, Duncan R (2004) Reptilian reovirus utilizes a small type III protein with an external myristylated amino terminus to mediate cell-cell fusion. J Virol *78*:4342–4351. https://doi.org/10.1128/jvi.78.8.4342-4351.2004
Corcoran JA, Syvitski R, Top D, Epand RM, Epand RF, Jakeman D, Duncan R (2004) Myristoylation, a protruding loop, and structural plasticity are essential features of a nonenveloped virus fusion peptide motif. J Biol Chem *279*:51386–51394. https://doi.org/10.1074/jbc.M406990200
Crump CM, Bruun B, Bell S, Pomeranz LE, Minson T, Browne HM (2004) Alphaherpesvirus glycoprotein M causes the relocalization of plasma membrane proteins. J Gen Virol *85*:3517–3527. https://doi.org/10.1099/vir.0.80361-0
Csoma E, Deli T, Konya J, Csernoch L, Beck Z, Gergely L (2006) Human herpesvirus 6A decreases the susceptibility of macrophages to R5 variants of human immunodeficiency virus 1: possible role of RANTES and IL-8. Virus Res *121*:161–168. https://doi.org/10.1016/j.virusres.2006.05.007
Dargan DJ, Douglas E, Cunningham C, Jamieson F, Stanton RJ, Baluchova K, McSharry BP, Tomasec P, Emery VC, Percivalle E et al (2010) Sequential mutations associated with adaptation of human cytomegalovirus to growth in cell culture. J Gen Virol *91*:1535–1546. https://doi.org/10.1099/vir.0.018994-0
Davis-Poynter N, Bell S, Minson T, Browne H (1994) Analysis of the contributions of herpes simplex virus type 1 membrane proteins to the induction of cell-cell fusion. J Virol *68*:7586–7590. https://doi.org/10.1128/JVI.68.11.7586-7590.1994
Dawe S, Duncan R (2002) The S4 genome segment of baboon reovirus is bicistronic and encodes a novel fusion-associated small transmembrane protein. J Virol *76*:2131–2140. https://doi.org/10.1128/jvi.76.5.2131-2140.2002
Dawe S, Corcoran JA, Clancy EK, Salsman J, Duncan R (2005) Unusual topological arrangement of structural motifs in the baboon reovirus fusion-associated small transmembrane protein. J Virol *79*:6216–6226. https://doi.org/10.1128/JVI.79.10.6216-6226.2005

de Parseval N, Lazar V, Casella JF, Benit L, Heidmann T (2003) Survey of human genes of retroviral origin: identification and transcriptome of the genes with coding capacity for complete envelope proteins. J Virol *77*:10414–10422. https://doi.org/10.1128/jvi.77.19.10414-10422.2003
de Parseval N, Diop G, Blaise S, Helle F, Vasilescu A, Matsuda F, Heidmann T (2005) Comprehensive search for intra- and inter-specific sequence polymorphisms among coding envelope genes of retroviral origin found in the human genome: genes and pseudogenes. BMC Genomics *6*:117. https://doi.org/10.1186/1471-2164-6-117
Di Giovine P, Settembre EC, Bhargava AK, Luftig MA, Lou H, Cohen GH, Eisenberg RJ, Krummenacher C, Carfi A (2011) Structure of herpes simplex virus glycoprotein D bound to the human receptor nectin-1. PLoS Pathog *7*:e1002277. https://doi.org/10.1371/journal.ppat.1002277
Dimitrov DS, Willey RL, Sato H, Chang LJ, Blumenthal R, Martin MA (1993) Quantitation of human immunodeficiency virus type 1 infection kinetics. J Virol *67*:2182–2190
Dingwell KS, Brunetti CR, Hendricks RL, Tang Q, Tang M, Rainbow AJ, Johnson DC (1994) Herpes simplex virus glycoproteins E and I facilitate cell-to-cell spread in vivo and across junctions of cultured cells. J Virol *68*:834–845. https://doi.org/10.1128/JVI.68.2.834-845.1994
Doane F, Rhodes AJ, Ormsby HL (1955) Tissue culture techniques in the study of herpetic infections of the eye. Am J Ophthalmol *40*:189–193
Dominguez G, Dambaugh TR, Stamey FR, Dewhurst S, Inoue N, Pellett PE (1999) Human herpesvirus 6B genome sequence: coding content and comparison with human herpesvirus 6A. J Virol *73*:8040–8052. https://doi.org/10.1128/JVI.73.10.8040-8052.1999
Dubay JW, Roberts SJ, Brody B, Hunter E (1992a) Mutations in the leucine zipper of the human immunodeficiency virus type 1 transmembrane glycoprotein affect fusion and infectivity. J Virol *66*:4748–4756. https://doi.org/10.1128/JVI.66.8.4748-4756.1992
Dubay JW, Roberts SJ, Hahn BH, Hunter E (1992b) Truncation of the human immunodeficiency virus type 1 transmembrane glycoprotein cytoplasmic domain blocks virus infectivity. J Virol *66*:6616–6625. https://doi.org/10.1128/JVI.66.11.6616-6625.1992
Duelli D, Lazebnik Y (2007) Cell-to-cell fusion as a link between viruses and cancer. Nat Rev Cancer *7*:968–976. https://doi.org/10.1038/nrc2272
Duncan R (2019) Fusogenic reoviruses and their fusion-associated small ransmembrane (FAST) proteins. Annu Rev Virol *6*:341–363. https://doi.org/10.1146/annurev-virology-092818-015523
Dupressoir A, Marceau G, Vernochet C, Benit L, Kanellopoulos C, Sapin V, Heidmann T (2005) Syncytin-A and syncytin-B, two fusogenic placenta-specific murine envelope genes of retroviral origin conserved in Muridae. Proc Natl Acad Sci U S A *102*:725–730. https://doi.org/10.1073/pnas.0406509102
Dupressoir A, Vernochet C, Bawa O, Harper F, Pierron G, Opolon P, Heidmann T (2009) Syncytin-A knockout mice demonstrate the critical role in placentation of a fusogenic, endogenous retrovirus-derived, envelope gene. Proc Natl Acad Sci U S A *106*:12127–12132. https://doi.org/10.1073/pnas.0902925106
Duprex WP, McQuaid S, Hangartner L, Billeter MA, Rima BK (1999) Observation of measles virus cell-to-cell spread in astrocytoma cells by using a green fluorescent protein-expressing recombinant virus. J Virol *73*:9568–9575. https://doi.org/10.1128/JVI.73.11.9568-9575.1999
Ejercito PM, Kieff ED, Roizman B (1968) Characterization of herpes simplex virus strains differing in their effects on social behaviour of infected cells. J Gen Virol *2*:357–364. https://doi.org/10.1099/0022-1317-2-3-357
El Kasmi I, Lippe R (2015) Herpes simplex virus 1 gN partners with gM to modulate the viral fusion machinery. J Virol *89*:2313–2323. https://doi.org/10.1128/JVI.03041-14
Engel JP, Boyer EP, Goodman JL (1993) Two novel single amino acid syncytial mutations in the carboxy terminus of glycoprotein B of herpes simplex virus type 1 confer a unique pathogenic phenotype. Virology *192*:112–120. https://doi.org/10.1006/viro.1993.1013
Esnault C, Priet S, Ribet D, Vernochet C, Bruls T, Lavialle C, Weissenbach J, Heidmann T (2008) A placenta-specific receptor for the fusogenic, endogenous retrovirus-derived, human syncytin-2. Proc Natl Acad Sci U S A *105*:17532–17537. https://doi.org/10.1073/pnas.0807413105

Fan Q, Longnecker R, Connolly SA (2014) Substitution of herpes simplex virus 1 entry glycoproteins with those of saimiriine herpesvirus 1 reveals a gD-gH/gL functional interaction and a region within the gD profusion domain that is critical for fusion. J Virol *88*:6470–6482. https://doi.org/10.1128/JVI.00465-14

Fan Q, Longnecker R, Connolly SA (2015) A Functional Interaction between Herpes Simplex Virus 1 Glycoprotein gH/gL Domains I and II and gD Is Defined by Using Alphaherpesvirus gH and gL Chimeras. J Virol *89*:7159–7169. https://doi.org/10.1128/JVI.00740-15

Farhatullah S, Kaza S, Athmanathan S, Garg P, Reddy SB, Sharma S (2004) Diagnosis of herpes simplex virus-1 keratitis using Giemsa stain, immunofluorescence assay, and polymerase chain reaction assay on corneal scrapings. Br J Ophthalmol *88*:142–144. https://doi.org/10.1136/bjo.88.1.142

Feng H, Yi R, Wu S, Wang G, Sun R, Lin L, Zhu S, Nie Z, He Y, Wang S et al (2022) KAP1 positively modulates influenza A virus replication by interacting with PB2 and NS1 proteins in human lung epithelial cells. Viruses *14*:689. https://doi.org/10.3390/v14040689

Foster TP, Alvarez X, Kousoulas KG (2003) Plasma membrane topology of syncytial domains of herpes simplex virus type 1 glycoprotein K (gK): the UL20 protein enables cell surface localization of gK but not gK-mediated cell-to-cell fusion. J Virol *77*:499–510. https://doi.org/10.1128/jvi.77.1.499-510.2003

Foster TP, Melancon JM, Baines JD, Kousoulas KG (2004) The herpes simplex virus type 1 UL20 protein modulates membrane fusion events during cytoplasmic virion morphogenesis and virus-induced cell fusion. J Virol *78*:5347–5357. https://doi.org/10.1128/jvi.78.10.5347-5357.2004

Frank JA, Singh M, Cullen HB, Kirou RA, Benkaddour-Boumzaouad M, Cortes JL, Garcia Perez J, Coyne CB, Feschotte C (2022) Evolution and antiviral activity of a human protein of retroviral origin. Science *378*:422–428. https://doi.org/10.1126/science.abq7871

Frankel SS, Wenig BM, Burke AP, Mannan P, Thompson LD, Abbondanzo SL, Nelson AM, Pope M, Steinman RM (1996) Replication of HIV-1 in dendritic cell-derived syncytia at the mucosal surface of the adenoid. Science *272*:115–117. https://doi.org/10.1126/science.272.5258.115

Franks TJ, Chong PY, Chui P, Galvin JR, Lourens RM, Reid AH, Selbs E, McEvoy CP, Hayden CD, Fukuoka J et al (2003) Lung pathology of severe acute respiratory syndrome (SARS): a study of 8 autopsy cases from Singapore. Hum Pathol *34*:743–748. https://doi.org/10.1016/s0046-8177(03)00367-8

Fugier-Vivier I, Servet-Delprat C, Rivailler P, Rissoan MC, Liu YJ, Rabourdin-Combe C (1997) Measles virus suppresses cell-mediated immunity by interfering with the survival and functions of dendritic and T cells. J Exp Med *186*:813–823. https://doi.org/10.1084/jem.186.6.813

Gage PJ, Levine M, Glorioso JC (1993) Syncytium-inducing mutations localize to two discrete regions within the cytoplasmic domain of herpes simplex virus type 1 glycoprotein B. J Virol *67*:2191–2201. https://doi.org/10.1128/JVI.67.4.2191-2201.1993

Galitska G, Biolatti M, De Andrea M, Leone A, Coscia A, Bertolotti L, Ala U, Bertino E, Dell'Oste V, Landolfo S (2018) Biological relevance of Cytomegalovirus genetic variability in congenitally and postnatally infected children. J Clin Virol *108*:132–140. https://doi.org/10.1016/j.jcv.2018.09.019

Gerdts V, Beyer J, Lomniczi B, Mettenleiter TC (2000) Pseudorabies virus expressing bovine herpesvirus 1 glycoprotein B exhibits altered neurotropism and increased neurovirulence. J Virol *74*:817–827. https://doi.org/10.1128/jvi.74.2.817-827.2000

Gerna G, Baldanti F, Revello MG (2004) Pathogenesis of human cytomegalovirus infection and cellular targets. Hum Immunol *65*:381–386. https://doi.org/10.1016/j.humimm.2004.02.009

Gerna G, Percivalle E, Perez L, Lanzavecchia A, Lilleri D (2016) Monoclonal antibodies to different components of the human cytomegalovirus (HCMV) Pentamer gH/gL/pUL128L and Trimer gH/gL/gO as well as antibodies elicited during primary HCMV infection prevent epithelial cell syncytium formation. J Virol *90*:6216–6223. https://doi.org/10.1128/JVI.00121-16

Gimenez J, Montgiraud C, Pichon JP, Bonnaud B, Arsac M, Ruel K, Bouton O, Mallet F (2010) Custom human endogenous retroviruses dedicated microarray identifies self-induced HERV-W family elements reactivated in testicular cancer upon methylation control. Nucleic Acids Res *38*:2229–2246. https://doi.org/10.1093/nar/gkp1214

Gompels UA, Nicholas J, Lawrence G, Jones M, Thomson BJ, Martin ME, Efstathiou S, Craxton M, Macaulay HA (1995) The DNA sequence of human herpesvirus-6: structure, coding content, and genome evolution. Virology *209*:29–51. https://doi.org/10.1006/viro.1995.1228

Granelli-Piperno A, Pope M, Inaba K, Steinman RM (1995) Coexpression of NF-kappa B/Rel and Sp1 transcription factors in human immunodeficiency virus 1-induced, dendritic cell-T-cell syncytia. Proc Natl Acad Sci U S A *92*:10944–10948. https://doi.org/10.1073/pnas.92.24.10944

Gray A, Tokumaru T, Scott TF (1958) Different cytopathogenic effects observed in HeLa cells infected with herpes simplex virus. Arch Gesamte Virusforsch *8*:59–76. https://doi.org/10.1007/BF01242313

Greenwood AD, Ishida Y, O'Brien SP, Roca AL, Eiden MV (2018) Transmission, evolution, and endogenization: lessons learned from recent retroviral invasions. Microbiol Mol Biol Rev *82*:e00044-17. https://doi.org/10.1128/MMBR.00044-17

Griffiths DJ (2001) Endogenous retroviruses in the human genome sequence. Genome Biol *2*:REVIEWS1017. https://doi.org/10.1186/gb-2001-2-6-reviews1017

Grivel JC, Ito Y, Faga G, Santoro F, Shaheen F, Malnati MS, Fitzgerald W, Lusso P, Margolis L (2001) Suppression of CCR5- but not CXCR4-tropic HIV-1 in lymphoid tissue by human herpesvirus 6. Nat Med *7*:1232–1235. https://doi.org/10.1038/nm1101-1232

Haanes EJ, Nelson CM, Soule CL, Goodman JL (1994) The UL45 gene product is required for herpes simplex virus type 1 glycoprotein B-induced fusion. J Virol *68*:5825–5834. https://doi.org/10.1128/JVI.68.9.5825-5834.1994

Hahn AS, Kaufmann JK, Wies E, Naschberger E, Panteleev-Ivlev J, Schmidt K, Holzer A, Schmidt M, Chen J, Konig S et al (2012) The ephrin receptor tyrosine kinase A2 is a cellular receptor for Kaposi's sarcoma-associated herpesvirus. Nat Med *18*:961–966. https://doi.org/10.1038/nm.2805

Han J, Chadha P, Starkey JL, Wills JW (2012) Function of glycoprotein E of herpes simplex virus requires coordinated assembly of three tegument proteins on its cytoplasmic tail. Proc Natl Acad Sci U S A *109*:19798–19803. https://doi.org/10.1073/pnas.1212900109

Han M, Cantaloube-Ferrieu V, Xie M, Armani-Tourret M, Woottum M, Pages JC, Colin P, Lagane B, Benichou S (2022) HIV-1 cell-to-cell spread overcomes the virus entry block of non-macrophage-tropic strains in macrophages. PLoS Pathog *18*:e1010335. https://doi.org/10.1371/journal.ppat.1010335

Harbison C, Zhuang K, Gettie A, Blanchard J, Knight H, Didier P, Cheng-Mayer C, Westmoreland S (2014) Giant cell encephalitis and microglial infection with mucosally transmitted simian-human immunodeficiency virus SHIVSF162P3N in rhesus macaques. J Neurovirol *20*:62–72. https://doi.org/10.1007/s13365-013-0229-z

Harrison SC (2008) Viral membrane fusion. Nat Struct Mol Biol *15*:690–698. https://doi.org/10.1038/nsmb.1456

Hartley O, Klasse PJ, Sattentau QJ, Moore JP (2005) V3: HIV's switch-hitter. AIDS Res Hum Retrovir *21*:171–189. https://doi.org/10.1089/aid.2005.21.171

Harvey WT, Carabelli AM, Jackson B, Gupta RK, Thomson EC, Harrison EM, Ludden C, Reeve R, Rambaut A, Consortium C-GU et al (2021) SARS-CoV-2 variants, spike mutations and immune escape. Nat Rev Microbiol *19*:409–424. https://doi.org/10.1038/s41579-021-00573-0

Havranek KE, Jimenez AR, Acciani MD, Lay Mendoza MF, Reyes Ballista JM, Diaz DA, Brindley MA (2020) SARS-CoV-2 spike alterations enhance pseudoparticle titers and replication-competent VSV-SARS-CoV-2 virus. Viruses *12*:1465. https://doi.org/10.3390/v12121465

He J, McCarthy M, Zhou Y, Chandran B, Wood C (1996) Infection of primary human fetal astrocytes by human herpesvirus 6. J Virol *70*:1296–1300. https://doi.org/10.1128/JVI.70.2.1296-1300.1996

Heldwein EE (2016) gH/gL supercomplexes at early stages of herpesvirus entry. Curr Opin Virol *18*:1–8. https://doi.org/10.1016/j.coviro.2016.01.010

Heldwein EE, Lou H, Bender FC, Cohen GH, Eisenberg RJ, Harrison SC (2006) Crystal structure of glycoprotein B from herpes simplex virus 1. Science *313*:217–220. https://doi.org/10.1126/science.1126548

Herschke F, Plumet S, Duhen T, Azocar O, Druelle J, Laine D, Wild TF, Rabourdin-Combe C, Gerlier D, Valentin H (2007) Cell-cell fusion induced by measles virus amplifies the type I interferon response. J Virol *81*:12859–12871. https://doi.org/10.1128/JVI.00078-07

Hoffman TL, Doms RW (1999) HIV-1 envelope determinants for cell tropism and chemokine receptor use. Mol Membr Biol *16*:57–65

Hoggan MD, Roizman B (1959) The isolation and properties of a variant of Herpes simplex producing multinucleated giant cells in monolayer cultures in the presence of antibody. Am J Hyg *70*:208–219. https://doi.org/10.1093/oxfordjournals.aje.a120071

Hou YJ, Chiba S, Halfmann P, Ehre C, Kuroda M, Dinnon KH 3rd, Leist SR, Schafer A, Nakajima N, Takahashi K et al (2020) SARS-CoV-2 D614G variant exhibits efficient replication ex vivo and transmission in vivo. Science *370*:1464–1468. https://doi.org/10.1126/science.abe8499

Huang C, Wang Y, Li X, Ren L, Zhao J, Hu Y, Zhang L, Fan G, Xu J, Gu X et al (2020) Clinical features of patients infected with 2019 novel coronavirus in Wuhan, China. Lancet *395*:497–506. https://doi.org/10.1016/S0140-6736(20)30183-5

Hubner W, McNerney GP, Chen P, Dale BM, Gordon RE, Chuang FY, Li XD, Asmuth DM, Huser T, Chen BK (2009) Quantitative 3D video microscopy of HIV transfer across T cell virological synapses. Science *323*:1743–1747. https://doi.org/10.1126/science.1167525

Hutt-Fletcher LM (2007) Epstein-Barr virus entry. J Virol *81*:7825–7832. https://doi.org/10.1128/JVI.00445-07

Jackson CB, Farzan M, Chen B, Choe H (2022) Mechanisms of SARS-CoV-2 entry into cells. Nat Rev Mol Cell Biol *23*:3–20. https://doi.org/10.1038/s41580-021-00418-x

Januszeski MM, Cannon PM, Chen D, Rozenberg Y, Anderson WF (1997) Functional analysis of the cytoplasmic tail of Moloney murine leukemia virus envelope protein. J Virol *71*:3613–3619. https://doi.org/10.1128/JVI.71.5.3613-3619.1997

Jasirwan C, Furusawa Y, Tang H, Maeki T, Mori Y (2014) Human herpesvirus-6A gQ1 and gQ2 are critical for human CD46 usage. Microbiol Immunol *58*:22–30. https://doi.org/10.1111/1348-0421.12110

Jiang X, Zhang Z, Wang C, Ren H, Gao L, Peng H, Niu Z, Ren H, Huang H, Sun Q (2020) Bimodular effects of D614G mutation on the spike glycoprotein of SARS-CoV-2 enhance protein processing, membrane fusion, and viral infectivity. Signal Transduct Target Ther *5*:268. https://doi.org/10.1038/s41392-020-00392-4

Johnston PB, Dubay JW, Hunter E (1993) Truncations of the simian immunodeficiency virus transmembrane protein confer expanded virus host range by removing a block to virus entry into cells. J Virol *67*:3077–3086. https://doi.org/10.1128/JVI.67.6.3077-3086.1993

Jolly C, Sattentau QJ (2004) Retroviral spread by induction of virological synapses. Traffic *5*:643–650. https://doi.org/10.1111/j.1600-0854.2004.00209.x

Jolly C, Kashefi K, Hollinshead M, Sattentau QJ (2004) HIV-1 cell to cell transfer across an Env-induced, actin-dependent synapse. J Exp Med *199*:283–293. https://doi.org/10.1084/jem.20030648

Jolly C, Mitar I, Sattentau QJ (2007) Adhesion molecule interactions facilitate human immunodeficiency virus type 1-induced virological synapse formation between T cells. J Virol *81*:13916–13921. https://doi.org/10.1128/JVI.01585-07

Kanai Y, Kawagishi T, Sakai Y, Nouda R, Shimojima M, Saijo M, Matsuura Y, Kobayashi T (2019) Cell-cell fusion induced by reovirus FAST proteins enhances replication and pathogenicity of non-enveloped dsRNA viruses. PLoS Pathog *15*:e1007675. https://doi.org/10.1371/journal.ppat.1007675

Kielian M (2006) Class II virus membrane fusion proteins. Virology *344*:38–47. https://doi.org/10.1016/j.virol.2005.09.036

Kim IJ, Chouljenko VN, Walker JD, Kousoulas KG (2013) Herpes simplex virus 1 glycoprotein M and the membrane-associated protein UL11 are required for virus-induced cell fusion and efficient virus entry. J Virol *87*:8029–8037. https://doi.org/10.1128/JVI.01181-13

Kirschner AN, Omerovic J, Popov B, Longnecker R, Jardetzky TS (2006) Soluble Epstein-Barr virus glycoproteins gH, gL, and gp42 form a 1:1:1 stable complex that acts like soluble gp42 in B-cell fusion but not in epithelial cell fusion. J Virol *80*:9444–9454. https://doi.org/10.1128/JVI.00572-06

Kitsou K, Lagiou P, Magiorkinis G (2022) Human endogenous retroviruses in cancer: oncogenesis mechanisms and clinical implications. J Med Virol 95:e28350. https://doi.org/10.1002/jmv.28350

Koenig S, Gendelman HE, Orenstein JM, Dal Canto MC, Pezeshkpour GH, Yungbluth M, Janotta F, Aksamit A, Martin MA, Fauci AS (1986) Detection of AIDS virus in macrophages in brain tissue from AIDS patients with encephalopathy. Science *233*:1089–1093

LaBranche CC, Sauter MM, Haggarty BS, Vance PJ, Romano J, Hart TK, Bugelski PJ, Marsh M, Hoxie JA (1995) A single amino acid change in the cytoplasmic domain of the simian immunodeficiency virus transmembrane molecule increases envelope glycoprotein expression on infected cells. J Virol *69*:5217–5227. https://doi.org/10.1128/JVI.69.9.5217-5227.1995

Lan J, Ge J, Yu J, Shan S, Zhou H, Fan S, Zhang Q, Shi X, Wang Q, Zhang L, Wang X (2020) Structure of the SARS-CoV-2 spike receptor-binding domain bound to the ACE2 receptor. Nature *581*:215–220. https://doi.org/10.1038/s41586-020-2180-5

Larsson LI, Holck S, Christensen IJ (2007) Prognostic role of syncytin expression in breast cancer. Hum Pathol *38*:726–731. https://doi.org/10.1016/j.humpath.2006.10.018

Lavi E, Wang Q, Weiss SR, Gonatas NK (1996) Syncytia formation induced by coronavirus infection is associated with fragmentation and rearrangement of the Golgi apparatus. Virology *221*:325–334. https://doi.org/10.1006/viro.1996.0382

Law KM, Komarova NL, Yewdall AW, Lee RK, Herrera OL, Wodarz D, Chen BK (2016) In vivo HIV-1 cell-to-cell transmission promotes multicopy micro-compartmentalized infection. Cell Rep *15*:2771–2783. https://doi.org/10.1016/j.celrep.2016.05.059

Lee GT, Spear PG (1980) Viral and cellular factors that influence cell fusion induced by herpes simplex virus. Virology *107*:402–414. https://doi.org/10.1016/0042-6822(80)90307-4

Lefkowitz EJ, Dempsey DM, Hendrickson RC, Orton RJ, Siddell SG, Smith DB (2018) Virus taxonomy: the database of the International Committee on Taxonomy of Viruses (ICTV). Nucleic Acids Res *46*:D708–D717. https://doi.org/10.1093/nar/gkx932

Leroy H, Han M, Woottum M, Bracq L, Bouchet J, Xie M, Benichou S (2020) Virus-mediated cell-cell fusion. Int J Mol Sci *21*:9644. https://doi.org/10.3390/ijms21249644

Li Q, Spriggs MK, Kovats S, Turk SM, Comeau MR, Nepom B, Hutt-Fletcher LM (1997) Epstein-Barr virus uses HLA class II as a cofactor for infection of B lymphocytes. J Virol *71*:4657–4662. https://doi.org/10.1128/JVI.71.6.4657-4662.1997

Li W, Moore MJ, Vasilieva N, Sui J, Wong SK, Berne MA, Somasundaran M, Sullivan JL, Luzuriaga K, Greenough TC et al (2003) Angiotensin-converting enzyme 2 is a functional receptor for the SARS coronavirus. Nature *426*:450–454. https://doi.org/10.1038/nature02145

Li D, Liu Y, Lu Y, Gao S, Zhang L (2022) Palmitoylation of SARS-CoV-2 S protein is critical for S-mediated syncytia formation and virus entry. J Med Virol *94*:342–348. https://doi.org/10.1002/jmv.27339

Lifson JD, Feinberg MB, Reyes GR, Rabin L, Banapour B, Chakrabarti S, Moss B, Wong-Staal F, Steimer KS, Engleman EG (1986) Induction of CD4-dependent cell fusion by the HTLV-III/LAV envelope glycoprotein. Nature *323*:725–728. https://doi.org/10.1038/323725a0

Liu J, Zheng X, Tong Q, Li W, Wang B, Sutter K, Trilling M, Lu M, Dittmer U, Yang D (2020) Overlapping and discrete aspects of the pathology and pathogenesis of the emerging human pathogenic coronaviruses SARS-CoV, MERS-CoV, and 2019-nCoV. J Med Virol *92*:491–494. https://doi.org/10.1002/jmv.25709

Liu X, Wei L, Xu F, Zhao F, Huang Y, Fan Z, Mei S, Hu Y, Zhai L, Guo J et al (2022) SARS-CoV-2 spike protein-induced cell fusion activates the cGAS-STING pathway and the interferon response. Sci Signal *15*:eabg8744. https://doi.org/10.1126/scisignal.abg8744

Liu X, Liu Z, Wu Z, Ren J, Fan Y, Sun L, Cao G, Niu Y, Zhang B, Ji Q et al (2023) Resurrection of endogenous retroviruses during aging reinforces senescence. Cell 186(2):287–304.e26. https://doi.org/10.1016/j.cell.2022.12.017

Lu R, Zhao X, Li J, Niu P, Yang B, Wu H, Wang W, Song H, Huang B, Zhu N et al (2020) Genomic characterisation and epidemiology of 2019 novel coronavirus: implications for virus origins and receptor binding. Lancet *395*:565–574. https://doi.org/10.1016/S0140-6736(20)30251-8

Luo Z, Weiss SR (1998) Roles in cell-to-cell fusion of two conserved hydrophobic regions in the murine coronavirus spike protein. Virology *244*:483–494. https://doi.org/10.1006/viro.1998.9121

Luo Z, Matthews AM, Weiss SR (1999) Amino acid substitutions within the leucine zipper domain of the murine coronavirus spike protein cause defects in oligomerization and the ability to induce cell-to-cell fusion. J Virol *73*:8152–8159. https://doi.org/10.1128/JVI.73.10.8152-8159.1999

Malassine A, Handschuh K, Tsatsaris V, Gerbaud P, Cheynet V, Oriol G, Mallet F, Evain-Brion D (2005) Expression of HERV-W Env glycoprotein (syncytin) in the extravillous trophoblast of first trimester human placenta. Placenta *26*:556–562. https://doi.org/10.1016/j.placenta.2004.09.002

Marechal V, Prevost MC, Petit C, Perret E, Heard JM, Schwartz O (2001) Human immunodeficiency virus type 1 entry into macrophages mediated by macropinocytosis. J Virol *75*:11166–11177. https://doi.org/10.1128/JVI.75.22.11166-11177.2001

McBride CE, Li J, Machamer CE (2007) The cytoplasmic tail of the severe acute respiratory syndrome coronavirus spike protein contains a novel endoplasmic reticulum retrieval signal that binds COPI and promotes interaction with membrane protein. J Virol *81*:2418–2428. https://doi.org/10.1128/JVI.02146-06

McChesney MB, Miller CJ, Rota PA, Zhu YD, Antipa L, Lerche NW, Ahmed R, Bellini WJ (1997) Experimental measles. I. Pathogenesis in the normal and the immunized host. Virology *233*:74–84. https://doi.org/10.1006/viro.1997.8576

Melikyan GB, Markosyan RM, Brener SA, Rozenberg Y, Cohen FS (2000) Role of the cytoplasmic tail of ecotropic moloney murine leukemia virus Env protein in fusion pore formation. J Virol *74*:447–455. https://doi.org/10.1128/jvi.74.1.447-455.2000

Meng B, Kemp SA, Papa G, Datir R, Ferreira I, Marelli S, Harvey WT, Lytras S, Mohamed A, Gallo G et al (2021) Recurrent emergence of SARS-CoV-2 spike deletion H69/V70 and its role in the Alpha variant B.1.1.7. Cell Rep *35*:109292. https://doi.org/10.1016/j.celrep.2021.109292

Mi S, Lee X, Li X, Veldman GM, Finnerty H, Racie L, LaVallie E, Tang XY, Edouard P, Howes S et al (2000) Syncytin is a captive retroviral envelope protein involved in human placental morphogenesis. Nature *403*:785–789. https://doi.org/10.1038/35001608

Mizusawa H, Hirano A, Llena JF, Kato T (1987) Nuclear bridges in multinucleated giant cells associated with primary lymphoma of the brain in acquired immune deficiency syndrome (AIDS). Acta Neuropathol *75*:23–26. https://doi.org/10.1007/BF00686788

Mlcochova P, Kemp SA, Dhar MS, Papa G, Meng B, Ferreira I, Datir R, Collier DA, Albecka A, Singh S et al (2021) SARS-CoV-2 B.1.617.2 Delta variant replication and immune evasion. Nature *599*:114–119. https://doi.org/10.1038/s41586-021-03944-y

Moll M, Klenk HD, Maisner A (2002) Importance of the cytoplasmic tails of the measles virus glycoproteins for fusogenic activity and the generation of recombinant measles viruses. J Virol *76*:7174–7186. https://doi.org/10.1128/jvi.76.14.7174-7186.2002

Moller AM, Delaisse JM, Soe K (2017) Osteoclast fusion: time-lapse reveals involvement of CD47 and syncytin-1 at different stages of nuclearity. J Cell Physiol *232*:1396–1403. https://doi.org/10.1002/jcp.25633

Montgomery RI, Warner MS, Lum BJ, Spear PG (1996) Herpes simplex virus-1 entry into cells mediated by a novel member of the TNF/NGF receptor family. Cell *87*:427–436. https://doi.org/10.1016/s0092-8674(00)81363-x

Mori Y, Seya T, Huang HL, Akkapaiboon P, Dhepakson P, Yamanishi K (2002) Human herpesvirus 6 variant A but not variant B induces fusion from without in a variety of human cells through a human herpesvirus 6 entry receptor, CD46. J Virol *76*:6750–6761. https://doi.org/10.1128/jvi.76.13.6750-6761.2002

Mori Y, Yang X, Akkapaiboon P, Okuno T, Yamanishi K (2003) Human herpesvirus 6 variant A glycoprotein H-glycoprotein L-glycoprotein Q complex associates with human CD46. J Virol *77*:4992–4999. https://doi.org/10.1128/jvi.77.8.4992-4999.2003

Mori Y, Akkapaiboon P, Yonemoto S, Koike M, Takemoto M, Sadaoka T, Sasamoto Y, Konishi S, Uchiyama Y, Yamanishi K (2004) Discovery of a second form of tripartite complex containing gH-gL of human herpesvirus 6 and observations on CD46. J Virol *78*:4609–4616. https://doi.org/10.1128/jvi.78.9.4609-4616.2004

Mothes W, Sherer NM, Jin J, Zhong P (2010) Virus cell-to-cell transmission. J Virol *84*:8360–8368. https://doi.org/10.1128/JVI.00443-10

Muggeridge MI, Grantham ML, Johnson FB (2004) Identification of syncytial mutations in a clinical isolate of herpes simplex virus 2. Virology *328*:244–253. https://doi.org/10.1016/j.virol.2004.07.027

Mullen MM, Haan KM, Longnecker R, Jardetzky TS (2002) Structure of the Epstein-Barr virus gp42 protein bound to the MHC class II receptor HLA-DR1. Mol Cell *9*:375–385. https://doi.org/10.1016/s1097-2765(02)00465-3

Murooka TT, Deruaz M, Marangoni F, Vrbanac VD, Seung E, von Andrian UH, Tager AM, Luster AD, Mempel TR (2012) HIV-infected T cells are migratory vehicles for viral dissemination. Nature *490*:283–287. https://doi.org/10.1038/nature11398

Murrell I, Tomasec P, Wilkie GS, Dargan DJ, Davison AJ, Stanton RJ (2013) Impact of sequence variation in the UL128 locus on production of human cytomegalovirus in fibroblast and epithelial cells. J Virol *87*:10489–10500. https://doi.org/10.1128/JVI.01546-13

Murrell I, Bedford C, Ladell K, Miners KL, Price DA, Tomasec P, Wilkinson GWG, Stanton RJ (2017) The pentameric complex drives immunologically covert cell-cell transmission of wild-type human cytomegalovirus. Proc Natl Acad Sci U S A *114*:6104–6109. https://doi.org/10.1073/pnas.1704809114

Navarro D, Paz P, Tugizov S, Topp K, La Vail J, Pereira L (1993) Glycoprotein B of human cytomegalovirus promotes virion penetration into cells, transmission of infection from cell to cell, and fusion of infected cells. Virology *197*:143–158. https://doi.org/10.1006/viro.1993.1575

Ng DL, Al Hosani F, Keating MK, Gerber SI, Jones TL, Metcalfe MG, Tong S, Tao Y, Alami NN, Haynes LM et al (2016) Clinicopathologic, immunohistochemical, and ultrastructural findings of a fatal case of middle east respiratory syndrome coronavirus infection in the United Arab Emirates, April 2014. Am J Pathol *186*:652–658. https://doi.org/10.1016/j.ajpath.2015.10.024

Nguyen CC, Kamil JP (2018) Pathogen at the gates: human cytomegalovirus entry and cell tropism. Viruses *10*:704. https://doi.org/10.3390/v10120704

Nicholls JM, Poon LL, Lee KC, Ng WF, Lai ST, Leung CY, Chu CM, Hui PK, Mak KL, Lim W et al (2003) Lung pathology of fatal severe acute respiratory syndrome. Lancet *361*:1773–1778. https://doi.org/10.1016/s0140-6736(03)13413-7

Nishimura M, Mori Y (2019) Entry of betaherpesviruses. Adv Virus Res *104*:283–312. https://doi.org/10.1016/bs.aivir.2019.05.005

Ogle BM, Cascalho M, Platt JL (2005) Biological implications of cell fusion. Nat Rev Mol Cell Biol *6*:567–575. https://doi.org/10.1038/nrm1678

Oliveira NM, Satija H, Kouwenhoven IA, Eiden MV (2007) Changes in viral protein function that accompany retroviral endogenization. Proc Natl Acad Sci U S A *104*:17506–17511. https://doi.org/10.1073/pnas.0704313104

Ozono S, Zhang Y, Ode H, Sano K, Tan TS, Imai K, Miyoshi K, Kishigami S, Ueno T, Iwatani Y et al (2021) SARS-CoV-2 D614G spike mutation increases entry efficiency with enhanced ACE2-binding affinity. Nat Commun *12*:848. https://doi.org/10.1038/s41467-021-21118-2

Paterson DA, Dyer AP, Milne RS, Sevilla-Reyes E, Gompels UA (2002) A role for human cytomegalovirus glycoprotein O (gO) in cell fusion and a new hypervariable locus. Virology *293*:281–294. https://doi.org/10.1006/viro.2001.1274

Pedro KD, Henderson AJ, Agosto LM (2019) Mechanisms of HIV-1 cell-to-cell transmission and the establishment of the latent reservoir. Virus Res *265*:115–121. https://doi.org/10.1016/j.virusres.2019.03.014

Peng X, Pan J, Gong R, Liu Y, Kang S, Feng H, Qiu G, Guo D, Tien P, Xiao G (2007) Functional characterization of syncytin-A, a newly murine endogenous virus envelope protein. Implication for its fusion mechanism. J Biol Chem *282*:381–389. https://doi.org/10.1074/jbc.M606353200

Peng Q, Zhou R, Liu N, Wang H, Xu H, Zhao M, Yang D, Au KK, Huang H, Liu L, Chen Z (2022) Naturally occurring spike mutations influence the infectivity and immunogenicity of SARS-CoV-2. Cell Mol Immunol *19*:1302–1310. https://doi.org/10.1038/s41423-022-00924-8

Petit CM, Chouljenko VN, Iyer A, Colgrove R, Farzan M, Knipe DM, Kousoulas KG (2007) Palmitoylation of the cysteine-rich endodomain of the SARS-coronavirus spike glycoprotein is important for spike-mediated cell fusion. Virology *360*:264–274. https://doi.org/10.1016/j.virol.2006.10.034

Phillips DM (1994) The role of cell-to-cell transmission in HIV infection. AIDS *8*:719–731. https://doi.org/10.1097/00002030-199406000-00001

Podbilewicz B (2014) Virus and cell fusion mechanisms. Annu Rev Cell Dev Biol *30*:111–139. https://doi.org/10.1146/annurev-cellbio-101512-122422

Pontow SE, Heyden NV, Wei S, Ratner L (2004) Actin cytoskeletal reorganizations and coreceptor-mediated activation of rac during human immunodeficiency virus-induced cell fusion. J Virol *78*:7138–7147. https://doi.org/10.1128/JVI.78.13.7138-7147.2004

Poste G (1970) Virus-induced polykaryocytosis and the mechanism of cell fusion. Adv Virus Res *16*:303–356. https://doi.org/10.1016/s0065-3527(08)60026-3

Poste G (1972) Mechanisms of virus-induced cell fusion. Int Rev Cytol *33*:157–252. https://doi.org/10.1016/s0074-7696(08)61451-5

Postler TS, Desrosiers RC (2013) The tale of the long tail: the cytoplasmic domain of HIV-1 gp41. J Virol *87*:2–15. https://doi.org/10.1128/JVI.02053-12

Pritt BS, Aubry MC (2017) Histopathology of viral infections of the lung. Semin Diagn Pathol *34*:510–517. https://doi.org/10.1053/j.semdp.2017.06.005

Rajah MM, Hubert M, Bishop E, Saunders N, Robinot R, Grzelak L, Planas D, Dufloo J, Gellenoncourt S, Bongers A et al (2021) SARS-CoV-2 alpha, beta, and delta variants display enhanced Spike-mediated syncytia formation. EMBO J *40*:e108944. https://doi.org/10.15252/embj.2021108944

Rajah MM, Bernier A, Buchrieser J, Schwartz O (2022) The mechanism and consequences of SARS-CoV-2 spike-mediated fusion and syncytia formation. J Mol Biol *434*:167280. https://doi.org/10.1016/j.jmb.2021.167280

Rauch DA, Rodriguez N, Roller RJ (2000) Mutations in herpes simplex virus glycoprotein D distinguish entry of free virus from cell-cell spread. J Virol *74*:11437–11446. https://doi.org/10.1128/jvi.74.24.11437-11446.2000

Read GS, Person S, Keller PM (1980) Genetic studies of cell fusion induced by herpes simplex virus type 1. J Virol *35*:105–113. https://doi.org/10.1128/JVI.35.1.105-113.1980

Read J, Clancy EK, Sarker M, de Antueno R, Langelaan DN, Parmar HB, Shin K, Rainey JK, Duncan R (2015) Reovirus FAST proteins drive pore formation and syncytiogenesis using a novel helix-loop-helix fusion-inducing lipid packing sensor. PLoS Pathog *11*:e1004962. https://doi.org/10.1371/journal.ppat.1004962

Ren Y, Bell S, Zenner HL, Lau SK, Crump CM (2012) Glycoprotein M is important for the efficient incorporation of glycoprotein H-L into herpes simplex virus type 1 particles. J Gen Virol *93*:319–329. https://doi.org/10.1099/vir.0.035444-0

Reuter N, Kropff B, Schneiderbanger JK, Alt M, Krawczyk A, Sinzger C, Winkler TH, Britt WJ, Mach M, Thomas M (2020) Cell fusion induced by a fusion-active form of human

cytomegalovirus glycoprotein B (gB) is inhibited by antibodies directed at antigenic domain 5 in the ectodomain of gB. J Virol *94*:e01276-20. https://doi.org/10.1128/JVI.01276-20
Ritter GD Jr, Mulligan MJ, Lydy SL, Compans RW (1993) Cell fusion activity of the simian immunodeficiency virus envelope protein is modulated by the intracytoplasmic domain. Virology *197*:255–264. https://doi.org/10.1006/viro.1993.1586
Rotola A, Di Luca D, Cassai E, Ricotta D, Giulio A, Turano A, Caruso A, Muneretto C (2000) Human herpesvirus 6 infects and replicates in aortic endothelium. J Clin Microbiol *38*:3135–3136. https://doi.org/10.1128/JCM.38.8.3135-3136.2000
Rowell JF, Stanhope PE, Siliciano RF (1995) Endocytosis of endogenously synthesized HIV-1 envelope protein. Mechanism and role in processing for association with class II MHC. J Immunol *155*:473–488
Rudnicka D, Feldmann J, Porrot F, Wietgrefe S, Guadagnini S, Prevost MC, Estaquier J, Haase AT, Sol-Foulon N, Schwartz O (2009) Simultaneous cell-to-cell transmission of human immunodeficiency virus to multiple targets through polysynapses. J Virol *83*:6234–6246. https://doi.org/10.1128/JVI.00282-09
Ryckman BJ, Jarvis MA, Drummond DD, Nelson JA, Johnson DC (2006) Human cytomegalovirus entry into epithelial and endothelial cells depends on genes UL128 to UL150 and occurs by endocytosis and low-pH fusion. J Virol *80*:710–722. https://doi.org/10.1128/JVI.80.2.710-722.2006
Sadasivan J, Singh M, Sarma JD (2017) Cytoplasmic tail of coronavirus spike protein has intracellular targeting signals. J Biosci *42*:231–244. https://doi.org/10.1007/s12038-017-9676-7
Saito A, Irie T, Suzuki R, Maemura T, Nasser H, Uriu K, Kosugi Y, Shirakawa K, Sadamasu K, Kimura I et al (2022) Enhanced fusogenicity and pathogenicity of SARS-CoV-2 Delta P681R mutation. Nature *602*:300–306. https://doi.org/10.1038/s41586-021-04266-9
Salahuddin SZ, Ablashi DV, Markham PD, Josephs SF, Sturzenegger S, Kaplan M, Halligan G, Biberfeld P, Wong-Staal F, Kramarsky B et al (1986) Isolation of a new virus, HBLV, in patients with lymphoproliferative disorders. Science *234*:596–601. https://doi.org/10.1126/science.2876520
Salsman J, Top D, Boutilier J, Duncan R (2005) Extensive syncytium formation mediated by the reovirus FAST proteins triggers apoptosis-induced membrane instability. J Virol *79*:8090–8100. https://doi.org/10.1128/JVI.79.13.8090-8100.2005
Salsman J, Top D, Barry C, Duncan R (2008) A virus-encoded cell-cell fusion machine dependent on surrogate adhesins. PLoS Pathog *4*:e1000016. https://doi.org/10.1371/journal.ppat.1000016
Santoro F, Kennedy PE, Locatelli G, Malnati MS, Berger EA, Lusso P (1999) CD46 is a cellular receptor for human herpesvirus 6. Cell *99*:817–827. https://doi.org/10.1016/s0092-8674(00)81678-5
Sarfo A, Starkey J, Mellinger E, Zhang D, Chadha P, Carmichael J, Wills JW (2017) The UL21 tegument protein of Herpes Simplex Virus 1 is differentially required for the syncytial phenotype. J Virol *91*:e01161-17. https://doi.org/10.1128/JVI.01161-17
Sathiyamoorthy K, Chen J, Longnecker R, Jardetzky TS (2017) The COMPLEXity in herpesvirus entry. Curr Opin Virol *24*:97–104. https://doi.org/10.1016/j.coviro.2017.04.006
Sato H, Orenstein J, Dimitrov D, Martin M (1992) Cell-to-cell spread of HIV-1 occurs within minutes and may not involve the participation of virus particles. Virology *186*:712–724
Satoh T, Arii J, Suenaga T, Wang J, Kogure A, Uehori J, Arase N, Shiratori I, Tanaka S, Kawaguchi Y et al (2008) PILRalpha is a herpes simplex virus-1 entry coreceptor that associates with glycoprotein B. Cell *132*:935–944. https://doi.org/10.1016/j.cell.2008.01.043
Sattentau Q (2008) Avoiding the void: cell-to-cell spread of human viruses. Nat Rev Microbiol *6*:815–826. https://doi.org/10.1038/nrmicro1972
Sauter MM, Pelchen-Matthews A, Bron R, Marsh M, LaBranche CC, Vance PJ, Romano J, Haggarty BS, Hart TK, Lee WM, Hoxie JA (1996) An internalization signal in the simian immunodeficiency virus transmembrane protein cytoplasmic domain modulates expression of envelope glycoproteins on the cell surface. J Cell Biol *132*:795–811. https://doi.org/10.1083/jcb.132.5.795

Sawitzky D (1997) Transmission, species specificity, and pathogenicity of Aujeszky's disease virus. Arch Virol Suppl *13*:201–206. https://doi.org/10.1007/978-3-7091-6534-8_19
Scherer WF, Syverton JT (1954) The viral range in vitro of a malignant human epithelial cell (strain HeLa, Gel). I. Multiplication of herpes simplex, pseudorabies, and vaccinia viruses. Am J Pathol *30*:1057–1073
Schmidt N, Domingues P, Golebiowski F, Patzina C, Tatham MH, Hay RT, Hale BG (2019) An influenza virus-triggered SUMO switch orchestrates co-opted endogenous retroviruses to stimulate host antiviral immunity. Proc Natl Acad Sci U S A *116*:17399–17408. https://doi.org/10.1073/pnas.1907031116
Schuitemaker H, Kootstra NA, Groenink M, De Goede RE, Miedema F, Tersmette M (1992) Differential tropism of clinical HIV-1 isolates for primary monocytes and promonocytic cell lines. AIDS Res Hum Retrovir *8*:1679–1682. https://doi.org/10.1089/aid.1992.8.1679
Schultz EP, Lanchy JM, Ellerbeck EE, Ryckman BJ (2015) Scanning mutagenesis of human cytomegalovirus glycoprotein gH/gL. J Virol *90*:2294–2305. https://doi.org/10.1128/JVI.01875-15
Scott TF, Mc LD (1959) Cellular responses to infection with strains of herpes simplex virus. Ann N Y Acad Sci *81*:118–128. https://doi.org/10.1111/j.1749-6632.1959.tb49300.x
Shieh MT, Spear PG (1994) Herpesvirus-induced cell fusion that is dependent on cell surface heparan sulfate or soluble heparin. J Virol *68*:1224–1228. https://doi.org/10.1128/JVI.68.2.1224-1228.1994
Shmulevitz M, Duncan R (2000) A new class of fusion-associated small transmembrane (FAST) proteins encoded by the non-enveloped fusogenic reoviruses. EMBO J *19*:902–912. https://doi.org/10.1093/emboj/19.5.902
Shmulevitz M, Salsman J, Duncan R (2003) Palmitoylation, membrane-proximal basic residues, and transmembrane glycine residues in the reovirus p10 protein are essential for syncytium formation. J Virol *77*:9769–9779. https://doi.org/10.1128/jvi.77.18.9769-9779.2003
Sigal A, Kim JT, Balazs AB, Dekel E, Mayo A, Milo R, Baltimore D (2011) Cell-to-cell spread of HIV permits ongoing replication despite antiretroviral therapy. Nature *477*:95–98. https://doi.org/10.1038/nature10347
Simmons G, McKnight A, Takeuchi Y, Hoshino H, Clapham PR (1995) Cell-to-cell fusion, but not virus entry in macrophages by T-cell line tropic HIV-1 strains: a V3 loop-determined restriction. Virology *209*:696–700. https://doi.org/10.1006/viro.1995.1307
Sinzger C, Schmidt K, Knapp J, Kahl M, Beck R, Waldman J, Hebart H, Einsele H, Jahn G (1999) Modification of human cytomegalovirus tropism through propagation in vitro is associated with changes in the viral genome. J Gen Virol *80(Pt 11)*:2867–2877. https://doi.org/10.1099/0022-1317-80-11-2867
Soe K, Andersen TL, Hobolt-Pedersen AS, Bjerregaard B, Larsson LI, Delaisse JM (2011) Involvement of human endogenous retroviral syncytin-1 in human osteoclast fusion. Bone *48*:837–846. https://doi.org/10.1016/j.bone.2010.11.011
Spear PG (2004) Herpes simplex virus: receptors and ligands for cell entry. Cell Microbiol *6*:401–410. https://doi.org/10.1111/j.1462-5822.2004.00389.x
Spear PG, Longnecker R (2003) Herpesvirus entry: an update. J Virol *77*:10179–10185. https://doi.org/10.1128/jvi.77.19.10179-10185.2003
Stampfer SD, Heldwein EE (2013) Stuck in the middle: structural insights into the role of the gH/gL heterodimer in herpesvirus entry. Curr Opin Virol *3*:13–19. https://doi.org/10.1016/j.coviro.2012.10.005
Sugimoto J, Sugimoto M, Bernstein H, Jinno Y, Schust D (2013) A novel human endogenous retroviral protein inhibits cell-cell fusion. Sci Rep *3*:1462. https://doi.org/10.1038/srep01462
Sylwester A, Wessels D, Anderson SA, Warren RQ, Shutt DC, Kennedy RC, Soll DR (1993) HIV-induced syncytia of a T cell line form single giant pseudopods and are motile. J Cell Sci *106(Pt 3)*:941–953. https://doi.org/10.1242/jcs.106.3.941
Symeonides M, Murooka TT, Bellfy LN, Roy NH, Mempel TR, Thali M (2015) HIV-1-induced small T cell syncytia can transfer virus particles to target cells through transient contacts. Viruses *7*:6590–6603. https://doi.org/10.3390/v7122959

Takeuchi K, Miyajima N, Nagata N, Takeda M, Tashiro M (2003) Wild-type measles virus induces large syncytium formation in primary human small airway epithelial cells by a SLAM(CD150)-independent mechanism. Virus Res *94*:11–16
Tanaka Y, Suenaga T, Matsumoto M, Seya T, Arase H (2013) Herpesvirus 6 glycoproteins B (gB), gH, gL, and gQ are necessary and sufficient for cell-to-cell fusion. J Virol *87*:10900–10903. https://doi.org/10.1128/JVI.01427-13
Tang H, Mori Y (2015) Determinants of human CD134 essential for entry of human herpesvirus 6B. J Virol *89*:10125–10129. https://doi.org/10.1128/JVI.01606-15
Tang H, Serada S, Kawabata A, Ota M, Hayashi E, Naka T, Yamanishi K, Mori Y (2013) CD134 is a cellular receptor specific for human herpesvirus-6B entry. Proc Natl Acad Sci U S A *110*:9096–9099. https://doi.org/10.1073/pnas.1305187110
Tang H, Wang J, Mahmoud NF, Mori Y (2014) Detailed study of the interaction between human herpesvirus 6B glycoprotein complex and its cellular receptor, human CD134. J Virol *88*:10875–10882. https://doi.org/10.1128/JVI.01447-14
Tang J, Frascaroli G, Lebbink RJ, Ostermann E, Brune W (2019) Human cytomegalovirus glycoprotein B variants affect viral entry, cell fusion, and genome stability. Proc Natl Acad Sci U S A *116*:18021–18030. https://doi.org/10.1073/pnas.1907447116
Terry-Allison T, Montgomery RI, Whitbeck JC, Xu R, Cohen GH, Eisenberg RJ, Spear PG (1998) HveA (herpesvirus entry mediator A), a coreceptor for herpes simplex virus entry, also participates in virus-induced cell fusion. J Virol *72*:5802–5810. https://doi.org/10.1128/JVI.72.7.5802-5810.1998
Terry-Allison T, Montgomery RI, Warner MS, Geraghty RJ, Spear PG (2001) Contributions of gD receptors and glycosaminoglycan sulfation to cell fusion mediated by herpes simplex virus 1. Virus Res *74*:39–45. https://doi.org/10.1016/s0168-1702(00)00244-6
Thorp EB, Boscarino JA, Logan HL, Goletz JT, Gallagher TM (2006) Palmitoylations on murine coronavirus spike proteins are essential for virion assembly and infectivity. J Virol *80*:1280–1289. https://doi.org/10.1128/JVI.80.3.1280-1289.2006
Tiwari V, O'Donnell C, Copeland RJ, Scarlett T, Liu J, Shukla D (2007) Soluble 3-O-sulfated heparan sulfate can trigger herpes simplex virus type 1 entry into resistant Chinese hamster ovary (CHO-K1) cells. J Gen Virol *88*:1075–1079. https://doi.org/10.1099/vir.0.82476-0
Tognon M, Manservigi R, Cavrini V, Campadelli-Fiume G (1984) Characterization of a herpes simplex virus type 1 mutant resistant to benzhydrazone, a selective inhibitor of herpesvirus glycosylation. Proc Natl Acad Sci U S A *81*:2440–2443. https://doi.org/10.1073/pnas.81.8.2440
Top D, Barry C, Racine T, Ellis CL, Duncan R (2009) Enhanced fusion pore expansion mediated by the trans-acting Endodomain of the reovirus FAST proteins. PLoS Pathog *5*:e1000331. https://doi.org/10.1371/journal.ppat.1000331
Trkola A, Gordon C, Matthews J, Maxwell E, Ketas T, Czaplewski L, Proudfoot AE, Moore JP (1999) The CC-chemokine RANTES increases the attachment of human immunodeficiency virus type 1 to target cells via glycosaminoglycans and also activates a signal transduction pathway that enhances viral infectivity. J Virol *73*:6370–6379. https://doi.org/10.1128/JVI.73.8.6370-6379.1999
Tsukamoto Y, Kotani T, Kohama K, Sakuma S, Sasaki F (1999) Nuclear proliferation in syncytia during avian reovirus replication. Can J Vet Res *63*:282–283
Tugizov S, Navarro D, Paz P, Wang Y, Qadri I, Pereira L (1994) Function of human cytomegalovirus glycoprotein B: syncytium formation in cells constitutively expressing gB is blocked by virus-neutralizing antibodies. Virology *201*:263–276. https://doi.org/10.1006/viro.1994.1291
Turner A, Bruun B, Minson T, Browne H (1998) Glycoproteins gB, gD, and gHgL of herpes simplex virus type 1 are necessary and sufficient to mediate membrane fusion in a Cos cell transfection system. J Virol *72*:873–875. https://doi.org/10.1128/JVI.72.1.873-875.1998
Van Goethem E, Poincloux R, Gauffre F, Maridonneau-Parini I, Le Cabec V (2010) Matrix architecture dictates three-dimensional migration modes of human macrophages: differential involvement of proteases and podosome-like structures. J Immunol *184*:1049–1061. https://doi.org/10.4049/jimmunol.0902223

Vanarsdall AL, Johnson DC (2012) Human cytomegalovirus entry into cells. Curr Opin Virol *2*:37–42. https://doi.org/10.1016/j.coviro.2012.01.001

Vanarsdall AL, Ryckman BJ, Chase MC, Johnson DC (2008) Human cytomegalovirus glycoproteins gB and gH/gL mediate epithelial cell-cell fusion when expressed either in cis or in trans. J Virol *82*:11837–11850. https://doi.org/10.1128/JVI.01623-08

Vanarsdall AL, Chase MC, Johnson DC (2011) Human cytomegalovirus glycoprotein gO complexes with gH/gL, promoting interference with viral entry into human fibroblasts but not entry into epithelial cells. J Virol *85*:11638–11645. https://doi.org/10.1128/JVI.05659-11

Vargiu L, Rodriguez-Tome P, Sperber GO, Cadeddu M, Grandi N, Blikstad V, Tramontano E, Blomberg J (2016) Classification and characterization of human endogenous retroviruses; mosaic forms are common. Retrovirology *13*:7. https://doi.org/10.1186/s12977-015-0232-y

Vicandi B, Jimenez-Heffernan JA, Lopez-Ferrer P, Patron M, Gamallo C, Colmenero C, Viguer JM (1999) HIV-1 (p24)-positive multinucleated giant cells in HIV-associated lymphoepithelial lesion of the parotid gland. A report of two cases. Acta Cytol *43*:247–251. https://doi.org/10.1159/000330987

Waldman WJ, Sneddon JM, Stephens RE, Roberts WH (1989) Enhanced endothelial cytopathogenicity induced by a cytomegalovirus strain propagated in endothelial cells. J Med Virol *28*:223–230. https://doi.org/10.1002/jmv.1890280405

Weed DJ, Nicola AV (2017) Herpes simplex virus membrane fusion. Adv Anat Embryol Cell Biol *223*:29–47. https://doi.org/10.1007/978-3-319-53168-7_2

Wheeler CE (1960) Herpes simplex virus. Characteristics of a strain which produces unusually large multinucleated giat cells in tissue culture. Arch Dermatol *82*:391–399. https://doi.org/10.1001/archderm.1960.01580030085011

Wilen CB, Tilton JC, Doms RW (2012) HIV: cell binding and entry. Cold Spring Harb Perspect Med *2*:a006866. https://doi.org/10.1101/cshperspect.a006866

Wilkinson GW, Davison AJ, Tomasec P, Fielding CA, Aicheler R, Murrell I, Seirafian S, Wang EC, Weekes M, Lehner PJ et al (2015) Human cytomegalovirus: taking the strain. Med Microbiol Immunol *204*:273–284. https://doi.org/10.1007/s00430-015-0411-4

Wille PT, Wisner TW, Ryckman B, Johnson DC (2013) Human cytomegalovirus (HCMV) glycoprotein gB promotes virus entry in trans acting as the viral fusion protein rather than as a receptor-binding protein. mBio *4*:e00332-00313. https://doi.org/10.1128/mBio.00332-13

Wilson DW, Davis-Poynter N, Minson AC (1994) Mutations in the cytoplasmic tail of herpes simplex virus glycoprotein H suppress cell fusion by a syncytial strain. J Virol *68*:6985–6993. https://doi.org/10.1128/JVI.68.11.6985-6993.1994

Wu CA, Shanley JD (1998) Chronic infection of human umbilical vein endothelial cells by human herpesvirus-6. J Gen Virol *79(Pt 5)*:1247–1256. https://doi.org/10.1099/0022-1317-79-5-1247

Xia S, Yan L, Xu W, Agrawal AS, Algaissi A, Tseng CK, Wang Q, Du L, Tan W, Wilson IA et al (2019) A pan-coronavirus fusion inhibitor targeting the HR1 domain of human coronavirus spike. Sci Adv *5*:eaav4580. https://doi.org/10.1126/sciadv.aav4580

Xia S, Liu M, Wang C, Xu W, Lan Q, Feng S, Qi F, Bao L, Du L, Liu S et al (2020) Inhibition of SARS-CoV-2 (previously 2019-nCoV) infection by a highly potent pan-coronavirus fusion inhibitor targeting its spike protein that harbors a high capacity to mediate membrane fusion. Cell Res *30*:343–355. https://doi.org/10.1038/s41422-020-0305-x

Xie M, Leroy H, Mascarau R, Woottum M, Dupont M, Ciccone C, Schmitt A, Raynaud-Messina B, Verollet C, Bouchet J et al (2019) Cell-to-Cell spreading of HIV-1 in myeloid target cells escapes SAMHD1 restriction. mBio *10*:e02457-19. https://doi.org/10.1128/mBio.02457-19

Yang C, Compans RW (1997) Analysis of the murine leukemia virus R peptide: delineation of the molecular determinants which are important for its fusion inhibition activity. J Virol *71*:8490–8496. https://doi.org/10.1128/JVI.71.11.8490-8496.1997

Zhen Z, Bradel-Tretheway B, Sumagin S, Bidlack JM, Dewhurst S (2005) The human herpesvirus 6 G protein-coupled receptor homolog U51 positively regulates virus replication and enhances cell-cell fusion in vitro. J Virol *79*:11914–11924. https://doi.org/10.1128/JVI.79.18.11914-11924.2005

Zhong P, Agosto LM, Munro JB, Mothes W (2013) Cell-to-cell transmission of viruses. Curr Opin Virol *3*:44–50. https://doi.org/10.1016/j.coviro.2012.11.004

Zhou P, Yang XL, Wang XG, Hu B, Zhang L, Zhang W, Si HR, Zhu Y, Li B, Huang CL et al (2020) A pneumonia outbreak associated with a new coronavirus of probable bat origin. Nature *579*:270–273. https://doi.org/10.1038/s41586-020-2012-7

Zingler K, Littman DR (1993) Truncation of the cytoplasmic domain of the simian immunodeficiency virus envelope glycoprotein increases env incorporation into particles and fusogenicity and infectivity. J Virol *67*:2824–2831. https://doi.org/10.1128/JVI.67.5.2824-2831.1993

Chapter 15
HIV-1 Induced Cell-to-Cell Fusion or Syncytium Formation

Tobias Starling and Sergi Padilla-Parra

Abstract HIV-1 cell-free infection has been thoroughly investigated; however, its relevance and importance in vitro are questionable. Cell-cell transmission is now thought to be the dominant mode of transmission within the host; however precise molecular details remain elusive. The considerable potency of cell-cell transmission hinges upon its ability to hijack and manipulate host immunological function to target uninfected cells, along with overcoming restriction factors and increasing the speed of latent pool formation. Another question of relevance is virus induced cell-cell fusion and how this process is regulated. How often HIV-1 induces the formation of syncytia? Is cell-cell function a potential process for HIV-1 transmission? These questions are discussed and reviewed together with a description of the most common ways of HIV-1 entry and transinfection.

Abbreviations

cART Combined antiretroviral therapy (cART)
Env Envelope glycoprotein HIV-1
HIV-1 Human immunodeficiency virus 1
HSP Heparin sulfate proteoglycans
LFMC Lymphocyte fused macrophage cell
MNG Multinucleated giant cell
NNRTI Non-nucleoside reverse transcriptase inhibitors
nAb Neutralizing antibody
SAMHD1 HD-domain-containing protein 1
TCR T cell receptor

T. Starling
Department of Infectious Diseases, King's College London, Faculty of Life Sciences & Medicine, London, UK

S. Padilla-Parra (✉)
Department of Infectious Diseases, King's College London, Faculty of Life Sciences & Medicine, London, UK

Randall Division of Cell and Molecular Biophysics, King's College London, London, UK
e-mail: sergio.padilla_parra@kcl.ac.uk

M. Kloc, A. Uosef (eds.), *Syncytia: Origin, Structure, and Functions*, Results and Problems in Cell Differentiation 71,
https://doi.org/10.1007/978-3-031-37936-9_15

TNT Tunnelling nanotubes
VCC Virus-containing compartment
vDNA Viral DNA

15.1 Introduction

The human immunodeficiency virus (HIV) replicates most efficiently in activated CD4+ T cells, the depletion of which leads to patients succumbing to canonical opportunistic infections, which characterize acquired immunodeficiency syndrome (AIDS). Since its identification, HIV's worldwide burden has exponentially expanded. In 2015, global projections estimated nearly 37 million patients to be infected with HIV1, contributing to at least 1.5 million deaths in 20,101, and depleting up to 20% of healthcare resources in significantly affected countries. To date, there is no cure or vaccine for HIV infection. The only available treatment is combination antiretroviral therapy (cART), which effectively halts a patient's progression to AIDS, barring the presence of drug-resistant strains. Despite the advent of cART, its use to control HIV infection is crippled by costs and notorious side effects, such as lipodystrophy and hepatotoxicity with protease inhibitors, and psychosis and mania with nucleoside (NRTIs) and non-nucleoside (NNRTIs) reverse transcriptase inhibitors. Furthermore, drug resistance is becoming an ever-increasing concern in treating HIV patients. It has been estimated that up to 10% of patients will be resistant to three classes of cART regimens within 10 years of initiating therapy. Moreover, with the expanding prevalence of drug-resistant strains, it is common for patients initiating therapy to already harbor drug-resistant virus. Therefore, it is imperative to develop novel therapeutics that target conserved, vital steps in HIV's replicative cycle to halt replication, CD4+ T cell depletion, and restore immunoprotective function while simultaneously avoiding the resurgence of resistant strains.

15.2 HIV-1 Entry, Cellular Targets, and Tropism

HIV-1 tropism is exclusively restricted to cells expressing the CD4 receptor, explaining the rapid decline of CD4+ immune cells such as CD4+ T helper cells. Immune cells of myeloid lineage, macrophages, monocytes, and dendritic cells (DCs) are also susceptible to HIV-1 infection. HIV-1 envelope glycoprotein (Env) is a heterotrimer of gp120 which sequentially binds the primary receptor (CD4) to cause a conformational change and the formation of the chemokine coreceptor binding site (CCR5 and CXCR4) (Ayouba et al. 2008). Co-receptor binding likely then stabilizes this conformation, culminating in gp120 dissociation triggering irreversible refolding of gp41 (a subunit of cleaved Env) and membrane fusion. CCR5 is the coreceptor of choice for transmission between patients and initial propagation, while CXCR4 mutants appear later in infection and are correlated with accelerated disease progression.

Macrophages express CD4 in lower levels than T cells resulting in macrophage-specific HIV-1 variants (M-tropic). These variants exhibit higher CD4 binding affinities through greater exposure of the CD4 binding site, but are more susceptible to antibody-mediated neutralization (Dale et al. 2011). While macrophages express CXCR4, it appears that HIV-1 is incapable of infecting macrophages using this coreceptor. It is possible that CXCR4 lab strains that are continuously cultured in T cell lines develop a greater dependence on T cell factors resulting in the CCR5 restriction seen in vitro but why the primary CXCR4 viruses remain productive. Macrophages are phenotypically highly heterogeneous creating a dramatic diversity in susceptibility to infection between macrophages even in the same tissue. Macrophages are phenotypically highly diverse which dramatically alters their susceptibility to HIV-1 infection.

Intestinal and alveolar macrophages are relatively resistant to HIV-1 while vaginal and rectal macrophages are relatively permissive. The original model of HIV-1 infection involves a CD4+ target cell becoming exposed to a freely diffusing HIV-1 virion (Cell-free) within the bloodstream. The probability of binding is principally dictated by viral concentration and fluid phase diffusion (Fig. 15.1). Following HIV-1 contact with a target cell, non-specific attachment to integrins, C-type lectins, and heparin sulfate proteoglycans (HSP) hold the virus in close proximity to the specific cellular receptors (Berger et al. 1999). Cell-free studies have demonstrated the fundamental differences in the molecular stoichiometries of the two HIV-1 strains. While X4 strains require two Env, R5 strains only need one Env to form the prefusion complex (Iliopoulou et al. 2018). Further studies have provided insights into Env and host receptor clustering, recruitment, and their neutralization (Fig. 15.1).

15.3 Cell-Cell HIV-1 Transmission in Macrophages

While the T cell virological synapse is well understood, cell-cell transmission between other cell types, macrophages remain understudied due to their complexity and questionable in vitro relevance. However, the relatively recent discovery of long-lived, self-renewing (independent of hematopoiesis) yolk sac-derived macrophages has reinvigorated research into the importance of macrophages during HIV-1 infection and as a latent pool. Due to macrophage low expression of entry receptors and high expression of HIV-1 restriction factors (SAMHD, tetherin, TRIM5α and APOBEC3G), cell-free HIV-1 infection of macrophage is highly inefficient, particularly early in infection by transmitter/founder viruses. However, like the VS, cell-cell HIV-1 spread between macrophages can dramatically enhance infection and propagation.

Due to macrophages low expression of entry receptors and high expression of HIV-1 restriction factors (SAMHD, tetherin, TRIM5α and APOBEC3G), cell-free

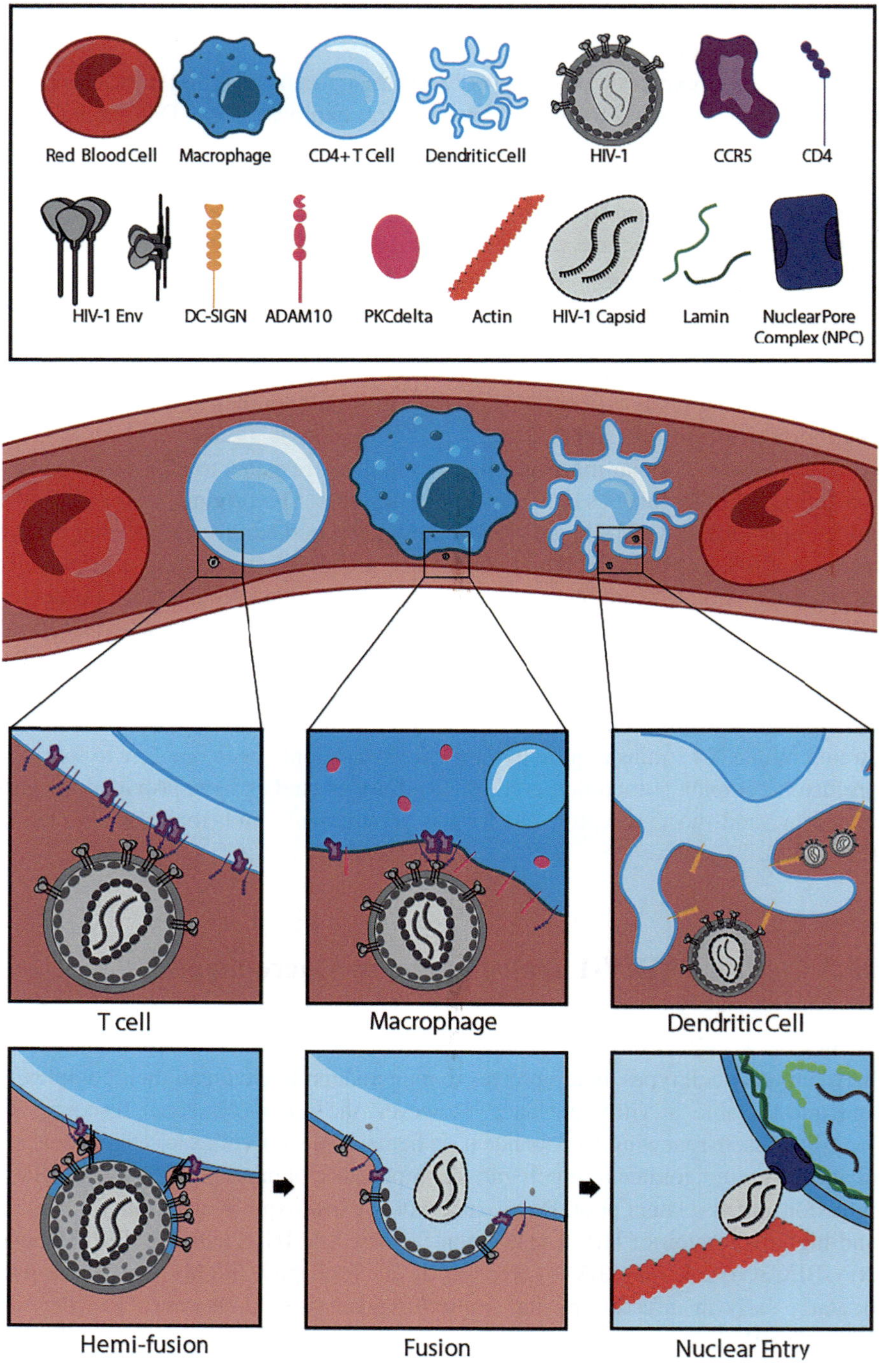

Fig. 15.1 Cartoon depicting single HIV-1 virus entry and fusion

HIV-1 infection of macrophage is highly inefficient, particularly early in infection with transmitter/founder viruses (Parrish et al. 2012). However, like the virological synapse, cell-cell HIV-1 spread between macrophages can dramatically enhance infection and propagation. Unlike T cells, macrophages are APCs and thus can form stable intracellular (ICAM-1/LFA-1) connections limiting their need for Env surface expression. Due to this reduced reliance on Env surface expression, these cell-cell contacts are often referred to as the infectious synapse (IS), and no similar structure has been observed between T cells. Despite infectious synapse formation being Env independent, infection remains Env dependent, indicting fusion is critical for all known methods of HIV-1 infection (Fig. 15.2b). In macrophages, Env surface expression and viral budding is low (~5%); instead, the virus is sequestered into a protective virus containing compartment (VCC). The formation of the VCC is poorly understood but is potentially an intrinsic property of macrophages as non-viral particles have been observed in similar compartments. The morphology of the VCC is highly dynamic; however it is always tightly linked to the surface (20 nm) and often linked to the plasma membrane. This potentially maintains its neutral pH while being narrow enough to prevent neutralizing antibody (nAb) access. It is speculated that local changes in phosphatidylinositol 4,5-bisphosphate (PIP_2), fatty acid chains, membrane fluidity, or lipid phases cause this affinity for gag binding and viral budding to occur within the VCC and not at the surface (Gobeil et al. 2013). The macrophage VCC is highly advantageous as it reduces the expression of surface Env which could be detected by the immune system, protects the virions from nAbs, and stores them for prolonged periods of time.

15.4 Phagocytosis and HIV-1 Entry

Macrophages are phagocytic cells, with multiple bacteria, fungi, and viruses exploiting this to gain entry via intracellular compartments by a "Trojan horse" strategy. While most viruses use the acidification of the endosome to trigger conformational changes for entry, there is no evidence of this for HIV-1. Macrophages preferentially engulf HIV-1 infected T cells up to 50 times more by an Env-independent mechanism, which also appears to be independent of apoptotic signalling (Gobeil et al. 2013).

While most phagocytosed T cells are degraded after engulfment, some remain in non-degradative compartments for up to 6 days; however their viability is unknown. While HIV-1 can use Tat interaction with Cdc42, and Nef interaction with AP-1 to inhibit phagocytosis, neither mechanism has been linked to HIV-1 escape from these phagocytosed T cells. Whether HIV-1 escapes immediately prior to T cell engulfment or from within an endocytic compartment, Env-mediated fusion is still essential. This method removes virion exposure to bNabs, fluid phase diffusion, and may drive virus and host receptor clustering similar to that of the VS. This phagocytic route even allows weakly M-tropic T/F viruses to productively infect macrophages before M-tropic variants develop leading to the suggestion that significant

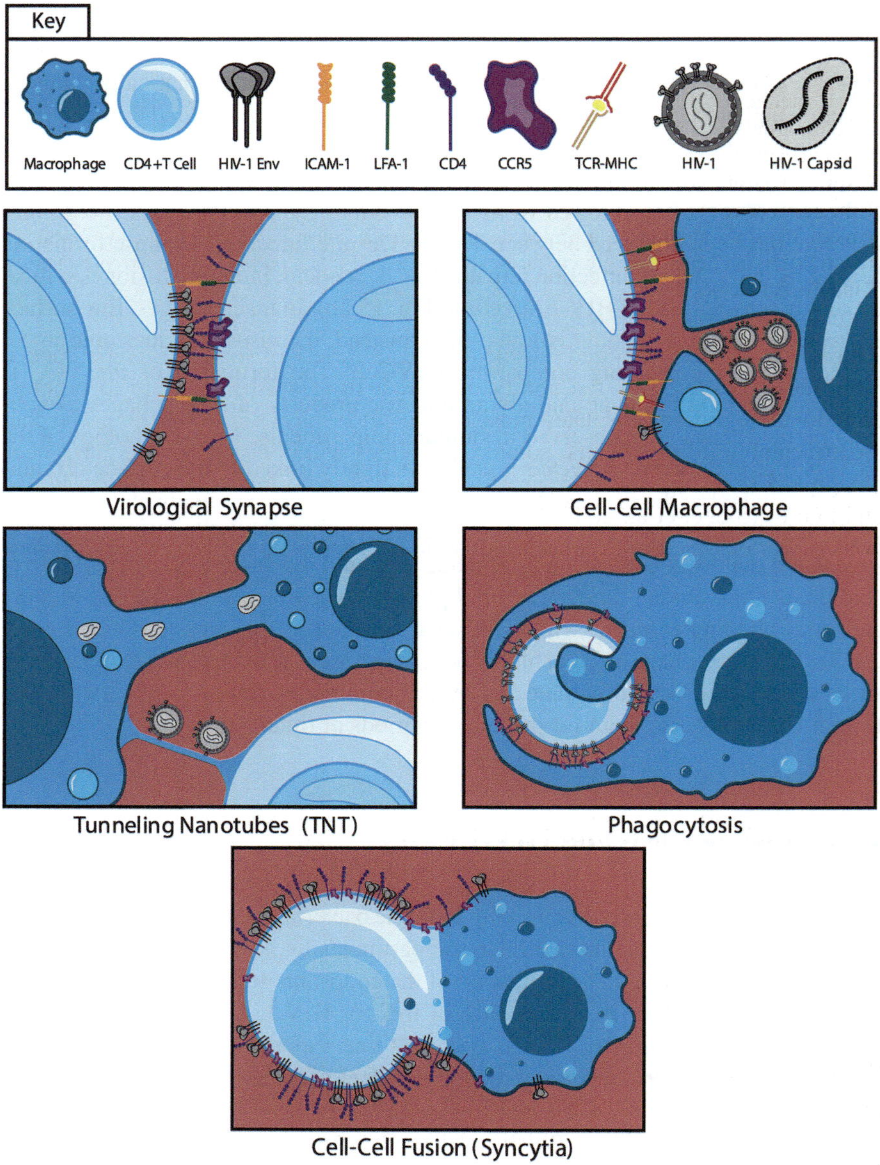

Fig. 15.2 Cartoon depicting cell to cell transinfection and cell-cell fusion for the first steps of HIV-1 infection

macrophage infection may occur earlier in disease pathogenesis than previously thought. It remains possible that infection via phagocytosis is not actually productive, with fusion either occurring at the plasma membrane before phagocytosis, or that HIV-1 nucleic acids, vDNA from the engulfed T- cell and viral proteins, provide a positive readout while not being truly infected.

15.5 Cell-Cell Fusion in HIV-1 Entry

Cell-free HIV-1 infection is the most studied in vitro; however cell-cell HIV-1 infection is considerably more infectious (at least ten times) and is likely the dominant mode of transmission, particularly in T cell-rich regions such as the lymph nodes (Yang et al. 2006). Cell-cell transmission was first described between T cells (infected and target). In infected T cells, HIV-1 expresses high levels of Env on the cell surface for viral budding. These surface Env can bind to CD4 on a target cell, along with endogenous integrin interactions holding the cells in close proximity for prolonged periods of time known as the virological synapse (Sloan et al. 2013) (VS) (Fig. 15.1c). This reduces one of the major limiting factors of HIV-1 entry, being the fluid phase diffusion distance, which is highly beneficial given the frequency of spontaneous gp120 dissociation (Bourgeois et al. 2006). Cell-cell contact rapidly reorganizes the cytoskeleton, viral proteins, and host receptors on both cells to this site of contact. On the infected cell, this is characterized by the localization of Gag to the intracellular surface and Env above. On the uninfected cell, integrins (LFA-1, ICAM-1), tetraspanins, and receptors (CD4/CXCR4/CCR5) also polarize to this contact site. The virological synapse dramatically reduces fluid phase diffusion distance, increases viral MOI at the site, increases the concentration of vital entry receptors, and spatially restricts the accessibility of antibodies and cells to create a highly advantageous system of viral spread. Env-CD4 cell-cell engagement relocates F-actin at the VS, potentially providing a less obstructed path for viral transfer to the nucleus. Cell-cell contact also appears to aid in T cell receptor (TCR) clustering and activation which triggers other downstream factors such as NF- κB.

HIV-1 targets T cells and macrophages that express CD4 (main receptor) and CCR5 or CXCR4 as coreceptors. The envelope glycoprotein oligomerizes as a trimer of dimers (cleaved gp160 into gp41 and gp120, which remain non-covalently associated). The gp120 subunit interacts asymmetrically with one CD4 molecule, and this interaction induces a series of conformational changes (opening Env). Cell-cell fusion between T cells (one infected and one being the target cell) was proposed to be an infection route (Leroy et al. 2020). Formation of T cell syncytia has been, however, controversial since other research showed no evidence for T cells to undergo cell-cell fusion employing primary CD4+ T cells (Sourisseau et al. 2007). The authors suggested that T cell syncytia formation might be artifacts coming from immortalized cell lines only observed in vitro and restricted to X4 tropic viruses (Moore and Ho 1995). This argument is against some observation of small T cell multinuclear cells found in vivo in lymph nodes from HIV-1 infected patients (Orenstein 2000). These small T cell syncytia could establish interactions with target T cells and macrophages and facilitate transinfection. Other HIV-1 target cells such as dendritic cells and macrophages also from syncytia in vivo. In conclusion cell-cell fusion might be an alternative way of productive infection in T cells, myeloid cells, and macrophages to avoid post-fusion restriction factors (e.g., HD-domain-containing protein 1 (SAMHD1) (Xie et al. 2019).

Cell-cell fusion (syncytia) is thought to occur due to infected T cells expressing high levels of surface Env that engage with CD4 and coreceptor on the macrophage. This high level of expression and interaction then facilitates fusion of the two membranes and the formation of a large multinucleated giant cell (MNG) or lymphocyte fused macrophage cell (LFMC). Why the T cells fuse instead of being phagocytosed is not understood but may relate to the duration of infection. Recently infected T cells (<3 days) may not expose "eat me" signals but would express surface Env allowing for fusion without phagocytosis. Following heterotypic fusion, massive Env and Gag transfer occurs, along with T cell membrane markers (CD3 and CD2) and intrinsic T cell factors (Lck, cyclin/CDK, and dNTP). This fusion to a LFMC is functionally important as it produces a macrophage that is now able to fuse with other macrophages. LMFCs provide another mechanism for macrophage to propagate HIV-1 by the more infectious cell-cell route. This creates LMFCs with enhanced lifetime and, surprisingly, faster migration than their mononuclear counterparts. While small LMFCs (<5 nuclei) have been observed in the LN of HIV-1 humanized mice and infected patients, the larger LMFCs (>5 nuclei) have only been observed in immortalized cell culture. Like cell-cell spread by phagocytosis, cell fusion may also represent a method of weakly M-tropic T/F viruses being capable of infecting macrophages early in infection. Finally, it has also been shown that HIV-1 employs a two-step cell-cell fusion approach for virus transfer from T cells to macrophages, DCs, and osteoclasts (Frankel et al. 1996); altogether these modes of dissemination by cell-cell fusion for HIV-1 might represent a very efficient mode of transinfection and dissemination in vivo.

15.6 Tunnelling Nanotubes and HIV-1

Tunnelling nanotubes (TNT) are used for cell-cell communication between homotypic and heterotypic cells. TNTs naturally facilitate calcium flux, protein, organelle, and genetic material exchange between cells up to 100 μm away (long TNTs). The exchange of these materials by the TNTs can be regulated by the presence of a gap junction, found at the tips of TNTs, with HIV being shown to manipulate connexin 43 (Cx43) at the gap junction to aid in infection. Unlike in T cells, TNT formation can be induced in macrophages by HIV-1 NEF interaction with M-Sec. HIV-1 particles have been shown to surf along the TNTs, with antibodies against Env and CD4 being capable of preventing fusion; however, HIV-1 containing vesicles (VCC or exosomes) have also been shown to be transported across thick TNTs (Fig. 15.2c). It is unknown if HIV-1 particles or vesicles are transported through TNTs or surfing top the TNT which could provide an opportunity for super resolution live cell imaging techniques.

It was shown that HIV-1 infection of primary macrophages induced the formation of TNT which correlated with viral replication (Eugenin et al. 2009). In this report, Eugenin and co-workers showed HIV-1 particles within TNT structures. These results should be supported with in vivo studies and represent a good hypothesis for yet another way of transmission that has received limited attention over the past decades.

15.7 Limitations: Biological Knowledge Gaps

The virological synapse is a complex system involving Env on the infected cell and then the CD4 primary receptor and CXCR4 coreceptor on the uninfected cell. Right at the virological synapse and the infectious synapse, other receptor interactions are also vital including ICAMs, CD3, Gag, integrins (including CD47), and potentially interactions with the other coreceptor. New technological developments are needed to examine with single molecule accuracy these interactions in line with efficient transinfection. This is not only valid for the microscopy approaches but also for labelling and use of in vitro and in vivo samples employing endogenous amounts of receptors and coreceptors.

Due to macrophage phenotype being inextricably linked to their environment, efforts have been made to replicate this for more biologically relevant study. Rat and bovine collagen (91.1% identity for the alpha I chains and 87.4% for the alpha 2 chains) has been used to replicate macrophages' physiological environment and found cell-free infection was even less efficient than previously thought. Innate interactions with collagen were found to reduce infectivity 20-fold by a mechanism-independent of Env shedding and viral aggregation. Similar reductions in infectivity were not seen in cell-cell transmission methods and were even less restricted as cell density increases further highlighting its importance in vitro.

15.8 Conclusion

The complete understanding of the mechanism of both single virus fusion and the process of transinfection is still not known. Also, how the process of virological synapse formation is regulated and how cell-cell fusion contributes to HIV-1 infection is not completely understood. Here we have highlighted several key facts that contribute to understand HIV-1 entry; but new technologies are needed to understand what all these diverse entry mechanisms have in common and how they are regulated.

Acknowledgments We thank all members of the Padilla-Parra lab for valuable discussions and criticism of the paper. This work has been supported by the European Research Council (ERC-2019-CoG-863869 FUSION to S.P.-P.).

Author Contributions Conceptualization, S.P.-P.; writing—original draft preparation, T.S.; writing—review and editing, T.S. and S.P.-P.; visualization, T.S.;

supervision, S.P.-P.; project administration, S.P.-P.; funding acquisition, S.P.-P. All authors have read and agreed to the published version of the paper.

Competing Interests The authors declare no competing interests.

References

Ayouba A et al (2008) Distinct efficacy of HIV-I entry inhibitors to prevent cell-to-cell transfer of R5 and X4 viruses across a human placental trophoblast barrier in a reconstitution model in vitro. Retrovirology 5:1–18

Berger EA, Murphy PM, Farber JM (1999) Chemokine receptors as HIV-1 coreceptors: roles in viral entry, tropism, and disease. Annu Rev Immunol 17:657–700

Bourgeois R et al (2006) Association between disruption of CD4 receptor dimerization and increased human immunodeficiency virus type 1 entry. Retrovirology 3:31

Dale BM et al (2011) Cell-to-cell transfer of HIV-1 via Virological synapses leads to endosomal Virion maturation that activates viral membrane fusion. Cell Host Microbe 10(6):551–562

Eugenin EA, Gaskill PJ, Berman JW (2009) Tunneling nanotubes (TNT): a potential mechanism for intercellular HIV trafficking. Commun Integr Biol 2(3):243–244

Frankel SS et al (1996) Replication of HIV-1 in dendritic cell-derived syncytia at the mucosal surface of the adenoid. Science 272(5258):115–117

Gobeil LA, Lodge R, Tremblay MJ (2013) Macropinocytosis-like HIV-1 internalization in macrophages is CCR5 dependent and leads to efficient but delayed degradation in endosomal compartments. J Virol 87(2):735–745

Iliopoulou M et al (2018) A dynamic three-step mechanism drives the HIV-1 pre-fusion reaction. Nat Struct Mol Biol 25(9):814–822

Leroy H et al (2020) Virus-Mediated Cell-Cell Fusion. Int J Mol Sci 21(24):9644

Moore JP, Ho DD (1995) HIV-1 neutralization: the consequences of viral adaptation to growth on transformed T cells. AIDS 9(Suppl A):S117–S136

Orenstein JM (2000) In vivo cytolysis and fusion of human immunodeficiency virus type 1-infected lymphocytes in lymphoid tissue. J Infect Dis 182(1):338–342

Parrish NF et al (2012) Transmitted/founder and chronic subtype C HIV-1 use CD4 and CCR5 receptors with equal efficiency and are not inhibited by blocking the integrin alpha 4 beta 7. PLoS Pathog 8(5):16

Sloan RD et al (2013) Productive entry of HIV-1 during cell-to-cell transmission via dynamin-dependent endocytosis. J Virol 87(14):8110–8123

Sourisseau M et al (2007) Inefficient human immunodeficiency virus replication in mobile lymphocytes. J Virol 81(2):1000–1012

Xie M et al (2019) Cell-to-cell spreading of HIV-1 in myeloid target cells escapes SAMHD1 restriction. MBio 10(6):e02457-19

Yang XZ et al (2006) Subunit stoichiometry of human immunodeficiency virus type 1 envelope glycoprotein trimers during virus entry into host cells. J Virol 80(9):4388–4395

Chapter 16
Relevance of the Entry by Fusion at the Cytoplasmic Membrane vs. Fusion After Endocytosis in the HIV and SARS-Cov-2 Infections

Leonor Huerta, Alejandro Gamboa-Meraz, and Pablo Samuel Estrada-Ochoa

Abstract HIV-1 and SARS-Cov-2 fuse at the cell surface or at endosomal compartments for entry into target cells; entry at the cell surface associates to productive infection, whereas endocytosis of low pH-independent viruses may lead to virus inactivation, slow replication, or alternatively, to productive infection. Endocytosis and fusion at the cell surface are conditioned by cell type-specific restriction factors and the presence of enzymes required for activation of the viral fusogen. Whereas fusion with the plasma membrane is considered the main pathway to productive infection of low pH-independent entry viruses, endocytosis is also productive and may be the main route of the highly efficient cell-to-cell dissemination of viruses. Alternative receptors, membrane cofactors, and the presence of enzymes processing the fusion protein at the cell membrane, determine the balance between fusion and endocytosis in specific target cells. Characterization of the mode of entry in particular cell culture conditions is desirable to better assess the effect of neutralizing and blocking agents and their mechanism of action. Whatever the pathway of

L. Huerta (✉)
Instituto de Investigaciones Biomédicas, Universidad Nacional Autónoma de México, Ciudad Universitaria, CDMX, Mexico
e-mail: leonorhh@biomedicas.unam.mx

A. Gamboa-Meraz
Instituto de Investigaciones Biomédicas, Universidad Nacional Autónoma de México, Ciudad Universitaria, CDMX, Mexico

Posgrado en Ciencias Bioquímicas, Universidad Nacional Autónoma de México, Ciudad de México, México

P. S. Estrada-Ochoa
Instituto de Investigaciones Biomédicas, Universidad Nacional Autónoma de México, Ciudad Universitaria, CDMX, Mexico

Facultad de Ciencias, Universidad Autónoma del Estado de México, Ciudad de México, México

M. Kloc, A. Uosef (eds.), *Syncytia: Origin, Structure, and Functions*, Results and Problems in Cell Differentiation 71,
https://doi.org/10.1007/978-3-031-37936-9_16

virus internalization, production of the viral proteins into the cells can lead to the expression of the viral fusion protein on the cell surface; if this protein is able to induce membrane fusion at physiological pH, it promotes the fusion of the infected cell with surrounding uninfected cells, leading to the formation of syncytia or heterokaryons. Importantly, particular membrane proteins and lipids act as cofactors to support fusion. Virus-induced cell-cell fusion leads to efficient virus replication into fused cells, cell death, inflammation, and severe disease.

16.1 Introduction

The human immunodeficiency virus (HIV) and the severe acute respiratory syndrome-coronavirus 2 (SARS-Cov-2) are constituted by an RNA chain surrounded by a lipid envelope acquired from the membrane of producer cells during virus budding. The viral membrane contains cellular glycosylated lipids and proteins which allow the attachment of the viral particles to the negatively charged surface of new target cells, as well as viral envelope proteins mediating specific high affinity receptor interactions and membrane fusion for virus entry. Membrane fusion induced by enveloped viruses proceeds initially by the close apposition of the virus and cell membranes, a process mediated by adhesion molecules. Upon interaction with specific receptors, viral envelope proteins undergo conformational changes that induce lipid perturbations leading to the fusion of the virus and cell membranes. The HIV and SARS-Cov-2 envelope proteins have similar structural organization and induce fusion by analogous mechanisms.

Fusion of the viral membrane can occur either with the cytoplasmic membrane at the cell surface, or with endosomal membranes after virus endocytosis. Fusion at the cell surface or at the endosomal compartments allows the entry of the viral capsid into the cytoplasm. The rate of productive infection seems to be higher when the virus entry takes place at the cell surface. HIV and SARS-Cov-2 use both routes for entry into the target cells, in a manner dependent on the virus strain and proteins acting as cofactors at the cell membrane. The relevance of the mode of entry has been recently emphasized by the observation that entry of SARS-Cov-2 by fusion at the cell surface associates with extensive and pathogenic virus replication in human tracheobronchial epithelial cells and pneumocytes in the lung, whereas the entry of variants through endocytosis leads to high virus replication almost exclusively in bronchial and upper respiratory tissues. Therefore, variants of the virus differ in their preference for one or the other pathway to enter target cells.

16.2 Fusion at the Cytoplasmic Membrane *vs.* Endocytosis in the SARS-Cov-2 Infection

SARS-Cov-2 replicates in double membranous structures known as "replication organelles" in the cytoplasm of infected cells. Structural proteins and genomic RNA are then directed to the endoplasmic reticulum-Golgi intermediate compartment (ERGIC) for virus assembly, glycosylation, and budding. The structural proteins spike (S), membrane (M), and envelope (E) are incorporated into the virion membrane. Like HIV Env, the S protein is cleaved by a furin or furin-like enzyme into the S1 and S2 subunits in the Golgi apparatus. S2 is anchored to the virus membrane through its transmembrane domain, whereas S1 remains non-covalently attached to it. Assembled viruses bud into the ERGIC lumen and are directed to the plasma membrane via the secretory pathway, or via deacidified lysosomes for release to the extracellular medium. Notably, lysosome deacidification limits virus degradation and impairs antigen presentation, thus avoiding the induction of an efficient immune response [Wang et al. 2021]. For a new infection, S1 establishes high affinity interaction with the virus receptor at the surface of target cells, the angiotensin-converting enzyme 2 (ACE2). Binding to ACE2 induces the shedding of S1 and the exposure of an additional site in S2, the S2' cleavage site. Cleavage at S2' is performed by the adyacent transmembrane protease serine 2 (TMPRSS2), and allows conformational changes leading to shedding of S1 and deployment of the S2 structure, favoring the insertion of its hydrophobic fusion peptide into the opposite cell membrane to promote virus-cell membrane fusion [Jackson et al. 2022].

The site of virus entry (plasma membrane or endocytic vesicles) is largely determined by the level of expression of TMPRSS2, which is expressed at the surface of the plasma membrane but not in the endosome. So, when insufficient levels of TMPRSS2 are expressed, S2' is not cleaved and the ACE2-bound virus is internalized by endocytosis; once in the endosome, the S2' cleavage is performed by endosomal cathepsins, which require an acidic pH to function. Activation of the S2 subunit then induces fusion with the endosomal membrane to release the viral capsid into the cytoplasm (Fig. 16.1).

The Omicron variant (B.1.1.529) is the main cause of the current pandemic in the world. Since its discovery in South Africa, Omicron caused concern due to the possible impact of its multiple mutations on the virus infective capacity, transmissibility, and immune system evasion. The SARS-Cov-2 S protein has 1273 amino acids, and its receptor-binding domain (RBD) comprises the 319–541 residues. A smaller region, known as the receptor binding motif (RBM), is located in residues 437–507. Of the 15 mutations carried by the Omicron variant in the RBD, 10 are located in the RBM. By comparison, the Delta variant (B.1.617.2) has only 2 or 4 amino acid mutations in the RBD. Given the high inter-individual Omicron transmissibility, concern aroused about the effect of mutations on its affinity for ACE2. Some studies suggested that the RBD of Omicron had weaker or similar binding affinity for ACE2 than the Delta variant [Wu et al. 2022; Han et al. 2022]. However, using biolayer interferometry or surface plasmon resonance to measure the binding of the RBD to

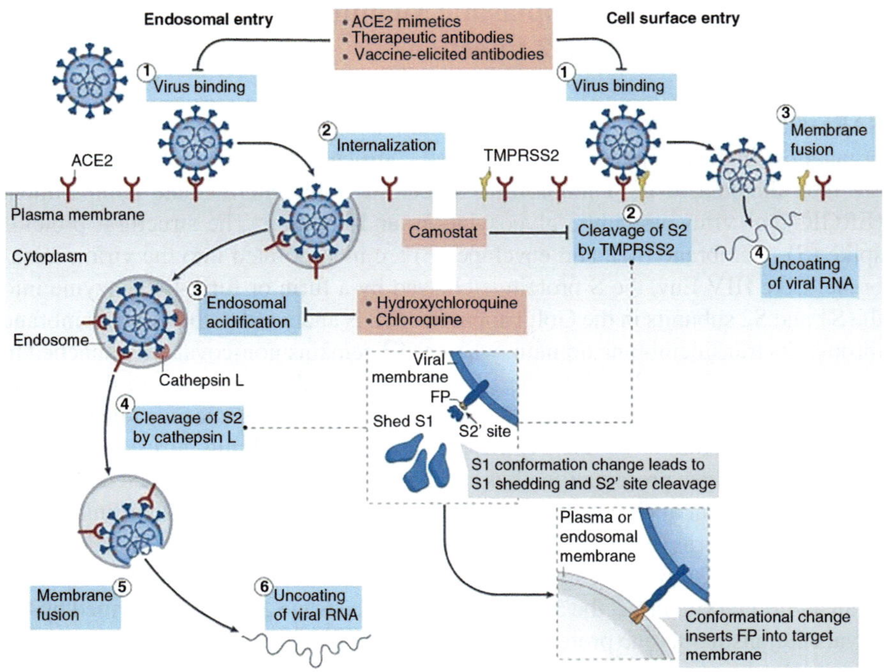

Fig. 16.1 Receptor-mediated entry of SARS-Cov-2 through the endosomal pathway or at the cell surface. Pink boxes show classic inhibitors of steps. Variants of the virus use different pathways of entry depending on the expression level of the TMPRRS2 in different tissues and exhibit different replication rates [Qu et al., 2023]. Analogous routes have been described for HIV, except that its fusion protein (Env) does not require enzymatic processing to induce fusion at the cell surface (see text). *(Reproduced from* Jackson et al., 2022*)*

human and mouse ACE2 receptor, it has been found that Omicron RBD binds to ACE2 with enhanced affinity compared to Wuhan-HU-1 and Delta RBD's [Meng et al. 2022; Cameroni et al. 2022].

Notably, Omicron harbors mutations near or in the furin cleavage site at the S1/S2 junction (H655Y, N679K, and P681H) that substantially reduce the intracellular processing by furin, and so, by TMPRSS2 at the S2' site on the plasma membrane. As a result, Omicron tends to use the endosomal entry pathway mediated by cathepsin L/B, rather than the plasma membrane entry pathway mediated by TMPRSS2. The H655Y mutation in Omicron is a major determinant of the intrinsic low fusogenicity with the plasma membrane and enhanced dependance on endosomal entry of this variant [Qu et al. 2023]. Accordingly, in contrast with previous variants, Omicron does not induce syncytia in spite of its high affinity for ACE2. Thus, the inability of the Omicron S protein to acquire a fusogenic conformation at the plasma membrane prevents it from inducing fusion between cells, thus limiting its pathogenic potential despite its high transmissibility [Meng et al. 2022] (Fig. 16.2).

Isolation of SARS-CoV-2 was initially carried out in China and other countries using the epithelial cell line Vero E6 from the kidney of the African monkey

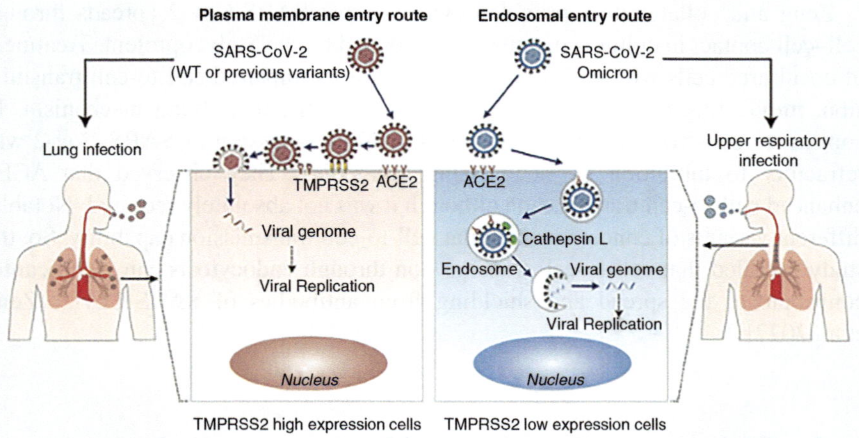

Fig. 16.2 Entry routes and pathogenesis of SASR-CoV-2 Omicron and previous variants. **Left:** Previous variants mainly infect lung epithelial cells, which are TMPRSS2-high expressing cells, and enter host cells by the plasma membrane route. Virus first binds to ACE2 and then to TMPRSS2 to be cleaved at the S protein. Then, the S2 protein anchors to the target cell membrane and mediates fusion of the virus with the cell membrane to release the viral genome into the cytoplasm. **Right:** Due to mutations at the furin cleavage site, Omicron S protein cannot be cleaved by TMPRSS2 and enter host cells by the endosomal route. Omicron mainly infects the upper airway epithelial cells, which express low levels of TMPRSS2. The virus–ACE2 complex is internalized via endocytosis into the endosomes, where S protein is cleaved by cathepsins. Then the S2 protein anchors to the endosomal membrane, and the viral and endosomal membranes are fused together to form a pore and release the viral genome into the cytoplasm. *(Reproduced from* Fan et al., 2022*)*

Chlorocebus aethiops. These cells express normal to moderate levels of ACE2 and TMPRSS2 proteins in their membrane. Research using this cell line showed that endocytosis was the entry route. Because of this, at that time it was considered that direct fusion of the viral membrane with the cell membrane was not a relevant form of entry of SARS-CoV-2 into target cells. However, the histopathological analysis of lung tissue of individuals at Wuhan diagnosed with COVID-19 after surgery due to lung cancer, showed the presence of multinucleated cells along with other inflammation features [Tian et al. 2020], indicating the possibility that the Wuhan strain harbor an S protein able to induce fusion at physiological pH. Lau and cols. finally demonstrated that deletion or point mutation at the S1/S2 junction seemed in Vero E6-derived viruses were not present in the original patient viruses [Lau et al. 2020]. Therefore, the original virus could enter the cells directly at the cell membrane level. Later, it was shown that different cell lines infected with SARS-Cov-2 express the S protein at their surface and can fuse with ACE2 and TMPRSS2-positive neighboring cells, and that the low levels of TMPRSS2 in Vero E6 cells had provoked the selection in vitro of viruses lacking the furin cleavage site [Klimstra et al. 2020]. Thus, Vero E6-derived viruses could enter the cells only through the endocytic pathway [Hoffmann et al. 2020].

Zeng and collaborators provided evidence that SARS-Cov-2 spreads through cell–cell contact in cultures in a manner mediated by the S glycoprotein. Treatment of cocultured cells with endosomal entry inhibitors impaired cell-to-cell transmission, implicating endosomal membrane fusion as the underlying mechanism. In contrast to cell-free virus infection, cell-to-cell transmission of SARS-Cov-2 was refractory to inhibition by neutralizing antibodies. They observed that ACE2 enhanced cell-to-cell transmission although it was not absolutely required. Notably, different variants of concern had similar cell-to-cell transmission capability. So, the study revealed that cell-to-cell transmission through endocytosis can significantly contribute to the spread and shielding from antibodies of SARS-Cov-2 [Zeng et al. 2022].

16.3 Fusion at the Cytoplasmic Membrane *vs.* Endocytosis in HIV Entry

The HIV-1 envelope protein (Env) is a receptor-activated molecular complex composed of the surface gp120 and the transmembrane gp41 subunits, organized in heterotrimers. The gp160 precursor of Env is 856 amino acids long and is synthetized in the endoplasmic reticulum membrane, where it trimerizes. After transport to the Golgi, the gp160 precursor is modified by complex glycans and cleaved by a cellular furin protease into the gp120 and gp41 subunits, which remain non-covalently bound. The gp41 subunit contains an ectodomain, a transmembrane domain, and a long cytoplasmic tail that mediates intracellular trafficking and interaction with the matrix HIV Gag protein for incorporation of Env into virions. The heterotrimer is folded in a metastable inactive state, which potential energy is used to induce membrane fusion. After exit from the ER, HIV-1 Env is transported to the Golgi complex and then to the trans-Golgi network (TGN) from where it follows the secretory pathway to the plasma membrane. Once incorporated into viruses, binding of one Env protomer to CD4 at the target cell membrane initiates the relaxing of the Env structure in a sequence leading to the binding of the other two Env protomers to CD4, a process that is followed by the extension of the heptad repeat coiled coil regions of gp41 to form a structure known as the pre-hairpin intermediate. At this stage, Env is able to interact with the coreceptor (CCR5) at the target cell. Coreceptor binding promotes the insertion of the hydrophobic fusion peptide of gp41 into the target cell membrane and the formation of a stable gp41 six-helix bundle that drives the fusion of the viral and cell membranes [Wang et al. 2020].

Early analyses on the entry mechanisms of HIV-1 into lymphocytes monitored the redistribution of lipid dyes labeling the virus and cellular membranes, as well as the movement of viral proteins and cytoplasmic probes during the first hours of contact between viruses and cells. Results pointed to a rapid kinetics of membrane fusion and entry. The virus entry and synthesis of viral proteins were not inhibited by modulators of the endosomal pH at concentrations which blocked the entry of several other viruses known to use the endocytic pathway, and no evidence of endocytosis was observed [Stein et al. 1987; Cavrois et al. 2004]. These observations prompted the assumption that HIV-1 enter cells through fusion with the cytoplasmic membrane. However, other studies using cell fractionation demonstrated that 50% of HIV-1 virions enter T cells by cytoplasm membrane fusion and 50% enter through endocytosis [Schaeffer et al. 2004]. Then, Schaeffer and collaborators, using GFP-Vpr-labeled viruses, demonstrated that each pathway was upregulated when the other one was inhibited. In both cases, internalization of the virus required CD4. It was observed that inhibition of endosome acidification increased virus replication, and it was proposed that endocytosed viruses can be rescued from degradation to allow fusion with the endosome membrane, or they may be recycled to the cytoplasmic membrane for entry. Since HIV-1 fusion at the cytoplasmic membrane was improved in the presence of endosome inhibitors, authors warned that endocytosis inhibition may disproportionately increase virion entry by cytoplasmic membrane fusion and lead to increased infectivity and disease, showing that a balance between fusion at the cytoplasmic membrane and endocytosis can take place [Schaeffer et al. 2004].

Other authors performed detailed analysis providing biochemical and imaging evidence of productive infection after entry by endocytosis. Images provided by high-speed tridimensional video microscopy of infected cells harboring a viral matrix *gag* gene coupled to green fluorescence protein (called HIV Gag-iGFP), showed a detailed picture and quantitative assessment of contact between cells and cell-to-cell transmission of virus. In this study, 24% of Jurkat cells expressing HIV Gag-iGFP formed stable adhesions with PHA-activated primary CD4 T cells in 4 h; 80% of these cells showed Gag accumulations at the contact site in approximately 82 min. Both formation of contacts and Gag accumulation required the expression of the viral Env protein. The virus was transmitted almost exclusively in such areas of contact, and transmission was dependent of CD4 and HIV co-receptors. Importantly, the transferred virus was observed inside multivesicular compartments in the target cells, suggesting endocytic entry. No evidence of syncytia formation was found. The expression of the fluorescent Gag product in the newly infected cells was indicative that infection through cell-to-cell transfer by endocytosis was productive [Hubner et al. 2009].

It is generally accepted that the rate of productive infection is lower for endocytosis than for entry at the cell surface [Fackler and Peterlin 2000]; however, endocytosis is the main route of entry during the cell-to-cell virus transfer at the virological synapse. The rate of infection with cell-associated HIV at site of close contact can be 23 orders of magnitude greater than infection by cell-free virus [Dimitrov et al. 1993]. Thus, the transfer of HIV through cell-cell contacts and endocytosis may be highly relevant for virus dissemination since the high efficiency of cell-to-cell virus transfer may compensate for the low rate of productive infection. In addition, this kind of entry may reduce the time of virus exposure to antibodies and other inhibitors. Melikyan and collaborators have proposed that binding of HIV particles to the cell surface is followed by conformational changes of gp120 that allow activation of the gp41 fusogenic protein, leading to hemifusion, i.e., the fusion between the external lipid layers of virus and plasma membranes. Complete fusion with formation of the fusion pore would take place at endocytic compartments [Melikyan 2014].

16.4 Expression of the Viral Fusion Protein at the Cell Membrane Is Required for Membrane Fusion and Syncytia Formation

Whatever the route of virus internalization into cells, the expression of the viral envelope protein at the cytoplasmic membrane enables the cells to fuse with surrounding non-infected cells at physiological pH, leading to syncytia (homologous fusion) or heterokaryons (heterologous fusion) [Rivera-Toledo et al. 2011; Ogle et al. 2005]. Fused cells are observed during both HIV and SARS-Cov-2 infections. Cell-cell fusion may lead to efficient virus production by syncytia and heterokaryons, cell death, and inflammation, so it is often associated with severe disease in both SARS-Cov-2 and HIV infections [Bussani et al. 2020; Xu et al. 2020; Blaak et al. 2000; Connell et al. 2020; Casado et al. 2018] (Fig. 16.3). Thus, factors associated to the expression of functional viral envelope proteins at the cell surface are relevant to the study of cell processes underlying pathology.

In HIV-1 infected cells, proteolytically processed Env trimers exit from the ER to be transported to the Golgi complex and then to the trans-Golgi network (TGN). From here, Env is trafficked via the secretory pathway to the plasma membrane, where it is incorporated into nascent virus particles, or alternatively is rapidly endocytosed [Egan et al. 1996]. Two motifs in the cytoplasmic tail of gp41 promote efficient internalization, since they can act autonomously to mediate clathrin-dependent endocytosis: the highly conserved and membrane proximal YSPL (YxxL) motif at position 712 [Boge et al. 1998], and a C-terminal dileucine sequence [Byland et al. 2007]. Internalization is also mediated by the retromer protein complex, a member of the endosomal sorting machinery associated with retrograde transport back to the Golgi complex. The retromer binds directly to the gp41 cytoplasmic tail. Recent studies propose that Env traffics through the endosomal recycling compartment (ERC), in a manner dependent on the Rab11-family interacting protein 1C (FIP1C), as Rab14 is required for Env incorporation in HIV-1 particles. Thus, the endocytic trafficking of HIV-1 Env may serve to both sequester excess viral Env epitope display on the plasma membrane, and provide a route for recycling back of Env to sites of virus assembly [Hoffman et al. 2022; Qi et al. 2013]. As HIV assembles and buds at the cytoplasmic membrane, the mature envelope HIV protein is normally expressed at the cell surface during the virus replication cycle. In contrast, SARS-Cov-2 assembles and buds into membranous replication organelles and reaches the extracellular space through exocytosis. So, additional mechanisms are required to allow the SARS-Cov-2 S protein to be retained at the plasma membrane to enable fusion with other cells, as observed for HIV.

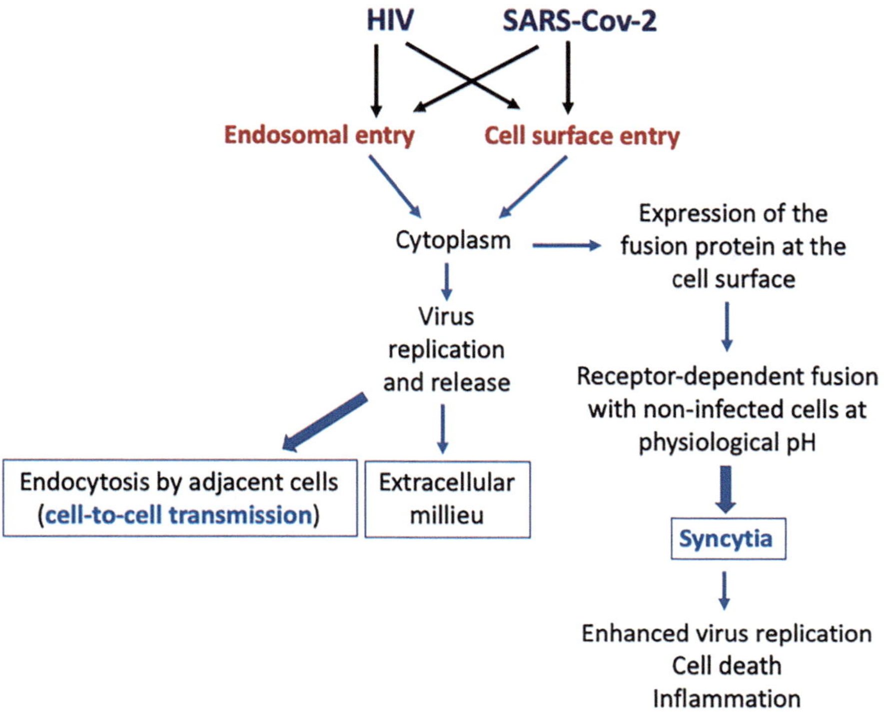

Fig. 16.3 Entry of HIV and SARS-CoV-2 by fusion with the endosomal or the cytoplasmic membrane leads to virus replication. Infected cells can transmit the virus to adjacent cells at sites of close contact through endocytosis. Expression of the viral fusogen at the cell surface and its activity at physiological pH leads to fusion with adjacent non-infected cells to induce syncytia formation. Besides specific receptors, host membrane protein and lipid cofactors are determinant for the induction of membrane fusion by viral proteins (see text)

The expression of the SARS-Cov-2 S protein at the cell membrane is required for its interaction with neighboring ACE2-positive non-infected cells and promotion of cell-cell fusion [Buchrieser et al. 2020]. However, S is naturally retained by cell and viral factors in intracellular membranes, mainly in ER, ERGIC, and Golgi apparatus, so, additional mechanisms should explain its expression at the cell membrane. During the infection cycle of SARS-Cov-2, structural viral proteins, envelope (E), membrane (M), nucleocapsid (N), and spike (S) are translocated into the endoplasmic reticulum (ER) and then trafficked through the ER-Golgi intermediate compartment (ERGIC) ultimately reaching the Golgi apparatus [Rajah et al. 2022]. The co-expression of M and E with the S protein allows maturation and assembly of virions, preventing the expression of S at the cytoplasm membrane and syncytia formation [Boson et al. 2021]. Sites at the transmembrane (TM) and cytoplasmic domains (CTD) interact with cellular factors and other viral structural proteins that facilitate retention of S in intracellular membranes. It has been described that S is continuously recycled in ER-ERGIC-Golgi through the interaction between the

CTD KxHxx motif with the host cytoplasmic coat protein complexes COPI and COPII [Rajah et al. 2022]. Interaction of the KxHxx motif (residues 1269–1273) with COPI allows the retrograde transport of S from the Golgi apparatus to the ER, augmenting its interactions with the M and E viral proteins, promoting the S retention in the ER [Boson et al. 2021]. Cattin-Ortolá and collaborators have described three features of S that promote its accumulation on the plasma membrane: the COPI-binding site is suboptimal, which allows some S protein to escape the Golgi apparatus; a region containing di-acidic COPII binding motifs directs efficient exit of S from the ER; and the S protein is not efficiently endocytosed back due to lack of a tyrosine-containing motif. The exact interaction between COPII and the cytoplasmic tail domain motif for di-acidic ER export (residues 1237–1254) is still unclear; this region contains eight cysteines and only one charged residue, so it binds non-specifically to the S CTD. So, disruption or inefficient interactions between the CTD domain and COPI and COPII provide S with the opportunity for escape to the cytoplasmic membrane [Cattin-Ortolá et al. 2021]. Further studies should provide a deeper insight into specific interactions between these protein complexes and adaptor proteins with the CTD motif of the spike protein that contribute to syncytia formation.

16.5 Membrane Cofactors Involved in Virus-Dependent Membrane Fusion

Besides specific receptors and enzymes processing viral proteins, entry of viruses through endocytosis or through the cell surface is mediated by a variety of host cell membrane molecules that influence the efficacy of specific binding and fusion of the HIV and SARS-Cov-2 viruses in different tissues. The best-known mechanisms of action of cellular proteins are the promotion of the attachment of viruses to the plasma membrane surface, the alteration of the cell membrane fluidity to impede or promote fusion, and the regulation of endocytosis. Some of these molecules are expressed ubiquitously.

The SARS-Cov-2 S protein interacts with both heparan sulfate and ACE2 through its RBD, and heparan sulfate can independently bind the S protein in vitro [Clausen et al. 2020]. In addition, several C-type lectin receptors (DC-SIGN, L-SIGN, LSECtin, ASGR1, and CLEC10A) and the Tweety family member 2 protein (TTYH2), a probable chloride channel, were identified as glycan-dependent binding sites of the S protein of SARS-Cov-2 [Lu et al. 2021]. A sialic acid-binding immunoglobulin-like lectin 1, (SIGLEC1) also known as CD169, sialoadhesin, or Siglec-1, has also been identified as a binding receptor that improves ACE2-dependent infection. A SIGLEC1-blocking antibody inhibited infection of HEK293T cells, supporting its role as a cofactor of SARS-Cov-2 [Lempp et al. 2021]. Likewise, the expression of members of the costimulatory T cell, immunoglobulin, and mucin proteins TIM (TIM-1 and TIM-4) and TAM (AXL) improves the binding of SARS-Cov-2 to cells, facilitate the internalization of fluorescence-labeled virions, and increases ACE2-dependent infection [Bohan et al. 2021]. Binding of the virus to proteins of the TIM family occurs through virion associated phosphatidylserine [Jemielity et al. 2013]. Finally, the lymphocyte function-associated antigen 1 (LFA-1), which is expressed in multiple leukocytes, was proposed as an ACE2-independent entry factor for SARS-Cov-2 in T cells [Shen et al. 2022].

Neuropilin-1 (NRP1) is a protein expressed on the surface membrane that promotes infection of SARS-Cov-2. Furin cleavage of the C terminus of the S1 protein generates an amino acid sequence (682RRAR685) that conforms a R/K]XX[R/K motif, termed the "C-end rule." NRP1 directly binds CendR motifs and significantly enhances the infectivity of SARS-Cov-2, [Cantuti-Castelvetri et al. 2020; Daly et al. 2020], through internalization of the virus by an endocytic process resembling micropinocytosis [Teesalu et al. 2009]. So, a natural deletion of the S1/S2 furin cleavage contributes to attenuate pathogenicity in animal models of human disease (Lau et al. 2020). Also, one group identified a receiver panel for SARS-Cov-2 with various binding properties, functions, and tissue distributions. From the panel of identified membrane proteins that interact with SARS-Cov-2 S, two, ASGR1 and KREMEN1, functioned independently of ACE2, directly mediating the entry of the virus both in vitro and in vivo [Gu et al. 2022]. The functional mechanism of these putative receptors still deserves analysis.

The expression of IFITM's (interferon-induced transmembrane proteins) is linked to the activation of signaling pathways that ultimately promote the expression of interferon-stimulated genes. It is thought that IFITMs prevent viral entry due to the alteration of mechanical and physical properties of the cell membrane, such as its curvature, rigidity, and composition of the cell's membrane bilayer [Barad et al. 2015]. As for HIV [Beitari et al. 2020], IFITM3 has been shown to inhibit SARS-Cov-2 infection independently of S-palmitoylation, thus having a restriction mechanism different from action at the cell membrane [Shi et al. 2021]. IFITM3 amphipathic helix was required for virus restriction, and mutation of residues within the IFITM3 endocytosis-promoting motif converted human IFITM3 into an enhancer of SARS-CoV-2 infection and cell-to-cell fusion promoter. Overexpression of TMPRSS2, which increases plasma membrane fusion, attenuated IFITM3 restriction and converted amphipatic helix mutants into infection enhancers. On the other hand, a recent study has shown that endogenous expression of IFITM2 and/or IFITM3 is critical for efficient replication of SARS-CoV-2 and other human coronaviruses as SARS-CoV-1 and hCoV-OC43, whereas overexpression of IFITIMs inhibits replication of these and other viruses (MERS-, NL63-and 229E-hCoVs). These results suggest that IFITM's favor the entry of ACE2-tropic SARS-CoVs and their overexpression cause artificial inhibitory effects [Xie et al. 2023]. From these reports, it can be noted that IFITIM's can alter the balance between fusion at the plasma membrane and endocytosis.

Similar to IFITM's, lymphocyte antigen 6 family member E (LY6E) promotes RNA virus infection from several viral families and modulates HIV-1 infection in a way that depends on the level of CD4 expression in target cells. However, LY6E was found to restrict human coronavirus infection [Zhao et al. 2020].

Transmembrane proteins favoring HIV fusion, different from receptors and coreceptors, have also been described, including ICAM-1, LFA-1, dynamin-2, and alpha4beta7 integrin. On the other hand, other proteins inhibiting fusion, and thus recognized as restriction factors, have been well recognized, such as tetraspanins CD9 and CD81, EWI-2, PSGL-1, CD137 (BST-2/tetherin), IFITM1, IFITM3, SERINC3, SERINC5, and MARCH8 [reviewed in Ruiz-Rivera et al. 2021].

A different mechanism that may increase membrane fusogenicity relates to the rise of intracellular Ca2+ induced by the S protein of SARS-Cov-2 and the Env protein of HIV-1. Ca2+ transiently activates fundamental proteins in cell physiology, such as a family of chloride channels termed TMEM16, which have a scramblase activity that naturally translocate phospholipids from the internal leaflets to the outer leaflets of the cell [Braga et al. 2021]. Exposure through translocation of certain phospholipids, such as phosphatidylserine (PS), triggers cell-cell fusion events [Whitlock and Chernomordik 2021]. Thus, non-apoptotic exposure of PS may influence fusion events and the recruitment of spike in the cell membrane. HIV-1-induced PS redistribution depends on Ca2+ signaling triggered by Env-coreceptor interactions and also involves the lipid scramblase TMEM16F. Externalized PS strongly promotes Env-mediated membrane fusion. Blocking externalized PS or suppressing TMEM16F inhibited Env-mediated fusion. These findings suggest that cell-surface PS acts as an important cofactor that facilitates receptor-dependent merger of viral and cell membranes and infection [Zaitseva et al. 2017].

The association of the transmembrane domain of the S protein, the membrane proximal regions, and the CTD, with certain lipids at the cell membrane such as cholesterol, allows the exposure of the S protein in the outer leaflet of the membrane [Sanders et al. 2021]. It is suggested that the presence of aromatic amino acid residues found in the TM S domain may be crucial for interactions with specific types of cholesterol in the cell membrane via lipidic raft-independent mechanisms [Sanders et al. 2021], ultimately favoring fusion events [Corver et al. 2009]. Post-transcriptional palmitoylation with lipidic moieties in cysteine amino acid residues present in the CTD S domain might also be fundamental for fusion, as well as the cysteine overall content in the CTD [McBride and Machamer 2010]. Punctual mutations that result in an amino acid substitution of cysteines to alanines in the CTD negatively alter cell-cell fusion [Sanders et al. 2021]. These observations lead to the proposal that membrane lipid-spike protein interactions are necessary to drive cell-cell fusion.

Thus, fusogenicity is dependent on the presence of adhesion molecules, alternative receptors, and the activity of promoting or restriction factors at the cytoplasmic or endosomal membranes of target cells. Importantly, once in the cytoplasm, replication of viruses is strikingly regulated by mechanisms termed together "intrinsic immunity." Mechanism developed by viruses to evade from these restrictions has been described. This aspect is out of the scope of this chapter and can be reviewed elsewhere [Majdoul and Compton 2022].

16.6 Conclusions

Receptor-mediated fusion with the endosomal membrane or the cytoplasmic membrane are alternative pathways for entry of HIV and SARS-CoV-2 into target cells. In addition to specific receptors, other membrane components such as adhesion molecules or alternative receptors, as well as restriction factors and particular lipids, promote or significantly limit the fusogenic activity of viral proteins. Fusion of virus particles at the cytoplasmic membrane allows the direct entry of the virus into the cytoplasm for replication, whereas endocytosis is slower and may lead to decreased or efficient replication. Endocytosis is a relevant mechanism of virus dissemination at sites of close contact between cells. The selection of an in vitro system for the study of virus replication and the analysis of neutralizing agents should consider the determination of the balance between the two entry routes and the role of membrane cofactors.

Acknowledgments The authors thank the Programa de Apoyo a Proyectos de Investigación e Innovación Tecnológica (PAPIIT) of Universidad Nacional Autónoma de México (UNAM) (grant AG200623 to LH) and Programa Nuevas Alternativas contra Enfermedades Infecciosas (NUATEI) of Instituto de Investigaciones Biomédicas, UNAM. AG-M is a post-graduate student of the Posgrado en Ciencias Bioquímicas, Universidad Nacional Autónoma de México, and receives a scholarship from the Consejo Nacional de Ciencia y Tecnología de México (CV 1184388).

References

Barad BA, Echols N, Wang RY-R, Cheng Y, DiMaio F, Adams PD, Fraser JS (2015) EMRinger: side chain–directed model and map validation for 3D cryo-electron microscopy. Nat Methods 12(10):943–946

Beitari S, Pan Q, Finzi A, Liang C (2020) Differential pressures of SERINC5 and IFITM3 on HIV-1 envelope glycoprotein over the course of HIV-1 infection. J Virol 94(16):e00514–20

Blaak H, van't Wout AB, Brouwer M, Hooibrink B, Hovenkamp E, Schuitemaker H (2000) In vivo HIV-1 infection of CD45RA+ CD4+ T cells is established primarily by syncytium-inducing variants and correlates with the rate of CD4+Tcell decline. Proc Natl Acad Sci 97(3):1269–1274

Boge M, Wyss S, Bonifacino JS, Thali M (1998) A membrane-proximal tyrosine-based signal mediates internalization of the HIV-1 envelope glycoprotein via interaction with the AP-2 clathrin adaptor. J Biol Chem 273(25):15773–15778

Bohan D, Van Ert H, Ruggio N, Rogers KJ, Badreddine M, Aguilar Briseño JA, Elliff JM, Rojas Chavez RA, Gao B, Stokowy T et al (2021) Phosphatidylserine receptors enhance SARS-Cov-2 infection. PLoS Pathog 17(11):e1009743

Boson B, Legros V, Zhou B, Siret E, Mathieu C, Cosset F-L, Lavillette D, Denolly S (2021) The SARS-Cov-2 envelope and membrane proteins modulate maturation and retention of the spike protein, allowing assembly of virus-like particles. J Biol Chem 296

Braga L, Ali H, Secco I, Chiavacci E, Neves G, Goldhill D, Penn R, Jimenez-Guardeño JM, Ortega-Prieto AM, Bussani R et al (2021) Drugs that inhibit TMEM16 proteins block SARS-Cov-2 spike-induced syncytia. Nature 594(7861):88–93

Buchrieser J, Dufloo J, Hubert M, Monel B, Planas D, Rajah MM, Planchais C, Porrot F, Guivel-Benhassine F, Van der Werf S et al (2020) Syncytia formation by SARS-Cov-2-infected cells. EMBO J 39(23):e106267

Bussani R, Schneider E, Zentilin L, Collesi C, Ali H, Braga L, Volpe MC, Colliva A, Zanconati F, Berlot G et al (2020) Persistence of viral RNA, pneumocyte syncytia and thrombosis are hallmarks of advanced COVID-19 pathology. EBioMedicine 61:103104

Byland R, Vance PJ, Hoxie JA, Marsh M (2007) A conserved dileucine motif mediates clathrin and AP-2–dependent endocytosis of the HIV-1 envelope protein. Mol Biol Cell 18(2):414–425

Cameroni E, Bowen JE, Rosen LE, Saliba C, Zepeda SK, Culap K et al (2022) Broadly neutralizing antibodies overcome SARS-CoV-2 omicron antigenic shift. Nature 602(7898):664–670

Casado C, Marrero-Herna´ndez S, Ma´rquez-Arce D, Pernas M, Marfil S, Borra's-Grañana F, Olivares I, Cabrera-Rodr´ıguez R, Valera M-S, de Armas-Rillo L, Lemey P, Blanco J, Valenzuela-Fernández A, Lopez-Galíndez C (2018) Viral characteristics associated with the clinical nonprogressor phenotype are inherited by viruses from a cluster of HIV-1 elite controllers. MBio 9(2):e02338–17

Cantuti-Castelvetri L, Ojha R, Pedro LD, Djannatian M, Franz J, Kuivanen S, van der Meer F, Kallio K, Kaya T, Anastasina M, Smura T, Levanov L, Szirovicza L, Tobi A, Kallio-Kokko H, Österlund P, Joensuu M, Meunier FA, Butcher SJ, Winkler MS, Mollenhauer B, Helenius A, Gokce O, Teesalu T, Hepojoki J, Vapalahti O, Stadelmann C, Balistreri G, Simons M (2020) Neuropilin-1 facilitates SARS-CoV-2 cell entry and infectivity Another host factor for SARSCoV-2. Science 370(6518):856–860. https://doi.org/10.1126/science.abd2985

Cattin-Ortolá J, Welch LG, Maslen SL, Papa G, James LC, Munro S (2021) Sequences in the cytoplasmic tail of SARS-Cov-2 spike facilitate expression at the cell surface and syncytia formation. Nat Commun 12(1):5333

Cavrois M, Neidleman J, Bigos M, Greene WC (2004) Fluorescence resonance energy transfer-based HIV-1 virion fusion assay. Flow Cytometry Protoc 263:333–343

Clausen TM, Sandoval DR, Spliid CB, Pihl J, Perrett HR, Painter CD, Narayanan A, Majowicz SA, Kwong EM, McVicar RN et al (2020) SARS-Cov-2 infection depends on cellular heparan sulfate and ACE2. Cell 183(4):1043–1057

Connell BJ, Hermans LE, Wensing AM, Schellens I, Schipper PJ, van Ham PM, de Jong DT, Otto S, Mathe T, Moraba R et al (2020) Immune activation correlates with and predicts CXCR4 co-receptor tropism switch in HIV-1 infection. Sci Rep 10(1):15866

Corver J, Broer R, Van Kasteren P, Spaan W (2009) Mutagenesis of the transmembrane domain of the SARS coronavirus spike glycoprotein: refinement of the requirements for SARS coronavirus cell entry. Virol J 6:1–13

Daly JL, Simonetti B, Klein K, Chen K-E, Williamson MK, Antón-Plágaro C, Shoemark DK, Simón-Gracia L, Bauer M, Hollandi R, Greber UF, Horvath P, Sessions RB, Helenius A, Hiscox JA, Teesalu T, Matthews DA, Davidson AD, Collins BM, Cullen PJ, Yamauchi Y (2020) Neuropilin-1 is a host factor for SARS-CoV-2 infection Another host factor for SARS-CoV-2. Science 370(6518):861–865. https://doi.org/10.1126/science.abd3072

Dimitrov D, Willey R, Sato H, Chang L-J, Blumenthal R, Martin M (1993) Quantitation of human immunodeficiency virus type 1 infection kinetics. J Virol 67(4):2182–2190

Egan MA, Carruth LM, Rowell JF, Yu X, Siliciano RF (1996) Human immunodeficiency virus type 1 envelope protein endocytosis mediated by a highly conserved intrinsic internalization signal in the cytoplasmic domain of gp41 is suppressed in the presence of the pr55gag precursor protein. J Virol 70(10):6547–6556

Fackler OT, Peterlin BM (2000) Endocytic entry of HIV-1. Curr Biol 10(16):1005–1008

Fan Y, Li X, Zhang L et al (2022) SARS-CoV-2 omicron variant: recent progress and future perspectives. Signal Transduct Target Ther 7:141

Gu Y, Cao J, Zhang X, Gao H, Wang Y, Wang J, He J, Jiang X, Zhang J, Shen G et al (2022) Receptome profiling identifies KREMEN1 and ASGR1 as alternative functional receptors of SARS-Cov-2. Cell Res 32(1):24–37

Han P, Li L, Liu S, Wang Q, Zhang D, Xu Z, Han P, Li X, Peng Q, Su C et al (2022) Receptor binding and complex structures of human ACE2 to spike RBD from omicron and delta SARS-Cov-2. Cell 185(4):630–640

Hoffman HK, Aguilar RS, Clark AR, Groves NS, Pezeshkian N, Bruns MM, van Engelenburg SB (2022) Endocytosed HIV-1 envelope glycoprotein traffics to Rab14+ late endosomes and lysosomes to regulate surface levels in T-cell lines. J Virol 96(14):e00767–22

Hoffmann M, Kleine-Weber H, Schroeder S, Krüger N, Herrler T, Erichsen S, Schiergens TS, Herrler G, Wu N-H, Nitsche A, Müller MA, Drosten C, Pöhlmann S (2020) SARS-Cov-2 cell entry depends on ACE2 and TMPRSS2 and is blocked by a clinically proven protease inhibitor. Cell 181(2):271–280.e8

Hubner W, McNerney GP, Chen P, Dale BM, Gordon RE, Chuang FY, Li X-D, Asmuth DM, Huser T, Chen BK (2009) Quantitative 3D video microscopy of HIV transfer across T cell virological synapses. Science 323(5922):1743–1747

Jackson CB, Farzan M, Chen B, Choe H (2022) Mechanisms of SARS-Cov-2 entry into cells. Nat Rev Mol Cell Biol 23(1):3–20

Jemielity S, Wang JJ, Chan YK, Ahmed AA, Li W, Monahan S, Bu X, Farzan M, Freeman GJ, Umetsu DT et al (2013) TIM-family proteins promote infection of multiple enveloped viruses through virion-associated phosphatidylserine. PLoS Pathog 9(3):e1003232

Klimstra WB, Tilston-Lunel NL, Nambulli S, Boslett J, McMillen CM, Gilliland T, Dunn MD, Sun C, Wheeler SE, Wells A et al (2020) SARS-Cov-2 growth, furin-cleavage-site adaptation and neutralization using serum from acutely infected hospitalized COVID-19 patients. J Gen Virol 101(11):1156

Lempp FA, Soriaga LB, Montiel-Ruiz M, Benigni F, Noack J, Park Y-J, Bianchi S, Walls AC, Bowen JE, Zhou J et al (2021) Lectins enhance SARS-Cov-2 infection and influence neutralizing antibodies. Nature 598(7880):342–347

Lu Q, Liu J, Zhao S, Castro MFG, Laurent-Rolle M, Dong J, Ran X, Damani-Yokota P, Tang H, Karakousi T et al (2021) SARS-Cov-2 exacerbates proinflammatory responses in myeloid cells through C-type lectin receptors and Tweety family member 2. Immunity 54(6):1304–1319

Majdoul S, Compton AA (2022) Lessons in self-defence: inhibition of virus entry by intrinsic immunity. Nat Rev Immunol 22(6):339–352

McBride CE, Machamer CE (2010) Palmitoylation of SARS-Cov S protein is necessary for partitioning into detergent-resistant membranes and cell–cell fusion but not interaction with m protein. Virology 405(1):139–148

Melikyan GB (2014) HIV entry: a game of hide-and-fuse? Curr Opin Virol 4:1–7

Meng B, Abdullahi A, Ferreira IATM et al (2022) Altered TMPRSS2 usage by SARS-CoV-2 omicron impacts infectivity and fusogenicity. Nature 603:706–714

Ogle BM, Cascalho M, Platt JL (2005) Biological implications of cell fusion. Nat Rev Mol Cell Biol 6(7):567–575. https://doi.org/10.1038/nrm1678

Qi M, Williams JA, Chu H, Chen X, Wang J-J, Ding L, Akhirome E, Wen X, Lapierre LA, Goldenring JR et al (2013) Rab11-FIP1C and Rab14 direct plasma membrane sorting and particle incorporation of the HIV-1 envelope glycoprotein complex. PLoS Pathog 9(4):e1003278

Qu P, Evans JP, Kurhade C, Zeng C, Zheng Y-M, Xu K, Shi P-Y, Xie X, Liu S-L (2023) Determinants and mechanisms of the low fusogenicity and high dependence on endosomal entry of omicron subvariants. MBio 14(1):e03176–22

Rajah MM, Bernier A, Buchrieser J, Schwartz O (2022) The mechanism and consequences of SARS-Cov-2 spike-mediated fusion and syncytia formation. J Mol Biol 434(6):167280

Rivera-Toledo E, Huerta L, Larralde C, Lamoyi E (2011) Quantitative and phenotypic analyses of lymphocyte–monocyte heterokaryons induced by the HIV envelope proteins: significant loss of lymphoid markers. Exp Mol Pathol 90(2):157–166. https://doi.org/10.1016/j.yexmp.2010.11.006

Ruiz-Rivera MB, Gómez-Icazbalceta G, Lamoyi E, Huerta L (2021) Host membrane proteins in the HIV-induced membrane fusion: role in pathogenesis and therapeutic potential of autoantibodies. Curr Opin Pharmacol 60:241–248

Sanders DW, Jumper CC, Ackerman PJ, Bracha D, Donlic A, Kim H, Kenney D, Castello-Serrano I, Suzuki S, Tamura T et al (2021) SARS-Cov-2 requires cholesterol for viral entry and pathological syncytia formation. elife 10:e65962

Schaeffer E, Soros VB, Greene WC (2004) Compensatory link between fusion and endocytosis of human immunodeficiency virus type 1 in human CD4 T lymphocytes. J Virol 78(3):1375–1383

Shen X-R, Geng R, Li Q, Chen Y, Li S-F, Wang Q, Min J, Yang Y, Li B, Jiang R-D et al (2022) ACE2-independent infection of T lymphocytes by SARS-Cov-2. Signal Transduct Target Ther 7(1):83

Shi G, Kenney AD, Kudryashova E, Zani A, Zhang L, Lai KK, Hall-Stoodley L, Robinson RT, Kudryashov DS, Compton AA et al (2021) Opposing activities of IFITM proteins in SARS-Cov-2 infection. EMBO J 40(3):e106501

Stein BS, Gowda SD, Lifson JD, Penhallow RC, Bensch KG, Engleman EG (1987) Ph-independent HIV entry into CD4-positive T cells via virus envelope fusion to the plasma membrane. Cell 49(5):659–668

Teesalu T, Sugahara KN, Kotamraju VR, Ruoslahti E (2009) C-end rule peptides mediate neuropilin-1-dependent cell vascular and tissue penetration. Proc Natl Acad Sci 106(38):16157–16162

Tian S, Hu W, Niu L, Liu H, Xu H, Xiao S et al (2020) Novel coronavirus (COVID-19) pneumonia in two patients with lung cancer. J Thoracic Oncol 15(5):700–704

Wang Q, Finzi A, Sodroski J (2020) The conformational states of the HIV-1 envelope glycoproteins. Trends Microbiol 28(8):655–667

Wang X, Melino G, Shi Y (2021) Actively or passively deacidified lysosomes push β-coronavirus egress. Cell Death Dis 12(3):235

Whitlock JM, Chernomordik LV (2021) Flagging fusion: phosphatidylserine signaling in cell–cell fusion. J Biol Chem 296:100411

Wu L, Zhou L, Mo M, Liu T, Wu C, Gong C, Lu K, Gong L, Zhu W, Xu Z (2022) SARS-Cov-2 omicron RBD shows weaker binding affinity than the currently dominant delta variant to human ACE2. Signal Transduct Target Ther 7(1):8

Xie Q, Bozzo CP, Eiben L, Noettger S, Kmiec D, Nchioua R, Niemeyer D, Volcic M, Lee JH, Zech F, Sparrer KMJ, Drosten C, Kirchhoff F (2023) Endogenous IFITMs boost SARS-coronavirus 1 and 2 replication whereas overexpression inhibits infection by relocalizing ACE2. iScience 26(4):106395

Xu Z, Shi L, Wang Y, Zhang J, Huang L, Zhang C, Liu S, Zhao P, Liu H, Zhu L, Tai Y, Bai C, Gao T, Song J, Xia P, Dong J, Zhao J, Wang F-S (2020) Pathological findings of COVID-19 associated with acute respiratory distress syndrome. Lancet Respir Med 8:420–422

Zaitseva E, Zaitsev E, Melikov K, Arakelyan A, Marin M, Villasmil R, Margolis LB, Melikyan GB, Chernomordik LV (2017) Fusion stage of HIV-1 entry depends on virus-induced cell surface exposure of phosphatidylserine. Cell Host Microbe 22(1):99–110

Zeng C, Evans JP, King T, Zheng Y-M, Oltz EM, Whelan SP, Saif LJ, Peeples ME, Liu S-L (2022) SARS-Cov-2 spreads through cell-to-cell transmission. Proc Natl Acad Sci 119(1):e2111400119

Zhao X, Zheng S, Chen D, Zheng M, Li X, Li G, Lin H, Chang J, Zeng H, Guo J-T (2020) Ly6e restricts entry of human coronaviruses, including currently pandemic SARS-Cov-2. J Virol 94(18):e00562–20

Chapter 17
Mathematical Modeling of Virus-Mediated Syncytia Formation: Past Successes and Future Directions

Hana M. Dobrovolny

Abstract Many viruses have the ability to cause cells to fuse into large multi-nucleated cells, known as syncytia. While the existence of syncytia has long been known and its importance in helping spread viral infection within a host has been understood, few mathematical models have incorporated syncytia formation or examined its role in viral dynamics. This review examines mathematical models that have incorporated virus-mediated cell fusion and the insights they have provided on how syncytia can change the time course of an infection. While the modeling efforts are limited, they show promise in helping us understand the consequences of syncytia formation if future modeling efforts can be coupled with appropriate experimental efforts to help validate the models.

17.1 Introduction

Viruses have developed the ability to infect a wide variety of hosts, ranging from bacteria to plants to animals (Abdelsattar et al. 2022; Bandin and Dopazo 2011; Roossinck 2019; Truyen et al. 1995). Viruses can also jump across species, showing great flexibility and adaptability to different host environments (Martinez-Turino et al. 2021; McLeish et al. 2019; Nova 2021; Parrish et al. 2008; Versoza and Pfeifer 2022). In humans, viruses can infect a variety of cells in different organs and are responsible for a number of illnesses (Chauhan et al. 2020). The broad variety of viral targets and the variety of environments encountered by viruses has led to viruses developing a number of strategies for transmission within a host.

Most commonly, viruses are known to leave the host cell, travel through the extracellular environment, typically via diffusion, but also sometimes getting caught

H. M. Dobrovolny (✉)
Department of Physics & Astronomy, Texas Christian University, Fort Worth, TX, USA
e-mail: h.dobrovolny@tcu.edu

M. Kloc, A. Uosef (eds.), *Syncytia: Origin, Structure, and Functions*, Results and Problems in Cell Differentiation 71,
https://doi.org/10.1007/978-3-031-37936-9_17

in fluid flows in the body or in extracellular vesicles (Bello-Morales et al. 2020; Huang et al. 2011; Johnson et al. 2022; Kouwaki and Okamoto 2017), to land on a nearby, but not necessarily adjacent, cell. This process, known as cell-free transmission, allows viruses to infect cells that can be some distance from the originating host cell. During cell-free transmission, viruses need to interact with cell surface receptors to gain entry into the cell and must also somehow free themselves from the surface of the original host cell (Ahmed et al. 2019; Albornoz et al. 2016; Laureti et al. 2018; Navaratnarajah et al. 2020). They are also exposed to possible dangers in the extracellular environment, such as components of the immune response (Viejo-Borbolla et al. 2017). These processes can hinder effective spread of the virus, so it's not surprising that viruses have developed other modes of transmission that avoid some of these potential dangers.

Viruses can also travel more directly from one cell to another via cell-to-cell transmission (Mothes et al. 2010; Zhong et al. 2013). In cell-to-cell transmission, viruses either create or take advantage of nano-tunnels between cells to travel directly between cells (Labudova 2020; Panasiuk et al. 2018). This allows viruses to avoid the difficulties of cell entry and exit, as well as avoiding the dangers of the extracellular environment. The drawback of cell-to-cell transmission, however, is that viruses can only spread to cells that neighbor the originating cell, potentially slowing spread of the virus (Ge et al. 2021).

There is a third method of transmission that some viruses use to spread between cells. Some viruses are known to cause cells to fuse into large multi-nucleated cells called syncytia (Leroy et al. 2020). Syncytia formation allows viruses to access the resources of another cell without venturing into extracellular space, so it has all the benefits (and drawbacks) of cell-to-cell transmission (Labudova 2020; Cifuentes-Munoz and Dutch 2019). However, the energy and complexity required to completely fuse two cells seems to be much larger than that needed for cell-to-cell transmission (Dittmar et al. 2021). Yet viruses as diverse as varicella-zoster virus, rotavirus, human immunodeficiency virus, coronaviruses, measles, and respiratory syncytial virus have been observed to form syncytia (Ayata et al. 2007; Buchrieser et al. 2020; Diller et al. 2019; Gonzalez-Reyes et al. 2001; Hu et al. 2020; Porotto et al. 2019; Qian et al. 2013; Symeonides et al. 2015; Wang et al. 2017). So what possible advantage does this mode of transmission confer?

It is possible that syncytia formation is an unintended consequence of the fusion proteins of some viruses. Fusion proteins are the viral surface proteins that bind to cell surface receptors and allow the virus to enter the cell by fusing the viral membrane with the cell membrane (Ci et al. 2016; Harrison 2015). During the viral replication process, fusion proteins are also expressed on the surface of infected cells before viruses start to bud off the cell surface. Since the fusion protein is meant to bind to cell surface receptors, fusion proteins on the cell membrane will naturally attempt to bind to the cell surface receptors of neighboring cells, causing the two cells to fuse (Hernandez et al. 1996; Ward et al. 1995). In this scenario, cell fusion is an accidental by-product of the viral mechanism for cell entry. However, there is some evidence that syncytia formation is an important mechanism for spread of the virus between cells. Studies using a number of different viruses have shown that viral strains that have a reduced capacity to cause cell-cell fusion also have reduced viral titers and slower replication during in vitro infections (Fukushima et al. 2019; Matsuyama and Taguchi 2000; Zsak et al. 1992). These studies suggest that even if the origin of syncytia formation was accidental, at least some viruses have come to rely on this mode of transmission as an integral part of how they spread.

Mathematical models can potentially provide insight into possible benefits and drawbacks of syncytia formation by allowing us to test different hypotheses about the role of syncytia in the infection process. It is still not known whether cells in syncytia produce the same amount of virus as infected cells not in syncytia. It is also not known if cells in syncytia have the same average infectious lifespan as infected cells not in syncytia. While the average size and number of syncytia has been measured for some infections, it is not known how either of these factors changes the course of the infection (Ayala-Breton et al. 2014; Gagliardi et al. 2017; Lopez-Balderas et al. 2007). Additionally, mathematical models can investigate the extent of protection provided by syncytia formation from extracellular immune responses and antivirals. These are all scenarios that can be manipulated and tested using mathematical models. While mathematical models have been used to study other aspects of a number of viral infections, such as emergence of drug resistance, coinfections, and the effect of specific mutations, their use in studying the role of syncytia formation is still limited (Dobrovolny and Beauchemin 2017; Holder et al. 2011; Paradis et al. 2015; Pinilla et al. 2012; Pinky and Dobrovolny 2016; Pinky and Dobrovolny 2017; Pinky et al. 2019).

In this chapter, we review the application of mathematical models to study syncytia formation and what we have learned about the role of syncytia formation on the course of an infection from mathematical modeling. We then discuss possible future modeling efforts that could further our understanding of syncytia formation during viral infections.

17.2 Mathematical Models of Virus-Mediated Cell Fusion

There are currently only a handful of mathematical models that incorporate some aspect of syncytia formation (Alzahrani et al. 2020; Amidei and Dobrovolny 2022; Bajzer et al. 2008; Biesecker 2010; Dingli et al. 2009; Jacobsen and Pilyugin 2015; Jessie and Dobrovolny 2021; Mulampaka and Dixit 2011). All the models are based on ordinary differential equations (ODEs), which lack the ability to capture any spatial dependence that is inherent in syncytia-forming infections, or on partial differential equations (PDEs), which have some ability to incorporate spatial dependence. While neither type of model can fully capture the spatial heterogeneity of syncytia formation, these models are providing important insight into our understanding of syncytia formation.

17.2.1 General Role in Viral Infection

To date, there has only been one mathematical model of syncytia formation during viral infections that explored the range of potential infection outcomes as properties of the syncytial cells were varied. The model, proposed by Jessie and Dobrovolny, incorporates both cell-free transmission and syncytia formation (Jessie and Dobrovolny 2021),

$$\frac{dT}{dt} = -\beta TV - \gamma T(I+S)$$
$$\frac{dI}{dt} = \beta TV - \gamma I(T+2I+S) - \delta I$$
$$\frac{dS}{dt} = \gamma T(2I+S) + \gamma I(2I+S) - r_\delta \delta S$$
$$\frac{dV}{dt} = pI + r_p pS - cV.$$

In the model, target cells, T, can be infected by free virus, V, at rate β or they can fuse with already infected cells, I, or syncytia, S, at rate γ. Infected cells can fuse with uninfected target cells, infected cells, or syncytia, to transition from single infected cells to cells in syncytia. The model assumes that single infected cells and cells in syncytia can have different viral production rates, p for single infected cells, $r_p p$ for cells in syncytia, and different infectious lifespans, $1/\delta$ for single infected cells and $1/r_\delta \delta$ for cells in syncytia. All free virus is cleared at rate c.

The model allows for transmission of the infection through syncytia formation if

$$\frac{\gamma T_0}{r_\delta \delta} > 1.$$

Note that it is possible for the infection to spread through fusion only (Fig. 17.1) without any free virus. In this scenario, the initial infected cells simply start fusing with neighboring cells to form syncytia. Since there is no free virus, there are no new single infected cells during the course of the infection, so we only see a growing number of syncytia that eventually decays as the syncytia die. If the cells in syncytia do not produce virus, then there will be no virus detected in swabs from the patient. While the model allows this type of spread as a possibility, it is not known whether any virus actually spreads in this manner.

The more realistic scenario is that a virus has the ability to transmit via both cell-free transmission and syncytia formation. In this case, the model predicts that the effect of syncytia formation starts to influence the time course of the infection when

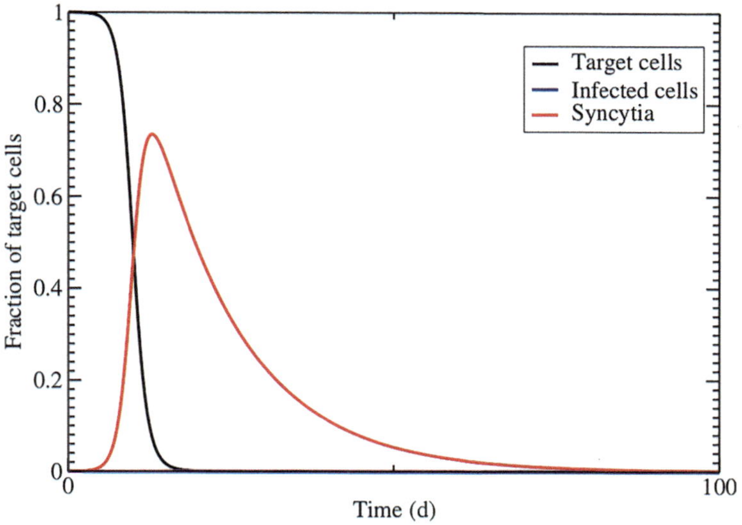

Fig. 17.1 Infection spreading through cell fusion only. The target cells are fused into existing syncytia that eventually die off. There are no single infected cells and no free virus in this infection

$$\frac{\gamma r_p T_0}{r_\delta \delta} \approx 1,$$

which comes from setting the contributions of syncytia and single infected cells to the basic reproduction number equal. This condition involves all the syncytia properties considered in this model. The fusion rate, γ, determines how many syncytia are formed during the infection, so if this is small, then there are few syncytia and their effect on the infection time course is small. The relative viral production rate, r_p, describes the amount of free virus produced by syncytia, so if this is low, the syncytia do not affect the time course of the viral infection since they are not contributing much virus. Finally, $1/r_\delta \delta$ is the lifespan of the syncytia, and if this is small, the syncytia don't survive long enough to have an effect. While all three of these factors contribute to the condition determining when the role of syncytia is important, each parameter has a different effect on the time course of the infection. Figure 17.2 shows how changing each of these parameters affects the time of viral peak (top row) and infection duration (bottom row). While changing the fusion rate and relative viral production rate lead to monotonic decreases in the time of viral peak, changing the relative syncytia lifespan leads to a minimal time of viral peak when the syncytia and single infectious cells contribute equally to spread of the infection. The infection duration decreases when the fusion rate increases but increases when the relative viral production rate increases and shows a clear minimal value as relative syncytia lifespan changes.

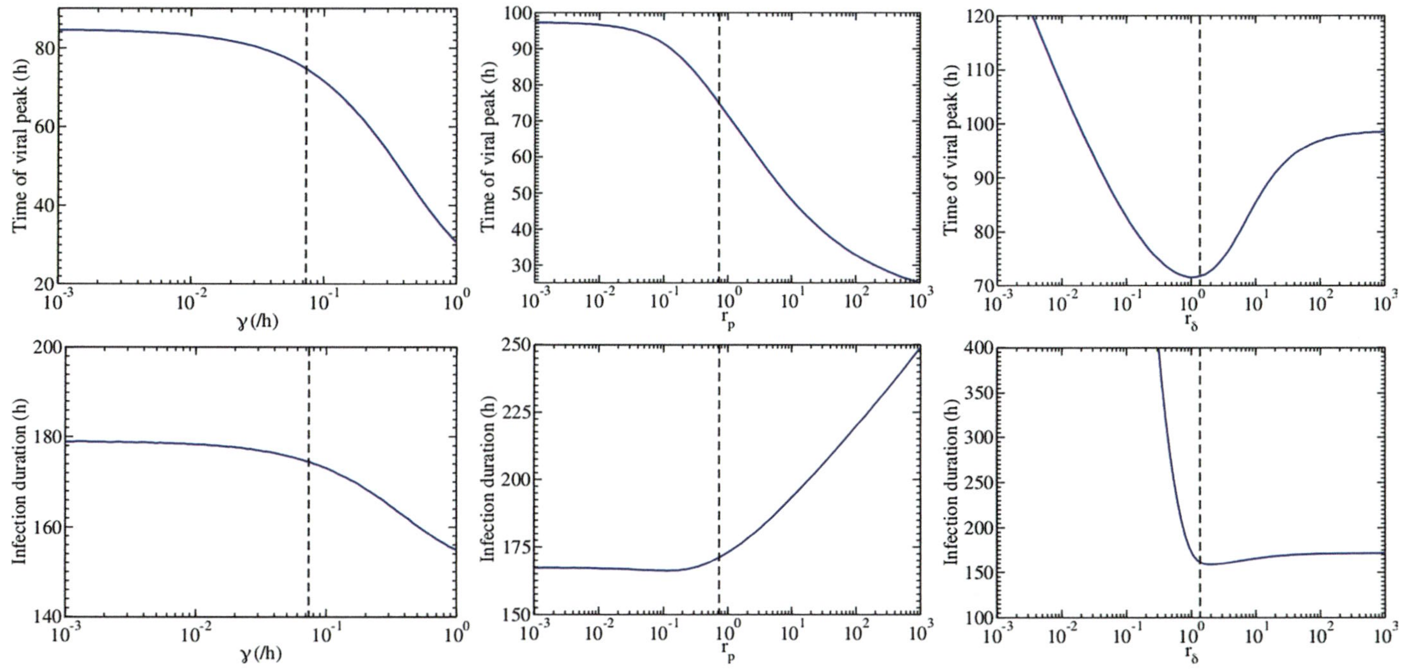

Fig. 17.2 Effect of syncytia characteristics on viral time course. We show how the syncytia formation rate, γ, (left column), relative syncytia viral production rate, r_p (center column), and relative syncytia death rate, r_δ (right column) affect the time of peak viral load (top row) and the infection duration (bottom row). The vertical dashed line indicates where $\frac{\gamma r_p T_0}{r_\delta \delta} = 1$

17.2.2 Cell-Cell Fusion Assay

Virus-mediated syncytia formation is sometimes studied using cell-cell fusion assays (Jenkinson et al. 2003). These assays use donor cells, cells transfected to express the viral fusion protein, and acceptor cells, cells expressing the cell surface proteins, to study virus-mediated syncytia formation without the presence of the virus. Each cell type is stained with a different dye, so that they fluoresce in different colors. Once the cells fuse, the two dyes mix causing syncytia to fluoresce in a third color. This simplified system is easier to model with some of the proposed models being analytically solvable.

The simplest such model considers only the three different cell types, donor (D), acceptor (A), and syncytia (S) with the fusion interactions represented in the model (Amidei and Dobrovolny 2022),

$$\frac{dD}{dt} = -\gamma DA - \gamma SD$$
$$\frac{dA}{dt} = -\gamma DA - \gamma SA$$
$$\frac{dS}{dt} = 2\gamma DA + \gamma SA + \gamma SD.$$

Here, donor and acceptor cells can fuse with each other or to existing syncytia. All fusion events are assumed to occur at the same rate. Under the assumption that the total number of cells is fixed ($D + A + S = N$), this system can be solved analytically to yield expressions for the donor and acceptor cells,

$$D = \frac{ND_0}{D_0 + (N - D_0)e^{\gamma Nt}}$$
$$A = \frac{NA_0}{A_0 + (N - A_0)e^{\gamma Nt}},$$

Figure 17.3 shows predicted time courses for this model under different initial conditions and different fusion rates.

An asymmetric version of this model was also considered by Amidei and Dobrovolny (Amidei and Dobrovolny 2022),

$$\frac{dD}{dt} = -\gamma DA$$
$$\frac{dA}{dt} = -\gamma DA - \gamma SA$$
$$\frac{dS}{dt} = 2\gamma DA + \gamma SA,$$

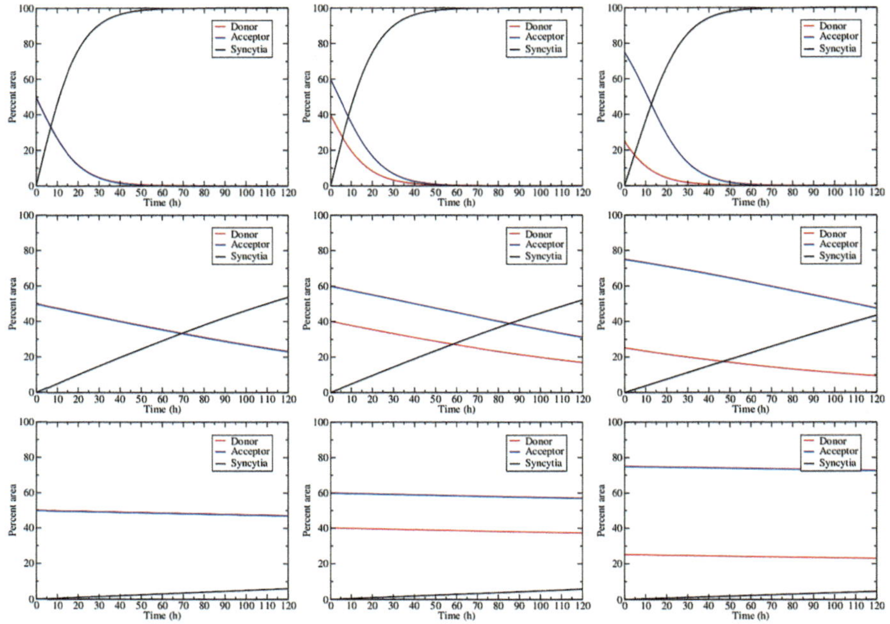

Fig. 17.3 Behavior of the symmetric model of cell-cell fusion. Figures show the percent area covered by donor cells (red), acceptor cells (blue), and syncytia (black) for fusion rates of $1{\times}10^{-3}$ h (top row), $1{\times}10^{-4}$ h (center row), and $1{\times}10^{-5}$ h (bottom row) with an initial 50/50 distribution of donor/acceptor (left column), an initial 40/60 distribution of donor/acceptor (center column), or an initial 25/75 distribution of donor/acceptor (right column)

where only acceptor cells are allowed to bind to existing syncytia. The biological motivation for this type of asymmetry is unclear, but breaking the donor/acceptor symmetry allows for a model that produces syncytia time courses that depend more strongly on the initial ratio of donor to acceptor (see Fig. 17.4). This flexibility produces better fits to experimental data, although breaking the symmetry in favor of the donors should allow for equally good fits to experimental data (Amidei and Dobrovolny 2022). This model can also be solved analytically to yield,

$$D = \frac{D_0}{N}\left[N - A_0 + A_{0e}^{-\gamma Nt}\right]$$

$$A = \frac{NA_0}{A_0 + \left(N - A_0\right)e^{\gamma Nt}}.$$

Figure 17.4 shows predicted time courses for this model under different initial conditions and different fusion rates.

A more detailed mathematical model of cell-cell fusion was used to study syncytia formation of human immunodeficiency virus (HIV) (Mulampaka and Dixit 2011). The model describes binding of the viral envelop protein gp120 to the CD4

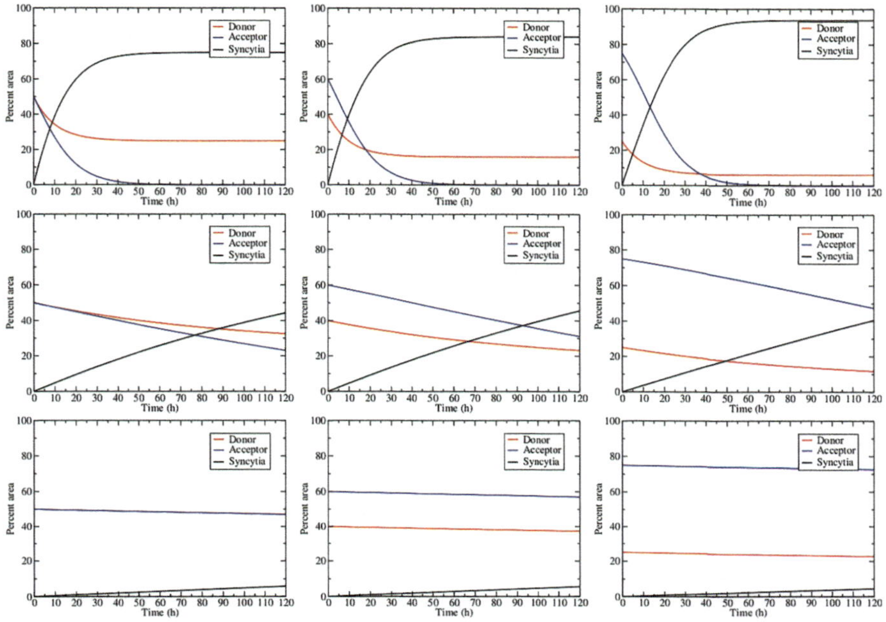

Fig. 17.4 Behavior of the asymmetric model of cell-cell fusion. Figures show the percent area covered by donor cells (red), acceptor cells (blue), and syncytia (black) for fusion rates of 1×10^{-3} h (top row), 1×10^{-4} h (center row), and 1×10^{-5} h (bottom row) with an initial 50/50 distribution of donor/acceptor (left column), an initial 40/60 distribution of donor/acceptor (center column), or an initial 25/75 distribution of donor/acceptor (right column)

CCR5 cell receptor—a necessary step in the fusion process. The basic model has an equilibrium solution of

$$GC = \frac{1}{2}\left[G_0 + C_0 + \frac{1}{K} - \sqrt{\left(G_0 + C_0 + \frac{1}{K}\right)^2 - 4G_0C_0}\right],$$

where GC is the density of gp120-CCR5 complexes, G_0 is the initial density of gp120 proteins, C_0 is the initial density of CCR5 receptors, and $K = k_{on}/k_{off}$ is the ratio of binding rate (k_{on}) and dissociation rate (k_{off}). Figure 17.5 shows the dependence of the equilibrium density of gp120-CCR5 on the initial densities of gp120 proteins, CCR5 receptors, and K. The equilibrium concentration shows an initial linear dependence on both initial quantities until a saturation point is reached. If there are low amounts of either the protein or the receptor, this will be a limiting factor and will determine the final number of complexes. The equilibrium concentration of complexes rises very steeply at very low values of K (either low binding rates or high dissociation rates), reaching the saturation value very quickly.

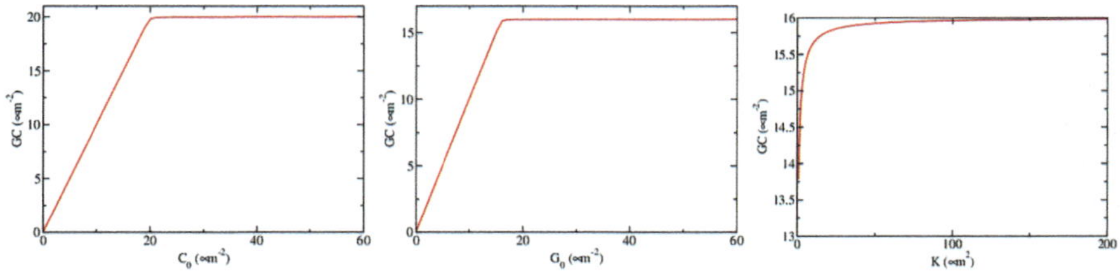

Fig. 17.5 Dependence of equilibrium density of gp120-CCR5 on (left) the initial density of CCR5 receptors, (center) the initial density of gp120 proteins, and (right) the ratio of binding rate to dissociation rate

17.2.3 Oncolytic Viruses

There is particular interest in use of syncytia forming viruses for oncolytic virotherapy (use of viruses to eradicate cancer) since there is some evidence that fusogenic viruses are more adept at eradicating cancer cells (Burton 2019; Ebert et al. 2004). This hypothesis is an ideal example of an idea that can be tested through the use of mathematical modeling. Several models have been developed to attempt to study the role of syncytia formation in oncolytic virotherapy (Alzahrani et al. 2020; Bajzer et al. 2008; Biesecker 2010; Dingli et al. 2009; Jacobsen and Pilyugin 2015).

Initial modeling of syncytia formation in oncolytic viruses used ordinary differential equation models. The first oncolytic virus model that incorporated syncytia formation was proposed by (Bajzer et al. 2008),

$$\frac{dC}{dt} = rC\left(1 - \frac{(C+I)^{\epsilon}}{K^{\epsilon}}\right) - \kappa CV - \rho CI$$

$$\frac{dI}{dt} = \kappa CV - \delta I$$

$$\frac{dV}{dt} = \alpha I - \omega V - \kappa CV.$$

Uninfected cancer cell (C) replication is described by a generalized logistic model with growth rate r, carrying capacity K, and shape parameter ϵ. The uninfected cancer cells can be infected by free virus at rate κ or can fuse with infected cells at rate ρ. Note that the *only* role played by syncytia in this model is to remove uninfected cancer cells from the system—any uninfected cancer cell that fuses with an infected cell is assumed to immediately die. There is a slight imbalance in the model in that the infected cell that fused with the uninfected cancer cell is not also assumed to die (there is no $-\rho CI$ term in the differential equation for I). The syncytia do not produce virus, nor do they take up any space or resources. Infected cells produce virus at rate α and die at rate δ. Free virus loses infectivity at rate ω.

Mathematical analysis of the model finds that the model has three fixed points. There are two disease-free equilibria; one with the tumor eliminated ($C^* = 0$, $I^* = 0$, $V^* = 0$) and one where the tumor remains after therapy ($C^*=K$, $I^* = 0$, $V^* = 0$). Unfortunately, the tumor eradication fixed point is unstable and cannot actually be reached. The third equilibrium is a chronic infection, meaning the patient is left with a tumor that does not grow, but also does not shrink, and the viral infection remains. The chronic infection equilibrium is shown to undergo a Hopf bifurcation, meaning that the model exhibits oscillations as it settles down to the fixed point (Fig. 17.6 (top row)). The syncytia fusion rate appears in the expression for the equilibrium number of infected cells and virus, so syncytia formation in this model helps reduce the final level of infection when there is a controlled tumor.

The model was further investigated by Biesecker et al., who used it to determine optimal treatment regimens (Biesecker 2010). The analysis found that one or two dose therapy was optimal—delivering small doses of virus periodically over a longer time frame is often insufficient to establish an infection that can completely eradicate the tumor. Along a similar vein, they found that small tumors were more difficult to treat than large tumors, again because a small number of uninfected

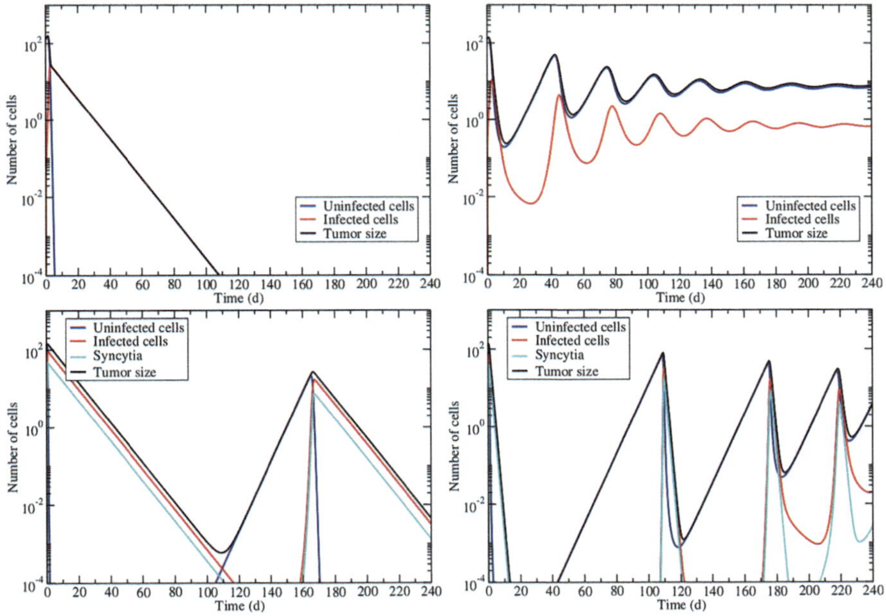

Fig. 17.6 Possible outcomes predicted by the models of (top) Bajzer et al. and (bottom) Dingli et al. (Bajzer et al. 2008; Dingli et al. 2009). We use the same initial conditions and parameter values (taken from Bajzer et al.) to allow for direct comparison of the two models, which differ only in how syncytia play a role in the infection. (top left) The Bajzer model predicts a fairly quick eradication of the tumor while (bottom right) the Dingli model with the same parameter values predicts large oscillations. (top right) The Bajzer model predicts decaying oscillations leading to a controlled tumor while (bottom right) the same parameters in the Dingli model predict much larger oscillations

cancer cells make it difficult to establish a long-lasting infection. It is unclear what role syncytia play in determining optimal treatment since this was not directly investigated in the study. However, since the role of syncytia in this model is to remove uninfected cancer cells without having them produce virus, the syncytia might be contributing to the difficulty in eradicating small tumors.

A later study by Dingli et al. extended the role of syncytia in the Bajzer model (Dingli et al. 2009),

$$\frac{dC}{dt} = rC\left(1 - \frac{(C+I+S)^{\epsilon}}{K^{\epsilon}}\right) - \kappa CV - \rho CI$$

$$\frac{dI}{dt} = \kappa CV + \lambda \rho CI - \delta I$$

$$\frac{dS}{dt} = (1-\lambda)\rho CI - \delta S$$

$$\frac{dV}{dt} = \alpha(I+S) - \omega V - \kappa CV.$$

Note that syncytia now comprise their own compartment in the model (S). They also take up space and resources, so they affect the ability of the tumor to spread. Perhaps more importantly, they produce virus and can contribute to perpetuating the infection.

The model was fit to experimental data from xenograft KAS-6/1 tumors in mice treated with measles virus (Myers et al. 2007). The model is capable of replicating a number of different treatment outcomes including tumor eradication, tumor control (the tumor size does not increase or decrease), and continued tumor growth, similar to the Bajzer model. Since the models are so similar, it is possible to use them to compare how the differing assumptions about the role of syncytia change model predictions. Figure 17.6 (top row) shows two possible outcomes of the Bajzer model as viral production rate and infected cell death rate are changed. Figure 17.6 (bottom row) shows the Dingli model predictions using the same parameter values. The ability of syncytia to produce virus and play a more active role in the infection (Dingli model) causes more extreme oscillations in tumor size.

These first syncytia-forming oncolytic virus models lack any spatial dependence since they consist of ordinary differential equations. Jacobsen et al. addressed this issue by building a model framework that incorporates at least some spatial dependence (Jacobsen and Pilyugin 2015). The model assumes a radially symmetric tumor with a boundary moving at velocity $u(r;t)$. Under this assumption, the partial integro-differential equations describing the system are

$$\frac{\partial c}{\partial t}+\frac{1}{r^2}\frac{\partial}{\partial r}\left(r^2uc\right)=\lambda c-\frac{\beta c}{\left|I_{r_c}\right|}\int_{I_{r_c}}v(s,t)ds-\frac{\rho c}{\left|I_{2r_c}\right|}\int_{I_{2r_c}}\left[i(s,t)+z(s,t)\right]ds$$

$$\frac{\partial i}{\partial t}+\frac{1}{r^2}\frac{\partial}{\partial r}\left(r^2ui\right)=\frac{\beta c}{\left|I_{r_c}\right|}\int_{I_{r_c}}v(s,t)ds-\left(\rho(c+i+z)+\delta\right)i$$

$$\frac{\partial z}{\partial t}+\frac{1}{r^2}\frac{\partial}{\partial r}\left(r^2uz\right)=\frac{\rho c}{\left|I_{2r_c}\right|}\int_{I_{2r_c}}\left[i(s,t)+z(s,t)\right]ds+\rho(c+i+z)-\mu z$$

$$\frac{\partial v}{\partial t}-\kappa\frac{1}{r^2}\frac{\partial}{\partial r}\left(r^2\frac{\partial v}{\partial r}\right)=\frac{N\delta}{\left|J_{r_c}\right|}\int_{I_{r_c}}\left[r_c^2-(r-s)^2\right]ids+\frac{\alpha}{I_{r_c}}\int_{I_{r_c}}\left[i(s,t)+z(s,t)\right]ds-\gamma v,$$

where $J_{r_c}=\int_{I_{r_c}}\left[r_c^2-(r-s)^2\right]ds$ and $I_{r_c}=\left(\max\left[0,r-r_c\right],\min\left[R,r+r_c\right]\right)$. Uninfected cancer cells c replicate with growth rate λ. They can be infected by free virus at rate β or incorporated into syncytia at rate ρ. The infected cancer cells i can also be incorporated into larger syncytia at rate ρ or they can die at rate δ. Syncytia, z, die at rate μ. Syncytia and infected cells produce virus at rate α, but only singly infected cells are assumed to have the ability to burst and release virus. The virus diffuses among the cells, while the positions of the cells themselves change with the moving boundary.

Under the assumption of spatial homogeneity, mathematical analysis results in the rather elegant result that the rate of tumor growth or decay is given by

$$F=\frac{1}{3}\left[(\lambda+\mu)c*+(\mu-\delta)i*-\mu\right],$$

where c^* and i^* are the fixed point values of the spatially homogeneous system. Unfortunately, this result suggests that it is not possible, at least in the spatially homogeneous case, to simultaneously eliminate the tumor along with the infection itself. Figure 17.7 shows combinations of c^* and i^* that result in a growth rate of zero for the parameters used in (Jacobsen and Pilyugin 2015). When the infected cells go to zero, we still have about half of the tumor cells remaining. The best that can be done in the spatially homogeneous system is to have a controlled tumor, one that doesn't grow or shrink, with a chronic viral infection.

However, real tumors are not spatially homogeneous, so more realistic scenarios were assessed using numerical simulations of the model assuming an initial injection of virus into the center of the tumor. In this case, it was found that the tumor radius could go to zero, i.e., that the cancer could be cured if the fusion rate was large. Interestingly, simulations showed that cure of the tumor was also possible when fusion rate was low if the viral burst size during cell lysis was also low. At low viral burst size, intermediate values of fusion rate did not result in cure of the tumor.

Another model, proposed by Alzahrani et al., explores the hypothesis that movement of syncytia helps spread the viral infection throughout a tumor, thus increasing

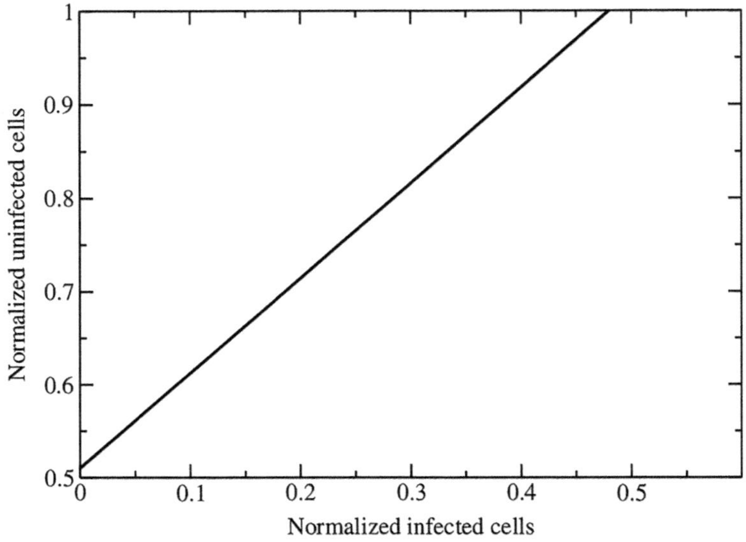

Fig. 17.7 Stable tumor sizes of the Jacobsen oncolytic virus model. Using values of $\lambda = 0.48$ /d, $\mu = 0.5$ /d, and $\delta = 1.5$ /d, we plot the normalized fixed point values of c^* and i^* that result in a growth rate, F of 0

the efficacy of oncolytic virus treatment (Alzahrani et al. 2020). This assumption is based on experimental studies that indicate that syncytia can use pseudopods to move (Sylwester et al. 1998; Sylwester et al. 1993). The model incorporates two types of motion for both cells and virus: diffusion from high concentrations to low concentrations, and haptotaxis determined by the extracellular matrix (ECM). Alzahrani et al. used a series of three mathematical models to systematically investigate the role of syncytia motion. The base model assumes stationary syncytia,

$$\frac{\partial c}{\partial t} = D_c \nabla^2 c - \eta_c \nabla \cdot (c \nabla u) + \mu_1 c(1-c) - \rho c v - \kappa c i$$

$$\frac{\partial i}{\partial t} = D_i \nabla^2 i - \eta_i \nabla \cdot (i \nabla u) + \rho c v + \rho_0 \kappa c i - \delta_i i$$

$$\frac{\partial s}{\partial t} = (1-\rho_0)\kappa c i - \delta_s s$$

$$\frac{\partial u}{\partial t} = -u(\alpha_c c + \alpha_i i + \alpha_s s) + \mu_2 u(1-u-c-i-s)$$

$$\frac{\partial v}{\partial t} = D_v \nabla^2 v - \eta_v \nabla \cdot (v \nabla u) + b_i i + b_s s - \rho c v - \delta_v v,$$

where c represents the uninfected cancer cells, i represents the infected cancer cells, s represents the syncytia, u is the ECM, and v is the virus. In this version, uninfected and infected cancer cells, as well as virus, move via diffusion (with diffusion coefficients D_c, D_i, and D_v) and through haptotactic movement towards higher concentrations of ECM (with rates η_c, η_i, and η_v). Uninfected cancer cells are assumed to replicate via a logistic model with growth rate μ_1. Uninfected cancer cells can be infected via free virus at rate ρ or can fuse with infected cells at rate κ. Infected cells and syncytia produce virus at rates b_i and b_s, and die at rates δ_i and δ_s, respectively. The ECM does not move, but is assumed to be remodeled in response to cell density. The authors compared the predictions of this model to a model where syncytia could move, both through diffusion or haptotactic motion, described by the modified syncytia equation

$$\frac{\partial s}{\partial t} = D_s \nabla^2 s - \eta_s \nabla \cdot (s \nabla u) + (1 - \rho_0) \kappa c i - \delta_s s.$$

The study finds that a virus' fusogenicity has a larger impact on its ability to clear a tumor than its motility. Both diffusion and haptotactic motion had little effect on the tumor invasion area, but changes in ρ_0, a measure of the fusogenicity of the virus, led to substantial changes in the tumor invasion area.

17.3 Future Directions

17.3.1 Experiments

Experiments are crucial for refining and improving mathematical modeling efforts. Without experimental data to validate or refute different mathematical hypotheses, modelers cannot construct models that accurately reproduce reality. In the case of syncytia formation, experimental observation of some of the most fundamental questions is still lacking. For example, the model presented by Jessie and Dobrovolny (Sect. 2.1) explores ranges of the amount of virus produced by syncytia and the lifespan of syncytia simply because measurements of these quantities do not yet exist for any virus (Jessie and Dobrovolny 2021). Viral production has been measured from single cells, so it seems that it should be feasible to measure production from individual syncytia (Timm and Yin 2012). Questions that need to be answered include whether syncytia produce more or less virus than the corresponding number of singly infected cells and whether the production rate varies with the size of the syncytium. It should also be reasonably straightforward to measure the infectious lifespan of syncytia and determine whether there is a size dependence for this quantity as well.

The cell-cell fusion assay described in Sect. 2.2 is one experimental assay that has been developed to study syncytia formation. Recent modeling work used data from this assay to estimate the fusion rate, although questions remain about the usefulness of this measurement when it comes to in vivo infections (Amidei and Dobrovolny 2022). Cell-cell fusion assays typically use cell types different from those actually infected by viruses in humans. For example, recent cell-cell fusion assays studying SARS-CoV-2 fusion used Vero cells, 293 T cells, HeLa cells, TZM cells, A549 cells, and U2OS cells, none of which are the typical cells targeted in the human respiratory tract (Buchrieser et al. 2020; Braga et al. 2021; Cheng et al. 2021; Essalmani et al. 2022; Hoernich et al. 2021; Jocher et al. 2022; Liu et al. 2022; Papa et al. 2021; Rajah et al. 2021; Ren et al. 2021; Sanders et al. 2021; Theuerkauf et al. 2021; Wang et al. 2021). One experiment examined the effect of having different cells as donors or acceptors, and analysis showed that the fusion rate remained the same (Amidei and Dobrovolny 2022; Papa et al. 2021). However, this experiment was limited to swapping the chosen donor and acceptor cells, which happened to be Vero and 293 cells, so it's not clear whether this result will extend to other cell types. There are also many other host factors that come into play when syncytia formation occurs in vivo that are not present in the cell-cell assays, so the correlation between in vitro findings and actual infections is not clear (New et al. 2021; Takemoto et al. 2022). One step in helping to bridge this gap is the development of organoids that are being used to study infection processes, including syncytia formation (Porotto et al. 2019; Clevers 2020; Ekanger et al. 2022; Mykytyn et al. 2021; Shpichka et al. 2022).

There is also a need to develop more detailed studies of some of the spatial aspects of syncytia. Since syncytia are spatially extended objects, the question of whether their surface properties are homogeneous or heterogeneous can impact the spatial dynamics of viral spread. There has been some study of the spatial distribution of both viral surface proteins and cell surface receptors for single infected cells (Gershon et al. 1979; Leary and Todd 1977; Sadzotdelvaux et al. 1992; Zimmerberg et al. 2004). While the distribution of cell surface receptors appears to be fairly homogeneous over the surface of a single cell before infection, one study found that in single infected cells, cell surface receptors redistribute themselves to aggregate in clumps on the cell surface after infection (Gershon et al. 1979; Leary and Todd 1977). Another study noted that viral surface proteins are also not homogeneously distributed about the surface of a single infected cell (Sadzotdelvaux et al. 1992). Does a similar effect occur on the surface of syncytia? There is an additional complication for syncytia since uninfected cells can fuse and become part of existing syncytia. How long does the surface of this newly added cell differ from the surface of the rest of the syncytium? Similar questions can be asked about the production of virus from syncytia—is it homogeneously distributed around the cell surface or is viral production highly localized? Additionally, single infected cells can burst to release a large amount of virus at once, which is known to alter the viral time course as compared to continuous viral production (Gilchrist et al. 2004). Lytic production for singly infected cells was incorporated into the Jacobsen model, but they did not include a similar lytic mechanism for syncytia based on a study that suggests that the mechanism of death for syncytia differs from that of singly infected cells (Bateman et al. 2002). More recent studies, however, suggest that syncytia can undergo apoptosis similar to singly infected cells, so viral bursts might need to be considered for syncytia as well (Hoffmann et al. 2008; Lin et al. 2010; Salsman et al. 2005).

Finally, there is a need to better understand the similarities and differences between virus-cell fusion and cell-cell fusion. While cell-cell fusion arises because viral surface proteins appear on the surface of infected cells, there is some evidence that the attachment and fusion processes differ between virus-cell and cell-cell fusion (Connolly and Lamb 2006; Fan et al. 2021). One mathematical study suggests that the redistribution of lipid molecules during fusion will differ for virus-cell fusion and cell-cell fusion simply due to the size difference between viruses and cells, although this doesn't appear to have been experimentally verified even though it appears to be experimentally feasible to measure such redistribution (Lowy et al. 1995; Rubin and Chen 1990). However, there are some experiments that suggest different fusion mechanisms for virus-cell and cell-cell fusion, or that different proteases appear to be responsible for facilitating cell-cell or virus-cell fusion or that antivirals have differential effects on virus-cell and cell-cell fusion (de Haan et al. 2004; Ghosh et al. 2000; Gombold et al. 1993; Konopka et al. 1995; Pleskoff et al. 1995; Schmid et al. 2000; Simmons et al. 2011). Understanding how virus-cell fusion and cell-cell fusion differ will help answer the question of whether syncytia formation is an accidental by-product of expression of viral surface proteins on the cell surface or whether syncytia formation evolved as a strategy to help evade extracellular immune responses.

17.3.2 Models

While some of the oncolytic virus models incorporate some spatial dynamics, they do not fully capture the full spatial heterogeneity of syncytia formation. The use of partial differential equations assumes continuity of the quantities being modeled, but this is not the case for cells. Cells are discrete objects, so the most realistic models need to take this into account. Models that simulate individual cells to study their collective behavior are known as agent-based models (ABMs). ABMs are ideal for capturing the spatially heterogeneous and discrete process of syncytia formation. While agent-based models have been used to simulate viral infections, they largely assume virus spreads through cell-free transmission, so none have incorporated syncytia formation (Alvarado et al. 2018; Beauchemin et al. 2005; Goyal and Murray 2016; Itakura et al. 2010; Tong et al. 2015; Wasik et al. 2014; Whitman et al. 2020; Wodarz et al. 2014). However, an ABM that includes syncytia formation has been created for models of bone formation, so incorporating syncytia formation into viral ABMs should be feasible (Van Scoy et al. 2017). The recent advance of implementing a viral ABM on graphical processing units (GPUs) allows for full-scale simulation of in vitro experiments in minutes on a desktop computer, so should also be able to replicate cell-cell fusion assays or organoid experiments in reasonable amounts of time (Fain and Dobrovolny 2022).

Another possible use of mathematical models is to help understand the possible benefits of syncytia formation. It is thought that syncytia formation protects the virus from extracellular threats such as many components of the immune response

(Labudova 2020; Cifuentes-Munoz and Dutch 2019). Mathematical models are an ideal vehicle for testing these types of hypotheses. For example, a model incorporating both syncytia formation and an antibody response, a common immune component already incorporated into many ODE viral models, could lead to insights on how much syncytia formation is needed to effectively evade antibodies (Ciupe 2015; Dobrovolny et al. 2013; Nikin-Beers et al. 2015). Similarly, a common application of viral dynamics models is to study the effect of different antivirals, so this could easily be incorporated into models that also include syncytia formation in order to investigate whether syncytia formation helps the virus elude the antiviral (Zitzmann and Kaderali 2018). Models that incorporate antivirals will also be useful for optimizing treatment doses and timing to quickly eliminate infections (Koizumi and Iwami 2014; Koizumi et al. 2017). Many newer antivirals target fusion proteins of various viruses, preventing the virus from fusing with the cell (Lan et al. 2022; Van Den Bergh et al. 2022). Antivirals that prevent fusion should not only prevent fusion of virions with the cell surface, but will also likely prevent cell-cell fusion during the infection, so might prove to be more effective than antivirals that target other parts of the viral replication cycle. In order to accurately model these types of antivirals, syncytia formation should also be included in the models in order to capture all the effects of the antiviral.

However, if models are to be used to make predictions, we must ensure that they are properly validated. While some of the oncolytic virus models have been fit to experimental data, it's not clear that models without syncytia formation would not have been able to replicate the same data. In order to properly validate models that include syncytia formation, we will need measurements of the time course of the syncytia. Currently, viral models are validated using measurements of viral time courses, or in the case of oncolytic virus treatment, measurement of tumor size. It is known that these limited measurements lead to problems with parameter identifiability (González-Parra et al. 2018; Miao et al. 2011) making it difficult to assess the reliability of the model itself and to determine any limitations.

17.4 Conclusions

The application of mathematical models within host viral dynamics started about 30 years ago and has led to a number of important advances, but there are still aspects of viral infections that have had limited mathematical modeling efforts (Beauchemin and Handel 2011; Perelson 2021). The formation of syncytia and its role in viral infections can be easily incorporated into mathematical models, but there have been few efforts to use mathematical modeling to improve our understanding of syncytia. This review describes these early efforts and what we have learned from them with the hope that both modeling and experimental efforts can be combined to help improve our understanding of this mechanism of viral spread.

References

Abdelsattar A, Dawoud A, Makky S, Nofal R, Aziz R, El-Shibiny A (2022) Bacteriophages: from isolation to application. Curr Pharmacol Biotech 23(3):337–360. https://doi.org/10.2174/1389201022666210426092002

Ahmed I, Akram Z, Iqbal HM, Munn AL (2019) The regulation of endosomal sorting complex required for transport and accessory proteins in multivesicular body sorting and enveloped viral budding — an overview. Intl J Biol Macromol 127:1–11. https://doi.org/10.1016/j.ijbiomac.2019.01.015

Albornoz A, Hoffmann AB, Lozach PY, Tischler ND (2016) Early bunyavirus-host cell interactions. Viruses 8(5):143. https://doi.org/10.3390/v8050143

Alvarado, A., Corrales, R., Leal, M.J., De la Ossa, A., Mora, R., Arroyo, M., Gomez, A., Calderon, A., Arias-Arias, J.L., Cellular-level characterization of dengue and Zika virus infection using multiagent simulation. In: 2018 IEEE International Work Conference on Bioinspired Intelligence (IWOBI), pp. 1–6. IEEE, San Carlos (2018). https://doi.org/10.1109/IWOBI.2018.8464219

Alzahrani T, Eftimie R, Trucu D (2020) Multiscale moving boundary modelling of cancer interactions with a fusogenic oncolytic virus: the impact of syncytia dynamics. Math Biosci 323:108296. https://doi.org/10.1016/j.mbs.2019.108296

Amidei A, Dobrovolny HM (2022) Estimation of virus-mediated cell fusion rate of sars-cov-2. Virol 575:91–100

Ayala-Breton C, Russell LO, Russell SJ, Peng KW (2014) Faster replication and higher expression levels of viral glycoproteins give the vesicular stomatitis virus/measles virus hybrid vsv-fh a growth advantage over measles virus. J Virol 88(15):8332–8339. https://doi.org/10.1128/JVI.03823-13

Ayata M, Shingai M, Ning X, Matsumoto M, Seya T, Otani S, Seto T, Ohgimoto S, Ogura H (2007) Effect of the alterations in the fusion protein of measles virus isolated from brains of patients with subacute sclerosing panencephalitis on syncytium formation. Virol Res 130(1–2):260–268. https://doi.org/10.1016/j.vir-usres.2007.07.017

Bajzer Z, Carr T, Kresimir Josic J, Russell S (2008) David Dingli: modeling of cancer virotherapy with recombinant measles viruses. J Theor Biol 252(1):109–122. https://doi.org/10.1016/j.jtbi.2008.01.016

Bandin I, Dopazo CP (2011) Host range, host specificity and hypothesized host shift events among viruses of lower vertebrates. Vet Res 42:67. https://doi.org/10.1186/1297-9716-42-67

Bateman A, Harrington K, Kottke T, Ahmed A, Melcher AA, Gough MJ, Linardakis E, Riddle D, Dietz A, Lohse CM, Strome S, Peterson T, Simari R, Vile RG (2002) Viral fusogenic membrane glycoproteins kill solid tumor cells by nonapoptotic mechanisms that promote cross presentation of tumor antigens by dendritic cells. Cancer Res 62(22):6566–6578

Beauchemin CA, Handel A (2011) A review of mathematical models of influenza A infections within a host or cell culture: lessons learned and challenges ahead. BMC Public Health 11(S1):S7–S22. www.biomedcentral.com/1471-2458-11-S1-S7

Beauchemin C, Samuel J, Tuszynski J (2005) A simple cellular automaton model for influenza A viral infections. J Theor Biol 232(2):223–234. https://doi.org/10.1016/j.jtbi.2004.08.001

Bello-Morales R, Ripa I, Antonio Lopez-Guerrero J (2020) Extracellular vesicles in viral spread and antiviral response. Viruses 12(6):623. https://doi.org/10.3390/v12060623

Biesecker M (2010) Jung-Han Kimn, Huitian Lu, David Dingli, Zeljko Bajzer: optimization of virotherapy for cancer. Bull Math Biol 72(2):469–489. https://doi.org/10.1007/s11538-009-9456-0

Braga L, Ali H, Secco I, Chiavacci E, Neves G, Goldhill D, Penn R, Jimenez-Guardeno JM, Ortega-Prieto AM, Bussani R, Cannata A, Rizzari G, Collesi C, Schneider E, Arosio D, Shah AM, Barclay WS, Malim MH, Burrone J, Giacca M (2021) Drugs that inhibit TMEM16 proteins block SARS-CoV-2 spike-induced syncytia. Nature 594(7861):88. https://doi.org/10.1038/s41586-021-03491-6

Buchrieser J, Dufloo J, Hubert M, Monel B, Planas D, Rajah MM, Planchais C, Porrot F, Guivel-Benhassine F, der Werf SV, Casartelli N, Mouquet H, Bruel T, Schwartz O (2020) Syncytia

formation by SARS-CoV-2 infected cells. EMBO J 39(23):e106267. https://doi.org/10.15252/embj.2020106267
Burton C (2019) Eric Bartee: syncytia formation in oncolytic virotherapy. Mol Ther 15:131–139. https://doi.org/10.1016/j.omto.2019.09.006
Chauhan RP, Dessie ZG, Noreddin A, El Zowalaty ME (2020) Systematic review of important viral diseases in Africa in light of the 'one health' concept. Pathogens 9(4):301. https://doi.org/10.3390/pathogens9040301
Cheng Y, Ma J, Wang H, Wang X, Hu Z, Li H, Zhang H, Liu X (2021) Co-infection of influenza A virus and SARS-CoV-2: A retrospective cohort study. J Med Virol 93(5):2947–2954. https://doi.org/10.1002/jmv.26817
Ci YL, Xu CM, Shi L (2016) The structure and function of viral fusion protein. Prog Biochem Biophys 43(9):850–857. https://doi.org/10.16476/j.pibb.2016.0055
Cifuentes-Munoz N, Dutch RE (2019) To assemble or not to assemble: the changing rules of pneumovirus transmission. Virus Res 265:68–73. https://doi.org/10.1016/j.virusres.2019.03.002
Ciupe SM (2015) Mathematical model of multivalent virus-antibody complex formation in humans following acute and chronic HIV infections. J Math Biol 71(3):513–532. https://doi.org/10.1007/s00285-014-0826-3
Clevers H (2020) COVID-19: organoids go viral. Nature Rev Mol Cell Biol 21(7):355–356. https://doi.org/10.1038/s41580-020-0258-4
Connolly SA, Lamb RA (2006) Paramyxovirus fusion: real-time measurement of parainfluenza virus 5 virus-cell fusion. Virol 355(2):203–212. https://doi.org/10.1016/j.virol.2006.07.021
de Haan C, Stadler K, Godeke G, Bosch BJ (2004) PJM Rottier: cleavage inhibition of the murine coronavirus spike protein by a furin-like enzyme affects cell-cell but not virus-cell fusion. J Virol 78(11):6048–6054. https://doi.org/10.1128/JVI.78.11.6048-6054.2004
Diller JR, Parrington HM, Patton JT, Ogden KM (2019) Rotavirus species B encodes a functional fusion-associated small transmembrane protein. J Virol 93(20):e00813–e00819. https://doi.org/10.1128/JVI.00813-19
Dingli D, Offord C, Myers R, Peng K-W, Carr TW, Josic K, Russell SJ, Bajzer Z (2009) Dynamics of multiple myeloma tumor therapy with a recombinant measles virus. Cancer Gene Ther 16(12):873–882. https://doi.org/10.1038/cgt.2009.40
Dittmar T, Weiler J, Luo T, Hass R (2021) Cell-cell fusion mediated by viruses and HERV-derived fusogens in cancer initiation and progression. Cancers 13(21):5363. https://doi.org/10.3390/cancers13215363
Dobrovolny HM, Beauchemin CA (2017) Modelling the emergence of influenza drug resistance: the roles of surface proteins, the immune response and antiviral mechanisms. PLoS One 12(7):e0180582. https://doi.org/10.1371/journal.pone.0180582
Dobrovolny HM, Reddy MB, Kamal MA, Rayner CR, Beauchemin CA (2013) Assessing mathematical models of influenza infections using features of the immune response. PLoS One 8(2):e57088. https://doi.org/10.1371/journal.pone.0057088
Ebert O, Shinozaki K, Kournioti C, Park MS, Garcia-Sastre A (2004) SLC woo: syncytia induction enhances the oncolytic potential of vesicular stomatitis virus in virotherapy for cancer. Cancer Res 64(9):3265–3270. https://doi.org/10.1158/0008-5472.CAN-03-3753
Ekanger CT, Zhou F, Bohan D, Lotsberg ML, Ramnefjell M, Hoareau L, Rosland GV, Lu N, Aanerud M, Gaertner F, Salminen PR, Bentsen M, Halvorsen T, Raeder H, Akslen LA, Langeland N, Cox R, Maury W, Stuhr LEB, Lorens JB, Engelsen AS (2022) Human organotypic airway and lung organoid cells of bronchiolar and alveolar differentiation are permissive to infection by influenza and SARS-CoV-2 respiratory virus. Front Cell Infect Microbiol 12:841447. https://doi.org/10.3389/fcimb.2022.841447
Essalmani R, Jain J, Susan-Resiga D, Andreo U, Evagelidis A, Mouna Derbali R, Huynh DN, Dallaire F, Laporte M, Delpal A, Sutto-Ortiz P, Coutard B, Mapa C, Wilcoxen K, Decroly E, Nq Pham T, Cohen EA, Seidah N (2022) Distinctive roles of furin and TMPRSS2 in SARS-CoV-2 infectivity. J Virol 96(8):e00128-22. https://doi.org/10.1128/jvi.00128-22

Fain B, Dobrovolny HM (2022) GPU acceleration and data fitting: agent-based models of viral infections can now be parameterized in hours. J Comp Sci 61:101662. https://doi.org/10.1016/j.jocs.2022.101662
Fan Q, Longnecker R, Connolly AS (2021) Herpes simplex virus glycoprotein B mutations define structural sites in domain I, the membrane proximal region, and the cytodomain that regulate entry. J Virol 95(2):e01050-21. https://doi.org/10.1128/JVI.01050-21
Fukushima K, Takahashi T, Takaguchi M, Ito S, Suzuki C, Agarikuchi T, Kurebayashi Y, Minami A, Suzuki T (2019) A I131V substitution in the fusion glycoprotein of human parainfluenza virus type 1 enhances syncytium formation and virus growth. Biol Pharmaceut Bull 42(5):827–832. https://doi.org/10.1248/bpb.b18-00714
Gagliardi TB, Criado MF, Proenca-Modena JL, Saranzo AM, Iwamoto MA, de Paula FE, Cardoso RS, Delcaro LS, Silva ML, Camara AA, Arruda E (2017) Syncytia induction by clinical isolates of human respiratory syncytial virus A. Intervirol 60(1–2):56–60. https://doi.org/10.1159/000480014
Ge Q, Wang X, Rong L (2021) A delayed reaction-diffusion viral infection model with nonlinear incidences and cell-to-cell transmission. Int J Biomath 14(8):2150100. https://doi.org/10.1142/S179352452150100X
Gershon N, Demsey A, Stackpole C (1979) Analysis of local order in the spatial distribution of cell-surface molecular assemblies. Exp Cell Res 122(1):115–126. https://doi.org/10.1016/0014-4827(79)90566-4
Ghosh J, Peisajovich SG, Shai Y (2000) Sendai virus internal fusion peptide: structural and functional characterization and a plausible mode of viral entry inhibition. Biochemist 39(38):11581–11592. https://doi.org/10.1021/bi0005963
Gilchrist M, Coombs D, Perelson AS (2004) Optimizing within-host viral fitness: infected cell lifespan and virion production rate. J Theor Biol 229(2):281–288. https://doi.org/10.1016/j.jtbi.2004.04.015
Gombold J, Hingley S, Weiss S (1993) Fusion-defective mutants of mouse hepatitis virus-A59 contain a mutation in the spike protein cleavage signal. J Virol 67(8):4504–4512. https://doi.org/10.1128/JVI.67.8.4504-4512.1993
González-Parra G, Dobrovolny HM, Aranda DF, Chen-Charpentier B, Roja RAG (2018) Quantifying rotavirus kinetics in the REH tumor cell line using in vitro data. Virus Res 244:53–63. https://doi.org/10.1016/j.virusres.2017.09.023
Gonzalez-Reyes L, Ruiz-Arguello M, Garcia-Barreno B, Calder L, Lopez J, Albar J, Skehel J, Wiley D, Melero J (2001) Cleavage of the human respiratory syncytial virus fusion protein at two distinct sites is required for activation of membrane fusion. Proc Natl Acad Sci U S A 98(17):9859–9864. https://doi.org/10.1073/pnas.151098198
Goyal A, Murray JM (2016) Modelling the impact of cell-to-cell transmission in hepatitis B virus. PLoS One 11(8):e0161978. https://doi.org/10.1371/journal.pone.0161978
Harrison SC (2015) Viral membrane fusion. Virol 479:498–507. https://doi.org/10.1016/j.virol.2015.03.043
Hernandez L, Hoffman L, Wolfsberg T, White J (1996) Virus-cell and cell-cell fusion. Ann Rev Cell Dev Biol 12:627–661. https://doi.org/10.1146/annurev.cellbio.12.1.627
Hoernich BF, Grosskopf AK, Schlagowski S, Tenbusch M, Kleine-Weber H, Neipel F, Stahl-Hennig C, Hahn AS (2021) SARS-CoV-2 and SARS-CoV spike-mediated cell-cell fusion differ in their requirements for receptor expression and proteolytic activation. J Virol 95(9):e00002-21. https://doi.org/10.1128/JVI.00002-21
Hoffmann D, Grunwald T, Bayer W, Wildner O (2008) Immune-mediated anti- neoplastic effect of intratumoral RSV envelope glycoprotein expression is related to apoptotic death of tumor cells but not to the size of syncytia. World J Gastroenterol 14(12):1842–1850. https://doi.org/10.3748/wjg.14.1842
Holder BP, Simon P, Liao LE, Abed Y, Bouhy X, Beauchemin CA, Boivin G (2011) As- sessing the in vitro fitness of an oseltamivir-resistant seasonal A/H1N1 influenza strain using a mathematical model. PLoS One 6(3):e14767. https://doi.org/10.1371/journal.pone.0014767

Hu M, Bogoyevitch MA, Jans DA (2020) Impact of respiratory syncytial virus on host func- tions: implications for antiviral strategies. Physiol Rev 100(4):1527–1594. https://doi.org/10.1152/physrev.00030.2019

Huang F, Dempsey C, Chona D, Suh J (2011) Quantitative nanoparticle tracking: applications to nanomedicine. Nanomed 6(4):693–700. https://doi.org/10.2217/NNM.11.42

Itakura J, Kurosaki M, Itakura Y, Maekawa S, Asahina Y, Izumi N, Enomoto N (2010) Reproducibility and usability of chronic virus infection model using agent-based simulation; comparing with a mathematical model. Biosys 99(1):70–78. https://doi.org/10.1016/j.biosystems.2009.09.001

Jacobsen K, Pilyugin SS (2015) Analysis of a mathematical model for tumor therapy with a fusogenic oncolytic virus. Math Biosci 270:169–182. https://doi.org/10.1016/j.mbs.2015.02.009

Jenkinson S, McCoy D, Kerner S, Ferris R, Lawrence W, Fox T, Smith C (2003) Development of a high-throughput viral-free assay for the measurement of CCR5-mediated HIV/cell fusion. Receptors Channels 9(2):117–123. https://doi.org/10.1080/10606820308248

Jessie B, Dobrovolny H (2021) The role of syncytia during viral infections. J Theor Biol 525:110749. https://doi.org/10.1016/j.jtbi.2021.110749

Jocher G, Vincent Grass K, Tschirner S, Riepler L, Breimann S, Kaya T, Oelsner M, Sabri Hamad M, Hofmann L, Blobel CP, Schmidt-Weber CB, Gokce O, Jakwerth CA, Trimpert J, Kimpel J, Pichlmair A, Lich-tenthaler SF (2022) ADAM10 and ADAM17 promote SARS-CoV-2 cell entry and spike protein- mediated lung cell fusion. EMBO Rep 23(6):e54305. https://doi.org/10.15252/embr.202154305

Johnson C, Exell J, Lin Y, Jonathan Aguilar D, Welsher K (2022) Capturing the start point of the virus-cell interaction with high-speed 3D single-virus tracking. Nature Meth 19:1642–1652. https://doi.org/10.1038/s41592-022-01672-3

Koizumi Y, Iwami S (2014) Mathematical modeling of multi-drugs therapy: a challenge for determining the optimal combinations of antiviral drugs. Theor Biol Med Model 11:41. https://doi.org/10.1186/1742-4682-11-41

Koizumi Y, Ohashi H, Nakajima S, Tanaka Y, Wakita T, Perelson AS, Iwami S, Watashi K (2017) Quantifying antiviral activity optimizes drug combinations against hepatitis C virus infection. Proc Natl Acad Sci U S A 114(8):1922–1927. https://doi.org/10.1073/pnas.1610197114

Konopka K, Pretzer E, Duzgunes N (1995) Differential effects of a hydrophobic tripeptide on human-immunodeficiency-virus type-1 (HIV-1)-induced syncytium formation and viral infectivity. Biophys Biochem Res Comm 208(1):75–81. https://doi.org/10.1006/bbrc.1995.1307

Kouwaki T, Okamoto M (2017) Hirotake Tsukamoto, Yoshimi Fukushima, Hiroyuki Oshiumi: extracellular vesicles deliver host and virus rna and regulate innate immune response. Intl J Mol Sci 18(3):666. https://doi.org/10.3390/ijms18030666

Labudova M (2020) Cell-to-cell transport in viral families: faster than usual. Acta Virol 64(2):154–166. https://doi.org/10.4149/av_2020_205

Lan Q, Wang L, Jiao F, Lu L, Xia S, Jiang S (2022) Pan-coronavirus fusion inhibitors to combat COVID-19 and other emerging coronavirus infectious diseases. J Med Virol 95(1):e28143. https://doi.org/10.1002/jmv.28143

Laureti M, Narayanan D, Rodriguez-Andres J, Fazakerley JK, Kedzierski L (2018) Flavivirus receptors: diversity, identity, and cell entry. Front Immunol 9:2180. https://doi.org/10.3389/fimmu.2018.02180

Leary J, Todd P (1977) Laser cytophotometric detection of variations in spatial distribution of concanavalin-A cell surface receptors following viral infection. J Histochem Cytochem 25(7):908–912. https://doi.org/10.1177/25.7.197163

Leroy H, Han M, Woottum M, Bracq L, Bouchet J, Xie M, Benichou S (2020) Virus-mediated cell-cell fusion. Int J Mol Sci 21(24):9644. https://doi.org/10.3390/ijms21249644

Lin EH, Salon C, Brambilla E, Lavillette D, Szecsi J, Cosset F-L, Coll J-L (2010) Fusogenic membrane glycoproteins induce syncytia formation and death in vitro and in vivo: a potential therapy agent for lung cancer. Cancer Gene Ther 17(4):256–265. 10.1038/cgt.2009.74

Liu X, Wei L, Fengwen X, Fei Zhao Y, Huang FZ, Mei S, Hu Y, Zhai L, Guo J, Zheng A, Cen S, Liang C, Guo F (2022) SARS-CoV-2 spike protein-induced cell fusion activates the cGAS-STING pathway and the interferon response. Sci Signal 15(729):eabg8744. 10.1126/scisignal.abg8744

Lopez-Balderas N, Huerta L, Villarreal C, Rivera-Toledo E, Sandoval G, Larralde C, Lamoyi E (2007) In vitro cell fusion between CD4(+) and HIV-1 Env(+) T cells generates a diversity of syncytia varying in total number, size and cellular content. Virus Res 123(2):138–146. https://doi.org/10.1016/j.virusres.2006.08.009

Lowy R, Sarkar D, Whitnall M, Blumenthal R (1995) Differences in dispersion of influenza-virus lipids and proteins during fusion. Exp Cell Res 216(2):411–421. https://doi.org/10.1006/excr.1995.1052

Martinez-Turino S, Calvo M, Bedoya LC, Zhao M, Garcia JA (2021) Virus host jumping can be boosted by adaptation to a bridge plant species. Microorganisms 9(4):805. https://doi.org/10.3390/microorganisms9040805

Matsuyama S, Taguchi F (2000) Impaired entry of soluble receptor-resistant mutants of mouse hepatitis virus into cells expressing MHVR2 receptor. Virology 273(1):80–89. https://doi.org/10.1006/viro.2000.0409

McLeish MJ, Fraile A, Garcia-Arena F (2019) Evolution of plant-virus interactions: host range and virus emergence. Curr Opin Virol 34:50–55. https://doi.org/10.1016/j.coviro.2018.12.003

Miao H, Xia X, Perelson AS, Wu H (2011) On identifiability of nonlinear ODE models and applications in viral dynamics. SIAM Rev 53(1):3–39. https://doi.org/10.1137/090757009

Mothes W, Sherer NM, Jin J, Zhong P (2010) Virus cell-to-cell transmission. J Virol 84(17):8360–8368. https://doi.org/10.1128/JVI.00443-10

Mulampaka SN, Dixit NM (2011) Estimating the threshold surface density of Gp120-CCR5 complexes necessary for HIV-1 envelope-mediated cell-cell fusion. PLoS One 6(5):e19941. https://doi.org/10.1371/journal.pone.0019941

Myers R, Greiner SM, Harvey ME, Griesmann G, JKuffel M, Buhrow SA, Reid JM, Fed-erspiel M, Ames MM, Dingli D (2007) Preclinical pharmacology and toxicology of intravenous MV-NIS, an oncolytic measles virus administered with or without cyclophosphamide. Clin Pharmacol Therapeut 82(6):700–710. https://doi.org/10.1038/sj.clpt.6100409

Mykytyn AZ, Breugem TI, Riesebosch S, Schipper D, van den Doel PB, Rottier RJ, Lamers MM, Haagmans BL (2021) SARS-CoV-2 entry into human airway organoids is serine protease-mediated and facilitated by the multibasic cleavage site. eLife 10:e64508. https://doi.org/10.7554/eLife.64508

Navaratnarajah CK, Generous AR, Yousaf I, Cattaneo R (2020) Receptor-mediated cell entry of paramyxoviruses: mechanisms, and consequences for tropism and pathogenesis. J Biol Chem 295(9):2771–2786. https://doi.org/10.1074/jbc.REV119.009961

New C, Lee Z-Y, Sen Tan K, Huee-Ping Wong A, Yun Wang D (2021) Thai Tran: Tetraspanins: host factors in viral infections. Int J Mol Sci 22(21):11609. https://doi.org/10.3390/ijms222111609

Nikin-Beers R, Ciupe M, S. (2015) The role of antibody in enhancing dengue virus infection. Math Biosci 263:83–92. https://doi.org/10.1016/j.mbs.2015.02.004

Nova N (2021) Cross-species transmission of coronaviruses in humans and domestic mammals, what are the ecological mechanisms driving transmission, spillover, and disease emergence? Front Public Health 9:717941. https://doi.org/10.3389/fpubh.2021.717941

Panasiuk M, Rychlowski M, Derewonko N, Bienkowska-Szewczyk K (2018) Tunneling nanotubes as a novel route of cell-to-cell spread of herpesviruses. J Virol 92(10):e00090-18. https://doi.org/10.1128/JVI.00090-18

Papa G, Mallery DL, Albecka A, Welch LG, Cattin-Ortola J, Luptak J, Paul D, McMahon HT, Goodfellow IG, Carter A, Munro S, James LC (2021) Furin cleavage of SARS-CoV-2 spike promotes but is not essential for infection and cell-cell fusion. PLoS Pathog 17(1):e1009246. https://doi.org/10.1371/journal.ppat.1009246

Paradis E, Pinilla L, Holder B, Abed Y, Boivin G, Beauchemin C (2015) Impact of the H275Y and I223V mutations in the neuraminidase of the 2009 pandemic influenza virus in vitro and

evaluating experimental reproducibility. PLoS One 10(5):e0126115. https://doi.org/10.1371/journal.pone.0126115
Parrish CR, Holmes EC, Morens DM, Park EC, Burke DS, Calisher CH, Laughlin CA, Saif LJ, Daszak P (2008) Cross-species virus transmission and the emergence of new epidemic diseases. Microbiol Mol Biol Rev 72(3):457. https://doi.org/10.1128/MMBR.00004-08
Perelson AS (2021) Ruian Ke: mechanistic modeling of SARS-CoV-2 and other infectious diseases and the effects of therapeutics. Clin Pharmacol Ther 109(4):829–840. https://doi.org/10.1002/cpt.2160
Pinilla LT, Holder BP, Abed Y, Boivin G, Beauchemin CAA (2012) The H275Y neuraminidase mutation of the pandemic A/H1N1 influenza virus lengthens the eclipse phase and reduces viral output of infected cells, potentially compromising fitness in ferrets. J Virol 86(19):10651–10660. https://doi.org/10.1128/JVI.07244-11
Pinky L, Dobrovolny HM (2016) Coinfections of the respiratory tract: viral competition for resources. PLoS One 11(5):e0155589. https://doi.org/10.1371/journal.pone.0155589
Pinky L, Dobrovolny HM (2017) The impact of cell regeneration on the dynamics of viral coinfection. Chaos 27(6):063109. https://doi.org/10.1063/1.4985276
Pinky L, González-Parra G, Dobrovolny HM (2019) Superinfection and cell regeneration can lead to chronic viral coinfections. J Theor Biol 466:24–38. https://doi.org/10.1016/j.jtbi.2019.01.011
Pleskoff O, Seman M, Alizon M (1995) Amphotericin B derivative blocks human immunodeficiency virus type 1 entry after CD4 binding: effect on virus-cell fusion but not on cell-cell fusion. J Virol 69(1):570–574. https://doi.org/10.1128/JVI.69.1.570-574.1995
Porotto M, Ferren M, Chen YW, Siu Y, Makhsous N, Rima B, Briese T, Greninger A, Snoeck HW, Moscona A (2019) Authentic modeling of human respiratory virus infection in human pluripotent stem cell-derived lung organoids. MBio 10(3):e00723-19. https://doi.org/10.1128/mBio.00723-19
Qian Z, Dominguez SR, Holmes KV (2013) Role of the spike glycoprotein of human middle east respiratory syndrome coronavirus (MERS-CoV) in virus entry and syncytia formation. PLoS One 8(10):e76469. https://doi.org/10.1371/journal.pone.0076469
Rajah MM, Hubert M, Bishop E, Saunders N, Robinot R, Grzelak L, Planas D, Dufloo J, Gellenoncourt S, Bongers A, Zivaljic M, Planchais C, Guivel-Benhassine F, Porrot F, Mouquet H, Chakrabarti LA, Buchrieser J, Schwartz O (2021) SARS-CoV-2 alpha, beta, and delta variants display enhanced spike-mediated syncytia formation. EMBO J 40(24):e108944. https://doi.org/10.15252/embj.2021108944
Ren H, Ma C, Peng H, Zhang B, Zhou L, Su Y, Gao X, Huang H (2021) Micronucleus production, activation of DNA damage response and cGAS-STING signaling in syncytia induced by SARS-CoV-2 infection. Biol Direct 16(1):20. https://doi.org/10.1186/s13062-021-00305-7
Roossinck MJ (2019) Viruses in the phytobiome. Curr Opin Virol 37:72–76. https://doi.org/10.1016/j.coviro.2019.06.008
Rubin R, Chen Y (1990) Diffusion and redistribution of lipid-like molecules between membranes in virus-cell and cell-cell fusion systems. Biophys J 58(5):1157–1167. https://doi.org/10.1016/S0006-3495(90)82457-7
Sadzotdelvaux C, Marc P, Lebon L, Mervillelouis M, Piette J, Rentier B (1992) Antibodies to varicella-zoster virus modulate antigen distribution but fail to induce viral persistence in vitro. J Virol 66(12):7499–7504. https://doi.org/10.1128/JVI.66.12.7499-7504.1992
Salsman J, Top D, Boutilier J, Duncan R (2005) Extensive syncytium formation mediated by the reovirus fast proteins triggers apoptosis-induced membrane instability. J Virol 79(13):8090–8100. https://doi.org/10.1128/JVI.79.13.8090-8100.2005
Sanders DW, Jumper CC, Ackerman PJ, Bracha D, Donlic A, Kim H, Kenney D, Castello-Serrano I, Suzuki S, Tamura T, Tavares AH, Saeed M, Holehouse AS, Ploss A, Levental I, Douam F, Padera RF, Levy BD, Brangwynne CP (2021) SARS-CoV-2 requires cholesterol for viral entry and pathological syncytia formation. eLife 10:e65962. https://doi.org/10.7554/eLife.65962
Schmid E, Zurbriggen A, Gassen U, Rima B, ter Meulen V, Schneider-Schaulies J (2000) Antibodies to CD9, a tetraspan transmembrane protein, inhibit canine distemper virus-induced

cell-cell fusion but not virus-cell fusion. J Virol 74(16):554–7561. https://doi.org/10.1128/JVI.74.16.7554-7561.2000

Shpichka A, Bikmulina P, Peshkova M, Heydari Z, Kosheleva N, Vosough M, Timashev P (2022) Organoids in modelling infectious diseases. Drug Discov Today 27(1):223–233. https://doi.org/10.1016/j.drudis.2021.08.005

Simmons G, Bertram S, Glowacka I, Steffen I, Chaipan C, Agudelo J, Kai L, Rennekamp AJ, Hofmann H, Bates P, Pohlmann S (2011) Different host cell proteases activate the SARS-coronavirus spike-protein for cell-cell and virus-cell fusion. Virol 413(2):265–274. 10.1016/j.virol.2011.02.020

Sylwester A, Wessels D, Anderson S, Warren R, Shutt D, Kennedy R, Soll D (1993) HIV-induced sycytia of a T-cell line form single giant pseudopods and are motile. J Cell Sci 106(3):941–953

Sylwester A, Daniels K, Soll D (1998) The invasive and destructive behavior of HIV-induced T cell syncytia on collagen and endothelium. J Leukoc Biol 63(2):233–244. https://doi.org/10.1002/jlb.63.2.233

Symeonides M, Murooka TT, Bellfy LN, Roy NH, Mempel TR, Thali M (2015) HIV-1-induced small T cell syncytia can transfer virus particles to target cells through transient contacts. Viruses 7(12):6590–6603. https://doi.org/10.3390/v7122959

Takemoto R, Suzuki T, Hashiguchi T, Yanagi Y, Shirogane Y (2022) Short-stalk isoforms of CADM1 and CADM2 trigger neuropathogenic measles virus-mediated membrane fusion by interacting with the viral hemagglutinin. J Virol 96(3):e01949-21. https://doi.org/10.1128/jvi.01949-21

Theuerkauf SA, Michels A, Vanessa Riechert J, Maier T, Flory E, Klaus Cichutek J, Buchholz C (2021) Quantitative assays reveal cell fusion at minimal levels of SARS-CoV-2 spike protein and fusion from without. iScience 24(3):102170. https://doi.org/10.1016/j.isci.2021.102170

Timm A, Yin J (2012) Kinetics of virus production from single cells. Virol 424:11–17. https://doi.org/10.1016/j.virol.2011.12.005

Tong X, Chen J, Miao H, Li T, Zhang L (2015) Development of an agent-based model (ABM) to simulate the immune system and integration of a regression method to estimate the key ABM parameters by fitting the experimental data. PLoS One 10(11):e0141295. https://doi.org/10.1371/journal.pone.0141295

Truyen U, Parrish C, Harder T, Kaaden O (1995) There is nothing permanent except change — the emergence of new virus diseases. Vet Microbiol 43(2–3):103–122. https://doi.org/10.1016/0378-1135(95)92531-F

Van Den Bergh A, Bailly B, Guillon P, von Itzstein M, Dirr L (2022) Antiviral strategies against human metapneumovirus: targeting the fusion protein. Antivir Res 207:105405. https://doi.org/10.1016/j.antiviral.2022.105405

Van Scoy GK, George EL, Asantewaa FO, Kerns L, Saunders MM, Prieto-Langarica A (2017) A cellular automata model of bone formation. Math Biosci 286:58–64. https://doi.org/10.1016/j.mbs.2017.02.001

Versoza CJ, Pfeifer SP (2022) Computational prediction of bacteriophage host ranges. Microorganisms 10(1):149. https://doi.org/10.3390/microorganisms10010149

Viejo-Borbolla A, Schildberg FA, Burgert H-G (2017) Mechanisms of extracellular immunomodulation mediated by infectious agents. J Immunol Res 2017:5107527. https://doi.org/10.1155/2017/5107527

Wang W, Yang L, Huang X, Fu W, Pan D, Cai L, Ye J, Liu J, Xia N, Cheng T, Zhu H (2017) Outer nuclear membrane fusion of adjacent nuclei in varicella-zoster virus-induced syncytia. Virol 512:34–38. https://doi.org/10.1016/j.virol.2017.09.002

Wang X, Chen CH, Badeti S, Cho JH, Naghizadeh A, Wang Z, Liu D (2021) Deletion of ER-retention motif on SARS-CoV-2 spike protein reduces cell hybrid during cell-cell fusion. Cell Biosci 11(1):114. https://doi.org/10.1186/s13578-021-00626-0

Ward C, Paterson R, Lamb R (1995) Mutants of the paramyxovirus SV5 fusion protein — regulated and extensive syncytium formation. Virol 209(1):242–249. https://doi.org/10.1006/viro.1995.1250

Wasik S, Jackowiak P, Figlerowicz M, Blazewicz J (2014) Multi-agent model of hepatitis C virus infection. Artif Intell Med 60(2):123–131. https://doi.org/10.1016/j.artmed.2013.11.001
Whitman J, Dhanji A, Hayot F, Sealfon SC, Jayaprakash C (2020) Spatiotemporal dynamics of host-virus competition: a model study of influenza A. J Theor Biol 484:110026. https://doi.org/10.1016/j.jtbi.2019.110026
Wodarz D, Chan CN, Trinite B, Komarova NL, Levy DN (2014) On the laws of virus spread through cell populations. J Virol 88(22):13240–13248. https://doi.org/10.1128/JVI.02096-14
Zhong P, Agosto LM, Munro JB, Mothes W (2013) Cell-to-cell transmission of viruses. Curr Opin Virol 3(1):44–50. https://doi.org/10.1016/j.coviro.2012.11.004
Zimmerberg J, Kumar M, Verma A, Farrington J, Roth M, Kenworthy A, Hess S (2004) Studying spatial distributions of influenza hemagglutinin on the plasma membrane of fibroblasts: a work in progress. Macromol Symp 219:17–23. https://doi.org/10.1002/masy.200550103
Zitzmann C, Kaderali L (2018) Mathematical analysis of viral replication dynamics and antiviral treatment strategies: from basic models to age-based multi-scale modeling. Front Microbiol 9:1546. https://doi.org/10.3389/fmicb.2018.01546
Zsak L, Zuckermann F, Sugg N, Benporat T (1992) Glycoprotein-gi of pseudorabies virus promotes cell-fusion and virus spread via direct cell-to-cell transmission. J Virol 66(4):2316–2325. https://doi.org/10.1128/JVI.66.4.2316-2325.1992

Chapter 18
Syncytium Induced by Plant-Parasitic Nematodes

Mateusz Matuszkiewicz and Mirosław Sobczak

Abstract Plant-parasitic nematodes from the genera *Globodera*, *Heterodera* (cyst-forming nematodes), and *Meloidogyne* (root-knot nematodes) are notorious and serious pests of crops. They cause tremendous economic losses between US $80 and 358 billion a year. Nematodes infect the roots of plants and induce the formation of specialised feeding structures (syncytium and giant cells, respectively) that nourish juveniles and adults of the nematodes. The specialised secretory glands enable nematodes to synthesise and secrete effectors that facilitate migration through root tissues and alter the morphogenetic programme of host cells. The formation of feeding sites is associated with the suppression of plant defence responses and deep reprogramming of the development and metabolism of plant cells.

In this chapter, we focus on syncytia induced by the sedentary cyst-forming nematodes and provide an overview of ultrastructural changes that occur in the host roots during syncytium formation in conjunction with the most important molecular changes during compatible and incompatible plant responses to infection with nematodes.

18.1 Introduction

Nematodes are one of the most numerous and widespread groups of invertebrate animals. Phylum Nematoda encompasses more than 27,000 described species, and the total number of species is estimated at more than one million. Thanks to their

---

M. Matuszkiewicz (✉)
Department of Plant Genetics, Breeding and Biotechnology, Institute of Biology, Warsaw University of Life Sciences (WULS-SGGW), Warsaw, Poland
e-mail: mateusz_matuszkiewicz@sggw.edu.pl

M. Sobczak
Department of Botany, Institute of Biology, Warsaw University of Life Sciences (WULS-SGGW), Warsaw, Poland

M. Kloc, A. Uosef (eds.), *Syncytia: Origin, Structure, and Functions*, Results and Problems in Cell Differentiation 71,
https://doi.org/10.1007/978-3-031-37936-9_18

great adaptability, nematodes are found all over the Earth. Although they are very conservative in body organisation and morphology, they inhabit a variety of habitats and occupy very different niches in virtually all climatic zones. About 5000 known species can be found in different soil environments, and the densities of their populations vary between 60,000 and 30 million individuals per square meter (Wasilewska 1979). Most of them are free-living species that feed on bacteria, fungi, or algae, but there are also species parasitizing animals or plants (Kikuchi et al. 2017). Plant-parasitic nematodes (PPNs) are among the most dangerous pests in modern agriculture, and they are parasites of virtually all crops, i.e. *Solanaceae* (tomato, potato, pepper, tobacco), *Fabaceae* (soybean, chickpea, bean, pea, alfalfa), *Malvaceae* (cotton), *Amaranthaceae* (beet), *Musaceae* (banana), *Brassicaceae* (cabbage and relatives), and *Poaceae* (rice, wheat, corn). The value of economic losses caused by them is estimated worldwide to be between US $80 billion (Sasser and Freckman 1987) and US $358 billion (Abd-Elgawad and Askary 2015) per year.

The classical systematics of nematodes has long been controversial, as it was based on the anatomical and morphological features and the type of food they take (Blaxter et al. 1998). Through the use of comparative genomics, Mitreva et al. (2005) distinguished five classes: Dorylaimia (I), Enoplia (II), Chromadorea/Spirurina (III), Tylenchina (IV), Rhabditina (V) across the phylum Nematoda. More precise taxonomic classification based on phylogenetic marker *18S rRNA* gene sequence allowed to distinguish 12 different clades (van Megen et al. 2009; Kikuchi et al. 2017). This classification indicates that ability to parasitize plants appeared several times during the evolution of nematodes. PPNs are found in clade 1-Triplonchida (class II according to Mitreva et al. (2005)), clade 2-Dorylaimida (accordingly class I), clade 10b-Aphelenchoidea (class IV), and clade 12-Tylenchida (class IV). It is worth to note that classes I and V encompass mostly dangerous parasites of animals and humans with few plant-parasitic species, whereas the most of plant parasites belongs to class IV Tylenchina (clade 12-Tylenchida) (Mitreva et al. 2005).

A common feature of all PPNs is the presence of solid U-shaped odontostyle or hollow stylet, which are used to withdraw cellular content from parasitized plant cells. However, except for the clade 12-Tylenchida parasitic behaviour of PPNs is very simple and practically restricted only to migratory ectoparasitism when the parasite remains motile during its whole life and feeds on rhizodermal cells or cortical cells below rhizodermis reachable to their mouth stylet (Sijmons et al. 1994; Wyss 1997). Probably PPNs have evolved from fungivorous soil-borne nematodes. The occurrence of thin fungal hyphae surrounding large and rich in nutrients plant roots seems to be a reasonable basis for such a transition. However, in the case of plant roots, the nutrients are not evenly distributed across their tissues. The outer cell layers, rhizodermis, and cortex are composed of thin-walled and strongly vacuolated cells containing only limited amounts of cytoplasm and nutrients. The bulk of nutrients is located in the central part of the root, called a vascular cylinder, where conductive tissues are located (Fig. 18.1a, d, g). This pattern of nutrient distribution may explain a tendency observed among lineages of PPNs to change their parasitic behaviour from ectoparasitism into endoparasitism. On the one hand, it provides

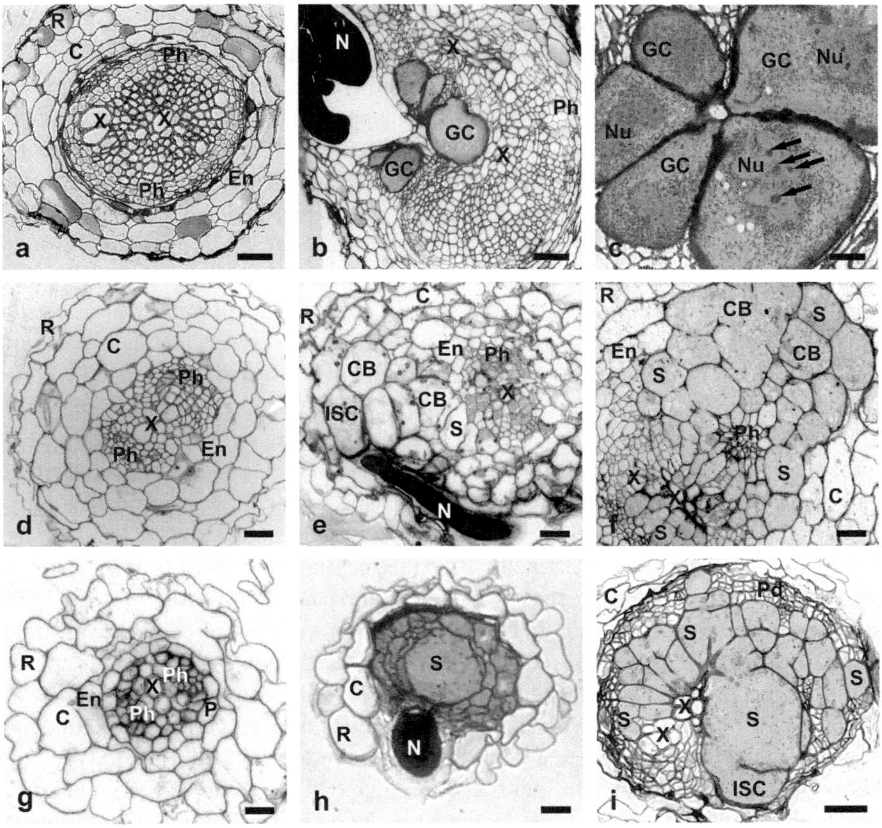

Fig. 18.1 Anatomy of feeding sites induced by sedentary plant-parasitic nematodes. Light microscopy micrographs of cross sections of giant cells induced by *Meloidogyne hapla* in sunflower (*Helianthus annuus*) root (**a**–**c**), and syncytia induced by *Globodera rostochiensis* in tomato (*Solanum lycopersicum*) root (**d**–**f**), and syncytia induced by *Heterodera schachtii* in *Arabidopsis thaliana* root (**g**–**i**). (**a**) uninfected sunflower root; (**b**) group of giant cells next to the nematode head; (**c**) group of giant cells above the nematode head. Arrows point to nucleoli; (**d**) uninfected tomato root; (**e**) syncytium at 7 dpi next to the nematode head; (**f**) syncytium at 14 dpi above the nematode head; (**g**) uninfected Arabidopsis root; (**h**) syncytium at 2 dpi next to the nematode head; (**i**) syncytium at 14 dpi above the nematode head. Abbreviations: C, cortex; CB, cortex bridge; En, endodermis; GC, giant cell; ISC, initial syncytial cell; N, nematode; Nu, nucleus; P, pericycle; Pd, periderm; Ph, phloem; R, rhizodermis; S, syncytium; X, xylem. Scale bars: 100 μm (**b**); 50 μm (**a**); 20 μm (**c**–**i**) (figure **a**–**c** *courtesy Kamila Chalamońska and Andrzej Skwiercz*)

access to cells rich in nutrients, and on the other hand, hiding inside the root tissues protects from predators. In the next step, the parasite will benefit from the restriction of its motility and thus energy expenses for migration among or inside root cells due to the switch from migratory into the sedentary mode of life. However, such transition demands an ability to induce development of specialised feeding cells, which will create a sink for plant nutrients towards the parasite.

Induction of feeding structures was achieved by some species from the clade Tylenchida, which developed the most advanced and sophisticated interactions with their hosts (Wyss 1997). Surprisingly, for unknown reasons, only nematode species parasitizing roots were able to develop permanent feeding sites whereas nematodes preferring above- and below-ground shoots remained migratory endoparasites (Palomares-Rius et al. 2017). However astonishingly, under specific artificial conditions, the sedentary root-parasitic nematodes can infect and induce their feeding sites also in leaves and stems (Miller and DiEdwardo 1962; Golinowski et al. 1996). The first permanent nematode feeding sites were single cells or groups of discreet, slightly enlarged root cortical cells, which were constantly explored by sessile parasitic nematodes (Hussey et al. 1992). From these cells evolved giant cells and syncytia. The former are hypertrophied cells with a large protoplast and reduced volume of the vacuole (Fig. 18.1b, c). They contain hypertrophied polyploid nuclei with enlarged nucleoli and increased numbers of plastids, mitochondria, and endoplasmic reticulum (ER). Giant cells induced by some nematode species contain a single enlarged polyploid nucleus whereas the other species achieve multinuclear polyploidy via endomitosis and mitosis without cytokinesis (Mundo-Ocampo and Baldwin 1983; Huang and Maggenti 1969). Syncytia are formed by the fusion of protoplasts of neighbouring cells stimulated by the parasite followed by partial dissolution of cell walls between them leading to the formation of cell wall openings (Grundler et al. 1998; Fig. 18.1e, f, h, i). This way, the multinuclear stage is also achieved, and the nuclei become polyploid due to endomitoses. Less evolutionary advanced PPNs induce syncytia and single or groups of giant cells among root cortex cells. Parasites being more evolutionary advanced developed tools to force the barrier of endodermis and to enter the vascular cylinder where they induce the development of nurse cells in direct contact with conductive tissues (Sijmons et al. 1994; Palomares-Rius et al. 2017). The sedentary mode of parasitism and feeding sites with direct access to plant resources allowed this group of PPNs to increase their reproduction rates several hundred times in comparison to migratory ectoparasites feeding on rhizodermal cells (Cohn and Spiegel 1991).

In agricultural practice, depending on crop species and type of agricultural yield, the most damaging are usually nematodes belonging to the group of sedentary endoparasites. They do not create a uniform group and differ greatly in plant species preferences, type of induced permanent feeding site, infection habits, and developmental and reproductive features. The two most detrimental groups of sedentary PPNs are root-knot nematodes (RKNs) with genus *Meloidogyne* which upon infection induce the development of groups of giant cells inside the vascular cylinder surrounded by hyperplastically dividing parenchymatic cells leading to the formation of swellings on the roots (galls, root-knots) (Fig. 18.1b, c), and cyst-forming nematodes (CNs) with leading genera *Globodera* and *Heterodera* which induce the development of syncytia located predominantly inside the vascular cylinder (Fig. 18.1e, f, h, i). The name of this group of PPNs originates from eggs-protecting cysts which are formed from the bodies of dead females. RKNs usually have a wide range of possible plant hosts, sometimes exceeding several hundreds of species. In contrast, CNs have a narrower host range sometimes restricted to a few closely related plant species. However, also some genera of migratory nematodes, i.e., *Pratylenchus* and *Radopholus*, can create serious problems on some crops (Jones et al. 2013).

Due to different morphogenetic pathways modulated at cellular and molecular levels by RKNs and CNs in plants during infection, and the establishment and development of their feeding sites, later in this chapter, we will focus mainly on syncytia induced by CNs to highlight developmental and molecular changes related to their parasitism, feeding site induction, and development. For readers interested in the structure, development, and functioning of nematode-induced giant cells, we recommend the following papers: Huang and Maggenti (1969), Paulson and Webster (1970), Fattah and Webster (1984), de Almeida et al. (2004), Jagdale et al. (2021), and Rutter et al. (2022).

18.2 Plant-Parasitic Nematodes

The life cycle of sedentary PPN is relatively uniform. It starts from an egg deposited inside the body of a dead female in the case of CNs. The first stage juvenile (J1) develops inside the eggshell and moults into the second stage juvenile (J2). The J2s may remain dormant inside the eggshell for different time, even for more than 20 years in the case of potato cyst nematode (*G. pallida*). Hatching of J2s is usually induced by exudates from host plant's roots. Using chemosensory organs juveniles migrate towards host roots. Infective J2s of CNs have large and robust stylets, thus they rely more on the physical disruption of cell walls than on the cell wall-degrading enzymes. They usually enter rhizodermis cells making several punctures with stylet in cell walls and then break them with head movements and stylet thrusts. They enter the cell and keep migrating intracellularly towards the vascular cylinder mechanically breaking the walls of cortical and endodermal cells (Fig. 18.1e, h). This savage behaviour changes rapidly into subtle and exploratory when the anterior part of the nematode body enters the vascular cylinder (Fig. 18.2a). The J2 carefully pierces procambial, pericyclic, and parenchymatic cells of the vascular cylinder and awaits their reaction. If the reaction is adverse, the cell is destroyed and the next cell is probed. This examination lasts until the properly-reacting cell is found. The molecular and cytological features of "proper reaction" are still unknown but getting this signal the J2 remains motionless with its stylet inserted into selected cell for several hours. During this period the density of granules in nematode's secretory glands decreases, which suggests that the sedentary J2 is releasing a cocktail of effectors to the selected cell. Afterward, the stylet is retracted and reinserted, and the metacorporal pump starts to pulse indicating the onset of food withdrawal (Wyss and Zunke 1986). Infective J2 of CNs selects only a single cell which becomes turned into the initial syncytial cell (ISC), and later, neighbouring meristematic and/or parenchymatic cells of the vascular cylinder become modified and fuse with the ISC via local dissolutions of cell walls, thus, forming a syncytium (Golinowski et al. 1996; Fig. 18.2a).

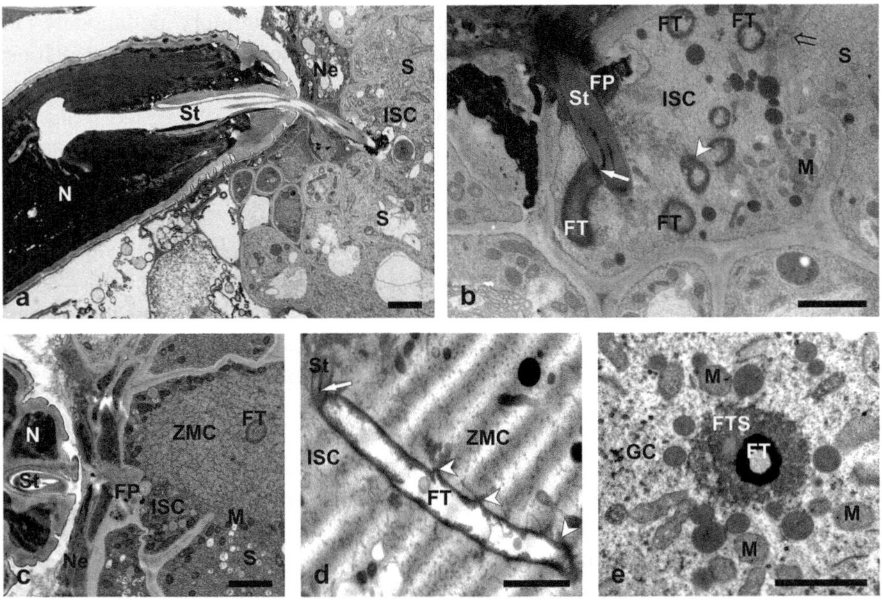

Fig. 18.2 Structures related to parasitism of sedentary plant-parasitic nematodes. Transmission electron microscopy micrographs of syncytia induced by *Heterodera schachtii* in *Arabidopsis thaliana* root (**a**–**d**) and giant cell induced by *Meloidogyne hapla* in sunflower (*Helianthus annuus*) root (**e**). (**a**) longitudinal section of nematode head and stylet inserted into the initial syncytial cell of 2 dpi syncytium; (**b**) section of 4 dpi syncytium with a feeding tube attached to the stylet orifice (arrow). The connection of feeding tubes with the endoplasmic reticulum is marked with the arrowhead. Double tail arrow points to widened plasmodesmata; (**c**) section of 4 dpi syncytium showing zone of modified cytoplasm in the initial syncytial cell; (**d**) section of 4 dpi syncytium with the longitudinal section of feeding tube attached to the stylet orifice (arrow). Arrowheads point to connections of endoplasmic reticulum with feeding tube wall; (**e**) cross-sectioned feeding tube inside the giant cell (*courtesy Kamila Chalamońska and Andrzej Skwiercz*). Abbreviations: FP, feeding plug; FT, feeding tube; FTS, feeding tube tubular structure; ISC, initial syncytial cell; M, mitochondrion; N, nematode; Ne, necrosis; S, syncytium; St, stylet; ZMC, zone of modified cytoplasm. Scale bars: 2 μm

After selection and induction of the ISC, the J2 loses its locomotive abilities and the induced feeding site becomes its only source of nutrients for all successive developmental stages. Sedentary J2 is still not sexually differentiated. It feeds for several days and then moults to sedentary and actively feeding J3, which exhibits the first anatomical features of sexual differentiation (Wyss 1992). There are numerous indications that the sex of sedentary PPNs is regulated epigenetically by the amount and composition of nutrients withdrawn by the J2 (Müller et al. 1982; Sobczak et al. 1997; Anjam et al. 2020). J3 feeds actively on the feeding site and moults to be J4. The male J4 undergoes successive moult, without a food uptake, to adult vermiform male, which breaks the cuticle of former stages and leaves the roots. The female J4 has another period of food uptake which ends with a moult to mature female. The female remains sedentary, and its body becomes round, citron-, or pear-shaped. CNs reproduce usually by amphimixis, and after fertilisation, the female deposits eggs inside its own body, which becomes a protective cyst, or outside in the gelatinous sacs (Raski 1950).

18.3 Structure of Syncytia in Susceptible Plants

After selection and induction of the ISC, the infective J2 withdraws and reinserts the stylet into the same cell (Wyss and Zunke 1986; Fig. 18.2a, b). Granules produced in nematode oesophageal glands move via glands ducts towards the stylet and they are injected into the selected cell through the stylet channel. Probably some constituents of nematode secretions spread inside the cytoplasm of the ISC and modify its gene expression whereas some other constituents interact with the cytoplasm of the ISC and form structures called feeding tubes (Rebois 1980; Rumpenhorst 1984; Endo 1991; Sobczak et al. 1999; Fig. 18.2b–d). There is a general agreement that feeding tubes serve as a kind of molecular sieve for the withdrawal of nutrients from modified plant cells (Eves-van den Akker et al. 2015). Like the stylet or odontostyle, the feeding tubes are formed by all microscopically analysed PPNs so far. In most of PPNs, the feeding tubes consist of a strongly electron-dense wall and electron translucent lumen, as found in the CNs (Fig. 18.2d). The recent feeding tube is attached to a stylet orifice (Fig. 18.2b, d) and surrounded by a zone of modified cytoplasm, which contains only short cisternae of endoplasmic reticulum and is devoid of all other organelles (Wyss 1992; Fig. 18.2c, d). Older feeding tubes are detached from the stylet tip, and in different stages of degradation, are present in various regions of syncytial cytoplasm (Sobczak et al. 1999; Fig. 18.2b). However, in the case of *Meloidogyne* species, this structure is additionally surrounded by a complex network called a fine tubular structure (Hussey and Mims 1991; Miyashita and Koga 2017; Fig. 18.2e), which is interconnected with tubules of endoplasmic reticulum (Miyashita and Koga 2017). In contrast, in feeding tubes of CNs such connections are relatively rare and have interconnections with cisternae of the endoplasmic reticulum (Sobczak et al. 1999; Fig. 18.2d).

After the release of gland secretions and formation of the feeding tube, the metacorporal pump of the nematode starts to pulse indicating the onset of active food withdrawal from the feeding site. The phase of active food uptake lasts for 2–4 h and then the juvenile withdraws the stylet and reinserts it again at the same place. The forward movement of gland granules and the formation of a new feeding tube takes place again. This feeding cycle occurs in all developmental stages of actively feeding sedentary nematodes (Wyss 1992). Parasitism of CNs also involves another specific structure at the interface between the nematode and plant cell wall called the feeding plug (Endo 1978; Sobczak et al. 1999; Fig. 18.2b, c). It is a kind of peg embedded in the syncytial cell wall that fills the space between the outer cell wall of the syncytium and the nematode head during later developmental stages. The stylet of CNs is inserted into the syncytial cytoplasm only through the feeding plug. The plug is probably composed of callose and functions as a kind of seal filling the space between the stylet and cell wall to protect syncytial cytoplasm leakage (Endo 1978; Sobczak et al. 1999) or acts as a glue providing durable adhesion of the parasite to its feeding site (Eves-van den Akker et al. 2015).

Arabidopsis thaliana has emerged as basic model plant in plant biology research, and fortunately, it is also a host for the beet cyst nematode (*H. schachtii*). Thus, the Arabidopsis–CNs interactions have been studied in detail (Sijmons et al. 1991). *H. schachtii*, a common parasite of beet and numerous brassicaceous plants (Raski 1950), selects the ISC preferentially among procambial cells located between xylem and phloem conductive elements (Golinowski et al. 1996; Figs. 18.1h and 18.2a, c). If these cells degrade or respond in an improper way, the juveniles select their ISCs among pericyclic cells (Sobczak et al. 1997; Fig. 18.2b). Syncytia induced in procambium expand further along the root axis by incorporation of modified procambial and parenchymatic cells located between conductive bundles and they develop an extensive interface with sieve tubes transporting assimilates and xylem vessels transporting water (Fig. 18.3a, c, i). It allows the formation of large and extensive syncytia full of nourishments and water, which can support development of nematode females (Golinowski et al. 1996). In contrast, syncytia induced in the pericycle are located on the outskirts of the vascular cylinder and they have a limited contact interface with conductive elements. They preferentially incorporate enlarged pericyclic cells, thus, close to the nematode head, the syncytia have a half-collar-like shape on cross sections (Fig. 18.2b). At some distance from the ISC, also procambial cells become incorporated, and syncytium enters the centre of the vascular cylinder. However, the degree of hypertrophy of incorporated cells is lower than in syncytia induced in the procambium. Being smaller and less effective, they are usually able to support only the development of male nematodes (Sobczak et al. 1997), which, during the entire development, have 29 times lower nutritional demands than females (Müller et al. 1982).

Other CNs species such as potato cyst nematodes (*G. rostochiensis* and *G. pallida*) select their ISCs usually among root cortical parenchyma cells or endodermis cells (Fig. 18.1e, f). In their case, the first syncytial elements are derived from cortical cells which fuse with the ISC forming a so-called cortex bridge extending towards the vascular cylinder (Sembdner 1963; Jones and Northcote 1972; Sobczak et al. 2005; Fig. 18.1e, f). When the "cortex bridge" reaches the vascular cylinder, it expands along the root by incorporation of procambial, cambial, and vascular parenchyma cells (Jones and Northcote 1972; Sobczak et al. 2005; Fig. 18.1f).

The ISC and neighbouring procambial cell enlarge. The amount of their cytoplasm increases whereas the volume of the vacuole decreases (Figs. 18.2a, b and 18.3a). About 24 h after ISC selection the first cell wall openings are formed by the widening of plasmodesmata between the ISC and neighbouring cells (Figs. 18.2b and 18.3b). The process of incorporation of new syncytial elements via incorporation of modified procambial and parenchymatic cells progresses and syncytium expands inside the vascular cylinder usually in acro- and basipetal directions from the ISC (Golinowski et al. 1996; Sobczak et al. 1997; Sobczak and Golinowski 2011).

The formation of syncytium is associated with numerous ultrastructural changes in incorporated cells. The most pronounced feature is the extreme hypertrophy of

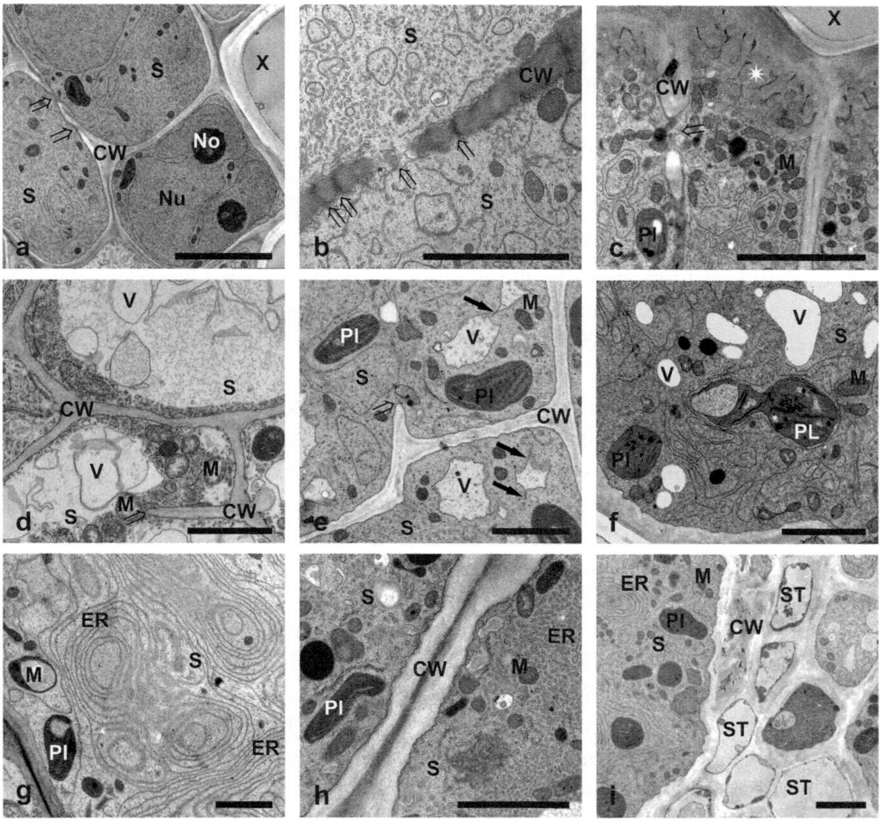

Fig. 18.3 Ultrastructural features of syncytia induced by *Heterodera schachtii* in roots of *Arabidopsis thaliana*. Transmission electron microscopy micrographs of syncytia at 4 dpi (**a**, **b**, **d**), 7 dpi (**e**–**h**), and 14 dpi (**c**, **i**). (**a**) terminal syncytial elements at leading edges of syncytium with small cell wall openings (double tail arrows); (**b**) cell wall openings (double tail arrows) formed by widening of plasmodesmata; (**c**) cell wall ingrowths (asterisks) formed at syncytial wall facing xylem vessels and cell wall opening (double tail arrow); (**d**) de-differentiation of the central vacuole in recently incorporated syncytial elements. Double tail arrow points to cell wall stub; (**e**) formation of small vacuoles in syncytial elements by dilation of endoplasmic reticulum cisternae (arrows). Double tail arrow points to cell wall stub; (**f**) portion of syncytium with dividing plastid; (**g**) extensive concentric arrays of smooth endoplasmic reticulum cisternae; (**h**) arrays of tubular endoplasmic reticulum; (**i**) new sieve tubes differentiated next to the syncytium. Abbreviations: CW, cell wall; ER, endoplasmic reticulum; M, mitochondrion; No, nucleolus; Nu, nucleus; Pl, plastid; S, syncytium; ST, sieve tube; X, xylem; V, vacuole. Scale bars: 5 μm (**a**–**c**); 2 μm (**d**–**i**)

syncytial elements (Fig. 18.1e, f, h, i) when compared to non-modified parenchymatic cells of the vascular cylinder. Accordingly, outer syncytial cell walls thicken to resist the increased osmotic pressure, which reaches 10 kPa in syncytial elements of *H. schachtii* syncytium induced in roots of *A. thaliana*, in comparison to 2 and 4 kPa in cortex and vascular parenchyma cells, respectively (Böckenhoff and Grundler 1994). Using monoclonal antibodies, Davies et al. (2012) showed that the structural architecture of syncytial walls is unique and different from neighbouring cell walls. The syncytial cell walls contain cellulose, xyloglucan, heteromannan, heavily methyl-esterified homogalacturonan, and arabinan, whereas galactan is absent. These cell wall components suggest that the syncytial cell wall is highly flexible to withstand internal pressure and turgor fluctuations during nematode feeding (Davies et al. 2012).

The syncytium expands via successive incorporation of modified surrounding cells via local cell wall dissolutions, which facilitate protoplasts fusion. First cell wall dissolutions form via widening of plasmodesmata between the ISC and neighbouring cells (Figs. 18.2b and 18.3b). At later stages of syncytium development, the cell wall openings form de novo by local digestion of syncytial walls (Grundler et al. 1998; Fig. 18.3a, c, e). 3-D reconstructions of syncytial walls show very regular arrangement of these cell wall openings, which are separated by remnants of non-digested cell wall, resembling the Parthenon pillars (Ohtsu et al. 2017). The widest and most numerous cell wall openings are in the old part of syncytium next to the nematode head (Fig. 18.1f, h, i). In young parts of the syncytium (remote from the nematode head) the cell wall openings are few and narrow.

With the growing nutritional demands of J3s, female J4s, and adults, the elaborated systems of cell wall ingrowths (protuberances) composed of cellulose and other polysaccharides form on the inner side of the outer syncytial wall facing conductive elements of xylem (Fig. 18.3c). This system is characteristic of transfer cells (cells specialised in short-distance transport of nutrients or water) and increases the surface of the xylem/syncytium exchange interface (Jones and Northcote 1972; Golinowski et al. 1996). The system of cell wall ingrowths is weakly developed, or even absent, in syncytia associated with male J3 (Sobczak et al. 1997). To provide enough nutrients, additional sieve tube elements differentiate next to the outer syncytial wall (Fig. 18.3i).

Syncytial cytoplasm undergoes extensive changes and re-organisation. The amount of cytoplasm in syncytial elements increases, whereas the vacuole volume decreases. The large central vacuole typical for plant cells is de-differentiated into numerous small vacuoles unevenly distributed within the syncytium (Fig. 18.3d–f). New vacuoles and vesicles also form via widening of endoplasmic reticulum cisternae (Baranowski et al. 2019; Fig. 18.3e). The number of ribosomes increases indicating high levels of protein synthesis, and the cytoskeleton undergoes extensive re-organisation (de Almeida Engler et al. 2004; Różańska et al. 2018). The system of endoplasmic reticulum expands. The cisternae of rough endoplasmic reticulum often form circular swirls (Fig. 18.3g). The tubules of smooth reticulum are well-developed and occur in the regions devoid of rough reticulum cisternae (Fig. 18.3h). The number of mitochondria and plastids increases (Fig. 18.3c). Mitochondria have a typical structure (Fig. 18.3c–i), but sometimes they acquire cup-like or coiled outlines (Fig. 18.3g). Plastids are round or elongated (Fig. 18.3c–i) and often become irregular (Fig. 18.3h), cup-like (Fig. 18.3g), or constricted (Fig. 18.3f). In contrast to typical root plastids, they usually contain relatively well-developed system of thylakoids, making them a bit like chloroplasts. They do not contain starch grains when the associated juvenile withdraws food, but they deposit large starch grains during nematode moults. The volume of syncytial nuclei also increases due to endoreduplication without cytokinesis (de Almeida Engler et al. 2012). Nuclei are irregular and amoeboid (Fig. 18.3a). The amount of heterochromatin decreases. The nucleolus is highly enlarged, which in combination with increased numbers of free cytoplasmic and reticulum-bound ribosomes indicates high translational activity in the syncytium (Fig. 18.3a).

The syncytium remains functional as long as the associated nematode withdraws nutrients and injects its secretions into the syncytial cytoplasm. When nematode stops feeding, i.e., when J3 male moults into non-feeding J4, or an adult female dies, the syncytium degenerates. It indicates that continuous stimulation and effectors delivery by the nematode are necessary for maintaining functional syncytium in plant roots (Sobczak et al. 1997).

18.4 Molecular Basis of Nematode–Plant Interaction and Susceptibility Genes

Changes in plant anatomy and cell ultrastructure occurring during syncytium development clearly indicate that PPNs can influence plant's morphogenetic pathways and fine tune metabolic processes for their own benefits. Because plants do not possess genes which exclusively profit the nematode, the nematode modulates genes necessary for plant development and functioning. These genes, which expression is induced, silenced, or modulated by infective juveniles, are called susceptibility genes. There is an extensive search for such genes because the modulation of their expression may be a useful tool in breeding plants with reduced susceptibility to PPNs. Two approaches are used in a search for plant nematode susceptibility genes and their proteins. The most extensively used approach uses new-generation sequencing methods to identify transcripts whose abundance is reduced or elevated in nematode feeding sites in comparison to uninfected control roots (Siddique et al. 2014, 2015; Różańska et al. 2018; Radakovic et al. 2018). The second approach focuses on the identification of putative nematode effectors and their plant protein interactors (Hewezi et al. 2015; Habash et al. 2017; Mitchum and Liu 2022). Overexpression or silencing of identified susceptibility genes was shown to reduce the susceptibility of mutants or overexpressors, which is reflected in lower numbers of developing females that deposit lower numbers of eggs, and induce smaller syncytia. Also, at cytological and anatomical levels, the differences between syncytia induced in control and genetically modified plants are rather quantitative than qualitative. Genetically modified plants have lower hypertrophy of syncytial elements and lower number and size of cell wall openings. Their cell wall ingrowths are maldeveloped or absent, and the syncytial cytoplasm is more electron translucent (possibly poorly supplied with nutrients) (Siddique et al. 2014, 2015; Różańska et al. 2018). Below we provide brief descriptions of some CNs susceptibility genes.

Early responses to nematode infection are associated with the breakdown of the plant's mechanical primary defence barrier constituted by the cell wall. During infection, plant NADPH oxidases are activated and there is an increased synthesis of superoxide anion radical, which forms H_2O_2 through the action of superoxide dismutase (SOD). H_2O_2 acts as a signalling molecule that triggers defence response. Alterations in reactive oxygen species (ROS) homeostasis can lead to the accumulation of toxic by-products of oxygen metabolism affecting plant cell metabolism and inducing programmed cell death (PCD). Activation of metabolic pathways associated with the biosynthesis of phytoalexins (involvement of $^1O_2$ in ROS production) and lignin (H_2O_2 as a reaction substrate) to seal the cell wall was described (Siddique and Grundler 2018). Siddique et al. (2015) found that Arabidopsis mutants impaired in NADPH oxidases (*Rboh*) expression were more susceptible (allowed development of higher numbers of females and males of the nematode) to *H. schachtii* than wild-type plants, but double knock-out mutant of *RbohD* and *RbohF* (*rbohD/F*) was less susceptible to CNs infection than wild-type plants, suggesting that Rboh-mediated ROS plays a role in promoting infection. Moreover, *H. schachtii* parasitism induces changes in ROS homeostasis also in the above-ground parts of plants where accumulation of SOD and H_2O_2 was found during early stages of syncytium development (Labudda et al. 2018). Additionally, activities of several antioxidative enzymes (SOD, catalase, and glutathione-S-transferase) were higher at 7 and 15 days post inoculation (dpi). This clearly shows that the development of syncytium in the host roots has a constitutive impact on ROS homeostasis in whole infected plant. Similarly, novel effector gene (*HsPDI*) from *H. schachtii*, with disulfide isomerase domain, was postulated to be involved in ROS detoxification, which is necessary for nematode protection during infection, migration, and syncytium development (Habash et al. 2017).

Nitrogen metabolism and scavenging of reactive nitrogen species (RNS) is also crucial for proper development of syncytium. These pathways were significantly altered during CNs parasitism. Good examples are 3 and 15 dpi syncytia induced by *H. schachtii* in *A. thaliana* roots, where nitrate reductase (NIA) activity and expression of *NIA2* gene were higher than in uninfected plants (Labudda et al. 2020).

The stress hormones salicylic acid (SA) and jasmonic acid (JA) act as signalling molecules in the plant defence responses to a range of pathogens. Biotrophic parasites show higher sensitivity to SA-based defence response. JA and ethylene (ET) are mediators of the defence response to necrotrophs and wounding (e.g. during grazing of insects). However, this clear separation seems to be too simplistic, as the action of SA, JA, and ET is strongly influenced by other phytohormones, ROS, and RNS (Gheysen and Mitchum 2011). When PPNs induce feeding sites in plant roots, massive reprogramming of expression of SA- and JA-related host genes occurs. Analyses of feeding site transcriptomes from various plant species clearly indicate strong activation of genes encoding pathogenesis-related proteins (PR) associated with activation of the SA synthesis pathway. Expression levels of the *PR-1* gene in the rosette, and *PR-2* and *PR-5* genes in infested Arabidopsis roots were elevated at the early stages of *H. schachtii* infestation. Moreover, Arabidopsis mutants with reduced content of SA (*sid2-1* and *pad4-1*) and mutants impaired in SA signal transduction (*npr1-2* and *npr1-3*) were more susceptible to *H. schachtii* (Wubben et al. 2008).

JA is responsible for triggering an early defence response during juvenile migration and the ISC selection stages. Application of JA methyl ester (methyl-JA; MeJA) to Arabidopsis leaves resulted in decreased susceptibility to *H. schachtii* infection (Kammerhofer et al. 2015). In contrast, Arabidopsis mutants with impaired JA biosynthesis (*dde2* and *lox6*) were more susceptible to the same nematode compared to control plants (Kammerhofer et al. 2015). The involvement of JA was studied by analysis of the transcriptome of syncytium induced by *H. glycines* in soybean roots where massive suppression of JA signalling was found (Ithal et al. 2007).

The role of ET in the plant response to PPNs is indisputable, however, unambiguous interpretation of its role is difficult. In the first line, the response is species-specific and depends on the developmental stage of syncytium. Secondly, it becomes complicated due to ET interaction with auxins, SA, and JA. ET modifies "attractiveness" of plant roots to CNs (Wubben et al. 2001; Hu et al. 2017) as they are more likely to attack plants that have higher levels of ET. The infection-stimulating effect is also obtained after treatment with ET precursors (Wubben et al. 2001). Analysis of Arabidopsis mutants with impaired ET synthesis pathway showed that this hormone plays a key role in the early stages of syncytium induction by *H. schachtii*. The ISCs were selected in roots of *ein3-1* and *ein2-1* mutants almost twice faster as in roots of wild-type plants, 4 h versus 8 h, respectively. The use of confocal laser scanning microscopy techniques together with ET-inducible reporter lines (*pACS6::NLS-3xVenus* and *pPR4::NLS-3xVenus*) confirmed the involvement of ET in the early stages of syncytium development. Therefore, modification of ET homeostasis may be one of the key strategies leading to the reproductive success of the CNs (Marhavý et al. 2019). Different susceptibility of ET mutants may depend on the type of signalling pathway that ET mutant deactivates. The canonical ET signalling pathway inhibits SA-dependent defence response, thus increasing plant susceptibility to infection with *H. schachtii*. In contrast, the second pathway involving ethylene receptor 1 (ETR1) and inhibition of cytokinin-related signalling pathways decreases plant susceptibility (Piya et al. 2018).

It is worth to stress that most of the plant–nematode interactions experiments is carried out in the in vitro culture. Closed culture containers with limited ventilation during experimental assays may change Arabidopsis mutants' phenotype due to ET and CO_2 accumulation (Matuszkiewicz et al. 2019). Plants cultured in well aerated containers were more susceptible when compared to plants grown in air-tight sealed plates. Therefore, choosing the proper sealing technique during in vitro assays is essential but neglected factor, which may mask the true phenotype of plant response to nematode infection.

Due to the developmental reprogramming of roots and cells incorporated into syncytia, levels of other phytohormones as auxins, gibberellins, cytokinins, and abscisic acid (ABA), which are transported in xylem and phloem sap, change upon nematode infection. Possibly, the nematode is capable to alter the plant's perception of these phytohormones by interfering with the direction of flow in conductive elements during feeding site development or to modulate their signalling (Gheysen and

Mitchum 2011, 2019). Using the expression of synthetic auxin-responsive *DR5* promoter in transgenic Arabidopsis plants infected with *H. schachtii*, it was shown that the increase in auxin levels in the developing syncytium is due to local biosynthesis and accumulation of this hormone (Goverse et al. 2000). The direction of auxin transport around syncytium region is also altered (Grunewald et al. 2009). Analyses of the expression of *PIN* genes encoding auxin transport proteins showed up-regulation of *PIN3* and *PIN4* expression during the early stages of syncytium development, thus auxin was transported to cells next to syncytium. At this time point expression of the *PIN1* gene coding auxin transporter responsible for auxin efflux from the cell was inhibited in a syncytium. Recently, two *H. schachtii* effectors were linked to the modification of auxin transport (Gheysen and Mitchum 2019). Effector 19C07 interacts with LAX3 protein that is responsible for auxin influx into the syncytium and induction of genes involved in cell wall degradation (Lee et al. 2011). Effector 10A07 interacts with the IAA16 protein thereby competing with selected transcription factors called auxin response factors (ARF) such as ARF5-9 and ARF19 (Hewezi et al. 2015). Another connection between auxin and ROS signalling in plant–nematode interaction is the discovery that auxin transporter WALLS ARE THIN1 (WAT1) is a downstream target of Rboh-mediated ROS. Transcriptomic analysis showed strong down-regulation of *WAT1* in *rbohD/F* mutant upon cyst nematode infection, but not in RKNs parasitism (Chopra et al. 2021). Moreover, WAT1 is crucial for the host indole metabolism via changing the levels of indole-3-acetic acid (auxin; IAA) in a syncytium. The role of auxin homeostasis in plant–nematode interactions has been extensively discussed in a review by Oosterbeek et al. (2021).

Cytokinins are hormones involved in regulation of plant cell divisions, stimulation of cell volumetric growth, induction of shoot development and branching, and organ senescence by altering nutrient distribution. Interestingly, De Meutter et al. (2003) characterised cytokinins in the secretions of CNs and RKNs. The involvement of cytokinins in syncytium development was also confirmed by identification of nematode genes responsible for cytokinins biosynthesis (isopentenyl transferase; *HsIPT*) (Siddique et al. 2015). Silencing of this gene in infective juveniles of *H. schachtii* resulted in decreased virulence of treated juveniles reflected in reduced numbers of developing females that were also smaller and induced smaller syncytia. Moreover, genes related to cytokinins biosynthesis, catabolism, and signalling played important but divergent roles during infection and parasitism of CNs and RKNs (Dowd et al. 2017).

18.5 Resistance Response

During evolution, plants have developed mechanisms of resistance to protect their resources. As described above, the susceptible interaction of plants with parasitic nematodes is complex and multidimensional. It depends on internal factors such as the physiological state of the host as well as external causes such as the virulence and severity of the pathogen, density of inoculum, and environmental conditions. The focus of researchers is on genetic determinants of the pathogen and host, and the resulting molecular mechanisms of plant–pathogen interaction. However, thanks to the multi-level nature of the interaction, plants have a wide range of responses to avoid or restrict pests. Pathogens, on the other hand, also constantly "armour up" to break down plants' defence mechanisms (Jones and Dangl 2006; Dangl and Jones 2019). The evolutionary "arms race" between plant defence mechanisms and pathogen virulence is best illustrated by the zig-zag model (Fig. 18.4) which is universal and applies to viral, bacterial, fungal, or herbivorous pests (Jones and Dangl 2006).

Once the nematode enters the root, a mechanism recognising pathogen-associated molecular patterns (PAMPs) and damage-associated molecular patterns (DAMPs)

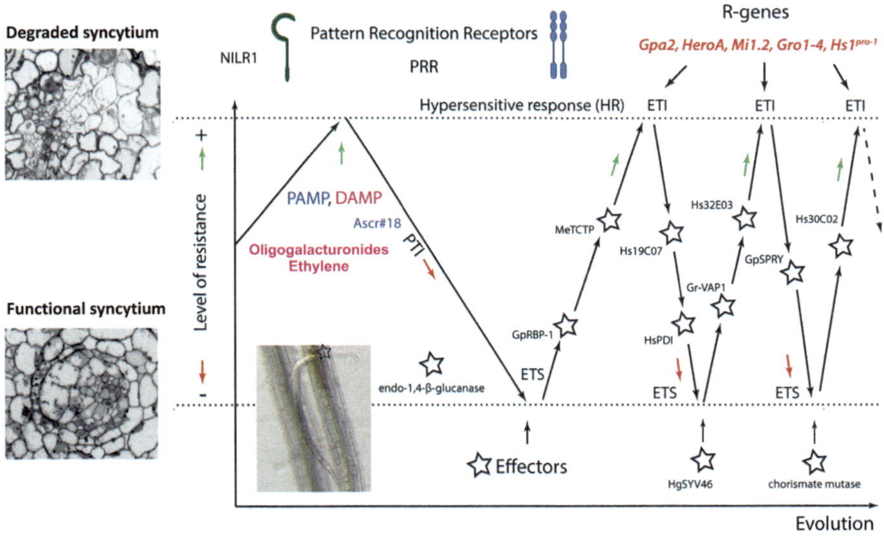

Fig. 18.4 Zig-zag model of plant response to parasitic nematode attack. The diagram depicts the concept of an "arms race" between plants and parasitic nematodes during evolution. The process consists of four phases. Phase I—activation of mechanism recognising pathogen-associated molecular patterns (PAMPs) and damage-associated molecular patterns (DAMPs) after root invasion. Plasma membrane located pattern recognising receptors (PRR), schematically depicted as a dimer, recognise DAMP and PAMP molecules. To date, only one PAMP, ascaroside 18 (Ascr#18), has been characterised. The PRR recognising Ascr#18 is not known. The only known PRR recognising plant-parasitic nematode pattern is NIRL1, but the recognised pattern is unknown. Phase II—nematodes have "produced" effectors, emblematically marked with asterisks, to break the immunity induced by PAMP factors (pattern-triggered immunity; PTI). The phenomenon of effectors breaking PTI resistance mechanisms is called effector-triggered susceptibility (ETS). Phase III—some effectors become recognised by NB-LRR/NLR immunoreceptors also called immune R proteins. The immunity produced in response to the effector is called effector-triggered immunity (ETI). Hypersensitive response (HR) involves the activation of programmed plant cell death at the site of infection and occurs during both, PTI and ETI responses. Phase IV—coevolution of new effectors of parasitic nematodes and resistance proteins (R) of plants. Please note that the names of genes and effectors are given only as examples (based on the original scheme of Smant and Jones (2011), modified)

derived from the debris of cells destroyed during root invasion is triggered (Fig. 18.4). Specific pattern recognition receptors (PRRs) located in the plasma membrane of plant cells recognise DAMP and PAMP signals. Molecular patterns are characteristic of pathogenic and non-pathogenic groups of organisms, i.e. bacteria (elongation factor EF-Tu, flagellin) and fungi (chitin, glucans). Recognition of the patterns leads to the activation of a cascade of signals that stimulates plant immune mechanisms eliciting PAMP-triggered immunity (PTI). So far, ascaroside, a pheromone being a glycolipid and containing ascarose, is the only PAMP characterised in PPNs (Manosalva et al. 2015). Among all ascarosides of PPNs, Ascr#18 is the most abundant. It stimulates the plant immune system by the activation of the mitogen-activated protein kinase cascade, increasing expression of PTI marker genes, and activating SA and JA signalling pathways. In addition, treatment of tomato, potato, barley, and Arabidopsis with Ascr#18 increased their resistance level to CNs and RKNs. This shows that the mechanism of PAMPs recognition is conserved in monocotyledonous and dicotyledonous plants. However, the specific plant PRR of Ascr#18 is still unknown.

Mendy et al. (2017) used "NemaWater", water in which hatched infective juveniles of CNs or RKNs were soaked for a longer time, to identify the first PRR receptor in Arabidopsis that recognises nematode-associated molecular patterns (NAMPs). NIRL1 (nematode-induced leucine-rich repeat receptor-like kinase) is responsible for the induction of plant defence responses upon nematode infection. Moreover, knock-out mutants of this gene are hypersensitive to parasitic nematodes. Its action depends on activity of the positive regulator of the BAK1 (BRI1-associated receptor kinase 1), which is a BRI1 (brassinosteroid insensitive 1) co-receptor involved in defence processes against pathogenic microorganisms. In addition, the extracellular receptor domain of NIRL1 is highly conserved in monocotyledonous and dicotyledonous plants. Different plant species (i.e. tomato, sugar beet, rice, and tobacco) treated with "NemaWater" activated PTI pathways with a burst of ROS. Arabidopsis plants treated with "NemaWater" were less susceptible to CNs, RKNs, and the virulent *Pseudomonas syringae* pv. tomato strain DC3000. The specific molecular pattern recognised by the NIRL1 receptor remains unknown (Mendy et al. 2017).

During migration, parasitic nematodes destroy cell walls via mechanical damage caused by the stylet, nematode body movements and by the action of secreted cell wall-degrading enzymes. This leads to the release of DAMPs and subsequent activation of PTI. An example of DAMPs released by CNs attack are oligogalacturonides. CNs secrete polygalacturonases (PGs) to degrade cell walls during infection. Arabidopsis has two genes coding for polygalacturonase inhibitors (*PGIP1* and *PGIP2*). Their action inhibits complete degradation of pectin to galacturonic acid by PGs, but oligogalacturonides with varying degrees of polymerisation are released (Benedetti et al. 2015). Spatio-temporal analysis of *PGIP1* and *PGIP2* expression during different CNs infection stages pointed out the superior role of *PGIP1* during plant–nematode interaction. Moreover loss-of-function mutant of *PGIP1* was hypersusceptible to CNs (Shah et al. 2017). A recent publication indicates that also ET may function as a DAMP due to its small-molecule size, gaseous nature, and the fact that even pinpoint root damage caused by the nematode stylet induces expression of ET-related genes that may impede selection of the ISC (Marhavý et al. 2019).

Trying to avoid recognition by plant defence mechanisms, parasites have evolved effectors blocking PTI resistance mechanisms (Fig. 18.4). This phenomenon has been called effector-triggered susceptibility (ETS) (Jones and Dangl 2006; Ngou et al. 2022). Sedentary PPNs synthesise chemical compounds in their oesophageal glands (subventral and dorsal), which are injected into the host cell protoplast via the hollow nematode stylet. On the one hand, these secretions are aimed at breaking the PTI mechanisms, and on the other hand, they are responsible for transforming the plant cells into functional feeding sites. A cocktail of nematode-derived effectors also affects the expression of plant genes involved in cell wall modification processes, changes in signal transduction pathways, and phytohormone homeostasis. The inhibition of expression of defence response genes is an important element of PTI collapse triggered by the PPNs. One of the first effector proteins characterised in CNs secretions was endo-1,4-β-glucanase, responsible for cellulose hydrolysis. The presence of this enzyme suggests that CNs may use it to weaken plant cell walls during migration, and additionally, injected into the ISC it may be involved in the formation of first cell wall openings between the ISC and neighbouring cells (Smant et al. 1998). A few years later 30C02 effector from *H. schachtii* was characterised to affect the expression of plant β-1,3-endoglucanase (Hamamouch et al. 2012). They both demonstrate the importance of the control of cell wall-related processes during migration, ISC selection, syncytium induction and development.

Another well-studied protein effector found in secretions of both, CNs and RKNs, is chorismate mutase (CM; EC 5.4.99.5). This enzyme is involved in the shikimate pathway and catalyses the conversion of chorismate into prephenate, which is substrate for phenylalanine, tyrosine, tryptophan, and precursors of secondary metabolites like indole derivatives (auxin) synthesis. The secretion of CM into the cytosol of host cells can modify the morphogenetic programme of root cells and inhibit plant defence responses (Lambert et al. 1999; Jones et al. 2003; Vanholme et al. 2009).

An example of effectors found so far only in CNs is the numerous and diverse group of SPRYSEC proteins (SP1a and RYanodine receptor—SPRY domain containing proteins) (Mei et al. 2015; Diaz-Granados et al. 2016). Their action is aimed at inhibition of plant defence responses. The GpRBP-1 (*G. pallida* effector containing the SPRY domain) breaks the resistance against white potato cyst nematode (*G. pallida*) provided by potato *Gpa2* resistance gene (Sacco et al. 2009). Another two SPRYSEC effectors derived from *G. pallida* (GpSPRY-12N3 and GpSPRY-33H17) only weaken the effect of *Gpa2*. However, several SPRYSEC effectors from *G. rostochiensis*, SPRYSEC-4, -5, -8, -18, and -19, were shown to suppress plant defence responses during infection of potato cyst nematode (Ali et al. 2015). Recently, the mechanism of action of GpSPRY-412-2 effector, which selectively attenuates the action of the *Gpa2* resistance gene was described. GpSPRY-412-2 binds to the potato cytoplasmic linker protein (CLIP)-associated protein (CLASP) and thus affects the stability and growth of microtubules. Changes in cytoskeleton polarity and organisation are necessary for syncytium induction and appropriate development (de Almeida Engler et al. 2004; Mei et al. 2018).

Another interesting group of effectors secreted by PPNs belongs to CLAVATA3/Endosperm Surrounding Region (CLE) like proteins, which imitate plant regulatory peptides. The transcript of the first nematode effector belonging to this "ligand-mimicry" group, HgSYV46, was identified in the cDNA library from the oesophageal glands of soybean cyst nematode (*H. glycines*). The HgSYV46 peptide contains a conserved 14 amino acid motif at its C-terminus, found among others in plant peptide CLAVATA3 (CLV3), which activates the CLV1/CLV2 receptor complex and inhibits WUSCHEL expression, thereby interfering in cell differentiation. The mechanism of action of the CLE effectors family may involve the induction of the CLE signalling pathway in the roots, although it is a stem-specific pathway, to promote syncytium formation (Wang et al. 2005).

Most of the effectors belongs to large gene families, but there are also virulence factors encoded by single genes. An example is the 30D08 effector isolated from *H. glycines* and *H. schachtii*. It interacts with Arabidopsis auxiliary spliceosomal protein SMU2. Most likely, this interaction affects pre-mRNA splicing and alters the expression of host genes involved in syncytium formation (Verma et al. 2018). But a yet higher level of complication of plant–nematode interactions was achieved with the characterisation of the mechanism of action of the beet cyst nematode-derived effector Hs32E03, a homologue of soybean cyst nematode effector Hg32E03 (Gao et al. 2003). It appeared to be the first nematode effector acting through epigenetic regulation. In Arabidopsis it binds to histone deacetylase (HDT1) and alters the histone acetylation profile, thus affecting chromatin containing *rDNA* fragments and further altering expression profiles of *rRNA* genes, which are important for syncytium development as the proliferation of ribosomes is one of the ultrastructural features of syncytial cytoplasm (Vijayapalani et al. 2018).

To defend themselves against pests, plants had evolved a specific mechanism recognising and inhibiting or deactivating nematode effectors. It is achieved by NB-LRR/NLR immunoreceptors, commonly called resistance (R) proteins (Table 18.1). Plant immunity triggered in response to effectors is called effector-triggered immunity (ETI). Upon effector recognition, the pathogen or parasite again becomes incapable of successfully infecting the plant, and the only way for it to gain the upper hand again is the production of new effectors which will break the established first level of ETI (Fig. 18.4). There are different concepts of interaction between R proteins and effectors. R proteins have an evolutionarily conserved structure consisting of nucleotide binding (NB) sequence and leucine-rich tandem repeats (LRR) motifs. The N-terminus of the protein contains the domain responsible for protein–protein interaction. Two main classes of R proteins are distinguished depending on the type of N-terminal domain. The first class has a coiled-coil (CC) domain, whereas the second has a Toll interleukin-1 receptor domain (TIR) that has similarity to the cytoplasmic portion of *Drosophila melanogaster* and human Toll interleukin-1 receptor protein. Several plant genes for PPNs resistance have been mapped and at least partly characterised (Kaloshian et al. 2011). The best analysed plant resistance genes are *Mi1.2*, *Mi-9,* and *Hero-A* from tomato, *H1*,

Table 18.1 List of the best-known plant resistance genes against PPNs

Crop plant	Resistance source	*R* gene	Plant-parasitic nematode (pathotype)	R-protein type	Reference
Potato	*Solanum tuberosum* sp. *andigena*	*H1*	*Globodera rostochiensis* (Ro1, Ro4)	CC-NB-LRR	Bakker et al. (2004)
	S. spegazzini	*Gro1-4*	*G. rostochiensis* (Ro1, Ro5)	TIR-NB-LRR	Paal et al. (2004)
	S. tuberosum sp. *andigena*	*Gpa2*	*G. pallida* (Pa2)	CC-NB-LRR	van der Voort et al. (1997)
Tomato	*S. pimpinellifolium*	*Hero-A*	*G. rostochiensis* (Ro1, Ro3, Ro5) *G. pallida* (Pa2, Pa3)	CC-NB-LRR	Ernst et al. (2002)
	S. peruvianum	*Mi1.2*	*Meloidogyne incognita* *M. javanica* *M. arenaria*	CC-NB-LRR	Milligan et al. (1998)
	S. arcanum	*Mi-9*	*M. incognita*	CC-NB-LRR	Jablonska et al. (2007)
Sweet pepper	*Capsicum annuum*	*CaMi*	*M. incognita*	CC-NB-LRR	Chen et al. (2007)
Sugar beet	*Beta procumbens*	*Hs1^pro-1^*	*Heterodera schachtii*	LRR-TM	Cai et al. (1997)
Soybean	*Glycine max*	*Rhg1*	*H. glycines*	Non-canonical *R*-gene	Kandoth et al. (2011)
		Rhg4			Liu et al. (2012); Matthews et al. (2013)
Wheat	*Aegilops tauschii*	*Cre1*	*H. avenae*	CC-NB-LRR	De Majnik et al. (2003)
	Triticum aestivum	*Cre3*	*H. avenae*	CC-NB-LRR	De Majnik et al. (2003)

Gpa2, and *Gro1-4* from potato, *CaMi* from pepper, *Rhg1,* and *Rhg4* from soybean, and *Cre1* and *Cre3* from wheat. Resistance to different species of RKNs is provided by *Mi1.2, Mi-9,* and *CaMi* genes, whereas resistance to different species of *Globodera* and *Heterodera* is granted by *Hero-A*, *H1*, *Gpa2*, *Gro1-4*, $Hs1^{pro\text{-}1}$, *Rhg1*, *Rhg4*, *Cre1,* and *Cre3* genes. Among these genes only *Gro1-4* belongs to the TIR-NB-LRR family, whereas genes listed above represent mostly CC-NB-LRR family of *R* genes, except of $Hs1^{pro\text{-}1}$, *Rhg1* and *Rhg4,* which do not encode R proteins with the canonical NB-LRR motif. $Hs1^{pro\text{-}1}$ encodes a protein rich in LRR with a transmembrane domain that has little similarity to other described R proteins. *Rhg1* and *Rhg4*, on the other hand, encode proteins with extracellular LRR domain, transmembrane domain, and cytosolic serine/threonine kinase domain (Kaloshian et al. 2011).

The classical model of plants resistant response to pathogens was proposed by Flor and it is known as "gene-for-gene" hypothesis. It assumes that a plant is resistant to a pathogen when it has a dominant resistance gene that matches to avirulence gene in pathogen (Flor 1942). Nowadays, some of *R* genes and pathogens effectors are known, but detailed analyses of interactions between them indicated some inconsistences, thus additional elements were added. "Guard model" postulates that interaction between immunity and avirulence proteins is coordinated by additional host proteins. "Decoy model" infers that effectors are recognised and bound by host proteins that resemble the target host protein for particular avirulence protein. In all models, binding of the effector to an *R* gene product (directly, or via guard or decoy) leads to activation of the R protein and further to induction of the plants' defence response (Kourelis and van der Hoorn 2018).

One of the best elaborated resistance models for plant–nematode interactions is potato resistance gene *Gpa2* and *G. pallida* effector Gp-RBP1. Upon recognition of the effector by the Gpa2 R protein and RanGTPase-activating protein (RanGAP2), a defence mechanism leading to syncytium degradation is activated. The presence of proline at position 187 in the SPRY domain of the Gp-RBP1 effector is essential for proper recognition by Gpa2. Some virulent *G. pallida* pathotypes have a mutation at this position, making them unrecognisable for the R protein, which cannot bind the effector. Moreover, the RanGAP2 protein interacts with the CC domain of the Gpa2 protein, which may suggest that RanGAP2–Gpa2 interactions are necessary for proper recognition of the SPRY domain of the Gp-RBP1 effector (Tameling and Baulcombe 2007; Sacco et al. 2009).

"Guard model" is well-described for interaction between tomato Mi-1.2 R protein, which inhibits RKNs parasitism, and creates signalling complex with chaperone proteins HSP90 (heat shock protein 90) and SGT1 (suppressor of G2 allele of skp1), and guard protein RME1 (resistance to *Meloidogyne* spp.), which participates in the response to RKNs effector(s). Conformational changes of RME1 protein, resulting from interaction with RKNs effector(s), lead to the binding of ATP molecules to the Mi1.2 R protein. Later ATP hydrolysis modifies the Mi1.2 protein and thus activates defence mechanisms (Bhattarai et al. 2007). Additionally, effector(s) interact also with the immune protein chaperone complex HSP90/SGT1/RAR1 and via it induce NB-LRR receptor-dependent response. It is also possible that this complex may induce receptor folding and formation of intramolecular interactions, controlling receptor levels in the cell, and regulating their intracellular mobility (Kadota and Shirasu 2012).

Another "guard/decoy" model was described for tomato and potato cyst nematode, *G. rostochiensis*. Gr-VAP1 (venom allergen-like protein) is secreted by the nematode into its feeding site where it interacts with $Rcr3^{pim}$ protein, which is papain-like cysteine protease (PLCP). The interaction between Gr-VAP1 and $Rcr3^{pim}$ is recognised by membrane receptor Cf-2 R protein which activates defence mechanisms. It has to be stressed that Cf-2 is originally an R protein providing resistance to the fungal pathogen *Cladosporium fulvum* (Lozano-Torres et al. 2012). Rcr3 is targeted also by other effectors produced by different pathogens, i.e., Cip1 from the bacterium *Pseudomonas syringae*, and several EPIC effectors from the oomycete *Phytophthora infestans* (Ilyas et al. 2015). It is postulated that the "guard/decoy" recognition model permits response to multiple pathogens more robustly and efficiently at lower metabolic costs for plants (van der Hoorn and Kamoun 2008).

18.6 Hypersensitive Response Activated by Plant-Parasitic Nematodes

One of the primary local plant defence reactions is the hypersensitive response (HR). It involves the activation of programmed plant cell death mechanisms at the site of infection and occurs in PTI and ETI responses. HR is a local reaction that limits the spread of the pathogen and protects the plant from further damage. It is assumed that there is a certain activation threshold, beyond which activation of cell death is observed (Coll et al. 2011). There is a consensus that HR shares biochemical, physiological, and molecular features with other types of PCD, as it is always associated with cytoplasm shrinkage and degradation, the release of cytochrome c from the mitochondria, electrolytes leakage, vacuole collapse, loss of water potential, chromatin condensation, and DNA fragmentation. Symptoms typical for autophagy such as the formation of various types of membranous structures, including autophagosomes, and increased activity of vacuolar processing enzymes (VPE) and proteases are also observed in the HR (Mur et al. 2008; Tang and Bassham 2018). However, despite symptoms characteristic of other types of cell death, HR is considered a separate type of PCD (Coll et al. 2011; Mur et al. 2008). It is also worth remembering that the cascade of processes activated in the HR is unique for particular plant–parasite interaction (Zheng et al. 2021). Under conditions where no activation of defence mechanisms is needed, inappropriate activation of the HR can lead to spontaneous cell death, which can have a detrimental impact on plant growth. This is why precise control of HR is so important. Under stress less conditions, *R* gene transcription is usually kept at a low level. In contrast, when a stress stimulus is received, the effective production or activation of R proteins is required. This process is controlled by the regulation of transcription factors, alternative splicing, and post-transcriptional modifications (Balint-Kurti 2019). Control of the HR is also provided at the level of protein accumulation and stability. Many plant NB-LRR receptors are likely tuned by chaperone proteins HSP90 and HSP70 and co-chaperons (auxiliary proteins), which can affect their stability, maturation, and proper functioning. Co-chaperone proteins in plants include RAR1 (required for MLA12 resistance 1) and SGT1, which interact with each other and with HSP90 to form the so-called HSP90/RAR1/SGT1 chaperone complex (Kadota and Shirasu 2012). The action of this complex affects the NB-LRR receptor-dependent response. In addition to interaction with the chaperone complex, the action of the R receptor complex may be associated with the formation of homo- and hetero-oligomers as shown for ZAR1 (HopZ-activated resistance 1) activation in Arabidopsis, where NLR complexes form homo-pentamer pores embedded in the cell membrane and thus initiating defence response in plants (Wang et al. 2019a, b). An extensive network of sensors and helper proteins responsible for triggering the HR has been successfully characterised in tomato (Baggs et al. 2017; Wu et al. 2017). Recently Derevnina et al. (2021) found *G. rostochiensis* effector, SPRYSEC15 which suppresses signalling mediated by helper NLRs—NRC2 and NRC3. The HR in plants resistant to PPNs can be switched on at three different phases of infection: (1) in the cortex and rhizodermis during penetration and migration of the nematode, e.g. tomato carrying *Mi-1.2 R* gene (Paulson and Webster 1972; Fig. 18.5a); (2) directly in feeding site after its induction and initial development, e.g. potato carrying *Gpa2* or *H3* from *S. tuberosum* ssp. *andigena* CPC2802 *R* gene (Varypatakis et al. 2020; Fig. 18.5b); and (3) in non-modified cells adjacent to developing feeding sites, e.g. tomato carrying *Hero-A R* gene (Sobczak et al. 2005; Kaloshian et al. 2011; Zheng et al. 2021; Fig. 18.5c). In the case of CNs syncytia, the HR most frequently occurs in cells next to syncytium. It leads to the formation of a physical barrier composed of dead cells, due to induced HR, between the syncytium and nutrients and water conducting elements. Degenerated cells inhibit syncytium enlargement and development by "cutting" off nutrient influx from neighbouring cells. This type of defence reaction is commonly called "delayed HR" or "male-based" resistance as usually the feeding site functions long enough to allow only the development of males, but too short to allow the development of females which have higher nutritional demands. Thus, the sex ratio is biased towards males and the PPNs reproduction rate is decreased. However, both CNs and RKNs have evolved effectors to inhibit HR. In the case of RKNs, the effector MeTCTP has been characterised, which suppresses PCD triggered by the pro-apoptotic protein BAX (Bcl-2-associated X; Zhuo et al. 2017) while cyst nematodes inhibit the HR rather in an indirect manner, as described for SPRYSEC effectors family (Diaz-Granados et al. 2016).

18.7 Concluding Remarks

Advances in multi-omics techniques used to study plant responses to infections with PPNs at genomic (i.e. Cotton et al. 2014; Eves-van den Akker et al. 2016; Masonbrink et al. 2019), transcriptomic (i.e. Szakasits et al. 2009; Matuszkiewicz et al. 2018; Siddique et al. 2022; Zhang et al. 2017), microtranscriptomic (i.e. Hewezi et al. 2008; Koter et al. 2018), proteomic (i.e. Hütten et al. 2015; Zhang et al. 2015; Filipecki et al. 2021), metabolomic (i.e. Hofmann et al. 2010; Assefa et al. 2021), and effectoromic (i.e. Lilley et al. 2018; Elashry et al. 2020; Maier et al. 2021) levels and the use of model plants, including crops, contribute to better understanding of the complexity of plant–nematode interactions. An ample example of a such multi-omics approach to identify susceptibility genes was published recently by Siddique et al. (2022) where authors studied the role of vitamin B5 in plant–nematode interaction. Their work was supported by assembling a high-quality genome of *H. schachtii* with the usage of two sequencing strategies—PacBio and

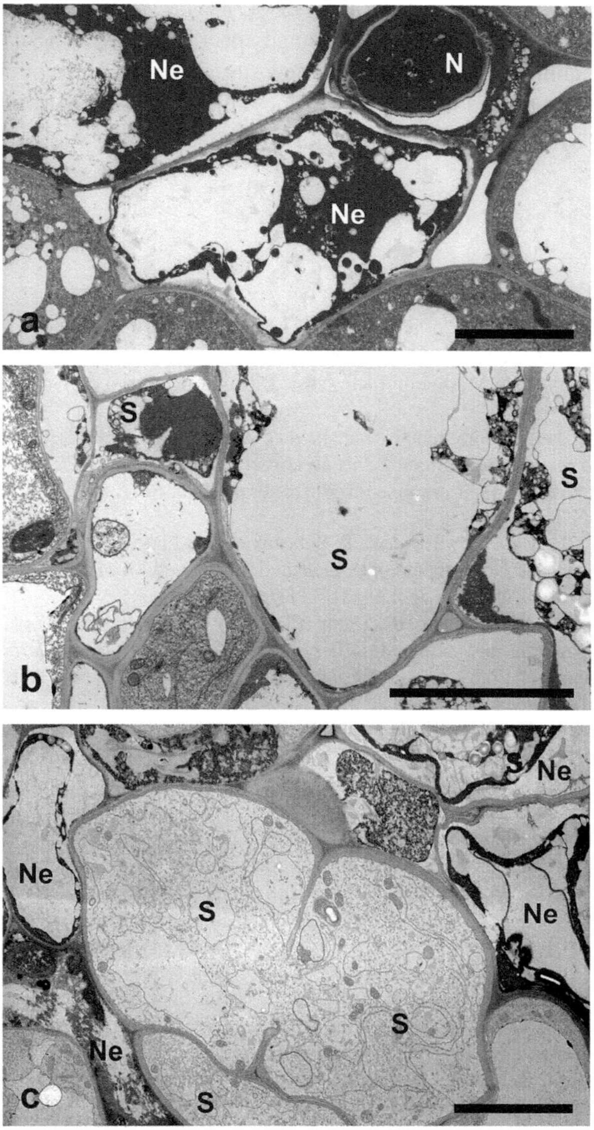

Fig. 18.5 Ultrastructural features of plant resistance reactions. Transmission electron microscopy micrographs of (**a**) resistant genotype of *Solanum sparsipilum* root infected with *Meloidogyne incognita* (2 dpi). Cell next to the nematode degrade during juvenile migration. (*courtesy Abou Bakari Kouassi*); (**b**) degenerated syncytium induced in tomato (*Solanum lycopersicum*) carrying *Cf-2* and *Rcr3*^{pim} genes by *Globodera rostochiensis* (7 dpi) (*courtesy Geert Smant*); (**c**) degenerated cells of the vascular cylinder next to syncytium induced by *Globodera rostochiensis* pathotype Ro1 in tomato (*Solanum lycopersicum*) genotype LA1792 (contains *Hero* resistance gene) roots at 4 dpi. Abbreviations: N, nematode; Ne, necrosis; S, syncytium. Scale bars: 5 μm (**a**–**c**)

Illumina. Moreover, transcriptomic analysis of *H. schachtii* and infected Arabidopsis roots was performed in a nematode life stage-specific manner. These applications of multi-omics techniques will help to find new susceptibility genes or metabolic pathways which could be crucial for proper syncytium development. Searching for new plant susceptibility factors or resistance enhancers could be a way to restrict these parasites.

References

Abd-Elgawad MMM, Askary TH (2015) Impact of phytonematodes on agriculture economy. In: Askary TH, Martinelli PRP (eds) Biocontrol agents of phytonematodes. CAB, Wallingford, pp 3–49

Ali S, Magne M, Chen S, Obradovic N, Jamshaid L, Wang X, Bélair G, Moffett P (2015) Analysis of *Globodera rostochiensis* effectors reveals conserved functions of SPRYSEC proteins in suppressing and eliciting plant immune responses. Front Plant Sci 6:623. https://doi.org/10.3389/fpls.2015.00623

Anjam MS, Shah SJ, Matera C, Różańska E, Sobczak M, Siddique S, Grundler FMW (2020) Host factors influence the sex of nematodes parasitizing roots of *Arabidopsis thaliana*. Plant Cell Environ 43:1160–1174. https://doi.org/10.1111/pce.13728

Assefa AD, Kim SH, Mani V, Ko HR, Hahn BS (2021) Metabolic analysis of the development of the plant-parasitic cyst nematodes *Heterodera schachtii* and *Heterodera trifolii* by capillary electrophoresis time-of-flight mass spectrometry. Int J Mol Sci 22:10488. https://doi.org/10.3390/ijms221910488

Baggs E, Dagdas G, Krasileva K (2017) NLR diversity, helpers and integrated domains: making sense of the NLR IDentity. Curr Opin Plant Biol 38:59–67. https://doi.org/10.1016/j.pbi.2017.04.012

Bakker E, Achenbach U, Bakker J, van Vliet J, Peleman J, Segers B, van der Heijden S, van der Linde P, Graveland R, Hutten R, van Eck H, Coppoolse E, van der Vossen E, Bakker J, Goverse A (2004) A high-resolution map of the *H1* locus harbouring resistance to the potato cyst nematode *Globodera rostochiensis*. Theor Appl Genet 109:146–152. https://doi.org/10.1007/s00122-004-1606-z

Balint-Kurti P (2019) The plant hypersensitive response: concepts, control and consequences. Mol Plant Pathol 20:1163–1178. https://doi.org/10.1111/mpp.12821

Baranowski Ł, Różańska E, Sańko-Sawczenko I, Matuszkiewicz M, Znojek E, Filipecki M, Grundler FMW, Sobczak M (2019) Arabidopsis tonoplast intrinsic protein and vacuolar H^+-adenosinetriphosphatase reflect vacuole dynamics during development of syncytia induced by the beet cyst nematode *Heterodera schachtii*. Protoplasma 256:419–429. https://doi.org/10.1007/s00709-018-1303-4

Benedetti M, Pontiggia D, Raggi S, Cheng Z, Scaloni F, Ferrari S, Ausubel FM, Cervone F, Lorenzo GD (2015) Plant immunity triggered by engineered *in vivo* release of oligogalacturonides, damage-associated molecular patterns. Proc Natl Acad Sci USA 112:5533–5538. https://doi.org/10.1073/pnas.1504154112

Bhattarai KK, Li Q, Liu Y, Dinesh-Kumar SP, Kaloshian I (2007) The *Mi-1*-mediated pest resistance requires Hsp90 and Sgt1. Plant Physiol 144:312–323. https://doi.org/10.1104/pp.107.097246

Blaxter ML, De Ley P, Garey JR, Liu LX, Scheldeman P, Vierstraete A, Vanfleteren JR, Mackey LY, Dorris M, Frisse LM, Vida JT, Thomas WK (1998) A molecular evolutionary framework for the phylum Nematoda. Nature 392:71–75. https://doi.org/10.1038/32160

Böckenhoff A, Grundler FMW (1994) Studies on the nutrient uptake by the beet cyst nematode *Heterodera schachtii* by *in situ* microinjection of fluorescent probes into the feeding structures in *Arabidopsis thaliana*. Parasitology 109:249–255. https://doi.org/10.1017/S003118200007637X

Cai D, Kleine M, Kifle S, Harloff J, Sandal N, Marcker A, Klein-Lankhorst M, Salentijn M, Lange W, Stiekema J, Wyss U, Grundler FM, Jung C (1997) Positional cloning of a gene for nematode resistance in sugar beet. Science 275:832–834. https://doi.org/10.1126/science.275.5301.832

Chen R, Li H, Zhang L, Zhang J, Xiao J, Ye Z (2007) *CaMi*, a root-knot nematode resistance gene from hot pepper (*Capsicum annuum* L.) confers nematode resistance in tomato. Plant Cell Rep 26:895–905. https://doi.org/10.1007/s00299-007-0304-0

Chopra D, Hasan MS, Matera C, Chitambo O, Mendy B, Mahlitz SV, Naz AA, Szumski S, Janakowski S, Sobczak M, Mithöfer A, Kyndt T, Grundler FMW, Siddique S (2021) Plant parasitic cyst nematodes redirect host indole metabolism via NADPH oxidase-mediated ROS to promote infection. New Phytol 232:318–331. https://doi.org/10.1111/nph.17559

Cohn E, Spiegel Y (1991) Root-nematode interactions. In: Weisel Y, Eshel A, Kafkafi U (eds) Plant roots; The hidden part. Marcel Dekker, New York, pp 789–805

Coll NS, Epple P, Dangl JL (2011) Programmed cell death in the plant immune system. Cell Death Differ 18:1247–1256. https://doi.org/10.1038/cdd.2011.37

Cotton JA, Lilley CJ, Jines LM, Kikuchi T, Reid AJ, Thorpe P, Tsai IJ, Beasley H, Blok V, Cock PJA, Eves-van den Akker S, Holroyd N, Hunt M, Mantelin S, Naghra H, Pain A, Palomares-Rius JE, Zarowiecki M, Berriman M, Jones JT, Urwin PE (2014) The genome and life-stage specific transcriptomes of *Globodera pallida* elucidate key aspects of plant parasitism by a cyst nematode. Genome Biol 15:R43. https://doi.org/10.1186/gb-2014-15-3-r43

Dangl JL, Jones JDG (2019) A pentangular plant inflammasome. Science 364:31–32. https://doi.org/10.1126/science.aax0174

Davies LJ, Lilley CJ, Knox JP, Urwin PE (2012) Syncytia formed by adult female *Heterodera schachtii* in *Arabidopsis thaliana* roots have a distinct cell wall molecular architecture. New Phytol 196:238–246. https://doi.org/10.1111/j.1469-8137.2012.04238.x

de Almeida Engler J, Van Poucke K, Karimi M, De Groodt R, Gheysen G, Engler G, Gheysen G (2004) Dynamic cytoskeleton rearrangements in giant cells and syncytia of nematode-infected roots. Plant J 38:12–26. https://doi.org/10.1111/j.1365-313X.2004.02019.x

de Almeida Engler J, Kyndt T, Vieira P, Van Cappelle E, Boudolf V, Sanchez V, Escobar C, De Veylder L, Engler G, Abad P, Gheysen G (2012) *CCS52* and *DEL1* genes are key components of the endocycle in nematode-induced feeding sites. Plant J 72:185–198. https://doi.org/10.1111/j.1365-313X.2012.05054.x

de Majnik J, Ogbonnaya FC, Moullet O, Lagudah ES (2003) The *Cre1* and *Cre3* nematode resistance genes are located at homeologous loci in the wheat genome. Mol Plant Microbe Interact 16:1129–1134. https://doi.org/10.1094/mpmi.2003.16.12.1129

De Meutter J, Tytgat T, Witters E, Gheysen G, Onckelen V, Gheysen G (2003) Identification of cytokinins produced by the plant parasitic nematodes *Heterodera schachtii* and *Meloidogyne incognita*. Mol Plant Pathol 4:271–277. https://doi.org/10.1046/j.1364-3703.2003.00176.x

Derevnina L, Contreras MP, Adachi H, Upson J, Vergara Cruces A, Xie R, Sklenar J, Menke FLH, Mugford ST, MacLean D, Ma W, Hogenhout SA, Goverse A, Maqbool A, Wu CH, Kamoun S (2021) Plant pathogens convergently evolved to counteract redundant nodes of an NLR immune receptor network. PLoS Biol 19:e3001136. https://doi.org/10.1371/journal.pbio.3001136

Diaz-Granados A, Petrescu AJ, Goverse A, Smant G (2016) SPRYSEC effectors: A versatile protein-binding platform to disrupt plant innate immunity. Front Plant Sci 7:1575. https://doi.org/10.3389/fpls.2016.01575

Dowd CD, Chronis D, Radakovic ZS, Siddique S, Schmülling T, Werner T, Kakimoto T, Grundler FMW, Mitchum MG (2017) Divergent expression of cytokinin biosynthesis, signaling and catabolism genes underlying differences in feeding sites induced by cyst and root-knot nematodes. Plant J 92:211–228. https://doi.org/10.1111/tpj.13647

Elashry AM, Habash SS, Vijayapalani P, Brocke-Ahmadinejad N, Blümel R, Seetharam A, Schoof H, Grundler FMW (2020) Transcriptome and parasitome analysis of beet cyst nematode *Heterodera schachtii*. Sci Rep 10:3315. https://doi.org/10.1038/s41598-020-60186-0

Endo BY (1978) Feeding plug formation in soybean roots infected with the soybean cyst nematode. Phytopathology 68:1022–1031

Endo BY (1991) Ultrastructure of initial responses of susceptible and resistant soybean roots to infection by *Heterodera glycines*. Rev Nématol 14:73–94

Ernst K, Kumar A, Kriseleit D, Dorothee-U K, Phillips MS, Ganal MW (2002) The broad-spectrum potato cyst nematode resistance gene (*Hero*) from tomato is the only member of a large gene family of NBS-LRR genes with an unusual amino acid repeat in the LRR region. Plant J 31:127–136. https://doi.org/10.1046/j.1365-313x.2002.01341.x

Eves-van den Akker S, Lilley CJ, Jones JT, Urwin PE (2015) Plant-parasitic nematode feeding tubes and plugs: new perspectives on function. Nematology 17:1–9. https://doi.org/10.1163/15685411-00002832

Eves-van den Akker S, Laetsch DR, Thorpe P, Lilley CJ, Danchin EG, Da Rocha M, Rancurel C, Holroyd NE, Cotton JA, Szitenberg A, Grenier E, Montarry J, Mimee B, Duceppe M-O, Boyes I, Marvin JMC, Jones LM, Yusup HB, Lafond-Napalme J, Esquibet M, Sabeh M, Rott M, Overmars H, Finkers-Tomczak A, Smant G, Koutsovoulos G, Blok V, Mantelin S, Cock PJA, Phillips W, Henrissat B, Urwin PE, Blaxter M, Jones JT (2016) The genome of the yellow potato cyst nematode, *Globodera rostochiensis*, reveals insights into the basis of parasitism and virulence. Genome Biol 17:124. https://doi.org/10.1186/s13059-016-0985-1

Fattah F, Webster JM (1984) Fine structure of the giant cells induced by *Meloidogyne javanica* in lima bean. Can J Bot 62:429–436

Filipecki M, Żurczak M, Matuszkiewicz M, Święcicka M, Kurek W, Olszewski J, Koter MD, Lamont D, Sobczak M (2021) Profiling the proteome of cyst nematode-induced syncytia on tomato roots. Int J Mol Sci 22:12147. https://doi.org/10.3390/ijms222212147

Flor HH (1942) Inheritance of pathogenicity in *Melampsora lini*. Phytopathology 32:653–669

Gao B, Allen R, Maier T, Davis EL, Baum TJ, Hussey RS (2003) The parasitome of the phytonematode *Heterodera glycines*. Mol Plant-Microbe Interact 16:720–726. https://doi.org/10.1094/mpmi.2003.16.8.720

Gheysen G, Mitchum MG (2011) How nematodes manipulate plant development pathways for infection. Curr Opin Plant Biol 14:415–421. https://doi.org/10.1016/j.pbi.2011.03.012

Gheysen G, Mitchum MG (2019) Phytoparasitic nematode control of plant hormone pathways. Plant Physiol 179:1212–1226. https://doi.org/10.1104/pp.18.01067

Golinowski W, Grundler FMW, Sobczak M (1996) Changes in the structure of *Arabidopsis thaliana* during female development of the plant-parasitic nematode *Heterodera schachtii*. Protoplasma 194:103–116. https://doi.org/10.1007/BF01273172

Goverse A, Overmars H, Engelbertink J, Schots A, Bakker J, Helder J (2000) Both induction and morphogenesis of cyst nematode feeding cells are mediated by auxin. Mol Plant Microbe Interact 13:1121–1129. https://doi.org/10.1094/MPMI.2000.13.10.1121

Grundler FMW, Sobczak M, Golinowski W (1998) Formation of wall openings in root cells of *Arabidopsis thaliana* following infection by the plant-parasitic nematode *Heterodera schachtii*. Eur J Plant Pathol 104:545–551

Grunewald W, Cannoot B, Friml J, Gheysen G (2009) Parasitic nematodes modulate PIN-mediated auxin transport to facilitate infection. PLoS Pathog 5:e1000266. https://doi.org/10.1371/journal.ppat.1000266

Habash SS, Sobczak M, Siddique S, Voigt B, Elashry A, Grundler FMW (2017) Identification and characterization of a putative protein disulfide isomerase (HsPDI) as an alleged effector of *Heterodera schachtii*. Sci Rep 7:13536. https://doi.org/10.1038/s41598-017-13418-9

Hamamouch N, Li C, Hewezi T, Baum TJ, Mitchum MG, Hussey RS, Vodkin LO, Davis EL (2012) The interaction of the novel 30C02 cyst nematode effector protein with a plant β-1,3-endoglucanase may suppress host defence to promote parasitism. J Exp Bot 63:3683–3695. https://doi.org/10.1093/jxb/ers058

Hewezi T, Howe P, Maier TR, Baum TJ (2008) *Arabidopsis* small RNAs and their targets during cyst nematode parasitism. Mol Plant Microbe Interact 21:1622–1634. https://doi.org/10.1094/mpmi-21-12-1622

Hewezi T, Juvale PS, Piya S, Maier TR, Rambani A, Rice JH, Mitchum MG, Davis EL, Hussey RS, Baum TJ (2015) The cyst nematode effector protein 10A07 targets and recruits host posttranslational machinery to mediate its nuclear trafficking and to promote parasitism in Arabidopsis. Plant Cell 27:891–907. https://doi.org/10.1105/tpc.114.135327

Hofmann J, El Ashry AN, Anwar S, Erban A, Kopka J, Grundler F (2010) Metabolic profiling reveals local and systemic responses of host plants to nematode parasitism. Plant J 62:1058–1071. https://doi.org/10.1111/j.1365-313X.2010.04217.x

Hu Y, You J, Li C, Williamson VM, Wang C (2017) Ethylene response pathway modulates attractiveness of plant roots to soybean cyst nematode *Heterodera glycines*. Sci Rep 7:41282. https://doi.org/10.1038/srep41282

Huang CS, Maggenti AR (1969) Mitotic aberrations and nuclear changes of developing giant cells in *Vicia faba* caused by root knot nematode, *Meloidogyne javanica*. Phytopathology 59:447–455

Hussey RS, Mims CW (1991) Ultrastructure of feeding tubes formed in giant-cells induced in plants by the root-knot nematode *Meloidogyne incognita*. Protoplasma 162:99–107

Hussey RS, Mims CW, Westcott SW III (1992) Ultrastructure of root cortical cells parasitized by the ring nematode *Criconemella xenoplax*. Protoplasma 167:55–65

Hütten M, Geukes M, Misas-Villamil JC, van der Hoorn RAL, Grundler FMW, Siddique S (2015) Activity profiling reveals changes in the diversity and activity of proteins in Arabidopsis roots in response to nematode infection. Plant Physiol Biochem 97:36–43. https://doi.org/10.1016/j.plaphy.2015.09.008

Ilyas M, Hörger AC, Bozkurt TO, van den Burg HA, Kaschani F, Kaiser M, Belhaj K, Smoker M, Joosten MHAJ, Kamoun S, van der Hoorn RAL (2015) Functional divergence of two secreted immune proteases of tomato. Curr Biol 25:2300–2306. https://doi.org/10.1016/j.cub.2015.07.030

Ithal N, Recknor J, Nettleton D, Maier T, Baum TJ, Mitchum MG (2007) Developmental transcript profiling of cyst nematode feeding cells in soybean roots. Mol Plant Microbe Interact 20:510–525. https://doi.org/10.1094/mpmi-20-5-0510

Jablonska B, Ammiraju JSS, Bhattarai KK, Mantelin S, Martinez de Ilarduya O, Roberts PA, Kaloshian I (2007) The *Mi-9* gene from *Solanum arcanum* conferring heat-stable resistance to root-knot nematodes is a homolog of *Mi-1*. Plant Physiol 143:1044–1054. https://doi.org/10.1104/pp.106.089615

Jagdale S, Rao U, Giri AP (2021) Effectors of root-knot nematodes: An arsenal for successful parasitism. Front Plant Sci 12:800030. https://doi.org/10.3389/fpls.2021.800030

Jones JDG, Dangl JL (2006) The plant immune system. Nature 444:323–329. https://doi.org/10.1038/nature05286

Jones MGK, Northcote DH (1972) Nematode-induced syncytium-a multinucleate transfer cell. J Cell Sci 10:789–809

Jones JT, Furlanetto C, Bakker E, Banks B, Blok V, Chen Q, Phillips M, Prior A (2003) Characterization of a chorismate mutase from the potato cyst nematode *Globodera pallida*. Mol Plant Pathol 4:43–50. https://doi.org/10.1046/j.1364-3703.2003.00140.x

Jones JT, Haegeman A, Danchin EGJ, Gaur HS, Helder J, Jones MGK, Kikuchi T, Manzanilla-López R, Palomares-Rius JE, Wesemael WML, Perry RN (2013) Top 10 plant-parasitic nematodes in molecular plant pathology. Mol Plant Pathol 14:946–961. https://doi.org/10.1111/mpp.12057

Kadota Y, Shirasu K (2012) The HSP90 complex of plants. Biochim Biophys Acta 1823:689–697. https://doi.org/10.1016/j.bbamcr.2011.09.016

Kaloshian I, Desmond OJ, Atamian HS (2011) Disease resistance-genes and defense responses during incompatible interactions. In: Jones J, Gheysen G, Fenoll C (eds) Genomics and molecular genetics of plant-nematode interactions. Springer, Dordrecht, pp 309–324. https://doi.org/10.1007/978-94-007-0434-3_15

Kammerhofer N, Radakovic Z, Regis JMA, Dobrev P, Vankova R, Grundler FMW, Siddique S, Hofmann J, Wieczorek K (2015) Role of stress-related hormones in plant defence during early infection of the cyst nematode *Heterodera schachtii* in Arabidopsis. New Phytol 207:778–789. https://doi.org/10.1111/nph.13395

Kandoth PK, Ithal N, Recknor J, Maier T, Nettleton D, Baum TJ, Mitchum MG (2011) The soybean *Rhg1* locus for resistance to the soybean cyst nematode *Heterodera glycines* regulates the expression of a large number of stress- and defense-related genes in degenerating feeding cells. Plant Physiol 155:1960–1975. https://doi.org/10.1104/pp.110.167536

Kikuchi T, Eves-van den Akker S, Jones JT (2017) Genome evolution of plant-parasitic nematodes. Annu Rev Phytopathol 55:333–354. https://doi.org/10.1146/annurev-phyto-080516-035434

Koter MD, Święcicka M, Matuszkiewicz M, Pacak A, Derebecka N, Filipecki M (2018) The miRNAome dynamics during developmental and metabolic reprogramming of tomato root infected with potato cyst nematode. Plant Sci 268:18–29. https://doi.org/10.1016/j.plantsci.2017.12.003

Kourelis J, van der Hoorn RAL (2018) Defended to the nines: 25 years of resistance gene cloning identifies nine mechanisms for R protein function. Plant Cell 30:285–299. https://doi.org/10.1105/tpc.17.00579

Labudda M, Różańska E, Czarnocka W, Sobczak M, Dzik JM (2018) Systemic changes in photosynthesis and reactive oxygen species homeostasis in shoots of *Arabidopsis thaliana* infected with the beet cyst nematode *Heterodera schachtii*. Mol Plant Pathol 19:1690–1704. https://doi.org/10.1111/mpp.12652

Labudda M, Różańska E, Muszyńska E, Marecka D, Głowienka M, Roliński P, Prabucka B (2020) *Heterodera schachtii* infection affects nitrogen metabolism in *Arabidopsis thaliana*. Plant Pathol 69:794–803. https://doi.org/10.1111/ppa.13152

Lambert KN, Allen KD, Sussex IM (1999) Cloning and characterization of an esophageal-gland-specific chorismate mutase from the phytoparasitic nematode *Meloidogyne javanica*. Mol Plant Microbe Interact 12:328–336. https://doi.org/10.1094/mpmi.1999.12.4.328

Lee C, Chronis D, Kenning C, Peret B, Hewezi T, Davis EL, Baum TJ, Hussey R, Bennett M, Mitchum MG (2011) The novel cyst nematode effector protein 19C07 interacts with the Arabidopsis auxin influx transporter LAX3 to control feeding site development. Plant Physiol 155:866–880. https://doi.org/10.1104/pp.110.167197

Lilley CJ, Maqbool A, Wu D, Yusup HB, Jones LM, Birch PRJ, Banfield MJ, Urwin PE, Eves-van den Akker S (2018) Effector gene birth in plant parasitic nematodes: Neofunctionalization of a housekeeping glutathione synthetase gene. PLoS Genet 14:e1007310. https://doi.org/10.1371/journal.pgen.1007310

Liu S, Kandoth PK, Warren SD, Yeckel G, Heinz R, Alden J, Yang C, Jamai A, El-Mellouki T, Juvale PS, Hill J, Baum TJ, Cianzio S, Whitham SA, Korkin D, Mitchum MG, Meksem K (2012) A soybean cyst nematode resistance gene points to a new mechanism of plant resistance to pathogens. Nature 492:256–260. https://doi.org/10.1038/nature11651

Lozano-Torres JL, Wilbers RHP, Gawronski P, Boshoven JC, Finkers-Tomczak A, Cordewener JHG, America AHP, Overmars HA, Van't Klooster JW, Baranowski L, Sobczak M, Ilyas M, van der Hoorn RAL, Schots A, de Wit PJGM, Bakker J, Goverse A, Smant G (2012) Dual disease resistance mediated by the immune receptor Cf-2 in tomato requires a common virulence target of a fungus and a nematode. Proc Natl Acad Sci USA 109:10119–10124. https://doi.org/10.1073/pnas.1202867109

Maier TR, Masonbrink RE, Vijayapalani P, Gardner M, Howland AD, Mitchum MG, Baum TJ (2021) Esophageal gland RNA-seq resource of a virulent and avirulent population of the soybean cyst nematode *Heterodera glycines*. Mol Plant Microbe Interact 34:1084–1087. https://doi.org/10.1094/MPMI-03-21-0051-A

Manosalva P, Manohar M, von Reuss SH, Chen S, Koch A, Kaplan F, Choe A, Micikas RJ, Wang X, Kogel KH, Sternberg PW, Williamson VM, Schroeder FC, Klessig DF (2015) Conserved nematode signalling molecules elicit plant defenses and pathogen resistance. Nat Commun 6:7795. https://doi.org/10.1038/ncomms8795

Marhavý P, Kurenda A, Siddique S, Dénervaud Tendon V, Zhou F, Holbein J, Hasan MS, Grundler FMW, Farmer EE, Geldner N (2019) Single-cell damage elicits regional, nematode-restricting ethylene responses in roots. EMBO J 38:e100972. https://doi.org/10.15252/embj.2018100972

Masonbrink R, Maier TR, Muppirala U, Seetharam AS, Lord E, Juvale PS, Schmutz J, Johnson NT, Korkin D, Mitchum MG, Mimee B, Eves-van den Akker S, Hudson M, Severin AKJ, Baum TJ (2019) The genome of the soybean cyst nematode (*Heterodera glycines*) reveals complex patterns of duplications involved in the evolution of parasitism genes. BMC Genomics 20:119. https://doi.org/10.1186/s12864-019-5485-8

Matthews BF, Beard H, MacDonald MH, Kabir S, Youssef RM, Hosseini P, Brewer E (2013) Engineered resistance and hypersusceptibility through functional metabolic studies of 100 genes in soybean to its major pathogen, the soybean cyst nematode. Planta 237:1337–1357. https://doi.org/10.1007/s00425-013-1840-1

Matuszkiewicz M, Sobczak M, Cabrera J, Escobar C, Karpiński S, Filipecki M (2018) The role of programmed cell death regulator LSD1 in nematode-induced syncytium formation. Front Plant Sci 9:314. https://doi.org/10.3389/fpls.2018.00314

Matuszkiewicz M, Koter MD, Filipecki M (2019) Limited ventilation causes stress and changes in Arabidopsis morphological, physiological and molecular phenotype during *in vitro* growth. Plant Physiol Biochem 135:554–562. https://doi.org/10.1016/j.plaphy.2018.11.003

Mei Y, Thorpe P, Guzha A, Haegeman A, Blok VC, MacKenzie K, Gheysen G, Jones JT, Mantelin S (2015) Only a small subset of the SPRY domain gene family in *Globodera pallida* is likely to encode effectors, two of which suppress host defences induced by the potato resistance gene *Gpa2*. Nematology 17:409–424. https://doi.org/10.1163/15685411-00002875

Mei Y, Wright KM, Haegeman A, Bauters L, Diaz-Granados A, Goverse A, Gheysen G, Jones JT, Mantelin S (2018) The *Globodera pallida* SPRYSEC effector GpSPRY-414-2 that suppresses plant defenses targets a regulatory component of the dynamic microtubule network. Front Plant Sci 9:1019. https://doi.org/10.3389/fpls.2018.01019

Mendy B, Wang'ombe MW, Radakovic ZS, Holbein J, Ilyas M, Chopra D, Holton N, Zipfel C, Grundler FMW, Siddique S (2017) Arabidopsis leucine-rich repeat receptor-like kinase NILR1 is required for induction of innate immunity to parasitic nematodes. PLoS Pathog 13:e1006284. https://doi.org/10.1371/journal.ppat.1006284

Miller HN, DiEdwardo AA (1962) Leaf galls on *Siderasis fuscata* caused by root-knot nematode *Meloidogyne incognita incognita*. Phytopathology 52:1070–1073

Milligan SB, Bodeau J, Yaghoobi J, Kaloshian I, Zabel P, Williamson VM (1998) The root knot nematode resistance gene *Mi* from tomato is a member of the leucine zipper, nucleotide binding, leucine-rich repeat family of plant genes. Plant Cell 10:1307–1319. https://doi.org/10.1105/tpc.10.8.1307

Mitchum M, Liu X (2022) Peptide effectors in phytonematode parasitism and beyond. Annu Rev Phytopathol 60:97–119. https://doi.org/10.1146/annurev-phyto-021621-115932

Mitreva M, Blaxter ML, Bird DM, McCarter JP (2005) Comparative genomics of nematodes. Trends Genet 21:573–581. https://doi.org/10.1016/j.tig.2005.08.003

Miyashita N, Koga H (2017) Three-dimensional ultrastructure of feeding tubes and interconnected endoplasmic reticulum in root-knot nematode-induced giant cells in rose balsam. Protoplasma 254:1941–1951. https://doi.org/10.1007/s00709-016-1072-x

Müller J, Rehbock K, Wyss U (1982) Growth of *Heterodera schachtii* with remarks on amounts of food consumed. Rev Nématol 4:227–234

Mundo-Ocampo M, Baldwin JG (1983) Host response to *Meloidodera* spp. (Heteroderidae). J Nematol 15:544–554

Mur LAJ, Kenton P, Lloyd AJ, Ougham H, Prats E (2008) The hypersensitive response; The centenary is upon us but how much do we know? J Exp Bot 59:501–520. https://doi.org/10.1093/jxb/erm239

Ngou BPM, Ding P, Jones JDG (2022) Thirty years of resistance: Zig-zag through the plant immune system. Plant Cell 34:1447–1478. https://doi.org/10.1093/plcell/koac041

Ohtsu M, Sato Y, Kurihara D, Suzaki T, Kawaguchi M, Maruyama D, Higashiyama T (2017) Spatiotemporal deep imaging of syncytium induced by the soybean cyst nematode *Heterodera glycines*. Protoplasma 254:2107–2115. https://doi.org/10.1007/s00709-017-1105-0

Oosterbeek M, Lozano-Torres JL, Bakker J, Goverse A (2021) Sedentary plant-parasitic nematodes alter auxin homeostasis via multiple strategies. Front Plant Sci 12:668548. https://doi.org/10.3389/fpls.2021.668548

Paal J, Henselewski H, Muth J, Meksem K, Menéndez CM, Salamini F, Ballvora A, Gebhardt C (2004) Molecular cloning of the potato *Gro1-4* gene conferring resistance to pathotype Ro1 of the root cyst nematode *Globodera rostochiensis*, based on a candidate gene approach. Plant J 38:285–297. https://doi.org/10.1111/j.1365-313X.2004.02047.x

Palomares-Rius JE, Escobar C, Cabrera J, Vovlas A, Castillo P (2017) Anatomical alterations in plant tissues induced by plant-parasitic nematodes. Front Plant Sci 8:1987. https://doi.org/10.3389/fpls.2017.01987

Paulson RE, Webster JM (1970) Giant cell formation in tomato roots caused by *Meloidogyne incognita* and *Meloidogyne hapla* (Nematoda) infection. A light and electron microscope study. Can J Bot 48:271–276

Paulson RE, Webster JM (1972) Ultrastructure of the hypersensitive reaction in roots of tomato, *Lycopersicon esculentum* L. to infection by the root-knot nematode, *Meloidogyne incognita*. Physiol Plant Pathol 2:227–234

Piya S, Binder BM, Hewezi T (2018) Canonical and noncanonical ethylene signaling pathways that regulate Arabidopsis susceptibility to the cyst nematode *Heterodera schachtii*. New Phytol 221:946–959. https://doi.org/10.1111/nph.15400

Radakovic ZS, Anjam MS, Escobar E, Chopra D, Cabrera J, Silva AC, Escobar C, Sobczak M, Grundler FMW, Siddique S (2018) Arabidopsis *HIPP27* is a host susceptibility gene for the beet cyst nematode *Heterodera schachtii*. Mol Plant Pathol 19:1917–1928. https://doi.org/10.1111/mpp.12668

Raski DJ (1950) The life history and morphology of the sugar-beet nematode, *Heterodera schachtii* Schmidt. Phytopathology 40:135–152

Rebois RV (1980) Ultrastructure of a feeding peg and tube associated with *Rotylenchulus reniformis* in cotton. Nematologica 26:396–405

Różańska E, Czarnocka W, Baranowski Ł, Mielecki J, de Almeida EJ, Sobczak M (2018) Expression of both Arabidopsis γ-tubulin genes is essential for development of a functional syncytium induced by *Heterodera schachtii*. Plant Cell Rep 37:1279–1292. https://doi.org/10.1007/s00299-018-2312-7

Rumpenhorst HJ (1984) Intracellular feeding tubes associated with sedentary plant parasitic nematodes. Nematologica 30:77–85

Rutter WB, Franco J, Gleason C (2022) Rooting out the mechanisms of root-knot nematode-plant interactions. Annu Rev Phytopathol 60:43–76. https://doi.org/10.1146/annurev-phyto-021621-120943

Sacco MA, Koropacka K, Grenier E, Jaubert MJ, Blanchard A, Goverse A, Smant G, Moffett P (2009) The cyst nematode SPRYSEC protein RBP-1 elicits Gpa2- and RanGAP2-dependent plant cell death. PLoS Pathog 5:e1000564. https://doi.org/10.1371/journal.ppat.1000564

Sasser JN, Freckman DW (1987) A world perspective on nematology: The role of the society. In: Veech JA, Dickson DW (eds) Vistas on nematology. Society of Nematologists, Hyattsville, pp 7–14

Sembdner G (1963) Anatomie der *Heterodera-rostochiensis*-Gallen an Tomatenwurzeln. Nematologica 9:55–64

Shah SJ, Anjam MS, Mendy B, Anwer MA, Habash SS, Lozano-Torres JL, Grundler FMW, Siddique S (2017) Damage-associated responses of the host contribute to defence against cyst nematodes but not root-knot nematodes. J Exp Bot 68:5949–5960. https://doi.org/10.1093/jxb/erx374

Siddique S, Grundler FMW (2018) Parasitic nematodes manipulate plant development to establish feeding sites. Curr Opin Microbiol 46:102–108. https://doi.org/10.1016/j.mib.2018.09.004

Siddique S, Matera C, Radakovic ZS, Hasan MS, Gutbrod P, Rozanska E, Sobczak M, Torres MA, Grundler FMW (2014) Parasitic worms stimulate host NADPH oxidases to produce reactive oxygen species that limit plant cell death and promote infection. Sci Signal 7:ra33. https://doi.org/10.1126/scisignal.2004777

Siddique S, Radakovic ZS, De La Torre CM, Chronis D, Novák O, Ramireddy E, Holbein J, Matera C, Hütten M, Gutbrod P, Anjam MS, Rozanska E, Habash S, Elashry A, Sobczak M, Kakimoto T, Strnad M, Schmülling T, Mitchum MG, Grundler FMW (2015) A parasitic nematode releases cytokinin that controls cell division and orchestrates feeding site formation in host plants. Proc Natl Acad Sci USA 112:12669–12674. https://doi.org/10.1073/pnas.1503657112

Siddique S, Radakovic Z, Hiltl C, Pellegrin C, Baum TJ, Beasley H, Bent AF, Chitambo O, Chopra D, Danchin EGJ, Grenier E, Habash SS, Hasan MS, Helder J, Hewezi T, Holbein J, Holterman M, Janakowski S, Koutsovoulos GD, Kranse OP, Lozano-Torres JL, Maier TR, Masonbrink RE, Mendy B, Riemer E, Sobczak M, Sonawala U, Sterken MG, Thorpe P, van Steenbrugge JJM, Zahid N, Grundler F, Eves-van den Akker S (2022) The genome and lifestage-specific transcriptomes of a plant-parasitic nematode and its host reveal susceptibility genes involved in trans-kingdom synthesis of vitamin B5. Nat Commun 13:6190. https://doi.org/10.1038/s41467-022-33769-w

Sijmons PC, Grundler FMW, von Mende N, Burrows PR, Wyss U (1991) *Arabidopsis thaliana* as a new model host for plant-parasitic nematodes. Plant J 1:245–254

Sijmons PC, Atkinson HJ, Wyss U (1994) Parasitic strategies of root nematodes and associated host cell responses. Annu Rev Phytopathol 32:235–259

Smant G, Jones J (2011) Suppression of plant defences by nematodes. In: Jones J, Gheysen G, Fenoll C (eds) Genomics and molecular genetics of plant-nematode interactions. Springer, Dordrecht, pp 273–286. https://doi.org/10.1007/978-94-007-0434-3_13

Smant G, Stokkermans JPWG, Yan Y, de Boer JM, Baum TJ, Wang X, Hussey RS, Gommers FJ, Henrissat B, Davis EL, Helder J, Schots A, Bakker J (1998) Endogenous cellulases in animals: Isolation of β-1,4-endoglucanase genes from two species of plant-parasitic cyst nematodes. Proc Natl Acad Sci USA 95:4906–4911. https://doi.org/10.1073/pnas.95.9.4906

Sobczak M, Golinowski W (2011) Cyst nematodes and syncytia. In: Jones J, Gheysen G, Fenoll C (eds) Genomics and molecular genetics of plant-nematode interactions. Springer, Dordrecht, pp 61–82. https://doi.org/10.1007/978-94-007-0434-3_4

Sobczak M, Golinowski W, Grundler FMW (1997) Changes in the structure of *Arabidopsis thaliana* roots induced during development of males of the plant parasitic nematode *Heterodera schachtii*. Eur J Plant Pathol 103:113–124. https://doi.org/10.1023/A:1008609409465

Sobczak M, Golinowski W, Grundler FMW (1999) Ultrastructure of feeding plugs and feeding tubes formed by *Heterodera schachtii*. Nematology 1:363–374

Sobczak M, Avrova A, Jupowicz J, Phillips MS, Ernst K, Kumar A (2005) Characterization of susceptibility and resistance responses to potato cyst nematode (*Globodera* spp.) infection of tomato lines in the absence and presence of the broad-spectrum nematode resistance *Hero* gene. Mol Plant Microbe Interact 18:158–168. https://doi.org/10.1094/mpmi-18-0158

Szakasits D, Heinen P, Wieczorek K, Hofmann J, Wagner F, Kreil DP, Sykacek P, Grundler FMW, Bohlmann H (2009) The transcriptome of syncytia induced by the cyst nematode *Heterodera schachtii* in Arabidopsis roots. Plant J 57:771–784. https://doi.org/10.1111/j.1365-313x.2008.03727.x

Tameling WIL, Baulcombe DC (2007) Physical association of the NB-LRR resistance protein Rx with a Ran GTPase-activating protein is required for extreme resistance to Potato virus X. Plant Cell 19:1682–1694. https://doi.org/10.1105/tpc.107.050880

Tang J, Bassham DC (2018) Autophagy in crop plants: what's new beyond Arabidopsis? Open Biol 8:180162. https://doi.org/10.1098/rsob.180162

van der Hoorn RAL, Kamoun S (2008) From guard to decoy: A new model for perception of plant pathogen effectors. Plant Cell 20:2009–2017. https://doi.org/10.1105/tpc.108.060194

van der Voort JR, Wolters P, Folkertsma R, Hutten R, van Zandvoort P, Vinke H, Kanyuka K, Bendahmane A, Jacobsen E, Janssen R, Bakker J (1997) Mapping of the cyst nematode resistance locus *Gpa2* in potato using a strategy based on comigrating AFLP markers. Theor Appl Genet 95:874–880. https://doi.org/10.1007/s001220050638

Vanholme B, Kast P, Haegeman A, Jacob J, Grunewald W, Gheysen G (2009) Structural and functional investigation of a secreted chorismate mutase from the plant-parasitic nematode *Heterodera schachtii* in the context of related enzymes from diverse origins. Mol Plant Pathol 10:189–200. https://doi.org/10.1111/j.1364-3703.2008.00521.x

van Megen H, Elsen S, Holterman M, Karssen G, Mooyman P, Bongers T, Holovachov O, Bakker J, Helder J (2009) A phylogenetic tree of nematodes based on about 1200 full-length small subunit ribosomal DNA sequences. Nematology 11:927–950. https://doi.org/10.1163/156854109X456862

Varypatakis K, Véronneau P-Y, Thorpe P, Cock PJA, Lim JT-Y, Armstrong MR, Janakowski S, Sobczak M, Hein I, Mimee B, Jones JT, Blok VC (2020) The genomic impact of selection for virulence against resistance in the potato cyst nematode, *Globodera pallida*. Genes 11:1429. https://doi.org/10.3390/genes11121429

Verma A, Lee C, Morriss S, Odu F, Kenning C, Rizzo N, Spollen WG, Lin M, McRae AG, Givan SA, Hewezi T, Hussey R, Davis EL, Baum TJ, Mitchum MG (2018) The novel cyst nematode effector protein 30D08 targets host nuclear functions to alter gene expression in feeding sites. New Phytol 219:697–713. https://doi.org/10.1111/nph.15179

Vijayapalani P, Hewezi T, Pontvianne F, Baum TJ (2018) An effector from the cyst nematode *Heterodera schachtii* derepresses host rRNA genes by altering histone acetylation. Plant Cell 30:2795–2812. https://doi.org/10.1105/tpc.18.00570

Wang X, Mitchum MG, Gao B, Li C, Diab H, Baum TJ, Hussey RS, Davis EL (2005) A parasitism gene from a plant-parasitic nematode with function similar to CLAVATA3/ESR (CLE) of *Arabidopsis thaliana*. Mol Plant Pathol 6:187–191. https://doi.org/10.1111/j.1364-3703.2005.00270.x

Wang J, Hu M, Wang J, Qi J, Han Z, Wang G, Qi Y, Wang HW, Zhou JM, Chai J (2019a) Reconstitution and structure of a plant NLR resistosome conferring immunity. Science 364:eaav5870. https://doi.org/10.1126/science.aav5870

Wang J, Wang J, Hu M, Wu S, Qi J, Wang G, Han Z, Qi Y, Gao N, Wang HW, Zhou JM, Chai J (2019b) Ligand-triggered allosteric ADP release primes a plant NLR complex. Science 364:eaav5868. https://doi.org/10.1126/science.aav5868

Wasilewska L (1979) Structure and function of soil nematode communities in natural ecosystems and agrocenoses. Pol Ecol Stud 5:97–146. https://agris.fao.org/agris-search/search.do?recordID=US201302822891

Wu CH, Abd-El-Haliem A, Bozkurt TO, Belhaj K, Terauchi R, Vossen JH, Kamoun S (2017) NLR network mediates immunity to diverse plant pathogens. Proc Natl Acad Sci USA 114:8113–8118. https://doi.org/10.1073/pnas.1702041114

Wubben MJ, Su H, Rodermel SR, Baum TJ (2001) Susceptibility to the sugar beet cyst nematode is modulated by ethylene signal transduction in *Arabidopsis thaliana*. Mol Plant Microbe Interact 14:1206–1212. https://doi.org/10.1094/MPMI.2001.14.10.1206

Wubben MJE, Jin J, Baum TJ (2008) Cyst nematode parasitism of *Arabidopsis thaliana* is inhibited by salicylic acid (SA) and elicits uncoupled SA-independent pathogenesis-related gene expression in roots. Mol Plant Microbe Interact 21:424–432. https://doi.org/10.1094/MPMI-21-4-0424

Wyss U (1992) Observations on the feeding behaviour of *Heterodera schachtii* throughout development, including events during moulting. Fundam Appl Nematol 15:75–89

Wyss U (1997) Root parasitic nematodes: An overview. In: Fenoll C, Grundler FMW, Ohl S (eds) Cellular and molecular aspects of plant-nematode interactions. Developments in plant pathology, vol 10. Springer, Dordrecht, pp 5–22

Wyss U, Zunke U (1986) Observations on the behaviour of second stage juveniles of *Heterodera schachtii* inside host roots. Rev Nématol 9:153–165

Zhang H, Deery MJ, Gannon L, Powers SJ, Lilley KS, Theodoulou FL (2015) Quantitative proteomics analysis of the Arg/N-end rule pathway of targeted degradation in Arabidopsis roots. Proteomics 15:2447–2457. https://doi.org/10.1002/pmic.201400530

Zhang H, Kjemtrup-Lovelace S, Li C, Luo Y, Chen LP, Song BH (2017) Comparative RNA-seq analysis uncovers a complex regulatory network for soybean cyst nematode resistance in wild soybean (*Glycine soja*). Sci Rep 7:9699. https://doi.org/10.1038/s41598-017-09945-0

Zheng Q, Putker V, Goverse A (2021) Molecular and cellular mechanisms involved in host-specific resistance to cyst nematodes in crops. Front Plant Sci 12:641582. https://doi.org/10.3389/fpls.2021.641582

Zhuo K, Chen J, Lin B, Wang J, Sun F, Hu L, Liao J (2017) A novel *Meloidogyne enterolobii* effector MeTCTP promotes parasitism by suppressing programmed cell death in host plants. Mol Plant Pathol 18:45–54. https://doi.org/10.1111/mpp.12374

Part V
Cell Fusion and Syncytia in Cancer

Chapter 19
Mechanisms of Cell Fusion in Cancer

Felicite K. Noubissi, Oluwatoyin V. Odubanjo, Brenda M. Ogle, and Paul B. Tchounwou

Abstract Cell–cell fusion is a normal physiological mechanism that requires a well-orchestrated regulation of intracellular and extracellular factors. Dysregulation of this process could lead to diseases such as osteoporosis, malformation of muscles, difficulties in pregnancy, and cancer. Extensive literature demonstrates that fusion occurs between cancer cells and other cell types to potentially promote cancer progression and metastasis. However, the mechanisms governing this process in cancer initiation, promotion, and progression are less well-studied. Fusogens involved in normal physiological processes such as syncytins and associated factors such as phosphatidylserine and annexins have been observed to be critical in cancer cell fusion as well. Some of the extracellular factors associated with cancer cell fusion include chronic inflammation and inflammatory cytokines, hypoxia, and viral infection. The interaction between these extracellular factors and cell's intrinsic factors potentially modulates actin dynamics to drive the fusion of cancer cells. In this review, we have discussed the different mechanisms that have been identified or postulated to drive cancer cell fusion.

F. K. Noubissi (✉) · O. V. Odubanjo · P. B. Tchounwou
Department of Biology, Jackson State University, Jackson, MS, USA

Research Centers in Minority Institutions (RCMI), Center for Health Disparity Research (RCMI-CHDR), Jackson State University, Jackson, MS, USA
e-mail: felicite.noubissi_kamdem@jsums.edu

B. M. Ogle
Department of Biomedical Engineering, University of Minnesota-Twin Cities, Minneapolis, MN, USA

Department of Pediatrics, University of Minnesota-Twin Cities, Minneapolis, MN, USA

Stem Cell Institute, University of Minnesota-Twin Cities, Minneapolis, MN, USA

M. Kloc, A. Uosef (eds.), *Syncytia: Origin, Structure, and Functions*, Results and Problems in Cell Differentiation 71,
https://doi.org/10.1007/978-3-031-37936-9_19

Abbreviations

293T	Human kidney cell
APC	Adenomatous polyposis coli
ASCT1	Alanine/serine/cysteine/threonine-preferring transporter 1
Bai1	Brain-specific angiogenesis inhibitor 1
Bai3	Brain-specific angiogenesis inhibitor 3
BCL9L	β-catenin/BCL9-Like
BMDCs	Bone marrow-derived cells
BRCA	Breast cancer gene 1
CSC	Cancer stem cells
CXCR-4	C-X-C chemokine receptor type 4
cyt	Cytosolic tail
DOCK 180	Dedicator of cytokinesis
ECM	Extracellular matrix
ELMO	Engulfment cell and mobility protein
EMT	Epithelial-mesenchymal transition
ENV	Envelop
ERK	Extracellular signal-regulated kinase
FBGC	Foreign body giant cell
FISH	Fluorescence in situ hybridization
FP	Fusion peptide
GCM1	Glial cells missing
GRP78	Glucose-regulating protein 78
HCC	Hepatocellular carcinoma
HERVs	Human endogenous retroviruses
HEY	Human ovarian carcinoma cell line
HIOECs	Human immortalized oral epithelial cells
HR	Heptad repeats
HS578T Hyg	Hygromycin resistant cancer cell lines
HUVEC	Human umbilical vein endothelial cells
ICAM1	Intercellular adhesion molecule 1
IDC	Invasive ductal carcinoma
IκB-α	Nuclear factor of kappa light polypeptide gene enhancer in B cells inhibitor alpha
MCF7	Human breast cancer cells
MDA-MB-231	Human breast adenocarcinoma cells
MMP9	Matrix metalloproteinase-9
MPC-1	Monocyte chemoattractant protein 1
MSC	Multipotent stromal/mesenchymal stem cell
NF-Kb	Nuclear factor kappa-light-chain-enhancer of activated B cells
OSCCs	Oral squamous carcinoma cells
PGCCs	Polyploid giant cancer cells
PHPP	Pre-hybrid preparation process

PtdSer	Phosphatidylserine
Rac1	Ras-related C3 botulinum toxin substrate 1
RBD	Receptor binding domain
RNA	Ribonucleic acid
RNAseq	RNA sequence
ROS	Reactive oxygen species
RSK	Ribosomal S6 kinase
SCC-9	Squamous cell carcinoma cells 9
shRNA	Small hairpin RNAs
siRNA	Small interfering RNA
SKOv3	Human ovarian cancer cell line
STAB2	Stabilin-2
SU	Surface unit
T47D	Human ductal breast epithelial tumor cell
TCF/LEF	T-cell factor/lymphoid enhancer factor
TCF4	T-cell factor 4
Tim4	T-cell membrane protein 4
TM	Transmembrane
tm	Transmembrane anchorage domain
TNF-α	Tumor necrosis factor alpha
TNFR1	TNF receptor 1
TRAF2	TNF receptor associated factor 2
VCAM-1	Vascular cell adhesion molecule 1
VEGF	Vascular endothelial growth factor
VLA-4	Very late activation antigen 4

19.1 Introduction

Cell–cell fusion is a fundamental requirement in numerous developmental and physiological processes in eukaryotes. These processes include fecundation, placentation, muscle growth, osteoclast differentiation in mammals, and tissue regeneration (Aguilar et al. 2013; Bastida-Ruiz et al. 2016; Seyfried and Huysentruyt 2013). However, cell fusion has also gained importance in cancer progression as a source of genetic instability, as well as a mechanism of drug resistance and metastasis (Kerbel et al. 1983a; Lu and Kang 2009; Seyfried and Huysentruyt 2013). In the early 1900s, Aichel proposed that metastatic cancer cells acquire leukocyte-like properties that allow them to migrate through the bloodstream by fusing with white blood cells (Aichel 1911). Numerous studies have since confirmed that cell fusion can contribute to cancer initiation, promotion, and progression (reviewed in Hass et al. 2021). Several in vitro and in vivo studies have shown that metastatic cells could result from the fusion of primary tumor cells with adjacent non-tumor cell types, including macrophages, fibroblasts, and stem cells (Chakraborty et al. 2004; Mandel et al. 2013; Pawelek 2014; Pawelek and Chakraborty 2008; Rachkovsky et al. 1998; Yang et al. 2015). To highlight a few, hybrids generated between weakly metastatic Cloudman S91 mouse melanoma cells and normal mouse macrophages, or human macrophages, showed a more aggressive phenotype than the parental cell line and induced aggressive metastases in a mouse model. In addition, the hybrids were highly heterogeneous (Rachkovsky et al. 1998). Prostate cancer cells were shown to fuse with stromal and skeletal muscle cells, whereas breast cancer cells fused with normal mammary gland cells and endothelial cells, resulting in hybrids with increased metastatic potential, proliferation rate, and drug resistance (Kerbel et al. 1983a, b). Li et al. showed that human hepatocellular carcinoma cells with low metastatic potential exhibited heterogeneity and significantly increased metastatic potential following fusion with mesenchymal/multipotent stem/stromal cells (MSCs) (Li et al. 2014). Wang et al. demonstrated that hybrid cells generated from the fusion of multiple myeloma cells and bone marrow-derived MSCs were highly heterogeneous and expressed stemness markers with increased resistance to drug treatment (Wang et al. 2018). Lung cancer cells were shown to fuse with MSCs to give rise to hybrids with enhanced metastatic capacity and characteristics of cancer stem cells by undergoing EMT (Zhang et al. 2019b). Chan et al. showed that adipose-derived stem cells fused with the breast cancer cell MDA-MB-231 spontaneously to produce breast cancer stem cells (CSC) expressing $CD44^{+}$ $CD24^{-/low}$ $EpCAM^{+}$. Furthermore, these hybrids exhibited a higher tumorigenic potential in xenograft mice compared to the parent tumorigenic triple-negative breast cancer cell line MDA-MB-231 (Chan et al. 2020). Fusion of mesenchymal stem/stromal cells and breast cancer cells (MDA-MB-231 Hyg and HS578T Hyg) was also shown by Dörnen et al. to lead to the formation of hybrid cells exhibiting a heterogeneous phenotype with stem cell characteristics (Dörnen et al. 2020). Chitwood et al. showed that breast tumor cells spontaneously fused with other cells in vivo to form hybrids. Those hybrids were found to be more prevalent in lung metastases than in the primary tumors supporting the role of cancer cell fusion in metastasis development (Chitwood et al. 2018). This was substantiated by the finding of Gast et al. who showed that cancer cell hybrids generated in vivo potentiate tumor heterogeneity and metastasis, and they are enriched in the circulation (Gast et al. 2018). Mezler et al. showed that spontaneous fusion between primary human MSC and human MDA-MB-231 breast cancer cells produced hybrids with elevated ability to develop a tumor and various distant organ metastases in a much shorter time than the parental breast cancer cells (Melzer, von der Ohe, and Hass 2018a). They also showed that even in the instance of dormancy observed with some hybrids to induce tumor initiation, tumor growth and formation of metastases in various organs occurred rapidly within about 10.5 days (Melzer et al. 2021). Yart et al. showed that fusion between two ovarian cancer cell lines SKOV3 and COV318 could generate ovarian polyploid giant cancer cells (PGCCs) with increased invasive capacity and those giant cells could also form ovarian tumors in ovo (Yart et al. 2020).

Although cancer cell fusion as a source of metastasis is more challenging to demonstrate in patients, there have been some indirect and direct links between cancer cell fusion and metastasis in the clinic. The discovery in cancer patients of circulating tumor cells expressing both carcinoma and leukocyte cell markers points to fusion events between bone marrow-derived cells and tumor cells (Andersen et al. 2007; Chakraborty et al. 2004; Yilmaz et al. 2005). A study used FISH to demonstrate that a renal cell carcinoma arising as secondary malignancy after the patient had undergone an allogeneic bone marrow transplant contained the donor's Y chromosome in the nuclei of the mother's tumor cells (Yilmaz et al. 2005). In further support, recent reports have used short tandem repeat length-polymorphism and forensic genetic techniques to show that a metastatic melanoma lesion in a patient arose from the fusion between a bone marrow-derived cell that the patient received as a transplant and a tumor cell (Lazova et al. 2011). Another transplant patient developed a nodular malignant melanoma on the upper back, which spread to an axillary sentinel lymph node. Analysis of both the primary tumor and the metastasis displayed both the patient and donor characteristics (LaBerge et al. 2017). More recently, LaBerge et al. traced hybrids from a primary melanoma to an axillary lymph node and to the brain metastasis of a transplant patient, therefore documenting the progression of the cancer from the primary tumor to the metastasis (LaBerge et al. 2021). Gast et al. analyzed circulating tumor cells in patients' peripheral blood and showed that the number of hybrids in the peripheral blood of human cancer patients correlated with disease stage and predicted overall survival (Gast et al. 2018) suggesting that those circulating hybrids have the potential of generating metastasis.

Even though fusion between cancer cells and bone marrow-derived cells or other cell types has been shown in many instances, the mechanisms driving this process are yet to be determined. A collaboration between cells' intrinsic components and their environmental factors seems to control this process. Some of the fusogens and proteins identified in normal physiological fusion processes have been observed to be critical in cancer cell fusion as well. These factors include HERV proteins (syncytin-1 and -2) (Bastida-Ruiz et al. 2016), the receptors ASCT1, ASCT2 (Bastida-Ruiz et al. 2016; Larsen et al. 2009; Larsson et al. 2007; Strick et al. 2007), phosphatidylserine (Sharma and Kanwar 2018), and annexins (Uygur et al. 2019). External factors associated with cancer cell fusion include chronic inflammation, inflammatory cytokines, hypoxia, and viral infection. In this review, we have highlighted the different mechanisms that have been identified or postulated to drive cancer cell fusion.

19.2 Mechanisms of Cancer Cell Fusion

The membrane that separates cells from the extracellular environment is made of a phospholipid bilayer. The fusion of two or more cells will require the morphological reconstruction of the cell membrane and the merging of the cell contents. The normal physiological mechanism of cell fusion, which is observed in fertilization, placentation, muscle differentiation, bone homeostasis, regeneration, and repair processes (Hass et al. 2021), is a highly regulated process by internal and external factors. A dysregulation of this process could lead to diseases such as osteoporosis

(Zhou et al. 2015), malformation of muscles (Buckingham 2006; Horsley and Pavlath 2004), difficulties in pregnancy (Chen et al. 2006; Langbein et al. 2008; Lee et al. 2001), and cancer (Bastida-Ruiz et al. 2016).

A series of five common steps have been proposed to take place during fusion events regardless of the cell type. These steps involve priming, chemotaxis, membrane adhesion, membrane fusion, and post-fusion resetting (Zhou and Platt 2011) . A pre-hybrid preparation process (PHPP) and a post-hybrid selection process have been identified as well (Hass et al. 2020; Melzer et al. 2020). The priming step enables the cells to become competent to fuse. During this process, adhesion molecules are expressed, the lipid composition of the cell membrane is modified with the translocation of inner-leaflet lipids such as phosphatidylserine and loss of the inhibitory state of the membrane due to the degradation of the extracellular matrix which allows cell migration and promotes cell–cell contact of the fusing cells (Zhou and Platt 2011). This close proximity is a requirement for cell fusion, and it triggers dehydration of the contact site, therefore reducing the hydration repulsion between the outer leaflets of the membranes. This reduction of repulsion allows the formation of a fusion stalk with only the outer leaflets of the membranes fused together (Chernomordik and Kozlov 2008; Gerbaud and Pidoux 2015; Pérot et al. 2011). The formation of a fusion stalk is followed by its radial expansion resulting in a hemifusion diaphragm with the inner leaflets still unaffected. The fusion of the inner leaflets will follow leading to a complete merge of the membranes and the opening of a pore through which the cell contents can blend completing the cell–cell fusion event (Pérot et al. 2011). The complex biochemical and thermodynamic processes that occur during the fusion of membranes are still not well understood. Studies have identified proteins or fusogens that play critical roles in this process and function to overcome the energy barrier to facilitate cell fusion (Brukman et al. 2019; Hernández and Podbilewicz 2017; Pérez-Vargas et al. 2014; Podbilewicz et al. 2006). In addition to fusogens and phosphatidylserine externalization, actin reorganization plays a critical role in cell–cell fusion (Brukman et al. 2019; Hernández and Podbilewicz 2017; Ogle et al. 2005; Whitlock and Chernomordik 2021). Indeed, inhibition of actin polymerization or lamellipodia and filopodia formation results in a drastic reduction in breast cancer cell fusion with MSCs (Melzer et al. 2019). After fusion, cells revert to the post-hybrid state and become non-fusogenic again. The factors and mechanisms involved in this change of phenotype are still not understood.

A well-orchestrated interaction between intracellular and external factors governs the mechanisms of cell fusion. Some of those factors that are responsible for mediating membrane fusion in normal physiological processes have been observed in cancer cells and seem to play similar roles there as well. Further insights might be gleaned from our understanding of intracellular fusion and the fusion of the envelope virus with the host cell.

19.2.1 Intrinsic Factors

Some fusogens involved in normal physiological processes have been identified. Syncytin-1 and syncytin-2 have been shown to drive fusion of cytotrophoblasts to form syncytiotrophoblasts (Chalbi et al. 2014; Huppertz et al. 2006; Malassiné et al. 2005; Muir et al. 2006; Pérez-Vargas et al. 2014). In myoblast fusion, the two fusogens, myomaker, and myomerger have been identified (Bi et al. 2017; Leikina et al. 2013; Millay and Olson 2013; Quinn et al. 2017). In fertilization, Izumo1 and Juno expressed in sperm and oocyte, respectively, were shown to be critical to the fusion of the sperm and oocyte (Bianchi and Wright 2015; Chalbi et al. 2014; Inoue et al. 2005; Kato et al. 2016). Some of these fusogens have been found in cancer cells suggesting their role in cancer cell fusion.

19.2.1.1 Syncytins, the Human Endogenous Retrovirus Envelope Genes (HERV env) and Cancer Cell Fusion

Syncytin-1 and -2 are human endogenous retrovirus envelope genes (HERV env) that express env proteins or fusogens involved in trophoblast fusion during placenta development (Blond et al. 1999; Sha et al. 2000). The syncytiotrophoblasts resulting from trophoblasts fusion are very invasive with high migratory capability (reviewed in Bastida-Ruiz et al. 2016). Syncytins have been demonstrated to be expressed or overexpressed in many cancers and cell lines (Bastida-Ruiz et al. 2016; Larsen et al. 2009; Larsson et al. 2007; Strick et al. 2007). The increase in the expression of syncytins in some cancer cells might be modulated by the higher number of chromosomes, chromosome mutation, or chromosomal rearrangement (Strick et al. 2011). Bjerregaard et al. demonstrated in their studies that syncytin was expressed in breast cancers and breast cancer cell lines. An inhibition of syncytin levels by using syncytin antisense oligonucleotide or the use of syncytin inhibitory peptides prevented fusion between breast cancer cells and endothelial cells (Bjerregaard et al. 2006). Syncytin antisense oligonucleotide could inhibit human myoblasts fusion as well (Bjerregard et al. 2014). Cell fusion was also inhibited in 293T cells when peptides designed to investigate the functions of the heptad repeat regions of the transmembrane (TM) subunit of syncytin were used (Chang et al. 2004). Using the siRNA approach, Strick et al. showed that silencing of syncytin-1 gene expression blocked cell–cell fusion of endometrial carcinoma (Strick et al. 2007). The syncytin receptors ASCT1 and ACST2 are upregulated in many cancers (Bastida-Ruiz et al. 2016; Strick et al. 2007) and in the cancer fusion partner cells such as endothelial cells which are breast cancer cell fusion partners (Bjerregaard et al. 2006). The syncytin receptor ASCT2 was found to be expressed in prostate cancer cells and to promote tumor growth and metastasis. Inhibition of its expression was shown to reduce tumor progression and metastasis in xenograft mice and was suggested as a target for therapy (Wang et al. 2015).

The mechanism by which syncytin-1 (HERV-W) promotes cancer cell fusion is still not well understood. However, syncytin-1 (HERV-W) resembles the retroviral envelope proteins, and the mechanism by which retroviral envelope proteins fuse cell membranes is well-studied. It is therefore expected that syncytin-1 (HERV-W) might employ the retroviral envelope proteins mechanism to promote cancer cell fusion.

Syncytin-1 is a trimeric glycoprotein composed of two units: a surface unit (SU) and a transmembrane unit (TM). The SU is composed of four main elements: (1) a receptor binding domain (RBD), (2) a furin cleavage site ($^{314-}$RNKR^{-317}), (3) 6 *N*-glycosylation sites, and (4) a $C_{\Phi\Phi}C$ ($^{186-}CX_2C^{-189}$) motif whereas the TM is composed of six main elements; (1) a fusion peptide (FP), (2) two heptad repeats (HR1 and HR2), (3) a transmembrane anchorage domain (tm), (4) a cytosolic tail (cyt), (5) 1 *N*-glycosylation site, and (6) a $^{397-}Cx_6C^{-407}$ domain (reviewed in Bastida-Ruiz et al. 2016). The protein is translated and folded in the endoplasmic reticulum. Both SU and TM are translated together but then separated by cleavage at the furin cleavage site of SU by proteases like the furin-convertase enzyme. Subsequently, a disulfide bond is formed between the $^{397-}Cx_6C^{-407}$ domain of TM and $^{186-}CX_2C^{-189}$ motif of SU (Gerbaud and Pidoux 2015; Pérot et al. 2011). During syncytin-driven cell–cell fusion, syncytin-1 in one cell recognizes and binds to its receptor (ASCT1 or ASCT2) on another cell. During this process, the receptor binding domain on SU recognizes its receptor on the target cell, and a conformational change in the trimeric structure of syncytin results in the separation of the SU unit and the TM unit by rupture of the disulfide bond. This produces a loop-to-helix movement of the syncytin-1 fusion peptide in the TM unit which triggers its projection toward the top of the glycoprotein, promoting its interaction with the target membrane, and its strong insertion into the membrane (Gerbaud and Pidoux 2015; Pérot et al. 2011). The tight connection of both membranes through the envelope protein allows the HR2 domain in the TM unit to fold back and interact with the HR1 domain, reversing the direction of the cell membrane and bringing both membranes in close proximity. The proximity of the membranes considerably reduces the free energy needed to overcome the barrier to merge (reviewed in Bastida-Ruiz et al. 2016). This mechanism of syncytin-dependent fusion might as well control fusion of cancer cells expressing this fusogen.

19.2.1.2 Phosphatidylserine

Phosphatidylserine (PtdSer) is an important component of the inner leaflet of eukaryotic cells. It connects signals between cells to drive processes such as apoptotic engulfment, synaptic pruning, and immune cells activation (reviewed in Whitlock and Chernomordik 2021). Its presence in the outer leaflet of membranes has been associated with all cell fusion mechanisms studied including myoblast fusion (van den Eijnde et al. 2001; Jeong and Conboy 2011), macrophage fusion (Faust et al. 2019; Helming et al. 2009), trophoblast fusion (Adler et al. 1995; Das et al. 2004; Žigon et al. 2015), and fertilization (Rival et al. 2019). PtdSer interacts with different receptors and binding proteins including Tim4, Bai1, Bai3, STAB2, Annexin 1, Annexin 5, and CD36 (reviewed in Whitlock and Chernomordik 2021). It was shown to interact with Bai1 and Bai3 to activate Elmo/Dock180/Rac signaling during myoblast fusion (Hamoud et al. 2014; Hochreiter-Hufford et al. 2013). PtdSer was also shown to bind to Bai1, Bai3, and CD36 during sperm and egg fusion and to signal through activation of the Elmo1/Rac1 pathway (Rival et al. 2019). The PtdSer receptor CD36 was demonstrated to contribute to the fusion of macrophages (Helming et al. 2009; Park et al. 2016) and monocytes with lung tumor stem cells to form highly invasive tumor-hybrid cells (Aguirre et al. 2020). PtdSer is selectively exposed to the surface of many cancer cells and not on normal cells when exposed to oxidative stress(Sharma and Kanwar 2018). Events such as inflammation, hypoxia, and oxidative stress are linked to the apoptotic pathways (Mohr et al. 2015) and promote PtdSer externalization (Brukman et al. 2019). This externalization of PtdSer could result in cell fusion. The externalization of PtdSer during cancer cell fusion may trigger syncytin-driven fusion by means of a mechanism similar to the PtdSer-dependent activation of HIV Env-mediated fusion. In this proposed mechanism, the localization of PtdSer in the outer leaflet of the membrane might regulate the reorganization of fusogenic proteins that are already present at the fusion site in preparation for fusion. It is also suggested that PtdSer binds to its specific proteins to direct the timing and location of fusion protein assembly (Whitlock and Chernomordik 2021).

A study showed that inhibition of apoptosis or inhibition of PtdSer prevented fusion between breast cancer cells and MSCs suggesting a mechanism of cancer cell fusion similar to that of myoblast fusion that involves PtdSer interaction with Bai1-dependent activation of the ELMO/Dock180/Rac1 signaling (Noubissi et al. 2015; Noubissi and Ogle 2016). High levels of the guanine nucleotide exchange factors ELMO and Dock180 have been observed in breast cancer cell lines (Li et al. 2013) and activation of the ELMO-Dock signaling pathway has been demonstrated to contribute to breast cancer metastasis (Abu-Thuraia et al. 2015; Li et al. 2013). ELMO1 is also a modifier of breast cancer risk for BRCA mutation carriers (Walker et al. 2010) and Rac1 was shown to be overexpressed in many breast diseases including breast carcinoma and breast metastases (Bid et al. 2013; Feng et al. 2014). This Rac1 overexpression correlates positively with the aggressive form of breast cancer (Abu-Thuraia et al. 2015; Li et al, 2013; Lu and Ravichandran 2006). Rac1's contribution to cancer metastasis seems to be controlled by episodes of hypoxia and re-oxygenation (Lee et al. 2015). Rac1 modulates cytoskeletal rearrangements which is critical for several cellular activities, such as phagocytosis, mesenchymal-like migration, axonal growth, adhesion and differentiation of multiple cell types, and reactive oxygen species (ROS)-mediated cell killing (reviewed in Heasman and Ridley 2008; Walker and Spurdle 2010). Rac1 also regulates signaling pathways that drive cellular growth and cell cycle regulation (Saci et al. 2011), the formation of cell–cell adhesions (Ehrlich et al. 2002), and the process of contact inhibition (Bosco et al. 2010), which are critical in malignant transformation. PtdSer-dependent activation of the ELMO/Dock180/Rac1 signaling might drive cancer development and progression via several mechanisms including Rac1-driven actin cytoskeleton reorganization which promotes cancer cell fusion.

In *Drosophila*, cell–cell fusion was demonstrated to involve the interaction between non-muscle Myosin II and actin network. During cell–cell fusion, Myosin II senses the mechanical strains in the actin network and is recruited to the fusogenic synapse where it accumulates and generates cortical tension necessary to resist podosome-like structure invasion, thereby promoting membrane juxtaposition and fusion (Kim et al. 2015). During osteoclast fusion, actin filaments were shown to polymerize to form thick actin filament bundles in the podosome field at the zone of contact of two cells and to recruit non-muscle myosin IIA to promote osteoclast fusion by generating protrusive forces against matrix (Takito et al. 2017). This mechanism of actin dynamics-dependent cell–cell fusion might operate similarly in cancer cell fusion as a result of the PtSer-dependent activation of the ELMO/Dock180/Rac1 signaling. In support, Rac1 was found to be expressed within the actin-rich region of the fusogenic domain of osteoclasts (Takito et al. 2017). Inhibition of proteins associated with lamellipodia and filopodia formation such as Arp2/3 and formin with a sub-lethal concentration of cytochalasin D resulted in a reduction in breast cancer cell fusion with MSCs as well (Melzer et al. 2019). This supports the role of actin polymerization in cancer cell fusion.

19.2.1.3 Annexins and Glucose-Regulated Protein 78 (GRP78)

Annexins are proteins that bind to PtdSer to drive different processes (Rosenbaum et al. 2011). PtdSer was shown to bind to Annexin A1 and A5 to promote myoblast (Leikina et al. 2013) and osteoclast (Verma et al. 2018) formation. Annexin A5 was also shown to be important in trophoblast fusion (Degrelle et al. 2017; di Simone et al. 2001). Together with syncytin-1, Annexin 5 has been implicated in prostate cancer cell fusion cocultured with muscle cells (Uygur et al. 2019). The presence of muscle cells significantly increased the concentrations of interleukins 4 and 13 which induced high levels of syncytin-1 and Annexin A5 in prostate cancer cells to facilitate cell fusion. However, inhibition of Annexin A5 in prostate cancer cells or muscle cells using siRNA reduced the fusion of prostate cancer cells. In addition, high levels of Annexin A5 were also observed in human prostate cancer tissues compared to non-malignant tissues (Uygur et al. 2019).

GRP78 is a well-known endoplasmic reticulum chaperone commonly used as a marker for stress. However, it was also found to be expressed in other compartments of the cell such as the cytoplasm, mitochondria, the nucleus, and the cell surface (reviewed in Ni et al. 2011). GRP78 expression as a cell surface receptor was observed to be selectively expressed in various cancer cells as a cell surface receptor (Li and Lee 2006; Ni et al. 2011). In prostate cancer cells, GRP78 interaction with its ligand α2-macroglobulin (α2M) was shown to activate anti-apoptotic pathways driving cell proliferation such as ERK1/2, p38 MAPK, and PI3K-dependent signaling and survival pathways such as Akt and NF-kB signaling cascade (Misra et al.

2006). Activation of these pathways results in tumor progression, metastasis, and resistance to therapy (Ni et al. 2011). ERK1/2 and PKA have been shown to induce an increase in syncytin and trophoblast fusion (Delidaki et al. 2011). Recently, Ruiz et al. used the human placental cell line BeWo to demonstrate that binding of α2M to GRP78 activates ERK1/2 and CREB, leading to the stimulation of the unfolded protein response (UPR)-driven increase in cell fusion (Bastida-Ruiz et al. 2020). It is therefore possible that α2M and GRP78 interaction in cancer cells activates the ERK1/2 signaling to induce cancer cell fusion.

19.2.2 Extracellular Factors

19.2.2.1 Inflammation, Inflammatory Cytokines, and Signaling Pathways

Inflammation and inflammatory cytokines have been shown to induce cell fusion (Davies et al. 2009; Melzer et al. 2018b; Weiler and Dittmar 2019; Yan et al. 2017). Davies et al. showed that in an inflammatory mouse model, intestinal inflammation and epithelial proliferation promoted bone marrow-derived cells (BMDCs) fusion with intestinal progenitors (Davies et al. 2009).

In cancer, Yang et al. demonstrated that the inflammatory cytokine TNF-α enhanced fusion between squamous cell carcinoma cells 9 (SCC-9) and human umbilical vein endothelial cells (HUVEC). This enhancement in fusion was paralleled with an upregulation of syncytin-1 in SCC-9 and its receptor ASCT-2 in HUVEC. Activation of the Wnt/β-catenin signal pathway by TNF-α also upregulated syncytin-1 expression and fusion between SCC-9 and HUVEC cells showing the contribution of the Wnt/β-catenin signaling in cancer cell fusion (Yan et al. 2017). Song et al. also showed that inflammation and inflammatory cytokines triggered fusion of oral cancer cells with endothelial cells (Song et al. 2012). They demonstrated that the addition of exogenous TNF-α induced a threefold increase in fusion between human oral squamous cell carcinoma cell lines and human umbilical vein endothelial cells. They detected the integrin called very late activation antigen 4 (VLA-4, α4β1) in human oral squamous cell carcinoma cell lines and in most of the oral squamous cell carcinoma specimens whereas the VLA-4 ligand the vascular cell adhesion molecule 1 (VCAM-1) was detected in the vascular endothelium of oral squamous cell carcinoma. TNF-α cell treatment upregulated VCAM-1 expression and its inhibition with a specific antibody or inhibition of VLA-4 prevented cancer-endothelial adhesion and fusion. This suggests that TNF-α triggers the fusion of cancer cells with endothelial cells by a mechanism that signals through VCAM-1/VLA-1 pathway (Song et al. 2012). The overexpression of VCAM-1 in the inflammatory tumor microenvironment has been linked to tumor development and progression. The integrin VLA-4 α4β1, on the other hand, is expressed in many different cancers (reviewed in Song et al. 2012). These integrins might well facilitate cell–cell adhesion which will facilitate fusion of cancer cells.

TNF-α has also been shown to induce fusion of breast cancer cells MDA-MD-435 with breast epithelial cells M13SV1(Mohr et al. 2015; Weiler et al. 2018; Weiler and Dittmar 2019) by a mechanism that activates the NF-κB signaling and its target genes. This TNF-α-induced fusion was inhibited using minocycline (Weiler et al. 2018). Studies of the mechanisms by which minocycline impairs TNF-α-induced fusion showed that it prevents the interaction between the TNF-α receptor TNFR1 and its associated protein TRAF2 and selectively inhibits IκB-α phosphorylation and NF-κB-p65 activation in the non-cancerous M13SV1 cells resulting in the reduction in the expression of the matrix metalloproteinase-9 (MMP9) and the intercellular adhesion molecule 1 (ICAM1) (Weiler and Dittmar 2019). Downregulation of the NF-kB pathway reduced TNF-α-dependent cell fusion (Weiler and Dittmar 2019) suggesting a critical role for NF-κB activation in TNF-α-driven cell fusion.

Skokos et al. showed that TNF-α and the matrix metalloproteinase-9 (MMP-9) were important in macrophage fusion as well (Skokos et al. 2011). They demonstrated that fusion events were significantly reduced in monocyte chemoattractant protein 1 (MPC-1) null mice and MPC-1 null macrophages. MPC-1 is one of the key chemokines required for foreign body giant cell (FBGC) formation in the foreign body response by the immune system. The reduction in macrophage fusion was associated with a decrease in the levels of MMP-9 and TNF-α production. An abnormal subcellular redistribution of E-cadherin and β-catenin during fusion was also observed in these MPC-1-null macrophages. The addition of exogenous TNF-α to MCP1-null macrophages restored fusion, which was paralleled with the restoration of MMP-9 expression and redistribution of E-cadherin (Skokos et al. 2011). MMP-9 directly degrades extracellular matrix (ECM) proteins and activates cytokines and chemokines to regulate tissue remodeling (Yabluchanskiy et al. 2013). MMP-9 probably cleaves E-cadherin to stimulate membrane–membrane apposition and induces normal and cancer cell–cell fusion.

β-catenin/E-cadherin complexes seem to be critical for sperm-oocyte adhesion as well. In this scenario, after adhesion, β-catenin is rapidly ubiquitinated and degraded in the proteasomes to allow membrane fusion between sperm-oocyte. The redistribution or degradation of E-cadherin and β-catenin complexes may change the lipid composition in the cell membrane, which will then provide microenvironments where cell fusion occurs (Takezawa et al. 2011). In cytotrophoblast fusion, E-cadherin was detectable at the cell–cell contact sites of cell aggregates, but its level was significantly reduced in fusing cytotrophoblasts suggesting its relocation after cell–cell adhesion.

A disintegrin and metalloproteinase 10 (ADAM10) was shown to cleave E-cadherin and affect E-cadherin and β-catenin subcellular localization in epithelial cells adhesion, migration, and proliferation (Maretzky et al. 2005). ADAM12 which controls cell fusion in myoblasts was found to induce E-cadherin ectodomain shedding to promote trophoblast fusion as well through remodeling of intercellular boundaries which includes disorganization of adherens junctions (Aghababaei et al.

2015). This shows that metalloproteinases play an important role in extracellular remodeling during cell fusion.

Matsuura et al. showed that the β-catenin/BCL9-Like (BCL9L)/T-cell factor 4 (TCF4) signaling directly targets the glial cells missing 1 (GCM1)/syncytin pathway and thereby regulates the fusion of human choriocarcinoma cells(Matsuura et al. 2011). This β-catenin/GCM1/syncytin pathway-dependent cell fusion might be important in tissue regeneration and/or cancer development. In Min mice, which have a mutated adenomatous polyposis coli (APC) that constitutively activates the Wnt/β-catenin signaling, the frequency of fusion of intestinal cells with bone marrow-derived cells in the intestine is increased compared with the wild type (Rizvi et al. 2006). The APC ^{Min} mice are known to develop a high number of intestinal tumors (Hamilton et al. 2013). Fusion between intestinal cells and bone marrow-derived cells might contribute to the development and/or progression of those tumors. In breast cancer, Zhang et al. showed that fusion of macrophages with breast cancer cells produces hybrids with high proliferation, migration, and invasion capabilities modulated through activating epithelial-mesenchymal transition and Wnt/β-catenin signaling pathway (Zhang et al. 2019a).

TNF-α might create an environment conducive to cell–cell adhesion and after adhesion of cancer cells with other cells, E-cadherin is cleaved from the E-cadherin/β-catenin complexes by metalloproteinases (such as MMP-9, ADAM10, ADAM12). β-catenin is then free to translocate into the nucleus where it interacts with LEF/TCF to activate Wnt/β-catenin signaling and NF-κB pathway therefore promoting tumor development and metastasis.

19.2.2.2 Virus

Members of four classes of viruses have been shown to have the ability to induce the fusion of various cells in vitro and in vivo resulting in heterokaryon and hybrid cell formation. These viruses involve members of paramyxoviridae, retroviridae, coronaviridae, poxviridae, and reoviridae (Dittmar et al. 2021). Some of the viruses that infect human cells have been associated with cell fusion-driven cancer development (Duelli et al. 2005; Duelli and Lazebnik 2003, 2007) and progression (Ganem et al. 2007; Larizza and Schirrmacher 1984; Shackney et al. 1989). It is possible that by inducing cell fusion, viruses trigger chromosomal instability and abnormal gene expression (Duelli and Lazebnik 2003; Ogle et al. 2004), which might lead to cancer development (Duelli et al. 2005; Duelli and Lazebnik 2007) and/or cancer metastasis (Ganem et al. 2007; Larizza and Schirrmacher 1984; Shackney et al. 1989).

The mechanisms by which viruses cause cell fusion are still not clear. Two mechanisms have been postulated including: (1) viral infection-driven fusogens activation and (2) virus serving as a bridge between two cells leading to their fusion (Duelli and Lazebnik 2007). As fusogens activator, viral infection of a cell stimulates the expression of fusogenic proteins by the cell at different locations including the cell membrane, and the cell can subsequently fuse with any other cell expressing a receptor for the fusogenic protein. The most studied fusogens activated by this mechanism involve the Env proteins (Duelli and Lazebnik 2007). They are expressed in normal tissues and in cancers. This fusogens activation mechanism of fusion is being extensively studied and considered in oncolytic virus therapy (Burton and Bartee 2019; Krabbe and Altomonte 2018; Liu and Kirn 2007). Since viruses infect many different cell types, these cells will potentially fuse to give rise to subsets of heterogeneous hybrids with unstable genome and abnormal gene expression which could lead to cancer (Duelli and Lazebnik 2007). In the second bridge builder mechanism, a viral particle serves as a bridge between two cells promoting their fusion. In this case, the cells are not infected by the virus. This mechanism in this case is less clear than the fusogens activation mechanism.

Although the direct evidence between viral infection and cancer cell fusion is challenging to prove, many indirect links have been shown between the expression of some human endogenous retroviruses (HERV) Env elements and cancer development and progression. HERVs are relics of ancient retroviral infections and account for about 8% of the human genome. They were believed for a long time to be silent passengers within our genomes. However, they have been found to be reactivated in many diseases especially the Env elements HERV-K HML-2 which is the most recent integration group with the least number of mutations. This Env element is not expressed in normal cells but has been observed to be reactivated in cancers and associated with cancer progression and poor outcome (Curty et al. 2020; Dervan et al. 2021). The HERV-K (HML-2) family encodes functional retroviral proteins and produces retrovirus-like particles (Grandi and Tramontano 2018).

Johanning et al. analyzed four HERV-K loci in different breast cancer subtypes (basal, Her2E, LumA, and LumB breast cancer subtypes) of 512 breast cancer patients with invasive ductal carcinoma (IDC) and found that HERV-K was expressed exclusively in the basal subtype of breast cancer and seemed associated with extremely poor prognosis and high frequencies of recurrence and metastasis (Johanning et al. 2017). Similarly, upregulation of HERV-K (HML-2) was observed in 84 hepatocellular carcinoma (HCC) tissues compared with adjacent normal tissues and significantly correlated with cancer progression and poor outcome. It was proposed as a novel candidate prognostic biomarker for HCC (Ma et al. 2016). Analysis of 106 pancreatic cancer patient sera showed that HERV-K viral RNA levels and anti-HERV-K antibody titers were significantly higher than in normal donor sera and played an important role in the progression of the disease (Li et al. 2017, 2019). HERV-K expression in melanoma tissues from 82 patients was also associated with recurrence and reduced disease-free survival of melanoma patients (Cardelli et al. 2020).

Huang et al. used the RNAi approach and HERV-K ENV monoclonal antibodies to demonstrate that proteins encoded by HERV-K can mediate intercellular fusion of melanoma cells, which may generate multinuclear cells and drive the evolution of genetic changes that provide growth and survival advantages (Huang et al. 2013). HERV-K reactivation in viral infected cells might result from its hypomethylation as HERV-K methylation was decreased in melanoma patients with clinical parameters associated with reduced disease-free survival (Cardelli et al. 2020). RNAseq studies showed that activation of HERV-K potentially drives cancer progression by contributing to the Ras-dependent signaling (Zhou et al. 2016) and involves ERK signaling in the RAS-ERK-RSK pathway (Lemaître et al. 2017; Li et al. 2017). HERV ENV proteins can also promote cancer via immune suppression (Gao et al. 2021). Inhibition of HERV-K using shRNA was shown to induce the reversion of breast cancer cells to a non-tumorigenic phenotype such as inhibition of cell proliferation, colony formation, and cell transformation, as well as inhibition of tumor formation and metastasis in vivo (Ma et al. 2016) suggesting that HERV-K could be targeted for therapy. Since HERV-K is selectively activated in cancer cells and not in normal cells and its activation correlates with the poor outcome of many cancers, it has been proposed as a candidate for diagnostics, prognostic, and cancer vaccines for immunotherapy.

The reactivation of HERV env elements such HERV-K HML2 env and syncytin-1 in cells by viral infections has been extensively reported in the literature and shown to be associated with disease progression including cancer (reviewed in Dittmar and Hass 2022). Although many studies have shown a link between syncytin-1 expression and cell fusion in cancer development and progression (reviewed in Dittmar et al. 2021), studies on the direct link between HERV-K reactivation and cell fusion in cancer are limited (Huang et al. 2013). Further studies will be required to elucidate the role of HERV env elements in cancer cell fusion and cancer development and progression.

19.2.2.3 Mediators of Cell Stress and Other Factors (Hypoxia, Radiotherapy, Chemotherapy, pH, Exosomes, Cellocytosis, Entosis)

Hypoxia is a phenotype of solid tumors in response to rapid cell and tissue growth. This hypoxic microenvironment puts cancer cells under stress and might promote fusion with cells of the stroma recruited to the tumor microenvironment. Interestingly, hypoxia has been shown to promote fusion between certain eukaryotic cell types such as bone marrow-derived mesenchymal stem cells and cardiomyocytes (Haneef et al. 2014), bone marrow-derived MSCs and myoblasts (Archacka et al. 2021), and between urine-derived stem cells and different types of liver cells (Hu et al. 2021). In cancer, hypoxia was also shown to stimulate a significant increase in fusion between MSCs and breast cancer cells T47D and MCF7 (Noubissi et al. 2015; Noubissi and Ogle 2016) and between oral squamous carcinoma cells (OSCCs) and human immortalized oral epithelial cells (HIOECs) (Huang et al. 2018). Morh et al. demonstrated that TNF-α together with hypoxia was a strong inducer of cell fusion in human MDA-MB-435 and MDA-MB-231 breast cancer cells (Mohr et al. 2015). Hypoxic conditions might promote cell fusion by mechanisms that involve the production of growth factors such as VEGF (Archacka et al. 2021), the upregulation of CXCR4 (Hu et al. 2021), the activation of the apoptotic pathway (Noubissi et al. 2015; Noubissi and Ogle 2016), and EMT mechanism (Huang et al. 2018).

Additional factors such as chemotherapy (Yart et al. 2020) and radiotherapy (Levin et al. 2010; Seyfried 2012) are other potential inducers of cancer cell fusion which can result in the formation of polyploid giant cancer cells (PGCCs). PGCCs are large, atypical cancer cells with multiple copies of DNAs. They form a special subpopulation of cancer cells that contribute to tumor heterogeneity. Their formation can be induced by stress factors such as hypoxia, radiotherapy, chemotherapy, and viruses (reviewed in Zhou et al. 2022). Although PGCCs can be formed through repeated incomplete mitotic events, studies using several cancer cell lines such as HEY, MDA-MB-231, and SKOv3 demonstrated that PGCCs formation can be induced by cell fusion (Zhang et al. 2014). Paclitaxel treatment of ovarian cancer cells was shown to induce ovarian PGCCs formation in vitro (Yart et al. 2022). There is evidence also that patients with radiation-treated cancer have a lower chance of survival due to the increased fusion of epithelial cells with macrophages (Seyfried 2012). Interestingly, advanced rectal cancer patients who received a combination of radiotherapy and chemotherapy (capecitabine alone or a combination of capecitabine and oxaliplatin) developed more PGCCs in their tumors than patients who did not receive the radiochemotherapy. Those PGCCs were more prevalent in metastases than in the primary tumors and associated with recurrent metastasis, chemoresistance, and poor prognosis. Consistent with these observations, the treatment of colon cancer cell lines LoVo and HCT116 with radiation or the chemotherapeutics capecitabine, oxaliplatin, or irinotecan induced the formation of PGCCs. These PGCCs produced many daughter cells that were resistant to further treatments and exhibited strong migration, invasion, and proliferation capabilities and mesenchymal phenotypes (Fei et al. 2019).

Additional factors that might contribute to cancer cell fusion include low pH, exosomes-dependent signaling (reviewed in Dittmar and Hass 2022; Hass et al. 2020; Howcroft et al. 2011; Uygur et al. 2019), cellocytosis (Vignery 2005), or entosis (Overholtzer et al. 2007). However, direct evidence of the role of these factors and processes in cancer cell fusion is yet to be demonstrated.

19.3 Conclusion

Although fusion has been extensively shown to happen between many different cancer cells and other cell types and to promote cancer progression and metastasis, the mechanisms of how fusion is initiated and terminated are still scarcely understood. Communication between cells intrinsic factors and extracellular factors appears to drive this process. Activation of fusogenic proteins (syncytins) in combination with phosphatidylserine externalization and cytoskeletal reorganization seems to be critical in cancer cell fusion. Inflammation/inflammatory cytokines, together with metalloproteinases, hypoxia, and apoptotic cells seem to support cancer cell fusion by converting cells from non-fusogenic states into fusogenic states or by increasing cell–cell contacts. Inflammation could be induced by viral infection which may at the same time reactivate fusogens such as HERV Env elements

(HERV-K HML2). Inflammation and inflammatory cytokines such as TNF-α also activate signaling pathways including the Wnt/β-catenin and NF-κB signal pathways that drive cancer progression. The non-apoptosis-dependent externalization of phosphatidylserine also activates different receptors which signal through specific pathways to induce fusion. Additional factors such as radiotherapy and chemotherapy have been implicated in cancer cell fusion as well. The roles of other factors such as pH, exosomes, entosis, and cellotosis in cancer cell fusion are yet to be demonstrated. Fusion promotes cancer progression by inducing chromosomal instability. Identification of additional fusogens or factors driving cancer cell fusion will help understand this process and potentially develop more effective drugs for cancer treatment. Indeed, the selective expression of HERV-K HML2 env in cancer cells has made it a potential candidate for diagnostics, prognostic, and cancer vaccines for immunotherapy. Although cell fusion seems deleterious in cancer, this mechanism is being studied for oncolytic virus therapy by fusing cancer cells with dendritic cells to elicit a strong immune response to eradicate cancer cells. However, further studies are required to understand the mechanisms of cancer cell fusion in order to determine how to prevent it or exploit it and reduce cancer progression and metastasis.

Acknowledgments This work was funded by the SID Freinkel Diversity Fellowship and the NIH/ NIMHD grant 5U54MD015929-03.

References

Abu-Thuraia A, Gauthier R, Chidiac R, Fukui Y, Screaton RA, Gratton J-P, Côté J-F (2015) Axl phosphorylates Elmo scaffold proteins to promote Rac activation and cell invasion. Mol Cell Biol 35(1). https://doi.org/10.1128/mcb.00764-14

Adler RR, Ng AK, Rote NS (1995) Monoclonal antiphosphatidylserine antibody inhibits intercellular fusion of the choriocarcinoma line, JAR. Biol Reprod 53(4). https://doi.org/10.1095/biolreprod53.4.905

Aghababaei M, Hogg K, Perdu S, Robinson WP, Beristain AG (2015) ADAM12-directed ectodomain shedding of E-cadherin potentiates trophoblast fusion. Cell Death Differ 22(12). https://doi.org/10.1038/cdd.2015.44

Aguilar PS, Baylies MK, Fleissner A, Helming L, Inoue N, Podbilewicz B, Wang H, Wong M (2013) Genetic basis of cell-cell fusion mechanisms. Trends Genet 29(7)

Aguirre LA, Montalbán-Hernández K, Avendaño-Ortiz J, Marín E, Lozano R, Toledano V, Sánchez-Maroto L, Terrón V, Valentín J, Pulido E, Casalvilla JC, Rubio C, Diekhorst L, Laso-García F, del Fresno C, Collazo-Lorduy A, Jiménez-Munarriz B, Gómez-Campelo P, Llanos-González E, Fernández-Velasco M, Rodríguez-Antolín C, de Diego RP, Cantero-Cid R, Hernádez-Jimenez E, Álvarez E, Rosas R, López-Ayllón BD, de Castro J, Wculek SK, Cubillos-Zapata C, de Cáceres II, Díaz-Agero P, Fernández MG, de Miguel MP, Sancho D, Schulte L, Perona R, Belda-Iniesta C, Boscá L, López-Collazo E (2020) Tumor stem cells fuse with monocytes to form highly invasive tumor-hybrid cells. OncoImmunology 9(1). https://doi.org/10.1080/2162402X.2020.1773204

Aichel O (1911) Vortrage Und Aufsatze Uber Entvickelungsmechanik Der Organismen. Wilhelm Engelmann, Leipzig

Andersen TL, Boissy P, Sondergaard TE, Kupisiewicz K, Plesner T, Rasmussen T, Haaber J, Kølvraa S, Delaissé JM (2007) Osteoclast nuclei of myeloma patients show chromosome translocations specific for the myeloma cell clone: a new type of cancer-host partnership? J Pathol 211(1). https://doi.org/10.1002/path.2078

Archacka K, Grabowska I, Mierzejewski B, Graffstein J, Górzyńska A, Krawczyk M, Różycka AM, Kalaszczyńska I, Muras G, Stremińska W, Jańczyk-Ilach K, Walczak P, Janowski M, Ciemerych MA, Brzoska E (2021) Hypoxia preconditioned bone marrow-derived mesenchymal stromal/stem cells enhance myoblast fusion and skeletal muscle regeneration. Stem Cell Res Ther 12(1). https://doi.org/10.1186/s13287-021-02530-3

Bastida-Ruiz D, van Hoesen K, Cohen M (2016) The dark side of cell fusion. Int J Mol Sci 17(5)

Bastida-Ruiz D, Wuillemin C, Pederencino A, Yaron M, Martinez B, de Tejada S, Pizzo V, Cohen M (2020) Activated A2-macroglobulin binding to cell surface GRP78 induces trophoblastic cell fusion. Sci Rep 10(1). https://doi.org/10.1038/s41598-020-66554-0

Bi P, Ramirez-Martinez A, Li H, Cannavino J, McAnally JR, Shelton JM, Sánchez-Ortiz E, Bassel-Duby R, Olson EN (2017) Control of muscle formation by the fusogenic micropeptide myomixer. Science 356(6335). https://doi.org/10.1126/science.aam9361

Bid HK, Roberts RD, Manchanda PK, Houghton PJ (2013) RAC1: An Emerging Therapeutic Option for Targeting Cancer Angiogenesis and Metastasis. Mol Cancer Ther 12(10):1925–1934. https://doi.org/10.1158/1535-7163.MCT-13-0164

Bianchi E, Wright GJ (2015) Cross-species fertilization: the hamster egg receptor, Juno, binds the human sperm ligand, Izumo1. Philos Trans Roy Soc B Biol Sci 370(1661). https://doi.org/10.1098/rstb.2014.0101

Bjerregaard B, Holck S, Christensen IJ, Larsson LI (2006) Syncytin is involved in breast cancer-endothelial cell fusions. Cell Mol Life Sci 63(16). https://doi.org/10.1007/s00018-006-6201-9

Bjerregard B, Ziomkiewicz I, Schulz A, Larsson LI (2014) Syncytin-1 in differentiating human myoblasts: relationship to caveolin-3 and myogenin. Cell Tissue Res 357(1). https://doi.org/10.1007/s00441-014-1930-9

Blond J-L, Besème F, Duret L, Bouton O, Bedin F, Perron H, Mandrand B, Mallet F (1999) Molecular characterization and placental expression of HERV-W, a new human endogenous retrovirus family. J Virol 73(2). https://doi.org/10.1128/jvi.73.2.1175-1185.1999

Bosco EE, Nakai Y, Hennigan RF, Ratner N, Zheng Y (2010) NF2-deficient cells depend on the Rac1-canonical Wnt signaling pathway to promote the loss of contact inhibition of proliferation. Oncogene 29(17). https://doi.org/10.1038/onc.2010.20

Brukman NG, Uygur B, Podbilewicz B, Chernomordik LV (2019) How cells fuse. J Cell Biol 218(5)

Buckingham M (2006) Myogenic progenitor cells and skeletal myogenesis in vertebrates. Curr Opin Genet Dev 16(5)

Burton C, Bartee E (2019) Syncytia formation in oncolytic virotherapy. Mol Ther Oncolytics 15:131–139

Cardelli M, van Doorn R, Larcher L, di Donato M, Piacenza F, Pierpaoli E, Giacconi R, Malavolta M, Rachakonda S, Gruis NA, Molven A, Andresen PA, Pjanova D, van den Oord JJ, Provinciali M, Nagore E, Kumar R (2020) Association of HERV-K and LINE-1 hypomethylation with reduced disease-free survival in melanoma patients. Epigenomics 12(19). https://doi.org/10.2217/epi-2020-0127

Chakraborty A, Lazova R, Davies S, Bäckvall H, Ponten F, Brash D, Pawelek J (2004) Donor DNA in a renal cell carcinoma metastasis from a bone marrow transplant recipient. Bone Marrow Transplant 34(2)

Chalbi M, Barraud-Lange V, Ravaux B, Howan K, Rodriguez N, Soule P, Ndzoudi A, Boucheix C, Rubinstein E, Wolf JP, Ziyyat A, Perez E, Pincet FDR, Gourier C (2014) Binding of sperm protein Izumo1 and its egg receptor Juno drives Cd9 accumulation in the intercellular contact area prior to fusion during mammalian fertilization. Development (Cambridge) 141(19). https://doi.org/10.1242/dev.111534

Chan YW, So C, Yau KL, Chiu KC, Wang X, Chan FL, Tsang SY (2020) Adipose-derived stem cells and cancer cells fuse to generate cancer stem cell-like cells with increased tumorigenicity. J Cell Physiol 235(10). https://doi.org/10.1002/jcp.29574

Chang C, Chen PT, Chang GD, Huang CJ, Chen H (2004) Functional characterization of the placental fusogenic membrane protein syncytin. Biol Reprod 71(6). https://doi.org/10.1095/biolreprod.104.033340

Chen CP, Wang KG, Chen CY, Yu C, Chuang HC, Chen H (2006) Altered placental syncytin and its receptor ASCT2 expression in placental development and pre-eclampsia. BJOG 113(2). https://doi.org/10.1111/j.1471-0528.2005.00843.x

Chernomordik LV, Kozlov MM (2008) Mechanics of membrane fusion. Nat Struct Mol Biol 15(7)

Chitwood CA, Dietzsch C, Jacobs G, McArdle T, Freeman BT, Banga A, Noubissi FK, Ogle BM (2018) Breast tumor cell hybrids form spontaneously in vivo and contribute to breast tumor metastases. APL Bioeng 2(3). https://doi.org/10.1063/1.5024744

Curty G, Marston JL, de Mulder Rougvie M, Leal FE, Nixon DF, Soares MA (2020) Human endogenous retrovirus K in cancer: a potential biomarker and immunotherapeutic target. Viruses 12(7)

Das M, Xu B, Lin L, Chakrabarti S, Shivaswamy V, Rote NS (2004) Phosphatidylserine efflux and intercellular fusion in a BeWo model of human villous cytotrophoblast. Placenta 25(5). https://doi.org/10.1016/j.placenta.2003.11.004

Davies PS, Powell AE, Swain JR, Wong MH (2009) Inflammation and proliferation act together to mediate intestinal cell fusion. PLoS One 4(8). https://doi.org/10.1371/journal.pone.0006530

Degrelle SA, Gerbaud P, Leconte L, Ferreira F, Pidoux G (2017) Annexin-A5 organized in 2D-network at the plasmalemma eases human trophoblast fusion. Sci Rep 7. https://doi.org/10.1038/srep42173

Delidaki M, Gu M, Hein A, Vatish M, Grammatopoulos DK (2011) Interplay of CAMP and MAPK pathways in HCG secretion and fusogenic gene expression in a trophoblast cell line. Mol Cell Endocrinol 332(1–2). https://doi.org/10.1016/j.mce.2010.10.013

Dervan E, Bhattacharyya DD, McAuliffe JD, Khan FH, Glynn SA (2021) Ancient adversary—HERV-K (HML-2) in cancer. Front Oncol 11:658489

di Simone N, Castellani R, Caliandro D, Caruso A (2001) Monoclonal anti-annexin V antibody inhibits trophoblast gonadotropin secretion and induces syncytiotrophoblast apoptosis. Biol Reprod 65(6). https://doi.org/10.1095/biolreprod65.6.1766

Dittmar T, Hass R (2022) Extracellular events involved in cancer cell–cell fusion. Int J Mol Sci 23(24):16071. https://doi.org/10.3390/ijms232416071

Dittmar T, Weiler J, Luo T, Hass R (2021) Cell-cell fusion mediated by viruses and Herv-derived fusogens in cancer initiation and progression. Cancer 13(21)

Dörnen J, Myklebost O, Dittmar T (2020) Cell fusion of mesenchymal stem/stromal cells and breast cancer cells leads to the formation of hybrid cells exhibiting diverse and individual (stem cell) characteristics. Int J Mol Sci 21(24). https://doi.org/10.3390/ijms21249636

Duelli D, Lazebnik Y (2003) Cell fusion: a hidden enemy? Cancer Cell 3(5)

Duelli D, Lazebnik Y (2007) Cell-to-cell fusion as a link between viruses and cancer. Nat Rev Cancer 7(12)

Duelli DM, Hearn S, Myers MP, Lazebnik Y (2005) A primate virus generates transformed human cells by fusion. J Cell Biol 171(3). https://doi.org/10.1083/jcb.200507069

Ehrlich JS, Hansen MDH, Nelson WJ (2002) Spatio-temporal regulation of Rac1 localization and lamellipodia dynamics during epithelial cell-cell adhesion. Dev Cell 3(2). https://doi.org/10.1016/S1534-5807(02)00216-2

Faust JJ, Balabiyev A, Heddleston JM, Podolnikova NP, Page Baluch D, Chew TL, Ugarova TP (2019) An actin-based protrusion originating from a podosome-enriched region initiates macrophage fusion. Mol Biol Cell 30(17). https://doi.org/10.1091/mbc.E19-01-0009

Fei F, Zhang M, Li B, Zhao L, Wang H, Liu L, Li Y, Ding P, Yanjun G, Zhang X, Jiang T, Zhu S, Zhang S (2019) Formation of polyploid giant cancer cells involves in the prognostic value

of neoadjuvant chemoradiation in locally advanced rectal cancer. J Oncol 2019. https://doi.org/10.1155/2019/2316436
Feng H, Hu B, Vuori K, Sarkaria JN, Furnari FB, Cavenee WK, Cheng SY (2014) EGFRvIII Stimulates Glioma Growth and Invasion through PKA-Dependent Serine Phosphorylation of Dock180. Oncogene 33(19):2504–2512. https://doi.org/10.1038/onc.2013.198
Ganem NJ, Storchova Z, Pellman D (2007) Tetraploidy, aneuploidy and cancer. Curr Opin Genet Dev 17(2)
Gao Y, Xiao Fang Y, Chen T (2021) Human endogenous retroviruses in cancer: expression, regulation and function (review). Oncol Lett 21(2)
Gast CE, Silk AD, Zarour L, Riegler L, Burkhart JG, Gustafson KT, Parappilly MS, Roh-Johnson M, Goodman JR, Olson B, Schmidt M, Swain JR, Davies PS, Shasthri V, Iizuka S, Flynn P, Watson S, Korkola J, Courtneidge SA, Fischer JM, Jaboin J, Billingsley KG, Lopez CD, Burchard J, Gray J, Coussens LM, Sheppard BC, Wong MH (2018) Cell fusion potentiates tumor heterogeneity and reveals circulating hybrid cells that correlate with stage and survival. Sci Adv 4(9). https://doi.org/10.1126/sciadv.aat7828
Gerbaud P, Pidoux G (2015) Review: an overview of molecular events occurring in human trophoblast fusion. Placenta 36(S1)
Grandi N, Tramontano E (2018) HERV envelope proteins: physiological role and pathogenic potential in cancer and autoimmunity. Front Microbiol 9:462
Hamilton KE, Noubissi FK, Katti PS, Hahn CM, Davey SR, Lundsmith ET, Klein-Szanto AJ, Rhim AD, Spiegelman VS, Rustgi AK (2013) IMP1 promotes tumor growth, dissemination and a tumor-initiating cell phenotype in colorectal cancer cell xenografts. Carcinogenesis 34(11):2647–2654. https://doi.org/10.1093/carcin/bgt217
Hamoud N, Tran V, Croteau LP, Kania A, Côté JF (2014) G-protein coupled receptor BAI3 promotes myoblast fusion in vertebrates. Proc Natl Acad Sci U S A 111(10). https://doi.org/10.1073/pnas.1313886111
Haneef K, Naeem N, Khan I, Iqbal HAA, Kabir N, Jamall S, Zahid M, Salim A (2014) Conditioned medium enhances the fusion capability of rat bone marrow mesenchymal stem cells and cardiomyocytes. Mol Biol Rep 41(5). https://doi.org/10.1007/s11033-014-3170-1
Hass R, von der Ohe J, Ungefroren H (2020) Impact of the tumor microenvironment on tumor heterogeneity and consequences for cancer cell plasticity and stemness. Cancer 12(12)
Hass R, von der Ohe J, Dittmar T (2021) Hybrid formation and fusion of cancer cells in vitro and in vivo. Cancer 13(17)
Heasman SJ, Ridley AJ (2008) Mammalian Rho GTPases: new insights into their functions from in vivo studies. Nat Rev Mol Cell Biol 9(9)
Helming L, Winter J, Gordon S (2009) The scavenger receptor CD36 plays a role in cytokine-induced macrophage fusion. J Cell Sci 122(4). https://doi.org/10.1242/jcs.037200
Hernández JM, Podbilewicz B (2017) The hallmarks of cell-cell fusion. Development (Cambridge) 144(24)
Hochreiter-Hufford AE, Lee CS, Kinchen JM, Sokolowski JD, Arandjelovic S, Call JA, Klibanov AL, Yan Z, Mandell JW, Ravichandran KS (2013) Phosphatidylserine receptor BAI1 and apoptotic cells as new promoters of myoblast fusion. Nature 497(7448). https://doi.org/10.1038/nature12135
Horsley V, Pavlath GK (2004) Forming a multinucleated cell: molecules that regulate myoblast fusion. Cells Tissues Organs 176(1–3):67–78
Howcroft TK, Zhang HG, Dhodapkar M, Mohla S (2011) Vesicle transfer and cell fusion: emerging concepts of cell-cell communication in the tumor microenvironment. Cancer Biol Ther 12(3):159–164
Hu C, He Y, Liu D, Zhao L, Fang S, Tan B, Dong S, Wang Y, He T, Bi Y (2021) Hypoxia preconditioning promotes the proliferation and migration of human urine-derived stem cells in chronically injured liver of mice by upregulating CXCR4. Stem Cells Dev 30(10). https://doi.org/10.1089/scd.2021.0008

Huang G, Li Z, Wan X, Wang Y, Dong J (2013) Human endogenous retroviral K element encodes fusogenic activity in melanoma cells. J Carcinog 12. https://doi.org/10.4103/1477-3163.109032

Huang CM, Yan TL, Zhi X, Wang M, Zhou XC, Jiang EH, Liu K, Shao Z, Shang ZJ (2018) Hypoxia enhances fusion of oral squamous carcinoma cells and epithelial cells partly via the epithelial-mesenchymal transition of epithelial cells. Biomed Res Int 2018. https://doi.org/10.1155/2018/5015203

Huppertz B, Bartz C, Kokozidou M (2006) Trophoblast fusion: fusogenic proteins, syncytins and ADAMs, and other prerequisites for syncytial fusion. Micron 37(6)

Inoue N, Ikawa M, Isotani A, Okabe M (2005) The immunoglobulin superfamily protein Izumo is required for sperm to fuse with eggs. Nature 434(7030). https://doi.org/10.1038/nature03362

Jeong J, Conboy IM (2011) Phosphatidylserine directly and positively regulates fusion of myoblasts into myotubes. Biochem Biophys Res Commun 414(1). https://doi.org/10.1016/j.bbrc.2011.08.128

Johanning GL, Malouf GG, Zheng X, Esteva FJ, Weinstein JN, Wang-Johanning F, Xiaoping S (2017) Expression of human endogenous retrovirus-K is strongly associated with the basal-like breast cancer phenotype. Sci Rep 7. https://doi.org/10.1038/srep41960

Kato K, Satouh Y, Nishimasu H, Kurabayashi A, Morita J, Fujihara Y, Oji A, Ishitani R, Ikawa M, Nureki O (2016) Structural and functional insights into IZUMO1 recognition by JUNO in mammalian fertilization. Nat Commun 7. https://doi.org/10.1038/ncomms12198

Kerbel RS, Lagarde AE, Dennis JW, Donaghue TP (1983a) Spontaneous fusion in vivo between normal host and tumor cells: possible contribution to tumor progression and metastasis studied with a lectin-resistant mutant tumor. Mol Cell Biol 3(4). https://doi.org/10.1128/mcb.3.4.523-538.1983

Kerbel RS, Tartamella-Biondo R, Cefalu W (1983b) Mutagen treatment as a means for selecting immunogenic variants from otherwise poorly immunogenic malignant murine tumors. Cancer Res 43(1)

Kim JH, Ren Y, Ng WP, Li S, Son S, Kee YS, Zhang S, Zhang G, Fletcher DA, Robinson DN, Chen EH (2015) Mechanical tension drives cell membrane fusion. Dev Cell 32(5). https://doi.org/10.1016/j.devcel.2015.01.005

Krabbe T, Altomonte J (2018) Fusogenic viruses in oncolytic immunotherapy. Cancer 10(7)

LaBerge GS, Duvall E, Grasmick Z, Haedicke K, Pawelek J (2017) A melanoma lymph node metastasis with a donor-patient hybrid genome following bone marrow transplantation: a second case of leucocyte-tumor cell hybridization in cancer metastasis. PLoS One 12(2). https://doi.org/10.1371/journal.pone.0168581

LaBerge G, Duvall E, Grasmick Z, Haedicke K, Galan A, Pawelek J (2021) A melanoma patient with macrophage-cancer cell hybrids in the primary tumor, a lymph node metastasis and a brain metastasis. Cancer Genet 256–257. https://doi.org/10.1016/j.cancergen.2021.05.009

Langbein M, Strick R, Strissel PL, Vogt N, Parsch H, Beckmann MW, Schild RL (2008) Impaired cytotrophoblast cell-cell fusion is associated with reduced syncytin and increased apoptosis in patients with placental dysfunction. Mol Reprod Dev 75(1). https://doi.org/10.1002/mrd.20729

Larizza L, Schirrmacher V (1984) Somatic cell fusion as a source of genetic rearrangement leading to metastatic variants. Cancer Metastasis Rev 3(3). https://doi.org/10.1007/BF00048385

Larsen JM, Christensen IJ, Nielsen HJ, Hansen U, Bjerregaard B, Talts JF, Larsson LI (2009) Syncytin immunoreactivity in colorectal cancer: potential prognostic impact. Cancer Lett 280(1). https://doi.org/10.1016/j.canlet.2009.02.008

Larsson LI, Holck S, Christensen IJ (2007) Prognostic role of syncytin expression in breast cancer. Hum Pathol 38(5). https://doi.org/10.1016/j.humpath.2006.10.018

Lazova R, Chakraborty AK, Pawelek JM (2011) Cancer cell fusion with myeloid cells: implications for energy metabolism in malignant hybrids. In: Larsson L-I (ed) Cell fusions: regulation and control. Springer, Dordrecht

Lee JW, Ryu YK, Ji YH, Kang JH, Moon EY (2015) Hypoxia/Reoxygenation-Experienced Cancer Cell Migration and Metastasis Are Regulated by Rap1-and Rac1-GTPase Activation via the Expression of Thymosin Beta-4. Oncotarget 6(12). https://doi.org/10.18632/oncotarget.3218

Lee X, Keith JC, Stumm N, Moutsatsos I, McCoy JM, Crum CP, Genest D, Chin D, Ehrenfels C, Pijnenborg R, van Assche FA, Mi S (2001) Downregulation of placental syncytin expression and abnormal protein localization in pre-eclampsia. Placenta 22(10). https://doi.org/10.1053/plac.2001.0722

Leikina E, Melikov K, Sanyal S, Verma SK, Eun B, Gebert C, Pfeifer K, Lizunov VA, Kozlov MM, Chernomordik LV (2013) Extracellular annexins and dynamin are important for sequential steps in myoblast fusion. J Cell Biol 200(1). https://doi.org/10.1083/jcb.201207012

Lemaître C, Tsang J, Bireau C, Heidmann T, Dewannieux M (2017) A human endogenous retrovirus-derived gene that can contribute to oncogenesis by activating the ERK pathway and inducing migration and invasion. PLoS Pathog 13(6). https://doi.org/10.1371/journal.ppat.1006451

Levin TG, Powell AE, Davies PS, Silk AD, Dismuke AD, Anderson EC, Swain JR, Wong MH (2010) Characterization of the intestinal cancer stem cell marker CD166 in the human and mouse gastrointestinal tract. Gastroenterology 139(6). https://doi.org/10.1053/j.gastro.2010.08.053

Li J, Lee A (2006) Stress induction of GRP78/BiP and its role in cancer. Curr Mol Med 6(1). https://doi.org/10.2174/156652406775574523

Li H, Yang L, Hui F, Yan J, Wang Y, Guo H, Hao X, Xuehua X, Jin T, Zhang N (2013) Association between Gαi2 and ELMO1/Dock180 connects chemokine signalling with Rac activation and metastasis. Nat Commun 4. https://doi.org/10.1038/ncomms2680

Li H, Feng Z, Tsang TC, Tang T, Jia X, He X, Pennington ME, Badowski MS, Liu AKM, Chen D, Harris DT, Martinez J, Meade-Tollin LC (2014) Fusion of HepG2 cells with mesenchymal stem cells increases cancer-associated and malignant properties: an in vivo metastasis model. Oncol Rep 32(2). https://doi.org/10.3892/or.2014.3264

Li M, Radvanyi L, Yin B, Li J, Chivukula R, Lin K, Yue L, Shen JJ, Chang DZ, Li D, Johanning GL, Wang-Johanning F (2017) Downregulation of human endogenous retrovirus type K (HERV-K) viral Env RNA in pancreatic cancer cells decreases cell proliferation and tumor growth. Clin Cancer Res 23(19). https://doi.org/10.1158/1078-0432.CCR-17-0001

Li M, Radvanyi L, Yin B, Rycaj K, Li J, Chivukula R, Lin K, Yue L, Shen JJ, Chang DZ, Li D, Johanning GL, Wang-Johanning F (2019) Correction: downregulation of human endogenous retrovirus type K (HERV-K) viral Env RNA in pancreatic cancer cells decreases cell proliferation and tumor growth. Clin Cancer Res 25(9). https://doi.org/10.1158/1078-0432.ccr-19-0700

Liu TC, Kirn D (2007) Systemic efficacy with oncolytic virus therapeutics: clinical proof-of-concept and future directions. Cancer Res 67(2)

Lu X, Kang Y (2009) Efficient acquisition of dual metastasis organotropism to bone and lung through stable spontaneous fusion between MDA-MB-231 variants. Proc Natl Acad Sci U S A 106(23). https://doi.org/10.1073/pnas.0900108106

Lu M, Ravichandran KS (2006) Dock180-ELMO cooperation in Rac activation. Methods Enzymol 406:388–402

Ma W, Hong Z, Liu H, Xi Chen L, Ding ZL, Zhou F, Yuan Y (2016) Human endogenous retroviruses-k (HML-2) expression is correlated with prognosis and progress of hepatocellular carcinoma. Biomed Res Int 2016. https://doi.org/10.1155/2016/8201642

Malassiné A, Handschuh K, Tsatsaris V, Gerbaud P, Cheynet V, Oriol G, Mallet F, Evain-Brion D (2005) Expression of HERV-W Env glycoprotein (syncytin) in the extravillous trophoblast of first trimester human placenta. Placenta 26(7). https://doi.org/10.1016/j.placenta.2004.09.002

Mandel K, Yang Y, Schambach A, Glage S, Otte A, Hass R (2013) Mesenchymal stem cells directly interact with breast cancer cells and promote tumor cell growth in vitro and in vivo. Stem Cells Dev 22(23). https://doi.org/10.1089/scd.2013.0249

Maretzky T, Reiss K, Ludwig A, Buchholz J, Scholz F, Proksch E, de Strooper B, Hartmann D, Saftig P (2005) ADAM10 mediates E-cadherin shedding and regulates epithelial cell-cell adhesion, migration, and β-catenin translocation. Proc Natl Acad Sci U S A 102(26). https://doi.org/10.1073/pnas.0500918102

Matsuura K, Jigami T, Taniue K, Morishita Y, Adachi S, Senda T, Nonaka A, Aburatani H, Nakamura T, Akiyama T (2011) Identification of a link between Wnt/I 2-catenin signalling and the cell fusion pathway. Nat Commun 2(1). https://doi.org/10.1038/ncomms1551

Melzer C, von der Ohe J, Hass R (2018a) Enhanced metastatic capacity of breast cancer cells after interaction and hybrid formation with mesenchymal stroma/stem cells (MSC). Cell Commun Signal 16(1). https://doi.org/10.1186/s12964-018-0215-4

Melzer C, von der Ohe J, Hass R (2018b) MSC stimulate ovarian tumor growth during intercellular communication but reduce tumorigenicity after fusion with ovarian cancer cells. Cell Commun Signal 16(1). https://doi.org/10.1186/s12964-018-0279-1

Melzer C, von der Ohe J, Hass R (2019) Involvement of actin cytoskeletal components in breast cancer cell fusion with human mesenchymal stroma/stem-like cells. Int J Mol Sci 20(4). https://doi.org/10.3390/ijms20040876

Melzer C, von der Ohe J, Hass R (2020) Altered tumor plasticity after different cancer cell fusions with MSC. Int J Mol Sci 21(21)

Melzer C, von der Ohe J, Luo T, Hass R (2021) Spontaneous fusion of Msc with breast cancer cells can generate tumor dormancy. Int J Mol Sci 22(11). https://doi.org/10.3390/ijms22115930

Millay DP, Olson EN (2013) Making muscle or mitochondria by selective splicing of PGC-1α. Cell Metab 17(1)

Misra UK, Deedwania R, Pizzo SV (2006) Activation and cross-talk between Akt, NF-KB, and unfolded protein response signaling in 1-LN prostate cancer cells consequent to ligation of cell surface-associated GRP78. J Biol Chem 281(19). https://doi.org/10.1074/jbc.M511694200

Mohr M, Tosun S, Arnold WH, Edenhofer F, Zänker KS, Dittmar T (2015) Quantification of cell fusion events human breast cancer cells and breast epithelial cells using a Cre-LoxP-based double fluorescence reporter system. Cell Mol Life Sci 72(19). https://doi.org/10.1007/s00018-015-1910-6

Muir A, Lever AML, Moffett A (2006) Human endogenous retrovirus-W envelope (Syncytin) is expressed in both villous and extravillous trophoblast populations. J Gen Virol 87(7). https://doi.org/10.1099/vir.0.81412-0

Ni M, Zhang Y, Lee AS (2011) Beyond the endoplasmic reticulum: atypical GRP78 in cell viability, signalling and therapeutic targeting. Biochem J 434(2)

Noubissi FK, Ogle BM (2016) Cancer cell fusion: mechanisms slowly unravel. Int J Mol Sci 17(9)

Noubissi FK, Harkness T, Alexander CM, Ogle BM (2015) Apoptosis-induced cancer cell fusion: a mechanism of breast cancer metastasis. FASEB J 29(9). https://doi.org/10.1096/fj.15-271098

Ogle BM, Butters KA, Plummer TB, Ring KR, Knudsen BE, Litzow MR, Cascalho M, Platt JL (2004) Spontaneous fusion of cells between species yields transdifferentiation and retroviral transfer in vivo. FASEB J 18(3). https://doi.org/10.1096/fj.03-0962fje

Ogle BM, Cascalho M, Platt JL (2005) Biological implications of cell fusion. Nat Rev Mol Cell Biol 6(7)

Overholtzer M, Mailleux AA, Mouneimne G, Normand G, Schnitt SJ, King RW, Cibas ES, Brugge JS (2007) A nonapoptotic cell death process, entosis, that occurs by cell-in-cell invasion. Cell 131(5). https://doi.org/10.1016/j.cell.2007.10.040

Pajcini KV, Pomerantz JH, Alkan O, Doyonnas R, Blau HM (2008) Myoblasts and macrophages share molecular components that contribute to cell-cell fusion. J Cell Biol 180(5). https://doi.org/10.1083/jcb.200707191

Park D, Ravichandran KS (2010) Emerging roles of brain-specific angiogenesis inhibitor 1. Adv Exp Med Biol 706. https://doi.org/10.1007/978-1-4419-7913-1_15

Park SY, Yun Y, Lim JS, Kim MJ, Kim SY, Kim JE, Kim IS (2016) Stabilin-2 modulates the efficiency of myoblast fusion during myogenic differentiation and muscle regeneration. Nat Commun 7. https://doi.org/10.1038/ncomms10871

Pawelek JM (2014) Fusion of bone marrow-derived cells with cancer cells: metastasis as a secondary disease in cancer. Chin J Cancer 33(3). https://doi.org/10.5732/cjc.013.10243

Pawelek JM, Chakraborty AK (2008) Chapter 10 The cancer cell-leukocyte fusion theory of metastasis. Adv Cancer Res 101:397–444

Pérez-Vargas J, Krey T, Valansi C, Avinoam O, Haouz A, Jamin M, Raveh-Barak H, Podbilewicz B, Rey FA (2014) Structural basis of eukaryotic cell-cell fusion. Cell 157(2). https://doi.org/10.1016/j.cell.2014.02.020

Pérot P, Montgiraud C, Lavillette D, Mallet F (2011) A comparative portrait of retroviral fusogens and syncytins. In: Larsson L-I (ed) Cell fusions: regulation and control. Springer, Dordrecht

Podbilewicz B, Leikina E, Sapir A, Valansi C, Suissa M, Shemer G, Chernomordik LV (2006) The C. elegans developmental fusogen EFF-1 mediates homotypic fusion in heterologous cells and in vivo. Dev Cell 11(4). https://doi.org/10.1016/j.devcel.2006.09.004

Quinn ME, Goh Q, Kurosaka M, Gamage DG, Petrany MJ, Prasad V, Millay DP (2017) Myomerger induces fusion of non-fusogenic cells and is required for skeletal muscle development. Nat Commun 8. https://doi.org/10.1038/ncomms15665

Rachkovsky M, Sodi S, Chakraborty A, Avissar Y, Bolognia J, McNiff JM, Platt J, Bermudes D, Pawelek J (1998) Melanoma x macrophage hybrids with enhanced metastatic potential. Clin Exp Metastasis 16(4). https://doi.org/10.1023/A:1006557228604

Ravichandran KS (2010) Find-me and eat-me signals in apoptotic cell clearance: progress and conundrums. J Exp Med 207(9)

Rival CM, Wenhao X, Shankman LS, Morioka S, Arandjelovic S, Lee CS, Wheeler KM, Smith RP, Haney LB, Isakson BE, Purcell S, Lysiak JJ, Ravichandran KS (2019) Phosphatidylserine on viable sperm and phagocytic machinery in oocytes regulate mammalian fertilization. Nat Commun 10(1). https://doi.org/10.1038/s41467-019-12406-z

Rizvi AZ, Swain JR, Davies PS, Bailey AS, Decker AD, Willenbring H, Grompe M, Fleming WH, Wong MH (2006) Bone marrow-derived cells fuse with normal and transformed intestinal stem cells. Proc Natl Acad Sci U S A 103(16). https://doi.org/10.1073/pnas.0508593103

Rosenbaum S, Kreft S, Etich J, Frie C, Stermann J, Grskovic I, Frey B, Mielenz D, Pöschl E, Gaipl U, Paulsson M, Brachvogel B (2011) Identification of novel binding partners (annexins) for the cell death signal phosphatidylserine and definition of their recognition motif. J Biol Chem 286(7). https://doi.org/10.1074/jbc.M110.193086

Saci A, Cantley LC, Carpenter CL (2011) Rac1 regulates the activity of MTORC1 and MTORC2 and controls cellular size. Mol Cell 42(1). https://doi.org/10.1016/j.molcel.2011.03.017

Seyfried TN (2012) Cancer as a metabolic disease: on the origin, management and prevention of cancer. Wiley, Hoboken

Seyfried TN, Huysentruyt LC (2013) On the origin of cancer metastasis. Crit Rev Oncog 18(1–2). https://doi.org/10.1615/CritRevOncog.v18.i1-2.40

Sha M, Lee X, Li XP, Veldman GM, Finnerty H, Racie L, LaVallie E, Tang XY, Edouard P, Howes S, Keith JC, McCoy JM (2000) Syncytin is a captive retroviral envelope protein involved in human placental morphogenesis. Nature 403(6771). https://doi.org/10.1038/35001608

Shackney SE, Smith CA, Miller BW, Burholt DR, Murtha K, Pollice AA, Giles HR, Ketterer DM (1989) Model for the genetic evolution of human solid tumors. Cancer Res 49(12)

Sharma B, Kanwar SS (2018) Phosphatidylserine: a cancer cell targeting biomarker. Semin Cancer Biol 52:17–25

Skokos EA, Charokopos A, Khan K, Wanjala J, Kyriakides TR (2011) Lack of TNF-α-induced MMP-9 production and abnormal E-cadherin redistribution associated with compromised fusion in MCP-1-null macrophages. Am J Pathol 178(5). https://doi.org/10.1016/j.ajpath.2011.01.045

Song K, Zhu F, Zhang HZ, Shang ZJ (2012) Tumor necrosis factor-α enhanced fusions between oral squamous cell carcinoma cells and endothelial cells via VCAM-1/VLA-4 pathway. Exp Cell Res 318(14). https://doi.org/10.1016/j.yexcr.2012.05.022

Strick R, Ackermann S, Langbein M, Swiatek J, Schubert SW, Hashemolhosseini S, Koscheck T, Fasching PA, Schild RL, Beckmann MW, Strissel PL (2007) Proliferation and cell-cell fusion of endometrial carcinoma are induced by the human endogenous retroviral syncytin-1 and regulated by TGF-β. J Mol Med 85(1). https://doi.org/10.1007/s00109-006-0104-y

Strick R, Beckmann MW, Strissel PL (2011) Cell-cell fusions and human endogenous retroviruses in cancer. In: Larsson L-I (ed) Cell fusions: regulation and control. Springer, Dordrecht

Takezawa Y, Yoshida K, Miyado K, Sato M, Nakamura A, Kawano N, Sakakibara K, Kondo T, Harada Y, Ohnami N, Kanai S, Miyado M, Saito H, Takahashi Y, Akutsu H, Umezawa A (2011) β-catenin is a molecular switch that regulates transition of cell-cell adhesion to fusion. Sci Rep 1. https://doi.org/10.1038/srep00068

Takito J, Otsuka H, Inoue S, Kawashima T, Nakamura M (2017) Symmetrical retrograde actin flow in the actin fusion structure is involved in osteoclast fusion. Biol Open 6(7). https://doi.org/10.1242/bio.025460

Uygur B, Leikina E, Melikov K, Villasmil R, Verma SK, Vary CPH, Chernomordik. LV (2019) Interactions with muscle cells boost fusion, stemness, and drug resistance of prostate cancer cells. Mol Cancer Res 17(3). https://doi.org/10.1158/1541-7786.MCR-18-0500

van den Eijnde SM, van den Hoff MJB, Reutelingsperger CPM, van Heerde WL, Henfling MER, Vermeij-Keers C, Schutte B, Borgers M, Ramaekers FCS (2001) Transient expression of phosphatidylserine at cell-cell contact areas is required for myotube formation. J Cell Sci 114(20). https://doi.org/10.1242/jcs.114.20.3631

Verma SK, Leikina E, Melikov K, Gebert C, Kram V, Young MF, Uygur B, Chernomordik LV (2018) Cell-surface phosphatidylserine regulates osteoclast precursor fusion. J Biol Chem 293(1). https://doi.org/10.1074/jbc.M117.809681

Vignery A (2005) Macrophage fusion: are somatic and cancer cells possible partners? Trends Cell Biol 15(4). https://doi.org/10.1016/j.tcb.2005.02.008

Walker LC, Spurdle AB (2010) Prioritizing candidate genetic modifiers of BRCA1 and BRCA2 using a combinatorial analysis of global expression and polymorphism association studies of breast cancer. Methods Mol Biol (Clifton, N.J.) 653:23–34

Wang Q, Hardie RA, Hoy AJ, van Geldermalsen M, Gao D, Fazli L, Sadowski MC, Balaban S, Schreuder M, Nagarajah R, Wong JJL, Metierre C, Pinello N, Otte NJ, Lehman ML, Gleave M, Nelson CC, Bailey CG, Ritchie W, Rasko JE, Holst J (2015) Targeting ASCT2-mediated glutamine uptake blocks prostate cancer growth and tumour development. J Pathol 236(3). https://doi.org/10.1002/path.4518

Wang Z, Yuan Y, Zhang L, Min Z, Zhou D, Sun Y, Wang P, Songguang J, Jun L, Jinxiang F (2018) Impact of cell fusion in myeloma marrow microenvironment on tumor progression. Oncotarget 9(57). https://doi.org/10.18632/oncotarget.25742

Weiler J, Dittmar T (2019) Minocycline impairs TNF-α-induced cell fusion of M13SV1-Cre cells with MDA-MB-435-PFDR1 cells by suppressing NF-KB transcriptional activity and its induction of target-gene expression of fusion-relevant factors. Cell Commun Signal 17(1). https://doi.org/10.1186/s12964-019-0384-9

Weiler J, Mohr M, Zänker KS, Dittmar T (2018) Matrix metalloproteinase-9 (MMP9) is involved in the TNF-α-induced fusion of human M13SV1-Cre breast epithelial cells and human MDA-MB-435-PFDR1 cancer cells. Cell Commun Signal 16(1). https://doi.org/10.1186/s12964-018-0226-1

Whitlock JM, Chernomordik LV (2021) Flagging fusion: phosphatidylserine signaling in cell–cell fusion. J Biol Chem 296:100411

Yabluchanskiy A, Ma Y, Iyer RP, Hall ME, Lindsey ML (2013) Matrix metalloproteinase-9: many shades of function in cardiovascular disease. Physiology 28(6)

Yan TL, Wang M, Zhi X, Huang CM, Zhou XC, Jiang EH, Zhao XP, Song Y, Song K, Shao Z, Liu K, Shang ZJ (2017) Up-regulation of syncytin-1 contributes to TNF-α-enhanced fusion between OSCC and HUVECs partly via Wnt/β-catenin-dependent pathway. Sci Rep 7. https://doi.org/10.1038/srep40983

Yang Y, Otte A, Hass R (2015) Human mesenchymal stroma/stem cells exchange membrane proteins and alter functionality during interaction with different tumor cell lines. Stem Cells Dev 24(10). https://doi.org/10.1089/scd.2014.0413

Yart L, Bastida-Ruiz D, Wuillemin C, Ribaux P, Allard M, Dietrich PY, Petignat P, Cohen M (2020) Linking endoplasmic reticulum stress to polyploidy in ovarian cancer cells. BioRxiv. https://doi.org/10.1101/2020.01.24.918029

Yart L, Bastida-Ruiz D, Allard M, Dietrich PY, Petignat P, Cohen M (2022) Linking unfolded protein response to ovarian cancer cell fusion. BMC Cancer 22(1). https://doi.org/10.1186/s12885-022-09648-4

Yilmaz Y, Lazova R, Qumsiyeh M, Cooper D, Pawelek J (2005) Donor Y chromosome in renal carcinoma cells of a female BMT recipient: visualization of putative BMT—tumor hybrids by FISH [2]. Bone Marrow Transplant 35(10)

Zhang S, Mercado-Uribe I, Xing Z, Sun B, Kuang J, Liu J (2014) Generation of cancer stem-like cells through the formation of polyploid giant cancer cells. Oncogene 33(1). https://doi.org/10.1038/onc.2013.96

Zhang LN, Huang YH, Zhao L (2019a) Fusion of macrophages promotes breast cancer cell proliferation, migration and invasion through activating epithelial-mesenchymal transition and Wnt/β-catenin signaling pathway. Arch Biochem Biophys 676. https://doi.org/10.1016/j.abb.2019.108137

Zhang LN, Kong CF, Zhao D, Cong XL, Wang SS, Ma L, Huang YH (2019b) Fusion with mesenchymal stem cells differentially affects tumorigenic and metastatic abilities of lung cancer cells. J Cell Physiol 234(4). https://doi.org/10.1002/jcp.27011

Zhou X, Platt JL (2011) Molecular and cellular mechanisms of mammalian cell fusion. Adv Exp Med Biol 713. https://doi.org/10.1007/978-94-007-0763-4_4

Zhou Y, Mohan A, Moore DC, Lin L, Zhou FL, Cao J, Qian W, Qin YX, Reginato AM, Ehrlich MG, Yang W (2015) SHP2 regulates osteoclastogenesis by promoting preosteoclast fusion. FASEB J 29(5)

Zhou F, Li M, Wei Y, Lin K, Yue L, Shen J, Johanning GL, Wang-Johanning F (2016) Activation of HERV-K Env protein is essential for tumorigenesis and metastasis of breast cancer cells. Oncotarget 7(51). https://doi.org/10.18632/oncotarget.11455

Zhou X, Zhou M, Zheng M, Tian S, Yang X, Ning Y, Li Y, Zhang S (2022) Polyploid giant cancer cells and cancer progression. Front Cell Dev Biol 10:1017588

Žigon P, Pirkmajer KP, Tomšič M, Kveder T, Božič B, Šemrl SS, Čučnik S, Ambrožič A (2015) Anti-phosphatidylserine/prothrombin antibodies are associated with adverse pregnancy outcomes. J Immunol Res. https://doi.org/10.1155/2015/975704

Chapter 20
Cell Fusion and Syncytia Formation in Cancer

Mareike Sieler and Thomas Dittmar

Abstract The natural phenomenon of cell–cell fusion does not only take place in physiological processes, such as placentation, myogenesis, or osteoclastogenesis, but also in pathophysiological processes, such as cancer. More than a century ago postulated, today the hypothesis that the fusion of cancer cells with normal cells leads to the formation of cancer hybrid cells with altered properties is in scientific consensus. Some studies that have investigated the mechanisms and conditions for the fusion of cancer cells with other cells, as well as studies that have characterized the resulting cancer hybrid cells, are presented in this review. Hypoxia and the cytokine TNFα, for example, have been found to promote cell fusion. In addition, it has been found that both the protein Syncytin-1, which normally plays a role in placentation, and phosphatidylserine signaling on the cell membrane are involved in the fusion of cancer cells with other cells. In human cancer, cancer hybrid cells were detected not only in the primary tumor, but also in the circulation of patients as so-called circulating hybrid cells, where they often correlated with a worse outcome. Although some data are available, the questions of how and especially why cancer cells fuse with other cells are still not fully answered.

Abbreviations

ASCT2	Alanine serine cysteine transporter 2
BiFC	Bimolecular fluorescence complementation
BMDC	Bone marrow-derived cell
BMT	Bone marrow transplantation
CAF	Cancer-associated fibroblast
CCC	Circulating cancer cell

M. Sieler (✉) · T. Dittmar
Institute of Immunology, Center for Biomedical Education and Research (ZBAF), University of Witten/Herdecke, Witten, Germany
e-mail: mareike.sieler@uni-wh.de; Thomas.Dittmar@uni-wh.de

M. Kloc, A. Uosef (eds.), *Syncytia: Origin, Structure, and Functions*, Results and Problems in Cell Differentiation 71,
https://doi.org/10.1007/978-3-031-37936-9_20

CHC	Circulating hybrid cell
CRLM	Colorectal liver metastasis
CS/IC	Cancer stem/initiating cell
CTC	Circulating tumor cell
DC-STAMP	Dendritic cell-specific transmembrane protein
DSP	Dual split protein
DSS	Disease-specific survival
EAC	Esophageal adenocarcinoma
EC	Endometrial carcinoma
FDR	Fluorescence double reporter
g-aCTC	Giant atypical circulating tumor cell
GFP	Green fluorescent protein
GPI	Glycosylphosphatidylinositol
HCC	Hepatocellular carcinoma
HE	Hematoxylin-eosin
HERV	Human endogenous retrovirus
HIF-1α	Hypoxia inducible factors-1α
HST/PR	Heterokaryon-to-synkaryon transition/ploidy reduction
HTR2B	5-hydroxytryptamine receptor 2B
HUVEC	Human umbilical vein endothelial cell
IHC	Immunohistochemistry
iMSC	Immortalized mesenchymal stem cell
mRNA	Messenger ribonucleic acid
MSC	Mesenchymal stem cell
NAT	Neoadjuvant therapy
NK cells	Natural killer cells
NSCLC	Non-small cell lung cancer
PANK/KRT/CK	Pan cytokeratin
pCR	Pathologic complete response
PGCC	Polyploid giant cancer cell
PHSP	Post-hybrid selection process
PS	Phosphatidylserine
RAC	Rectal adenocarcinoma
RANKL	Receptor activator of NF-κB ligand
RFP	Red fluorescent protein
RT-PCR	Real-time polymerase chain reaction
SCC-9	Squamous cell carcinoma 9
shRNA	Small hairpin ribonucleic acid
siRNA	Small interfering ribonucleic acid
STR	Short tandem repeat
TAM	Tumor-associated macrophage
TMEM16	Transmembrane member 16
UCC	Urothelial cell carcinoma

20.1 Introduction

Cell fusion is a biological mechanism crucial for physiological processes, such as fertilization and placentation, wound healing, and tissue regeneration as well as myogenesis and osteoclastogenesis (Aguilar et al. 2013; Dittmar and Zänker 2011a; Zhou and Platt 2011; Brukman et al. 2019; Blond et al. 1999). In contrast, it is known that cell fusion plays a role in pathophysiological processes, like viral infections or tumor development (Dittmar and Zänker 2011b; Weiler and Dittmar 2019a; Dittmar et al. 2021; Manjunath et al. 2020a). Although the fusion of cells with other cells seems to be easy at first glance, there are many proteins and signaling molecules necessary to regulate cell fusion and until now the whole process is mechanistically not fully understood (Zhou and Platt 2011; Ogle et al. 2005). Cells which are known for their fusogenicity, like cytotrophoblasts (Vargas et al. 2009; Song et al. 2021a; Msheik et al. 2019; Gauster et al. 2009) or myoblasts (Lehka and Redowicz 2020; Eigler et al. 2021; Pircher et al. 2022), are not fusogenic per se and need to enter a profusion state first (Podbilewicz 2014). Then, after the energy consuming actual fusion step, the resulting hybrid cell either stays in a multinucleated state, like syncytiotrophoblasts and myoblasts, or needs to undergo a post-hybrid selection process (PHSP) to achieve genomic stability and a structured cell metabolism to survive (Hass et al. 2021a; Hernandez and Podbilewicz 2017). Cell–cell fusion in a cancer context is believed to result in cancer hybrid cells exposing new malignant properties, like a higher tumorigenic or metastatic potential, an enhanced drug and radiation resistance as well as cancer recurrence (Lindström et al. 2017; Powell et al. 2011; Rachkovsky et al. 1998; Dittmar et al. 2009, 2011; He et al. 2015; Melzer et al. 2018a; Wang et al. 2012; Gast et al. 2018; Pawelek 2014; Wang et al. 2018; Clawson et al. 2015; Lartigue et al. 2020; Rubio et al. 2022; Merle et al. 2021; Yart et al. 2022; Montalban-Hernandez et al. 2022). In this review the current knowledge about the mechanisms of cell–cell fusion, which proteins and mediators are involved and the properties of the resulting multinucleated cells, called syncytia, is discussed.

20.2 How Does Cell–Cell Fusion Work and Which Proteins/ Mediators Are Involved?

Cell–cell fusion is a strictly regulated and energy-consuming process which can be divided into different steps, called the hallmarks of cell–cell fusion (Podbilewicz 2014; Hernandez and Podbilewicz 2017). Since the membranes of two cells are strongly repelled by hydrostatic forces there are several intermediate states that need to be overcome, namely the dehydration of the membrane lipid's polar head groups, the promotion of a hemifusion stalk, and the opening and expansion of a pore between the fusing cells. The soluble contents of the cells are mixed, and a hybrid cell is formed, which can further fuse with other cells and build up syncytia.

Different triggers are capable of getting cells into a fusion competent state and inducing cell–cell fusion. Inflammation and inflammatory cytokines are known mediators for cell–cell fusion (Melzer et al. 2018b; Yan et al. 2017; Davies et al. 2009; Weiler and Dittmar 2019b; Song et al. 2012; Mohr et al. 2015; Skokos et al. 2011; Hotokezaka et al. 2007) and are found in the environment of tumors, since they are assumed as wounds that do not heal (Dvorak 1986, 2015; Ribatti and Tamma 2018; Balkwill and Mantovani 2001). In addition, viruses are known to induce cell–cell fusion by the expression of virus-specific proteins, like *env* elements, in the targeted cell (Duelli and Lazebnik 2007). Those proteins have fusogenic properties of which advantage is taken in the development of oncolytic viruses (Burton and Bartee 2019; Matveeva and Shabalina 2020; Krabbe and Altomonte 2018; Del Papa et al. 2021). The effects of pH and different ions in cell–cell fusion are also discussed, since they have an impact on phospholipid scramblases (Liang and Yang 2021). Another trigger of cell fusion is hypoxia, which also is a characteristic feature of the tumor microenvironment (Petrova et al. 2018; Emami Nejad et al. 2021; Li et al. 2021; Eltzschig and Carmeliet 2011; Tinganelli and Durante 2020). It is known that hypoxia activates the transcription factor hypoxia-inducible factors-1α (HIF-1α), which induces the transcription of hypoxia-associated proteins (Petrova et al. 2018; Emami Nejad et al. 2021; Li et al. 2021; Eltzschig and Carmeliet 2011; Tinganelli and Durante 2020). Several studies showed that hypoxia has different impacts on the protein expression and cell–cell fusion of different kind of cells (Yart et al. 2022; Archacka et al. 2021; Huang et al. 2018; Kudo et al. 2003), but what exactly happens and how fusogenic proteins are (in-)activated remain to be elucidated. As already mentioned, after getting the cells in a pro-fusogenic state the fusion process needs to be catalyzed by viral or intracellular proteins to overcome the energetic barrier between the membranes (Brukman et al. 2019). For example, for the fusion of gametes in mammals, the conserved immunoglobulin superfamily member Izumo1 in sperms (Inoue et al. 2005) as well as the glycosylphosphatidylinositol (GPI)-anchored Izumo1 receptor Juno in eggs (Bianchi et al. 2014) is crucial. Studies showed that human infertility can be caused by female antibodies against Izumo1 or mutations in Juno (Clark and Naz 2013; Yu et al. 2018). Another well-known protein playing a role in the reproduction of mammals is the human endogenous retroviral (HERV) envelope protein Syncytin-1 (ERV-W), which mediates the fusion of villous cytotrophoblasts to form syncytiotrophoblasts, serving as fetomaternal barrier, during pregnancy (Potgens et al. 2004; Bolze et al. 2017; Soygur and Sati 2016; Muir et al. 2006; Malassine et al. 2005; Mi et al. 2000). A decreased expression of Syncytin-1 in pregnancy can lead to fetal growth restriction (Wang et al. 2022) or preeclampsia (Ruebner et al. 2013; Huang et al. 2014; Knerr et al. 2004). For the regulation of Syncytin-1, an antagonist protein, named Suppressyn, which can bind to Syncytin's receptor alanine serine cysteine transporter (ASCT) 2 and therefore inhibit trophoblast syncytialization, was found (Sugimoto et al. 2013, 2019). Likewise, Syncytin-2 (ERV-FRD), another member of the HERV family, was also shown to play a role in the development of a functional placenta (Vargas et al. 2009; Chen et al. 2008; Soygur et al. 2016). HERV elements in general make up 8% of the human genome and although the majority of them have become inactive due to mutations or silencing (Xue et al. 2020; Garcia-Montojo et al. 2018; Gao et al. 2021; Dervan et al. 2021; de Parseval and Heidmann 2005; Curty et al. 2020), viral infections can increase or reactivate, for example, the expression of Syncytins (Marston et al. 2021; Uleri et al. 2014; Mameli et al. 2012; Liu et al. 2017) or the most transcriptionally active HERV element HERV K, which was found to be expressed in different cancers (Dai et al. 2018; Zhou et al. 2016; Li et al. 2017). Furthermore, a variety of intracellular proteins are known to mediate cell–cell fusion crucial for building up and maintenance of body tissues. The fusion of myoblasts to form muscle fibers depends on the two muscle-specific proteins Myomaker and Myomerger (Myomixer/Minion) (Millay et al. 2013; Bi et al. 2017; Gamage et al. 2017; Quinn et al. 2017; Leikina et al. 2018; Zhang et al. 2017); the expression of the both proteins on fibroblasts is sufficient to induce cell–cell fusion. Additionally, osteoclasts resulting from Receptor Activator of NF-κB Ligand (RANKL)-stimulated cell–cell fusion of monocyte progenitors are crucial for bone remodeling during bone development and maintenance since they are capable of resorbing bone (Levaot et al. 2015). It was shown that the formation of multinucleated osteoclasts leads to more efficient bone resorption compared to bone resorption by mononucleated cells (Lees and Heersche 1999). Proteins involved in cell–cell fusion of monocyte progenitors to form osteoclasts are dendritic cell-specific transmembrane protein (DC-STAMP), CD36, CD44, CD47, and CD200, whose expression decreases as the osteoclast gains nuclei (Soe 2020; Helming and Gordon 2009; Helming et al. 2009; Fang et al. 2020; Vignery 2005). For further fusion of multinucleated osteoclasts, Syncytin-1 was found to be involved (Soe et al. 2011; Moller et al. 2017). A dysregulation of osteoclast fusion was connected to osteopetrosis (Yagi et al. 2005) or osteoporosis (Mizuno et al. 1998). In addition to fusogenic proteins, cell fusion depends on the reorganization of the actin skeleton, proteases, chemokines, and cytokines (Aguilar et al. 2013; Brukman et al. 2019; Dittmar et al. 2021; Podbilewicz 2014; Hass et al. 2021a, b; Hernandez and Podbilewicz 2017; Helming and Gordon 2009; Abmayr and Pavlath 2012; Martens and McMahon 2008; Petrany and Millay 2019) as well as the crosstalk with plasma membrane phospholipids and their binding proteins, like phosphatidylserine (PS) and PS-binding annexins (Aguilar et al. 2013; Zhou and Platt 2011; Brukman et al. 2019; Ogle et al. 2005; Hernandez and Podbilewicz 2017; Liang and Yang 2021; Whitlock and Chernomordik 2021; Whitlock et al. 2018; Zhang et al. 2020, 2022; Sharma and Kanwar 2018; Melzer et al. 2019a; Bevers and Williamson 2016). Viable cells show an asymmetric distribution of phospholipids between the inner and outer leaflets of the membrane, whereas PS is usually located in the inner leaflet of the membrane and its translocation to the outer leaflet is associated with apoptosis (Martin et al. 1995). Translocation of PS to the outer leaflet is mediated by phospholipid scramblases, such as members of the transmembrane member 16 (TMEM16) family (Suzuki et al. 2010, 2013; Gyobu et al. 2017). Targeting of TMEM16 family members or their receptors leads to failure in fusions taking place in myogenesis, osteoclastogenesis, and syncytialization as well as a diminished entry of HIV into host cells and syncytia formation induced by SARS-CoV-2 spike protein (Zhang et al. 2020; Verma et al. 2018; Braga et al. 2021; Zaitseva et al. 2017;

Hochreiter-Hufford et al. 2013). In addition, a mutation in TMEMF16 causing its inactivity is related to a rare bleeding disease, named Scott syndrome (Suzuki et al. 2010). In summary, cell–cell fusion is dependent not only on fusogenic proteins, but also on multiple factors, including further proteins, conditions given by the microenvironment of the cells and the composition of the cell membrane.

20.3 Does Cell–Cell Fusion Naturally Occur in Cancer?

The German physician Otto Aichel already proposed in 1911 that the formation of malignant tumors could be attributed to the fusion of tumor cells with infiltrating leukocytes (Aichel 1911). He also assumed that cell–cell fusion could be an explanation for aneuploidy and that the altered chromosomal content could lead to a metastatic phenotype in the hybrid cell. Indeed, in current studies ploidy abnormalities of cells are assumed as a hallmark of cancer and are found in approximately 30 % of human tumors (Matsumoto et al. 2021; Zack et al. 2013; Bielski et al. 2018; Quinton et al. 2021; Dornen et al. 2020a). In addition, the existence of cancer hybrid cells with altered metastatic capability, drug resistance, and cancer stem/initiating cell (CS/IC) characteristics was proven in a plethora of in vitro and vivo studies (for review, see Dittmar et al. 2021; Manjunath et al. 2020a; Hass et al. 2021a, b; Wang et al. 2021). Until today, it is not clear how the fusion process of cancer cells with other cells is directed and why cancer cells fuse at all. Physiological fusion processes are tightly regulated and depend on the expression of specific fusogens and their affiliated receptors, while in cancer cells the distribution of fusogens is highly heterogeneous. Likewise, the fusion rate or the amount of detected cancer hybrid cells, respectively, varied between 0.0066 and 6.5% in in vitro studies (Miroshnychenko et al. 2021; Wakeling et al. 1994; Fortuna et al. 1989; Lu and Kang 2009; Yan et al. 2016) and between 0.5% and up to 51% in vivo (Powell et al. 2011; Gast et al. 2018; Miroshnychenko et al. 2021; Yan et al. 2016; Melzer et al. 2019b; Ramakrishnan et al. 2013; Rizvi et al. 2006). The values of these studies should only be compared with caution, since they differ not only in the experimental setup and use of different models and cancer types, but also in the read-out method as well as in the time point for analysis. A late time point for detection might not refer to the fusion rate of cancer cells but show the proliferation of stable hybrid cells. One possible influence on the formation of cancer hybrid cells may be chemotherapy by favoring the formation of therapy-resistant hybrid cells or promoting the proliferation of already existing hybrid cells. In a study published by Yan and colleagues, it was shown that the use of a chemotherapeutic agent in a xenograft mouse experiment can almost double the amount of tumor hybrid cells in a solid tumor (Yan et al. 2016). In addition, they observed that after the administration of the chemotherapeutic agent, a large number of hybrid cells were found in the outer layer of the tumor, probably because more chemotherapeutic agent comes into contact with the cells in this area compared to inside the tumor (Yan et al. 2016). Since the chemotherapeutic agent in this study was given to the mice only once and this does not correspond to the real treatment of cancer patients, it cannot be said in general that chemotherapy promotes the development of cancer hybrid cells. Likewise, in several studies it was observed that treatment with chemotherapeutics or irradiation led to a (dose-dependent) formation of polyploid giant cancer cells (PGCCs), which can arise from cancer-cell fusion (Lin et al. 2019; Coward and Harding 2014; Puig et al. 2008; Kaur et al. 2015; Mirzayans et al. 2017; Song et al. 2021b). In addition, Rizvi et al. found out that irradiation can induce the fusion of bone-marrow-derived cells (BMDCs) with the irradiated tissue, but the hybrid cells in this study failed to initiate tumor growth (Rizvi et al. 2006). However, it was also shown that irradiation therapy can dissect tumor cells, leading to circulating tumor cells (CTCs) (Tinganelli and Durante 2020). In most of the studies, it was not analyzed whether those circulating cells surviving in the body and generating metastasis were originated by cancer-cell fusion. Another type of cancer cells, the CS/ICs, being able to generate a tumor when transplanted into an immune deficient animal and often showing chemo- and radiation-therapy resistance, are identified by a $CD44^{+}/CD24^{-low}$ or $CD133^{+}$ phenotype (Dittmar et al. 2009; Li et al. 2008; Meirelles et al. 2012; Zielske et al. 2011). Although it is often not characterized in studies whether the CS/ICs arose from cell fusion, it is interesting to keep in mind that CD44 is not only a marker for CS/ICs, but also mediates the homotypic fusion of macrophages, which would support the suggestion that CS/ICs could arise from the fusion of cancer cells with other cells (Fang et al. 2020; Vignery 2005; Dittmar 2022).

20.3.1 Syncytin-1 Contributes to Cancer-Cell Fusion and Progression

Little is known about the role of fusogenic proteins in cancer-cell fusion, but it is likely that dysregulation of previously characterized fusogenic proteins may play a role in the formation of cancer hybrid cells and syncytia formation. One prominent fusogenic protein already mentioned before is Syncytin-1, which under physiological conditions is crucial for placental development and binds to ASCT2 (Grandi and Tramontano 2018; Blond et al. 2000). A plethora of studies showed that Syncytin-1 expression was increased in cancers, for example in endometrial carcinoma (EC) (Liu et al. 2019; Strissel et al. 2012; Strick et al. 2007) and breast cancer (Bjerregaard et al. 2006; Larsson et al. 2007), but also in T-cell lymphoma (Maliniemi et al. 2013), leukemia (Sun et al. 2010), urothelial cell carcinoma (Yu et al. 2014), non-small cell lung cancer (Fu et al. 2021; Li et al. 2019), colorectal cancer (Montalban-Hernandez et al. 2022; Larsen et al. 2009), prostate cancer (Uygur et al. 2019), and hepatocellular carcinoma (Zhou et al. 2021). Furthermore, in vitro studies showed that the expression of Syncytin-1 is increased in PGCCs resulting from coculture of the human colorectal cancer cell lines HCT166 and LoVo and stimulation with cobalt chloride ($CoCl_2$) to mimic hypoxia (Fei et al. 2019). Likewise, Syncytin-1 expression was increased by TNFα stimulation of squamous cell carcinoma 9

(SCC-9) cells and human umbilical vein endothelial cells (HUVEC) and led to an increased fusion of the cells (Yan et al. 2017). In contrast, the sole expression of Syncytin-1 in cancer cells was not sufficient to induce a high rate of cell–cell fusion as compared to the fusion rate of cytotrophoblasts. Indeed, it was seen in different studies using immunohistochemistry (IHC) to characterize breast cancer, lung cancer, colorectal cancer, and endometrial cancer that Syncytin-1 expression was heterogenous among the cancer cells (Yan et al. 2017; Liu et al. 2019; Strick et al. 2007; Bjerregaard et al. 2006; Fu et al. 2021), indicating that Syncytin-1 expression is only increased in a specific compartment of the tumor or in specific cells, respectively. To add, Syncytin-1 was mainly found in the cytoplasm and not in the plasma membrane of the cancer cells (Yan et al. 2017; Liu et al. 2019; Strick et al. 2007; Bjerregaard et al. 2006; Fu et al. 2021; Fei et al. 2019) where it needs to be located to mediate cell–cell fusion and what consequently could explain the low fusion rate of the cells. Nevertheless, there are studies showing that Syncytin-1 is not only upregulated in cancer, but also plays a role in the fusion of cancer cells as well as in the progression of cancer (Yan et al. 2017; Liu et al. 2019; Strick et al. 2007; Bjerregaard et al. 2006; Larsson et al. 2007; Yu et al. 2014; Fu et al. 2021; Larsen et al. 2009; Uygur et al. 2019; Fei et al. 2019; Chignola et al. 2019; Benesova et al. 2017). In their study, Uygur et al. cocultured prostate cancer cells with primary human myocytes and observed an increase of Syncytin-1 as well as Annexin 5 expression after the coculturing period and resulting hybrid cells with CS/IC characteristics (Uygur et al. 2019). When using a Syncytin-1 inhibitory peptide or knocking down Syncytin-1 with small hairpin ribonucleic acid (shRNA) less multinucleated cells could be observed in the coculturing assay. In addition, human prostate tissue was characterized for its Syncytin-1 expression in normal, benign, and malignant human prostate tissue samples via tissue microarrays and it was found that the Syncytin-1 expression correlated with malignancy of the tissue (Uygur et al. 2019). Bjerregaard and colleagues characterized the breast cancer cell lines MCF-7 and MDA-MB-231 as well as samples from breast cancer patients (Bjerregaard et al. 2006). By real-time polymerase chain reaction (RT-PCR) and IHC the expression of Syncytin-1 in the breast cancer cell lines and the breast cancer specimens was proven. Fusion between MCF-7 cells and either HUVECs or CPAE cells was observed after 24 h of coculturing and was significantly decreased when MCF-7 cells were either treated with a phosphorothioate -protected Syncytin antisense oligonucleotide, which reduced Syncytin-1 messenger ribonucleic acid (mRNA), or when a Syncytin-1 inhibitory peptide was used (Bjerregaard et al. 2006). Also working with samples from breast cancer patients, Larsson et al. observed the expression of Syncytin-1 in more than a third of the specimens by using IHC (Larsson et al. 2007). Interestingly, in their study they could identify Syncytin-1 expression as an independent prognostic indicator of increased recurrence-free survival (Larsson et al. 2007). In contrast to the study by Larsson et al., Yu and colleagues were able to show that the overexpression of Syncytin-1 correlated with the degree and aggressiveness of urothelial cell carcinoma (UCC) and enhanced proliferation as well as viability of the cancer cells (Yu et al. 2014). When they xenografted SV-HUC-1 cells that normally cannot initiate tumors but had been transfected with a Syncytin-1 overexpression vector for the experiment, a tumor developed in nude mice which also showed signs of fused cells in hematoxylin-eosin (HE) staining (Yu et al. 2014). In samples from UCC patients they found a mutation in the promoter region of Syncytin-1 which enables the proto-oncogenic transcription factor c-Myb to bind and initiate Syncytin-1 expression in this kind of cancer (Yu et al. 2014). Like Yu et al. in UCC, Strick and colleagues demonstrated by RT-PCR, Northern, and Western Blots that Syncytin-1 expression correlated with tumor malignancy and metastasis in endometrial carcinoma (EC) (Strick et al. 2007). Using HE staining, they were also able to observe syncytia in EC biopsies. The EC cell line RL95-2 was used as an in vitro model and the expression of Syncytin-1 was decreased using small interfering RNAs (siRNAs). Normally, cell fusion can be induced in these cells by forskolin, which could be significantly decreased in siRNA treated cells (Strick et al. 2007). Similarly, Liu et al. also found in EC a correlation between Syncytin-1 expression and clinical stages of the tumor and overall survival (Liu et al. 2019). In in vitro studies using EC cell lines they observed that the overexpression of Syncytin-1 led to an increase in proliferation, cell migration and invasion, and G2/M phase transition in the cell cycle, meaning that Syncytin-1 promoted cells to enter mitosis (Liu et al. 2019). In human colorectal tumor tissues, Fei et al. detected Syncytin-1 expression by IHC which again correlated with stage and metastasis formation of the cancer (Fei et al. 2019). Larsen et al. also worked on colorectal cancer and observed heterogenous Syncytin-1 expression among patient samples, which correlated negatively with the overall survival (Larsen et al. 2009). Likewise, the expression of Syncytin-1 in non-small cell lung cancer (NSCLC) was proven by Fu et al. by IHC and they observed that patients with lower Syncytin-1 expression survived more than 5 years after tumor treatment (Fu et al. 2021). They also had a look on the methylation and therefore epigenetic regulation of the 5′-LTR in the HERVW gene and found less methylation what equals less inactivation in NSCLC tissues (Fu et al. 2021). A similar discovery was made by Benesova et al. when they analyzed the methylation of the Syncytin-1 promoter region on seminoma tissue (Benesova et al. 2017). Recently, Zhou et al. found in hepatocellular carcinoma (HCC), similar to the other studies mentioned, that Syncytin-1 expression correlated with a poor prognosis and they declared it to be a prognostic factor to predict vascular invasion, metastasis, and tumor size (Zhou et al. 2021). The group also observed that the phosphorylation and therefore activation of the kinases MEK1/2 and ERK1/2 correlated linearly to the Syncytin-1 expression in HCC (Zhou et al. 2021).

20.3.2 Phosphatidylserine as Possible Cell Fusion Mediator in Cancer

The role of the phospholipid PS in cell fusion of cancer cells with other cells is under investigation. It is known that cancer cells often display PS on their surface (Whitlock and Chernomordik 2021; Sharma and Kanwar 2018). In the already mentioned study from Uygur et al. not only an increase in Syncytin-1 but also in Annexin 5 expression was observed in cancer hybrid cells (Uygur et al. 2019), whereas the latter is a known PS-binding protein playing a role in myoblast, osteoclast, and trophoblast fusion (Whitlock and Chernomordik 2021). Knockdown of Annexin 5 via siRNAs impaired the formation of multinucleated fusion cells in this study (Uygur et al. 2019), illustrating the need for PS in cell fusion. Another well-known appearance of PS in the outer leaflet of the plasma membrane is during apoptosis (Martin et al. 1995) and in a study by Noubissi et al. proof was given that apoptotic cells can induce cancer-cell fusion (Noubissi et al. 2015). The fusion of MSCs and T47D or MCF7 human breast cancer cells was significantly increased by addition of apoptotic cells to the cell coculture and was impaired when the caspase inhibitor Z-VAD-FMK was added (Noubissi et al. 2015). Unfortunately, since there were no PS masking or Annexin targeting experiments in this study, one element is missing to complete the circle of the role of PS in cell fusion. It is also not known whether the apoptotic cells themselves fused with other cells or whether they only induced the fusion between neighboring cells. The fusion of apoptotic cells with other cells resulting in viable hybrid cells would need a termination of apoptotic processes in an early stage and studies need to investigate whether this is possible. Also, the role of apoptotic bodies needs to be further investigated. Data of Hochreiter-Hufford and colleagues showed that the activation of the PS receptor BAI1 and apoptosis promoted myoblast fusion (Hochreiter-Hufford et al. 2013). As in the study by Noubissi et al., fusion could be prevented by blocking apoptosis and could be increased again by adding apoptotic cells. In this study they could see that apoptotic cells did not directly fuse with normal cells but rather induced the fusion of non-apoptotic cells (Hochreiter-Hufford et al. 2013).

20.4 How Do Tumor Hybrid Cells Survive and How Does Fusion Alter Them?

20.4.1 The Post-hybrid Selection Process (PHSP)

The fusion of two or more cells normally leads to a mechanism called cellular senescence, which is a stable and irreversible growth arrest playing an important role in tumor suppression and tissue repair (Gal and Krizhanovsky 2014; Chuprin et al. 2013; Bojko et al. 2020). In this perspective, the fusion of cancer cells with neighboring cancer cells is investigated to empower oncolytic viruses, which replicate in and kill cancer cells specifically (Del Papa et al. 2021; Fu et al. 2003; Nakamura et al. 2004; Higuchi et al. 2000; Ackermann et al. 2020; Lin et al. 2010; Allen et al. 2004). By arming those viruses with fusogenic membrane proteins, they can spread more efficiently throughout the cancer mass and enhance cancer cell–cell fusion, which results in the formation of non-viable syncytia (Krabbe and Altomonte 2018; Salsman et al. 2005; Guedan et al. 2012; Le Boeuf et al. 2017). In addition, dying syncytial cells can release immunologically active syncytiosomes, which can load specific melanoma tumor antigens in dendritic cells and therefore stimulate the immune system to target cancer cells (Bateman et al. 2002). But how can cancer hybrid cells circumvent this destiny? In normal and in neoplastic tissues the occurrence of cell–cell fusion is a rare process, but even rarer is the survival of resulting hybrid cells. Since hybrid cells exhibit both aneuploidy and genomic instability, they need to restabilize their chromosomal imbalances by PHSP, which plays a crucial role in determining the fate of normal hybrid cells and the malignant properties of cancer hybrid cells (Hass et al. 2021a; Hass 2020; Chunduri and Storchova 2019; Melzer et al. 2020; Holland and Cleveland 2009). Knowing how these cells survive can help to prevent the formation of CS/ICs, the increased growth of tumors or the formation of metastasis (Hass et al. 2021a). A proposed initial step in PHSP is the heterokaryon-to-synkaryon transition (HST)/ploidy reductions (PR) step which means the segregation of chromosomes in dividing tumor hybrid cells. Since after the fusion of two or more (cancer) cells there are also several centrosomes in the resulting hybrid cell, chromosomal missegregation is occurring, leading to aneuploidy, micronuclei formation, neosis and/or chromothripsis by bi-, tri, or even multipolar divisions of the cell (Hass et al. 2021a; Song et al. 2021b; Chunduri and Storchova 2019; Holland and Cleveland 2009; Duncan et al. 2009; Ganem et al. 2009; Godinho et al. 2009; Passerini et al. 2016; Duncan et al. 2010). The distribution of the segregated chromosomes to daughter cells is random and cannot be predicted. These serious changes in the genomic content of the cells are the main cause for the poor survival of hybrid cells, since genes coding for crucial cellular processes such as metabolism, proliferation, signaling, or DNA replication are eliminated during PSHP and HST/PR. Cells are responding to this stress with autophagy, protein aggregation, further DNA damage or cell cycle arrest resulting in senescence or apoptosis (Chunduri and Storchova 2019; Passerini et al. 2016; Jonas et al. 2013; Barr et al. 2017; Sheltzer 2013; Stefani and Dobson 2003; Oromendia et al. 2012). And even if the cancer hybrid cells have successfully undergone HST/PR, this does not guarantee their continued survival. The hybrid cells are further selected until the previously unstable genome stabilizes to some extent but remains aneuploid. This further selection has been named autocatalytic karyotype evolution (Hass et al. 2021a; Li et al. 2000) and goes on until a chromosomal stabilized subclone arises, able to survive. As already described before, cancer hybrid cells are existing and were found in a variety of human cancer patient samples, where in general they were found to contribute to tumor heterogeneity and worsening the patient prognosis. With a mathematical model Miroshnychenko and colleagues demonstrated that along the well-known mutations found in tumors cell fusion can increase tumor-tissue plasticity, promote tumor progression and metastasis, although cell fusion rarely occurs and hybrid cells are even less likely to survive the PHSP process (Miroshnychenko et al. 2021). However, also other factors supporting the survival of the hybrid cells are existent. Do et al. used HeLa cells with low expression levels of p53, a transcription factor with pro-apoptotic properties in their study and let them fuse with each other (Do et al. 2017). Fused cells mainly went into cell cycle arrest or died as levels of p53 increased again after fusion. In surviving hybrid clones increased levels of the protein Survivin were found in the cytoplasm, whereas knockdown of Survivin led to a decreased survival rate of stable hybrid cell clones again. In addition, Athanassiadou et al. showed that an increased Survivin expression in breast carcinomas was correlated to a worse prognosis (Athanassiadou et al. 2011).

20.4.2 Stable Cancer Hybrid Cells Show Altered Properties Within the Tumor

Cancer hybrid cells which survived PHSP show a random new phenotype, which cannot be predicted. Concerning the tumorigenicity of cancer hybrid cells only a few studies exist showing that tumorigenicity of cancer hybrid cells has remained the same or even decreased in comparison with parental cancer cells (Wang et al. 2012; Miroshnychenko et al. 2021; Melzer et al. 2018c), which would be favorable for patients. Unfortunately, many more studies exist showing that tumorigenicity and the metastatic potential have increased or even newly developed in cancer hybrid cells compared to parental cells (Rachkovsky et al. 1998; He et al. 2015; Melzer et al. 2018a, b; Gast et al. 2018; Lartigue et al. 2020; Merle et al. 2021; Miroshnychenko et al. 2021; Zhou et al. 2015; Delespaul et al. 2019, 2020; Hass et al. 2019; Xue et al. 2015; Zhang et al. 2019; Yan et al. 2015; Xu et al. 2014; Tal et al. 2019; Sun et al. 2019; Sodi et al. 1998; Rappa et al. 2012; Miller et al. 1989; Mi et al. 2012; Luo et al. 2016; Duelli et al. 2007; Ding et al. 2012). It should be emphasized that even homotypic fusion of two cancer cells can decisively alter the properties of the resulting hybrid cell, as seen in a study where the fusion of B16-F10 melanoma cells leads to hybrid cells with enhanced metastatic potential in comparison with the parental cells (Mi et al. 2012). So, cancer hybrid cells can gather new properties by cell fusion, but how do they become metastasis or recurrence associated CS/ICs? It is possible that CS/ICs result from fusion of non-CS/ICs with normal cells and that the resulting hybrid cells have gained the new characteristics by HST/PR (Zhou et al. 2015; Delespaul et al. 2019; Duelli et al. 2007). Moreover, cancer cells could fuse with stem cells and gain CS/IC characteristics, which was shown in several studies (He et al. 2015; Ramakrishnan et al. 2013; Melzer et al. 2018c; Hass et al. 2019; Xue et al. 2015; Zhang et al. 2019; Gauck et al. 2017; Dornen et al. 2020b; Wei et al. 2014). As stated before, CS/ICs are identified by a $CD44^{+}/CD24^{-low}$ or $CD133^{+}$ phenotype (Dittmar et al. 2009; Li et al. 2008; Meirelles et al. 2012; Zielske et al. 2011) and also by identification of overall stemness markers such as Oct4, Nanog, Bmi1, Sox2, or ALDH1 (Xu et al. 2014). Cancer hybrid cells and especially CS/ICs are also associated with an increased resistance against chemotherapeutics, which implicates that those cells are less sensitive or can even survive chemotherapeutic treatments (Lindström et al. 2017; Dittmar et al. 2011; Kaur et al. 2015; Uygur et al. 2019; Liu et al. 2021; Mirzayans and Murray 2020). Cancer hybrid cells resulting from spontaneous in vitro fusion of macrophages with MCF7 cells showed radiation resistance with a higher survival rate, less DNA damage, and less heterogeneity in DNA damage compared to parental cells (Lindström et al. 2017). Less DNA heterogeneity could result from a PHSP with an enhanced DNA-repair capacity in the cancer hybrid cells. Concerning the treatment of cancer hybrid cells with different chemotherapeutic reagents, Dittmar et al. could show that cancer hybrid cell clones from the epithelial cell line M13SV1 and the breast cancer cell lines HS578T or MDA-MB-435, respectively, obtained by spontaneous in vitro cell fusion and antibiotic double selection showed different reactivity toward doxorubicin, 5-fluorouracil, etoposide, or paclitaxel (Dittmar et al. 2011), which probably is attributed to different PHSP. Likewise, a paclitaxel resistance was observed in cancer hybrid cells from head and neck squamous cell carcinoma and mesenchymal stem cells (MSCs) (Liu et al. 2021). In a recent study by Montalbán-Hernández and colleagues it was shown that cancer hybrid cells resulting from fusion of colorectal cancer cells and monocytes found in vitro and in patients' blood are able to evade the immune system by expression of SIGLEC5, which suppresses T-cell activation (Montalban-Hernandez et al. 2022). It was also shown that the count of tumor hybrid cells in patients' blood correlated positively with SIGLEC5 levels (Montalban-Hernandez et al. 2022). Similarly, Aguirre and colleagues used NSCLC cell lines and monocytes to generate cancer hybrid cells, which were able to avoid immune control by inhibiting natural killer (NK) cells via expression of HLA class I members, which resulted in less perforin generation as well as less cytotoxicity of the NK cells (Aguirre et al. 2020). In addition, also T-cell proliferation was reduced by cancer hybrid cells (Aguirre et al. 2020). The role of NK cell inactivation especially in metastasis was recently discussed (Chan and Ewald 2022) and since cancer hybrid cells play a major role in metastasis generation, they may be also responsible for the effects on immune cells.

20.5 How Can Tumor Hybrid Cells Be Detected In Vitro and In Vivo?

20.5.1 Detection of Cancer Hybrid Cells In Vitro and In Vivo in an Experimental Setup

In experimental studies tumor hybrid cells are detectable in vivo and in vitro by overlapping fusion markers like fluorescent reporter genes, antibiotic resistances, or lineage characteristics received from parental cells and thus, a plethora of studies using modifies cancer cells and transgenic mouse models are existent showing either homotypic cancer cell fusion or heterotypic cancer cell fusion with other cells, such as MSCs or macrophages (Lindström et al. 2017; Powell et al. 2011; Rachkovsky et al. 1998; Dittmar et al. 2009, 2011; Melzer et al. 2018a–c, 2019b; Wang et al. 2012; Gast et al. 2018; Lartigue et al. 2020; Mohr et al. 2015; Miroshnychenko et al. 2021; Wakeling et al. 1994; Lu and Kang 2009; Yan et al.

2016; Ramakrishnan et al. 2013; Strick et al. 2007; Noubissi et al. 2015; Hass 2020; Delespaul et al. 2019, 2020; Zhang et al. 2019; Xu et al. 2014; Tal et al. 2019; Rappa et al. 2012; Luo et al. 2016; Duelli et al. 2007; Ding et al. 2012; Dornen et al. 2020b; Aguirre et al. 2020; Wang et al. 2020; Kerbel et al. 1983). For example, Dörnen et al. used immortalized MSCs (iMSCs) that showed resistance to puromycin and cocultured them with either the hygromycin resistant breast cancer cell line HS578T Hyg or MDA-MB-231 Hyg to receive double antibiotic resistant cancer hybrid cells clones (Dornen et al. 2020b). The cancer hybrid cells clones were also characterized by short tandem repeat (STR) analysis to verify their fusogenic origin and showed new properties, indicating that cell fusion can lead to tumor heterogeneity (Dornen et al. 2020b). Likewise, the dual antibiotic selection method was used to generate hybrid cells from coculturing of breast cancer cells and breast epithelial cells (Dittmar et al. 2009, 2011; Fahlbusch et al. 2020) as well as of two bone and lung-tropic sublines of the before mentioned breast cancer cell line MDA-MB-231 (Lu and Kang 2009). To study the impact of ERVW-1 (Syncytin-1) expression on the cell fusion of normal human diploid fibroblasts (IMR-90 cells) and cancer cells, Chuprin et al. started their experiments by generating IMR-90 cells overexpressing ERVW-1 and in addition either the green fluorescent protein (GFP) or mCherry (Chuprin et al. 2013). The coculturing of these cells led to fused multinuclear cells with overlapping expression of both fluorescent proteins, from which the group concluded that ERVW-1 expression induces cell fusion (Chuprin et al. 2013). They also observed that hybrid cells underwent cellular senescence mediated by p53, a transcription and tumor suppression factor (Chuprin et al. 2013). In other studies, antibiotic resistance was used concurrently with fluorescent reporter genes, such as in the study by Wang et al. (2020). They used the human prostate cancer cell line LNCaP, being resistant to G418 and expressing the red fluorescent protein (RFP) AsRed2, and the human prostate stromal cell line HPS-15, being resistant to puromycin and expressing GFP (Wang et al. 2020). After coculturing the cells for several weeks, cancer hybrid cells were isolated by dual antibiotic treatment and showed dual fluorescence as well as genomic heterogeneity among the cancer hybrid clones. In addition, they were also analyzed by STR to prove the origin through the fusion of the parental cells (Wang et al. 2020). A similar approach for coculturing was done with different breast cancer cell lines and cancer-associated fibroblasts (CAFs) by Miroshnychenko and colleagues (Miroshnychenko et al. 2021). Lineage characteristics of M2-macrophages, namely $CD163^+/CD45^+$, were used to characterize cancer hybrid cells in a coculture experiment with GFP-labeled MCF-7 human breast cancer cells by Lindström and colleagues (Lindström et al. 2017). Using flow cytometry after the coculturing the cells, cancer hybrid cells were identified as $GFP^+/CD163^+/CD45^+$ cells and were further characterized for their radiation resistance, as described before (Lindström et al. 2017). In a study by Aguirre et al. the lung cancer cell lines H460/H460^{GFP} and A549 were cultured in cancer stem cell medium and used for rising cancer hybrid cells by fusion with human monocytes isolated from buffy coats (Aguirre et al. 2020). The hybrid cells were characterized by the expression of either GFP^+CD14^+ or $PANK^+CD14^+$, whereas CD14 is a specific marker for monocytes and PANK (Pan-Cytokeratin) is a cancer-epithelial-cytokeratin marker. To detect the earliest fusion time point in these cells, the group used the vital colorants DID and DIO on the monocytes and the H460-CSCs and could see double positive $DID^{+}DIO^{+}$ fusion cells in about 30 min of coculture (Aguirre et al. 2020). Melzer et al. cocultured MSCs labeled with GFP and the human benign breast cancer line MCF10A labeled with the RFP mCherry and did time lapse imaging to analyze the time needed for a fusion event, which turned out to be less than five minutes (Melzer et al. 2018b). In another in vivo study by Melzer et al. GFP-labeled MSCs (MSC^{GFP}) and the breast cancer cell line MDA-MB-231-labeled with mCherry (MDA-MB-231^{cherry}) were cocultured, two cancer hybrid cells were isolated by flow cytometry and injected subcutaneously in NOD/scid mice in comparison with the parental MDA-MB-231^{GFP} breast cancer cell line (Melzer et al. 2018a). A significantly elevated tumor growth as well as multiple metastases in distant organs was observed after a shorter time in comparison with the parental breast cancer cell line in the mice (Melzer et al. 2018a). Similar approaches were used by Gast et al. when cancer hybrid cells from GFP-tagged murine macrophages and H2B-RFP-labeled mouse colon carcinoma cells MC38 were analyzed concerning formation of metastasis in mice (Gast et al. 2018). Another possibility to detect fusion events is the use of a Cre recombinase in combination with recombinable vectors (Mohr et al. 2015; Tal et al. 2019; Sprangers et al. 2012). For example, Mohr et al. used so-called fluorescence double reporter (FDR) vectors which code for a loxP-flanked RFP cassette followed by GFP and which can be recombined by Cre recombinase leading to a fluorescence switch from red to green and therefore display a fusion event (Mohr et al. 2015). An approach described either as bimolecular fluorescence complementation (BiFC) (Noubissi et al. 2015) or as dual split protein (DSP) assay (Ishikawa et al. 2012) based on the expression of only a half of a fluorescent protein in each of the cells to be fused was used in in vitro studies. With all these approaches, however, it must be remembered that other mechanisms can resemble cancer-cell fusion, like cell cannibalism, entosis, emperipolesis, or cytophagocytosis, that forms cell-in-cell structures with DNA exchange (Fais and Overholtzer 2018; Overholtzer and Brugge 2008; Xia et al. 2008; Janssen and Medema 2011; Wang et al. 2019). Moreover, genes or mRNAs of fusion markers can be transferred via extracellular vesicles or by horizontal/lateral gene transfer and thereby alter the phenotype of the recipient cell (Yanez-Mo et al. 2015; Colombo et al. 2014; van Niel et al. 2018; Mittelbrunn and Sanchez-Madrid 2012) and tumor cells can shed vesicles even larger than exosomes, called ectosomes, and are known to produce blebs more often (Mittelbrunn and Sanchez-Madrid 2012). All of those naturally occurring processes can lead to false positive results in the setup for the detection of cell fusion events. Also, when using a stably expressed Cre recombinase, it should be noted that Cre can be cytotoxic and is altering the genome of possible cancer hybrid cells, because of active recombinase recognition sites in the mammalian genome (Thyagarajan et al. 2000; Schmidt-Supprian and Rajewsky 2007). To verify the fusogenic origin of the cancer hybrid cells, for example, karyotypization or STR analysis needs to be done, which is limited to heterotypic cancer hybrid cells only, since hybrids arising from homotypic fusion events show the same genetic background (Weiler and Dittmar 2019a).

20.5.2 Detection of Cancer Hybrid Cells In Vivo in Human Cancers

The detection of tumor hybrid cells outside of an experimental setting and in the context of human disease is challenging due to the lack of predetermined markers and the unknown fusion partners. Nevertheless, several studies have been able to demonstrate the fusion between cancer and normal cells in human cancers, and thus the existence of cancer hybrid cells, using a variety of markers (Manjunath et al. 2020a, b; Gast et al. 2018; Clawson et al. 2015; Montalban-Hernandez et al. 2022; Ramakrishnan et al. 2013; Larsson et al. 2007; Shabo et al. 2008, 2009, 2013, 2015; Lazova et al. 2013; Yilmaz et al. 2005; Clawson et al. 2012, 2017; Andersen et al. 2007, 2010; Chakraborty et al. 2004; Kurgyis et al. 2016). On a genomic level, Yilmaz et al. could detect tumor hybrid cells in a female patient with a primary renal cell carcinoma, who received a bone marrow transplantation (BMT) from a male donor in the past, by detection of the Y chromosome (Yilmaz et al. 2005). Similar findings were reported for patients with pancreatic ductal adenocarcinoma, lung cancer, renal carcinoma, and head and neck squamous carcinoma who received a sex-mismatched BMT in a study by Gast et al. (2018). Also, after BMT, Lazova et al. showed both alleles, from donor and patient, in tumor cells and in associated metastases by performing STR genotyping (Lazova et al. 2013). In order to characterize more precisely which normal cells specifically underwent heterogenous cell fusion with cancer cells, a commonly used approach is to look for non-cancer specific epitopes of the normal fusion partner, such as epithelial, macrophage, or hematopoietic antigens (Dittmar and Zänker 2011b; Ramakrishnan et al. 2013; Shabo et al. 2008, 2013; Manjunath et al. 2020b; Clawson et al. 2012). Studies by Shabo et al. concentrated on detecting tumor hybrid cells (probably) originating from the fusion of breast cancer or colorectal cancer cells with tumor-associated macrophages (TAMs) (Dittmar and Zänker 2011b; Shabo et al. 2008, 2013). These macrophages of the M2 type are promoting tumor progression and present specific antigens like CD14, CD68, CD163, MAC387, or DAP12, which were also found in the cancer hybrid cells (Dittmar and Zänker 2011b). In detail, in breast cancer tissues from 133 patients the macrophage-specific antigen CD163 was found in 48% of the samples, whereas MAC387 was found in 12% (Shabo et al. 2008). The expression of CD163 correlated positively with the occurrence of distant metastasis and shorter survival of the patients (Shabo et al. 2008), which was also seen in a study analyzing rectal cancer (Shabo et al. 2009). The expression of DAP12 was found in 66% of the analyzed breast cancer samples and was also associated with a high tumor grade and liver and skeletal metastases (Shabo et al. 2013). Likewise, Ramakrishnan and colleagues demonstrated that the pan-hematopoietic marker CD45 is expressed in up to 23.9% of human epithelial ovarian carcinoma cells, suggesting the existence of fusion cells (Ramakrishnan et al. 2013).

20.5.3 Detection of Circulating Hybrid Cells (CHCs) in the Blood of Human Cancer Patients

In addition to the possibility of studying CHCs in the primary solid cancer, there are also circulating cancer/tumor cells (CCCs/CTCs) and, amongst these, circulating hybrid cells (CHCs), which have been detected in studies and are moving further into the focus of research as liquid biomarkers for early tumor detection (Gast et al. 2018; Clawson et al. 2012, 2015, 2017; Manjunath et al. 2020b; Dietz et al. 2021; Walker et al. 2021; Sutton et al. 2022; Deng et al. 2022; Pereira-Veiga et al. 2022; Sulaiman et al. 2022; Lopresti et al. 2022; Menyailo et al. 2022; Ruano et al. 2022; Hu et al. 2022; Parappilly et al. 2022). In the already mentioned study by Gast et al. CHCs were detected in the circulation of female cancer patients who received a sex-mismatched BMT by coexpression of CD45 and EPCAM as well as the Y chromosome amongst others (Gast et al. 2018). The amount of CHCs in the circulation correlated with stage and survival of the patients (Gast et al. 2018). Studies by Clawson and colleagues identified CHCs in patients with melanoma (Clawson et al. 2012, 2015) or pancreatic ductal adenocarcinoma (Clawson et al. 2017). Originally CTCs were analyzed in the blood of melanoma patients, but when staining for pan-cytokeratin (KRT) and the leukocyte marker CD45 50% double positive cells were found, indicating that a fusion event between cancer cells and leukocytes happened, which was also seen in the blood from colorectal and pancreatic cancer patients (Clawson et al. 2012). CHCs from melanoma patients showed morphological characteristics from macrophages, high ploidy or aneuploidy, and expression of macrophage markers CD14 and CD 68 as well as markers specific for M2 macrophages CD163, CD204, and CD206 in combination with melanocytic specific markers ALCAM and MLANA (Clawson et al. 2015). When transplanted subcutaneously in nude mice, the hybrid cells disseminated and produced metastasis (Clawson et al. 2015). A similar CHC immunophenotype and DNA content was detected in samples from patients with pancreatic ductal adenocarcinoma which also resulted from cancer-cell fusion with M2 macrophages (Clawson et al. 2017). In addition, single cell RNASeq revealed high levels of expression of various metastasis-related markers, such as MIF, CD44, CD74, and CXCR4 and indeed the CHCs were able to induce metastases when transplanted into mice (Clawson et al. 2017). CHCs were also found in the blood of most NSCLC patients (76.5%), staining positive for both, the epithelial markers cytokeratin (CK) and EpCAM and the myeloid/macrophage markers CD14 and CD45, and showing more than one nucleus (Manjunath et al. 2020b). It was found that the count and the size of (giant) CHCs correlated with tumor stage and overall survival in NSCLC patients (Manjunath et al. 2020b). A plethora of 14 cancer types was characterized for the occurrence of CHCs from fusion of cancer cells (epithelial or tissue specific markers) with macrophages (CD45) by Dietz et al. (2021). CHCs were found in all cancer types, even in glioblastoma, which is reported to rarely disseminate outside the central nervous system (Dietz et al. 2021). The same working group did research on whether CHCs are suitable liquid biomarkers during treatment of gastrointestinal cancers (Walker et al. 2021). Samples from patients with rectal adenocarcinoma (RAC), esophageal adenocarcinoma (EAC), and colorectal liver metastasis (CRLM) were characterized concerning CHCs (macrophage-specific staining of CD45; epithelial cell staining of pan-cytokeratin (CK)), starting with the response of RAC and EAC patients to neoadjuvant therapy (NAT) (Walker et al. 2021). Significantly less CHCs were found in the patients after resection and NAT with a pathologic complete response (pCR) to NAT (4.7 ± 4 vs. 31 ± 19.9 CHCs in RAC patients; 3.8 ± 2.1 vs. 21 ± 21.2 CHCs in EAC patients), whereas patients with an incomplete response to NAT showed comparable levels of CHCs like before treatment (20 ± 18.1 in RAC and EAC patients combined) and NAT non-responsive patients had an increased occurrence of CHCs (47 ± 20.9 CHCs in RAC and EAC patients combined) (Walker et al. 2021). In EAC patients more CHCs were found in lymph node positive samples and the overall number of CHCs correlated with disease-specific survival (DSS), indicating that regardless of the degree of pathologic response lower levels of CHCs in EAC may improve patients' survival (Walker et al. 2021). The amount of CHCs in patients with CRLM correlated with treatment success and with recurrence and spreading of further metastasis in patients (Walker et al. 2021). In a recent study, CHCs were detected in EC patients by staining for CD45+/Pan-Cytokeratin+/EpCam+/CD31+ cells, which showed a macrophage-like morphology and polyploidy (Sulaiman et al. 2022). The CHCs were characterized morphologically, for example whether they were tiny or giant, and were then compared to pathological parameters of the EC. This led the authors to the conclusion that the presence of CHCs correlated neither with stage nor with grade of the disease (Sulaiman et al. 2022). It should be noted that the authors did not quantify CHCs in this study, which was done in the studies described previously. Also recently published was a study by Lopresti and colleagues who characterized, as termed by them, atypical circulating tumor cells in the blood of patients with metastatic breast cancer (Lopresti et al. 2022). They divided the CTCs into three different subsets, whereas the fraction called giant atypical CTCs (g-aCTCs) resembles CHCs, which were found in 46% of the patients and were characterized to have an enlarged size with multiple nuclei as well as a hybrid epithelial-mesenchymal phenotype, but were negative for CD45 (Lopresti et al. 2022). Nevertheless, the occurrence of g-aCTCs was correlated with overall and progression-free survival of the patients (Lopresti et al. 2022). Ruano et al. analyzed CHCs in the blood of patients with high-grade serous ovarian cancer by collecting them either by size or by antigen expression. The cells were characterized by CD45 expression and aneuploidy of chromosome 8, whereas the former had no prognostic value, but ploidies of the chromosome correlated with a shortened lifespan (Ruano et al. 2022). To analyze CHCs in uveal melanoma, Parappilly and colleagues used immunohistochemical staining for the common melanocytic marker gp100, for the 5-hydroxytryptamine receptor 2B (HTR2B) upregulated in high-risk uveal melanoma, and for the leukocyte specific antigen CD45 (Parappilly et al. 2022). They thus demonstrated that CHCs outnumber CTCs and are prognostic markers for 3-year progression-free survival (Parappilly et al. 2022). In sum, these data indicate that cancer cells expressing specific epitopes from non-cancerous cells were identified in human cancers and serve as a proof of the cancer hybrid cells existence. However, it must be kept in mind that the expression of these epitopes can also originate by the genomic instability of the primary cancer and that markers on cancer hybrid cells can get lost, as seen, for example, in a study by Melzer et al., where the Y chromosome of the parental cell was not detectable in the cancer hybrid cell anymore (Melzer et al. 2018a). Furthermore, the detection of homotypic cell fusions in human cancers turns out to be much more difficult because the cells have an identical genetic background and no specific epitopes to prove a fusion event exist. Thus, the establishment of potential fusion markers in human cancers would be helpful not only for a better characterization of the tumor but also for a more accurate diagnosis and treatment of patients.

20.6 Conclusions

Cell–cell fusion is a crucial process in development and maintenance of different organs and tissues in mammals (Inoue et al. 2005; Bianchi et al. 2014; Potgens et al. 2004; Bolze et al. 2017; Soygur and Sati 2016; Muir et al. 2006; Millay et al. 2013; Bi et al. 2017; Gamage et al. 2017; Quinn et al. 2017; Leikina et al. 2018; Zhang et al. 2017), whereas in a tumor context it leads to an increase of heterogeneity in the tumor and its regulation, if any is existent, is not understood (Matsumoto et al. 2021; Zack et al. 2013; Bielski et al. 2018; Quinton et al. 2021; Dornen et al. 2020a). Still ongoing research has shown that there are some conditions and proteins that can promote or induce cell fusions in the tumor microenvironment, such as hypoxia (Petrova et al. 2018; Emami Nejad et al. 2021; Li et al. 2021; Eltzschig and Carmeliet 2011; Tinganelli and Durante 2020), the presence of inflammatory cytokines (Melzer et al. 2018b; Yan et al. 2017; Davies et al. 2009; Weiler and Dittmar 2019b; Song et al. 2012; Mohr et al. 2015; Skokos et al. 2011; Hotokezaka et al. 2007) or the expression of Syncytin-1 (Liu et al. 2019; Strissel et al. 2012; Strick et al. 2007; Bjerregaard et al. 2006; Larsson et al. 2007; Maliniemi et al. 2013; Sun et al. 2010; Yu et al. 2014; Fu et al. 2021; Li et al. 2019; Larsen et al. 2009; Uygur et al. 2019; Zhou et al. 2021). A plethora of studies investigated the fusion process itself and showed with different experimental setups that cancer cells are indeed able to fuse with normal cells, like macrophages, to form cancer hybrid cells (Lindström et al. 2017; Powell et al. 2011; Rachkovsky et al. 1998; Dittmar et al. 2009, 2011; Melzer et al. 2018a–c, 2019b; Wang et al. 2012; Gast et al. 2018; Lartigue et al. 2020; Mohr et al. 2015; Miroshnychenko et al. 2021; Wakeling et al. 1994; Lu and Kang 2009; Yan et al. 2016; Ramakrishnan et al. 2013; Strick et al. 2007; Noubissi et al. 2015; Hass 2020; Delespaul et al. 2019, 2020; Zhang et al. 2019; Xu et al. 2014; Tal et al. 2019; Rappa et al. 2012; Luo et al. 2016; Duelli et al. 2007; Ding et al. 2012; Dornen et al. 2020b; Aguirre et al. 2020; Wang et al. 2020; Kerbel et al. 1983). In addition, the hypothesis that cancer-cell fusions take place in human cancers, initially doubted, is gaining more and more scientific consensus. Cancer hybrid cells were detected by the expression of cancer-specific as well as non-cancer-specific epitopes in solid tumors and in the circulation of cancer patients as circulating hybrid cells (CHCs) and were further characterized (Gast et al. 2018; Clawson et al. 2012, 2015 2017; Manjunath et al. 2020b; Dietz et al. 2021; Walker et al. 2021; Sutton et al. 2022; Deng et al. 2022; Pereira-Veiga et al. 2022; Sulaiman et al. 2022; Lopresti et al. 2022; Menyailo et al. 2022; Ruano et al. 2022; Hu et al. 2022; Parappilly et al. 2022). Most of those hybrid cells were aneuploid and gathered new malignant properties, such as a higher metastatic potential or resistance to chemotherapy (Dittmar et al. 2021; Manjunath et al. 2020a, b; Hass et al. 2021a, b; Gast et al. 2018; Clawson et al. 2012, 2015, 2017; Wang et al. 2021; Dietz et al. 2021; Walker et al. 2021; Sutton et al. 2022; Deng et al. 2022; Pereira-Veiga et al. 2022; Sulaiman et al. 2022; Lopresti et al. 2022; Menyailo et al. 2022; Ruano et al. 2022; Hu et al. 2022; Parappilly et al. 2022). Since the detection of cancer hybrid cells is challenging, due to their genomic instability and potential loss of markers, there needs to be more research done into reliable markers for cancer hybrid cells. Furthermore, especially CHCs were found to be prognostic markers for progression-free and overall survival (Gast et al. 2018; Manjunath et al. 2020b; Walker et al. 2021; Lopresti et al. 2022; Parappilly et al. 2022) of cancer patients what may be explained with a higher probability for metastases. It is discussed whether it would be possible to stop the adhesion of CHCs and therefore prevent the invasion and metastasizing of distant organs (Xie et al. 2022), which would be a benefit for cancer patients. In sum, the fusion of cancer cells with normal cells is proven to result in cancer hybrid cells, which can worsen the prognosis of cancer patients. Future studies should focus on how to stop the cancer-cell fusion or least the seeding of distant metastasis by cancer hybrid cells.

References

Abmayr SM, Pavlath GK (2012) Myoblast fusion: lessons from flies and mice. Development 139:641–656

Ackermann M, Verleden SE, Kuehnel M, Haverich A, Welte T, Laenger F et al (2020) Pulmonary vascular endothelialitis, thrombosis, and angiogenesis in Covid-19. N Engl J Med 383(2):120–128

Aguilar PS, Baylies MK, Fleissner A, Helming L, Inoue N, Podbilewicz B et al (2013) Genetic basis of cell-cell fusion mechanisms. Trends Genet 29:427–437

Aguirre LA, Montalban-Hernandez K, Avendano-Ortiz J, Marin E, Lozano R, Toledano V et al (2020) Tumor stem cells fuse with monocytes to form highly invasive tumor-hybrid cells. Oncoimmunology 9:1773204

Aichel O (1911) Über Zellverschmelzung mit qualitativ abnormer Chromosomenverteilung als Ursache der Geschwulstbildung. In: Vorträge und Aufsätze über Entwicklungsmechanik der Organismen; 8

Allen C, McDonald C, Giannini C, Peng KW, Rosales G, Russell SJ et al (2004) Adenoviral vectors expressing fusogenic membrane glycoproteins activated via matrix metalloproteinase cleavable linkers have significant antitumor potential in the gene therapy of gliomas. J Gene Med 6:1216–1227

Andersen TL, Boissy P, Sondergaard TE, Kupisiewicz K, Plesner T, Rasmussen T et al (2007) Osteoclast nuclei of myeloma patients show chromosome translocations specific for the myeloma cell clone: a new type of cancer-host partnership? J Pathol 211:10–17

Andersen TL, Soe K, Sondergaard TE, Plesner T, Delaisse JM (2010) Myeloma cell-induced disruption of bone remodelling compartments leads to osteolytic lesions and generation of osteoclast-myeloma hybrid cells. Br J Haematol 148:551–561

Archacka K, Grabowska I, Mierzejewski B, Graffstein J, Gorzynska A, Krawczyk M et al (2021) Hypoxia preconditioned bone marrow-derived mesenchymal stromal/stem cells enhance myoblast fusion and skeletal muscle regeneration. Stem Cell Res Ther 12:448

Athanassiadou AM, Patsouris E, Tsipis A, Gonidi M, Athanassiadou P (2011) The significance of Survivin and Nectin-4 expression in the prognosis of breast carcinoma. Folia Histochem Cytobiol 49:26–33

Balkwill F, Mantovani A (2001) Inflammation and cancer: back to Virchow? Lancet 357:539–545

Barr AR, Cooper S, Heldt FS, Butera F, Stoy H, Mansfeld J et al (2017) DNA damage during S-phase mediates the proliferation-quiescence decision in the subsequent G1 via p21 expression. Nat Commun 8:14728

Bateman AR, Harrington KJ, Kottke T, Ahmed A, Melcher AA, Gough MJ et al (2002) Viral fusogenic membrane glycoproteins kill solid tumor cells by nonapoptotic mechanisms that promote cross presentation of tumor antigens by dendritic cells. Cancer Res 62:6566–6578

Benesova M, Trejbalova K, Kovarova D, Vernerova Z, Hron T, Kucerova D et al (2017) DNA hypomethylation and aberrant expression of the human endogenous retrovirus ERVWE1/syncytin-1 in seminomas. Retrovirology 14:20

Bevers EM, Williamson PL (2016) Getting to the outer leaflet: physiology of phosphatidylserine exposure at the plasma membrane. Physiol Rev 96:605–645

Bi P, Ramirez-Martinez A, Li H, Cannavino J, McAnally JR, Shelton JM et al (2017) Control of muscle formation by the fusogenic micropeptide myomixer. Science 356:323–327

Bianchi E, Doe B, Goulding D, Wright GJ (2014) Juno is the egg Izumo receptor and is essential for mammalian fertilization. Nature 508:483–487

Bielski CM, Zehir A, Penson AV, Donoghue MTA, Chatila W, Armenia J et al (2018) Genome doubling shapes the evolution and prognosis of advanced cancers. Nat Genet 50:1189–1195

Bjerregaard B, Holck S, Christensen IJ, Larsson LI (2006) Syncytin is involved in breast cancer-endothelial cell fusions. Cell Mol Life Sci 63:1906–1911

Blond JL, Besème F, Duret L, Bouton O, Bedin F, Perron H et al (1999) Molecular characterization and placental expression of HERV-W, a new human endogenous retrovirus family. J Virol 73:1175–1185

Blond J-L, Lavillette D, Cheynet V, Bouton O, Oriol G, Chapel-Fernandes S et al (2000) An envelope glycoprotein of the human endogenous retrovirus HERV-W is expressed in the human placenta and fuses cells expressing the type D mammalian retrovirus receptor. J Virol 74:3321–3329

Bojko A, Staniak K, Czarnecka-Herok J, Sunderland P, Dudkowska M, Sliwinska MA et al (2020) Improved autophagic flux in escapers from doxorubicin-induced senescence/polyploidy of breast cancer cells. Int J Mol Sci 21:6084

Bolze PA, Mommert M, Mallet F (2017) Contribution of syncytins and other endogenous retroviral envelopes to human placenta pathologies. Prog Mol Biol Transl Sci 145:111–162

Braga L, Ali H, Secco I, Chiavacci E, Neves G, Goldhill D et al (2021) Drugs that inhibit TMEM16 proteins block SARS-CoV-2 spike-induced syncytia. Nature 594:88–93

Brukman NG, Uygur B, Podbilewicz B, Chernomordik LV (2019) How cells fuse. J Cell Biol 218:1436–1451

Burton C, Bartee E (2019) Syncytia formation in oncolytic virotherapy. Mol Ther Oncolytics 15:131–139

Chakraborty A, Lazova R, Davies S, Backvall H, Ponten F, Brash D et al (2004) Donor DNA in a renal cell carcinoma metastasis from a bone marrow transplant recipient. Bone Marrow Transplant 34:183–186

Chan IS, Ewald AJ (2022) The changing role of natural killer cells in cancer metastasis. J Clin Invest 132(6):e143762

Chen CP, Chen LF, Yang SR, Chen CY, Ko CC, Chang GD et al (2008) Functional characterization of the human placental fusogenic membrane protein syncytin 2. Biol Reprod 79:815–823
Chignola R, Sega M, Molesini B, Baruzzi A, Stella S, Milotti E (2019) Collective radioresistance of T47D breast carcinoma cells is mediated by a Syncytin-1 homologous protein. PLoS One 14:e0206713
Chunduri NK, Storchova Z (2019) The diverse consequences of aneuploidy. Nat Cell Biol 21:54–62
Chuprin A, Gal H, Biron-Shental T, Biran A, Amiel A, Rozenblatt S et al (2013) Cell fusion induced by ERVWE1 or measles virus causes cellular senescence. Genes Dev 27:2356–2366
Clark S, Naz RK (2013) Presence and incidence of izumo antibodies in sera of immunoinfertile women and men. Am J Reprod Immunol 69:256–263
Clawson GA, Kimchi E, Patrick SD, Xin P, Harouaka R, Zheng S et al (2012) Circulating tumor cells in melanoma patients. PLoS One 7:e41052
Clawson GA, Matters GL, Xin P, Imamura-Kawasawa Y, Du Z, Thiboutot DM et al (2015) Macrophage-tumor cell fusions from peripheral blood of melanoma patients. PLoS One 10:e0134320
Clawson GA, Matters GL, Xin P, McGovern C, Wafula E, dePamphilis C et al (2017) "Stealth dissemination" of macrophage-tumor cell fusions cultured from blood of patients with pancreatic ductal adenocarcinoma. PLoS One 12:e0184451
Colombo M, Raposo G, Thery C (2014) Biogenesis, secretion, and intercellular interactions of exosomes and other extracellular vesicles. Annu Rev Cell Dev Biol 30:255–289
Coward J, Harding A (2014) Size does matter: why polyploid tumor cells are critical drug targets in the war on cancer. Front Oncol 4:123
Curty G, Marston JL, de Mulder RM, Leal FE, Nixon DF, Soares MA (2020) Human endogenous retrovirus K in cancer: a potential biomarker and immunotherapeutic target. Viruses 12(7):726
Dai L, Del Valle L, Miley W, Whitby D, Ochoa AC, Flemington EK et al (2018) Transactivation of human endogenous retrovirus K (HERV-K) by KSHV promotes Kaposi's sarcoma development. Oncogene 37:4534–4545
Davies PS, Powell AE, Swain JR, Wong MH (2009) Inflammation and proliferation act together to mediate intestinal cell fusion. PLoS One 4:e6530
Delespaul L, Merle C, Lesluyes T, Lagarde P, Le Guellec S, Perot G et al (2019) Fusion-mediated chromosomal instability promotes aneuploidy patterns that resemble human tumors. Oncogene 38:6083–6094
Delespaul L, Gelabert C, Lesluyes T, Le Guellec S, Perot G, Leroy L et al (2020) Cell-cell fusion of mesenchymal cells with distinct differentiations triggers genomic and transcriptomic remodelling toward tumour aggressiveness. Sci Rep 10:21634
Del Papa J, Clarkin RG, Parks RJ (2021) Use of cell fusion proteins to enhance adenoviral vector efficacy as an anti-cancer therapeutic. Cancer Gene Ther 28:745–756
Deng Z, Wu S, Wang Y, Shi D (2022) Circulating tumor cell isolation for cancer diagnosis and prognosis. EBioMedicine 83:104237
de Parseval N, Heidmann T (2005) Human endogenous retroviruses: from infectious elements to human genes. Cytogenet Genome Res 110:318–332
Dervan E, Bhattacharyya DD, McAuliffe JD, Khan FH, Glynn SA (2021) Ancient adversary—HERV-K (HML-2) in cancer. Front Oncol 11:658489
Dietz MS, Sutton TL, Walker BS, Gast CE, Zarour L, Sengupta SK et al (2021) Relevance of circulating hybrid cells as a non-invasive biomarker for myriad solid tumors. Sci Rep 11:13630
Ding J, Jin W, Chen C, Shao Z, Wu J (2012) Tumor associated macrophage x cancer cell hybrids may acquire cancer stem cell properties in breast cancer. PLoS One 7:e41942
Dittmar T (2022) Generation of cancer stem/initiating cells by cell-cell fusion. Int J Mol Sci 23(9):4514
Dittmar T, Zänker KS (2011a) Cell fusion in health and disease; Volume I: Cell fusion in health. Springer, Dordrecht
Dittmar T, Zänker KS (2011b) Cell fusion in health and disease; Volume II: Cell fusion in disease. Springer, Dordrecht

Dittmar T, Nagler C, Schwitalla S, Reith G, Niggemann B, Zanker KS (2009) Recurrence cancer stem cells-made by cell fusion? Med Hypotheses 73:542–547
Dittmar T, Schwitalla S, Seidel J, Haverkampf S, Reith G, Meyer-Staeckling S et al (2011) Characterization of hybrid cells derived from spontaneous fusion events between breast epithelial cells exhibiting stem-like characteristics and breast cancer cells. Clin Exp Metastasis 28:75–90
Dittmar T, Weiler J, Luo T, Hass R (2021) Cell-cell fusion mediated by viruses and HERV-derived fusogens in cancer initiation and progression. Cancers (Basel) 13(21):5363
Do M, Kwak IH, Ahn JH, Lee IJ, Lee JH (2017) Survivin protects fused cancer cells from cell death. BMB Rep 50:361–366
Dornen J, Sieler M, Weiler J, Keil S, Dittmar T (2020a) Cell fusion-mediated tissue regeneration as an inducer of polyploidy and aneuploidy. Int J Mol Sci 21:1811
Dornen J, Myklebost O, Dittmar T (2020b) Cell fusion of mesenchymal stem/stromal cells and breast cancer cells leads to the formation of hybrid cells exhibiting diverse and individual (stem cell) characteristics. Int J Mol Sci 21(24):9636
Duelli D, Lazebnik Y (2007) Cell-to-cell fusion as a link between viruses and cancer. Nat Rev Cancer 7:968–976
Duelli DM, Padilla-Nash HM, Berman D, Murphy KM, Ried T, Lazebnik Y (2007) A virus causes cancer by inducing massive chromosomal instability through cell fusion. Curr Biol 17:431–437
Duncan AW, Hickey RD, Paulk NK, Culberson AJ, Olson SB, Finegold MJ et al (2009) Ploidy reductions in murine fusion-derived hepatocytes. PLoS Genet 5:e1000385
Duncan AW, Taylor MH, Hickey RD, Hanlon Newell AE, Lenzi ML, Olson SB et al (2010) The ploidy conveyor of mature hepatocytes as a source of genetic variation. Nature 467:707–710
Dvorak HF (1986) Tumors: wounds that do not heal. N Engl J Med 315:1650–1659
Dvorak HF (2015) Tumors: wounds that do not heal-redux. Cancer Immunol Res 3:1–11
Eigler T, Zarfati G, Amzallag E, Sinha S, Segev N, Zabary Y et al (2021) ERK1/2 inhibition promotes robust myotube growth via CaMKII activation resulting in myoblast-to-myotube fusion. Dev Cell 56:3349–3363.e3346
Eltzschig HK, Carmeliet P (2011) Hypoxia and inflammation. N Engl J Med 364:656–665
Emami Nejad A, Najafgholian S, Rostami A, Sistani A, Shojaeifar S, Esparvarinha M et al (2021) The role of hypoxia in the tumor microenvironment and development of cancer stem cell: a novel approach to developing treatment. Cancer Cell Int 21:62
Fahlbusch SS, Keil S, Epplen JT, Zanker KS, Dittmar T (2020) Comparison of hybrid clones derived from human breast epithelial cells and three different cancer cell lines regarding in vitro cancer stem/ initiating cell properties. BMC Cancer 20:446
Fais S, Overholtzer M (2018) Cell-in-cell phenomena in cancer. Nat Rev Cancer 18:758–766
Fang JY, Yang Z, Han B (2020) Switch of macrophage fusion competency by 3D matrices. Sci Rep 10:10348
Fei F, Li C, Wang X, Du J, Liu K, Li B et al (2019) Syncytin 1, CD9, and CD47 regulating cell fusion to form PGCCs associated with cAMP/PKA and JNK signaling pathway. Cancer Med 8:3047–3058
Fortuna MB, Dewey MJ, Furmanski P (1989) Cell fusion in tumor development and progression: occurrence of cell fusion in primary methylcholanthrene-induced tumorigenesis. Int J Cancer 44:731–737
Fu X, Tao L, Jin A, Vile R, Brenner MK, Zhang X (2003) Expression of a fusogenic membrane glycoprotein by an oncolytic herpes simplex virus potentiates the viral antitumor effect. Mol Ther 7:748–754
Fu Y, Zhuang X, Xia X, Li X, Xiao K, Liu X (2021) Correlation between promoter hypomethylation and increased expression of Syncytin-1 in non-small cell lung cancer. Int J Gen Med 14:957–965
Gal H, Krizhanovsky V (2014) Cell fusion induced senescence. Aging (Albany NY) 6:353–354

Gamage DG, Leikina E, Quinn ME, Ratinov A, Chernomordik LV, Millay DP (2017) Insights into the localization and function of myomaker during myoblast fusion. J Biol Chem 292:17272–17289

Ganem NJ, Godinho SA, Pellman D (2009) A mechanism linking extra centrosomes to chromosomal instability. Nature 460:278–282

Gao Y, Yu XF, Chen T (2021) Human endogenous retroviruses in cancer: Expression, regulation and function. Oncol Lett 21:121

Garcia-Montojo M, Doucet-O'Hare T, Henderson L, Nath A (2018) Human endogenous retrovirus-K (HML-2): a comprehensive review. Crit Rev Microbiol 44:715–738

Gast CE, Silk AD, Zarour L, Riegler L, Burkhart JG, Gustafson KT et al (2018) Cell fusion potentiates tumor heterogeneity and reveals circulating hybrid cells that correlate with stage and survival. Sci Adv 4:eaat7828

Gauck D, Keil S, Niggemann B, Zanker KS, Dittmar T (2017) Hybrid clone cells derived from human breast epithelial cells and human breast cancer cells exhibit properties of cancer stem/initiating cells. BMC Cancer 17:515

Gauster M, Moser G, Orendi K, Huppertz B (2009) Factors involved in regulating trophoblast fusion: potential role in the development of preeclampsia. Placenta 30 Suppl A:S49–S54

Godinho SA, Kwon M, Pellman D (2009) Centrosomes and cancer: how cancer cells divide with too many centrosomes. Cancer Metastasis Rev 28:85–98

Grandi N, Tramontano E (2018) HERV envelope proteins: physiological role and pathogenic potential in cancer and autoimmunity. Front Microbiol 9:462

Guedan S, Grases D, Rojas JJ, Gros A, Vilardell F, Vile R et al (2012) GALV expression enhances the therapeutic efficacy of an oncolytic adenovirus by inducing cell fusion and enhancing virus distribution. Gene Ther 19:1048–1057

Gyobu S, Ishihara K, Suzuki J, Segawa K, Nagata S (2017) Characterization of the scrambling domain of the TMEM16 family. Proc Natl Acad Sci U S A 114:6274–6279

Hass R (2020) Role of MSC in the tumor microenvironment. Cancers (Basel) 12(8):2107

Hass R, von der Ohe J, Ungefroren H (2019) Potential role of MSC/cancer cell fusion and EMT for breast cancer stem cell formation. Cancers (Basel) 11:1432

Hass R, von der Ohe J, Dittmar T (2021a) Cancer cell fusion and post-hybrid selection process (PHSP). Cancers (Basel) 13(18):4636

Hass R, von der Ohe J, Dittmar T (2021b) Hybrid formation and fusion of cancer cells in vitro and in vivo. Cancers (Basel) 13(17):4496

He X, Li B, Shao Y, Zhao N, Hsu Y, Zhang Z et al (2015) Cell fusion between gastric epithelial cells and mesenchymal stem cells results in epithelial-to-mesenchymal transition and malignant transformation. BMC Cancer 15:24

Helming L, Gordon S (2009) Molecular mediators of macrophage fusion. Trends Cell Biol 19:514–522

Helming L, Winter J, Gordon S (2009) The scavenger receptor CD36 plays a role in cytokine-induced macrophage fusion. J Cell Sci 122:453–459

Hernandez JM, Podbilewicz B (2017) The hallmarks of cell-cell fusion. Development 144:4481–4495

Higuchi H, Bronk SF, Bateman A, Harrington K, Vile RG, Gores GJ (2000) Viral fusogenic membrane glycoprotein expression causes syncytia formation with bioenergetic cell death: implications for gene therapy. Cancer Res 60:6396–6402

Hochreiter-Hufford AE, Lee CS, Kinchen JM, Sokolowski JD, Arandjelovic S, Call JA et al (2013) Phosphatidylserine receptor BAI1 and apoptotic cells as new promoters of myoblast fusion. Nature 497:263–267

Holland AJ, Cleveland DW (2009) Boveri revisited: chromosomal instability, aneuploidy and tumorigenesis. Nat Rev Mol Cell Biol 10:478–487

Hotokezaka H, Sakai E, Ohara N, Hotokezaka Y, Gonzales C, Matsuo K et al (2007) Molecular analysis of RANKL-independent cell fusion of osteoclast-like cells induced by TNF-alpha, lipopolysaccharide, or peptidoglycan. J Cell Biochem 101:122–134

Hu B, Gong Y, Wang Y, Xie J, Cheng J, Huang Q (2022) Comprehensive atlas of circulating rare cells detected by SE-iFISH and image scanning platform in patients with various diseases. Front Oncol 12:821454

Huang Q, Chen H, Wang F, Brost BC, Li J, Gao Y et al (2014) Reduced syncytin-1 expression in choriocarcinoma BeWo cells activates the calpain1-AIF-mediated apoptosis, implication for preeclampsia. Cell Mol Life Sci 71:3151–3164

Huang CM, Yan TL, Xu Z, Wang M, Zhou XC, Jiang EH et al (2018) Hypoxia enhances fusion of oral squamous carcinoma cells and epithelial cells partly via the epithelial-mesenchymal transition of epithelial cells. Biomed Res Int 2018:5015203

Inoue N, Ikawa M, Isotani A, Okabe M (2005) The immunoglobulin superfamily protein Izumo is required for sperm to fuse with eggs. Nature 434:234–238

Ishikawa H, Meng F, Kondo N, Iwamoto A, Matsuda Z (2012) Generation of a dual-functional split-reporter protein for monitoring membrane fusion using self-associating split GFP. Protein Eng Des Sel 25:813–820

Janssen A, Medema RH (2011) Entosis: aneuploidy by invasion. Nat Cell Biol 13:199–201

Jonas K, Liu J, Chien P, Laub MT (2013) Proteotoxic stress induces a cell-cycle arrest by stimulating Lon to degrade the replication initiator DnaA. Cell 154:623–636

Kaur E, Rajendra J, Jadhav S, Shridhar E, Goda JS, Moiyadi A et al (2015) Radiation-induced homotypic cell fusions of innately resistant glioblastoma cells mediate their sustained survival and recurrence. Carcinogenesis 36:685–695

Kerbel RS, Lagarde AE, Dennis JW, Donaghue TP (1983) Spontaneous fusion in vivo between normal host and tumor cells: possible contribution to tumor progression and metastasis studied with a lectin-resistant mutant tumor. Mol Cell Biol 3:523–538

Knerr I, Huppertz B, Weigel C, Dotsch J, Wich C, Schild RL et al (2004) Endogenous retroviral syncytin: compilation of experimental research on syncytin and its possible role in normal and disturbed human placentogenesis. Mol Hum Reprod 10:581–588

Krabbe T, Altomonte J (2018) Fusogenic viruses in oncolytic immunotherapy. Cancers (Basel) 10(7):216

Kudo Y, Boyd CA, Sargent IL, Redman CW (2003) Hypoxia alters expression and function of syncytin and its receptor during trophoblast cell fusion of human placental BeWo cells: implications for impaired trophoblast syncytialisation in pre-eclampsia. Biochim Biophys Acta 1638:63–71

Kurgyis Z, Kemeny LV, Buknicz T, Groma G, Olah J, Jakab A et al (2016) Melanoma-derived BRAF(V600E) mutation in peritumoral stromal cells: implications for in vivo cell fusion. Int J Mol Sci 17(6):980

Larsen JM, Christensen IJ, Nielsen HJ, Hansen U, Bjerregaard B, Talts JF et al (2009) Syncytin immunoreactivity in colorectal cancer: potential prognostic impact. Cancer Lett 280:44–49

Larsson LI, Holck S, Christensen IJ (2007) Prognostic role of syncytin expression in breast cancer. Hum Pathol 38:726–731

Lartigue L, Merle C, Lagarde P, Delespaul L, Lesluyes T, Le Guellec S et al (2020) Genome remodeling upon mesenchymal tumor cell fusion contributes to tumor progression and metastatic spread. Oncogene 39:4198–4211

Lazova R, Laberge GS, Duvall E, Spoelstra N, Klump V, Sznol M et al (2013) A melanoma brain metastasis with a donor-patient hybrid genome following bone marrow transplantation: first evidence for fusion in human cancer. PLoS One 8:e66731

Le Boeuf F, Gebremeskel S, McMullen N, He H, Greenshields AL, Hoskin DW et al (2017) Reovirus FAST protein enhances vesicular stomatitis virus oncolytic virotherapy in primary and metastatic tumor models. Mol Ther Oncolytics 6:80–89

Lees RL, Heersche JN (1999) Macrophage colony stimulating factor increases bone resorption in dispersed osteoclast cultures by increasing osteoclast size. J Bone Miner Res 14:937–945

Lehka L, Redowicz MJ (2020) Mechanisms regulating myoblast fusion: A multilevel interplay. Semin Cell Dev Biol 104:81–92

Leikina E, Gamage DG, Prasad V, Goykhberg J, Crowe M, Diao J et al (2018) Myomaker and myomerger work independently to control distinct steps of membrane remodeling during myoblast fusion. Dev Cell 46:767–780 e767

Levaot N, Ottolenghi A, Mann M, Guterman-Ram G, Kam Z, Geiger B (2015) Osteoclast fusion is initiated by a small subset of RANKL-stimulated monocyte progenitors, which can fuse to RANKL-unstimulated progenitors. Bone 79:21–28

Li R, Sonik A, Stindl R, Rasnick D, Duesberg P (2000) Aneuploidy vs. gene mutation hypothesis of cancer: recent study claims mutation but is found to support aneuploidy. Proc Natl Acad Sci U S A 97:3236–3241

Li X, Lewis MT, Huang J, Gutierrez C, Osborne CK, Wu MF et al (2008) Intrinsic resistance of tumorigenic breast cancer cells to chemotherapy. J Natl Cancer Inst 100:672–679

Li M, Radvanyi L, Yin B, Rycaj K, Li J, Chivukula R et al (2017) Downregulation of human endogenous retrovirus type K (HERV-K) viral env RNA in pancreatic cancer cells decreases cell proliferation and tumor growth. Clin Cancer Res 23:5892–5911

Li X, Fu Y, Xia X, Zhang X, Xiao K, Zhuang X et al (2019) Knockdown of SP1/Syncytin1 axis inhibits the proliferation and metastasis through the AKT and ERK1/2 signaling pathways in non-small cell lung cancer. Cancer Med 8:5750–5759

Li Y, Zhao L, Li XF (2021) Hypoxia and the Tumor Microenvironment. Technol Cancer Res Treat 20:15330338211036304

Liang P, Yang H (2021) Molecular underpinning of intracellular pH regulation on TMEM16F. J Gen Physiol 153

Lin EH, Salon C, Brambilla E, Lavillette D, Szecsi J, Cosset FL et al (2010) Fusogenic membrane glycoproteins induce syncytia formation and death in vitro and in vivo: a potential therapy agent for lung cancer. Cancer Gene Ther 17:256–265

Lin KC, Torga G, Sun Y, Axelrod R, Pienta KJ, Sturm JC et al (2019) The role of heterogeneous environment and docetaxel gradient in the emergence of polyploid, mesenchymal and resistant prostate cancer cells. Clin Exp Metastasis 36:97–108

Lindström A, Midtbö K, Arnesson LG, Garvin S, Shabo I (2017) Fusion between M2-macrophages and cancer cells results in a subpopulation of radioresistant cells with enhanced DNA-repair capacity. Oncotarget 8:51370–51386

Liu C, Liu L, Wang X, Liu Y, Wang M, Zhu F (2017) HBV X Protein induces overexpression of HERV-W env through NF-kappaB in HepG2 cells. Virus Genes 53:797–806

Liu C, Xu J, Wen F, Yang F, Li X, Geng D et al (2019) Upregulation of syncytin-1 promotes invasion and metastasis by activating epithelial-mesenchymal transition-related pathway in endometrial carcinoma. Onco Targets Ther 12:31–40

Liu C, Billet S, Choudhury D, Cheng R, Haldar S, Fernandez A et al (2021) Bone marrow mesenchymal stem cells interact with head and neck squamous cell carcinoma cells to promote cancer progression and drug resistance. Neoplasia 23:118–128

Lopresti A, Acquaviva C, Boudin L, Finetti P, Garnier S, Aulas A et al (2022) Identification of atypical circulating tumor cells with prognostic value in metastatic breast cancer patients. Cancers (Basel) 14(4):932

Lu X, Kang Y (2009) Efficient acquisition of dual metastasis organotropism to bone and lung through stable spontaneous fusion between MDA-MB-231 variants. Proc Natl Acad Sci U S A 106:9385–9390

Luo F, Liu T, Wang J, Li J, Ma P, Ding H et al (2016) Bone marrow mesenchymal stem cells participate in prostate carcinogenesis and promote growth of prostate cancer by cell fusion in vivo. Oncotarget 7:30924–30934

Malassine A, Handschuh K, Tsatsaris V, Gerbaud P, Cheynet V, Oriol G et al (2005) Expression of HERV-W Env glycoprotein (syncytin) in the extravillous trophoblast of first trimester human placenta. Placenta 26:556–562

Maliniemi P, Vincendeau M, Mayer J, Frank O, Hahtola S, Karenko L et al (2013) Expression of human endogenous retrovirus-w including syncytin-1 in cutaneous T-cell lymphoma. PLoS One 8:e76281

Mameli G, Poddighe L, Mei A, Uleri E, Sotgiu S, Serra C et al (2012) Expression and activation by Epstein Barr virus of human endogenous retroviruses-W in blood cells and astrocytes: inference for multiple sclerosis. PLoS One 7:e44991

Manjunath Y, Porciani D, Mitchem JB, Suvilesh KN, Avella DM, Kimchi ET et al (2020a) Tumor-cell-macrophage fusion cells as liquid biomarkers and tumor enhancers in cancer. Int J Mol Sci 21(5):1872

Manjunath Y, Mitchem JB, Suvilesh KN, Avella DM, Kimchi ET, Staveley-O'Carroll KF et al (2020b) Circulating giant tumor-macrophage fusion cells are independent prognosticators in patients with NSCLC. J Thorac Oncol 15(9):1460–1471

Marston JL, Greenig M, Singh M, Bendall ML, Duarte RRR, Feschotte C et al (2021) SARS-CoV-2 infection mediates differential expression of human endogenous retroviruses and long interspersed nuclear elements. JCI Insight 6:e147170

Martens S, McMahon HT (2008) Mechanisms of membrane fusion: disparate players and common principles. Nat Rev Mol Cell Biol 9:543–556

Martin SJ, Reutelingsperger CP, McGahon AJ, Rader JA, van Schie RC, LaFace DM et al (1995) Early redistribution of plasma membrane phosphatidylserine is a general feature of apoptosis regardless of the initiating stimulus: inhibition by overexpression of Bcl-2 and Abl. J Exp Med 182:1545–1556

Matsumoto T, Wakefield L, Peters A, Peto M, Spellman P, Grompe M (2021) Proliferative polyploid cells give rise to tumors via ploidy reduction. Nat Commun 12:646

Matveeva OV, Shabalina SA (2020) Prospects for using expression patterns of paramyxovirus receptors as biomarkers for oncolytic virotherapy. Cancers (Basel) 12(12):3659

Meirelles K, Benedict LA, Dombkowski D, Pepin D, Preffer FI, Teixeira J et al (2012) Human ovarian cancer stem/progenitor cells are stimulated by doxorubicin but inhibited by Mullerian inhibiting substance. Proc Natl Acad Sci U S A 109:2358–2363

Melzer C, von der Ohe J, Hass R (2018a) Enhanced metastatic capacity of breast cancer cells after interaction and hybrid formation with mesenchymal stroma/stem cells (MSC). Cell Commun Signal 16:2

Melzer C, von der Ohe J, Hass R (2018b) In vitro fusion of normal and neoplastic breast epithelial cells with human mesenchymal stroma/stem cells partially involves tumor necrosis factor receptor signaling. Stem Cells 36:977–989

Melzer C, von der Ohe J, Hass R (2018c) MSC stimulate ovarian tumor growth during intercellular communication but reduce tumorigenicity after fusion with ovarian cancer cells. Cell Commun Signal 16:67

Melzer C, von der Ohe J, Hass R (2019a) Involvement of actin cytoskeletal components in breast cancer cell fusion with human mesenchymal stroma/stem-like cells. Int J Mol Sci 20(4):876

Melzer C, von der Ohe J, Hass R (2019b) In vivo cell fusion between mesenchymal stroma/stem-like cells and breast cancer cells. Cancers (Basel) 11(2):185

Melzer C, Ohe JV, Hass R (2020) Altered tumor plasticity after different cancer cell fusions with MSC. Int J Mol Sci 21(21):8347

Menyailo ME, Zainullina VR, Khozyainova AA, Tashireva LA, Zolotareva SY, Gerashchenko TS et al (2022) Heterogeneity of circulating epithelial cells in breast cancer at single-cell resolution: identifying tumor and hybrid cells. Adv Biol (Weinh) 7(2):e2200206

Merle C, Lagarde P, Lartigue L, Chibon F (2021) Acquisition of cancer stem cell capacities after spontaneous cell fusion. BMC Cancer 21:241

Mi S, Lee X, Li X-p, Veldman GM, Finnerty H, Racie L et al (2000) Syncytin is a captive retroviral envelope protein involved in human placental morphogenesis. Nature 403:785–789

Mi R, Pan C, Bian X, Song L, Tian W, Cao F et al (2012) Fusion between tumor cells enhances melanoma metastatic potential. J Cancer Res Clin Oncol 138:1651–1658

Millay DP, O'Rourke JR, Sutherland LB, Bezprozvannaya S, Shelton JM, Bassel-Duby R et al (2013) Myomaker is a membrane activator of myoblast fusion and muscle formation. Nature 499:301–305

Miller FR, Mohamed AN, McEachern D (1989) Production of a more aggressive tumor cell variant by spontaneous fusion of two mouse tumor subpopulations. Cancer Res 49:4316–4321

Miroshnychenko D, Baratchart E, Ferrall-Fairbanks MC, Velde RV, Laurie MA, Bui MM et al (2021) Spontaneous cell fusions as a mechanism of parasexual recombination in tumour cell populations. Nat Ecol Evol 5:379–391

Mirzayans R, Murray D (2020) Intratumor heterogeneity and therapy resistance: contributions of dormancy, apoptosis reversal (anastasis) and cell fusion to disease recurrence. Int J Mol Sci 21(4):1308

Mirzayans R, Andrais B, Scott A, Wang YW, Kumar P, Murray D (2017) Multinucleated giant cancer cells produced in response to ionizing radiation retain viability and replicate their genome. Int J Mol Sci 18(2):360

Mittelbrunn M, Sanchez-Madrid F (2012) Intercellular communication: diverse structures for exchange of genetic information. Nat Rev Mol Cell Biol 13:328–335

Mizuno A, Amizuka N, Irie K, Murakami A, Fujise N, Kanno T et al (1998) Severe osteoporosis in mice lacking osteoclastogenesis inhibitory factor/osteoprotegerin. Biochem Biophys Res Commun 247:610–615

Mohr M, Tosun S, Arnold WH, Edenhofer F, Zanker KS, Dittmar T (2015) Quantification of cell fusion events human breast cancer cells and breast epithelial cells using a Cre-LoxP-based double fluorescence reporter system. Cell Mol Life Sci 72:3769–3782

Moller AM, Delaisse JM, Soe K (2017) Osteoclast fusion: time-lapse reveals involvement of CD47 and Syncytin-1 at different stages of nuclearity. J Cell Physiol 232:1396–1403

Montalban-Hernandez K, Cantero-Cid R, Casalvilla-Duenas JC, Avendano-Ortiz J, Marin E, Lozano-Rodriguez R et al (2022) Colorectal cancer stem cells fuse with monocytes to form tumour hybrid cells with the ability to migrate and evade the immune system. Cancers (Basel) 14(14):3445

Msheik H, El Hayek S, Bari MF, Azar J, Abou-Kheir W, Kobeissy F et al (2019) Transcriptomic profiling of trophoblast fusion using BeWo and JEG-3 cell lines. Mol Hum Reprod 25:811–824

Muir A, Lever AML, Moffett A (2006) Human endogenous retrovirus-W envelope (syncytin) is expressed in both villous and extravillous trophoblast populations. J Gen Virol 87:2067–2071

Nakamura T, Peng KW, Vongpunsawad S, Harvey M, Mizuguchi H, Hayakawa T et al (2004) Antibody-targeted cell fusion. Nat Biotechnol 22:331–336

Noubissi FK, Harkness T, Alexander CM, Ogle BM (2015) Apoptosis-induced cancer cell fusion: a mechanism of breast cancer metastasis. FASEB J 29:4036–4045

Ogle BM, Cascalho M, Platt JL (2005) Biological implications of cell fusion. Nat Rev Mol Cell Biol 6:567–575

Oromendia AB, Dodgson SE, Amon A (2012) Aneuploidy causes proteotoxic stress in yeast. Genes Dev 26:2696–2708

Overholtzer M, Brugge JS (2008) The cell biology of cell-in-cell structures. Nat Rev Mol Cell Biol 9:796–809

Parappilly MS, Chin Y, Whalen RM, Anderson AN, Robinson TS, Strgar L et al (2022) Circulating neoplastic-immune hybrid cells predict metastatic progression in uveal melanoma. Cancers (Basel) 14(19):4617

Passerini V, Ozeri-Galai E, de Pagter MS, Donnelly N, Schmalbrock S, Kloosterman WP et al (2016) The presence of extra chromosomes leads to genomic instability. Nat Commun 7:10754

Pawelek JM (2014) Fusion of bone marrow-derived cells with cancer cells: metastasis as a secondary disease in cancer. Chin J Cancer 33:133–139

Pereira-Veiga T, Schneegans S, Pantel K, Wikman H (2022) Circulating tumor cell-blood cell crosstalk: Biology and clinical relevance. Cell Rep 40:111298

Petrany MJ, Millay DP (2019) Cell fusion: merging membranes and making muscle. Trends Cell Biol 29:964–973

Petrova V, Annicchiarico-Petruzzelli M, Melino G, Amelio I (2018) The hypoxic tumour microenvironment. Oncogenesis 7:10

Pircher T, Wackerhage H, Akova E, Bocker W, Aszodi A, Saller MM (2022) Fusion of normoxic- and hypoxic-preconditioned myoblasts leads to increased hypertrophy. Cells 11(6):1059
Podbilewicz B (2014) Virus and cell fusion mechanisms. Annu Rev Cell Dev Biol 30:111–139
Potgens AJ, Drewlo S, Kokozidou M, Kaufmann P (2004) Syncytin: the major regulator of trophoblast fusion? Recent developments and hypotheses on its action. Hum Reprod Update 10:487–496
Powell AE, Anderson EC, Davies PS, Silk AD, Pelz C, Impey S et al (2011) Fusion between Intestinal epithelial cells and macrophages in a cancer context results in nuclear reprogramming. Cancer Res 71:1497–1505
Puig PE, Guilly MN, Bouchot A, Droin N, Cathelin D, Bouyer F et al (2008) Tumor cells can escape DNA-damaging cisplatin through DNA endoreduplication and reversible polyploidy. Cell Biol Int 32:1031–1043
Quinn ME, Goh Q, Kurosaka M, Gamage DG, Petrany MJ, Prasad V et al (2017) Myomerger induces fusion of non-fusogenic cells and is required for skeletal muscle development. Nat Commun 8:15665
Quinton RJ, DiDomizio A, Vittoria MA, Kotynkova K, Ticas CJ, Patel S et al (2021) Whole-genome doubling confers unique genetic vulnerabilities on tumour cells. Nature 590:492–497
Rachkovsky M, Sodi S, Chakraborty A, Avissar Y, Bolognia J, McNiff JM et al (1998) Melanoma x macrophage hybrids with enhanced metastatic potential. Clin Exp Metastasis 16:299–312
Ramakrishnan M, Mathur SR, Mukhopadhyay A (2013) Fusion-derived epithelial cancer cells express hematopoietic markers and contribute to stem cell and migratory phenotype in ovarian carcinoma. Cancer Res 73:5360–5370
Rappa G, Mercapide J, Lorico A (2012) Spontaneous formation of tumorigenic hybrids between breast cancer and multipotent stromal cells is a source of tumor heterogeneity. Am J Pathol 180:2504–2515
Ribatti D, Tamma R (2018) A revisited concept. Tumors: wounds that do not heal. Crit Rev Oncol Hematol 128:65–69
Rizvi AZ, Swain JR, Davies PS, Bailey AS, Decker AD, Willenbring H et al (2006) Bone marrow-derived cells fuse with normal and transformed intestinal stem cells. Proc Natl Acad Sci U S A 103:6321–6325
Ruano APC, Gadelha Guimaraes AP, Braun AC, Flores B, Tariki MS, Abdallah EA et al (2022) Fusion cell markers in circulating tumor cells from patients with high-grade ovarian serous carcinoma. Int J Mol Sci 23(23):14687
Rubio C, Avendano-Ortiz J, Ruiz-Palomares R, Karaivanova V, Alberquilla O, Sanchez-Dominguez R et al (2022) Toward tumor fight and tumor microenvironment remodeling: PBA induces cell cycle arrest and reduces tumor hybrid cells' pluripotency in bladder cancer. Cancers (Basel) 14(2):287
Ruebner M, Strissel PL, Ekici AB, Stiegler E, Dammer U, Goecke TW et al (2013) Reduced syncytin-1 expression levels in placental syndromes correlates with epigenetic hypermethylation of the ERVW-1 promoter region. PLoS One 8:e56145
Salsman J, Top D, Boutilier J, Duncan R (2005) Extensive syncytium formation mediated by the reovirus FAST proteins triggers apoptosis-induced membrane instability. J Virol 79:8090–8100
Schmidt-Supprian M, Rajewsky K (2007) Vagaries of conditional gene targeting. Nature Immunology 8:665–668
Shabo I, Stal O, Olsson H, Dore S, Svanvik J (2008) Breast cancer expression of CD163, a macrophage scavenger receptor, is related to early distant recurrence and reduced patient survival. Int J Cancer 123:780–786
Shabo I, Olsson H, Sun XF, Svanvik J (2009) Expression of the macrophage antigen CD163 in rectal cancer cells is associated with early local recurrence and reduced survival time. Int J Cancer 125:1826–1831
Shabo I, Olsson H, Stal O, Svanvik J (2013) Breast cancer expression of DAP12 is associated with skeletal and liver metastases and poor survival. Clin Breast Cancer 13:371–377

Shabo I, Midtbo K, Andersson H, Akerlund E, Olsson H, Wegman P et al (2015) Macrophage traits in cancer cells are induced by macrophage-cancer cell fusion and cannot be explained by cellular interaction. BMC Cancer 15:922

Sharma B, Kanwar SS (2018) Phosphatidylserine: A cancer cell targeting biomarker. Semin Cancer Biol 52:17–25

Sheltzer JM (2013) A transcriptional and metabolic signature of primary aneuploidy is present in chromosomally unstable cancer cells and informs clinical prognosis. Cancer Res 73:6401–6412

Skokos EA, Charokopos A, Khan K, Wanjala J, Kyriakides TR (2011) Lack of TNF-alpha-induced MMP-9 production and abnormal E-cadherin redistribution associated with compromised fusion in MCP-1-null macrophages. Am J Pathol 178:2311–2321

Sodi SA, Chakraborty AK, Platt JT, Kolesnikova N, Rosemblat S, Keh-Yen A et al (1998) Melanoma x macrophage fusion hybrids acquire increased melanogenesis and metastatic potential: altered N-glycosylation as an underlying mechanism. Pigment Cell Res 11:299–309

Soe K (2020) Osteoclast fusion: physiological regulation of multinucleation through heterogeneity-potential implications for drug sensitivity. Int J Mol Sci 21(20):7717

Soe K, Andersen TL, Hobolt-Pedersen AS, Bjerregaard B, Larsson LI, Delaisse JM (2011) Involvement of human endogenous retroviral syncytin-1 in human osteoclast fusion. Bone 48:837–846

Song K, Zhu F, Zhang HZ, Shang ZJ (2012) Tumor necrosis factor-alpha enhanced fusions between oral squamous cell carcinoma cells and endothelial cells via VCAM-1/VLA-4 pathway. Exp Cell Res 318:1707–1715

Song H-L, Liu T-H, Wang Y-H, Li F-F, Ruan L-L, Adu-Gyamfi EA et al (2021a) Appropriate expression of P57kip2 drives trophoblast fusion via cell cycle arrest. Reproduction 161:633–644

Song Y, Zhao Y, Deng Z, Zhao R, Huang Q (2021b) Stress-induced polyploid giant cancer cells: unique way of formation and non-negligible characteristics. Front Oncol 11:724781

Soygur B, Sati L (2016) The role of syncytins in human reproduction and reproductive organ cancers. Reproduction 152:R167–R178

Soygur B, Sati L, Demir R (2016) Altered expression of human endogenous retroviruses syncytin-1, syncytin-2 and their receptors in human normal and gestational diabetic placenta. Histol Histopathol 31:1037–1047

Sprangers AJ, Freeman BT, Kouris NA, Ogle BM (2012) A Cre-Lox P recombination approach for the detection of cell fusion in vivo. J Vis Exp. https://doi.org/10.3791/3581

Stefani M, Dobson CM (2003) Protein aggregation and aggregate toxicity: new insights into protein folding, misfolding diseases and biological evolution. J Mol Med (Berl) 81:678–699

Strick R, Ackermann S, Langbein M, Swiatek J, Schubert SW, Hashemolhosseini S et al (2007) Proliferation and cell-cell fusion of endometrial carcinoma are induced by the human endogenous retroviral Syncytin-1 and regulated by TGF-beta. J Mol Med (Berl) 85:23–38

Strissel PL, Ruebner M, Thiel F, Wachter D, Ekici AB, Wolf F et al (2012) Reactivation of codogenic endogenous retroviral (ERV) envelope genes in human endometrial carcinoma and prestages: Emergence of new molecular targets. Oncotarget 3(10):1204–1219

Sugimoto J, Sugimoto M, Bernstein H, Jinno Y, Schust D (2013) A novel human endogenous retroviral protein inhibits cell-cell fusion. Sci Rep 3:1462

Sugimoto J, Schust DJ, Kinjo T, Aoki Y, Jinno Y, Kudo Y (2019) Suppressyn localization and dynamic expression patterns in primary human tissues support a physiologic role in human placentation. Sci Rep 9:19502

Sulaiman R, De P, Aske JC, Lin X, Dale A, Vaselaar E et al (2022) Identification and morphological characterization of features of circulating cancer-associated macrophage-like cells (CAMLs) in endometrial cancers. Cancers (Basel) 14(19):4577

Sun Y, Ouyang DY, Pang W, Tu YQ, Li YY, Shen XM et al (2010) Expression of syncytin in leukemia and lymphoma cells. Leuk Res 34:1195–1202

Sun C, Dai X, Zhao D, Wang H, Rong X, Huang Q et al (2019) Mesenchymal stem cells promote glioma neovascularization in vivo by fusing with cancer stem cells. BMC Cancer 19:1240

Sutton TL, Patel RK, Anderson AN, Bowden SG, Whalen R, Giske NR et al (2022) Circulating cells with macrophage-like characteristics in cancer: the importance of circulating neoplastic-immune hybrid cells in cancer. Cancers (Basel) 14(16):3871
Suzuki J, Umeda M, Sims PJ, Nagata S (2010) Calcium-dependent phospholipid scrambling by TMEM16F. Nature 468:834–838
Suzuki J, Fujii T, Imao T, Ishihara K, Kuba H, Nagata S (2013) Calcium-dependent phospholipid scramblase activity of TMEM16 protein family members. J Biol Chem 288:13305–13316
Tal A, Tal R, Shaikh S, Gidicsin S, Mamillapalli R, Taylor HS (2019) Characterization of cell fusion in an experimental mouse model of endometriosis. Biol Reprod 100:390–397
Thyagarajan B, Guimarães MJ, Groth AC, Calos MP (2000) Mammalian genomes contain active recombinase recognition sites. Gene 244:47–54
Tinganelli W, Durante M (2020) Tumor hypoxia and circulating tumor cells. Int J Mol Sci 21:9592
Uleri E, Mei A, Mameli G, Poddighe L, Serra C, Dolei A (2014) HIV Tat acts on endogenous retroviruses of the W family and this occurs via Toll-like receptor 4: inference for neuroAIDS. AIDS 28:2659–2670
Uygur B, Leikina E, Melikov K, Villasmil R, Verma SK, Vary CPH et al (2019) Interactions with muscle cells boost fusion, stemness, and drug resistance of prostate cancer cells. Mol Cancer Res 17:806–820
van Niel G, D'Angelo G, Raposo G (2018) Shedding light on the cell biology of extracellular vesicles. Nat Rev Mol Cell Biol 19:213–228
Vargas A, Moreau J, Landry S, LeBellego F, Toufaily C, Rassart E et al (2009) Syncytin-2 plays an important role in the fusion of human trophoblast cells. J Mol Biol 392:301–318
Verma SK, Leikina E, Melikov K, Gebert C, Kram V, Young MF et al (2018) Cell-surface phosphatidylserine regulates osteoclast precursor fusion. J Biol Chem 293:254–270
Vignery A (2005) Macrophage fusion: the making of osteoclasts and giant cells. J Exp Med 202:337–340
Wakeling WF, Greetham J, Bennett DC (1994) Efficient spontaneous fusion between some co-cultured cells, especially murine melanoma cells. Cell Biol Int 18:207–210
Walker BS, Sutton TL, Zarour L, Hunter JG, Wood SG, Tsikitis VL et al (2021) Circulating hybrid cells: a novel liquid biomarker of treatment response in gastrointestinal cancers. Ann Surg Oncol 28:8567–8578
Wang R, Sun X, Wang CY, Hu P, Chu CY, Liu S et al (2012) Spontaneous cancer-stromal cell fusion as a mechanism of prostate cancer androgen-independent progression. PLoS One 7:e42653
Wang Z, Yuan Y, Zhang L, Min Z, Zhou D, Yu S et al (2018) Impact of cell fusion in myeloma marrow microenvironment on tumor progression. Oncotarget 9(57):30997–31006
Wang X, Li Y, Li J, Li L, Zhu H, Chen H et al (2019) Cell-in-Cell phenomenon and its relationship with tumor microenvironment and tumor progression: a review. Front Cell Dev Biol 7:311
Wang R, Lewis MS, Lyu J, Zhau HE, Pandol SJ, Chung LWK (2020) Cancer-stromal cell fusion as revealed by fluorescence protein tracking. Prostate 80:274–283
Wang HF, Xiang W, Xue BZ, Wang YH, Yi DY, Jiang XB et al (2021) Cell fusion in cancer hallmarks: Current research status and future indications. Oncol Lett 22:530
Wang YN, Ye Y, Zhou D, Guo ZW, Xiong Z, Gong XX et al (2022) The role of syncytin in placental angiogenesis and fetal growth. Front Cell Dev Biol 10:852561
Wei HJ, Nickoloff JA, Chen WH, Liu HY, Lo WC, Chang YT et al (2014) FOXF1 mediates mesenchymal stem cell fusion-induced reprogramming of lung cancer cells. Oncotarget 5:9514–9529
Weiler J, Dittmar T (2019a) Cell fusion in human cancer: the dark matter hypothesis. Cells 8(2):132
Weiler J, Dittmar T (2019b) Minocycline impairs TNF-alpha-induced cell fusion of M13SV1-Cre cells with MDA-MB-435-pFDR1 cells by suppressing NF-kappaB transcriptional activity and its induction of target-gene expression of fusion-relevant factors. Cell Commun Signal 17:71
Whitlock JM, Chernomordik LV (2021) Flagging fusion: Phosphatidylserine signaling in cell-cell fusion. J Biol Chem 296:100411
Whitlock JM, Yu K, Cui YY, Hartzell HC (2018) Anoctamin 5/TMEM16E facilitates muscle precursor cell fusion. J Gen Physiol 150:1498–1509

Xia P, Wang S, Guo Z, Yao X (2008) Emperipolesis, entosis and beyond: dance with fate. Cell Res 18:705–707
Xie X, Li Y, Lian S, Lu Y, Jia L (2022) Cancer metastasis chemoprevention prevents circulating tumour cells from germination. Signal Transduct Target Ther 7:341
Xu MH, Gao X, Luo D, Zhou XD, Xiong W, Liu GX (2014) EMT and acquisition of stem cell-like properties are involved in spontaneous formation of tumorigenic hybrids between lung cancer and bone marrow-derived mesenchymal stem cells. PLoS One 9:e87893
Xue J, Zhu Y, Sun Z, Ji R, Zhang X, Xu W et al (2015) Tumorigenic hybrids between mesenchymal stem cells and gastric cancer cells enhanced cancer proliferation, migration and stemness. BMC Cancer 15:793
Xue B, Sechi LA, Kelvin DJ (2020) Human endogenous retrovirus K (HML-2) in health and disease. Front Microbiol 11:1690
Yagi M, Miyamoto T, Sawatani Y, Iwamoto K, Hosogane N, Fujita N et al (2005) DC-STAMP is essential for cell-cell fusion in osteoclasts and foreign body giant cells. J Exp Med 202:345–351
Yan Y, Hsu Y, He X, Lu N, Wei W, Zhang Z et al (2015) Evidence of cell fusion in carcinogen-induced mice gastric carcinoma. Tumour Biol 36:5089–5094
Yan B, Wang J, Liu L (2016) Chemotherapy promotes tumour cell hybridization in vivo. Tumour Biol 37:5025–5030
Yan TL, Wang M, Xu Z, Huang CM, Zhou XC, Jiang EH et al (2017) Up-regulation of syncytin-1 contributes to TNF-alpha-enhanced fusion between OSCC and HUVECs partly via Wnt/beta-catenin-dependent pathway. Sci Rep 7:40983
Yanez-Mo M, Siljander PR, Andreu Z, Zavec AB, Borras FE, Buzas EI et al (2015) Biological properties of extracellular vesicles and their physiological functions. J Extracell Vesicles 4:27066
Yart L, Bastida-Ruiz D, Allard M, Dietrich PY, Petignat P, Cohen M (2022) Linking unfolded protein response to ovarian cancer cell fusion. BMC Cancer 22:622
Yilmaz Y, Lazova R, Qumsiyeh M, Cooper D, Pawelek J (2005) Donor Y chromosome in renal carcinoma cells of a female BMT recipient: visualization of putative BMT-tumor hybrids by FISH. Bone Marrow Transplant 35:1021–1024
Yu H, Liu T, Zhao Z, Chen Y, Zeng J, Liu S et al (2014) Mutations in 3′-long terminal repeat of HERV-W family in chromosome 7 upregulate syncytin-1 expression in urothelial cell carcinoma of the bladder through interacting with c-Myb. Oncogene 33:3947–3958
Yu M, Zhao H, Chen T, Tian Y, Li M, Wu K et al (2018) Mutational analysis of IZUMO1R in women with fertilization failure and polyspermy after in vitro fertilization. J Assist Reprod Genet 35:539–544
Zack TI, Schumacher SE, Carter SL, Cherniack AD, Saksena G, Tabak B et al (2013) Pan-cancer patterns of somatic copy number alteration. Nat Genet 45:1134–1140
Zaitseva E, Zaitsev E, Melikov K, Arakelyan A, Marin M, Villasmil R et al (2017) Fusion stage of HIV-1 entry depends on virus-induced cell surface exposure of phosphatidylserine. Cell Host Microbe 22:99–110 e117
Zhang Q, Vashisht AA, O'Rourke J, Corbel SY, Moran R, Romero A et al (2017) The microprotein minion controls cell fusion and muscle formation. Nat Commun 8:15664
Zhang LN, Kong CF, Zhao D, Cong XL, Wang SS, Ma L et al (2019) Fusion with mesenchymal stem cells differentially affects tumorigenic and metastatic abilities of lung cancer cells. J Cell Physiol 234:3570–3582
Zhang Y, Le T, Grabau R, Mohseni Z, Kim H, Natale DR et al (2020) TMEM16F phospholipid scramblase mediates trophoblast fusion and placental development. Sci Adv 6:eaba0310
Zhang Y, Liang P, Yang L, Shan KZ, Feng L, Chen Y et al (2022) Functional coupling between TRPV4 channel and TMEM16F modulates human trophoblast fusion. Elife 11:e78840
Zhou X, Platt JL (2011) Molecular and cellular mechanisms of mammalian cell fusion. Adv Exp Med Biol 713:33–64
Zhou X, Merchak K, Lee W, Grande JP, Cascalho M, Platt JL (2015) Cell fusion connects oncogenesis with tumor evolution. Am J Pathol 185:2049–2060

Zhou F, Li M, Wei Y, Lin K, Lu Y, Shen J et al (2016) Activation of HERV-K Env protein is essential for tumorigenesis and metastasis of breast cancer cells. Oncotarget 7:84093–84117
Zhou Y, Liu L, Liu Y, Zhou P, Yan Q, Yu H et al (2021) Implication of human endogenous retrovirus W family envelope in hepatocellular carcinoma promotes MEK/ERK-mediated metastatic invasiveness and doxorubicin resistance. Cell Death Discov 7:177
Zielske SP, Spalding AC, Wicha MS, Lawrence TS (2011) Ablation of breast cancer stem cells with radiation. Transl Oncol 4:227–233

Chapter 21
The Hallmarks of Circulating Hybrid Cells

Ranish K. Patel, Michael Parappilly, Shahrose Rahman, Issac R. Schwantes, Marisa Sewell, Nicole R. Giske, Riley M. Whalen, Naside Gozde Durmus, and Melissa H. Wong

Abstract While tumor metastases represent the primary driver of cancer-related mortality, our understanding of the mechanisms that underlie metastatic initiation and progression remains incomplete. Recent work identified a novel tumor-macrophage hybrid cell population, generated through the fusion between neoplastic and immune cells. These hybrid cells are detected in primary tumor tissue, peripheral blood, and in metastatic sites. In-depth analyses of hybrid cell biology indicate that they can exploit phenotypic properties of both parental tumor and immune cells, in order to intravasate into circulation, evade the immune response, and seed tumors at distant sites. Thus, it has become increasingly evident that the development and dissemination of tumor-immune hybrid cells play an intricate and fundamental role in the metastatic cascade and can provide invaluable information regarding tumor characteristics and patient prognostication. In this chapter, we review the current understanding of this novel hybrid cell population, the specific

Shahrose Rahman, Issac R. Schwantes, Marisa Sewell, Nicole R. Giske and Riley M. Whalen contributed equally with all other contributors.

R. K. Patel · S. Rahman · I. R. Schwantes · M. Sewell
Department of Surgery, Oregon Health & Science University, Portland, OR, USA

M. Parappilly · N. R. Giske · R. M. Whalen
Department of Cell, Developmental and Cancer Biology, Oregon Health & Science University, Portland, OR, USA

N. G. Durmus
Molecular Imaging Program at Stanford (MIPS), Department of Radiology, Stanford University School of Medicine, Palo Alto, CA, USA

M. H. Wong (✉)
Department of Cell, Developmental and Cancer Biology, Oregon Health & Science University, Portland, OR, USA

Knight Cancer Institute, Oregon Health & Science University, Portland, OR, USA
e-mail: wongme@ohsu.edu

M. Kloc, A. Uosef (eds.), *Syncytia: Origin, Structure, and Functions*, Results and Problems in Cell Differentiation 71,
https://doi.org/10.1007/978-3-031-37936-9_21

hallmarks of cancer that these cells exploit to promote cancer progression and metastasis, and discuss exciting new frontiers that remain to be explored.

Abbreviations

CHCs	Circulating hybrid cells
CTCs	Circulating tumor cells
EMT	Epithelial to mesenchymal transition
EpCAM	Epithelial cellular adhesion molecule
FISH	Fluorescence in situ hybridization
GFP	Green fluorescent protein
GVHD	Graft versus host disease
MET	Mesenchymal to epithelial transition
MMP	Matrix metalloproteinase
PBMCs	Peripheral blood mononuclear cells
RFP	Red fluorescent protein
TAMs	Tumor-associated macrophages
VEGF	Vascular endothelial growth factor

21.1 Introduction

Cellular fusion is a phenomenon by which cells from identical (homotypic) or distinct (heterotypic) lineages combine to form a single cell. Through fusion, the cellular cytoplasmic and nuclear contents of two or more cells merge, all within a single plasma membrane. The parent nuclei themselves may either fuse into a single nucleus, or remain distinct from one another resulting in either a mononuclear (synkaryotic) or a multinuclear (heterokaryotic) daughter cell. Synkaryotic syncytia are less common compared to heterokaryotic syncytia, as best evidenced by the vast abundance of heterokaryotic syncytia in normal human skeletal, cardiac, and smooth muscle, as well as osteoclasts and placental tissue.

Fusion is essential to a variety of physiologic processes, such as fertilization, organ development, immunity, and tissue regeneration and repair (Hass et al. 2021). It is a fundamental biologic mechanism by which cellular phenotypic and genotypic diversity are rapidly enhanced, rendering the newly generated hybrid cell distinct from the parents from which they are derived, and thus increases overall heterogeneity and tissue plasticity. In cancer, this normal biologic process is hijacked to contribute to tumor heterogeneity, tumorigenesis, metastasis, and therapeutic drug resistance (Jiang et al. 2019), thus highlighting the need to better understand how hybrid cells contribute to overall homeostasis or disease processes.

In 1911, German pathologist Otto Aichel first theorized that fusion between immune and cancer cells may play a central role in malignancy and metastasis (Aichel 1911). This theory was rooted in the pioneering work of Theodore and Marcella Boveri detailing the relationship between aberrant chromosome numbers and abnormal mitosis in sea urchin eggs experimentally fertilized with two sets of spermatozoa (Boveri 2008; Larsson 2010; Vande Woude and Klein 2009). Aichel postulated, that through leukocyte fusion, tumor cells could develop malignant phenotypes by exploiting quantitative and qualitative chromosomal differences in order to acquire motile properties inherent to immune cells. As a result, such fusion would result in an "entirely new cell having characteristics of both mother cells" (Aichel 1911; LaBerge et al. 2019). Almost a century later, the leukocyte fusion hypothesis re-emerged, and considerable work has since been conducted to identify immune-neoplastic hybrid cells and demonstrate their aneuploidy and greater malignant potential in both in vitro and in vivo models (Avilès et al. 1977; Chakraborty et al. 2000; Fortuna et al. 1989, 1990; Goldenberg 1968; Goldenberg et al. 1971, 1974; Mekler 1971; Pawelek 2005, 2007; Pawelek and Chakraborty 2008a, b; Kerbel et al. 1983).

While cellular mechanisms such as cytoplasmic bridges and exosome transfer can explain how tumor and immune cells transiently share limited genotypic and phenotypic information, a cellular fusion mechanism conveys larger scale and heritable genotypic and phenotypic alterations in the newly formed neoplastic-immune hybrid cell population. Of note, these hybrid cells are mononuclear, or synkaryotic, as a result of the both cytoplasmic and nuclear fusion of their two parent lineages. In vitro studies involving co-cultured murine colorectal cancer cells expressing nuclear red fluorescent protein (RFP) and macrophages expressing cytoplasmic green fluorescent protein (GFP) demonstrated real-time spontaneous cellular and nuclear fusion, generating RFP^+/GFP^+ daughter hybrid cells that persisted through multiple generations (Gast et al. 2018). Perhaps the strongest in vivo evidence comes from the identification of hybrid cells in female patients with prior sex-mismatched bone marrow transplants, who subsequently developed pancreatic ductal adenocarcinoma. These tumors were found to have intra-tumoral hybrid cells that expressed both Y-chromosome and specific epithelial tumor protein expression (Gast et al. 2018). Further, the Pawelek group reported similarly identified tumors, likely seeded by hybrid cells in patients who received bone marrow or allogeneic stem cell donations, where these tumors contained DNA from both the patient and donor (Chakraborty et al. 2004; LaBerge et al. 2017, 2021; Lazova et al. 2013; Yilmaz et al. 2005). Recently, heterotypic immune-neoplastic hybrid cells were described and identified in a variety of human malignancies (Clawson et al. 2012; Davies et al. 2009; Powell et al. 2011; Rizvi et al. 2006; Silk et al. 2013; Walker et al. 2021; Lazova et al. 2013; Duelli and Lazebnik 2003; LaBerge et al. 2017; Pawelek 2005; Pawelek and Chakraborty 2008b), and their role in cancer progression is increasingly intriguing.

Subsequent to their formation within the primary tumor tissue, hybrid cells gain the ability to intravasate into circulation, and thus they become circulating hybrid cells (CHCs) (Dietz et al. 2021; Gast et al. 2018). CHCs co-expressing the pan-leukocyte antigen CD45 and specific canonical tumor protein expression have been identified in peripheral blood specimens of patients across a myriad of solid tumor malignancies (Dietz et al. 2021; Allan et al. 2005; de Wit et al. 2018; Li et al. 2018; Liu et al. 2016; Manjunath et al. 2020; Nel et al. 2014; Parappilly et al. 2022; Toyoshima et al. 2015; Walker et al. 2021; Zhang et al. 2015). CHCs demonstrate great potential to be utilized as an important biomarker to predict survival, prognostication, relapse, and response to treatment.

The dissemination of neoplastic cells into circulation remains an important step within the larger metastatic cascade. Historically, CD45-negative, unfused circulating tumor cells (CTCs) have dominated study in the field of circulating neoplastic cells; however, these cells are difficult to identify owing to their relative rarity in peripheral blood (Dietz et al. 2021; Gast et al. 2018; Lin et al. 2021; Sutton et al. 2022). In contrast, CHCs have relative enrichment in circulation at an order of magnitude greater than CTCs and have been found to carry their own malignant potential. They promote tumor growth when isolated and injected intra-dermally, as well as seed distant tissue when injected into circulation (Gast et al. 2018; Clawson et al. 2012).

Hanahan and Weinberg first proposed "the Hallmarks of Cancer" (Hanahan and Weinberg 2000), which initially included six essential alterations in cell physiology that are fundamental to the development of malignancy, and have since expanded this list to include fourteen different characteristics (Hanahan 2022; Hanahan and Weinberg 2011). As we further establish the role that CHCs play in malignancy and metastasis, it has become increasingly apparent that these neoplastic-immune cells share many of the fundamental hallmarks of cancer, namely: *tumor-promoting inflammation, genomic instability and mutation, unlocking phenotypical plasticity, invasion and dissemination into vasculature, and avoiding immune destruction* (Fig. 21.1). Herein, we provide insight into how CHCs possess and exploit each of these fundamental properties, to promote tumor progression and metastasis.

21.2 Tumor-Promoting Inflammation

It is well-established that virtually every neoplastic lesion contains notable populations of immune cells, both from the innate and adaptive arms of the immune system (Hanahan and Weinberg 2011), and that these tumor-associated inflammatory conditions are key mediators of tumorigenesis and enhance malignant progression. The tumor microenvironment has a significant impact on the proliferative and metastatic capabilities of cancer cells (Hanahan and Weinberg 2011). Initially reported as far back as 1863 by Rudolf Virchow, chronic inflammation was identified as a predisposing event in the development of cancer, but no specific biological pathways were implicated in this finding for over a century (Virchow 1863). In 1986, neoplasms were shown to demonstrate the ability to induce similar effects in microenvironment as immune responses, initially identified as changes in endothelial permeability via the secretion of vascular endothelial growth factor (VEGF) (Senger et al. 1986). It is now understood that local inflammation provides an influx of bioactive molecules, including cytokines inducing proliferative signaling, enzymes which modify surrounding stroma to enhance angiogenesis, suppress the antitumor response and metastatic potential of disease, as well as other factors, which can induce de-differentiation of neoplastic cells (DeNardo et al. 2010; Parker et al.

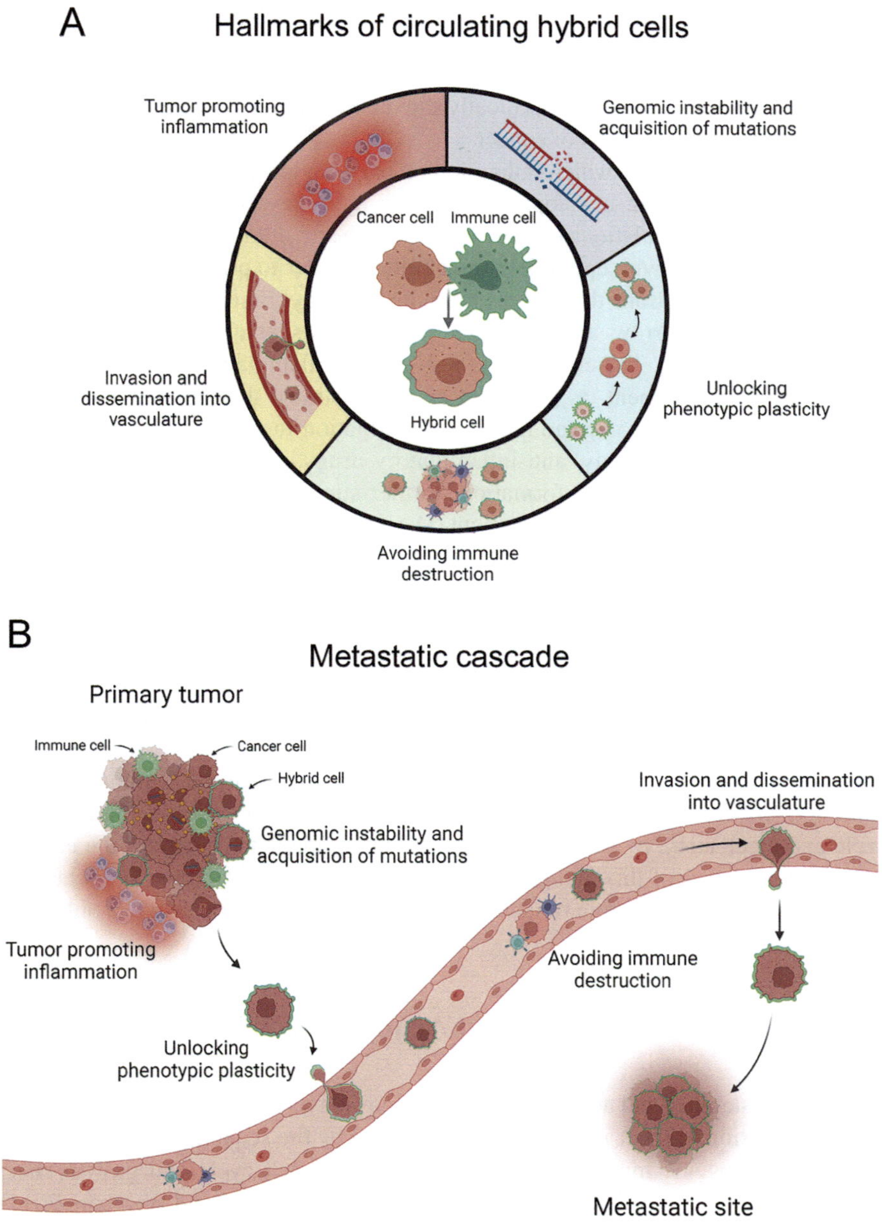

Fig. 21.1 (**a**) The hallmarks of circulating hybrid cells currently embody five of the characteristics from Hanahan and Weinberg's hallmarks of cancer (Hanahan and Weinberg 2011). Genomic instability and acquisition of mutations and invasion and dissemination into vasculature are considered core characteristics as they contribute to the functional capabilities necessary for tumor growth and progression. Tumor-promoting inflammation and unlocking phenotypic plasticity are considered enabling characteristics as they facilitate acquisition of the core characteristics. Lastly, avoiding immune destruction is an emerging characteristic, as it is not yet generalizable and fully validated in the hallmarks of cancer. (**b**) Illustration of the hallmark advantages conferred by circulating hybrid cells in the metastatic cascade. Figure created with BioRender.com

2015; Qian and Pollard 2010). Of significance, inflammation is evident at earliest stages of neoplasia and has been demonstrated to play a role in the ultimate development into fully invasive cancers (de Visser and Coussens 2006; Qian and Pollard 2010). The immune system, specifically the innate immune system, has significant impacts in the pathogenesis of cancer.

A key mechanism by which inflammation plays an integral role in malignancy is through the generation of neoplastic-immune hybrid cells. By their very nature, these hybrid cell populations rely on inflammation and immune infiltration to facilitate myeloid fusion with epithelial tumor cells, in order to produce hybrid progeny. Inflammatory conditions themselves promote tissue hybrid cell generation through fusion of bone marrow and intestinal cells (Davies et al. 2009; Pawelek and Chakraborty 2008b). The effects of tumor-related inflammation extend beyond the tumor microenvironment, as Davies et al. demonstrated that inflammatory conditions increased in vivo-derived hybrid cell formation in mouse models of colonic inflammation. Interestingly, anti-inflammatory drugs counteracted this effect and inhibited robust hybrid cell formation, further supporting inflammation as a key driving force in hybrid cell development (Davies et al. 2009). Chronic inflammation has additionally been shown to be a driving factor in CHC development in brain, muscle, liver, and heart muscles as well (Johansson et al. 2008; Nygren et al. 2008).

While it is clear that tumor-promoting inflammation is an enabling characteristic of hybrid cells, the mechanisms by which inflammation within the context of the tumor microenvironment facilitates the generation and dissemination of hybrids into circulation remain to be investigated. It is unclear what environmental cues promote neoplastic-immune hybridization, and further, what signaling pathways respond to these inflammatory conditions to promote cellular intravasation into circulation. Given that CHCs play a key role in the metastatic cascade, clarifying the role and mechanism by which tumor-related inflammation accelerates the creation and dissemination of this malignant cell population is of paramount importance in discovering new therapeutic targets for the management of aggressive neoplasms.

21.3 Genomic Instability and Acquisition of Mutations

Genomic instability and acquisition of mutations play a key role in cancer initiation and progression. Cancer is thought to be initiated by the sequential acquisition of a series of oncogenic mutations that allow bypass of cell cycle and growth check points to gain uncontrolled proliferative phenotypes (Hanahan and Weinberg 2011). While the majority of mutations are deleterious to a cell, the right combination of mutations in select genes has tumorigenic consequences. Established cancer cells often have increased rates of mutational acquisition and higher genomic instability (Hanahan and Weinberg 2011). This concept is known as genetic heterogeneity and is one way disease can evolve or become treatment resistant (Ben-David and Amon 2020).

Aneuploidy is an important source of genomic instability and mutations in cancer cells (Ben-David and Amon 2020). The addition and deletion of chromosomes in a cell changes gene copy number on a massive scale across many genes, which can amplify the effects of oncogenic gene mutations or reduce the effect of tumor suppressor genes. Overall, this causes aberrant gene expression and increased genetic heterogeneity, which can drive tumor initiation, evolution, and the potential for cancer cells to adapt to new conditions and become treatment resistant (Goldenberg et al. 2014; Wang et al. 2021). This is best evidenced by the fact that 90% of solid tumors are aneuploid and in many cancer types more complex aneuploidy, or increased karyotype diversity, is linked to worse patient outcomes and more aggressive disease. There is also evidence that increased chromosome missegregation, which is one cause of aneuploidy, is associated with higher rates of metastatic disease (Ben-David and Amon 2020).

There is a preponderance of evidence to suggest that neoplastic-immune hybrid cells are aneuploid. Silk et al. demonstrated that epithelial-hematopoietic fusion cells in the human gut were aneuploid through fluorescence in situ hybridization (FISH) analysis of intestinal samples of female patients with graft versus host disease (GVHD) that had received sex mis-matched peripheral stem cell or bone marrow transplants, where hybrid cells were identified by their XXY phenotype (Silk et al. 2013). Gast et al. cultured in vitro-derived murine tumor-macrophage hybrids in which the macrophages were derived from an XY mouse and fused with a XO murine cancer cell line. Downstream karyotype analysis on the resulting hybrid cells revealed that they were XXY. In addition, they also observed variable chromosome number outside the sex chromosomes, and that subsequent cell passaging resulted in chromosome loss (Gast et al. 2018). As expected, CHCs derived from intra-tumoral hybrids similarly demonstrate aneuploidy; Clawson et al. cultured CHCs isolated from peripheral blood of melanoma patients and identified variable ploidy using 3D rendered confocal imaging to measure DNA ploidy in hybrids (Clawson et al. 2015). When considering that these cell populations independently display aggressive malignant phenotypes, it stands to reason that a component of their acquired tumorigenicity could be derived from the resultant genomic instability.

Like most malignant cell types, CHCs demonstrate aneuploidy and thus have increased genetic heterogeneity and genomic instability. These characteristics are generally linked to poor patient outcomes and more aggressive disease and may contribute to the metastatic cascade, as higher numbers of CHCs in patient peripheral blood correlate with late-stage cancers or poor patient outcomes (Ben-David and Amon 2020; Gast et al. 2018; Parappilly et al. 2022; Walker et al. 2021; Manjunath et al. 2020). Further investigation is needed to better establish the specific novel phenotypic changes attributed by CHC aneuploidy, as this may shed light on how aneuploidy contributes to the ability of hybrid cells to disseminate into the peripheral blood and seed metastatic disease elsewhere in the body.

21.4 Unlocking Phenotypic Plasticity

Neoplastic-immune hybrid cells uniquely possess phenotypic plasticity, as they can alter their phenotypic expression in response to local stressors present within the tumor microenvironment that regulate either/or neoplastic and immune behavior. Due to the inherent plasticity conveyed by combining the genetic material of two distinct cell lineages, the act of fusion may confer adaptive advantages to the newly formed hybrid that allow its survival in heterogeneous and evolving microenvironments. It is possible that cell fusion may underlie a cell changing morphologic identity from an epithelial to a mesenchymal cell, such as the case in the developmental process of epithelial to mesenchymal transition (EMT) or mesenchymal to epithelial transition (MET). This transition may facilitate a tumor cell to lose its adhesive properties in order to escape the primary tumor, or migrate through peripheral blood untargeted by the immune system. Studies indicate that fusion between cancer cells and mesenchymal stroma/stem-like cells increases phenotypic plasticity in the resulting hybrid cell (Melzer et al. 2020, 2021). These studies provide strong evidence that tumor-macrophage cell fusion synergistically generates a relatively undifferentiated daughter hybrid cell with altered phenotypes that play an integral part in the metastatic cascade (Dietz et al. 2021; Gast et al. 2018).

One of the fundamental fusion partners is the macrophage, which is already well-documented to harbor the ability to change phenotypes dependent on microenvironmental cues. Macrophages are known to play a role in both tumor resolution, propagation and demonstrate plasticity in response to the tumor microenvironment (Leopold Wager and Wormley 2014; Malyshev and Malyshev 2015). Local chemotactic factors in the tumor microenvironment alter both the phenotypic expression of the macrophage and their function. For example, the classical pathway of macrophage activation results in the M1 macrophage which causes the release of proinflammatory cytokines including tissue necrosis factor-alpha, interleukin-6, and interferon gamma, to directly kill tumor cells. In contrast, the alternative pathway of macrophage activation results in the M2 macrophage, which breaks down the basement membrane, promotes angiogenesis, and allows for tumor invasion (Leopold Wager and Wormley 2014; Sutton et al. 2022). Plasticity inherently present in macrophages can subsequently be passed to hybrid cells derived from macrophages. This is best evidenced by the gained ability of otherwise polar, fixed epithelial tumor cells to express the mesenchymal motile, stem-like, invasive properties (i.e., EMT). A study by Ding et al. demonstrates that stem cell-like properties are present in macrophage-breast cancer hybrids. In this in vitro study, tumor-associated macrophages were fused to breast cancer cell lines. A subset of the resulting hybrids demonstrated increased expression of EMT-associated genes such as snail1 and snail 2, and exhibited enhanced migrative and invasive abilities (Ding et al. 2012). On the contrary, after hybrids enter circulation, they seed at distant sites and are able to return to a more epithelial phenotype and grow as distant epithelial metastases (i.e., MET), thus altering their phenotype to promote their spread and survival (Dietz et al. 2021; Gast et al. 2018; Powell et al. 2011).

Beyond macrophage plasticity, hybrid cells harbor phenotypes of less differentiated cells types, indicating that hybrid cell plasticity could also be a post-fusion event. For example, a cell with combined genomic contribution may be responsive to various environmental cues, including those that impact stem cell or differentiation states. Furthermore, CHCs have been demonstrated to harbor less differentiated, stem cell phenotypes. Dietz et al. examined hybrids from the peripheral blood of patients with triple negative breast cancer and determined that CHC populations predominantly harbored stem cell signatures ($CD44^{+}/CD24^{lo}$) when compared to CTCs from the same patient. These findings supported the undifferentiated status of hybrid cells, and thus, independent tumor-initiating potential (Dietz et al. 2021; Li et al. 2017). In addition, when primary tumor hybrid cells were isolated from a murine breast cancer model and injected into secondary recipient mice, the hybrid cells rapidly recapitulated tumor growth. Notably, 100-fold more unfused tumor cells were required to support tumor initiation (Dietz et al. 2021). Similarly, breast cancer hybrids from Ding et al. showed enhanced tumorigenicity and metastatic ability when injected into NOD/SCID mice (Ding et al. 2012). Through fusion, it is possible that CHCs independently gain undifferentiated stem-like plasticity that is fundamental to their oncogenic capacity.

Although evidence indicates that CHCs harbor phenotypic plasticity, much remains to be understood regarding the mechanisms by which the CHCs gain plasticity. For example, differentially expressed cellular mechanisms that drive mesenchymal characteristics of CHCs have yet to be revealed. Additionally, we do not yet understand how CHCs phenotypically behave like stem-like progenitor cells to support metastatic tumor growth. It is also possible that tumor fusion partners are already undifferentiated cells, however it is equally possible that hybrid cells may dedifferentiate as a consequence of fusion. Nevertheless, the phenotypic plasticity exhibited by circulating hybrids is paramount to their fundamental role within the tumor-metastasis axis.

21.5 Invasion and Dissemination into Vasculature

It is believed that greater than 90% of cancer patients who succumb to their disease die of complications from metastatic spread of cancer (Cancer Statistics Center 2020; Gupta and Massagué 2006). Metastatic tumors derive from primary tumor cells that can successfully navigate a number of hurdles, including invasion into surrounding tissue, and dissemination into the peripheral blood (Friedl and Wolf 2003), which are two critical hallmarks of cancer progression (Hanahan and Weinberg 2011). In order for tumor cells to invade, they lose cell–cell adhesive properties, breakdown then cross the basement membrane, and upregulate migratory phenotypes (Friedl and Wolf 2003). Pawelek et al. documented several instances where a hybrid cell likely seeded metastatic tumors in patients who had received a bone marrow transplant. This was evidenced by the metastatic tumor containing a large proportion of donor DNA when compared to the primary tumor (LaBerge et al. 2021; LaBerge et al. 2017). The case for hybrid cells seeding these tumors is strengthened by the notion that mechanisms underlying invasion and dissemination are varied, thus it is plausible that cell fusion imparts properties to neoplastic cells enabling them to metastasize.

Tumor cell invasion leads to disease expansion across tissue as a result of the cell's acquisition of migratory traits (Friedl and Wolf 2003) that are common attributes of immune cells. Emerging data clearly support this hypothesis. Hybrid cells demonstrate loss of adhesion from down regulation of cell adhesion molecules (e.g., E-cadherin, epithelial cellular adhesion molecule (EpCAM)), which allows them to move through tissue (Hanahan and Weinberg 2011; Skokos et al. 2011). Additionally, neoplastic cells at the leading-edge of the tumor and discrete immune cell populations such as macrophages harbor expression of proteases, including matrix metalloproteinases (MMPs), which can mediate breakdown of the extracellular matrix. Recruitment of bone marrow-derived cells through increased intra-tumoral oxysterol levels (Park et al. 2022) can provide neoplastic cells with macrophage partners that have MMP expression (Skokos et al. 2011), as tumor-immune hybrid cells harbor high expression of MMPs at macrophage levels (Mohr et al. 2015).

An additional way that hybrid cells may contribute to invasion is by gaining inherent migratory traits of macrophages, which are known to respond to sensory cues (Stramer and Mayor 2017). Consistent with this, Gast et al. identified acquired functional gene expression patterns that convey response to specific microenvironmental cues in neoplastic-macrophage hybrid cells (Gast et al. 2018). Specifically, macrophages express receptors that permit their migration toward paired ligands (e.g., colony-stimulating factor 1 receptor or stromal cell-derived factor-1). Gast et al. further demonstrated that gain of macrophage-associated receptors imparted ligand-mediated migration to the hybrid cells, which could be blocked by antibodies to these macrophage receptors (Gast et al. 2018). Furthermore, a study on glioblastoma-tumor associated macrophage hybrids found that hybrids were enriched from glioma invasion-associated genes in a subset of hybrids that also displayed increased invasiveness (Cao et al. 2019). Taken collectively, these data nicely demonstrate that aspects of cell invasion and migration could be attributed to the hybrid cells within the primary tumor. Hybrid cells have the potential to escape the primary site through the tumor vasculature and disseminate to develop metastases.

Dissemination is a critical process for metastatic spread of disease. A cancer cell must leave the primary tumor, enter peripheral circulation, and survive in this new environment as it travels to colonize new locations (Hanahan and Weinberg 2011). In order for neoplastic cells to disseminate into peripheral blood, they downregulate their epithelial adhesive properties (Peinado et al. 2004), and it is demonstrated that they take on mesenchymal features (Taube et al. 2010). Disseminated cancer-leukocyte hybrid cells have been shown to express mesenchymal features by Gast and colleagues (Gast et al. 2018). Additionally, Lu et al. demonstrated an acquisition of mesenchymal phenotypes in hybrid cells from lung cancer (Lu and Kang 2009). This point was also demonstrated by Clawson et al when they cultured macrophage-tumor hybrid cells isolated from peripheral blood samples of a melanoma patients. The cells they isolated co-expressed melanoma tumor markers and several immune markers and when these cells were transplanted subcutaneously into nude mice they produced metastatic tumors (Clawson et al. 2015). To this point, tumor-associated macrophages are known to promote tumor cell invasion and intravasation (Noy and Pollard 2014), and hybrid cells gain these mechanistic abilities by combining genetic information.

Disseminated CHCs have demonstrated their potential prognostic utility, as they carry information about the primary tumor from which they originated (Dietz et al. 2021; Sutton et al. 2019, 2022; Manjunath et al. 2020; Parappilly et al. 2022; Walker et al. 2021). It has been shown that CHCs are enriched in peripheral blood of cancer patients, outnumbering the population of CTCs, making them an ideal population for downstream analysis of tumor protein expression (Gast et al. 2018). This anecdotally may imply an enhanced ability of hybrid cells for survival in the blood, as the hybrid cell's immune identity permits cloaking from the immune system (Casanova-Acebes et al. 2021; Mohr et al. 2015). Ultimately, it is clear that neoplastic-immune hybrid cells have a remarkable ability to enter and survive in circulation. Given that these cells have been clearly shown to have tumorigenic properties, invasion and dissemination into vasculature is an enabling and foundational characteristic that is at the core of this cell population's role in the metastatic cascade.

21.6 Avoiding Immune Destruction

The avoidance of peripheral immune destruction of circulating tumor cells, and the identification of associated immunogenic targets for therapy, is a complex and important frontier in cancer biology (Mohme et al. 2017). Circulating hybrid cells in particular are an interesting component of this phenomenon. These cells are formed within the tumor microenvironment in an abundance of immune system components and demonstrate expression of markers specific to leukocytes, such as CD45. It is reasonable then to conclude that their ability to survive in the peripheral blood, similar to circulating tumor cells, likely depends on an immune privilege that allows them to avoid destruction.

A 2017 review by Mohme et al. postulates several mechanisms (and potential drug targets) for peripheral CTC survival. For example, there is some evidence to suggest that peripheral survival of circulating breast carcinoma cells may specifically depend on PD-L1 mediated Treg activation (Mohme et al. 2017). Similar investigations into CHCs are ongoing, though much of the current evidence supports development of a tumorigenic, immunologically active CHC phenotype in the primary tumor itself. One of the strongest links between peripheral CHC survival and tumorigenesis lies in its phenotypic similarities to the tumor-associated macrophage.

Tumor-associated macrophages, or TAMs, are found in the tumor microenvironment and play a key role in orchestrating tumor-associated inflammation (Mantovani et al. 2017). Similar to tissue-derived macrophages, they are stimulated by CSF-1, a macrophage-associated gene colony-stimulating factor involved in recruitment, differentiation, and cellular survival, and are extremely abundant in the tumor. The presence of CSF-1 is associated with several tumorigenic and immunosuppressive actions in the primary tumor itself. Strong evidence supporting the contribution of CSF-1 to peripheral CHC survival could lie in the expression of CSF-1R by CHC's (Gast et al. 2018). Though in vivo murine studies demonstrate that activation of TAMs by CSF-1 initially supports antitumor activity, eventually that response is abrogated by chemokines from the tumor and tumor stroma itself, diversifying the macrophage phenotype (Mantovani et al. 2017). TAMs progress to express an indirectly immunosuppressive phenotype, which aid in tumor growth, angiogenesis, and metastasis (Lin and Pollard 2004). Tumor metastasis in particular is directly related to CSF-1 production by primary tumor cells (Wyckoff et al. 2004). It is possible then that this expression also contributes to tumor escape and immune evasion of CHCs.

Transcriptomic analysis as well as in vitro studies has shown that CSF-1 production is induced after exposure to tumor-related CD8+ T cells, reflecting a conserved resistance mechanism by TAMs in the presence of adaptive immune cell activation (Neubert et al. 2018). This implies that a cell such as a CHC which is stimulated by CSF-1 may thrive in an environment where CSF-1 is abundant. Meaning, though TAMs themselves could be responsible for much of the tumorigenic properties of the tumor microenvironment, it could be that macrophage stimulating factors such as CSF-1 may be stimulating other members of the tumor microenvironment, such as CHCs, to produce the same overall result.

The expression of CSF-1R likely contributes to the CHC's enhanced migration and proliferation as compared to a non-hybrid neoplastic cell (Gast et al. 2018). It could be postulated then that the tumor microenvironment may simply select for cancer cells and CHCs responsive to CSF-1 (Mantovani et al. 2017). This opens an avenue for further understanding not only of CHC survival in the periphery, but of direct immunologic contributions by CHCs in the periphery and in distant tissues.

21.7 Impact of Cell Fusion in Cancer

From conception to organ development, and immunity, cellular fusion is an essential process for life. Through fusion, neoplastic-immune hybrid cells hijack the same mechanism that initiates life, in order to promote malignant progression fueled by dissemination into circulation. "The Hallmarks of Cancer" describe a set of fundamental alterations in cell physiology that are essential for the development of malignancy. As we learn more about the tumor hybrid cell population, it is increasingly apparent that these cells share and propagate of many of the fundamental hallmarks of cancer. As we learn more of the role cellular senescence plays in stimulating invasion and metastasis, senescent cells could play an important paracrine role in influencing cellular selection for hybridization, or it could be possible that tumor hybrid cells that transiently senesce confer therapeutic resistance. Given that CHCs independently harbor tumor-initiating properties (Clawson et al. 2015; Dietz et al. 2021; Gast et al. 2018), fusion itself could allow neoplastic cells to gain improved replicative immortality over their unfused CTC counterparts.

While CHCs represent an important avenue by which tumors may metastasize, when isolated from peripheral blood, they provide valuable clinical information in regard to prognostication and oncologic surveillance. For example, it has been observed that CHC levels decreased with therapy response and increased prior to clinical evidence of disease progression in rectal adenocarcinoma (Walker et al. 2021) Additionally, CHC levels also correlate with disease stage and survival in patients with pancreatic ductal adenocarcinoma, lung cancer, and uveal melanoma (Gast et al. 2018; Manjunath et al. 2020; Parappilly et al. 2022). Given that CHC levels track with disease burden, it is plausible that they could be developed as a biomarker to measure disease response to systemic therapies, to optimize and tailor therapy to improve survival outcomes. Additionally, CHCs isolated from peripheral blood have been demonstrated to harbor phenotypic and genotypic signatures of the tumors from which they are derived (Dietz et al. 2021; Parappilly et al. 2022; Walker et al. 2021), thus opening the door for this cell population to potentially be utilized as a "liquid biopsy" to reliably characterize the molecular character of neoplasms without facing the challenges of obtaining direct tissue samples through biopsy.

To facilitate this line of inquiry, novel and efficient approaches for CHC isolation must be developed. Further, since the biologic consequences of cell fusion are not fully appreciated as it pertains to gene and protein expression, devices that isolate cells in a label-free fashion (i.e., not dependent on knowing cell surface antigen expression) would yield the greatest impact on the field. One such area is magnetic levitation and sorting of neoplastic cell hybrids (Liang et al. 2022). This approach relies on the biophysical properties of the cell and leverages the principles of magnetic levitation to separate unique cell populations, like CHCs, from peripheral blood mononuclear cells (PBMCs) (Durmus et al. 2015), based on density differences. Due to the low intrinsic magnetic susceptibility of cells, the levitation process does not perturb cells nor affect their long-term viability or function, thus it is compatible with downstream assays (Durmus et al. 2015; Puluca et al. 2020; Tocchio et al. 2018). The potential of unique systems, such as magnetic levitation, promise to isolate all sub-types of CHCs (independent of their size and surface marker expression) with high purity and to break open our biologic knowledgebase of hybrid cell biology.

21.8 Prospectus

- Heterotypic cell fusion provides a mechanism for acquisition of phenotypes in a cell population that can differentially respond to its tissue microenvironment.
- Cell fusion may underlie alteration in a cell's identity, akin to gain in cellular plasticity.

- Cell–cell fusion remains an under-explored frontier in cancer biology.

References

Aichel O (1911) Über Zellverschmelzung mit qualitativ abnormer Chromosomenverteilung als Ursache der Geschwulstbildung, vol 13. W. Engelmann, Leipzig

Allan AL, Vantyghem SA, Tuck AB, Chambers AF, Chin-Yee IH, Keeney M (2005) Detection and quantification of circulating tumor cells in mouse models of human breast cancer using immunomagnetic enrichment and multiparameter flow cytometry. Cytometry A 65(1):4–14. https://doi.org/10.1002/cyto.a.20132

Avilès D, Jami J, Rousset JP, Ritz E (1977) Tumor x host cell hybrids in the mouse: chromosomes from the normal cell parent maintained in malignant hybrid tumors. J Natl Cancer Inst 58(5):1391–1399. https://doi.org/10.1093/jnci/58.5.1391

Ben-David U, Amon A (2020) Context is everything: aneuploidy in cancer. Nat Rev Genet 21(1):44–62. https://doi.org/10.1038/s41576-019-0171-x

Boveri T (2008) Concerning the origin of malignant tumours by Theodor Boveri. Translated and annotated by Henry Harris. J Cell Sci 121 Suppl 1:1–84. https://doi.org/10.1242/jcs.025742

Cancer Statistics Center (2020) American Cancer Society. http://cancerstatisticscenter.cancer.org. Accessed 1 Dec 2022

Cao MF, Chen L, Dang WQ, Zhang XC, Zhang X, Shi Y, Yao XH, Li Q, Zhu J, Lin Y, Liu S, Chen Q, Cui YH, Zhang X, Bian XW (2019) Hybrids by tumor-associated macrophages × glioblastoma cells entail nuclear reprogramming and glioblastoma invasion. Cancer Lett 442:445–452. https://doi.org/10.1016/j.canlet.2018.11.016

Casanova-Acebes M, Dalla E, Leader AM, LeBerichel J, Nikolic J, Morales BM, Brown M, Chang C, Troncoso L, Chen ST, Sastre-Perona A, Park MD, Tabachnikova A, Dhainaut M, Hamon P, Maier B, Sawai CM, Agulló-Pascual E, Schober M, Brown BD, Reizis B, Marron T, Kenigsberg E, Moussion C, Benaroch P, Aguirre-Ghiso JA, Merad M (2021) Tissue-resident macrophages provide a pro-tumorigenic niche to early NSCLC cells. Nature 595(7868):578–584. https://doi.org/10.1038/s41586-021-03651-8

Chakraborty AK, Sodi S, Rachkovsky M, Kolesnikova N, Platt JT, Bolognia JL, Pawelek JM (2000) A spontaneous murine melanoma lung metastasis comprised of host x tumor hybrids. Cancer Res 60(9):2512–2519

Chakraborty A, Lazova R, Davies S, Bäckvall H, Ponten F, Brash D, Pawelek J (2004) Donor DNA in a renal cell carcinoma metastasis from a bone marrow transplant recipient. Bone Marrow Transplant 34(2):183–186. https://doi.org/10.1038/sj.bmt.1704547

Clawson GA, Kimchi E, Patrick SD, Xin P, Harouaka R, Zheng S, Berg A, Schell T, Staveley-O'Carroll KF, Neves RI, Mosca PJ, Thiboutot D (2012) Circulating tumor cells in melanoma patients. PLoS One 7(7):e41052. https://doi.org/10.1371/journal.pone.0041052

Clawson GA, Matters GL, Xin P, Imamura-Kawasawa Y, Du Z, Thiboutot DM, Helm KF, Neves RI, Abraham T (2015) Macrophage-tumor cell fusions from peripheral blood of melanoma patients. PLoS One 10(8):e0134320. https://doi.org/10.1371/journal.pone.0134320

Davies PS, Powell AE, Swain JR, Wong MH (2009) Inflammation and proliferation act together to mediate intestinal cell fusion. PLoS One 4(8):e6530. https://doi.org/10.1371/journal.pone.0006530

DeNardo DG, Andreu P, Coussens LM (2010) Interactions between lymphocytes and myeloid cells regulate pro- versus anti-tumor immunity. Cancer Metastasis Rev 29(2):309–316. https://doi.org/10.1007/s10555-010-9223-6

de Visser KE, Coussens LM (2006) The inflammatory tumor microenvironment and its impact on cancer development. Contrib Microbiol 13:118–137. https://doi.org/10.1159/000092969

de Wit S, Zeune LL, Hiltermann TJN, Groen HJM, Dalum GV, Terstappen L (2018) Classification of cells in CTC-enriched samples by advanced image analysis. Cancers (Basel) 10(10). https://doi.org/10.3390/cancers10100377

Dietz MS, Sutton TL, Walker BS, Gast CE, Zarour L, Sengupta SK, Swain JR, Eng J, Parappilly M, Limbach K, Sattler A, Burlingame E, Chin Y, Gower A, Mira JLM, Sapre A, Chiu YJ, Clayburgh DR, Pommier SJ, Cetnar JP, Fischer JM, Jaboin JJ, Pommier RF, Sheppard BC, Tsikitis VL, Skalet AH, Mayo SC, Lopez CD, Gray JW, Mills GB, Mitri Z, Chang YH, Chin K, Wong MH (2021) Relevance of circulating hybrid cells as a non-invasive biomarker for myriad solid tumors. Sci Rep 11(1):13630. https://doi.org/10.1038/s41598-021-93053-7

Ding J, Jin W, Chen C, Shao Z, Wu J (2012) Tumor associated macrophage × cancer cell hybrids may acquire cancer stem cell properties in breast cancer. PLoS One 7(7):e41942. https://doi.org/10.1371/journal.pone.0041942

Duelli D, Lazebnik Y (2003) Cell fusion: a hidden enemy? Cancer Cell 3(5):445–448. https://doi.org/10.1016/s1535-6108(03)00114-4

Durmus NG, Tekin HC, Guven S, Sridhar K, Arslan Yildiz A, Calibasi G, Ghiran I, Davis RW, Steinmetz LM, Demirci U (2015) Magnetic levitation of single cells. Proc Natl Acad Sci U S A 112(28):E3661–E3668. https://doi.org/10.1073/pnas.1509250112

Fortuna MB, Dewey MJ, Furmanski P (1989) Cell fusion in tumor development and progression: occurrence of cell fusion in primary methylcholanthrene-induced tumorigenesis. Int J Cancer 44(4):731–737. https://doi.org/10.1002/ijc.2910440430

Fortuna MB, Dewey MJ, Furmanski P (1990) Enhanced lung colonization and tumorigenicity of fused cells isolated from primary MCA tumors. Cancer Lett 55(2):109–114. https://doi.org/10.1016/0304-3835(90)90019-t

Friedl P, Wolf K (2003) Tumour-cell invasion and migration: diversity and escape mechanisms. Nat Rev Cancer 3(5):362–374. https://doi.org/10.1038/nrc1075

Gast CE, Silk AD, Zarour L, Riegler L, Burkhart JG, Gustafson KT, Parappilly MS, Roh-Johnson M, Goodman JR, Olson B, Schmidt M, Swain JR, Davies PS, Shasthri V, Iizuka S, Flynn P, Watson S, Korkola J, Courtneidge SA, Fischer JM, Jaboin J, Billingsley KG, Lopez CD, Burchard J, Gray J, Coussens LM, Sheppard BC, Wong MH (2018) Cell fusion potentiates tumor heterogeneity and reveals circulating hybrid cells that correlate with stage and survival. Sci Adv 4(9):eaat7828. https://doi.org/10.1126/sciadv.aat7828

Goldenberg DM (1968) [On the progression of malignity: a hypothesis]. Klin Wochenschr 46(16):898–899. https://doi.org/10.1007/bf01746251

Goldenberg DM, Bhan RD, Pavia RA (1971) In vivo human-hamster somatic cell fusion indicated by glucose 6-phosphate dehydrogenase and lactate dehydrogenase profiles. Cancer Res 31(8):1148–1152

Goldenberg DM, Pavia RA, Tsao MC (1974) In vivo hybridisation of human tumour and normal hamster cells. Nature 250(5468):649–651. https://doi.org/10.1038/250649a0

Goldenberg DM, Rooney RJ, Loo M, Liu D, Chang CH (2014) In-vivo fusion of human cancer and hamster stromal cells permanently transduces and transcribes human DNA. PLoS One 9(9):e107927. https://doi.org/10.1371/journal.pone.0107927

Gupta GP, Massagué J (2006) Cancer metastasis: building a framework. Cell 127(4):679–695. https://doi.org/10.1016/j.cell.2006.11.001

Hanahan D (2022) Hallmarks of cancer: new dimensions. Cancer Discov 12(1):31–46. https://doi.org/10.1158/2159-8290.Cd-21-1059

Hanahan D, Weinberg RA (2000) The hallmarks of cancer. Cell 100(1):57–70. https://doi.org/10.1016/s0092-8674(00)81683-9

Hanahan D, Weinberg RA (2011) Hallmarks of cancer: the next generation. Cell 144(5):646–674. https://doi.org/10.1016/j.cell.2011.02.013

Hass R, von der Ohe J, Dittmar T (2021) Hybrid formation and fusion of cancer cells in vitro and in vivo. Cancers (Basel) 13(17). https://doi.org/10.3390/cancers13174496

Jiang E, Yan T, Xu Z, Shang Z (2019) Tumor microenvironment and cell fusion. Biomed Res Int 2019:5013592. https://doi.org/10.1155/2019/5013592

Johansson CB, Youssef S, Koleckar K, Holbrook C, Doyonnas R, Corbel SY, Steinman L, Rossi FM, Blau HM (2008) Extensive fusion of haematopoietic cells with Purkinje neurons in response to chronic inflammation. Nat Cell Biol 10(5):575–583. https://doi.org/10.1038/ncb1720

Kerbel RS, Lagarde AE, Dennis JW, Donaghue TP (1983) Spontaneous fusion in vivo between normal host and tumor cells: possible contribution to tumor progression and metastasis studied with a lectin-resistant mutant tumor. Mol Cell Biol 3(4):523–538. https://doi.org/10.1128/mcb.3.4.523-538.1983

LaBerge GS, Duvall E, Grasmick Z, Haedicke K, Pawelek J (2017) A melanoma lymph node metastasis with a donor-patient hybrid genome following bone marrow transplantation: a second case of leucocyte-tumor cell hybridization in cancer metastasis. PLoS One 12(2):e0168581. https://doi.org/10.1371/journal.pone.0168581

LaBerge GS, Duvall E, Haedicke K, Pawelek J (2019) Leukocyte- cancer cell fusion-genesis of a deadly journey. Cells 8(2). https://doi.org/10.3390/cells8020170

LaBerge G, Duvall E, Grasmick Z, Haedicke K, Galan A, Pawelek J (2021) A melanoma patient with macrophage-cancer cell hybrids in the primary tumor, a lymph node metastasis and a brain metastasis. Cancer Genet 256-257:162–164. https://doi.org/10.1016/j.cancergen.2021.05.009

Larsson LI (2010) Cell fusions: regulation and control. Springer, Dordrecht

Lazova R, LaBerge GS, Duvall E, Spoelstra N, Klump V, Sznol M, Cooper D, Spritz RA, Chang JT, Pawelek JM (2013) A melanoma brain metastasis with a donor-patient hybrid genome following bone marrow transplantation: first evidence for fusion in human cancer. PLoS One 8(6):e66731. https://doi.org/10.1371/journal.pone.0066731

Leopold Wager CM, Wormley FL Jr (2014) Classical versus alternative macrophage activation: the Ying and the Yang in host defense against pulmonary fungal infections. Mucosal Immunol 7(5):1023–1035. https://doi.org/10.1038/mi.2014.65

Li W, Ma H, Zhang J, Zhu L, Wang C, Yang Y (2017) Unraveling the roles of CD44/CD24 and ALDH1 as cancer stem cell markers in tumorigenesis and metastasis. Sci Rep 7(1):13856. https://doi.org/10.1038/s41598-017-14364-2

Li H, Meng QH, Noh H, Somaiah N, Torres KE, Xia X, Batth IS, Joseph CP, Liu M, Wang R, Li S (2018) Cell-surface vimentin-positive macrophage-like circulating tumor cells as a novel biomarker of metastatic gastrointestinal stromal tumors. Oncoimmunology 7(5):e1420450. https://doi.org/10.1080/2162402x.2017.1420450

Liang K, Yaman S, Patel RK, Parappilly MS, Walker BS, Wong MH, Durmus NG (2022) Magnetic levitation and sorting of neoplastic circulating cell hybrids. bioRxiv. https://doi.org/10.1101/2022.11.03.515127

Lin EY, Pollard JW (2004) Macrophages: modulators of breast cancer progression. Novartis Found Symp 256:158–168. Discussion 168–172, 259–169

Lin D, Shen L, Luo M, Zhang K, Li J, Yang Q, Zhu F, Zhou D, Zheng S, Chen Y, Zhou J (2021) Circulating tumor cells: biology and clinical significance. Signal Transduct Target Ther 6(1). https://doi.org/10.1038/s41392-021-00817-8

Liu Q, Liao Q, Zhao Y (2016) Myeloid-derived suppressor cells (MDSC) facilitate distant metastasis of malignancies by shielding circulating tumor cells (CTC) from immune surveillance. Med Hypotheses 87:34–39. https://doi.org/10.1016/j.mehy.2015.12.007

Lu X, Kang Y (2009) Efficient acquisition of dual metastasis organotropism to bone and lung through stable spontaneous fusion between MDA-MB-231 variants. Proc Natl Acad Sci U S A 106(23):9385–9390. https://doi.org/10.1073/pnas.0900108106

Malyshev I, Malyshev Y (2015) Current concept and update of the macrophage plasticity concept: intracellular mechanisms of reprogramming and M3 macrophage "switch" phenotype. Biomed Res Int 2015:341308. https://doi.org/10.1155/2015/341308

Manjunath Y, Mitchem JB, Suvilesh KN, Avella DM, Kimchi ET, Staveley-O'Carroll KF, Deroche CB, Pantel K, Li G, Kaifi JT (2020) Circulating giant tumor-macrophage fusion cells are independent prognosticators in patients with NSCLC. J Thorac Oncol 15(9):1460–1471. https://doi.org/10.1016/j.jtho.2020.04.034

Mantovani A, Marchesi F, Malesci A, Laghi L, Allavena P (2017) Tumour-associated macrophages as treatment targets in oncology. Nat Rev Clin Oncol 14(7):399–416. https://doi.org/10.1038/nrclinonc.2016.217

Mekler LB (1971) [Hybridization of transformed cells with lymphocytes as 1 of the probable causes of the progression leading to the development of metastatic malignant cells]. Vestn Akad Med Nauk SSSR 26(8):80–89

Melzer C, Ohe JV, Hass R (2020) Altered tumor plasticity after different cancer cell fusions with MSC. Int J Mol Sci 21(21). https://doi.org/10.3390/ijms21218347

Melzer C, Ohe JV, Luo T, Hass R (2021) Spontaneous fusion of MSC with breast cancer cells can generate tumor dormancy. Int J Mol Sci 22(11). https://doi.org/10.3390/ijms22115930

Mohme M, Riethdorf S, Pantel K (2017) Circulating and disseminated tumour cells—mechanisms of immune surveillance and escape. Nat Rev Clin Oncol 14(3):155–167. https://doi.org/10.1038/nrclinonc.2016.144

Mohr M, Tosun S, Arnold WH, Edenhofer F, Zänker KS, Dittmar T (2015) Quantification of cell fusion events human breast cancer cells and breast epithelial cells using a Cre-LoxP-based double fluorescence reporter system. Cell Mol Life Sci 72(19):3769–3782. https://doi.org/10.1007/s00018-015-1910-6

Nel I, Jehn U, Gauler T, Hoffmann AC (2014) Individual profiling of circulating tumor cell composition in patients with non-small cell lung cancer receiving platinum based treatment. Transl Lung Cancer Res 3(2):100–106. https://doi.org/10.3978/j.issn.2218-6751.2014.03.05

Neubert NJ, Schmittnaegel M, Bordry N, Nassiri S, Wald N, Martignier C, Tillé L, Homicsko K, Damsky W, Maby-El Hajjami H, Klaman I, Danenberg E, Ioannidou K, Kandalaft L, Coukos G, Hoves S, Ries CH, Fuertes Marraco SA, Foukas PG, De Palma M, Speiser DE (2018) T cell-induced CSF1 promotes melanoma resistance to PD1 blockade. Sci Transl Med 10(436). https://doi.org/10.1126/scitranslmed.aan3311

Noy R, Pollard JW (2014) Tumor-associated macrophages: from mechanisms to therapy. Immunity 41(1):49–61. https://doi.org/10.1016/j.immuni.2014.06.010

Nygren JM, Liuba K, Breitbach M, Stott S, Thorén L, Roell W, Geisen C, Sasse P, Kirik D, Björklund A, Nerlov C, Fleischmann BK, Jovinge S, Jacobsen SE (2008) Myeloid and lymphoid contribution to non-haematopoietic lineages through irradiation-induced heterotypic cell fusion. Nat Cell Biol 10(5):584–592. https://doi.org/10.1038/ncb1721

Parappilly MS, Chin Y, Whalen RM, Anderson AN, Robinson TS, Strgar L, Sutton TL, Conley P, Klocke C, Gibbs SL, Chang YH, Wu G, Wong MH, Skalet AH (2022) Circulating neoplastic-immune hybrid cells predict metastatic progression in uveal melanoma. Cancers (Basel) 14(19). https://doi.org/10.3390/cancers14194617

Park MD, Silvin A, Ginhoux F, Merad M (2022) Macrophages in health and disease. Cell 185(23):4259–4279. https://doi.org/10.1016/j.cell.2022.10.007

Parker KH, Beury DW, Ostrand-Rosenberg S (2015) Myeloid-derived suppressor cells: critical cells driving immune suppression in the tumor microenvironment. Adv Cancer Res 128:95–139. https://doi.org/10.1016/bs.acr.2015.04.002

Pawelek JM (2005) Tumour-cell fusion as a source of myeloid traits in cancer. Lancet Oncol 6(12):988–993. https://doi.org/10.1016/s1470-2045(05)70466-6

Pawelek JM (2007) Viewing malignant melanoma cells as macrophage-tumor hybrids. Cell Adh Migr 1(1):2–6

Pawelek JM, Chakraborty AK (2008a) The cancer cell—leukocyte fusion theory of metastasis. Adv Cancer Res 101:397–444. https://doi.org/10.1016/s0065-230x(08)00410-7

Pawelek JM, Chakraborty AK (2008b) Fusion of tumour cells with bone marrow-derived cells: a unifying explanation for metastasis. Nat Rev Cancer 8(5):377–386. https://doi.org/10.1038/nrc2371

Peinado H, Marin F, Cubillo E, Stark HJ, Fusenig N, Nieto MA, Cano A (2004) Snail and E47 repressors of E-cadherin induce distinct invasive and angiogenic properties in vivo. J Cell Sci 117(Pt 13):2827–2839. https://doi.org/10.1242/jcs.01145

Powell AE, Anderson EC, Davies PS, Silk AD, Pelz C, Impey S, Wong MH (2011) Fusion between Intestinal epithelial cells and macrophages in a cancer context results in nuclear reprogramming. Cancer Res 71(4):1497–1505. https://doi.org/10.1158/0008-5472.Can-10-3223

Puluca N, Durmus NG, Lee S, Belbachir N, Galdos FX, Ogut MG, Gupta R, Hirano KI, Krane M, Lange R, Wu JC, Wu SM, Demirci U (2020) Levitating cells to sort the fit and the fat. Adv Biosyst 4(6):e1900300. https://doi.org/10.1002/adbi.201900300

Qian BZ, Pollard JW (2010) Macrophage diversity enhances tumor progression and metastasis. Cell 141(1):39–51. https://doi.org/10.1016/j.cell.2010.03.014

Rizvi AZ, Swain JR, Davies PS, Bailey AS, Decker AD, Willenbring H, Grompe M, Fleming WH, Wong MH (2006) Bone marrow-derived cells fuse with normal and transformed intestinal stem cells. Proc Natl Acad Sci U S A 103(16):6321–6325. https://doi.org/10.1073/pnas.0508593103

Senger DR, Perruzzi CA, Feder J, Dvorak HF (1986) A highly conserved vascular permeability factor secreted by a variety of human and rodent tumor cell lines. Cancer Res 46(11):5629–5632

Silk AD, Gast CE, Davies PS, Fakhari FD, Vanderbeek GE, Mori M, Wong MH (2013) Fusion between hematopoietic and epithelial cells in adult human intestine. PLoS One 8(1):e55572. https://doi.org/10.1371/journal.pone.0055572

Skokos EA, Charokopos A, Khan K, Wanjala J, Kyriakides TR (2011) Lack of TNF-α-induced MMP-9 production and abnormal E-cadherin redistribution associated with compromised fusion in MCP-1-null macrophages. Am J Pathol 178(5):2311–2321. https://doi.org/10.1016/j.ajpath.2011.01.045

Stramer B, Mayor R (2017) Mechanisms and in vivo functions of contact inhibition of locomotion. Nat Rev Mol Cell Biol 18(1):43–55. https://doi.org/10.1038/nrm.2016.118

Sutton TL, Walker BS, Wong MH (2019) Circulating hybrid cells join the fray of circulating cellular biomarkers. Cell Mol Gastroenterol Hepatol 8(4):595–607. https://doi.org/10.1016/j.jcmgh.2019.07.002

Sutton TL, Patel RK, Anderson AN, Bowden SG, Whalen R, Giske NR, Wong MH (2022) Circulating cells with macrophage-like characteristics in cancer: the importance of circulating neoplastic-immune hybrid cells in cancer. Cancers (Basel) 14(16). https://doi.org/10.3390/cancers14163871

Taube JH, Herschkowitz JI, Komurov K, Zhou AY, Gupta S, Yang J, Hartwell K, Onder TT, Gupta PB, Evans KW, Hollier BG, Ram PT, Lander ES, Rosen JM, Weinberg RA, Mani SA (2010) Core epithelial-to-mesenchymal transition interactome gene-expression signature is associated with claudin-low and metaplastic breast cancer subtypes. Proc Natl Acad Sci U S A 107(35):15449–15454. https://doi.org/10.1073/pnas.1004900107

Tocchio A, Durmus NG, Sridhar K, Mani V, Coskun B, El Assal R, Demirci U (2018) Magnetically guided self-assembly and coding of 3D living architectures. Adv Mater 30(4). https://doi.org/10.1002/adma.201705034

Toyoshima K, Hayashi A, Kashiwagi M, Hayashi N, Iwatsuki M, Ishimoto T, Baba Y, Baba H, Ohta Y (2015) Analysis of circulating tumor cells derived from advanced gastric cancer. Int J Cancer 137(4):991–998. https://doi.org/10.1002/ijc.29455

Vande Woude GF, Klein G (2009) Advances in cancer research. Elsevier, San Diego

Virchow R (1863) Cellular pathology as based upon physiological and pathological histology. Dover, New York

Walker BS, Sutton TL, Zarour L, Hunter JG, Wood SG, Tsikitis VL, Herzig DO, Lopez CD, Chen EY, Mayo SC, Wong MH (2021) Circulating hybrid cells: a novel liquid biomarker of treatment response in gastrointestinal cancers. Ann Surg Oncol 28(13):8567–8578. https://doi.org/10.1245/s10434-021-10379-2

Wang HF, Xiang W, Xue BZ, Wang YH, Yi DY, Jiang XB, Zhao HY, Fu P (2021) Cell fusion in cancer hallmarks: Current research status and future indications. Oncol Lett 22(1):530. https://doi.org/10.3892/ol.2021.12791

Wyckoff J, Wang W, Lin EY, Wang Y, Pixley F, Stanley ER, Graf T, Pollard JW, Segall J, Condeelis J (2004) A paracrine loop between tumor cells and macrophages is required for tumor cell migration in mammary tumors. Cancer Res 64(19):7022–7029. https://doi.org/10.1158/0008-5472.Can-04-1449

Yilmaz Y, Lazova R, Qumsiyeh M, Cooper D, Pawelek J (2005) Donor Y chromosome in renal carcinoma cells of a female BMT recipient: visualization of putative BMT-tumor hybrids by FISH. Bone Marrow Transplant 35(10):1021–1024. https://doi.org/10.1038/sj.bmt.1704939

Zhang Y, Wang F, Ning N, Chen Q, Yang Z, Guo Y, Xu D, Zhang D, Zhan T, Cui W (2015) Patterns of circulating tumor cells identified by CEP8, CK and CD45 in pancreatic cancer. Int J Cancer 136(5):1228–1233. https://doi.org/10.1002/ijc.29070